Springer Series in
SOLID-STATE SCIENCES 169

Springer Series in
SOLID-STATE SCIENCES

Series Editors:
M. Cardona   P. Fulde   K. von Klitzing   R. Merlin   H.-J. Queisser   H. Störmer

The Springer Series in Solid-State Sciences consists of fundamental scientific books prepared by leading researchers in the field. They strive to communicate, in a systematic and comprehensive way, the basic principles as well as new developments in theoretical and experimental solid-state physics.

Please view available titles in *Springer Series in Solid-State Sciences*
on series homepage http://www.springer.com/series/682

Bernard Pajot
Bernard Clerjaud

Optical Absorption of Impurities and Defects in Semiconducting Crystals

Electronic Absorption
of Deep Centres and Vibrational Spectra

With 159 Figures

 Springer

Dr. Bernard Pajot
Dr. Bernard Clerjaud
Institut des NanoSciences de Paris
Université Pierre et Marie Curie
Boîte courier 840
4, place Jussieu, F-75252 Paris cedex 05, France
bernard.pajot@insp.jussieu.fr, bernard.clerjaud@insp.jussieu.fr

Series Editors:

Professor Dr., Dres. h. c. Manuel Cardona
Professor Dr., Dres. h. c. Peter Fulde*
Professor Dr., Dres. h. c. Klaus von Klitzing
Professor Dr., Dres. h. c. Hans-Joachim Queisser
Max-Planck-Institut für Festkörperforschung, Heisenbergstrasse 1, 70569 Stuttgart, Germany
* Max-Planck-Institut für Physik komplexer Systeme, Nöthnitzer Strasse 38
  01187 Dresden, Germany

Professor Dr. Roberto Merlin
Department of Physics, University of Michigan
450 Church Street, Ann Arbor, MI 48109-1040, USA

Professor Dr. Horst Störmer
Dept. Phys. and Dept. Appl. Physics, Columbia University, New York, NY 10027 and
Bell Labs., Lucent Technologies, Murray Hill, NJ 07974, USA

Springer Series in Solid-State Sciences   ISSN 0171-1873
ISBN 978-3-642-18017-0   ISBN 978-3-642-18018-7 (eBook)
DOI 10.1007/978-3-642-18018-7
Springer Heidelberg New York Dordrecht London

Library of Congress Control Number: 2012945420

Printed on acid-free paper

Springer is part of Springer Science+Business Media (www.springer.com)

Foreword

In 1956, Bardeen, Brattain, and Shockley were awarded the Nobel prize in Physics for the invention, some 9 years earlier, of the solid-state transistor. The aim had been to find a device that could replace the fragile thermionic valve, as used in the (mainframe) computers of the day. The direct descendants of that device, developed through improvements in materials, design, production techniques, and software, have revolutionized the way we live and work. Earlier technological advances had improved life by increasing the production of goods and the means of transport. In contrast, the electronic revolution has so far been concerned with the processing of data, its transmission, and its storage. The impact of this revolution has been enormous, too great to summarize concisely. But take a few simple examples, which would have been completely unpredictable at the time of the Bell Labs invention. Within 60 years of the basic invention, the power and cost of electronic components have reached the point where there is one subscription to a mobile phone network for every two people on the planet. Cheap handheld phones have given billions of people their first simple, instant means of communication. A single handheld "phone", more sophisticated but still not expensive, can be used to video-call anyone worldwide, to tell them your position to within a metre, to show them photographs from your locality. Or it can be used to keep up to date with the news, to watch a film, to order the weekly shopping, to control equipment at home... Sixty years after the potential of solid-state devices was established, a device the size of a thin book can carry a vast library; and a visually handicapped person can download a new book and hear it being read to them within minutes. High-speed electronics not only control much of our equipment today, but also allow life-like images to be generated for games. Picking a few isolated examples cannot, of course, illustrate the extent to which electronic systems underpin our lives. But in addition to the direct impact of modern electronics on the way we live, the unprecedented ability to store and process large amounts of information affects our entire social systems, and no doubt will continue to impact in completely unexpected ways on our lives.

Crucial to all the industries based on modern electronics have been developments in hardware. As the sizes of devices have shrunk, knowledge of the material has become increasingly crucial. To select the most appropriate semiconductor material for a particular application requires considerable knowledge of the materials themselves. But semiconductors are greatly affected by the impurities and other atomic-sized defects that they contain. Some impurities, such as the donors and acceptors, are essential for most applications of semiconductors; these impurities have been discussed by one of the authors in Vol. I of this set. But there is another, crucially important, class of defects: the "deep centres". They can be loosely characterized as having electronic states which are localized within one or two bond lengths, and electronic energy levels that are remote from the valence and conduction bands. Deep centres are formed by a wide range of effects: by radiation damage, by metallic contamination, by substitution of iso-electronic atoms for lattice atoms... Their deep energy states may trap electrons or holes, with beneficial or harmful results. The localized electronic orbitals have a strong interaction with nearby atoms, producing a rich variety of deformations of the defect from its geometric configuration. The same local bonding, or a defect atom's low mass, result in another fingerprint of a deep centre – its local vibrational mode. Deep centres may produce significant effects at very low concentrations, maybe at one per billion atoms of the semiconductor. Scientific techniques have to be applied that give quantitative information about the concentration, the chemical species, the molecular structure, the electronic and vibrational states, and the diffusion or thermal stability of the defects. A huge effort has been made worldwide, in universities, industrial laboratories, and government laboratories, making fundamental studies of the centres. For the basic scientists, the deep centres are a wonderful playground in which apparently very simple systems, consisting of one defect atom in an otherwise perfectly repetitive structure, can display a remarkable range of properties. Simply exciting the electrons, so that the distribution of the electron density changes, might make the defects adopt a completely different molecular structure. The systems being studied are simple enough that each observed phenomenon must have a unique cause - a classic field for reductionist science - but finding that cause has often been unexpectedly difficult.

Optical measurements have turned out to be a very powerful method of probing the properties of deep centres in semiconductors. Optical spectroscopy has the essential requirement that the spectral resolution is usually sufficiently good that a researcher can be confident that measurements are being made on one species of centre. The high spectral resolution permits the effects of small perturbations to be observed, so that isotope substitution can be used to identify the chemistry of the centre, strains can be applied to determine the symmetry of the centre, the nature of its vibrations and the orbital nature of the electronic states, magnetic fields can be used to identify electron spin states, and so on. Optical measurements can be calibrated to give, quickly and non-destructively, accurate values of the concentrations of the defects at the very low levels that are important. They can be used to enhance other techniques, as in optically detected magnetic resonance. Used in conjunction with independent experimental and theoretical techniques, optical measurements have been crucial in advancing our understanding of defects in semiconductors.

This volume gathers into one accessible work the results of the last four or five decades of intensive studies into the properties of deep centres found in the key semiconductor materials. The volume has been written by world experts, who themselves have made very significant contributions to the work. The many, carefully chosen, examples illustrate the level of understanding that has been reached through the application of optical techniques in the key semiconductor materials, making the book an essential read for those involved in, or starting to apply, these powerful techniques.

London, UK   Gordon Davies

Preface

The optical spectroscopy of impurities in bulk semiconductors and insulators is a field of research that has been actively investigated in the last 60 years, resulting in a huge amount of published results. With time, there has been an evolution of the spectroscopic techniques toward higher spectral resolutions necessary for the observation of important spectroscopic details. There has been also a parallel evolution of the materials studied toward a better control of their impurity content and of their isotopic composition, to the resurgence of interest for materials like diamond, silicon carbide, or zinc oxide, and to the study of new materials like the nitrides and the magnetic semiconductors. The growth of the interest for the properties of hydrogen in semiconductors has also been a constant of the last 30 years. Last, but not least, optical spectroscopy has also been determinant in the study of quasi-monoisotopic materials made of nearly a single isotope.

One of the goals of optical spectroscopy of defects and impurity centres is to help in the determination of their atomic structure (chemical composition and symmetry in the host lattice), and when possible, of their electronic properties, in order to better understand their other physical properties, for their eventual use in various processes, or for mere curiosity and understanding. This is generally possible for not too complicated and reasonably localized centres, known as point defects, as opposed to extended defects. This requires the identification of good-quality spectra, which can be facilitated when isotope shifts are observed, and when measurements under uniaxial mechanical stresses or magnetic fields can be performed. This is not in general sufficient and when applicable, additional methods involving the electronic and nuclear spins of the atoms of the centres (electron spin resonance or ESR, electron nuclear double resonance or ENDOR) must be used. Important information on the atomic structure and location of defect atoms can also be obtained from the results of channelling measurements of alpha particles along specific crystalline orientations and from the interpretation of extended X-rays fine structure (EXAFS) measurements. The concentrations of electrically active centres are obtained by electrical methods like Hall effect measurements, and some of the energy levels of these centres can be obtained by deep-level transient spectroscopy

(DLTS). At a more general level, impurity concentrations can also be obtained by secondary ions mass spectroscopy (SIMS).

The atomic structures and electronic and vibrational properties of impurities and defects have often been confirmed or predicted from ab initio calculations. Different methods associated with these calculations have been of a great help to elucidate the thermodynamic stability and the vibrational spectra of impurity centres, and when different possibilities existed, such calculations have allowed making a choice based on thermodynamic arguments.

Three methods are mainly used in optical spectroscopy of impurity centres in semiconductors and insulators. One is absorption spectroscopy and the other one is based on photoluminescence. In absorption spectroscopy, the spectral characteristics of the radiation transmitted or reflected by a sample is analysed. In usual photoluminescence experiments, an intense optical excitation beam, generally at photon energies larger than the band gap, is focussed on the sample, and radiative transitions emitted by the sample at lower energies are recorded. In Raman scattering experiment, the external excitation must be monochromatic and at low temperature, one records weak photoluminescence lines at energies equal to the difference between that of the excitation line and of transitions of the centres with specific selection rules (Stokes lines). Each of these methods has its specific advantages and limitations.

One goal of optical absorption spectroscopy, which is a non-destructive technique, is the routine analysis of semiconductor materials (impurity concentration and homogeneity). A prerequisite is the existence of calibration factors of the absorption. These calibrations are obtained from the correlation of absorption coefficients or integrated absorption of well- identified features with atomic or defects concentration, obtained by ESR, SIMS, activation analysis, or atomic absorption analysis.

This book is the second volume of an ensemble devoted to the presentation of the optical absorption of impurities and defects in semiconductors and insulators. The first one, entitled "Optical Absorption of Impurities and Defects in Semiconducting Crystals I. Hydrogen-Like Centres" dealt with the electronic spectroscopy of centres whose electronic excited states are close to the first electronic continua of the crystals (conduction and valence bands).

The present volume is complementary to the preceding one, which was focussed on electrically active dopants. This one deals with two types of absorption: the first one is electronic and due to transitions between levels generally separated by energies representing a notable fraction of the band gap of the semiconductors, and associated with so-called deep-level centres or more shortly deep centres. Depending on the value of the band gap of the material, these transitions can occur in the ultraviolet, visible, or near-infrared region. The second type of absorption studied in this book is due to the vibration of isolated impurity atoms or impurity complexes, and it generally occurs in the infrared when purely vibrational. One advantage of this latter type of absorption is that it is a form of solid-state molecular spectroscopy which is independent of the electrical activity of the centres. These two fields cover also parts of the authors' experimental work during many years and the presentation given here can look somewhat biased but we think, however, it reflects the situation in these domains, without pretending to be exhaustive.

Deep-level centres are defects or impurities that have energy levels generally located well below the conduction band minimum and well above the valence band maximum. They are found in the native state in crystals like diamond or aluminium oxide (Al_2O_3) and they are responsible for the different colours found for these crystals. They can also be produced in these crystals by thermal treatments at high temperature or by irradiation with high-energy particles, and this is used indeed to produce artificially coloured gemstones. In semiconductors, they are due to metallic impurities like transition metals and to defect centres produced by high-energy irradiation of various origins. Deep centres can play important roles in semiconductor devices as they often act as recombination centres for charge carriers and therefore affect minority carrier life times. Deep centres associated with some metal atoms can also be useful for obtaining semi-insulating materials that can serve as substrates for devices technology. For instance, state-of-the-art bulk "pure" gallium arsenide would be p-type because of residual carbon; the natural introduction during crystal growth of As_{Ga} anti-site defects, which act as double-donors, allows pinning the Fermi level around mid-gap and therefore obtaining materials with resistivity about 10^8 Ω cm at room temperature. In the same spirit, state-of-the-art "pure" indium phosphide is n-type because of residual sulphur, the main contaminant of phosphorus, and doping with iron, a mid-gap acceptor, allows InP substrates to be grown with high resistivity.

Electrical measurements, including space-charge techniques (DLTS), are indeed techniques privileged for their high sensitivity to investigate deep-level centres in semiconductors. However, they are not so efficient for identifying precisely the microscopic structure of the defects involved. Most of the centres with deep levels display electronic transitions which can be detected by absorption or photolumi-nescence spectroscopy, which help elucidate their physical properties and atomic structures.

In semiconducting and insulating crystals, centres involving foreign atoms (FAs) with masses significantly smaller than those of the atom(s) of the crystal can give vibrational modes at frequencies higher than those of the lattice vibrations. Such FAs can be residual impurities of the starting materials used in crystal growth (e.g. carbon in silicon), but also impurities or components of gases used during crystal growth or thermal processes (e.g. hydrogen, carbon, or nitrogen), or constituents or impurities of materials used to contain the molten material before crystal growth (e.g. boron, oxygen, or nitrogen). These modes do not propagate in the crystal and remain localized in the vicinity of the centres, hence their denomination of localized vibrational modes (LVMs), and when they produce a dipole moment, they can be observed in absorption. The role of the masses of the FAs in the frequencies of these LVMs implies the existence of isotope shifts of these modes when the FAs are not mono-isotopic, and this property is very useful for the chemical identification of the FAs. As for the electronic transitions, the concentration of centres can also be obtained from the intensities of the LVMs when a suitable calibration factor exists.

Chapter 1 starts with basic definitions necessary for the understanding of the subject, followed by sections summarizing different physical properties of the centres discussed later. It ends with indications of the terminology used in the description of the spectroscopic features, sometimes in relation to ESR spectra for the paramagnetic centres.

An overview of the bulk optical properties of the semiconducting crystals necessary to understand the spectroscopic properties of impurities and defects is provided in Chap. 2.

Chapter 3 describes the principles of absorption spectroscopy in solids, the main types of optical instruments used for this purpose, together with additional techniques to cool and compress samples, or to submit them to magnetic fields.

Electronic absorption of deep centres is discussed in Chap. 4. For this topic, we have been obliged to make a choice that we hope to be representative. A first category of centres includes some intrinsic and extrinsic radiation defects in silicon. Another one concerns the impurity centres and defects in diamond, with a special part devoted to nitrogen. For the third one, we consider EL2 in GaAs, and the last part is devoted to some transition metals in different semiconducting crystals.

The principles of the vibrational spectroscopy of impurity centres and defects are presented at the beginning of Chap. 5. We have also added some hints on the calculation methods used to predict the most stable atomic structures for defect centres and to obtain vibrational frequencies. The need to make a classification in the vibrational spectra of impurity and defects have led us to present in this chapter those related to substitutional FAs, isolated, paired, or complexed, after the general presentation. We have roughly followed the columns of the periodic table, with an inversion between carbon and boron because of the general conclusions which can be drawn from the results on carbon, especially in GaAs.

The vibrational spectra of interstitial atoms and related centres are presented in Chap. 6, with an important part related to oxygen. Hydrogen is nearly absent from this chapter as its vibrational properties are discussed in Chap. 8.

A large part of Chap. 7 is devoted to FAs sometimes classified as quasi-substitutional or off-centre, well represented by the oxygen-vacancy centre in silicon and germanium. We have also added in this chapter the features related to the C- and B-related irradiation defects.

Chapter 8 deals with the vibrational transitions of centre where hydrogen is involved. The centres first considered are the intrinsic defects containing one or several hydrogen atoms bonded to crystal atoms. A large part is devoted to the vibrational properties of the isolated H_2 molecule in III–V compounds and in elemental semiconductors. The spectroscopy of complexes of hydrogen with FAs is divided into two parts. One concerns the centres resulting from the interaction with isoelectronic and interstitial atoms. The other one deals with the centres formed when the electrical activity of donors and acceptors is passivated by hydrogen. We have tried to describe in the last part the spectroscopy of some hydrogen-related centres whose origin is not clear.

This book is primarily intended for students and scientists interested in the optical properties of semiconductors, but it should be also useful to scientists and engineers interested or involved in the characterization of semiconductors. For the understanding of the principles underlying the experimental data, an elementary knowledge of quantum mechanics and of group theory applied to solid-state physics and spectroscopy is required.

The writing of different parts of this book has benefited from oral or e-mail discussions with Brian Bech Nielsen, Jacques Chevallier, Paul Clauws, Alan Collins, Monique Combescot, Denis Côte, Gordon Davies, Stephan Estreicher, Robert Jones, Kurt Lassmann, Michael Stavola, Wihelm Ulrici, Hiroshi Yamada-Kaneta, and George Watkins. Gordon Davies deserves special thanks for having accepted to write the foreword and for many valuable suggestions.

We are also indebted to Vladimir Akhmetov, Horst Bettin, Naomi Fujita, Christian Julien, Ludmila Khirunenko, Edvard Lavrov, Matthew McCluskey, Vladimir Markevich, Leonid Murin, Roland Pässler, Adele Sassella, Wilhelm Ulrici, and Ralf Vogelgesang for kindly sending information or figures used in the book. Warm thanks are addressed to Georg Bosshart for providing a very good photograph of the Dresden Green diamond and information on the Allnatt diamond.

The help and the suggestions of Claude Naud for a substantial part of the spectroscopic results obtained in our laboratories are gratefully acknowledged.

We wish to thank Bernard Perrin, director of the Institut des NanoSciences de Paris (INSP), for allowing one of us (BP) to co-write this book within the frame of this Institute.

Finally, let Claus Ascheron be thanked for his patience during the preparation of the manuscript, Adelheid Duhm for her support in the editing phase and Deepa Gopalakrishnan for her help in correcting the proofs.

Paris, France   Bernard Pajot
July 2011   Bernard Clerjaud

Contents

1 Introduction 1
  1.1   Preamble and Basic Definitions 1
  1.2   Origins of Impurities and Defects 4
      1.2.1   Natural Occurrence 4
      1.2.2   Contamination 6
      1.2.3   Doping 8
      1.2.4   Growth, Thermal Treatments and Irradiation 9
           1.2.4.1   Growth and Thermal Treatments 9
           1.2.4.2   Radiation Defects 10
  1.3   Structural Properties 12
      1.3.1   Global Atomic Configurations 12
      1.3.2   Lattice Distortion and Metastability 16
  1.4   Physico-chemical Properties 18
      1.4.1   Solubilities 18
      1.4.2   Diffusion Coefficients 20
  1.5   Electrical Activity 22
      1.5.1   Donors and Acceptors 22
      1.5.2   Compensation 24
      1.5.3   Passivation 25
      1.5.4   Deep Levels and Negative-U Properties 26
      1.5.5   Transition Metals 28
      1.5.6   Attracting Isoelectronic Impurities 29
      1.5.7   Lattice Point Defects 30
  1.6   Optical Properties 31
      1.6.1   Electronic Studies 31
      1.6.2   Vibrations of Impurities and Defects 32
  1.7   Spin Effects 34
  1.8   Labelling of the Spectroscopic Signatures 35
  References 37

2   Bulk Optical Absorption... 43
   2.1   Introduction 43
   2.2   Lattice Absorption 45
         2.2.1   Fundamental Lattice Absorption 45
         2.2.2   Multi-phonon Absorption and Anharmonicity 51
   2.3   Electronic Absorption 56
         2.3.1   Energy Gap and Intrinsic Absorption 56
                 2.3.1.1   Band Structure 56
                 2.3.1.2   Band Parameters 60
                 2.3.1.3   Absorption Coefficients 65
                 2.3.1.4   Band Gap Dependences 67
                               Temperature.. 67
                               Hydrostatic Pressure.............................. 70
                               Magnetic Field 72
                               Isotopic Composition 72
         2.3.2   Excitons 74
         2.3.3   Electron–Hole Droplets (EHDs), Biexcitons,
               and Polyexcitons 79
   2.4   Conclusion 81
   References 81

3   Instrumental Methods for Absorption Spectroscopy in Solids 89
   3.1   Introduction 89
   3.2   Spectrometer–Detector Combinations 94
   3.3   Tunable Sources 99
   3.4   Filtering and Polarization.. 100
   3.5   Samples Conditioning 101
   3.6   Cooling the Samples... 102
   3.7   Compressing the Samples 105
         3.7.1   Uniaxial Stresses 105
         3.7.2   Hydrostatic Stresses... 107
   3.8   Magnetooptical Measurements....................................... 109
   References 110

4   Absorption of Deep Centres and Bound Excitons 113
   4.1   Radiation Defects in Silicon.. 115
         4.1.1   The Divacancy 116
         4.1.2   The Higher Order Bands (HOBs) 125
         4.1.3   Group V-related Defects and Li-related Defects............. 126
                 4.1.3.1   The Vacancy-Group V Pair........................ 126
                 4.1.3.2   Other Group V-related Defects.................... 127
                 4.1.3.3   Li-related Defects 128
         4.1.4   Carbon-related Bands 130
                 4.1.4.1   Interstitial Carbon (C_i) 130
                 4.1.4.2   The G Line (969 meV Line) and the
                          C_iSiC_s Defect 131

		4.1.4.3	The C and P Lines	134
		4.1.4.4	The $3942\,cm^{-1}$ Line Defect	137
4.2	Impurity Centres and Defects in Diamond			137
	4.2.1	Intrinsic Defects		139
		4.2.1.1	Self-interstitial-related Centres	139
		4.2.1.2	Vacancy-related Centres	140
		4.2.1.3	Brown Diamonds	143
	4.2.2	Nitrogen-related Centres		144
		4.2.2.1	Substitutional and Vacancy-associated Centres	144
			Substitutional Nitrogen (N_s)	145
			The Substitutional N Pair (N_{2s}) (A Aggregate or A Centre)	146
			VN_4 (B Aggregate or B Centre) and VN_3	146
			The NV Centre	147
			VN_2 and V_2N_4	148
		4.2.2.2	Other N-related centres	149
	4.2.3	Si-related Absorption		149
4.3	EL2 in GaAs			152
	4.3.1	EL2 as a Double Donor		152
	4.3.2	Photoinduced Metastability of EL2		152
	4.3.3	Optical Absorption		153
	4.3.4	Other Optical Properties		156
4.4	Transition Metal (TM) Impurities			158
	4.4.1	Introduction		158
	4.4.2	Generalities on TM Impurities		158
	4.4.3	Intracentre Transitions		160
	4.4.4	High- or Low-Spin Ground States?		165
	4.4.5	Isotope Effects		166
	4.4.6	Transitions Between TMs Energy Levels and Valence or Conduction Band		169
	4.4.7	Transitions Involving Hydrogen-Like Hole States		171
References				175

5 Vibrational Absorption of Substitutional Atoms and Related Centres 189
5.1	Introduction			189
5.2	Overview of Theoretical Methods			193
5.3	Vibrational Spectra of Specific Centres			194
	5.3.1	Li, Group-II Atoms, and $3d$ Transition Metals (TMs)		195
		5.3.1.1	Li	195
		5.3.1.2	Group-II Atoms	197
		5.3.1.3	TM Atoms	200
	5.3.2	C		201
		5.3.2.1	Silicon and Germanium	201
		5.3.2.2	III–V Compounds	203

	5.3.3	B and Al	210
		5.3.3.1 Group-IV Crystals	211
		5.3.3.2 Compound Crystals	212
	5.3.4	B Pairing with Donors	214
		5.3.4.1 Group-IV Crystals	214
		5.3.4.2 III–V Compounds	217
	5.3.5	Si	217
	5.3.6	N	221
		5.3.6.1 Diamond	221
		5.3.6.2 Other Group-IV Crystals	224
		5.3.6.3 Other Semiconductors	226
	5.3.7	P and As	227
	5.3.8	O, S, and Se	228
		5.3.8.1 III-V Compounds	228
		5.3.8.2 II–VI Compounds	229
	References		233

6 Vibrational Absorption of O and N Interstitial Atoms and Related Centres ... 243

	6.1	Bond-Centred O in Semiconductors	244
		6.1.1 Interstitial Oxygen in Silicon	244
		6.1.1.1 Isolated Oxygen (O_i)	244
		Frequencies	249
		Linewidths and Lifetimes	260
		Stress Effects and Reorientation	263
		O_i Perturbation by Ge and Sn Atoms	266
		Relative Intensities and Calibration	268
		6.1.1.2 The Dimer (O_{2i}) and the TDDs	271
		6.1.1.3 Oxygen Precipitation	276
		6.1.1.4 O_i Interaction with Other Foreign Atoms	278
		Carbon	278
		Hydrogen	279
		Lithium	282
		6.1.2 Interstitial Oxygen in Germanium	282
		6.1.2.1 Isolated Oxygen (O_i)	282
		Frequencies	286
		Linewidths and Lifetimes	298
		Reorientation and Calibration	300
		6.1.2.2 The Dimer (O_{2i}) and the TDDs	300
		6.1.2.3 Oxygen Precipitation	305
		6.1.3 Interstitial Oxygen in GaAs and GaP	306
	6.2	Interstitial-N-related Centres	308
		6.2.1 Diamond	308
		6.2.2 Silicon and Germanium	310
		6.2.2.1 The N_i-N_i Split Pair	310
		6.2.2.2 Complexes with Oxygen	312

6.3 Some Interstitial Centres in III–V Compounds...................... 315
6.4 Self-Interstitial-related Centres in Diamond......................... 316
References 316

**7 Vibrational Absorption of Quasi-substitutional Atoms
and Other Centres**.. 325
7.1 Quasi-substitutional O (VO) 325
   7.1.1 O-Vacancy Centres in Silicon 326
      7.1.1.1 Isolated VO 326
      7.1.1.2 V_nO_m Centres 332
      7.1.1.3 VO and Foreign Atoms 334
   7.1.2 VO in Germanium 337
   7.1.3 Off-centre Oxygen (O_{oc}) in GaAs 340
7.2 C-Related Irradiation Defects 346
   7.2.1 Interstitial Carbon (C_i), (IC_i) and (C_i, Sn_s) in Silicon 347
   7.2.2 The C_iSiC_s Defect in Silicon 348
   7.2.3 The Dicarbon nn Pair ($C_s - C_s$) in Silicon.................. 352
   7.2.4 The Carbon–Oxygen Centres in Silicon 353
      7.2.4.1 The C_iO_i Centre ["C" or C(3) Centre]............. 353
      7.2.4.2 The (IC_iO_i) Centre 355
   7.2.5 The C(1) Centre in GaAs and GaP 357
7.3 B-Related Irradiation Defects 359
   7.3.1 Silicon 359
   7.3.2 III–V Compounds 361
References 362

8 Vibrational Absorption Associated with Hydrogen 369
8.1 Introduction 369
8.2 Intrinsic Hydrogen Centres.. 370
   8.2.1 "Isolated" Hydrogen 370
      8.2.1.1 Diamond 370
      8.2.1.2 Silicon, Germanium, and SiGe Alloys............ 371
      8.2.1.3 Compound Semiconductors 375
   8.2.2 The Hydrogen Dimer H_2^* 376
   8.2.3 Hydrogen and Vacancies................................... 379
      8.2.3.1 Group IV Semiconductors......................... 379
      8.2.3.2 Compound Semiconductors 384
   8.2.4 Hydrogen and Self-Interstitials 388
8.3 Hydrogen Molecules... 389
   8.3.1 Reminder on Free Hydrogen Molecules 389
   8.3.2 Isolated Molecules in GaAs 390
   8.3.3 Isolated Molecules in Silicon.............................. 392
      8.3.3.1 Raman Spectroscopy 393
      8.3.3.2 Absorption Spectroscopy......................... 394
      8.3.3.3 Ortho–Para Conversion............................ 395
   8.3.4 Isolated Molecules in Germanium 396
   8.3.5 Isolated Molecules in ZnO................................ 396

8.3.6   Isolated Molecules in Gallium Nitride 397
8.3.7   Molecules Trapped by Oxygen in Silicon 399
8.3.8   Molecules in Platelets and Voids 399
8.4   Extrinsic Hydrogen Centres I 401
8.4.1   Isoelectronic Atoms... 402
8.4.1.1   Carbon in Silicon and Germanium................. 402
A Weakly Bound Centre 403
The CH_2^* Dimers.................................... 404
The Hydrogenated C_s Dicarbon Pair.............. 405
8.4.1.2   Nitrogen in III–V Compounds 406
GaP 407
GaAs 409
8.4.1.3 $III–V_{1-x}N_x$ Dilute Alloys 409
8.4.1.4   Discussion on (N, H) Complexes in
III–V Compounds and Their Alloys 411
8.4.2   Interaction with C_i in Silicon 412
8.5   Extrinsic Hydrogen Centres II 413
8.5.1   (Acceptor, Hydrogen) Centres.............................. 413
8.5.1.1   Elemental Semiconductors 413
8.5.1.2   Compound Semiconductors 418
III–V Compounds................................. 418
II–VI Compounds................................. 425
8.5.1.3   Stress-induced Reorientation...................... 427
8.5.2   (Donor, Hydrogen) Centres.................................. 430
8.5.2.1   Compound Semiconductors 431
Donors on Cation Sites 431
Donors on Anion Sites 433
8.5.2.2   Silicon 438
8.6   Hydrogen and Transition Metals (TMs) 440
8.6.1   Complexes with TMs in Silicon 440
8.6.2   Copper–Hydrogen Complexes in ZnO 442
8.6.3   TM-Hydrogen Complexes in III–V Compounds 443
8.7   Other H-Related Centres 445
8.7.1   Hydrogen-related LVMs in Diamond 445
8.7.2   (O, H) Centres in GaAs and GaP........................... 447
8.7.2.1   The $3300\,cm^{-1}$ Line in GaAs 447
8.7.2.2   Other (O,H) Centres in GaAs and GaP 448
8.7.3   "Hidden" Hydrogen in InP 449
8.7.4   Unidentified H-Related LVMs.............................. 450
References 451

A   Energy Units Used in Spectroscopy and Solid-State Physics 465
A.1   Values of Selected Physical Constants Recommended
by CODATA (2006) 466

B   Bravais Lattices, Symmetry, and Crystals 467
    B.1   The Reciprocal Lattice 469
    B.2   Lattice Planes and Miller Indices 470
    B.3   A Toolbox for Symmetry Groups 471
        B.3.1   The Abstract Groups 471
        B.3.2   The Symmetry Point Groups 472
        B.3.3   Representations and Basis Functions 473
        B.3.4   The Symmetry Space Groups 476
    B.4   Some Crystal Structures 478
        B.4.1   Cubic Structures 478
        B.4.2   Hexagonal Structures 480
    References 482

C   Optical Band Gaps and Crystal Structures of Some Insulators and Semiconductors 483

D   Table of Isotopes 485

E   Uniaxial Stress and Orientational Degeneracy 495
    References 499

Index 501

Notations and Symbols

We have tried to comply with the IUPAC recommendations, but when the same letter is used for too many meanings, we have diverged (e.g. k_B for the Boltzmann constant). When confusion with chemical symbols is possible, the abbreviations are generally in *italics*. Symbols in **bold characters** denote vectors.

With respect to the notation and symbols lists of Vol. 1 of *Optical Absorption of Impurities and Defects in Semiconducting Crystals*, amu in the acronyms list has been replaced by u in the symbols list, and the nuclear spin is noted here **I** instead of *I*.

Acronyms

2DLFM	2D low-frequency motion
3D	Three dimensional
A	Acceptor
AB	Antibonding or antibonded
AM	Average mass
a.u.	Atomic unit
BC	Bond-centred
BE	Bound exciton
BL	Bravais lattice
BS	Beam splitter
BWO	Backward wave oscillator
BZ	Brillouin zone
CB	Conduction band
CP	Critical point
CR	Cyclotron resonance
CT	Charge transfer
CVD	Chemical vapour deposition
CZ	Czochralski

D	Diffusion coefficient, donor
DAC	Diamond anvil cell
DAP	Donor–acceptor pair
DLTS	Deep-level transient spectroscopy
DoS	Density of states
DPA	Deformation potential approximation
DR	Dichroic ratio
ECR	Electron cyclotron resonance
EM(A)	Effective mass (approximation)
EMT	Effective-mass theory
ENDOR	Electron nuclear double resonance
ESR	Electron spin resonance
EXAFS	Extended X-ray absorption fine structure
FA	Foreign atom
FE	Free exciton
FEL	Free-electron laser
FT(S)	Fourier transform (spectrometer)
FWHM	Full width at half maximum
FZ	Float-zone or floating zone
GDMS	Glow discharge mass spectrometry
h-e	High-energy (applied to particles)
hfs	Hyperfine structure
HB	Horizontal Bridgman
HPHT	High pressure, high temperature
HVPE	Hydride vapour phase epitaxy
IA	Isoelectronic acceptor
IBE	Isoelectronic bound exciton
IMC	Isotopic mass coefficient
IR	Infrared
IR	Irreducible representation
IS	Isotope shift
J-T	Jahn–Teller
LA	Longitudinal acoustic
LEC	Liquid encapsulated Czochralski
LFE	Low-frequency excitation
LHeT	Liquid helium temperature
LNT	Liquid nitrogen temperature
LO	Longitudinal optic
LVM	Localized vibrational mode
M	Metal atom
MBE	Molecular beam epitaxy
MC	Momentum-conserving
MCD	Magnetic circular dichroism
MOCVD	Metal-organic chemical vapour deposition
MOVPE	Metal-organic vapour phase epitaxy

*nat*X	Element X with natural isotopic abundance
ND	Not detected
nn	Nearest neighbour
nnn	Next nearest neighbour (second nearest neighbour)
OS	Oscillator strength
PAC	Perturbed angular correlation
PL	Photoluminescence
ppba	Part per billion (atomic)
ppma	Part per million (atomic)
PTIS	Photo-thermal ionization spectroscopy
qmi	Quasi-monoisotopic
RF	Radiofrequency
RI	Relative intensity
RT	Room temperature
RTA	Rapid thermal annealing
SD	Shallow donor
SI	Semi-insulating
SIMS	Secondary ion mass spectroscopy
s-o	Spin–orbit
STD	Shallow thermal donor
STM	Scanning tunnelling microscopy
TA	Transverse acoustic
TD	Thermal donor
TDD	Thermal double donor
TEC	Thermal equilibrium conditions
TEM	Transmission electron microscopy
TM	Transition metal
TPA	Two-photon absorption
TO	Transverse optic
USTD	Ultrashallow thermal donor
VB	Valence band
VGF	Vertical gradient freeze
ZPL	Zero-phonon line (no-phonon line)

Symbols

a_0	Lattice parameter or lattice constant
a_0^*	Effective Bohr radius
A_i	Piezospectroscopic coefficient
A_I	Integrated absorption
B	Magnetic field flux density
B	Magnetic field
Ch	Chalcogen atom

d	Sample thickness
D_R	Dichroic ratio
E	Electric field strength, doubly degenerate irreducible representation
E	Electric field
E	Energy, identity operation
E_F	Fermi level
E_g	Band gap energy
E_{gx}	Excitonic band gap
E_i	Ionization energy
f_e or f_n	Electron or neutron fluence
g	g-factor
I	Self-interstitial atom, inversion operation
I	Nuclear spin
J	Rotational quantum number
k	Extinction coefficient
k	Electron or photon wave vector
K	Compensation ratio
K	Absorption coefficient
k_B	Boltzmann constant
$\bar{m}$	Reduced effective mass or reduced mass
m_e	Free electron mass
m_n	Electron effective mass
m_h	Hole effective mass
M	Magnetic quantum number
n	Refractive index, principal quantum number, neutron, integer
n	Electron or free carrier concentration, occupation number
N	Interference order
N_c	Conduction band effective density of state
N_v	Valence band effective density of state
N	Number per unit volume, integer
N_c	Critical concentration
P	Polarization, parity
p	Hole concentration
q	Phonon wave vector
R	Reflectance
R	Reflectivity, orientational degeneracy
R_H	Hall coefficient
R_∞^*	Effective Rydberg constant
S	Electronic spin
S	Huang–Rhys factor
T	Transmittance
T	Temperature, stress magnitude
V	Lattice vacancy
V_2	Divacancy
u	Unified atomic mass unit

v	Vibrational quantum number
[X]	Concentration of centre X per cm^3
α	Polarizability
β	Parameter
γ	Ratio of transverse and longitudinal effective masses, damping constant
γ_B	$\eta\omega_c / 2R_\infty^*$
Δ_{so}	Spin–orbit splitting or energy
Δ_{CF}	Crystal field energy
ε	Dielectric constant
ε_s	Static dielectric constant
ε	Strain
ϵ	Emissivity
η	Effective charge
λ	Wavelength
μ	Mobility, chemical potential
$\tilde{v}$	Wavenumber
ρ	Electrical resistivity
σ	Electrical conductivity
σ	Mechanical stress
σ	Absorption cross-section
τ	Lifetime
Φ	Work function
ω	Angular frequency, wavenumber
ω_c	Cyclotron frequency

Chapter 1
Introduction

1.1 Preamble and Basic Definitions

The properties of impurities and defects are treated globally in textbooks dealing with the physical properties of semiconductors (for instance, [1] or [2]). An overview on the optical measurements on point defects in semiconductors is given in the chapter by Davies [3], and the relations between their optical properties and their colours in books dealing with colour in general, like the one by Nassau [4]. More ancient textbooks were devoted to the optical properties of semiconductors, for instance, the book by Moss [5], or aimed toward special topics like the vibrational absorption of impurities and defects in some insulators and semiconductors [6]. Many book chapters and reviews in journals have also been devoted to specific aspects of the absorption of impurities and defects and they are duly referenced in this book.

In this monograph, we consider first the electronic absorption of some kind of impurities and defects giving deep levels in the band gap of semiconductors and of covalent or partially covalent insulators. The electronic absorption of hydrogen-like centres in these materials was presented in a first book subtitled "I. Hydrogen-like Centres" [7]. The second point considered in the present book is the vibrational absorption of impurities and defects, electrically active or not, in the same materials.

Semiconductors differ from metallic conductors by the existence, at low temperature, of a fully occupied electronic band (the valence band or VB) separated by an energy gap or band gap (E_g) from a higher energy band empty of electrons (the conduction band or CB). When E_g reduces to zero, like in mercury telluride, the materials are called semimetals. In metals, the highest occupied band is only partially filled with electrons so that the electrons in this band can be accelerated by an electric field, whatever small it is. From a chemical viewpoint, most of these semiconducting and insulating crystals are elements or compounds in which all the valence electrons are used to form covalent or partially covalent chemical bonds, leaving no extra electron for electrical conduction. This is the case for the diamond form of carbon, for silicon and germanium, for many crystals resulting from the

B. Pajot and B. Clerjaud, *Optical Absorption of Impurities and Defects in Semiconducting Crystals*, Springer Series in Solid-State Sciences 169,
DOI 10.1007/978-3-642-18018-7_1, © Springer-Verlag Berlin Heidelberg 2013

combination of group IIB or IIIA elements of the periodic table with group V or VI elements (the II–VI or III–V compounds) or for the partially ionic IB–VII (e.g. CuCl) compounds. In purely ionic insulators, like sodium chloride, electron capture from the electropositive element by the electronegative one produces ions with closed shells. Values of band gaps of different semiconductors and insulators are listed in Appendix C.

From the optical side, the difference between semiconductors and insulators lies in the value of E_g. The admitted boundary is usually set to 3 eV (see Appendix A for the energy units) and materials with E_g below this value are categorized as semiconductors, but crystals considered as semiconductors like the wurtzite forms of silicon carbide and gallium nitride have band gaps larger than 3 eV, so that this value is somewhat arbitrary. The translation into the electrical resistivity domain depends on the value of E_g, but also on the effective mass of the electrons and holes and on their mobilities, and the solution is not unique; moreover, the boundary is not clearly set. "Semi-insulating" silicon carbide 4H polytype samples with reported RT resistivities of the order of $10^{10}\,\Omega$ cm could constitute the electrical limit between semiconductors and insulators, but the definition of such a limit is finally of moderate importance. In the following, for simplification, "semiconductors and insulators" is generally replaced by "semiconductors".

In a category of materials known as Mott insulators, like MnO, CoO, or NiO, with band gaps of 4.8, 3.4, and 1.8 eV, respectively ([8], and references therein), the upper energy band made from 3d atomic states of the metal is partially occupied and metallic conduction should occur. The insulating behaviour of these compounds is ascribed to a strong intra-atomic Coulomb interaction, which results in the formation of a gap between the filled and empty 3d states [9].

In pure covalent or partially covalent semiconductor crystals, a free electron is created in the CB once sufficient energy has been provided to a VB electron to overcome the energy gap E_g. The required energy can be produced thermally under equilibrium at temperature T, by optical absorption of photons with energies $h\nu \geq E_g$, or by irradiation with electrons in the keV energy range. These processes leave in the VB a positively charged free "hole", noted h, which has no equivalent in metals, and whose absolute electric charge is the elementary charge. When free carriers can only be produced by the above processes, the semiconductors are said to be intrinsic. The electron–hole pair concentration n_i produced thermally in a semiconductor depends on the effective densities of states (DoS) N_c and N_v in the conduction and valence bands, which are temperature dependent, and on the band gap E_g. It can be expressed as:

$$n_i = (N_c N_v)^{1/2} \exp(-E_g/2k_B T). \tag{1.1}$$

In germanium, silicon, and GaAs, at room temperature (RT), n_i is 2×10^{13}, $\sim 10^{10}$, and 2×10^6 cm^{-3}, respectively. A consequence of the existence of an electronic band gap is that at sufficiently low temperature, the absorption of photon with energies below E_g by electronic processes does not occur in intrinsic semiconductors or insulators. Inversely, the photons with energies above E_g are strongly absorbed by optical transitions between the valence and conduction bands, to produce electron–hole pairs, and this absorption is called fundamental or intrinsic.

Compound semiconductor crystals show strong infrared absorption in some spectral regions at photon energies below E_g due to the vibrations of the atoms of the crystal lattice. In these regions, the lattice absorption can be so strong that the crystals are opaque at usual thicknesses. At energies below the lattice absorption region, the crystals become transparent again. In elemental crystals like diamond (C_{diam}) or silicon, this first-order vibration of the lattice atoms is not infrared-active (there is no dipole moment associated to the symmetric vibration of two identical atoms) so that pure crystals of this kind do not become opaque, but they do show weaker absorption bands due to combinations of vibration modes of the crystal lattice.

Extrinsic semiconductors are materials containing foreign atoms (FAs) or atomic impurity centres that can release electrons in the CB or trap an electron from the VB with energies smaller than E_g (from neutrality conservation, trapping an electron from the VB is equivalent to the release of a positive hole in this otherwise full band). These centres can be inadvertently present in the material or introduced deliberately by doping, and the term extrinsic refers to the electrical conductivity of such materials. The electron-releasing entities are called donors and the electron-accepting ones acceptors. When the majority of the impurities or dopants in a material is of the donor (acceptor) type, the material is termed n-type (p-type) and the electrical conduction comes from electrons (holes). In doped semiconductors with $E_g \gtrsim 0.6\,\text{eV}$, the intrinsic free-carrier concentration n_i can usually be neglected at RT compared to the extrinsic one. In these semiconductors, when the energy required to release a free carrier from the dominant donor or acceptor (the ionization energy) is comparable to the RT thermal energy ($\sim$26 meV), a measurement of the RT resistivity $\rho = (ne\mu)^{-1}$, where μ is the electrical mobility of the free carrier considered, gives a representative value of the concentration n of the dominant donor or acceptor.

For a semiconductor, the intrinsic resistivity $\rho_i(T)$ at a given temperature is determined by the concentration n_i of electron–hole pairs given by expression (1.1). For germanium, ρ_i is $\sim$50 Ω cm at RT, but it rises to $\sim 3 \times 10^5\,\Omega$ cm for silicon and to $\sim 10^8\,\Omega$ cm for GaAs. Above a temperature depending on the value of E_g, the concentration of electrons–holes pairs produced thermally by direct excitation through the band gap in extrinsic materials can become comparable to the extrinsic carrier concentration and the semiconductor is said to go into the intrinsic regime. The presence of free electrons produces at RT a Drude-type continuous optical absorption increasing as λ^2, where λ is the wavelength of the radiation. The wavelength dependence of the free-hole absorption is not as simple. It must be pointed out that several residual impurities present in semiconductors are not electrically active and cannot be detected by electrical methods, so that the term intrinsic cannot be systematically taken as a synonym for high purity.

For some values of the donor or acceptor concentrations depending on E_g, the free-carrier absorption can be so large that the material becomes opaque in the whole spectral range. For still higher dopant concentrations, a transition to a quasi-metallic state (the metal–insulator transition) occurs.

When temperature is reduced, the free carriers in extrinsic materials are normally re-trapped by the ionized donors or acceptors centres, increasing the resistivity of the materials. The free carriers can also be trapped irreversibly by deep centres like transition metal (TM) atoms, native defects, or irradiation defects. This trapping can produce materials with RT resistivity close to the intrinsic one, and the spectroscopic properties of some of these deep centres are presented in Chap. 4.

A large number of semiconducting and insulating crystals are known which are used in various technologies and for pure and applied research, and most of them are grown artificially. All these crystals are made from or contain an element with more than one isotope. For specific applications, some of these crystals, like germanium, have been grown with a substantial enrichment in only one isotope, which can reach nearly 100%, and they are referred to in this book as quasimonoisotopic (qmi) crystals. Insulating crystals, either native or artificial are also used in jewellery, where their optical properties are of utmost importance.

1.2 Origins of Impurities and Defects

1.2.1 Natural Occurrence

Some insulating and semiconducting crystals, like diamond, various forms of silicon dioxide or corundum (Al_2O_3) are found in the native state. In some cases, the crystals are transparent (their band gap lies in the UV), but they can also come with different hues. One of the origins of these colours is the presence of FAs in the crystal lattice. These atoms originate from the presence of the corresponding chemical elements in the earth mantle during the formation of the crystals. One example is diamond, where the presence of a significant nitrogen concentration is responsible for the yellow hues of many natural diamonds. The type of natural diamonds and, by extension, of artificial ones is first determined formally by their overall nitrogen content as type I diamonds containing typically more than 1 ppma ($\sim 1.8 \times 10^{17}\,cm^{-3}$) of N atoms and type II diamonds for the other crystals. This labelling is further refined by additional lettering bringing supplementary information. Thus, the rarely found natural type Ib diamonds contain only nitrogen atoms replacing carbon atoms (substitutional nitrogen), with an absorption onset toward higher energies at 2.2 eV ($\sim 564\,nm$), in the blue region of the spectrum, so that the complementary yellow colour, the canary yellow of the gemmologists, is preferentially perceived. The more common yellow type Ia diamonds contain, besides the so-called A aggregate (a pair of substitutional N atoms) and B aggregate [a missing atom (vacancy V) surrounded by four substitutional N atoms] the so-called N3 centre (VN_3) and the $(VN_2)^0$ centre. These latter centres have absorption lines in the visible (see Sect. 4.2) and their presence is responsible for the yellow colouration known as Cape yellow of these diamonds (Fig. 1.1).

All these N-containing diamonds are insulators, with resistivities in the $10^{15}\,\Omega\,cm$ range, because the electronic levels introduced by nitrogen are

Fig. 1.1 The Allnatt
diamond, weighing 101.29 ct
(20.26 g), was cut from a
stone assumed to come from
South Africa. It is presently
owned by the SIBA
Corporation. This diamond is
of the Ia type (G. Bosshart,
private communication)

near midgap. N-lean diamonds are known as type II diamonds, and the purest
ones are the colourless type IIa diamonds. The presence of boron in diamond
produces a blue coloration due to the substitutional B acceptor and these natural
type IIb diamonds are semiconducting, with resistivities as low as 5 Ω cm for a few
natural stones. A discussion of the electronic absorption of type IIb semiconducting
diamonds can be found in [7].

Blue-grey insulating diamonds have also been extracted from the Argyle mine in
Australia and this colour has been attributed to high concentrations of hydrogen and
of N-containing defects, together with the presence of Ni-related centres [10]. The
colour of many natural diamonds is also affected by the plastic deformation they
underwent during their growth and rise to the earth surface.

In a way similar to diamond, pure corundum (Al_2O_3) is colourless, but the
presence of Cr^{3+} ions replacing Al^{3+} (typically a few percent) turn it into the
red ruby while the presence of Fe^{2+} and Ti^{4+} ions gives the blue sapphire. Stable
lattice defects producing colouration of the crystals can also be introduced in
otherwise transparent minerals, like alkali halides, alkaline-earth[1] halides, or zircon
($ZrSiO_4$), and in diamond, by natural radioactivity. When crystals, like zircon,
contain radioactive isotopes, the defects inducing colour are evenly distributed in the
crystal, but when colour is due to external radioactivity, it is generally located in an
outer region of the crystal corresponding to the penetration depth of the particles. In
diamond, natural and artificial irradiations by neutrons, electrons, or γ-rays produce
more or less intense green colouration of the stones due to the presence of vacancies
and vacancy-related defects, and such irradiations are taken for responsible of the
colour of natural diamonds like the so-called Dresden Green diamond shown in
Fig. 4.12 [11]. These colour changes can arise from the selective absorption of light
by impurities or defects in the crystal at the corresponding energies or from the

---

[1]The alkaline earths are the atoms of column IIA of the periodic table (Be, Mg, Ca, Sr).

absence in the energy spectrum reflected by the crystal of a spectral domain which is absorbed by the crystal (see for instance, [4]).

1.2.2 Contamination

The contamination by impurities of artificially grown materials can have several origins and some of them have been reviewed in [7] in relation to hydrogen-like (or effective-mass) centres. For bulk crystals, the impurities can come from their presence in the polycrystalline starting material used, as for carbon in silicon [12], from the growth atmosphere and/or, when a crucible is used, from chemical elements of the crucible containing the molten material. For instance, most of the silicon crystals used in the electronic industry are grown from a melt contained in a silica crucible by dipping a monocrystalline silicon seed just below the melt and slowly pulling it while silicon solidifies as a crystal at the seed bottom. This growth method, known as the Czochralski (CZ) method, after the Polish metallurgist Jan Czochralski [13], introduces in the silicon crystal a rather large concentration $(\sim 10^{18}\,\mathrm{cm}^{-3}$ or 20 ppma) of electrically inactive interstitial oxygen (O_i), originating from the partial dissolution or etching of silica by molten silicon ([14, and references therein). The purest crucible-grown crystals are probably the undoped Ge crystals grown from a silica crucible in a hydrogen atmosphere by the CZ method, for which the overall bulk impurity concentrations (mainly Si, O and H) are in the $10^{14}\,\mathrm{cm}^{-3}$ range [$\sim$2 atomic parts per billion (ppba)]. In the case of germanium, the much lower O contamination is due to its melting temperature (937°C) compared to 1414°C for silicon, and to the smaller affinity of O for germanium. Severe O contamination ($\sim$0.02%) also occurs in the high-pressure growth of GaN from a gallium solution containing dissolved nitrogen, probably due to the etching by ammonia of O-containing components of the apparatus. An indirect source of contamination arises when crystals containing an element with a high vapour pressure, like P or As, are grown by the CZ method: to prevent the evaporation of the volatile element, a compound with a low vapour pressure and a low miscibility with the melt of the crystal to be grown, the encapsulant, is placed at the top of the polycrystalline charge to be melted. Once molten, this encapsulant makes a tight seal between the molten material and the atmosphere of the furnace (usually nitrogen). This growth method is known as the liquid encapsulation Czochralski (LEC) method and Fig. 1.2 shows a schematic of a LEC crystal grower.

The price to pay is the unavoidable introduction in the crystals grown by this method of a small amount of atoms of the molten encapsulant: to grow GaAs and InP crystals, the encapsulant used is wetted boron oxide (B_2O_3). In addition to the introduction of B and O impurities, at high temperature, the water added to B_2O_3 to prevent sticking between the encapsulant and the crystal dissociates and hydrogen is introduced in the crystal [15].

A large concentration of hydrogen is also introduced in III–V layers grown by dissociation of organometallic compounds in the vapour phase epitaxial process ([16], and references therein).

Fig. 1.2 Schematic view of a liquid encapsulated Czochralski (LEC) crystal pulling machine showing the principle of the method. It is used to grow compound crystals like GaAs, where one of the constituents of the melt has a high vapour pressure. The rotation directions of the crystal holder and of the pyrolitic BN crucible are opposite. In the Czochralski technique used to grow CZ silicon, there is no encapsulant and the crucible material is silica

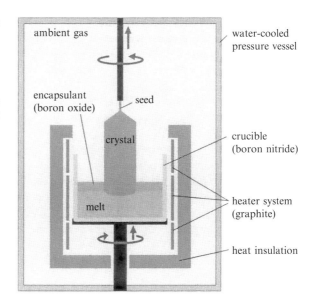

The contamination introduced by melting polycrystalline charges in a crucible at high temperature has led to the development of crucibleless growth methods for some crystals like silicon, when needed with a low O content for particular applications. In these methods, a monocrystalline seed is mounted at the bottom or at the top of a polycrystalline charge and the polycrystalline region in contact with the seed is melted by a contactless technique (RF field or halogen lamp furnace). The melted region is prevented to flow from capillarity forces alone and it is displaced upward or downward by moving the RF coil, leaving a monocrystalline region. This method, known as float-zone (FZ) growth method, was invented independently by several scientists ([17], and references therein). A schematic FZ setup for silicon crystal growth is shown in Fig. 1.1 of [14].

The growth of crystals with lower melting points and low reactivity has been obtained by the Bridgman method, named after the American physicist Percy Williams Bridgman, in an elongated crucible held horizontally, with the monocrystalline seed at one end of the crucible [horizontal Bridgman (HB) method]. The principle of the monocrystalline growth is to displace the molten zone from the seed region throughout the crucible length. The Bridgman method is also used in a variant where a sealed crucible is displaced vertically in a temperature gradient (vertical gradient freeze method), as for the growth of CdTe monocrystals [18]. With these methods, contamination of the crystals often occurs from the ambient gas.

Besides the nitrogen contamination due to the pollution of the carrier gas, the diamond films obtained by chemical vapour deposition (CVD) are usually contaminated with silicon. This can originate from the plasma etching of the silica walls of the reactor and from the commonly used silicon substrates [19].

Metallic contamination by TMs and Cu occurs in many semiconductors because of the high diffusion coefficients of these elements. This contamination can have many origins including the initial purity of the materials, chemical etching, electrical contacts, mechanical contacts with metallic parts or metallic constituents of heating resistances in thermal treatments. For technological reasons, this contamination has been widely studied in silicon CZ wafers, and also in germanium, in relation to nuclear detectors. It remains usually at a low level, but it can be detected by very sensitive methods like deep-levels transient spectroscopy (DLTS), an electrical method, or, when the TMs complex with shallow dopants, by photo-thermal ionization spectroscopy (PTIS), which is related to optical spectroscopy. In the metal-solvent method of growth of synthetic diamond, graphite is first dissolved into a molten metal solvent like nickel, cobalt, or iron under high pressure and high temperature (HPHT) above the eutectic melting point of the solvent–carbon diagram, in a (P, T) region of the phase diagram of carbon where diamond is the stable phase (5.5 GPa and 1400°C are typical values). As more and more graphite is dissolved, the molten metal becomes supersaturated and small diamond crystals start to nucleate and begin to grow. Substitutional Ni and Co have been shown to be incorporated in diamond by this so-called HPHT growth method when nickel or cobalt is one of the solvent metals [20].

1.2.3 Doping

Doping consists in introducing into the crystal FAs called dopants which usually lower the resistivity of the intrinsic material and convert it into a well-characterized n-type or p-type material. However, doping of compounds semiconductors with TMs introducing deep levels in the band gap has also been used for optical purposes. The introduction of dopant atoms in III–V compounds is discussed in [21] and some aspects of doping are also discussed in [7]. Standard doping of bulk semiconductor crystals is achieved by adding either to the solid charge or directly to the melt a selected amount of the dopant element or of a crystal/dopant alloy. This method works for many dopants in group IV and in III–V semiconductors. However, the introduction of a FA in a crystal at a given concentration is determined primarily by its solubility. Moreover, it is not possible to dope any crystal with any impurity.

The possibility to use the transmutation of atoms of the crystals by thermal neutrons to dope semiconductors was first demonstrated by Cleland et al. [22]. A brief review of the possibilities of this method (neutron transmutation doping or NTD) and of its limits is given in [7]. A consequence of this doping method is also the introduction in the crystals of lattice defects produced by fast neutrons present in the neutron beam. These defects must be removed by annealing before using the semiconductor for the production of devices, and more is said below of these defects.

Doping can also be achieved by implantation in the crystals of dopant ions at energies typically below 100 keV, followed by thermal or optical annealing in order to remove lattice defects and produce dopant diffusion in the crystal. During the growth of epilayer by vapour-phase epitaxy, doping is also possible by adding to the gaseous flux organic molecules containing the dopant atom.

In some cases, the spectroscopic study of specific dopants and FAs with several isotopes has been facilitated by doping with a specific isotope or with an isotopic mixture enriched with one isotope.

1.2.4 Growth, Thermal Treatments and Irradiation

1.2.4.1 Growth and Thermal Treatments

The growth of crystalline samples or their annealing at very high temperature produces a steady-state concentration of elementary defects. The simplest ones result from the thermal ejection of a crystal atom from its crystal sites into an interstitial location. These self-interstitial atoms (I) leave an empty crystal site or vacancy (V). A slow cooling-down allows for recombination of the interstitial atoms into the empty sites, but a relatively fast cooling-down or a quenching allows the most stable of these defects to survive and/or to agglomerate at RT. This is the mechanism put forward to explain the origin of point defects produced during the growth of some III–V and II–VI compound crystals under stoichiometric conditions, where anion and cation vacancies, as V_{Cd} or V_{Te} in CdTe, can be produced, as the cooling-down of the solidified fraction is not an equilibrium process. In CZ silicon crystals, depending on the pulling rate and on the thermal gradient at the solid/liquid interface, excess vacancies or interstitials are present. They are mobile and some V–I pairs can recombine, but in regions presenting a vacancy excess, the vacancies agglomerate to form macroscopic voids which can be present in CZ, and also FZ silicon crystals at concentrations $\sim 10^4$–$10^7\,\mathrm{cm}^{-3}$ [23, 24]. CZ void-free silicon crystals can be grown by adjusting the pulling rate and the temperature gradient at the solid/melt interface ([25], and references therein). In regions with excess of interstitials, extended defects can be produced, like series of dislocation loops, best known as striations or swirls, which are decorated by oxygen in CZ silicon. In GaAs crystals grown by the HB or LEC methods, and also in epitaxial layers grown by CVD, several native defects are present, including the EL2 centre [26, 27].

The main native defects in III–V and II–VI compounds are vacancies and atoms in antisites. For instance, the As antisite (As_{Ga}) as well as the As vacancy (V_{As}) are residual defects in LEC-grown GaAs crystals [27]. ZnO is a material whose electrical properties are determined by native lattice defects: the presence of interstitial Zn correlated with that of O vacancies (V_O) seems to be responsible for the n-type electrical conductivity of many crystals, but in high-resistivity crystals obtained by hydrothermal growth, the dominant defect is V_{Zn} [28].

Last, but not least, annealing of CZ silicon or O-containing germanium in the 350–500°C temperature range produces several O-related electrically active centres known as thermal donors (TDs), and their atomic structures differ as a function of the annealing duration. One category, which involves only O and Si or Ge atoms in the cores of the centres are thermal double donors (TDDs) that can bind two extra electrons [29, 30]. In CZ silicon containing N or H, short-time annealing in the 300–600°C range produces also donors known as shallow thermal donors

Fig. 1.3 Atomic structure of the VO centre in silicon. The O atom is bonded to two nearest neighbours of the vacancy and a bond is reconstructed between the two other nearest neighbours. There are six equivalent orientations for this centre about the central vacancy

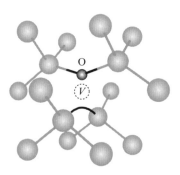

(STDs) [31]. The electronic spectra of these donor centres are discussed in Chap. 6 of [7] and their vibrational spectra in Chap. 6 of the present volume.

1.2.4.2 Radiation Defects

The radiation defects in semiconductors started to be investigated in the 1950s to elucidate their effects on semiconductor devices exposed to high-energy (h-e) particles, including specifically semiconductor detectors for h-e particles, and later in processes like NTD. Defects can be deliberately introduced in semiconductor and insulator crystals by irradiation with h-e photons (γ rays or X-rays), elementary particles, or ions. Irradiation with γ rays, h-e electrons and fast neutrons, protons or α particles (He nuclei) has been adopted for the deliberate production and study of lattice defects in covalent and partially covalent materials. Combinations of irradiations and thermal treatments are also used to colour gemstones by introducing lattice defects or complexes with absorption lines or bands in the visible region of the spectrum.

The penetration depth of the particles depends on their charge, on their energy, on their mass, and on the atomic number of the crystal target. The charged particles interact mainly with the electron cloud of the atoms or ions of the crystal and for the same energy, their penetration depth is much smaller than that of neutrons, which interact mainly with nuclei.

For irradiation with h-e nucleons, the primary defects produced depend on the energy, charge, and mass of the particles. In covalent crystals, the simplest primary defects produced by γ-ray or electron irradiation are I and V.

In silicon, vacancies produced by electron irradiation near 10 K start diffusing near 150–180 K in p-type and near 70–80 K in n-type material. During their diffusion, they can interact with impurities and defects. For instance, in silicon, two vacancies can combine to form a divacancy[2] usually noted V_2; vacancies can also get trapped by dopant atoms to give V-dopant pairs, or with an interstitial O atom (see [33]) to form the oxygen-vacancy centre (VO), represented in Fig. 1.3, which is stable up to $\sim$300°C.

---

[2]The divacancy is also produced as a primary defect in neutron-irradiated silicon [32].

Fig. 1.4 Schematic diagram
of the hexavacancy (V_6) in
silicon. The missing atoms
are the *large light balls*
showing the location of the
vacancy sites in a puckered
hexagonal configuration.
(**a**) Unrelaxed. (**b**) With six
reconstructed bonds. V_6
displays a trigonal symmetry
about the vertical $\langle 111 \rangle$
direction [38]

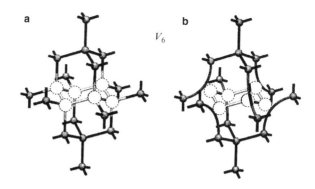

VO, also known in silicon as the A centre, can be also considered as a limiting case of substitutional oxygen when the O atom comes close to the vacancy site [34]. The vibrational properties of this centre in silicon and germanium are discussed in Sect. 7.1.

Besides V_2, which is stable up to $\sim$200°C, irradiation of silicon with fast neutrons can also produce polyvacancies, and indirect evidence for the formation of trivacancy (V_3) has recently been given [35]. Calculations have predicted that the hexavacancy (V_6) is a very stable defect in silicon [36, 37] and it is shown in Fig. 1.4.

An unattributed photoluminescence (PL) line observed a long time ago in heat-treated neutron-irradiated silicon has been ascribed to V_6 [38]. Vacancies have been mainly studied in diamond and silicon after their production by electron irradiation, but there is a difference between the two crystals: in diamond, V is stable up to $\sim$600°C while in silicon it is only stable below RT. Many other complexes involving lattice defects are known in semiconductors and insulators.

Protons are hydrogen nuclei so that irradiation with h-e protons introduces not only lattice defects but also hydrogen in the irradiated crystal. This hydrogen can "decorate" lattice defects produced by the implantation by forming H bonds whose vibrations can be detected by optical spectroscopy. The fully hydrogenated vacancy VH$_4$ where the four dangling bonds of a vacancy are saturated by a H atom is such an example [39]. A combination of proton implantation and thermal annealing of CZ or FZ silicon has also been shown to produce shallow donors [40]. Hydrogen can also be introduced in a region near from the semiconductor surface by hydrogen plasma treatments as long as the semiconductor surface does not suffer excessive plasma etching.

1.3 Structural Properties

1.3.1 Global Atomic Configurations

In crystals, FAs can take simple configurations, but depending on their concentration, diffusion coefficient, or chemical properties and also on the presence of different kind of impurities or of lattice defects, more complex situations can be found. Aside from indirect information like electrical measurements or X-ray diffraction, methods such as optical spectroscopy under uniaxial stress or a magnetic field, electron spin resonance, channelling, positron annihilation, or extended X-rays absorption fine structure (EXAFS) can bring more detailed results on the location and atomic structure of impurities and defects in crystals. We describe here the simplest atomic configurations. In the course of the chapters, more complicated configurations will also be met and discussed.

In the simplest case, a FA can replace an atom of the crystal at a regular lattice site. It becomes then a substitutional impurity or dopant, noted with index s in the general case (e.g. B_s). In covalent or partially covalent crystals, the main relevant parameters for the possible location of a FA on a substitutional site are its ability to form chemical bonds with its nearest neighbours (nns), the strengths of these bonds, and the difference between the atomic radius of the FA and that (or those) of the crystal atoms. When a crystal is made up of two different elements, like one with the sphalerite or wurtzite structure, a substitutional FA can in principle occupy two different lattice sites. Sometimes, depending on its concentration and on the growth conditions, the same atom can occupy in fact two different lattice sites. This kind of amphoteric behaviour occurs, for instance, in GaAs, a cubic crystal with sphalerite structure, where, depending on the growth conditions, a Si atom can occupy either a cation site (Si_{Ga}) where it is a donor or an anion site (Si_{As}) where it is an acceptor.[3] This duality is not a general rule, however, and, for instance, the doping or contamination of GaAs with C produces only C_{As}. In heteropolar semiconductors, each kind of atoms of the crystal occupies one sublattice (the group III and the group V sublattices in III–V compounds). For different reasons, it can occur that some group V (III) atoms get located on group III (V) sublattice and these already-mentioned antisite atoms can be considered as "internal" impurity atoms. Similarly, a foreign group III atom can occupy a group V site (B in GaAs, for instance) and act as an "external" antisite. The location of FAs at substitutional sites is very common among semiconductors and insulators. This does not necessarily mean that the FA takes the exact equilibrium position of the atom it replaces as, depending on the radius and valence of the FA, lattice distortion can occur that will be discussed later. Besides well-defined crystal structures like sphalerite and wurtzite where there is only one substitutional site for each kind of atom, there

---

[3]With reference to chemistry, when considering a binary compound, the site of the metal atom is called the cation site and the other one the anion site.

are more complicated structures like the polytypes found, for instance, in the SiC compound, where there can be several kinds of substitutional sites differing by the symmetry of their *nns* (see Appendix B).

Small FAs tend to occupy interstitial sites, but some of them can also locate in the same material on a substitutional site, as lithium in GaAs. Figure 1.5 shows possible interstitial locations of isolated FAs in a III–V compound with sphalerite structure.

In the T_i sites, the impurity is located at a tetrahedral interstitial site, where it is weakly bonded to the crystal lattice. In compounds with the sphalerite or wurtzite structure, in which there are two different substitutional sites, there are also two different T_i sites: one where the interstitial atom is nearer from atoms of one sublattice and another where it is nearer from atoms of the other sublattice (T_i III and T_i V of Fig. 1.5). As the electronic densities at these two sites are different, the interstitial atoms occupy preferably one of these sites. This T_i location is also met for Li atoms in silicon and germanium, and as there is no chemical bond between the Li and Si or Ge atoms, the $2s$ valence electron of Li has a low binding energy, making interstitial Li (Li_i) a shallow donor in these semiconductors. In the interstitial bond-centred (BC) location, bonding must be rearranged to allow the FA to form chemical bonds with its neighbours. Besides a rather small size of the

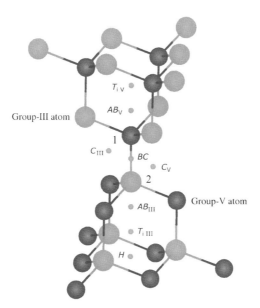

Fig. 1.5 High-symmetry interstitial sites (*small spheres*) in a III–V sphalerite lattice oriented along a $\langle 111 \rangle$ vertical axis (the simple substitutional sites of the crystal atoms are not indicated). The bond-centred (BC), antibonding (AB), tetrahedral interstitial (T_i) and hexagonal (H) sites are located along the $\langle 111 \rangle$ axis. The T_i and AB sites are noted according to the atoms closest to these sites. The C site, midway between two next nearest neighbours along a $\langle 110 \rangle$ axis, is noted according to these atoms. The M site (not shown) is midway between two adjacent C_{III} and C_V sites and also midway between a BC site and a H site

atom, this latter location also implies a strong affinity between the foreign and lattice atoms. The paradigm of such a structure is isolated interstitial oxygen (O_i) in silicon (Fig. 1.6). Low-temperature vibrational absorption spectroscopy shows that in the static configuration, the Si–O–Si geometry is puckered, with an apex angle of $\sim 162°$ (see Fig. 6.1) but the mixed rotation-tunnelling motion of O_i in a plane perpendicular to the $Si_c \cdots Si_e$ axis leads to a dynamical configuration where the probability of presence of O_i is maximum on the $Si_c \cdots Si_e$ axis, as shown in Fig. 1.6.

The above example introduces a more general property of impurities and impurity complexes in crystals, namely, orientational degeneracy. In the linear configuration of Fig. 1.6, the four equivalent $\langle 111 \rangle$ orientations of the Si–O–Si_c structure around the central Si_c atom are equally populated and display the same vibrational properties: this structure displays therefore a fourfold orientational degeneracy. For centres with noncubic symmetry in cubic crystals, orientational degeneracy is the rule and it can be extrapolated to other crystal structures. A value of the energy required to jump from one orientation to an equivalent one (the reorientation energy) can be of interest to determine some physical properties of the centre, such as the diffusion coefficient, and it is shown later in the book how it can be measured.

An interstitial atom in an antibonding (*AB*) site is bonded to its nearest neighbour lattice atom. This location is often found for H in complexes involving a donor atom and it results in the relaxation of the local lattice bonding. Special interstitial structures like the di-interstitial configuration are discussed later in this book. Incidentally, Fig. 1.5 emphasizes the ternary symmetry of the sphalerite lattice along a $\langle 111 \rangle$ direction. It allows to see the analogy with the wurtzite structure, where the $\langle 111 \rangle$ direction is replaced by the *c*-axis direction (Appendix B). In the $\langle 111 \rangle$ direction, the sphalerite lattice is made of alternate layers of atoms of the two sublattices. It can also be seen in Fig. 1.5 that the stacking sequence is a period of three layers of the same kind (the so-called *ABC* sequence).

Pairing between identical or different FAs is also found in semiconductors and insulators, depending on factors, which can be inter-related, and several pairing configurations can be met. We limit ourselves here to the description of the simplest

Fig. 1.6 Dynamical location of isolated interstitial oxygen in a silicon crystal (see text). The two Si atoms bonded to O are represented in the same plane as their neighbours because of the repulsion induced by the O atom. There are four equivalent orientations of O_i in the crystal around the central Si_c atom

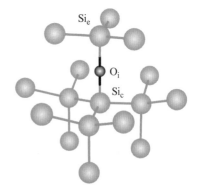

ones, and others will be eventually considered later. In elemental semiconductors, the simplest kind of pairing concerns impurities on *nn* substitutional sites or *nn* substitutional and interstitial sites, and the atoms of the pair are generally different. Another important kind of pairing found in silicon concerns two identical or different FAs located on equivalent distorted interstitial sites. This is illustrated by the location of nitrogen in N-containing silicon and germanium because of the relatively small size of the N atom compared to the lattice atoms, and also because of the high strength of the N_2 molecular bond. In these crystals, most of nitrogen goes in the form of a nitrogen split pair depicted schematically in Fig. 1.7. In this configuration, with C_{2h} symmetry, the two N atoms are located at equivalent distorted interstitial sites and they are bonded to their three *nn* Si atoms. They are separated from each other by two quasi-substitutional Si atoms (atoms 4 and 5 in Fig. 1.7), to which they are bonded. This trivalent bonding is the natural one of N in many molecules, and it makes this N_i–N_i pair, known as a split pair, electrically inactive.

Complexes similar to this N_i–N_i pair, where one of the N atoms is replaced by a C or an O atom, have been identified in silicon and they will be discussed in due time. A second atomic geometry has also been investigated by Jones et al. [41] for the nitrogen split di-interstitial, based on a model proposed as a building block of the platelets in diamonds assuming that they were made of nitrogen, called the Humble model [42]. This latter configuration, with C_{2v} symmetry and consisting of two [011] *nnn* split interstitials, is shown in Fig. 1.8. For N_i–N_i in silicon, the energy of this structure was found to be 0.9 eV larger than the one of Fig. 1.7, but it is found to be the most stable configuration for the diamagnetic self-di-interstitial in diamond [43].

In compound semiconductors, the same kind of pairing as in elemental semi-conductors can be found, with the addition of pairs of identical or different *nnn*

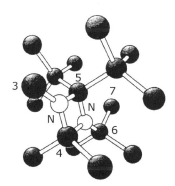

Fig. 1.7 Model of the nitrogen split pair (N_i–N_i) in the silicon crystal. In the perfect crystal, the Si atoms 3, 4, 5, 6, and 7 form a zigzag chain along a $\langle 110 \rangle$ direction in a $\{110\}$ plane. The introduction of the two N atom leads to the breaking of the Si–Si bonds between atoms (3, 4), (4, 5), and (5, 6), and to the bonding of one N to atoms 3, 4, and 5 and of the other N to atoms 4, 5, and 6 (after [41])

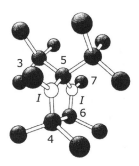

Fig. 1.8 Humble model of a split di-interstitial structure in the diamond lattice. The introduction of two interstitial atoms *I* (in *white*) leads to the breaking of the bonds between atoms (3, 4), (4, 5), (5, 6), and (6, 7), and to the rebonding between atoms 4 and 6. In this model, the two *I* atoms are bonded to five lattice atoms (after [41])

atoms on the same sublattice. For instance, in GaP, N is an isoelectronic FA with a relatively high solubility and at concentrations larger than $\sim 10^{17}\,\mathrm{cm}^{-3}$, it can first form the so-called NN_1 pair due to *nnn* N_P atoms and when increasing [N] to more distant NN pairs, whose spectroscopic properties were investigated by Thomas and Hopfield [44].

Pairing can be due to the interaction between atoms of opposite type, like for the donor–acceptor substitutional *nn* pairs found at high dopant concentrations in silicon [45] or *nnn* B_P–chalcogen pairs in GaP [46]. Another kind of pairing is a mixed one between a substitutional acceptor atom (actually a negative ion) and a positively charged interstitial atom. The $Li_i^+ B_s^-$ pair is an example of such a configuration, but many other pairs involving interstitial TM ions also exist. The mobility of interstitial atoms produced by electron irradiation can also result in pairing: in electron-irradiated silicon containing carbon, evidence of the presence of a mixed $C_i C_s$ pair has been obtained, related to the difference of electronic charge of the two atoms. A limiting case of pairing is that of the interstitial hydrogen molecule found in different semiconductors after hydrogen-plasma treatment, which is a nearly free rotator ([47], and references therein).

1.3.2 *Lattice Distortion and Metastability*

FAs in a crystal can induce a local distortion of the lattice. When they are substitutional, this is caused by the difference between their atomic radii and those of the atoms they replace, and also by their chemical affinity with the surrounding atoms. According to Vegard's law,[4] substitutional atoms having a smaller (larger) atomic

---

[4] Vegard's law is an empirical rule which holds that an approximate linear relation exists between the crystal lattice parameter of an alloy and the concentration of its constituent elements ([48]. See also [49]).

radius than the atom they replace should produce a uniform lattice contraction (expansion) of the crystal proportional to their concentration. With reference to the unperturbed lattice parameter a_0 of cubic crystal, the change Δa of the lattice parameter produced by a concentration N_f of FAs can be expressed as:

$$\frac{\Delta a}{a_0} = \beta_f N_f, \tag{1.2}$$

where β_f is the lattice contraction or expansion coefficient. For substitutional impurities in covalent or partially covalent cubic crystals, the sign and an order of magnitude of β_f can be obtained by replacing Δa by the difference between the covalent radii of the impurity and of the atom of the host crystal it replaces, a by the intrinsic atomic separation in the host crystal and N_f by the number of available sites for impurity sites per unit volume [50]. The value of this coefficient for substitutional C in silicon, calculated from the atomic radii, is $\beta_{C(calc)} = -6.85 \times 10^{-24}\,\mathrm{cm^3\,atom^{-1}}$. For dopants with a covalent radius showing a large difference with that of the atom it replaces, like Tl or Bi in silicon, or P in diamond, this distortion puts a limitation to their solubility. Another kind of local distortion met for substitutional impurities is a lowering of symmetry like the one for isolated N in silicon or diamond, where this atom is displaced along a N–X bond (X is an atom of the crystal) along a $\langle 111 \rangle$ direction. A local distortion can also reduce the symmetry of a centre through the Jahn–Teller effect, as for the atomic vacancy in silicon: this defect should normally display tetrahedral symmetry, but it is lowered to D_{2d}, and this can be detected in the paramagnetic states by the dependence of the ESR spectra on the magnetic field orientation (see for instance [3] of appendix B, p. 467, of this reference). Lattice distortion related to the bond lengths can also occur for an interstitial atom strongly bonded to atoms of the crystal in the *BC* configuration of Fig. 1.6. In this particular case, if the structure remains linear, the two *nn* atoms of the crystal can be pushed out from their equilibrium positions when the lengths of the new bonds exceed the equilibrium *nn* separation.

When the local effect of distortion and the impurity concentration are large, a difference in the average lattice parameter as a function of the impurity concentration can be measured with appropriate X-ray diffraction techniques. In silicon, values of $\beta_C = -6.9 \times 10^{-24}\,\mathrm{cm^3\,atom^{-1}}$ and $\beta_O = 4.4 \times 10^{-24}\,\mathrm{cm^3\,atom^{-1}}$ have been measured for C_s and O_i, respectively [51, 52]. One can note incidentally the good agreement between the measured value of β_C and the value predicted from Vegard's law. With $[O_i] \sim 10^{18}\,\mathrm{cm^{-3}}$ found in most CZ silicon crystals, the value of β_O corresponds to a relative increase of the lattice parameter of 4.4×10^{-6} compared to high-purity FZ silicon.

The lattice distortions induced by substitutional impurities can also be measured locally from the distance between an impurity atom and its nearest neighbours using EXAFS [53]. The results of the EXAFS experiments require sensible interpretations as they do not necessarily follow simple rules like the addition of the covalent radii of the elements involved [54, 55]. Local volume changes of group V and group VI donor atoms in silicon have been obtained indirectly from a comparison between the measured spacings of the electronic absorption lines of these donors with calculated values [56]. Interesting conclusions concerning the change of colour of ruby as a function of the chromium concentrations have been also drawn from EXAFS measurements [57]. Global lattice expansion or contraction can be also measured, for instance, by X-ray diffraction, in doped layers epitaxied on an undoped substrate of the same material from the positive or negative interface stresses, depending on the atomic radius of the doping atom with respect to that of the atom it replaces. In some cases, first-principle calculations have given a good insight of the local distortion induced by a foreign atom [58, 59].

For centres with different charge states (they will be fully discussed in the next section), the distortion can be modified by a change of the electronic density in the vicinity of the centre. The consequence is that a change of the charge state of a centre can produce a local lattice relaxation. It is usual to describe the electronic energy states of these centres as a function of configuration coordinates. When a change of the charge state induces lattice relaxation, the equilibrium configuration coordinates differ in the two states (see Fig. 1.9). It should be noted that, within the same global charge state of a centre, due to differences in electronic densities, lattice relaxation can also occur between the ground state and excited states, with the same consequences regarding the equilibrium configuration coordinates.

A limiting case of distortion is the occurrence of a second atomic configuration of a centre in the same charge state. The idea of this possibility was not obvious at first sight, but experimental results including optical spectroscopy results have led to admit this situation. When two such non-degenerate atomic configurations of a centre coexist, the one with the lowest energy is the stable one and the second is said to be metastable. There is an energy barrier between the two configurations and its value determines the temperature domain of the metastability. The corresponding centre is often said to be bistable. An example of such a bistable centre is the B_s–Si_i (BI) pair, produced in B-doped silicon by electron irradiation at low temperature, which is discussed in Sect. 7.1 in relation with its vibrational absorption.

The change of configuration of a centre induced by its transition into a metastable state produces a lattice distortion which can result in a macroscopic volume change. Transient effects due to the photocreation of electron–hole pairs in n-type GaP and SI GaAs have been attributed to this effect [60].

1.4   Physico-chemical Properties

1.4.1   Solubilities

In many cases, impurities and dopants are introduced in the molten phase, in which they have a definite solubility N_{sol-l}. In the solid phase, near from the melting point, the solubility N_{sol-s} decreases with respect to the liquid phase and the ratio N_{sol-s}/N_{sol-l} is the segregation (or distribution) coefficient, which is usually less than unity. The solubility of impurities in crystals can be considered, in most cases, as the maximum concentration of isolated FAs which can be introduced in the crystal before precipitation, formation of cluster of a mixed compound (e.g. SiC in C-doped silicon) or of an alloy. The solubility of an impurity is conditioned by characteristics such as its atomic radius, electronic structure, site(s) in the crystal, eventual binding energies with the atoms of the crystal, and tendencies to complex or to form pairs. As it generally requires energy to introduce an impurity in a crystal, solubility is a temperature-dependent (thermally activated) process characterized by an activation energy (the heat of solid solution), and for this reason, it is larger near the melting point of the crystal than at RT. When a solubility is mentioned, it is therefore mandatory to know to what temperature it corresponds. For solubilities measured near RT, one must distinguish between the equilibrium solubility, which correspond to a cooling down of the crystals after the introduction of the FAs under conditions close to thermodynamic equilibrium and non-equilibrium solubility. In the second case, the apparent solubility is larger than the equilibrium solubility and the crystal is oversaturated. This situation is met naturally for O_i in CZ silicon. There has been many studies of the solubility of O_i in silicon (see [61] for a review) and the equilibrium solubility $[O_i]_s$ between the melting point (1414°C) and 850°C can be reasonably represented by [62]:

$$[O_i]_s \, (\text{cm}^{-3}) = 9.0 \times 10^{22} \text{exp} \left[-1.52 \, (\text{eV}) / k_B T \right]. \tag{1.3}$$

Within these limits, the solubility calculated using expression (1.3) varies between 2.6×10^{18} and $1.4 \times 10^{16} \, \text{cm}^{-3}$. There is no exact value of the equilibrium solubility of O_i at RT, but it is expected to be lower than the value at 850°C. $[O_i]$ measured at RT in CZ silicon is none the less in the $10^{18} \, \text{cm}^{-3}$ range, showing an oversaturation of this material with O_i. Comparable values have been reported in O-doped germanium [63]. In silicon, nitrogen is not a residual impurity because its solubility is much smaller than that of the other group-V elements and of carbon and oxygen. One of the reasons for this can be the fact that its most stable configuration in silicon is the N pair presented above. As nitrogen doping improves the mechanical properties of silicon, its doping has been actively investigated. Non-equilibrium solubilities of dopants can also be deliberately reached after implantation by solid-phase-epitaxial regrowth, flash or laser anneals of the implanted zone. These fast annealing procedures produce a local out-of-equilibrium situation which is frozen at RT because of the very short cooling-down duration. The metastable solubilities obtained by such annealings can be one order of magnitude larger than the equilibrium solubilities [64]. The problem of the solubility of FAs in a crystal can be complicated by the fact that the same atom can sometimes occupy either interstitial or substitutional sites, like some TMs in silicon. In that case, the apparent solubility is larger for the interstitial location. Globally, the TMs are characterized by a solubility in the 10^{16}–$10^{17} \, \text{cm}^{-3}$ range and by diffusion coefficients significantly larger than those of the substitutional shallow donors and acceptors (see [65] for a review). The interstitial solubility of TMs and of group-IB elements in silicon, also depends on the concentration of substitutional acceptors in the material because, as mentioned before, they can form interstitial–substitutional pairs with these acceptors. This is also true from Li_i, which form pairs with substitutional acceptors, but there seems to be no consensus on the RT solubility of Li_i in FZ silicon. A value $\sim10^{16}\,cm^{-3}$ can be inferred from the conclusions of [66]. As a rule, the solubility of elements of groups II and VI in silicon decreases compared to that of elements of group III and V, with the notable exception of O and H. The solubility of C in III–V compounds has been thoroughly investigated because in these crystals, C is in some cases a pollutant and usually a p-type substitutional dopant with a rather large solubility limit (in the $10^{20}\,cm^{-3}$ range for GaAs). In silicon the solubility of substitutional C at the melting point is $\sim4 \times 10^{17}\,cm^{-3}$ [67] and it is larger in CZ silicon than in FZ silicon because the lattice contraction induced by C_s is compensated by the lattice dilatation induced by O_i. In crystals supersaturated with C, annealing can produce the precipitation of SiC microcrystals [68].

Besides the substitutional/interstitial location of the same FA, other centres can exist where more than one FA are involved, like the *nn* substitutional pairs for chalcogens in silicon or nitrogen in diamond so that in these cases, one must consider a global solubility of the FAs.

In most crystals supersaturated with substitutional or *BC* impurities, these atoms are usually immobile at RT because their diffusion coefficients are small at this temperature. However, when annealing is performed at relatively high temperatures where materials are still saturated with impurities, precipitation or formation of complexes involving impurities can take place because of their migration. This is the case in CZ silicon, where silica precipitates are produced during annealing at 800°C. Attempts to introduce high C concentration in silicon crystals, lead also to the already-mentioned precipitation of SiC microcrystals.

It is clear that a solubility limit can no longer be defined when the host crystal and the impurity are partially or fully miscible. This is the case with Ge in silicon, giving at high Ge concentrations Ge_xSi_{1-x} alloys, and also with most of the group III FAs in III–V compounds, like A*l* in GaAs giving Al_xGa_{1-x} As alloys.

1.4.2   Diffusion Coefficients

At the atomic scale, the diffusion of a FA in a crystal lattice can take place by different mechanisms, and the most common in silicon and germanium are the vacancy and interstitial mechanisms. The interstitial/substitutional or kick-out mechanism, which is an interstitial mechanism combined with the ejection of a lattice atom (self-interstitial) and its replacement by the dopant atom is also met for some atoms like Pt in silicon.

When the constant surface concentration of an impurity with diffusion coefficient D is N_{is}, its concentration $N_i(x, t)$ at depth x from the surface of a plane sample of thickness $d \gg x$ after a diffusion time t is given in the ideal case by:

$$N_{ix} = N_{is}\text{erfc}\frac{x}{2\sqrt{Dt}}, \tag{1.4}$$

where the complementary error function erfc $u = (1 - \text{erf}\, u) = 2/\sqrt{\pi} \int_x^\infty e^{-t^2} dt$. The error function erf u is tabulated p. 142 of the book by Runyan [69].

The temperature dependence of the diffusion coefficient is generally expressed as:

$$D(T) = D_0 \exp[-E_D/k_B T], \tag{1.5}$$

where E_D is an activation energy related to the diffusion mechanism. In Table 1.1 are listed values of D_0 and E_D for a few representative dopants and impurities in silicon. Values of D for many other FAs in silicon can be found in [78].

At a given temperature, in a cubic crystal, an order of magnitude of the atomic jump rate of a diffusing atom between two atomic sites (the number of jumps per unit time) is D/a_0^2, where a_0 is the lattice constant of the crystal. From the above value of D for the P atom in silicon at 1200°C, this jump rate is $\sim 400\,\text{s}^{-1}$ for the P atom. For interstitial atoms like Li_i or BC interstitial O, see Part IV of [79]. The value of the diffusion coefficients of impurities and dopants in semiconductors can be modified by the presence of compensating impurities or of crystal dislocations so that the interpretation of diffusion measurements requires some thought. This is also the case for the diffusion of hydrogen in silicon, whose values depend on the doping type and doping level, because of the possibility of pairing with the dopant atoms [80]. Finally, it must be mentioned that as the diffusing species can be ions, the diffusion coefficient can be modified by an electric field.

Table 1.1 Values of diffusion parameters of some representative foreign atoms in silicon

FA	References	D_0 (cm^{2}s^{-1})	E_D (eV)	D (cm^{2}s^{-1})
Al	[70]	1.8	3.2	2×10^{-11}(1200°C)
P	[71]	5.3	3.69	1.2×10^{-12}(1200°C)
S	[72]	0.047	1.8	3.3×10^{-8}(1200°C)
Li$_i$	[73]	2.5×10^{-3}	0.655	4.4×10^{-9}(300°C)
Cu$_i$	[74]	4.5×10^{-3}	0.39	1.7×10^{-6}(300°C)
Fe$_i$	[75]	9.5×10^{-4}	0.65	1.5×10^{-6}(900°C)
Ti$_i$	[76]	0.0145	1.79	1.1×10^{-8}(1200°C)
Pt	[65]	5.9	3.97	1.5×10^{-13}(1200°C)
O$_i$	[62]	0.13	2.53	2.9×10^{-10}(1200°C)
C	[77]	1.9	3.1	4.7×10^{-11}(1200°C)

D is calculated at the temperature indicated in parentheses using expression (1.5)

1.5   Electrical Activity

1.5.1   Donors and Acceptors

In a semiconductor or an insulator, a centre can be considered as electrically active if it can display more than one electronic charge state. This is the case when electrons or holes are either released from the centres with energies smaller than the band gap or trapped by the centres. This can occur for many FAs, but the substitutional FAs with one more (or less) valence electron than the atom they replace are of particular importance as their bonding with their nns leaves an extra electron not involved in the bonding or is completed by the borrowing of a valence electron of the crystal, leaving a positive hole. The FAs belonging to the first kind are called donors (D) and those belonging to the second kind, acceptors (A). The Coulomb interaction between the positive (negative) foreign ion and the extra electron (hole) has led to compare these centres with hydrogen-like (H-like) atoms. The main difference between these H-like atoms in semiconductors and in atomic physics comes in the former case from their embedding in a crystal matrix with static dielectric constant ε_s, from the mass of the electron, different from the mass m_e of free electron in vacuum, or from the "mass" of the positive hole. In a first step of the modelling of the properties of H-like centres, the relevant masses are replaced by scalar "effective" masses m^*_e or m^*_h, for electrons and holes, respectively. This is an oversimplification, but scalar values of the effective masses can be obtained from a modelling of the RT electrical measurements. The factor scaling the energy of these centres with the Lyman energy spectrum $E_{0n} = R_\infty/n^2$ of H in vacuum is $s = (m^*/m_e)/\varepsilon_s^2$, where m^* is the appropriate effective mass. The energy E_n of the effective-mass centre is thus $1.36 \times 10^4 \, s/n^2$ (meV), where n is the usual principal quantum number. This is the basis of the effective mass theory (EMT), which is presented in detail in Chap. 5 of [7]. Within this approximation, the ground state energy or level for the acceptors in silicon ($m^*_h \cong 0.6m_e$, $\varepsilon_s = 11.7$) is separated from the VB continuum by 60 meV, and for the donors in GaAs $\left(m^*_e \cong 0.07m_e, \, \varepsilon_s = 12.9\right)$ by 5.7 meV from the CB continuum. These values are orders of magnitude of the ionization energies of the shallowest of these centres, known as shallow centres, but the crude assumptions made cannot account for the effect of the chemical nature of the impurity on the ionization energies, which can be important for semiconductors like silicon. The technological importance of the shallow donors or acceptors is that they bind electrons or holes with energies comparable to the RT thermal energy ($\sim$26 meV). Thus, the carriers released at RT by these shallow centres act as a reservoir to control the electrical conductivity of the crystals. Under equilibrium, this release is a thermal process and as the electrons and holes are particles with non-integer spins, their energy distributions follow Fermi–Dirac statistics. At a given temperature T, the concentration of electrons and holes in the continua can be expressed as a function of the chemical potential μ of the semiconductor. In metal physics, the Fermi level E_F is the energy of the electron level whose occupancy probability is 1/2 and it has the same meaning as the more general chemical potential. The term "Fermi level" has been extrapolated from metal to semiconductor physics, despite the fact that in semiconductors, E_F lies in the band gap, where there is only a limited number of discrete allowed states. Here, to comply with the common use, we keep "Fermi level" which is at best a quasi-Fermi level.

At very low temperature, the concentration of free carriers in the continuum is negligible as they are trapped by the ionized impurities of the opposite charge and E_F is close to the energy level E_i of the dominant impurity. This level separates the band gap into two regions: one, between E_i and the relevant band continuum, taken as the energy origin and a second one for energies between E_i and the opposite band continuum. In energy diagrams for single donors (D) or acceptors (A), the zone contiguous to the opposite continuum is noted "$+$" for donors and "$-$" for acceptors as, when E_F lies in this zone, the centre is ionized (D^+ or A^-). In the same spirit, the second zone is noted "0" because when E_F lies in this zone, the centre is neutral (D^0 or A^0).

There are also FAs with two more (or less) valence electrons than the lattice atom they replace, which act as double donors (DD) or acceptors. In group IV semiconductors, this is, for instance, the case for the chalcogen atoms (S, Se, and Te) belonging to group VI of the periodic table, with three charge states DD^0, DD^+ and DD^{++}. $(DD)^0$ can be first ionized to give DD^+, with an ionization energy substantially higher than the one for single donors, and DD^+ can in turn be ionized with an energy approximately two times higher than the one for DD^0. In this aspect, they can be considered as some kind of He-like centres.

The H-like and He-like centres do not show much lattice relaxation between the neutral ground state and ionized states and the ionization energies measured electrically or spectroscopically are about the same. For deep centres, where a substantial lattice relaxation can occur between the ground and ionized states, the energies can be represented approximately in 1D by parabolas as a function of a general lattice (configuration) coordinate representing the lattice relaxation between the two states (Fig. 1.9). The threshold of the optical transitions (optical ionization energy E_{io}) takes place without lattice relaxation while the thermal ionization energy E_{ith} corresponds to an equilibrium configuration and Fig. 1.9 shows that in this particular situation, E_{ith} is smaller than E_{io}. The difference is the Franck–Condon shift. Such a diagram is called a Huang–Rhys diagram and it will be used later with some additions in the discussion of the coupling of the electronic transitions of impurities with phonon modes of the crystal.

These deep centres can still be considered as donors or acceptors, but in that case, a donor is better defined as a hole trap and an acceptor as an electron trap. The trapping and emission efficiencies of these centres are quantified by captures and emission cross-sections for electrons or holes. Some centres with capture cross-sections not too dissimilar for electrons and holes, as substitutional Ag or Au in silicon, are sometimes called amphoteric, by analogy with the chemical terminology. The term amphoteric has also been used to describe the same impurity in different lattice sites, where it can display either acceptor or donor properties (e.g. Si in GaAs). The separation between shallow and deep centres is an oversimplification as there are also centres in between.

Fig. 1.9 Configuration
coordinate Huang–Rhys
diagram of the electronic
energies of an impurity centre
whose lattice equilibrium
configurations in the ground
and ionized states are
represented by configuration
coordinates Q_{gr} and Q_{free}
with different values. The
thermal ionization energy E_{ith}
of such a centre is smaller
than the optical ionization
energy E_{io} by the
Franck–Condon energy E_{FC}

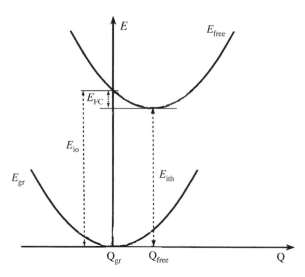

1.5.2   Compensation

In a real semiconductor, more than one kind of donor and acceptor impurities
are usually present at the same time, but to simplify, we start considering a
material containing only one kind of FAs of each type. The one with the highest
concentration N_{maj} is the majority impurity, which determines the electrical type
of the semiconductor and the other is the minority impurity with concentration
N_{min}. The net concentration of active centres able to contribute each a free carrier is
$N_{maj}-N_{min}$ and this comes from the annihilation of a concentration N_{min} of electron–
hole pairs. This situation is called compensation, and it can also arise from the
presence of centres in concentration N_{trap} which can trap carriers from the majority
impurity. The compensation ratio K is usually defined as the ratio N_{min}/N_{maj}.
When the intrinsic concentration of electrons and holes can be neglected, this
net concentration is close to the free carrier concentration measured when these
active centres are thermally ionized, or to the number of neutral centres which
can be spectroscopically detected at low temperature under thermal equilibrium.
Between the low-temperature region where the electron concentration n in a n-type
semiconductor is practically zero and the exhaustion region where it is $N_{maj}-N_{min}$,
the temperature dependence of the electron concentration n released in the CB by
the donor with ionization energy E_i is:

$$n = \frac{N_{maj} - N_{min}}{N_{min}} N_c e^{E_i/k_B T}, \qquad (1.6)$$

where N_c is the effective DoS in the CB. A similar equation holds for the hole
concentration p in the VB in a p-type semiconductor, by replacing N_c by the
effective DoS N_v in the VB. Expression (1.6) shows that for shallow impurities,

E_i can be derived from $n(T)$ and it can be obtained, for instance, from the temperature dependence of the Hall coefficient $R_H = -r/ne$ (the Hall factor $r = \langle \tau^2 \rangle / \langle \tau \rangle^2$ depends on the electron or hole scattering process through their lifetime τ, and in most semiconductors, it is close to $3\pi/8$). An alternative is a measurement of the energy absorption spectrum of the hydrogen-like impurities at low temperature, from which ionization energies can be extrapolated and this method is fully explained in [7].

Compensation reduces the concentration of active majority impurities, but it produces also additional impurity ions of both charges. These ions are the source of the so-called ionized impurity scattering for the majority free carriers and it reduces their lifetime. The electrical conductivity of a crystal is proportional to the number of free carriers and to their electrical mobility, which is in turn proportional to their lifetime. As a consequence, in the extrinsic regime, a high resistivity (or a low value of the carrier concentration measured directly from Hall effect) does not necessarily mean a high purity of the material.

We have mentioned the situation where a dopant atom can locate on two different sites where it behaves either like a donor or an acceptor. For some growth conditions, this possibility can produce what is known as self-compensation, and this occurs indeed for GaAs:Si. Another example of self-compensation is the doping of ZnO with Li: this results in a material with a relatively high resistivity, attributed to the occupancy with comparable probabilities by a Li atom of interstitial sites, where it acts as a donor, and of Zn sites, where it is an acceptor.

The actual compensation in a material is more complex than a simple balance between a majority impurity and a minority impurity as the material usually contains a combination of residual impurities, dopant and deep centres, whose concentrations must be estimated to know the actual degree of compensation in the material. As said before, compensation of majority impurities by adding opposite-type dopant or by the presence of deep centres leaves in the material charged ions, which reduce the lifetime of the free carriers. When the carrier lifetime in a given pure material is known, a lifetime measurement of an unknown sample of this material can inform on the degree of compensation of the sample.

1.5.3 Passivation

In the compensation process, there is only a change in the charge state of the impurity or dopant atom and it is temporarily reversible, for instance, by illumination of the crystal with band gap or above-band gap radiation. This illumination produces electrons and holes that are trapped by the ionized centres. This is of course an out-of-equilibrium situation, which prevails only during illumination.

When studying the interaction of hydrogen plasmas with crystalline silicon surfaces, it was discovered that hydrogen could penetrate in the bulk of the material and decrease its electrical conductivity [81, 82]. What could have been due to a compensation effect revealed itself as a passivation effect where hydrogen interacted chemically with the shallow acceptors in silicon to form electrically-inactive centres. This was reminiscent of older studies which had showed that hydrogen played a role in the passivation of deep centres at the Si/SiO_2 interfaces and later on bulk and interface defects in crystalline silicon, not to mention the role of hydrogen in amorphous silicon. A proof of this interaction with shallow acceptors in silicon was the observation of IR vibrational modes related to hydrogen–acceptor complexes. These complexes were electrically inactive so that they did not contribute to the ionized impurity scattering. This process has naturally been called passivation and it has been observed for many donors and acceptors in semiconductors (see, for instance, [39], and references therein). However, the interaction of hydrogen with impurities in semiconductor crystals is complex and in some cases it can turn electrically inactive impurities into electrically active complexes. Moreover, for double donors or acceptors, it can passivate partially the centre and turn a deep impurity into a shallow donor or acceptor complex. These points are discussed in Chap. 8.

1.5.4 Deep Levels and Negative-U Properties

Many impurities and most defects complexes give electrical levels deep in the band gap, whose excited states cannot be considered as EM states. When ionization of these centres in the continuum takes place, the measured thermal ionization energy is smaller than the optical one because of the Franck–Condon shift. The thermal ionization energies of these deep centres are usually measured by transient capacitance methods like deep-level transient spectroscopy (DLTS) rather than from the temperature dependence of the Hall effect because the former technique is easier to use and more adapted to the problem when deep levels are involved.

In a semiconductor, a neutral double donor (acceptor) atom can release two electrons (holes) into the conduction (valence) band. The ionization energy of the first particle is smaller than that of the second one as the energy of the centre binding two particles is smaller than the one when it binds only one particle. This can be understood considering the Coulomb repulsive energy between the two electrons or the two holes. In defect physics, the energy difference between the second particle and the first one is sometimes called the Hubbard correlation energy (noted U) and in the above example, it is positive. For a centre like a double donor, the band gap of the semiconductor is thus separated into three energy regions by the DD^+/DD^0 and DD^{++}/DD^+ ionization energies of the donor. With the same meaning as that for single donors, the energy region from the CB (E_c) to the DD^+/DD^0 level is noted "0" and the two other ones similarly "+" and "++" in order of increasing energy. For a double acceptor, the sequence would be "0", "−", and "−−" in order of increasing energy with respect to the VB (E_v). Such an ordering characterizes positive-U centres and most of the multicharged centres enter this category. For instance, the already-mentioned divacancy V_2, produced in silicon by electron or neutron irradiation, is stable up to about 250°C, but in an initially p-type material, it can a trap a hole, giving V_2^+. Inversely, in initially n-type material, it can trap one or two electrons to give V_2^- or V_2^{2-}. This defines three energy levels in the band gap: a V_2^{2-}/V_2^- "double acceptor" level, a V_2^-/V_2^0 acceptor level, and a V_2^0/V_2^+ "donor" level. The sequence of the charge states zones of V_2 in the silicon band gap is shown in Fig. 1.10. There is some spread in the values of the energy levels of V_2 in silicon and those given in Fig. 1.10 are those quoted by Svennson et al. [83]. More details on this defect are given in Chap. 4.

It turns out that for a few multicharged centres, the energy is larger with respect to the conduction (valence) band when the number of electrons (holes) bound to the centres is the largest. Following the suggestion of Anderson [84] applied to calchogenide glasses, this can be explained by the gain in energy provided by a large lattice relaxation between the two charge states. With the above convention, the correlation energy U is of these centres is negative and they are known as negative-U centres. Within the double donor model used above, the DD^0/DD^+ level should then be deeper in the band gap than the DD^+/DD^{++} level. In the case of negative-U properties, this ordering as function of energy is no longer possible and when two such DD^+ centres are present, their energy can be minimized by turning them into one DD^0 and one DD^{++} centres, which are the only stable charge states. It looks as if the two energy levels were replaced by a virtual DD^0/DD^{++} one, halfway between DD^0/DD^+ and DD^+/DD^{++}. An example of negative-U centre is off-centre oxygen (O_{oc}) in GaAs, also known as $V_{As}O$. This defect, produced when oxygen is present during the growth of GaAs crystals, contains an off-centre O atom bonded to two next-*nn* Ga atoms. It can be detected by the vibrational absorption of the Ga–O–Ga mode [85] and its absorption spectroscopy is presented in detail in Sect. 7.3. The two stable charge states of the centre are $V_{As}O^-$ and $V_{As}O^+$. As for other negative-U centres, the "forbidden" charge state, here $V_{As}O^0$, is metastable at low temperature and it can be produced by capture by $V_{As}O^+$ of electrons produced by photoionization of other defects. It has been thought for some time that the atomic configuration of this centre was the analogue in GaAs of the VO centre

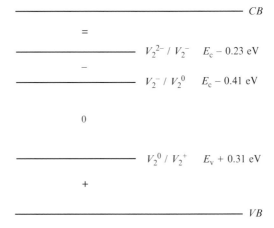

Fig. 1.10 Charge states sequence of the divacancy V_2 in silicon showing the charge state of V_2 when the Fermi level is located in a given energy zone of the band gap. E_c and E_v are the respective positions of the conduction band (*CB*) and valence band (*VB*). The values of the V_2^{2-} and V_2^- energy levels are those quoted by Svennson et al. [83]

in silicon or germanium, with O and V_{As} as the only ingredients. However, on the basis of total energy calculations [86] and of the absence of atomic reorientation, the two other neighbours of V_{As} are likely to be As_{Ga} antisite atoms instead of Ga atoms. Another example of negative-U centre is the one known as interstitial B in silicon discussed in Sect. 1.3.2 in relation with metastability [87, 88].

1.5.5 Transition Metals

TMs are elements with an incomplete d or f electron shell. The 3d TMs go from Sc to Cu, the 4d ones from Y to Ag, and the 5d from Hf to Au [elements of column IB (Cu, Ag and Au) are included in the TMs despite full d shells because of the presence of only one ns electron]. As the electrical properties of the TMs are rather different from those of the impurity elements of columns II, III, IV, V and VI of the periodic table, we present here some facts that are indirectly related to their electrical properties, but can clarify them, despite the fact that they have already been mentioned in this chapter. Compared to usual dopants, TMs are connected to the neighbouring lattice atoms by a rather strong short-range potential and they can therefore give deep levels in the band gap. In group IV crystals, some of them can be found indifferently at substitutional and interstitial sites, with different electrical properties, while others, like Fe, are found only at interstitial site. For instance, the diffusion coefficient is generally much larger at interstitial than substitutional locations, but it must be kept in mind that the measured diffusion coefficients of the TMs depend on several parameters, an important one being the initial doping of the material. The high diffusion coefficients of some TMs ions and their positive charge states result in the formation of pairs between the interstitial TMs and negative acceptor ions as well as with other substitutional impurities. This is a well-documented phenomenon, especially for Fe in silicon [89]. No general rules, however, exist concerning the diffusion coefficients of interstitial TMs in silicon as, although some elements like Fe and Cu show large diffusion coefficients, interstitial Ti, for instance, is a very slow diffuser (Table 1.1). In silicon, substitutional Au is amphoteric, with an acceptor level 0.61 eV above the *VB* and a deep donor level 0.79 eV below the *CB*. In III–V and II–VI crystals, the TMs replace the metal cation. The typical solubilities are in the 10^{16}–10^{18} cm^{-3} range, with some exceptions like Mn or Co, which have much larger solubilities or can even form definite compounds with group VI elements (the band gap of MnTe is ~1.3 eV). The III–V and II–VI alloys with Mn and Co belong to the category of diluted magnetic semiconductors. These materials offer perspectives in the domain of spintronics and their properties are actively investigated [90].

In a chemical description of the III–V and II–VI compounds, the metalloid atoms contribute five or six electrons to the crystal bonding and the metals three or two atoms, respectively. Consequently, the different charge states of the TMs are often described starting from an ionic description of the neutral TM atoms M omitting the electrons involved in the bonding. This is of course an approximation as the above crystals are partially covalent. In III–V and II–VI compounds, the neutral states are therefore noted M^{3+} and M^{2+}, respectively.

The following concerns III–V compounds, but it can be extrapolated, *mutatis mutandis*, to II–VI materials. In the same way as a neutral acceptor atom in silicon can be ionized by trapping an electron, a M^{3+} ion in a III–V compound can trap an electron to give M^{2+}. The energy change from M^{3+} to M^{2+} corresponds to the M^{2+}/M^{3+} acceptor level (M^-/M^0 in the usual shallow impurities description), but in contrast to the shallow impurities description, the level is considered as empty in the M^{3+} state. The same applies for the M^{4+}/M^{3+} (M^+/M^0) donor level. Depending on the initial availability of free electrons or holes in the material, several charge states of the same TM element can be detected: for instance, introducing vanadium in p-type GaP results in the observation of a deep V^{4+}/V^{3+} donor level at $E_v + 0.2\,eV$ while doing the same in n-type material allows a deep V^{3+}/V^{2+} acceptor level at $E_c - 0.58\,eV$ to be observed.

The absolute energy levels of impurities are related to vacuum instead of the valence or conduction band of the host crystal. These absolute values require knowing the *VB* energy of the crystal with respect to vacuum. This latter energy is the work function or photothreshold of the crystal, i.e. the energy Φ to photoionize an electron from the *VB* into vacuum (in GaP, for instance, Φ is $\sim6.0\,eV$, compared to 7.5 eV in ZnS and 5.2 eV in silicon). The absolute energy levels of the impurities are thus obtained by subtracting from Φ the energy levels related to the *VB*. It has been found that for a given TM in compound materials, the absolute energy levels for different host crystal with the same formal composition (i.e. III–V compounds or II–VI compounds) are roughly the same. For instance, the absolute V^{3+}/V^{2+} levels in GaP and GaAs are 4.3 and 4.2 eV, respectively. In semiconductors with a wide band gap like GaP or ZnSe, it is possible to find both the TMs donor and acceptor levels in the band gap, but when the gaps are smaller, only one level lies generally in the band gap. The other level is resonant with the valence band or the conduction band. This is the case with the V^{3+}/V^{2+} acceptor level in InP, which is resonant with the *CB*. This alignment of the TMs energy levels with respect to vacuum is very useful to predict such resonances.

1.5.6 Attracting Isoelectronic Impurities

When a FA from the same column of the periodic table replaces an atom of the crystal, it is in principle electrically inactive because both atoms have the same number of valence electrons. Such a FA is referred to as an isoelectronic atom, and in most cases, it is indeed electrically inactive. In a III–V compound, the substitution of an atom of the group III sublattice by another group III atom does not modify the electrical properties of the compound (at least as long as the foreign group III atom can be considered as an impurity and not as a component of an alloy). We have mentioned the role of the atomic potential in the immediate vicinity of the impurity site. It can occur that, for some isoelectronic impurities or for some electrically inactive complexes, the combination of the atomic potential with the potential produced by the local lattice distortion produces an overall electron- or hole-attractive potential. This potential can bind electrons or hole to the neutral centre with energies much larger than those for shallow electrically active acceptors or donors. In a semiconductor containing such isoelectronic centres, a free exciton (FE) (an electron–hole pair stabilized by Coulomb interaction) produced at low temperature by band gap illumination can therefore be trapped by these centres because of the preferential interaction with the electron (resp. hole) part of the exciton. The hole (electron) part of the exciton is then comparable to a hole (electron) bound to a negatively (positively) charged acceptor (donor) ion. This metastable entity is therefore called a pseudo-acceptor (resp. pseudo-donor). The study of excitons bound to isoelectronic centres in silicon and compound semiconductors (isoelectronic bound excitons or IBE) was actively investigated in the 1980s. In compound semiconductors, one of the best-studied electron-attractive (pseudo-acceptor) is probably N_P in GaP [44]. Isoelectronic substitutional oxygen can also play this role in some II–VI compounds ([91], and references therein). Bi at a P site in GaP and InP seems to be the best-documented hole attracting centres [92, 93]. In silicon, the potential near a C or Ge atom cannot bind an electron or a hole and electrically active isoelectronic centres are complexes like the Be pair at a Si site [94], or some (C, O) complexes in irradiated or annealed CZ silicon which are discussed in Sect. 6.1 of [7].

1.5.7 Lattice Point Defects

The simplest lattice point defect in covalent or mostly covalent crystals is the already-mentioned atomic vacancy V. In silicon it is stable only at low temperature and for a convenient study, it is produced by irradiation below typically 20 K, but in diamond (C_{diam}), it is stable up to ~870 K (~600°C). In n-type silicon, V can trap an electron to become V^-, which is paramagnetic, giving the Si-G2 ESR spectrum. V^- starts migrating at about 70 K. In p-type silicon, there is also an ESR evidence of V^+. The correlation between the charge states of V in silicon and the electronic levels in the band gap is far from simple. The position of the two donor levels V^0/V^+ and V^+/V^{++} are close to the valence band, but the study is complicated by the negative-U character of V^+. The positions of the V^-/V^0 and V^{2-}/V^- acceptor levels have not yet been established, except for the tentative conclusion that the V^{2-}/V^- level is deeper than 0.17 eV from the CB [95]. The electron or neutron irradiation of n- or p-type silicon near RT always lowers the electrical conductivity of the material. This indicates the production by irradiation of lattice defects and impurity-defect centres that act as traps for the donor electrons and acceptor holes. The creation of majority deep traps by h-e irradiation is not a general rule, however, and a counter-example is that of n-type germanium: under electron or neutron irradiation at RT, the electrical conductivity of this material starts first decreasing, but under larger fluences,[5] the conductivity goes p-type and it increases again with increasing fluences. This is attributed to the formation of a V_2–donor complex with acceptor levels ~ 0.1 eV for V_2–P, V_2–As, and V_2–Sb [96].

1.6   Optical Properties

The optical properties of impurities and defects are mainly studied by optical absorption, PL, or Raman scattering usually performed in reflexion geometry. Schematically, in the absorption measurements, the sample is inserted between a radiation source, which can be made monochromatic before or after the sample, and the transmission of the sample is measured as a function of the energy of the radiation by the appropriate radiation detector. In PL, the sample is illuminated by a relatively intense monochromatic radiation source and the radiation emitted by the sample analysed as a function of energy. In the Raman scattering measurements, the sample is excited at oblique incidence with a laser line at a given energy and the scattered radiation is usually analysed at 90° from the direction of the laser beam with the appropriate monochromator. In measurements performed at RT, the lines scattered at energies higher (anti-Stokes spectrum) and lower (Stokes spectrum) than that of the laser line can be detected, but in the Raman measurements performed at low temperature, only the Stokes spectrum is observed. In that latter case, the energies of the Raman transitions are the differences between the energy of the laser line and those of the Stokes spectrum.

The observed spectra can be continuous or discrete, but for identification or structural studies, the discrete spectra bring more information, especially when the lines are sharp. Precious information on the chemical nature and atomic structure of the centres is obtained from isotopic effects. For pure electronic transitions, called zero-phonon lines (ZPL), these isotopic effects are small.

1.6.1   Electronic Studies

Many impurity and defects centres are electrically active, displaying more than one charge state, and electronic absorption of one of these charge states is generally observed. Electronic absorption of the above-described H-like donor or acceptor centres when neutral is observed in the infrared at low temperature. It is charac-terized by a line spectrum converging with increasing energy into a continuous spectrum called the photoionization spectrum. Besides the ionization energies of the centres, the parameters for these spectra are those of the conduction and valence bands of the host crystal for the donors and the acceptors, respectively. For this

---

[5]The fluence of an irradiation is the total number of particles per unit area incident on the sample.

reason and in the absence of accidental resonances with the phonon spectrum of the host crystal, in a given material, the spectra of the different donor centres are very similar, the only difference being the energy region where the spectrum occurs, and this is also true for the spectra of the different acceptor centres. The spectroscopy of these centres, which can also include PL spectra, has been widely investigated and it is presented in details in [7].

The FEs can bind to neutral H-like centres with electron–hole binding energies E_{ex} slightly larger than that of the FEs. The difference is called the localization energy E_{loc} of the resulting bound exciton (BE). Excitons are created by laser illumination of a semiconductor sample at an energy larger than E_g and the study of their radiative recombination by PL has been an active field of the optical spectroscopy of semiconductors [97–99]. The excitons can recombine radiatively by emitting a photon at energy $E_{gx} = E_g - E_{ex}$. However, in indirect-gap semiconductors, conservation of the momentum of the weakly bound electron, comparable to that of a free electron, implies the creation of a lattice phonon of opposite momentum so that the energy of the photon emitted is $E_{gx} - E_{phon}$ where E_{phon} is the energy of the momentum-conserving phonon. In indirect-gap semiconductors, this phonon-assisted process has a significantly larger intensity than the zero-phonon intensity. Besides the phonon-assisted replicas, the recombination of excitons bound to complexes with internal vibration modes can take place with the excitation of some of these modes, producing what are known as vibronic sidebands.

In ionized deep centres, free electrons or holes produced by above-band gap excitation can be trapped in shallow excited states, from which they recombine radiatively to the ground state. This PL process is known as internal luminescence, and it provides a way to measure the ionization energy of the deep centre when the energy of the shallow state is known.

1.6.2 Vibrations of Impurities and Defects

The presence in a crystal lattice of FAs bonded to atoms of the crystal introduces perturbations in the lattice periodicity. These perturbations are due to local changes in the atom masses (in most cases), in the bond lengths and in the force constants between the atoms. When the mass differences and the difference with the crystal bonding are small, the main result is a perturbation of the one-phonon mode distribution of the perfect crystal, with additional features from the local disturbance. One then speaks of resonant modes, which can propagate at some distance of the FAs and give rise to IR absorption below the Raman frequency of the crystal (the maximum one-phonon frequency). The resonant modes have been mainly studied in elemental semiconductors as, contrary to compound materials, the former do not show one-phonon absorption and allow absorption measurements in the vicinity of the Raman frequency [100]. In some compound semiconductors where a phonon gap exists, some impurity vibrational modes can lie in this gap. They are then localized and are known as gap modes. When, for comparable or smaller FAs masses, the bonds with the FAs are stronger than the crystal bonds, the frequencies of the induced modes are above the Raman frequency and they do not propagate in the crystal. Such modes are known as localized vibrational modes (LVMs) and most of them are IR-active. These LVMs are observed whatever the electrical activity of the centres and their study is the only possible spectroscopic technique when the centres are electrically inactive. LVM spectroscopy is a sensitive tool to study the properties of the corresponding centres and there have been many reviews on the subject (see, for instance, [101]). Vibrational spectroscopy is based on the displacement of atomic masses and the LVMs display isotope shifts when the FAs have several natural isotopes with comparable concentrations, or whose concentrations can be adjusted. These isotope shifts either due to the FAs themselves or to their *nns*, are of a great help in the chemical identification of the FAs and in the determination of their local environment. The vibrational properties also lend themselves to many theoretical approaches which assist in determining the atomic form of the centre. It must be added that some of the centres showing LVMs in the near or medium IR can also present low-frequency quasi-rotational modes that give absorption in the far IR, and in this latter domain, phonon spectroscopy, whose principles are briefly described at the end of Sect. 3.1, has also been used. Vibrational absorption spectroscopy is a very general method which requires only a vibrational mode of appropriate symmetry for the existence of a dipole moment, and it allows the detection of electrically active as well as electrically inactive or compensated centres. In the absence of a dipole moment, a change of the polarizability of the bond allows the detection of the vibrational mode by Raman scattering. Electrically active FAs or defects display at least two charge states, and different charge states can change the frequencies of the LVMs. This can be seen clear in the case of centres with levels relatively deep in the band gap, where the bound electron or hole is localized on the centre, like for the oxygen-vacancy centre in silicon, but for centres where the bound carrier is more delocalized, the situation may not be so simple.

One drawback of vibrational spectroscopy is its slightly lower sensitivity compared to electronic absorption, thus requiring higher impurity concentrations (in the best cases, the sensitivity limit is however in the 10^{13}–10^{14} cm^{-3} range), and Raman scattering is still less sensitive.

Small foreign molecules, ions, or free radicals located at an interstitial or more rarely substitutional site, with much less vibrational interaction with the crystal, can also been encountered in semiconductors. The vibration frequencies of such centres are often comparable to the ones in vacuum and they can also display hindered rotation. Ions like OH$^-$ [102] or CN$^-$ [103] are often found in ionic crystals, where they replace an anion, and their dynamical properties can have similarities with the above centres or with molecules in rare gas matrices. We shall see that they can also be found at interstitial sites in compound semiconductors. Another example is molecular hydrogen in semiconductors, whose vibrational properties are presented in Sect. 8.3.

As already mentioned, the coupling between the electronic and vibrational motions (vibronic coupling) results sometimes in the observation of local modes replicas of PL lines. The study of these replicas, or vibronic transitions, can supplement the direct absorption measurements as some of the LVMs associated with the vibronic transitions are usually IR inactive and undetectable in pure vibrational absorption.

1.7   Spin Effects

Electron spin effects are observed for electrically active centres with an odd number of electrons. In charge states with an even number of electrons, the spins are generally paired off. There are however a few cases where a two-electron centre gives a resultant spin S $= 1$, with parallel individual spins. A centre in a charge state with nonzero spin is said to be paramagnetic. Such a centre interacts with an external magnetic field **B** through the magnetic dipole moment of the electron arising from the electron spin and from its angular momentum. For many centres, the angular momentum of the electron is quenched in the ground state so that one can only consider the spin. In a solid, the Zeeman term can then be expressed as [104]:

$$H_{Zee} = \mu_B g \mathbf{SB},$$

where μ_B is the Bohr magneton and g a symmetric tensor whose values g_1, g_2, and g_3 with respect to the principal axes (the g factors) of the g tensor are close to 2. The ground state of a centre with spin S $= 1/2$ is split by the magnetic field into a doublet with $M_S = +1/2$ and $-1/2$ separated by $\mu_B g B$ [for a magnetic field of 1 T and $g \sim 2$, this separation is ~ 30 GHz (~ 0.12 meV)] and a magnetic dipole transition can take place between the two components. We have seen that noncubic centres with different equivalent orientations in a cubic crystal present an orientational degeneracy. When these centres are paramagnetic, the doublet separation depends on the angle between the magnetic field and the main axis of the centres. In classical ESR experiments, the transition between the two levels is induced by the magnetic field of a microwave with a fixed frequency for a critical value of **B**. Practically, **B**, oriented along a high-symmetry axis of the crystal ($\langle 100 \rangle$, $\langle 111 \rangle$, or $\langle 110 \rangle$) is tuned in order to make the splittings of the centres with different orientations coincide with the microwave frequency and this is repeated for different orientations. The variation of the number of resonances for different orientations of **B** allows then the orientational degeneracy of the centre to be determined.

A paramagnetic atom with T_d symmetry should give only one resonance line, but when this atom has a nuclear spin, the electron and nuclear spins can couple by hyperfine interaction, and for a nuclear spin **I**, each electronic spin component splits into 2I $+ 1$ components giving the same number of $\Delta m_I = 0$ resonances. For instance, the ESR spectrum of tetrahedral interstitial Al (**I** $= 5/2$) produced by electron irradiation of Al-doped silicon is an isotropic sextuplet due to transitions between the six nuclear sublevels of each electronic-spin component ([104], and references therein). The electron spin of a centre can also interact with the nuclear spins of neighbouring atoms to give additional structures and this is clearly shown for ^{29}Si atoms ($\mathbf{I} = 1/2$) in Fig. 1.4 of the chapter by Watkins [104]. This shows that the analysis of the hyperfine structure of the ESR spectrum can inform on the chemical nature of the atoms involved in the paramagnetic centre. This can also occur for noncubic centres and the hyperfine structure is superimposed to the orientational structure.

For a given value of $\mathbf{B}$, the energies of the $\Delta m_I = 1$ transitions between the nuclear sublevels of a given electronic spin state are much lower than those between the electronic spin components. Information on the amplitude at different lattice sites in the vicinity of the centre of the wavefunction of the electron whose spin is responsible for the ESR spectrum was obtained by Feher [105] by monitoring the ESR spectrum as a function of the frequencies in the nuclear frequencies range and this technique was called electron nuclear double resonance (ENDOR). Improvements in the sensitivity of ESR can be obtained by using optical or electrical detection methods [106].

All the neutral single donors without d or f electrons have spin 1/2 while the double donors and acceptors have spin 0 in the ground state, but in some excited states, they have spin 1 and optically forbidden transitions between the singlet and triplet states have been observed. The spins of the neutral acceptors in the ground state depend on the electronic degeneracy of the valence band at its maximum. For silicon, the threefold degeneracy of the valence band results in a quasi-spin 3/2 of the acceptor ground state. In the high-spin configuration, the spin in the ground state of TMs with configurations d^n goes from 1/2 to 5/2 for configurations d^1 to d^5 decreasing from 2 to 1/2 from configurations d^6 to d^9.

Useful correlations have been often made between the ESR and optical spectra based on the similarities in the observation conditions, and these correlations have contributed identifying these spectra with specific centres.

1.8 Labelling of the Spectroscopic Signatures

The optical absorption, PL, ESR, and DLTS signatures of a multitude of centres have been reported in semiconductors and insulators, and many solutions to the problem of labelling and identifying these signatures have been used. In the few cases where the identification was established very early, as for the LVMs of interstitial oxygen in silicon, the signatures could be labelled by the atomic configuration of the centre. Such a situation is however met for only a few centres and practical solutions have had to be found.

For instance, a still used nomenclature of the first ZPLs observed in diamond, summarized in Table 3 of [107], was introduced by Clark et al. [108, 109], where prefixes N, GR, R, TR, and H followed by integers correspond to: Naturally occurring (centres occurring in natural diamonds), general radiation (centres induced in all diamonds by irradiation), radiation, type (II) radiation (centres not observed in type-I diamonds), and heat (centres produced by heat treatment preceded by irradiation), respectively.

At the beginning, unknown ESR spectra in silicon were labelled, for instance, by capital letters, like A and E by Watkins et al. [110]. In many cases, when an optical signature which could be correlated with the ESR spectrum was observed, the ESR labelling was kept for the optical labelling. This is illustrated by the Si-A centre in silicon (there are five different A centres in different semiconductor crystals!) named after the ESR spectrum, and characterized by a LVM at $12\,\mu$m, which was identified as the VO defect [111]. Later, in the ESR world, when an unidentified ESR spectrum was first observed, it was often the rule to label it by the initials of the laboratory, city, or country of the discoverers and by an integer corresponding to the order of discovery. With this method, there was no physical relation between the ESR label in different materials as a similarity in the labels did not mean necessarily a similarity in the nature of the centres: for instance, the C-P1 ESR spectrum in diamond, associated with substitutional nitrogen, was reported first by Smith et al. [112] from the IBM Research Laboratory located then at Poughkeepsie (NY). On the other hand, the Si-P1 ESR spectrum in silicon, first reported by Jung and Newell [113] from Purdue University, was later shown to be due to a non-planar five-vacancy cluster.

Levels found in DLTS measurements were also labelled for discrimination purposes, like the labellings of the EBN and ELN levels (N = 1, 2, 3, ...) of electron traps in GaAs by Martin et al. [114], where B was for Bell Telephone Laboratories and L for Laboratoires d'Électronique et de Physique appliquée (LEP), and from which, incidentally, emerged the EL2 centre. More generally, this led to the naming of spectra or lines by letters, acronyms, or energies of the transition. In many cases, even when the atomic structures of the centres have been properly identified, the old labellings can still be found in some publications. This labelling problem is not trivial and it has been seriously considered [115].

In this book, on the basis of the present knowledge, when possible the centres are labelled on the basis of their atomic structure, but the usual label is generally indicated. When the exact structure is not simple and when there exists an acronym for the centre, it has been used. When admitted labels are associated to transitions of a given centre and when there are no physical attributions to these transitions, we use also the current label.

Several lines or bands have also been denoted by their rounded wavelengths, wavenumbers, or energies, and there has been small changes in the quoted values when wavelength calibration improved, due mainly to the change from dispersive spectrometers needing external calibration to FTSs with built-in calibration. It can thus occur in this book that slightly different rounded values are found for the same band or line. As a rule, the values to take in consideration at a given temperature are those given in the tables, with eventually the proper reference; the unquoted spectral values in the tables are unpublished values of the authors.

Another point to consider is that the same centre can produce more than one spectroscopic feature and these features have to be identified adequately. This is often the case with the different vibrational modes of a given centre. When these modes are properly identified, they are generally labelled by the irreducible representation(s) of the symmetry point group of the centre to which they belong. Otherwise, they are denoted by their wavenumber.

References

1. M. Balkanski, R.F. Wallis, *Semiconductor Physics and Applications* (Oxford University Press, Oxford, 2000)
2. P.Y. Yu, M. Cardona, *Fundamentals of Semiconductors*, 3rd edn. (Springer, Berlin, 2001)
3. G. Davies, Optical measurements of point defects in semiconductors, in *Identification of Defects in Semiconductors, Semiconductors and Semimetals*, vol. 51B, ed. by M. Stavola (Academic, San Diego, 1999), pp. 1–92
4. K. Nassau, *The Physics and Chemistry of Color: The Fifteen Causes of Color*, 2nd edn. (Wiley-VCH, New York, 2001)
5. T.S. Moss, *Optical Properties of Semi-conductors* (Butterworths, London, 1969)
6. R.C. Newman, *Infrared Studies of Crystal Defects* (Taylor and Francis, London, 1973)
7. B. Pajot, *Optical Absorption of Impurities and Defects in Semiconducting Crystals. I. Hydrogen-Like Centres* (Springer, Berlin, 2010)
8. V.I. Anisimov, M.A. Korotin, E.Z. Kurmaev, Band-structure description of Mott insulators (NiO, MnO, FeO, CoO). J. Phys. Cond. Matter **2**, 3973–3987 (1990)
9. N.F. Mott, *Metal-Insulators Transitions* (Taylor and Francis, London, 1974)
10. K. Iakoubovskii, G.J. Adriaenssens, Optical characterization of natural Argyle diamonds. Diam. Relat. Mater. **11**, 125–131 (2002)
11. A.T. Collins, The detection of colour-enhanced and synthetic gem diamonds by optical spectroscopy. Diam. Relat. Mater. **12**, 1976–1983 (2003)
12. B.O. Kolbesen, A. Mühlbauer, Carbon in silicon: properties and impact on devices. Solid State Electron. **25**, 759–775 (1982)
13. J. Czochralski, Ein neues verfahren zur messung der kristallisationsgeschwindigkeit. Z. Phys. Chem. **92**, 219–221 (1918)
14. W. Lin, The incorporation of oxygen into silicon crystals, in *Oxygen in Silicon, Semiconductors and Semimetals*, vol. 42, ed. by F. Shimura (Academic, San Diego, 1994), pp. 9–52
15. W. Ulrici, F.M. Kiessling, P. Rudolph, The nitrogen-hydrogen-vacancy complex in GaAs. Phys. Stat. Sol. B **241**, 1281–1285 (2004)
16. V. Clerjaud, D. Côte, C. Naud, M. Gauneau, R. Chaplain, Unintentional hydrogen concentration in liquid encapsulation Czochralski grown III-V compounds. Appl. Phys. Lett. **59**, 2980–2982 (1991)
17. W. Keller, A. Mühlbauer, *Floating-Zone Silicon. Preparation and Properties of Solid State Materials*, vol. 5 (Marcel Dekker, New York, 1981)
18. B. Schaub, J. Gallet, A. Brunet-Jailly, B. Pelliciari, Preparation of cadmium telluride by a programmed solution growth technique. Rev. Phys. Appl. **12**, 147–150 (1977)
19. K. Iakoubovskii, G.J. Adriaenssens, M. Nesladek, Photochromism of vacancy-related centres in diamond. J. Phys. Condens. Matter **12**, 189–199 (2000)
20. A.T. Collins, Spectroscopy of defects and transition metals in diamond. Diam. Relat. Mater. **9**, 417–423 (2000)
21. E.F. Schubert, *Doping in III-V Compound Semiconductors* (Cambridge University Press, Cambridge, 2005)
22. J.W. Cleland, K. Lark-Horovitz, J.C. Pigg, Transmutation produced germanium semiconductors. Phys. Rev. **78**, 814–815 (1950)
23. M. Kato, T. Yoshida, Y. Ikeda, Y. Kitagawara, Transmission electron microscope observation of "IR scattering defects" in as-grown Czochralski Si crystals. Jpn. J. Appl. Phys. **35**, 5597–5601 (1996)

24. F. Spaepen, A. Eliat, New methods for determining the void content of silicon single crystals. IEEE Trans. Instrum. Meas. **48**, 230–232 (1999)
25. H. Yamada-Kaneta, T. Goto, Y. Nemoto, K. Sato, M. Hikin, Y. Saito, S. Nakamura, Vacancies in CZ silicon crystal observed by low-temperature ultrasonic measurements. Physica B **401–402**, 138–143 (2007)
26. G.M. Martin, Optical assessment of the main electron trap in bulk semi-insulating gallium arsenide. Appl. Phys. Lett. **39**, 747–748 (1981)
27. J.C. Bourgoin, H.J. von Bardeleben, D. Stiévenard, Native defects in gallium arsenide. J. Appl. Phys. **64**, R65–R91 (1988)
28. Z.Q. Chen, S. Yamamoto, M. Maekawa, A. Kawasuso, Postgrowth annealing of defects in ZnO studied by positron annihilation, x-ray diffraction, Rutherford backscattering, cathodo-luminescence and Hall measurements. J. Appl. Phys. **94**, 4807–4812 (2003)
29. P. Wagner, J. Hage, Thermal double donors in silicon. Appl. Phys. A **49**, 123–128 (1989)
30. P. Clauws, Oxygen-related defects in germanium. Mater. Sci. Eng. B **36**, 213–220 (1996)
31. C.A.J. Ammerlaan, Shallow thermal donors in c-Si, in *Properties of Crystalline Silicon, EMIS Data Reviews Series No. 20*, ed. by R. Hull (INSPEC, London, 1999), pp. 659–662
32. J.W. Corbett, G.D. Watkins, Silicon divacancy and its direct production by electron irradiation. Phys. Rev. Lett. **7**, 314–316 (1961)
33. G.D. Watkins, The lattice vacancy in silicon, in *Deep Centers in Semiconductors*, 2nd edn., ed. by S.T. Pantelides (Gordon and Breach, New York, 1992), pp. 177–213
34. G.G. DeLeo, W.B. Fowler, G.D. Watkins, Theory of off-center impurities in silicon: substitutional nitrogen and oxygen. Phys. Rev. B **29**, 3193–3207 (1984)
35. L.I. Murin, B.G. Svensson, J.L. Lindström, V.P. Markevich, C.A. Londos, Trivacancy-oxygen complex in silicon: local vibrational mode characterization. Physica B **404**, 4568–4571 (2009)
36. D.J. Chadi, K.J. Chang, Magic numbers for vacancy aggregation in crystalline Si. Phys. Rev. B **38**, 1523–1525 (1988)
37. S.K. Estreicher, J.L. Hastings, P.A. Fedders, The ring-hexavacancy in silicon: a stable and inactive defect. Appl. Phys. Lett. **70**, 432–434 (1997)
38. B. Hourahine, R. Jones, A.N. Safonov, S. Öberg, P.R. Briddon, S.K. Estreicher, Identification of the hexavacancy in silicon with the B^4_{80} optical center. Phys. Rev. B **61**, 12584–12597 (2000)
39. J. Chevallier, B. Pajot, Interaction of hydrogen with impurities and defects in semiconductors. Solid State Phenom. **85–86**, 203–284 (2002)
40. S.R. Wilson, W.M. Paulson, W.F. Krolikowski, D. Fathy, J.D. Gressett, A.H. Hamdi, F.D. McDaniel, Characterization of n-type layers formed in Si by ion implantation of hydrogen, in *Proceedings of Symposium on Ion Implantation and Ion Beam Processing of Materials*, ed. by G.K. Hubler, O.W. Holland, C.R. Clayton, C.W. White (North Holland, New York, 1984), pp. 287–292
41. R. Jones, S. Öberg, F. Berg Rasmussen, B. Bech Nielsen, Identification of the dominant nitrogen defect in silicon. Phys. Rev. Lett. **72**, 1882–1885 (1994)
42. P. Humble, D.F. Lynch, A. Olsen, Platelets defects in natural diamond. II. Determination of structure. Phil. Mag. **52**, 623–641 (1985)
43. J.P. Goss, B.J. Coomer, R. Jones, T.D. Shaw, P.R. Briddon, M. Rayson, S. Öberg, Self-interstitial aggregation in diamond. Phys. Rev. **63**, 195208/1–14 (2001)
44. D.G. Thomas, J.J. Hopfield, Isoelectronic traps due to nitrogen in gallium phosphide. Phys. Rev. **150**, 680–689 (1966)
45. R.C. Newman, R.S. Smith, Local mode absorption from boron complexes in silicon, in *Localized Excitations in Solids*, ed. by R.F. Wallis (Plenum, New York, 1968), pp. 177–184
46. S.R. Morrisson, R.C. Newman, F. Thompson, The behaviour of boron impurities in n-type gallium arsenide and gallium phosphide. J. Phys. C **7**, 633–644 (1974)
47. M. Hiller, E.V. Lavrov, J. Weber, Raman scattering study of H_2 in Si. Phys. Rev. B **74**, 235214/1–9 (2006)
48. L. Vegard, Die Konstitution der Mischkristalle und die Raumfüllung der Atome. Z. Phys. **5**, 17–26 (1921)

49. A.R. Denton, N.W. Ashcroft, Vegard's law. Phys. Rev. A **43**, 3161 (1991)
50. K.G. McQuhae, A.S. Brown, The lattice contraction coefficient of boron and phosphorus in silicon. Solid State Electron. **15**, 259–264 (1972)
51. Y. Takano, M. Maki, Diffusion of oxygen in silicon, in *Semiconductor Silicon 1973*, ed. by H.R. Huff, R.R. Burgess (The Electrochemical Society, Pennington, 1973), pp. 469–481
52. D. Windisch, P. Becker, Silicon lattice parameters as an absolute scale of length for high-precision measurements of fundamental constants. Phys. Stat. Sol. A **118**, 379–388 (1990)
53. J.E. Rowe, F. Sette, S.J. Pearton, J.M. Poate, Local structure of DX-like centers from extended x-ray absorption fine structure. Diffusion and Defect Data Part B. Solid State Phenom. **10**, 283–295 (1990)
54. K.L. Kavanagh, G.S. Gargill III, Lattice strain from substitutional Ga and from holes in heavily doped Si:Ga. Phys. Rev. B **45**, 3323–3331 (1992)
55. S. Wei, H. Oyonagi, H. Kawanami, T. Sakamoto, K. Tamura, N.L. Saini, K. Uosaki, Local structure of isovalent and heterovalent dilute impurities in Si crystal probed by fluorescence X-ray absorption fine structure. J. Appl. Phys. **82**, 4810–4815 (1997)
56. B. Pajot, A.M. Stoneham, A spectroscopic investigation of the lattice distortion at substitutional sites for group V and VI donors in silicon. J. Phys. C Solid State Phys. **20**, 5241–5252 (1987)
57. E. Gaudry, A. Kiratisin, P. Sainctavit, C. Brouder, F. Mauri, A. Ramos, A. Rogale, J. Goulon, Structural and electronic relaxation around substitutional Cr^{3+} and Fe^{3+} in corundum. Phys. Rev. B **67**, 094108 (2003)
58. M. Hakala, M.J. Puska, R.M. Nieminen, First-principle calculations of interstitial boron in silicon. Phys. Rev. B **61**, 8155–8161 (2000)
59. D. Sasireka, E. Palanyandi, K. Yakutti, Study of local lattice relaxation of substitutional impurities in silicon and germanium. Int. J. Quant. Chem. **99**, 142–152 (2004)
60. T. Sugiyama, K. Tanimura, N. Itoh, Direct measurement of transient macroscopic volume change induced by generation of electron-hole pairs in GaP and GaAs. Appl. Phys. Lett. **58**, 146–148 (1990)
61. B. Pajot, Solubility of O in silicon, in *Properties of Crystalline Silicon*, EMIS Data Reviews Series No. 20, ed. by R. Hull (INSPEC, London, 1999), pp. 488–491
62. J.C. Mikkelsen Jr., The diffusivity and solubility of oxygen in silicon. Mater. Res. Soc. Symp. Proc. **59**, 19–30 (1986)
63. W. Kaiser, C.D. Thurmond, Solubility of oxygen in germanium. J. Appl. Phys. **32**, 115–118 (1961)
64. R. Duffy, T. Dao, Y. Tamminga, K. van der Tak, F. Roozeboom, E. Augendre, Group III and V impurity solubilities in silicon due to laser, flash, and solid-phase-epitaxial-regrowth. Appl. Phys. Lett. **89**, 071915 (2006)
65. W. Schröter, M. Seibt, Solubility and diffusion of transition metal impurities in c-Si, in *Properties of Crystalline Silicon*, EMIS Data Reviews Series No. 20, ed. by R. Hull (INSPEC, London, 1999), pp. 543–560
66. R.C. Young, J.W. Westhead, J.C. Corelli, Interaction of Li and O with radiation-produced defects in Si. J. Appl. Phys. **40**, 271–278 (1969)
67. T. Nozaki, Y. Yatsurugi, N. Akiyama, Y. Endo, Y. Makide, Behaviour of light impurity elements in the production of semiconductor silicon. J. Radioanal. Chem. **19**, 109 (1974)
68. A.R. Bean, R.C. Newman, The solubility of carbon in pulled silicon crystals. J. Phys. Chem. Solids **32**, 1211–1219 (1971)
69. W.R. Runyan, *Silicon Semiconductor Technology, Texas Instruments Electronic Series* (McGraw Hill, New York, 1965)
70. W. Rosnowski, Aluminium diffusion into silicon in an open tube high vacuum system. J. Electrochem. Soc. **125**, 957–962 (1978)
71. J.S. Makris, B.J. Masters, Phosphorus isoconcentration diffusion studies in silicon. J. Electrochem. Soc. **120**, 1252–1255 (1973)
72. F. Rollert, N.A. Stolwijk, H. Mehrer, Diffusion of sulfur-35 into silicon using an elemental vapor source. Appl. Phys. Lett. **63**, 506–508 (1993)

73. S.J. Pearton, Alkali impurities (Na, K, Li) in c-Si, in *Properties of Crystalline Silicon, EMIS Data Reviews Series No. 20*, ed. by R. Hull (INSPEC, London, 1999), pp. 593–595
74. A. Mesli, T. Heiser, Interstitial defect reactions in silicon: the case of copper. *Defect and Diffusion Forum 131–132* (Trans Tech, Liechtenstein, 1996), p. 89
75. T. Isobe, H. Nakashima, K. Hashimoto, Diffusion coefficient of interstitial iron in silicon. Jpn. J. Appl. Phys. **28**, 1282–1283 (1989)
76. S. Hocine, D. Mathiot, Titanium diffusion in silicon. Appl. Phys. Lett. **53**, 1269–1271 (1988)
77. R.C. Newman, J. Wakefield, The diffusivity of carbon in silicon. J. Phys. Chem. Solids **19**, 230–234 (1961)
78. R. Hull (ed.), *Properties of Crystalline Silicon, EMIS Data Reviews Series No. 20* (INSPEC, London, 1999)
79. H. Mehrer, *Diffusion in Solids – Fundamentals, Methods, Materials, Diffusion-Controlled Processes* (Springer, Berlin, 2007)
80. C. Herring, N.M. Johnson, Hydrogen migration and solubility in silicon, in *Hydrogen in Semiconductors*, ed. by J.I. Pankove, N.M. Johnson, *Semiconductors and Semimetals*, ed. by R.K. Willardson, A.C. Beer (Academic, Boston, 1991), pp. 225–350
81. C.T. Sah, J.Y.C. Sun, J.J.T. Tzou, Deactivation of the boron acceptor in silicon by hydrogen. Appl. Phys. Lett. **43**, 204–206 (1983)
82. J.I. Pankove, D.E. Carlson, J.E. Berkeyheiser, R.O. Wance, Neutralization of shallow acceptor levels in silicon by atomic hydrogen. Phys. Rev. Lett. **51**, 2224–2225 (1983)
83. J.H. Svennson, B.G. Svennson, B. Monemar, Infrared absorption studies of the divacancy in silicon: new properties of the singly negative charge state. Phys. Rev. B **38**, 4192–4197 (1988)
84. P.W. Anderson, Model for the electronic structure of amorphous semiconductors. Phys. Rev. Lett. **34**, 953–955 (1975)
85. H.Ch. Alt, Experimental evidence for a negative-U center in gallium arsenide related to oxygen. Phys. Rev. Lett. **65**, 3421–3424 (1990)
86. M. Pesola, J. von Boehm, V. Sammalkorpi, T. Mattila, R.M. Nieminen, Microscopic structure of oxygen defects in gallium arsenide. Phys. Rev. B **60**, R16267–R16270 (1999)
87. J.R. Troxell, G.D. Watkins, Interstitial boron in silicon: a negative-U system. Phys. Rev. B **22**, 921–931 (1980)
88. R.D. Harris, J.L. Newton, G.D. Watkins, Negative-U defect: interstitial boron in silicon. Phys. Rev. B **36**, 1094–1104 (1987)
89. A.A. Istratov, H. Hielsmair, E.R. Weber, Iron and its complexes in silicon. Appl. Phys. A **69**, 13–44 (1999)
90. T. Dietl, A ten-year perspective on dilute magnetic semiconductors and oxides. Nat. Mater. **9**, 966–974 (2010)
91. K. Akimoto, H. Okuyama, M. Ikeda, Y. Mori, Isoelectronic oxygen in II-VI compounds. Appl. Phys. Lett. **60**, 91–93 (1992)
92. P.J. Dean, R.A. Faulkner, Zeeman effect and crystal-field splitting of excitons bound to isoelectronic bismuth in gallium phosphide. Phys. Rev. **185**, 1064–1067 (1969)
93. A.M. White, P.J. Dean, K.M. Fairhurst, W. Bardsley, B. Day, The Zeeman effect in the spectrum of exciton bound to isoelectronic bismuth in indium phosphide. J. Phys. C Solid State Phys. **7**, L35–L39 (1974)
94. M.L.W. Thewalt, D. Labrie, T. Timusk, The far infrared spectra of bound excitons in silicon. Solid State Commun. **53**, 1049–1054 (1985)
95. G.D. Watkins, Vacancies and interstitials and their interactions with other defects in silicon, in *Proceeding of Third International Symposium on Defects in Silicon, Electrochem. Soc. Symp. Proc.*, vol. 99–1, ed. by T. Abe, W.M. Bullis, S. Kobayashi, W. Lin, P. Wagner (Electrochemical Society, Pennington, 1999), pp. 38–52
96. T.V. Mashovets, V.V. Emtsev, Point defects in germanium, in *Inst. Phys. Conf. Ser. No. 23* (The Institute of Physics, Bristol, 1975), pp. 103–125
97. P.J. Dean, J.R. Haynes, W.F. Flood, New radiative recombination processes involving neutral donors and acceptors in silicon and germanium. Phys. Rev. **161**, 711–729 (1967)

98. G. Davies, The optical properties of luminescence centers in silicon. Phys. Rep. **176**, 83–188 (1989)

99. B. Monemar, U. Lindefelt, W.M. Chen, Electronic structure of bound excitons in semiconductors. Physica B & C **146**, 256–285 (1987)

100. J.F. Angress, A.R. Goodwin, S.D. Smith, A study of the vibrations of boron and phosphorus in silicon by infrared absorption. Proc. Roy. Soc. Lond. A **287**, 64–68 (1965)

101. M.D. McCluskey, Local vibrational modes of impurities in semiconductors. J. Appl. Phys. **87**, 3593–3617 (2000)

102. C.K. Chau, M.V. Klein, B. Wedding, Photon and phonon interactions with OH^- and OD^- in KCl. Phys. Rev. Lett. **17**, 521–525 (1966)

103. R.C. Spitzer, W.P. Ambrose, A.J. Sievers, Observation of persistent hole burning in the vibrational spectrum of CN in KBr. Phys. Rev. B **34**, 7307–7317 (1986)

104. G.D. Watkins, EPR and ENDOR studies of defects in semiconductors, in *Identification of Defects in Semiconductors, Semiconductors and Semimetals*, vol. 51A, ed. by M. Stavola (Academic, San Diego, 1998), pp. 1–43

105. G. Feher, Electron spin resonance experiments on donors in silicon. I. Electronic structure of donors by the electron nuclear double resonance technique. Phys. Rev. **114**, 1219–1244 (1959)

106. J.M. Spaeth, Magneto-optical and electrical detection of paramagnetic resonance in semiconductors, in *Identification of Defects in Semiconductors, Semiconductors and Semimetals*, vol. 51A, ed. by M. Stavola (Academic, San Diego, 1998), pp. 45–92

107. J. Walker, Optical absorption and luminescence in diamond. Rep. Prog. Phys. **42**, 1605–1659 (1979)

108. C.D. Clark, R.W. Ditchburn, H.B. Dyer, The absorption spectra of natural and irradiated diamond. Proc. R. Soc. Lond. A **234**, 363–381 (1956)

109. C.D. Clark, R.W. Ditchburn, H.B. Dyer, The absorption spectra of irradiated diamonds after heat treatment. Proc. Roy. Soc. Lond. A **237**, 75–89 (1956)

110. G.D. Watkins, J.W. Corbett, R.M. Walker, Spin resonance in electron irradiated silicon. J. Appl. Phys. **30**, 1198–1203 (1959)

111. J.W. Corbett, G.D. Watkins, R.M. Chrenko, R.S. McDonald, Defects in irradiated silicon. I. Infrared absorption of the Si-*A* center. Phys. Rev. **121**, 1015–1022 (1961)

112. W.V. Smith, P.P. Sorokin, I.L. Gelle, G.J. Lasher, Electron-spin resonance of nitrogen donors in diamond. Phys. Rev. **115**, 1546–1552 (1959)

113. W. Jung, G.S. Newell, Spin-1 centers in neutron-irradiated silicon. Phys. Rev. **132**, 648–662 (1963)

114. G.M. Martin, A. Mitonneau, A. Mircea, Electron traps in bulk and epitaxial GaAs crystals. Electron. Lett. **13**, 191–192 (1977)

115. F. Bridges, G. Davies, J. Robertson, A.M. Stoneham, The spectroscopy of crystal defects: a compendium of defect nomenclature. J. Phys. Cond. Matter **2**, 2875–2928 (1990)

Chapter 2
Bulk Optical Absorption

2.1 Introduction

Observations of the absorption of impurity centres in semiconductors and insulators are limited to the domains of transparency of these materials, which are in turn determined by their intrinsic electronic and vibrational absorptions. All these materials are characterized by an energy gap E_g above which incident photons are strongly absorbed to produce transitions from the valence band to the conduction band. This intrinsic absorption extends from the band gap E_g (see Appendix C) to the far UV and it precludes absorption measurements of impurities and defects in this spectral domain, but in this domain, the intrinsic absorption coefficient can be derived from reflectivity measurements (see [1]). A thorough treatment of the bulk absorption of semiconductors can be found in the books of Yu and Cardona [2] and Balkanski and Wallis [3]. At energies lower than E_g, intrinsic absorption of lattice phonon modes occurs in the mid-IR. In compound materials, one-phonon absorption takes place and this is a strong effect producing, for usual thicknesses, a local opacity of the materials at the energy concerned. At higher and lower energies, multiphonon sum or difference absorption processes involving two or three lattice phonons reduces the transparency of compound as well as elemental crystals, but their intensities decreases with temperature, allowing the measurement of the absorption of impurities and defects, and there are still large spectral domains of transparency without any intrinsic absorption.

The optical properties of semiconductors can be described by a complex dielectric constant whose real and imaginary parts are $n^2 - k^2$ and $2nk$, respectively, where n is the refractive index and k the extinction coefficient of the material. The refractive index is a quantity which is integrated over the whole spectral absorption of the material, and the contribution of the intrinsic absorption is predominant. In contrast, the extinction coefficient k is defined at a specific energy, and practically, one uses the absorption coefficient $K = 2\omega k/c$, where ω is the angular frequency or pulsation of the radiation. The reflectivity, also called reflection coefficient, of the

B. Pajot and B. Clerjaud, *Optical Absorption of Impurities and Defects in Semiconducting Crystals*, Springer Series in Solid-State Sciences 169, DOI 10.1007/978-3-642-18018-7_2, © Springer-Verlag Berlin Heidelberg 2013

intensity of an electromagnetic wave propagating in vacuo and incident normal to the plane polished surface of such crystals is:

$$R = \frac{(n-1)^2 + k^2}{(n+1)^2 + k^2} \qquad (2.1)$$

and at energies above E_g, R is close to unity because of the large value of k. Inversely, whenever k is small, R can be taken as $(n-1)^2/(n+1)^2$.

When the direction of unpolarized radiation incident on a nonabsorbing or weakly absorbing plane polished sample makes an angle i less than 90°, the reflected part of the beam is partially polarized perpendicularly to the plane of incidence, defined by the propagation vector of the radiation and the normal to the sample surface. For an angle of incidence i_B known as the Brewster angle and defined by tg $i_B = n$, where n is the refractive index of the sample for the incident radiation, defined as Brewster's incidence, the reflected beam is fully polarized perpendicularly to the plane of incidence (s-polarization) and it consists in the s-polarized part of the incident radiation. Thus, if a sample is illuminated at Brewster's incidence with radiation polarized parallel to the plane of incidence (p-polarization), there is no reflection loss at the interface. Properties of rotable polarizers for the near IR using polarized reflection on silicon mirrors at Brewster's incidence ($i_B \sim 73°$ for silicon) have been described in [4].

In the transparency regions, the transmission of unpolarized radiation as a function of the wavelength λ through a sample with plane parallel polished sides with reflectivity R produces a channelled spectrum consisting of sinusoidal equal-thickness interference fringes when the measurement is performed with a spectral bandpass smaller than the fringes spacing. For radiation incident normal to the surface of the sample, the theoretical contrast T_{max}/T_{min} of the transmission fringes is $(1+R)^2/(1-R)^2$. The transmission maxima and minima correspond to constructive and destructive interferences, respectively, between beams transmitted with increasing path differences. Under normal incidence, the path difference between two adjacent extrema is $2nd$, where d is the sample thickness. For adjacent transmission maximums at wavelengths λ_1 and $\lambda_2(\lambda_1 > \lambda_2)$, $2nd = N\lambda_1 = (N+1)\lambda_2$, where N is the interference order. From this, one derives $2nd = \lambda_1\lambda_2/(\lambda_1 - \lambda_2)$ and a value of n can be obtained by measuring the wavelengths of adjacent maximums [4], but it must be realized that a linear contribution of λ to the refractive index cannot be detected by this method unless similar measurements are performed at significantly longer wavelengths. When the spectral bandwidth used in spectroscopic measurements is larger than the spacing of the interference fringes, they are averaged out. However, the existence of multiple surface reflections must still be taken into account and, from the summation of a geometric series, the average transmittance T of a moderately absorbing sample under normal incidence can be calculated to be:

$$T = \frac{(1-R)^2 u}{1 - R^2 u^2}, \qquad (2.2)$$

where u $=\exp[-Kd]$, and similar expressions can be derived for the reflectance R and the absorbance A.

A value of n can also be obtained by measuring the dependence of the angle of incidence i of the intensity on the reflected part of a p-polarized incident beam. At Brewster's incidence, the contribution of p-polarized radiation to the reflected beam is zero and there is no reflection.

2.2 Lattice Absorption

2.2.1 Fundamental Lattice Absorption

Within the harmonic approximation, a crystal made of N atoms or ions can be seen as a set of 3N independent oscillators and the contribution to the total energy of a particular normal mode with angular frequency $\omega_s(\mathbf{q})$ is:

$$(\mathrm{n_{ks}} + 1/2)\hbar\omega_s\,(\mathbf{q})\,, \tag{2.3}$$

where $\mathrm{n_{ks}}$ can take any positive integer value or 0. The analogy with the normal modes of the radiation field in a cavity has led to the name "phonons" for the corresponding excitations of harmonic crystal, and this name is extended to situations involving anharmonicity.

The periodic pattern of displacement of the atoms about their equilibrium positions can be characterized by a wavelength λ, which can take any value between infinity (actually the average size of the crystal) and the lattice constant a_0 (for simplicity, we consider a cubic crystal). In a given direction of propagation of the lattice deformation, it is more convenient to use the propagation vector or wave vector $\mathbf{q}$, with amplitude $\mathbf{q} = 2\pi/\lambda$. This vector has the periodicity of the reciprocal lattice and the study of the $\mathbf{q}$-dependent physical quantities can be restricted to the first Brillouin zone (BZ) of the crystal. The fundamental energy spectrum of the crystal is determined by the angular frequency ω of the individual atoms or ions of crystals as a function of the propagation vector $\mathbf{q}$. These vibrational modes are multivalued functions of $\mathbf{q}$ that can be characterized by two kinds of dispersion curves. The ones with angular frequency (or energy) 0 at $\mathbf{q} = 0$ (long wavelengths) show a nearly linear behaviour near the origin and the proportionality coefficients are the sound velocities in the crystals. For this reason, they are called acoustic (A) branches and they correspond to neighbouring atoms vibrating in phase near $\mathbf{q} = 0$. The energy spectrum of crystals with more than one atom per unit cell displays also dispersion curves with a maximum of energy at $\mathbf{q} = 0$, corresponding to vibrational modes where neighbouring atoms have opposite displacements. When these atoms are different, the resulting first-order dipole moment gives rise to optical absorption and the corresponding dispersion curves are therefore called optic (O) branches. Along the main symmetry directions of the crystals, one distinguishes between longitudinal modes (LA and LO) where the atom displacements are parallel to **q**, and transverse modes (TA and TO) where they are perpendicular to **q**. For random orientations, the distinction between pure longitudinal or transverse modes is generally no longer valid. There are two transverse modes corresponding to the propagation of the atomic motion along mutually perpendicular axes and they can be degenerate or not, depending on the symmetry of the branches in the crystal considered. As already mentioned, because of the lattice periodicity of the crystals, the dispersion curves are studied for propagation vectors lying only in the first BZ of the crystal.

In elemental (homonuclear) crystals with cubic symmetry, the LO and TO branches are degenerate at **q** = 0 (the Γ point of the BZ) and the phonons at that point are noted O(Γ). The situation is different in compound crystals, where the energy of the LO branch is larger than that of the TO branch. The reason for this difference in compound crystals comes from the contribution of an electric field effect to the restoring forces, induced by the electric dipole moments due to the vibration of atoms of different kinds, and it can be shown that at **q** = 0:

$$\omega^2 \, (\text{LO}) = \frac{\varepsilon_s}{\varepsilon_\infty} \omega^2 \, (\text{TO}), \tag{2.4}$$

where ε_s is the dielectric constant at frequencies below that of the TO(Γ) mode, and ε_∞ the dielectric constant above that of the LO(Γ) mode. This expression is known as the Lyddane–Sachs–Teller relation [5].

The dispersion curves of phonons in gallium phosphide (sphalerite structure), represented in Fig. 2.1, show the separation between the LO and TO modes at the Γ point. The square of the ratio of their frequencies is ~1.20, not far from the ratio $\varepsilon_s / \varepsilon_\infty$ (1.22), in agreement with Lyddane–Sachs–Teller relation.

The abscissa correspond to the phonon wave vector, and at the boundary limits of the BZ it is ~π / a_0, where a_0 is the lattice parameter.

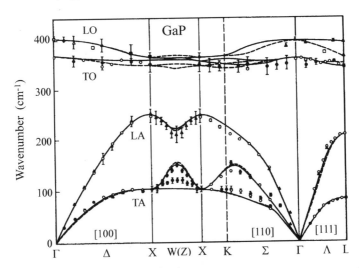

Fig. 2.1 Phonon dispersion curves of gallium phosphide along the main symmetry directions calculated with an 11-parameter rigid ion model. *Solid lines* are calculated to fit neutron diffraction data (*full symbols*) from [6], and the *dashed lines* to fit the data of [7] (after [8])

Table 2.1 RT energies (cm^{-1}) of zone-centre (Γ point) phonons in some cubic and hexagonal (wurtzite) semiconducting and insulating crystals with natural isotopic composition, compiled from literature

Cubic	TO	LO	Cubic	TO	LO	Hexagonal	TO(E_2)	TO(E_1)	LO(A_1)
C_{diam}	1332.4		GaSb	223.6	232.6	$2H$-SiC	764	799	968
SiC	796.2	972.2	InP	303.3	344.5	AlN	656	670	890
Si	520.8[c]		InAs	217.3	238.6	w-GaN[a]	569.2	560.0	739.3
Ge	301		InSb	179.1	190.4	InN	488	476	586
c-BN	1056	1306	MgO	402	718	ZnO	439	379	577
AlAs	361.7	403.7	c-ZnS	271	352	w-ZnS	274	274	352
AlSb	318.7	340.0	ZnSe	213	253	CdS[b]	256	243	305
GaN	552	739	ZnTe	177	207	w-CdSe	172	172	210
GaP	367.3	403.0	CdTe	140	169				
GaAs	268.3	291.8	CaF$_2$	261	482				

The values chosen are obtained from Raman scattering measurements. Slightly different values are reported from neutron scattering measurements. For the wurtzite-type crystal, the phonon symmetry is indicated in *brackets*

[a] At 10 K
[b] At 25 K
[c] [11]

In binary crystals with large differences between the masses of the two atoms (e.g. GaP, InP, or AlSb), the frequencies of the LA phonons at the BZ boundaries are significantly smaller than those of the TO phonons. This is clearly observed in Fig. 2.1 for GaP, and a frequency gap occurs, which extends between $\sim$250 and $330\,\text{cm}^{-1}$, referred to as the phonon gap. In InP, this gap extends from $\sim$168 to $313\,\text{cm}^{-1}$ [9], and in AlSb from $\sim$150 to $280\,\text{cm}^{-1}$ [10].

Table 2.1 gives values of the frequencies of optical phonons in some semiconducting and insulating materials with cubic and hexagonal structures. For the ones with cubic structure, these values are those of the zone-centre phonons. For the crystals with the wurtzite structure, the different phonons are usually denoted by the *IR*s of the C_{6v} point group and the strongest Raman lines in the usual scattering geometry are produced by the A_1 LO phonon at the zone centre along the c axis and by the E_2 TO folded phonon (see Table B.4 of Appendix B). When two structures of the same compound exist, the frequency of the E_1 zone-centre TO phonon of the wurtzite-type crystals is relatively close to the one of the TO(Γ) phonon.

In w-GaN, the comparison of phonon dispersion curves obtained by inelastic X-ray scattering with *ab initio* calculations and Raman scattering data have allowed to attribute two zone-centre phonons with B_1 symmetry, at 329 and $692\,\text{cm}^{-1}$, to silent modes, which are neither IR nor Raman active [12].

A temperature decrease produces generally a decrease of the lattice spacing and a corresponding moderate increase of the phonon modes frequencies.[1] The

---

[1] In a temperature domain below $\sim$50 K, a temperature decrease can result in an increase of the lattice spacing, as in silicon or in some sphalerite-type crystals (see for instance, [13]).

LO(Γ) phonon frequency of GaAs thus increases from 291.8 to 294.5 cm^{-1} between RT and LHeT. A hydrostatic pressure reduces also the lattice spacings of the crystals and one of the consequences is an increase of the phonon modes with pressure [14, 15]. For InP, the increases of the RT frequencies of the TO and LO modes between 0 and 9.6 GPa extrapolated from the data of [14] are $\sim$50 and 46 cm^{-1}, respectively. The variation of the phonon frequencies with isotopic composition has been measured in several semiconductors and the shifts observed can generally be accounted for by considering a virtual crystal with an averaged mass corresponding to the isotopic composition (the virtual crystal approximation or VCA). For instance, extensive results for C_{diam} were given by Hass et al. [16] and by Vogelgesang et al. [17], and they show a variation of the O(Γ) frequency from 1333 to $\sim$1282 cm^{-1} between quasimonoisotopic (qmi) $^{12}C_{diam}$ and $^{13}C_{diam}$ at RT. However, these results show also a departure from the VCA. A general account of the subject can be found in the review by Cardona and Thewalt [18]. The O(Γ) frequencies recently measured at RT in qmi ^{28}Si, ^{29}Si, and ^{30}Si are 521.4, 512.3, and 503.9 cm^{-1}, respectively, [11]. The low-temperature frequencies of some optical and acoustical lattice phonons can be obtained from the observation of phonon replicas on the low-energy side of zero-phonon (no-phonon) lines observed by PL.

The primitive cells of the nH and 3nR SiC polytypes contain n formula (Si–C) units and the unit cell of the polytypes along the c-axis is n times larger than that of cubic 3C SiC. The BZ of the corresponding polytype is thus reduced in the $\Gamma - L$ direction by a factor 1/n [19]. One then speaks of folded BZ and some of the folded acoustical phonons with non-zero frequencies at the zone centre are IR- and Raman-active. Their absorptions, with lines whose FWHMs is $\sim$0.03 cm^{-1} or less at LHeT, have been reported in the 160–260 cm^{-1} range for the 6H and 15R SiC polytypes ([20], and references therein).

The creation of an optical phonon by photon absorption requires the coupling of electromagnetic radiation with a dipole moment. For elemental crystals, the two neighbouring atoms are the same and there is no first-order dipole moment, hence no one-phonon absorption, but the opposite displacement of the atoms results in a change of the polarizability of the crystals. This change can be detected by Raman scattering with a frequency shift corresponding to the O(Γ) frequency, also known as the Raman frequency of the crystal. This is also true for compound crystals with the sphalerite [21] and wurtzite structure, and the Raman scattering of the LO and TO modes is detected in these crystals. In the crystals with the NaCl structure, however, no first-order Raman scattering of the optical modes is observed because of the inversion symmetry of this structure.

The one-phonon absorption can be measured in compound crystal films. From momentum conservation, the momentum of the created phonon must be opposite to that of the absorbed photon, and as the photon momentum is very small compared to the boundary limits of the BZ of the crystal, the energy of the created phonon can be taken as that at the Γ point of the BZ. At normal incidence, the photons can only couple with the TO(Γ) phonons.

The TO(Γ) mode absorption in compound crystals is very large and the refractive index of the crystal near TO frequencies becomes complex, with an imaginary part corresponding to absorption. The dispersion curves of Fig. 2.2 are obtained from a classical treatment of the dielectric constant [23], where γ results from a damping term added to the equation of motion of an anion and cation pair with given reduced mass and effective charge, vibrating at frequency ω_0 (corresponding to the TO mode) in a cubic crystal in the presence of a transverse electromagnetic field.

The high value of the extinction coefficient k results in a nearly metallic reflectivity in this spectral region. When the direction of the incident radiation beam is oblique with respect to the surface of the compound crystal film, it has been demonstrated that when the electric vector is polarized in the plane of incidence of the film, the absorption of the LO mode can also be observed [24]. This technique has been used to measure the frequencies of the TO and LO modes for several semiconductors and semiconducting alloys at LHeT [25].

The study of the one-phonon density of states of crystals has shown the existence of singularities corresponding to critical points (CPs) located within or at the surface of the BZs along particular directions (the BZs for the diamond and sphalerite structures are the same as the one shown in Fig. B.2 of Appendix B). They arise from the topology of the $\omega_t(\mathbf{q})$ dispersion curves, where index "t" refers to a given phonon branch. It can be shown that the density of vibrational state $g(\omega)$ can be written:

$$g(\omega) \propto \sum_t \int_{S_t(\omega)} \frac{dS_t}{\left| \nabla_\mathbf{q} \omega_t(\mathbf{q}) \right|},\tag{2.5}$$

where S_t is the surface in the BZ for which $\omega_t(\mathbf{q}) = \omega$. These CPs are those for which $\nabla_\mathbf{q}\omega(\mathbf{q}) = 0$ and for the diamond-like crystals, they correspond to points X, L, K, and W of Fig. B.2. They are defined by:

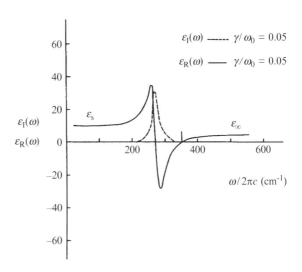

Fig. 2.2 Dispersions of ε_R and of ε_I near from the fundamental lattice absorption of cubic ZnS. The high-energy dispersion region of ε_R corresponds to ε_∞ and the low-energy one to ε_s. The peak of $\varepsilon_I(\omega)$ is at ω_0 (the TO frequency at 271 cm^{-1}), and the frequency at $\varepsilon_\infty = 0$ corresponds to the LO mode at 350 cm^{-1}. The curves are calculated for $\gamma/\omega_0 = 0.05$. For $\gamma = 0$, $\varepsilon_I(\omega)$ diverges at ω_0 (after [22, 23])

$$\mathbf{q}_X = (2\pi/a)\,(1,0,0)\,,$$

$$\mathbf{q}_L = (2\pi/a)\,(1/2,1/2,1/2)\,,$$

$$\mathbf{q}_K = (2\pi/a)\,(3/4,3/4,0)\,,$$ (2.6)

$$\mathbf{q}_W = (2\pi/a)\,(1,1/2,0)\,.$$

This analysis is based on topological considerations, but other CPs can emerge depending on the actual shape of the dispersion curves in the BZ [26]. Sometimes, the Γ point is included in the CPs, but when this is done, this can be only for the optical branches. We have mentioned the degeneracy of the TO and LO branches at the Γ point for the diamond structure. A similar topological degeneracy of the LA and LO branches at the X point [noted L(X)] exists also for this structure. For the 12-fold degenerate CP W of the surface of the BZ of Fig. B.2, which is the one to be considered for the group-IV cubic crystals, the LO and LA modes are degenerate and noted L(W), after [17].

For pure elemental semiconductors, the strong electronic absorption at energies above E_g produces a small nonlinear dispersion of the refractive index n: just below E_g, its room temperature (RT) value in diamond, silicon, and germanium, is 2.42, 3.55, and 4.06, respectively, and it steadily decreases to 2.387 [27], 3.42, and 3.98, respectively, at low frequencies. For these elemental crystals, the relative dielectric constant ε at energies below E_g is real in the whole spectral domain, and equal to n^2. The refractive index is isotropic for cubic crystals, but for crystals with one anisotropy axis, like those of the wurtzite type, the refractive index for the component of the electric field of the radiation parallel to this axis (n$_\parallel$) is slightly different from the one for the component perpendicular to this axis (n$_\perp$).

A consequence of the Kramers–Kronig relations (see for instance [28]) is that, for a semiconductor or an insulator, the static dielectric constant ε_s is:

$$\varepsilon_s = 1 + \frac{2}{\pi} \int_0^\infty 2\mathrm{nk}\frac{\mathrm{d}\omega}{\omega}.$$ (2.7)

This expression shows that a high value of ε_s or of the refractive index necessitates a large amount of absorption throughout the electromagnetic spectrum. This is the reason why crystals with a low value of E_g, for which the fundamental electronic absorption extends far in the infrared, display high values of the refractive indices. Values of n$_s$ and n$_\infty$ are given in Table 2.2. There can be small discrepancies between the values of n$_s$ and n$_\infty$ obtained by spectroscopic methods and the square root of the dielectric constants ε_s and ε_∞ because both present a small variation with energy due to dispersion. As far as possible the values for n$_s$ given in Table 2.2 are those at the lowest possible frequency. The values for n$_\infty$ are those in the spectral region where n stays reasonably constant, but dispersion results in an increase of its value with energy. For instance, the value of n$_\infty$ (GaAs) close to E_g is ~3.6 (see, for instance, [29]), about the same as that of n$_s$. For the cubic crystals, the frequencies of LO and TO modes are those at the Γ point of the BZ. For the wurtzite crystals, the LO mode corresponds to the Γ point.

Table 2.2 RT values of the LO and TO phonon frequencies (cm^{-1}) and average values of n_∞ (for $\tilde{v} > \tilde{v}$ (LO)) and of n_s (for $\tilde{v} < \tilde{v}$ (TO)) of some compound crystals

	$\tilde{v}$(LO), $\tilde{v}$(TO)	n_∞, n_s		$\tilde{v}$(LO), $\tilde{v}$(TO)	n_∞, n_s
MgO (c)	718, 402	1.7, 3.2	ZnTe (s)	207, 177	2.71, 3.11
c-BN (s)	1306, 1056	2.1, 2.7	GaP (s)	403, 367	3.02, 3.33
AlN (w)	670, 656	2.20, 2.9	AlAs (s)	404, 361	2.86, 3.2
w-ZnS	352, 274	2.26, 3.1	CdSe (w)	210, 172	2.5, 3.19
				211, 166	2.5, 3.05
c-ZnS	352, 271	2.3, 2.8	AlSb (s)	340, 319	3.14, 3.35
w-GaN	735, 570	2.4, 3.22	CdTe (s)	169, 140	2.6, 3.32
		2.3, 3.1			
c-GaN	739, 552	2.3, 3.1	GaAs (s)	292, 269	3.30, 3.62
ZnO (w)	576, 378	1.94, 2.96	InP (s)	345, 304	3.09, 3.54
	588, 410	1.9, 2.8			
2H-SiC (w)	968, 764	2.69, 2.71	GaSb (s)	233, 224	3.8, 4.0
ZnSe (s)	253, 213	2.3, 2.8	InAs (s)	239, 217	3.50, 3.89
CdS (w)	305, 256	2.31, 3.02	InSb (s)	190, 179	3.96, 4.22
AlP (s)	501, 439	2.75, 3.1			
3C-SiC (s)	972, 796	2.63, 3.12			

The crystal symmetry is indicated by (c), (s), and (w) for cubic, sphalerite, and wurtzite, respectively. In wurtzite-type crystals, the values for **E** $\parallel$ c and **E**$\perp c$, when known, are given in the first and second row, respectively

Usually, the low-temperature dielectric constant (or the refractive index) is slightly lower than the one at RT [30]. LHeT values of ε_s for group-IV crystals have also been obtained indirectly from a comparison between experimental and calculated line spacings of shallow donor impurities [31].

What has been presented above is based on an interaction of electrons or atoms with the electric field through a quadratic harmonic potential. When potentials including higher-order terms are used, the polarizations, electric dipole moment, and optical susceptibility include in turn higher-order terms whose properties are the base of nonlinear optics and anharmonic effects.

2.2.2 Multi-phonon Absorption and Anharmonicity

Higher-order lattice absorption or Raman scattering has been observed in elemental ([17, 32], and references therein) as well as in compound semiconducting and insulating crystals [33, 34]. Higher-order effects can arise from two mechanisms: (a) anharmonic coupling between phonons, arising from third and higher-order terms in the potential energy, and (b) second and higher-order terms in the electric moment. Fundamentally, these effects are similar to those leading to overtones, summation, or difference bands in molecular spectroscopy. In process (a), the anharmonic mechanism has been described [23] by the coupling of a photon with a TO phonon which couples in turn with two other phonons. The net result can be either the creation of two phonons (summation process) or the creation of one phonon and the annihilation of another one (difference process). The condition for process (a) to occur is the existence of a first-order dipole moment and it is therefore ruled out in elemental crystals. In process (b), where the first-order dipole moment can be zero, the photon couples directly with two phonons, the first one producing an asymmetry in the electronic charge distribution, which is then displaced by the second phonon. The phonons involved in both processes are the short-wavelength phonons for which the nearest-neighbour atomic motion is more asymmetric than the zone-centre phonons. As a result, an electric moment is produced that couples with the photon. The net result is the same as the one implying anharmonicity. The above description implies that both the anharmonicity and the effect of higher-order moment can be present in compound crystals while multi-phonon absorption of elemental crystals can only be explained by second-order dipole moment. The absorption due to the summation process is observed at energies above that of the zone-centre phonons while that due to the difference processes is observed below, in the far infrared.

The absorption coefficient for a multi-phonon combination can be written as the product of three terms. The first one is the matrix element of the coupling term between the phonons involved in the process. It is non-zero only for specific phonon combinations determined by selection rules derived from symmetry considerations. The second one describes the temperature dependence of the phonon populations and the third one is related to the phonons density of states. The IR and Raman selection rules for two- and three-phonon summation processes in the diamond and sphalerite structures have been derived by Birman [35]. It is found that in diamond-like crystals, the two-phonon combinations are usually IR active when they come from different branches [e.g. TO(X) + LO(X) or LO(L) + LA(L)], but that the overtones are IR forbidden (the situation for the W CP is more complicated). In sphalerite-like crystals, in addition, some acoustical or optical overtones are IR active [36,37]. The frequency of the 2TA(X) overtone of some cubic semiconducting compounds measured in absorption is given in Table 2.3.

Tables of the symmetry-allowed three-phonon combinations in the diamond and sphalerite structures can also be found in Birman's paper. In addition to the selection rules involving specific phonons at CPs of the BZ, the two phonons must have opposite momenta to comply with the total momentum conservation (the photon

Table 2.3 Frequency (cm^{-1}) of the 2TA(X) overtone of some cubic semiconducting compounds measured by absorption at LHeT

GaP	c-ZnS	GaAs	ZnSe	ZnTe	InP	InSb	CdTe
212[a]	182[a]	161[b]	141.5[a]	110[a]	136.5[a]	83[b]	71.3[a]

These frequencies are lower than the Raman frequency TO(Γ) and measurements at low temperature are necessary to limit the contribution of the multiphonon difference processes which are discussed below

[a]Slacks and Roberts [36], and references therein

[b]Koteles and Datars [33]

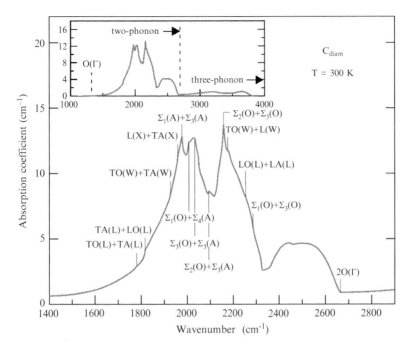

Fig. 2.3 RT absorption spectra of a natural type-IIa diamond showing the two-phonon and three-phonon features. The positions of the one-phonon Raman frequency and of its overtone, which are IR-inactive, are indicated as $O(\Gamma)$ and $2O(\Gamma)$, respectively. After [17]

momentum is neglected). The multi-phonon summation spectrum of diamond at RT is shown in Fig. 2.3. The letters in parentheses after the phonon types correspond to the CPs of the BZ of diamond (fcc lattice). The space group of diamond is O_h^7 and the point group symmetries of the CPs of this BZ are given in Table B.6 of Appendix B.

In silicon, the strongest two-phonon absorption features are the $TO(L) + TA(L)$ and $TO(X) + TA(X)$ summation bands, whose energies, deduced from Raman scattering experiments, are 603 and $611\,\mathrm{cm}^{-1}$, respectively ([38], and references therein), and they give a broad unresolved band with a peak absorption of $\sim 10\,\mathrm{cm}^{-1}$ at RT[2] [39]. The strongest features of the multiphonon absorption spectrum in natural and qmi silicon are given in Table 2.4.

In Sennikov et al.'s report [40], the frequencies of the $2TO(X) + TA(X)$ feature in the qmi ^{29}Si and ^{30}Si samples were not indicated because of the high concentration of interstitial oxygen in the samples available. From the shifts for the $TO(L) + TA(L)$ and $O(\Gamma)$ features with the isotopic Si masses, shifts of approximately -20

---

[2]The local mode of C_s at $605\,\mathrm{cm}^{-1}$ is superimposed on this two-phonon band.

Table 2.4 Attributions based on similarities in the frequencies of RT frequencies (cm^{-1}) of the main two- and three-phonon absorption features in $^{nat}Si(^{28.09}Si)$ and in quasi-monoisotopic Si (after [40])

Attribution	$^{nat}Si^a$	^{nat}Si	qmi ^{28}Si	qmi ^{29}Si	qmi ^{30}Si
LO(X) + TA(X)	563	566.4	569.1	559	548.1
TO(L) + TA(L)	608.5	610.8	612	601.6	591.5
LO(W) + LA(W)	743	739.1	741	728.3	715.6
TO(L) + LA(L)	823	819	817.8	804.3	791
TO(X) + LO(X)	890	~888	~890	873.5	859
TO(W) + TO(W)	956	~959	~962	944	930
TO(L) + O(Γ) + TA(L)	1124	1119	1122.4	NM	NM
TO(X) + 2LO(X)	1287^b	1299.5	~1302	1278.3	1253.6
3TO(W)	1434	1448.8	~1450	1425.1	1398.6

The CPs in brackets correspond to the BZ of diamond. The values of the second column are deduced from IR measurements with attributions including the CPs
[a] Balkanski and Nusimovici [41]
[b] Johnson and Loudon [42]

and $-40\,cm^{-1}$ with respect to qmi ^{28}Si are estimated for the so-called 2TO + TA combination in qmi ^{29}Si and ^{30}Si, respectively.

The temperature-dependent term represents the difference in the occupation numbers of the phonon states involved in the process. As phonons are bosons, the occupation number for a phonon of angular frequency ω at temperature T is given by Bose–Einstein statistics as

$$n(\omega, T) = [\exp(\hbar\omega/k_B T) - 1]^{-1}. \tag{2.8}$$

For two-phonon processes involving branches t and t' and phonons with wave vectors $\mathbf{q}$ and $-\mathbf{q}$, the temperature-dependent term is:

$$[(n_{qt} + 1)(n_{-qt'} + 1) - n_{qt}n_{-qt'}] \quad \text{(summation process)}$$

or

$$[n_{qt}(n_{-qt'} + 1) - (n_{qt} + 1)n_{-qt'}] \quad \text{(difference process)}.$$

Similar relationships can be obtained for three-phonon processes. At low temperature, $n(\omega, T)$ is much smaller than unity and the above expression for summation tends to unity, but it goes to zero for the difference processes, which are therefore not observed at low temperature. At higher temperature, the absorption intensity increases for both processes [39]. Multiphonon spectra of germanium at different

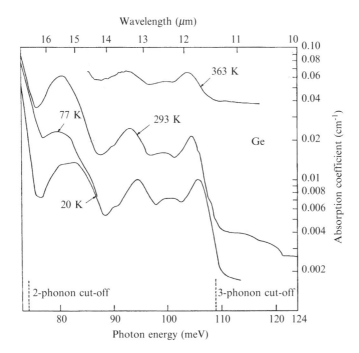

Fig. 2.4 High-energy multiphonon spectra of germanium at different temperatures (the Raman frequency is $301\,\mathrm{cm}^{-1}$ or $37.3\,\mathrm{meV}$). The spectrum extends from ~585 to $1000\,\mathrm{cm}^{-1}$. Note the logarithmic ordinate scale (after [43]). With permission from the Institute of Physics

temperatures are shown in Figs. 2.4 and 2.5, and they illustrate the temperature dependence of the summation and difference processes.

In Fig. 2.5, no conspicuous absorption is observed at the Raman frequency. We have mentioned the existence of CPs in the one-phonon density of states, but this can be measured only for compound crystals. The situation is different in multiphonon absorption because there, the high-frequency phonons of the BZ boundary are mostly involved (note that in three-phonon processes, the $\mathbf{q} = 0$ zone-centre phonons can also be involved without problem for momentum conservation). For two-phonon absorption, the density of states is proportional to an integral similar to the one of expression (2.5), but there, ω_t is replaced by the sum of the two frequencies ω_t and $\omega_{t'}$ of the phonons of the combination. Besides the trivial case where $\omega_t = \omega_{t'} = 0$, the condition $\nabla_{\mathbf{q}}[\omega_t(\mathbf{q}) + \omega_{t'}(\mathbf{q})] = 0$ is fulfilled when $\nabla_{\mathbf{q}}(\omega_t) = \nabla_{\mathbf{q}}(\omega_{t'}) = 0$ or when $\nabla_{\mathbf{q}}(\omega_t) = -\nabla_{\mathbf{q}}(\omega_{t'})$. The observed two-phonon absorption is the sum of the contributions of the possible two-phonon processes. As already mentioned, the multi-phonon absorption in compound crystals can arise both from anharmonicity and from induced dipole moments and it is stronger than in elemental crystals. For this reason absorption measurements of impurities and defects in the two-phonon spectral domain are more difficult in the compound crystals.

Fig. 2.5 Multiphonon spectra of germanium at 2 K and at RT. The vertical line gives the position of the Raman frequency $\omega(O(\Gamma))$. The part at the RHS is the two-phonon spectrum and the LHS to phonon difference processes. The spectrum extends in energy up to 62 meV. Note the logarithmic ordinate scale (after [44]). Reproduced with permission from the Physical Society of Japan

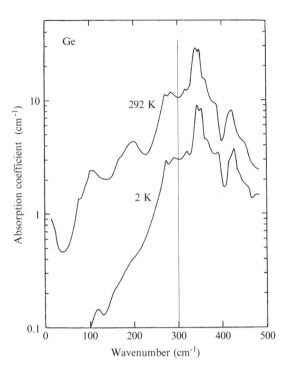

2.3   Electronic Absorption

2.3.1   Energy Gap and Intrinsic Absorption

2.3.1.1   Band Structure

It is usual to represent the electron energies in crystalline solids as a function of the wave vector $\mathbf{k}$ of the electron along directions of the reciprocal lattice in the first BZ. These energies are labelled $\mathcal{E}_b(\mathbf{k})$ where index b refers to a particular band and it can be shown that the mean velocity of an electron in the CB is $\mathbf{v}_c(\mathbf{k}) = \hbar^{-1}\nabla_{\mathbf{k}}\mathcal{E}_c(\mathbf{k})$. An effective-mass tensor $[\mathbf{M}^{-1}(\mathbf{k})]_{ij} = \mp\hbar^{-2}\partial^2\mathcal{E}_b(\mathbf{k})/\partial k_i k_j$ can be similarly derived, where the $-$ and $+$ signs refer to a band maximum (for holes) or to a band minimum (for electrons). The effective-mass tensor plays an important role in the spectroscopy of impurities in semiconductors, especially when a magnetic field is involved. The optical interband transitions take place between extrema of the valence and conduction bands. Near extrema $\mathcal{E}_c$ or $\mathcal{E}_v$ of the conduction or valence bands, it is possible to express the energy as:

$$\mathcal{E}_b(\mathbf{k}) = \mathcal{E}_b \pm \hbar^2 \left(k_1^2/2m_1 + k_2^2/2m_2 + k_3^2/2m_3 \right), \tag{2.9}$$

where $\mathcal{E}_b$ is $\mathcal{E}_c$ or $\mathcal{E}_v$. The $+$ and $-$ signs refer to the conduction and valence bands, respectively. The effective mass parameters m_i are of course different for the two bands. For the *CB*, the absolute energy minimum can occur at $\mathbf{k} = 0$, but also at $\mathbf{k} \neq 0$ while the *VB* maximum occurs mostly at $\mathbf{k} = 0$. The *VB* states at this maximum are related to the atomic bonding between the atoms of the crystal. For diamond, silicon, and germanium covalent crystals, the *VB* maxima correspond to p-like bonds in a threefold electronic degeneracy when electron spin is neglected. When spin–orbit (s–o) interaction is considered, the valence band edge splits into fourfold degenerate $p_{3/2}$-like states separated from twofold degenerate $p_{1/2}$-like states by the s–o splitting energy Δ_{so}. The constant energy surfaces about the extrema are ellipsoids or warped spheres specified by their principal axes $\mathbf{k}_i$, the three effective masses m_i, and the location in $\mathbf{k}$-space of the ellipsoids, which determine the symmetry of the ellipsoids and the orientational degeneracy of the extrema in $\mathbf{k}$-space. In the case of revolution symmetry, as for the *CB*s in group IV non-metallic crystals, there are only two effective-mass parameters, one longitudinal electron mass m_{nl} along the main axis of the ellipsoid and one transverse electron mass m_{nt} along the two perpendicular axes. For the onset of the optical absorption, one is generally concerned with the energies of the absolute maximum of the *VB* and absolute minimum of the *CB*, whose difference determines the values of the band gaps. In a semiconducting crystal, when these extrema have the same wave vector $\mathbf{k}$, the optical transitions between them are called direct and the crystal a direct-gap semiconductor. In all the non-metallic group IV crystals, the absolute minimum of the *CB* is at $\mathbf{k} \neq 0$ and it is material-dependent. In that case, the optical transitions between these two extrema imply a change in the electron momentum and they are forbidden to zeroth order. Indirect absorption can however take place, the difference in electron momentum being compensated by annihilation or creation of lattice phonons of the opposite momentum. Because of the above particularity, whenever the band gap of semiconductors is a relevant parameter, such materials are labelled as indirect-gap semiconductors.

When studying the optical properties related to the *VB*, the s–o coupling must be taken into account. This is not usually necessary for the properties of the *CB*, and Table 2.5 gives the correspondence between the *IR*s of the double group used above and those of the standard group used, for instance, by Cohen and Bergstresser [47] to label in the electronic band structure different symmetry points of the BZ we are concerned with here.

The group-IV materials are not the only indirect-gap cubic semiconductors. For instance, GaP and the III–V compounds with sphalerite structure involving A*l* are also indirect gap semiconductors. There is a difference between the two structures however: while the energy dispersion curves of diamond-type crystals are degenerate at the X point of the BZ, this degeneracy is lifted for the sphalerite structure into a lower band, X_1 and a higher one, X_3 (X_6 and X_7, respectively, when s–o interaction is considered) separated typically by an energy of $\sim 0.4\,\mathrm{eV}$. It turns out that the relative ordering of these two bands depends on the origin of the coordinate system, which can be chosen at a group-III or group-V site (the same reasoning holds also for the II–VI compounds) through a potential that differs for

Table 2.5 Correspondence between the *IR*s of some particular points of the electronic band structure of cubic crystals with space groups O_h^7 and T_d^2 with and without s–o coupling [45]

	Diamond structure (O_h^7)		Sphalerite structure (T_d^2)	
	Double group (s–o coupling)	Standard group no spin	Double group (s–o coupling)	Standard group no spin
VB	Γ_8^+ (4) Γ_7^+ (2)	Γ_5^+ (Γ_{25}') (3)	Γ_8 (4) Γ_7 (2)	$\Gamma_5(\Gamma_{15})$ (3)
CB	Γ_7^- (2) Γ_6^- (2) Γ_8^- (4) X_5 (4)	Γ_2^- (Γ_2') (1) $\Gamma_5^-(\Gamma_{15})$ (3) X_1 (2)	Γ_6 (2) Γ_7 (2) Γ_8 (4) X_6 (2), X_7 (2)	Γ_1 (1) $\Gamma_5(\Gamma_{15})$ (3) X_1 (1), X_3 (1)

The notation used in [46] and the dimensions of the *IR*s are given in brackets

the two sites [48]. When this potential on one site is attracting for electrons, the electron states on that site are the lowest and they belong to the *CB* minimum. For GaP, it has been found that this potential is positive and attracting for electrons on a P site while repulsive for electrons on a Ga site. As a consequence, an electron on the X_1 band is concentrated on a P site and on the X_3 band on a Ga site.

Logically, when the absolute conduction and valence bands extrema of a semiconductor lay both at the same value of **k**, the gap is said to be direct. III–V compounds like InP, GaAs, and InSb belong to this category, with extrema for **k** = 0. As already mentioned, in the crystals with the diamond or sphalerite structure, the electrons wave functions at the top of the *VB* at **k** = 0 are triply degenerate. They form a basis for a three-dimensional irreducible representation (*IR*) of the diamond (O_h) or sphalerite (T_d) symmetry point group. This *IR* is Γ_5^+ for diamond and Γ_5 for sphalerite (see Table 2.5). Under s–o interaction, the Γ_5^+ *VB* splits into the Γ_8^+ and Γ_7^+ bands (sometimes noted $^4\Gamma_8^+$ and $^2\Gamma_7^+$, respectively, by including the *VB* degeneracy $2j + 1$) separated by the s–o splitting Δ_{so}. For isolated atoms, Δ_{so} increases as $Z^{\sim 4}$ and a similar trend is observed in crystals. The calculated band structure of CdTe is shown in Fig. 2.6.

The intrinsic absorption of semiconductors and insulators is very strong and the value of E_g determines the visual aspect of intrinsic polished crystals (the visible spectrum extends from about 400 to 750 nm, that is for photons between 3.10 and 1.65 eV). In germanium and in other indirect-gap semiconductors, the intensity of the phonon-assisted indirect transitions from the *VB* to *CB* is smaller than that of the direct one. At RT, the absorption coefficient of germanium near 1.0 eV, in the direct-transition region, is $\sim 10^4$ cm^{-1}, compared to $\sim 10^2$ cm^{-1} at 0.78 eV. The crystals with $E_g < 1.65$ eV display a quasi-metallic aspect due to their intrinsic absorption in the visible region, correlated with a high reflectivity; those with E_g between 1.65 and 3.10 eV are transparent, with a colour depending on the value of E_g while those with $E_g > 3.10$ eV are colourless when pure. An estimation of the indirect absorption coefficient of silicon can be obtained from Fig. 2.7.

In most crystals, the symmetry of the valence band with the highest energy is $\Gamma_8^+(O_h)$ or $\Gamma_8(T_d)$. This is not the case for CuC*l*, where Γ_7 is about 60 meV

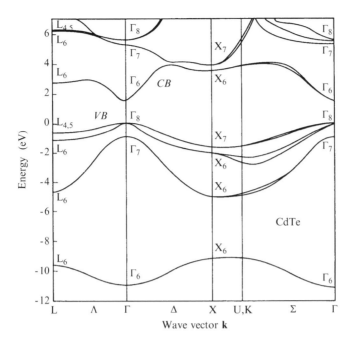

Fig. 2.6 Electronic band structure of direct band gap CdTe from a non-local pseudopotential calculation (after [49]). The experimental Γ_6–Γ_8 difference (E_g) is 1.53 eV and the VB spin–orbit splitting Γ_8–Γ_7 is 0.80 eV. The energy reference is the VB maximum at the Γ point. The location of the critical points in the BZ is shown in Fig. B.2

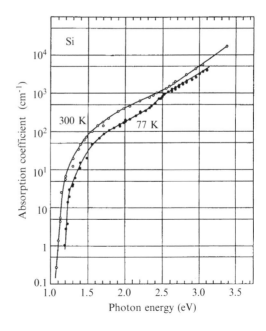

Fig. 2.7 Semi-logarithmic plot of the indirect intrinsic absorption of a silicon sample with a RT resistivity >20 Ωcm (after [50]). The direct band gap value is ~3.5 eV. This figure shows also the increase of E_g when temperature decreases. © 1955 by the American Physical Society

above Γ_8, due to the hybridization of the Cu $3d$ levels with the Cl $3p$ levels, because their energies are similar [51], and also for Cu$_2$O. Cuprous oxide is a direct-band-gap cubic semiconductor known for a very long time. Its BZ is a simple cube with O_h point group symmetry (see Appendix B), which is also the symmetry point group of the centre of the BZ, where the direct optical transitions take place. In this material, the order of the s–o split VBs is the inverse of the one in group-IV and III–V semiconductors ([52], and references therein). The absolute VB maximum of Cu$_2$O is the Γ_7^+ band, and the $\Gamma_7^+(VB) \rightarrow \Gamma_6^+(CB)$ band-gap transition is nominally parity-forbidden. A continuum absorption corresponding to this transition is however observed above 2.174 eV at LHeT, but it is much lower than the one above 2.304 eV, corresponding to the onset of allowed transitions between the s–o split $^4\Gamma_8^+$ VB and the same $^2\Gamma_6^+$ CB (see Fig. 2.9 of [53]). The admitted value of the LHeT band gap E_g of Cu$_2$O is thus 2.174 eV. The values of the $\Gamma_7^+(VB) - \Gamma_8^-(CB)$ and $\Gamma_8^+(VB) - \Gamma_8^-(CB)$ energy gaps are 2.624 and 2.755 eV, respectively, and the absolute value of the $\Gamma_8^+ - \Gamma_7^+$ s–o splitting Δ_{so} of the VB is 0.13 eV [54, 55]. It has been suggested that the minimum forbidden (intrinsic) band gap in Cu$_2$O derived from transport measurements was only $\sim$0.8 eV, close to the one of 0.78 eV calculated from first principles using the local-density approximation (LDA) [56]. However, the calculations within the LDA are known to give undervalued estimations of the band gap, and the experimental activation energies in this range, initially thought to correspond to the optically forbidden band gap are presumably due to the excitation of an electron to an impurity level. This suggestion is therefore unfounded. Inversely, a calculation of the band gap energy of Cu$_2$O using the periodic Hartree–Fock method and an a posteriori density functional correction [57] leads to an overestimation (9.7 eV) of the band-gap energy inherent to this calculation method.[3]

Choosing the z axis along $\mathbf{k}_{min}$ and taking the electron energy origin at $\mathbf{k}_{min}$, the energy E of a conduction electron near $\mathbf{k}_{min}$ for the indirect gap semiconductors of Table 2.6 is:

$$E = \frac{\hbar^2}{2}\left[(\mathbf{k}_z - \mathbf{k}_{min})^2/m_{nl} + \left(\mathbf{k}_x^2 + \mathbf{k}_y^2\right)/m_{nt}\right] \tag{2.10}$$

and the constant energy surfaces in the $\mathbf{k}$-space are prolate revolution ellipsoids with their main axis along z. The band structure of the sphalerite-type crystals is similar to that of the diamond-type crystals, with a few differences, however, but for most of them, the CB minimum is at $\mathbf{k} = 0$.

2.3.1.2 Band Parameters

In the vicinity of the VB maximum at $\mathbf{k} = 0$, the expressions for the constant-energy surfaces of the VB electrons in the highest-energy band of the diamond- or sphalerite-type crystals are usually given as functions of three parameters noted

---

[3] In [57], the correct volume and page number for Ref. 27 of this paper are 17 and 292, respectively.

Table 2.6 Selected band-structure parameters of indirect-band-gap cubic crystals

	C_{diam}[a]	$3C$-SiC	Si[a]	Ge[a]	GaP[a]	AlAs[a]	AlSb[a]
k_{min}	$0.76k(X)$	$k(X)$	$0.84k(X)$	$k(L)$	$k(X)$[b]	$k(X)$[b]	$k(X)$[b]
m_{nl}	1.7[c]	0.667[d]	0.9163	1.57	0.90[e]	1.1	1.8
m_{nt}	0.31[c] (0.36[f])	0.247[d]	0.1905	0.0807	0.25[e]	0.19	0.259
m_{hh}	1.08		0.54	0.35[g]	0.67[g]	1.022[g]	0.872[g]
m_{lh}	0.36	0.45	0.15	0.043[g]	0.17[g]	0.109[g]	0.091[g]
m_{so}	0.15		0.24	0.095	0.465		
Δ_{so} (meV)	6[h]–13[i]	14.4[j]	42.65[k]	295[l]	80	300	673
E_g (eV)	5.49	2.4	1.17	0.745	2.35	2.23	1.69
Direct gap (eV)	7.3	7.0	3.48	0.888[l]	2.90	3.13	2.38
γ_1	3.61[m]	2.8	4.28	13.3[n]	4.05	4.04	4.15
γ_2	0.09[m]	0.16	0.375	4.24[n]	0.49	0.78	1.01
γ_3	1.06[m]	0.65	1.45	5.69[n]	1.25	1.57	1.75

k_{min} denotes the wave-vector symmetry and modulus for the absolute CB minimum with respect to the critical points of the BZ. The electron and hole effective masses are in units of m_e. Δ_{so} is the s–o splitting of the VB. E_g is given at LHeT. The direct gap corresponds to $\Gamma_8(VB) - \Gamma_6(CB)$ for sphalerite and $\Gamma_8{}^+(VB) - \Gamma_7{}^-(CB)$ for diamond. The VB parameters γ_i are in units of $2m_e/\hbar^2$
[a]Madelung [8]
[b]See text
[c]Casanova et al. [64]
[d]Kaplan et al. [65]
[e]Oshikiri et al. [66]
[f]Nava et al. [67]
[g]Values near from $\mathbf{k} = 0$ along the [111] direction
[h]Rauch [68]
[i]Willatzen et al. [69]
[j]Willatzen et al. [70]
[k]Pajot [71]
[l]Li et al. [72], LHeT
[m]Reggiani et al. [73]
[n]Hensel and Suzuki [74]

A, B, and C [58]. These maxima are warped spheres in the $\mathbf{k}$-space given by:

$$E_{(3/2)\pm} = -A\mathbf{k}^2 \mp \left[B^2\mathbf{k}^4 + C^2 \left(k_x^2 k_y^2 + k_y^2 k_z^2 + k_z^2 k_x^2 \right) \right]^{1/2} \quad (2.11)$$

which can be seen as the sum of a spherical and a cubic contributions. $E_{(3/2)+}$, with a smaller energy dispersion (a corresponding larger mass) than $E_{(3/2)-}$, is the heavy-hole band (E_{hh}) and E_- the light-hole band (E_{lh}). The dispersion curves of the constant-energy surfaces for the holes of the VB split by s–o coupling are:

$$E_{(1/2)} = -\Delta_{so} + A\mathbf{k}^2,$$

and they are spheres in the $\mathbf{k}$-space.

New VB parameters γ_1, γ_2, and γ_3 were introduced by Luttinger [59] in his description of holes in the silicon VB. They are formally different from the ones in (2.11) as holes were considered instead of electrons, but it can be checked that:

$$\frac{\hbar^2}{2m_e}\gamma_1 = -A \quad \frac{\hbar^2}{m_e}\gamma_2 = -B \quad \frac{\hbar^2}{m_e}\gamma_3 = \sqrt{B^2 + C^2/3}.$$

In the practical case, these so-called Luttinger parameters are given in units of $\hbar^2/2m_e$ so that $\gamma_1 = -A$, $\gamma_2 = -B/2$, and $\gamma_3 = (1/2)\sqrt{B^2 + C^2/3}$. The Hamiltonian for holes in the upper *VB* of silicon using these parameters, known as the Luttinger's Hamiltonian H_L is:

$$H_L = \frac{1}{m_e}\left[\left(\gamma_1 + \frac{5}{2}\gamma_2\right)\frac{p^2}{2}\right.$$

$$\left. -\gamma_2\left(p_x^2 J_x^2 + p_y^2 J_y^2 + p_z^2 J_z^2\right) - 2\gamma_3(\{p_x p_y\}\{J_x J_y\} + \mathrm{cp})\right], \quad (2.12)$$

where the p_i are the components of the hole linear momentum and the J_i the components of the angular momentum operator corresponding to spin 3/2. The cyclic permutation is noted cp and $\{ab\}$ is $(ab + ba)/2$. This Hamiltonian is more convenient for diagonalization, especially in the presence of additional perturbations and it has also been used for a description of the upper *VB* of other cubic semiconductors. For sphalerite-type crystals with symmetry point group T_d, the Hamiltonian must include a term taking into account the asymmetry of these crystals with respect to inversion. This additional term is written:

$$H_A = \frac{2C}{\sqrt{3}\hbar}(p_x\{J_x, V_x\} + p_y\{J_y, V_y\} + p_z\{J_z, V_z\}), \quad (2.13)$$

where $V_x = J_y^2 - J_z^2$, $V_y = J_z^2 - J_x^2$, and $V_z = J_x^2 - J_y^2$. Parameter C in expression (2.13) is different from the *VB* parameter C in expression (2.11). Estimations of the values of parameter C of expression (2.13) for different semiconducting compounds have been calculated by Cardona et al. [60]. For InSb, C is $\sim 8.7 \times 10^{-8}$ meV cm and this value is in reasonable agreement with the one (9.3×10^{-8} meV cm) obtained by Pidgeon and Groves [61] from magneto-optical reflection measurements at 1.5 K. In many practical cases, the contribution of H_A is neglected.

Hamiltonian (2.12) is used to explain the principles of the calculation of the shallow acceptor levels in these crystals. When a magnetic field **B** is present, assumed to derive from a vector potential **A** through $\mathbf{B} = \nabla \times \mathbf{A}$ satisfying the condition $\nabla \cdot \mathbf{A} = 0$, the hole momentum π can be written, in SI units:

$$\pi = -i\hbar\nabla + eA$$

and an expression similar to (2.12) can be obtained for the EM Hamiltonian with the addition of field-dependent *VB* parameters κ and q introduced by Luttinger [59]. The term H_B linear in **B** added to Hamiltonian (2.12) can be written:

$$H_B = \mu_B(g_1 \mathbf{B} \cdot \mathbf{J} + g_2(\mathbf{B}_x J_x^3 + \mathbf{B}_y J_y^3 + \mathbf{B}_z J_z^3)), \qquad (2.14)$$

where parameters g_1 and g_2 [62] are called the g-factors of the top of the VB. These VB g-factors are related to the above VB parameters by $g_1 = 2\kappa$ and $g_2 = 2q$.

From the experimental side, the band-structure parameters are mainly determined from the cyclotron resonance (CR) spectra of electron and holes (see, for instance, [3]). Some of these parameters can also be obtained from the Zeeman splitting of electronic transitions of shallow impurities involving levels for which the electronic masses can be taken as those of free electrons or holes, or from free-carriers magnetoreflectivity. Averaged effective masses can also be deduced from Hall-effect measurements or from other transport measurements. Calculation methods that have been used to obtain band-structure parameters free from experimental input are the *ab initio* pseudopotential method, the k–p method, and combination of both. These theoretical methods are presented in Chap. 2 of the book by Yu and Cardona [2]. VB parameters at $\mathbf{k} = 0$ including κ and q have been calculated for several semiconductors with diamond and zinc-blende structures by Lawaetz [63].

Table 2.6 gives a few relevant band-structure parameters of group-IV and group-III–V crystals with indirect band gaps. The structure of the CB near from its minimum is generally simpler to model than that of the VB. The CB parameters are known therefore with a reasonable accuracy from the experimental data. For diamond, $m_{nt} = 0.31 m_e$ is deduced from the Zeeman splitting of $2p_{\pm 1}(P)$ in C_{diam} [64] and $m_{nl} = 1.7 m_e$ from the ratio $\gamma = m_{nt}/m_{nl}$.

This is not the case for the VB and there is still a significant uncertainty on the exact values of the VB parameters of diamond (see, for instance, [69]). For silicon and germanium, there is only a moderate dispersion of the values of these parameters. Grey tin (α-Sn) is a semimetal stable below 13°C, where the energy separation (0.14 eV) between Γ_8(v,c) and the conduction band minimum at L_6^+ is sometimes called the optical energy gap because it corresponds to the onset of a higher absorption [75], but the absorption coefficient of α-Sn at energies below this onset is already in the 10^4 cm^{-1} range. The VB s–o splitting of α-Sn is 0.8 eV.

The X point is on the Δ axis of the BZ with $\langle 100 \rangle$ orientation and the absolute energy minimum of the CB of the corresponding cubic crystals of Table 2.6 is sixfold degenerate in **k**-space. For germanium, it is only fourfold degenerate as the minimum is along the Λ axis with $\langle 111 \rangle$ orientation. For the Ge$_{1-x}$Si$_x$ alloys, this location is observed up to $x \sim 0.15$, where E_g is ~ 0.93 eV. For higher values of x, the band structure becomes similar to that of silicon. Values of the indirect excitonic band gap of Ge$_{1-x}$Si$_x$ alloys in the x range between 0 and 1 have been obtained from PL measurements by Weber and Alonso ([76], and references therein).

For GaP and AlSb, to quantify what has been mentioned before, the CB minima are also along the $\langle 100 \rangle$ directions, at $0.925 \mathbf{k}(X_1)$ and $0.90\mathbf{k}(X_1)$, respectively. The energy dispersion curve of these crystals shows a small maximum ΔE (see Table 2.6) at the X_1 point with respect to the two nearby minima and this

Table 2.7 Experimentally determined effective masses (in units of m_e) at $\mathbf{k} = 0$ extrema and *VB* Luttinger parameters for some direct-band-gap cubic semiconductors

	GaAs	GaSb	InP	InAs	InSb	ZnSe	ZnTe	CdTe
m_n	0.0662[a]	0.041	0.0793[b]	0.022[c]	0.0139[c]	0.13	0.122[d]	0.093
m_{hh}	0.53	0.8	0.58	0.4	0.42			0.84
m_{lh}	0.08	0.05	0.12	0.026	0.016			0.12
m_{so}	0.15		0.12	0.14				
E_g (eV)	1.519	0.811	1.424	0.418	0.2344	2.82	2.394	1.607
Δ_{so} (eV)	0.341	0.76	0.108	0.39	0.850[c]	0.40[c]	0.91	0.80
γ_1	6.98[e]	11.80[f]	6.28[f]	19.67[f]	35.65[g] (3.25)	3.77[f]	3.90[h]	5.30[i]
γ_2	2.25	4.03	2.08	8.37	15.70 (−1.3)	1.24	0.80	1.70
γ_3	2.9	5.26	2.76	9.29	16.97(0.0)	1.67	1.70	2.00

At this *CB* minimum, m_n can be considered as nearly isotropic. For germanium, m_n at $\mathbf{k} = 0$ is 0.038 [77]. For InSb, values of the effective Luttinger parameters are given in brackets. The Luttinger parameters of Lawaetz are calculated values. The values of E_g and Δ_{so} are the ones at LHeT

[a] Kozhevnikov et al. [78]
[b] Hopkins et al. [79]
[c] Madelung [8,49]
[d] Clerjaud et al. [80]
[e] Skolnick et al. [81]
[f] Lawaetz [63]
[h] Fröhlich et al. [82]
[i] Le-Si-Dang and Romestain [83]
[g] Yakunin [84]

configuration has been coined the camel's back. This situation is discussed in detail in [71].

The electron effective masses m_n at the *CB* minimum at $\mathbf{k} = 0$ are generally smaller than the ones for the $\mathbf{k} \neq 0$ *CB* minima, as can be judged from Table 2.7. The *VB* Luttinger parameters have been determined by many authors and we have been obliged to make a choice, biased in some cases by the values used in the most recent calculations of the shallow-acceptor levels. The situation is complicated by the fact that for semiconductors like InSb, where there is an interaction between the valence and the conduction bands, effective Luttinger parameters $\tilde{\gamma}_i$ have been defined by Pidgeon and Brown [85] as:

$$\tilde{\gamma}_1 = \gamma_1 - \frac{E_P}{3E_g} \text{ and } \tilde{\gamma}_1 = \gamma_1 - \frac{E_P}{6E_g} \text{ for } i = 2, 3,$$

where E_P, known as Kane energy and related to the *VB–CB* interaction, is of the order of 20 eV. For InSb, the parameters $\tilde{\gamma}_i$ are sometimes given as the Luttinger parameters and this can create some confusion.

The band structure of the $Al_x Ga_{1-x} As$ alloy depends on the value of x. The crossover from a direct-gap to an indirect-gap alloy takes place for $x \sim 0.4$.

For III–V alloys like $Al_x Ga_{1-x} As$ or $GaAs_{1-x} P_x$, the value of the band gap increases with x as E_g for AlAs and GaP are 2.15 and 2.27 eV, respectively, compared to 1.42 eV for GaAs (RT values). In $GaAs_{1-x} N_x$, despite values of 3.30 and ~ 3.4 eV for *c*-GaN and *w*-GaN, respectively, a decrease of the band gap has been measured in alloys with x up to ~ 0.15, where E_g is 1.03 eV [86], in qualitative agreement with theoretical results [87].

The structure of the *VB* maximum of the wurtzite-type crystals differs from that of the sphalerite-type crystals. This difference, which determines intrinsic optical features of these crystals, is due to the combination of the crystal field mentioned in Appendix B with the s–o coupling. When ignoring the s–o coupling and the crystal field, the *VB* maximum of wurtzite at $\mathbf{k} = 0$ is made of two degenerate bands associated with the one-dimensional Γ_1 and two-dimensional Γ_5 *IRs* of the C_{6v} symmetry point group. The combined effects of the crystal field and of the s–o coupling are to lift (1) the degeneracy between the Γ_1 and Γ_5 bands and (2) the intrinsic degeneracy of the Γ_5 band. In the double group representation of C_{6v} due to the introduction of spin, the *VB* maximum then corresponds to the Γ_9 *IR*, separated from two bands both corresponding to the *IR* Γ_7. These three *VBs* are usually noted A, B, and C in order of decreasing energy. The separations E_{AB} and E_{AC} between the $\Gamma_9(A)$, $\Gamma_7(B)$, and $\Gamma_7(C)$ bands as a function of the crystal field and s–o energy parameters Δ_{cf} and Δ_{so} have been calculated to be [88]:

$$E_{AB} = \frac{1}{2}(\Delta_{so} + \Delta_{cf}) - \left[\frac{1}{4}(\Delta_{so} + \Delta_{cf})^2 - \frac{2}{3}\Delta_{so}\Delta_{cf}\right]^{\frac{1}{2}}, \qquad (2.15a)$$

$$E_{AC} = \frac{1}{2}(\Delta_{so} + \Delta_{cf}) + \left[\frac{1}{4}(\Delta_{so} + \Delta_{cf})^2 - \frac{2}{3}\Delta_{so}\Delta_{cf}\right]^{\frac{1}{2}}. \qquad (2.15b)$$

Most wurtzite-type crystals are direct band-gap materials ($2H$-SiC is an exception) and interband transitions can take place between these three *VBs* and the Γ_7 *CB* minimum. These materials are anisotropic and this anisotropy reflects on the selection rules for the optical transitions and on the effective masses. The $\Gamma_9(A) \rightarrow \Gamma_7(CB)$ transitions are only allowed for $\mathbf{E} \perp c$ while the two $\Gamma_7(B,C) \rightarrow \Gamma_7(CB)$ transitions are allowed for both polarizations. However, the relative values of the transition matrix elements for the $\Gamma_7(B,C) \rightarrow \Gamma_7(CB)$ transitions can vary with the material. For instance, in *w*-GaN, the $\Gamma_7(B) \rightarrow \Gamma_7(CB)$ transition is predominantly allowed for $\mathbf{E} \perp c$ while the $\Gamma_7(C) \rightarrow \Gamma_7(CB)$ transition is predominantly allowed for $\mathbf{E} \parallel c$ [89]. Table 2.8 gives band structure parameters of representative materials with the wurtzite structure.

A few semiconductors have *VB* extrema at other points of the BZ, like the direct-gap lead chalcogenides (PbS, PbSe, PbTe), with rock salt structure, where the valence and conduction bands extrema are both located at the L point of the BZ.

2.3.1.3   Absorption Coefficients

For direct-gap semiconductors, assuming spherical effective masses m_n and for electrons at the *CB* minimum and m_h for holes at the *VB* maximum, the interband absorption coefficient $K(\omega)$ for allowed transitions can be shown (see, for instance, [28]) to be proportional to

Table 2.8 Selected band structure parameters of five compounds with the wurtzite structure

	w-CdSe	ZnO	w-GaN	w-ZnS	w-CdS
$\Gamma_9(A) - \Gamma_7(CB)$ (eV)	1.829	3.4370[a]	3.504	3.864	2.573
$\Gamma_9(A) - \Gamma_7(B)$ (meV)	-26	-9.5[a]	-6	-29	-16
$\Gamma_9(A) - \Gamma_7(C)$ (meV)	-1429	-49.8[a]	-43	-117	-70
Δ_{so} (meV)	416	16[a]	12	86	62
Δ_{cf} (meV)	39	43[a]	37.5	58	28
$m_{n\perp}$, $m_{n\parallel}$	0.12	0.24–0.28	0.19	0.28	0.20–0.25
$m_{h\perp}(A)$, $m_{h\parallel}(A)$	0.45, >1	0.59, 0.59	0.33, 2.03	0.48, 1.4	0.7, $\cong 5$
$m_{h\perp}(B)$, $m_{h\parallel}(B)$		0.59, 0.59	0.34, 1.25		
$m_{h\perp}(C)$, $m_{h\parallel}(C)$		0.35, 0.31	2.22, 0.15		

The energies for ZnO and GaN are given at LHeT and at 80 K for CdSe and CdS (the effective masses are expressed in units of m_e). The $\Gamma_9(A) - \Gamma_7(CB)$ difference is the energy gap E_g [49]
[a]Reynolds et al. [90]

$$|\mathbf{p}_{cv}|^2 \ (2\bar{m})^{3/2} \frac{(\hbar\omega - E_g)^{1/2}}{\hbar\omega}, \tag{2.16}$$

where $\mathbf{p}_{cv}$ is the momentum matrix element governing the transition probability between the valence and conduction bands and $\bar{m}$ the reduced effective mass $(m_n m_h)/(m_n + m_h)$. This energy dependence is closely followed in the vicinity of E_g by semiconductors like InSb. When the direct transitions between the valence and conduction bands are forbidden by a selection rule at $\mathbf{k} = 0$, the interband absorption coefficient is given by:

$$|\mathbf{p}_{cv}|^2 \ (2\bar{m})^{3/2} \frac{(\hbar\omega - E_g)^{3/2}}{\hbar\omega}. \tag{2.17}$$

The indirect transitions involve the creation (emission) or the annihilation (absorption) of a phonon for momentum conservation. It has been proposed that in the vicinity of the indirect band gap energy, the absorption coefficient with phonon annihilation was:

$$K_a \propto n(\omega_{ph}) \left[\hbar\omega - (E_g - \hbar\omega_{ph}) \right]^2, \tag{2.18}$$

where $n(\omega_{ph})$ in (2.18) is the occupation number for the annihilated phonon. Similarly, the absorption coefficient with phonon creation is

$$K_c \propto \left[1 + n(\omega_{ph}) \right] (\hbar\omega - E_g - \hbar\omega_{ph})^2, \tag{2.19}$$

where $n(\omega_{ph})$ in (2.19) is the occupation number for the created phonon. It follows that the indirect absorption should be proportional to the sum of expressions (2.18) and (2.19). The phonons ω_{ph} involved in the momentum-conserving (MC) process for the indirect band-gap absorption of semiconductors must have wave vectors $\mathbf{q}$ opposite to the electrons wave vectors $\mathbf{k}_{min}$ at the CB minimum given in Table 2.6. Values of the energies of MC phonons in germanium, silicon, and diamond are given in Table 2.10.

In indirect-gap semiconductors, this phonon-assisted electronic absorption is revealed by kinks in the vicinity of the electronic absorption edge. They are due to the different energies of the MC phonons involved as well as to the above-discussed different phonon processes (see [71], where the original results of [93] are presented). In indirect band-gap semiconductors, accurate determination of the band gap at low temperature relies mainly on the interpretation of the free exciton (FE) spectra.

The interband absorption of semiconductors produces free electrons and holes in the conduction and valence bands. These free carriers produce intrinsic photoconductivity above the band gap in adequate structures and several types of infrared photodetectors have been built on that principle [28].

When a semiconductor is illuminated with above band-gap radiation, excess electrons and holes are photo-created. They can form FEs or be trapped by ionized impurities, but their ultimate fate is their annihilation by thermal or radiative recombination. The FE case will be discussed in Sect. 2.3.2, but in direct band-gap semiconductors, electron–hole radiative recombination can also occur at an energy which can be very close to E_g if the pumping beam is kept at a low level. This can provide an accurate determination of E_g [94].

Band gap energies at RT and LHeT of different semiconductors and insulators are given in Appendix C.

2.3.1.4 Band Gap Dependences

The electronic band gaps are correlated with the cohesive energies of the materials and, for covalent crystals, with the atomic binding energies. This is why, for group-IV elements, the band gap decreases as the atomic number of the element increases. This rule is also followed by binary compounds with one element fixed and it allows for a very few exceptions like PbSe and PbTe with band gaps of 0.26 eV and 0.29 eV, respectively, at RT.

For elements or compounds with the same crystal structure, belonging to the same row of the periodic table, the band gap strongly opens with ionicity: E_g increases from 0.68 eV for covalent germanium to 1.42 and 2.67 eV for GaAs and ZnSe, respectively, and while α-Sn is a semimetal, InSb and CdTe are semiconductors with E_g of 0.18 and 1.53 eV, respectively (see also Appendix C). We consider below the effects of temperature, pressure, magnetic field, and isotopic composition on the value of the band gap.

Temperature

When temperature is lowered, the band gaps of semiconductors usually increase (see Appendix C). The role of the lattice vibrations on the temperature dependence of the band gap of semiconductors was treated by Fan [95], and comparisons were made with the experimental results for germanium and silicon known at that time, and since then, this effect has been actively investigated. The temperature dependence changes from relatively weak and apparently quadratic one at cryogenic temperatures to relatively strong one, approaching linear asymptotes at temperatures above the Debye temperature. Phenomenological relations describing this temperature dependence have been given ([96, 97], and references therein). Varshni [96] proposed an empirical expression:

$$E_g(T) = E_0 - \alpha T^2/(T + \beta), \tag{2.20}$$

where the band gap E_g can be direct or indirect, E_0 being the band gap at $0\,K$, and α and β constants to be determined from experimental results. An expression proposed by Pässler [98] is:

$$E_g(T) = E_0 - (\alpha\Theta_p/2)[\sqrt[p]{1 + (2T/\Theta_p)^p} - 1], \tag{2.21}$$

where α represents here the limit of the slope $S(T)$ of $-dE/dT$ when $T \to \infty$, Θ_p is a phonon temperature which can be considered as a substantial fraction of the maximum TA energy, and p is related to the phonon dispersion in the material. In these expressions, the value of E_0 can be slightly adjusted with reference to the experimental values at LHeT.

A fit of the temperature dependence of the energies at CPs of the germanium CB measured by ellipsometry in the visible region has been made by Viña et al. [99] assuming that the energy thresholds decrease with temperature proportionally to Bose–Einstein statistical factors for phonon emission plus absorption. This fit has also been tested by Pässler [98, 100] to model the temperature dependence of E_g of some semiconductors. The fitting expression is:

$$E_g(T) = E_B - a_B \left(1 + \frac{2}{\exp\left(\frac{\Theta}{T}\right) - 1}\right) \equiv E_0 - \frac{\alpha\Theta}{2}\left(\coth\left(\frac{\Theta}{2T}\right) - 1\right), \tag{2.22}$$

where $E_B - a_B$ is taken as E_0 and the effective phonon temperature as Θ. The last term of (2.22) results from a trivial algebraic manipulation, with α put equal to a_B/Θ.

The fit of the experimental values of $E_g(T)$ for GaAs [101] with expressions (2.20), (2.21), and (2.22) is represented in Fig. 2.8. It is seen there that in the high-temperature limit, expressions (2.20) and (2.22) both result in a linear variation of $E_g(T)$, but that in the low-temperature limit, Varshni's formula leads to a quadratic dependence of $E_g(T)$ while expression (2.22) produces a plateau.

The parameters used to fit the experimental points of Fig. 2.8 are given in Table 2.9.

The parameters of expression (2.21) obtained for many semiconducting and insulating crystals are listed in Table 2.1 of [100], and graphical comparisons

Fig. 2.8 Fitting by three different functions of the experimental values of $E_g(T)$ of GaAs (*black dots*). The *straight dotted line* is the high-temperature asymptote $E_0 - \alpha(T - \Theta/2)$ associated with expression (2.21). The *cross* marks the point $\Theta/2$ (112.8 K) where the asymptote crosses the E_0 level (1.5191 eV). The *inset* shows in more detail the variation of $E_g(T)$ produced with the three models in the low-temperature region (after [98])

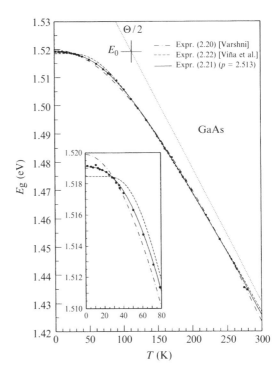

Table 2.9 Sets of parameters used in expressions (2.20), (2.21), and (2.22) to obtain the fit of the experimental values of E_g(T) of GaAs shown in Fig. 2.8

	E_0 (eV)	α (meV K^{-1})	β(K)	Θ (K)	p
Expression (2.20)	1.5199	0.7821	429.8		
Expression (2.21)	1.51909	0.4730		225.6	2.513
Expression (2.22)	1.5185	0.4513		221.7	
Expression (2.21)	1.519	0.472		230	2.44

The values of the last line for expression (2.21) are from [100]

between the experimental points and the fits with expressions (2.20), (2.21), and (2.22) for several materials are also given in this reference.

A few materials like lead sulphides or some copper halides are exceptions with a band gap *increasing* with temperature ([102], and references therein). A quantitative analysis of the temperature dependence of the energy gaps must take into account the electron–phonon interactions, which are the predominant contributions, and the thermal expansion effect. The effect of thermal expansion can be understood intuitively on the basis of the increase of the interatomic distances with temperature. A quantitative analysis of the electron–phonon contributions is more difficult and most calculations have been performed for direct band-gap structures [103], but calculations have also been performed for diamond-type semiconductors [104]. Multi-parameters calculations of the temperature dependence of band gaps in

Table 2.10 Energies (meV, cm^{-1} in brackets) of momentum-conserving phonons in transitions involving electrons at the *CB* minimum in germanium, silicon, and diamond

Ge[a]	Si[b]	C$_{diam}$[c]
TA(L): 7.9 (64)	TA(Δ): 18.4 (148)	TA(Δ): 87 (702)
LA(L): 26.3 (212)		
TO(L): 34.6 (279)	TO(Δ): 58.0 (468)	TO(Δ): 141 (1137)
LO(L): 30.7 (248)	LO(Δ): 56.3 (454)	LO(Δ): 163 (1315)

In germanium, this minimum lies at the L point of the surface of the BZ. In silicon and diamond, this minimum lies in the Δ direction (the $\langle 100 \rangle$ axis), at $0.84\mathbf{k}(X)$ and $0.76\mathbf{k}(X)$, where X is the intersection of the $\langle 100 \rangle$ axis with the surface of the BZ (see Table 2.6 and Fig. B.2)
[a] Johnson and Loudon [42]
[b] Vouk and Lightowlers [91] <20 K
[c] Collins et al. [92] LNT

semiconductors can be found in [97]. From the practical side, an increase of the absorption of CdTe near the RT band gap ($\sim$1.5 eV) has been correlated with 10.06-μm laser illumination [105]. It has been attributed to the temperature-induced shift of the band gap to lower energies generated by residual absorption of the crystal at 10.06 μm. The band-gap increase of silicon between RT and LHeT is $\sim$50 meV, and recent measurements at ultra-high resolution of the shift with temperature of the strongest B acceptor BE line of qmi ^{28}Si between 4.8 and 1.3 K show a band-gap increase of $\sim$1 GHz or 4 μeV in this temperature domain [106].

Hydrostatic Pressure

The positions of the energy bands are also pressure-dependent and the sign of the pressure-induced shift depends on the relative signs and magnitudes of the shifts of the *VB* and *CB*. As a rule the direct band gaps increase with pressure and the order of magnitude of the linear part of the increase for II–V and III–V compounds is 100 meV GPa^{-1} ([107], and references therein). In indirect-band-gap group-IV semiconductors, the situation is the opposite. For silicon, the indirect band gap decreases with pressure (-14.4 meV GPa^{-1}), and this is shown in Fig. 2.9.

In diamond, the inverse is observed, with a shift of $\sim +7$ meV GPa^{-1} ([109], and references therein). In germanium, the energy of the absolute minimum of the *BC* at the L point increases with pressure at a rate of $\sim$50 meV GPa^{-1} with respect to the *VB* maximum while the relative minimum along the Δ axis decreases at a rate of -15 meV GPa^{-1}. As the two minima are separated by $\sim$0.2 eV at atmospheric pressure, the crossover takes place for a pressure of $\sim$3 GPa [110]. It can be inferred from this that above 3 GPa, the band gap of germanium decreases with increasing pressure. For 3C-SiC, as for silicon, the shift is negative but nearly one order of magnitude lower (-1.9 meV GPa^{-1}) [111].

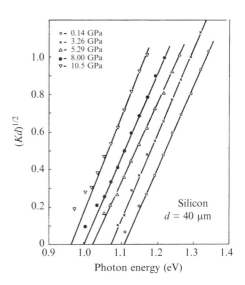

Fig. 2.9 Variation with hydrostatic pressure of the indirect band-gap absorption threshold of silicon at RT (10 kbar are taken as 9.81 GPa). K is the absorption coefficient and d the sample thickness. After Welber et al. [108]. © 1975 with permission from Elsevier

The increase of the direct band gap E_{dir} of germanium for pressures up to 12.3 GPa has been measured[4] at LHeT by Li et al. [72]. A least-square fit gives:

$$E_{dir} (eV) = 0.888\,(2) + 0.137\,(5)\,P - 0.0025\,(4)\,P^2,$$

while the s–o splitting Δ_{so} increases slowly with P according to:

$$\Delta_{so} (eV) = 0.294\,(5) + 0.0016\,(5)\,P,$$

where the pressure P is expressed in GPa.

At high hydrostatic pressures, the diamond-type lattice becomes unstable: for instance, the opacity of indirect-gap semiconductors silicon and germanium in the IR for pressures above about 12 GPa ($\sim$120 kbar) is attributed to a change from the cubic phase into the tetragonal β-Sn metallic phase. Diamond has been found to remain stable under pure hydrostatic pressure up to 140 GPa, but instabilities at comparable or higher pressures have also been reported [15].

In GaAs, the global sublinear increase of E_g at 7 GPa is $\sim$0.7 eV [112], but the energies of the Γ and L CB minimums increase with pressure while that of the X point decreases, with a crossover near 4 GPa, producing a change of the band-gap structure of this material from direct to indirect [113].

The decrease of the indirect band gaps of silicon and germanium under uniaxial stress along different axes has been measured indirectly by Bulthuis [114] at RT and the values lay between -50 and -100 meV GPa^{-1}.

---

[4]At this pressure, the semiconducting phase remained stable for several hours while the normal pressure for the phase transition at RT is 10.6 GPa (Menoni et al. Phys. Rev. B **34**, 362 (1986)).

Magnetic Field

In the presence of a magnetic field $\mathbf{B}$, the calculations of [115] have shown that the energy of electrons in metals becomes quantized in a plane perpendicular to the field, but remains continuous in the direction of the field. The result is an helical motion of the electrons in the plane perpendicular to $\mathbf{B}$ with the Landau energy $E_N = \hbar\omega_c(N + 1/2)$, when neglecting electron spin. In this expression, which can also be written $2\mu_B B(N + 1/2)$, $\omega_c = eB/m_e$ is the cyclotron pulsation and N can be 0 or a positive integer. In semiconductors, the band structure in the presence of a magnetic field becomes complicated in the direction perpendicular to the field as the continua of the valence and conduction bands split into different Landau levels ladders characterized by different total angular momenta J and spacings. In addition, the spin degeneracy of these Landau levels is removed (for s-type bands, the $J = 1/2$ level is split into sublevels with $M_J = +1/2$ and $-1/2$). When a degeneracy of the *VB* occurs for $\mathbf{k} = 0$, this degeneracy is also lifted by the magnetic field. The absorption coefficient for interband transitions in the presence of a magnetic field takes then the form:

$$K(\mathbf{B}, \omega) \propto \frac{1}{2}\hbar\omega_c \sum_l (\hbar\omega - E_N)^{-1/2}, \qquad (2.23)$$

where

$$E_N = E_g + (N + 1/2)\hbar\omega_c + \mu_B (g_c M_{Jc} - g_v M_{Jv}) \mathbf{B}. \qquad (2.24)$$

The effective mass involved in expression for the cyclotron frequency is a reduced mass $\bar{m}$ comparable to the one used for the interband transitions in expression (2.16) and for each band, $(N + 1/2)\hbar\omega_c$ is $1.1577 \times 10^{-1}(N + 1/2)B m_e/m^*$ (meV T^{-1}). The effective electron g-factors in the valence and conduction bands are g_c and g_v. The selection rules require that $\Delta N = 0$, whatever the polarization, with $\Delta M_J = 0$ for $\mathbf{B} \parallel \mathbf{E}$ and $\Delta M_J = \pm 1$ for $\mathbf{B} \perp \mathbf{E}$. The net result is that the onset of absorption is shifted to higher energies by $\hbar\omega_c/2$ and the absorption displays an oscillatory behaviour (see, for instance, [116]).

Isotopic Composition

Finally, a band-gap change can also occur when the isotopic composition of the crystal changes. There are two contributions to the change of the band gap with the isotopic mass. One is due to a volume change of the crystal due to the combined effects of the mass-dependence of the zero-point motion of the atoms, and to the anharmonicity of the potential. In silicon, for instance, the relative decrease $\Delta a/a$ of the lattice constant between qmi ^{28}Si and ^{30}Si has been measured to be 6×10^{-5} at 30 K [117]. In indirect-band-gap semiconductors, the related contribution to the band gap is positive when the relative volume change $\Delta V/V$ is positive, and it is given by Hayama et al. [118]:

$$\Delta E_g^V = \Xi \, \Delta V / V,$$

where Ξ is the deformation potential for the lowest indirect energy gap, which is $+1380\,\mathrm{meV}$ for silicon. For monoatomic semiconducting crystals, $\Delta V / V$ is negative when the change ΔM in the isotopic mass is positive (the zero-point frequency is smaller). The relative volume change $(3\Delta a / a)$ from qmi $^{28}\mathrm{Si}$ to $^{30}\mathrm{Si}$ at 30 K is -1.8×10^{-4}. For silicon, this yields a volume contribution of $-0.25\,\mathrm{meV}$ between $^{28}\mathrm{Si}$ and $^{30}\mathrm{Si}$. The second contribution is due to the change of the electron–phonon interaction, and this contribution is positive when ΔM is positive. In silicon, it is one order of magnitude larger ($+2.09\,\mathrm{meV}$ between $^{28}\mathrm{Si}$ and $^{30}\mathrm{Si}$) than the contribution of the volume change, giving for silicon at low temperature an overall isotopic mass coefficient (IMC) increase ΔE_g^M of $+0.92\,\mathrm{meV}\,\mathrm{u}^{-1}$ for the band gap of silicon.

Natural diamond is $^{12}\mathrm{C}_{0.989}{}^{13}\mathrm{C}_{0.011}$, but qmi $^{13}\mathrm{C}_{\mathrm{diam}}$ crystals have been grown and their physical properties investigated. In a 99% monoisotopic $^{13}\mathrm{C}_{\mathrm{diam}}$ sample, an increase of 13.6 meV of E_g – a relative increase of 0.25% – has been measured by comparison with $^{\mathrm{nat}}\mathrm{C}_{\mathrm{diam}}$ [119]. The major contribution to this upward shift has been attributed to the effect of the isotope change on the mean-square displacement of the crystal atoms in relation with the electron–phonon coupling. The other (positive) contribution[5] is the effect of the negative volume change (-4.5×10^{-4} deduced from [120]) between $^{12}\mathrm{C}_{\mathrm{diam}}$ and $^{13}\mathrm{C}_{\mathrm{diam}}$ due to the decrease of the zero-point vibration frequency [119]. Spectroscopic IS values of E_g have also been obtained in silicon from the ISs of excitons bound to shallow impurities (BEs) measured by PL at LHeT [118, 121] and a value of the IMC of $+0.98\,\mathrm{meV}\,\mathrm{u}^{-1}$ can be deduced from these results. For germanium, a value of the IMC of E_g of $+0.36\,\mathrm{meV}\,\mathrm{u}^{-1}$ is reported [122, 123].

In compound crystals, the value of the IMC generally differs for anion or cation isotopic substitution: for cubic ZnS, the S IMC measured keeping the Zn mass unchanged is $+0.40\,\mathrm{meV}\,\mathrm{u}^{-1}$ while the Zn IMC keeping $^{32}\mathrm{S}$ as the anion is $+0.79\,\mathrm{meV}\,\mathrm{u}^{-1}$ [124]. The sign of the IS can depend on the nature of the atom replaced: in CuCl, it has been observed that the direct band gap (3.399 eV at LHeT) increased by 364 μeV u^{-1} when increasing the mass of the Cl atom, but *decreased* by 76 μeV u^{-1} when increasing the mass of the Cu atom [51]. One might expect that a simple explanation could have been found from the usual band structure of many compound crystals, for which the upper *VB* corresponds to the valence electrons of the most electronegative element and the *CB* band to the valence electrons of the most positive element. However, for CuCl, the role of the phonon modes in the gap renormalization is determinant and it explains the above isotope effects as well as the increase of the band gap with temperature [51].

In nanocrystals with average radii typically below 10 nm, the band gap increases due to confinement. Calculations of the effect of confinement on the excitonic gap of

---

[5]In the expression Δ_2 for this contribution in [119], parameter a of expression (4) of this reference becomes negative because the stress considered is of the expansive type.

some II–VI and III–V direct-gap semiconductors have been performed by Tomasulo and Ramakrishna [125], and comparisons with the experimental situation are given for CdS and CdSe. The review by Yoffe [126] provides a good account of the optical properties of nanocrystals in compound semiconductors while the one by Sattler [127] is more general.

In semiconductors with small band gaps and small electron effective masses, a high concentration of n-type dopants produces a large accumulation of electrons in the *CB*. This can block the interband transitions with the lowest energies and efficient interband absorption takes places only at energies larger than E_g, producing an apparent increase of the band gap. The above explanation was provided independently by Burstein [128] and Moss [129] to explain the high-energy shift of the band gap observed in InSb with increasing free electron concentrations (from a value of 0.18 eV up to an apparent band-gap value of $\sim$0.6 eV for an electron concentration of $\sim 10^{19}$ cm^{-3}). This energy shift, coined the Burstein–Moss effect, has later been observed in PbS [130] and GaSb [131].

2.3.2 Excitons

In a semiconductor, a steady-state concentration of out-of-equilibrium electron–hole pairs can be produced near from its surface by continuous or pulsed illumination at energies higher than E_g or by irradiation with electrons in the $\sim$10–100 keV range, depending on the value of the band gap. When this is performed at low temperature, the Coulomb attraction between an electron not too distant from a hole can single out the pair as an exciton, and a steady-state concentration of excitons co-exists with free electrons and holes. Excitons can be seen as pseudo-hydrogenic atoms where the role of the positive ion is taken by the positive hole, and in a given material, they are characterized by the binding energy E_{ex} between the two particles and by a finite lifetime which depends on several parameters. These excitons are free to propagate as a whole in the crystal during their lifetimes, therefore, they are called FEs. Nonradiative recombination of FEs by phonon emission has a small probability because of the energies involved, the most efficient decay being electron–hole radiative recombination, which can be followed by PL measurements. The direct production of FEs can also be followed by absorption measurements.

The exciton mass to be considered in this centre-of-mass motion is the sum of the masses of the two particles. Their binding energy E_{ex} obviously depends on the effective masses of the particles, on the static dielectric constant of the crystal and on its ionicity, through free-carrier screening. The dissociation of these so-called Mott–Wannier excitons results in a free electron and a free hole. The energy required to produce directly such a pair is often referred to as the excitonic gap $E_{gx} = E_g - E_{ex}$. For a direct-gap semiconductor with spherical energy bands at $\mathbf{k} = 0$, the exciton levels can be fitted to a hydrogen-like series whose energies are given by:

$$E_{ex}(n) = R_{\infty\,eff}/n^2, \tag{2.25}$$

where $R_{\infty\text{eff}}$ is an effective Rydberg $R_\infty \bar{m}/\varepsilon_s^2$, where R_∞ is weighted by the reduced effective exciton mass $\bar{m}$, where $(\bar{m})^{-1} = (m_n)^{-1} + (m_h)^{-1}$, and by the static dielectric constant ε_s. By analogy with the H atom, the dimension of such an exciton can be defined by an effective Bohr radius a_0^* expressed as $\varepsilon_s \hbar^2/\bar{m}e^2$ in cgs units. It is typically of the order of 1–10 nm. The determination of the energies of the FEs in some semiconductors has allowed obtaining accurate values of E_g at LHeT.

For instance, the absorption of the FEs in direct-band-gap CdTe has been measured at 2 K [132] and at 4 K [133]. The energies of the transitions $E_g - E_{ex}(n)$ measured to create an exciton in the n = 1 and 2 states are very close in these two references (1.5965 and 1.6142 eV, respectively [132] and 1.5961 and 1.6037 eV, respectively [133]. The binding energy $E_{ex}(1)$, or simply E_{ex}, deduced from these two values of the transition energy using expression (2.25) are also very close (10.3 and 10.1 meV, respectively). Figure 2.10 shows the spectrum of the FE absorption in CdTe at LHeT.

The theoretical value of E_{ex} calculated by using the VB parameters of CdTe for the hole part of the exciton is 9.05 meV [132]. From the above-measured energies of the 1s transition of the FE and from its binding energy, an average value of 1.6065 eV is obtained for the band gap of CdTe at LHeT.

In indirect-gap semiconductors, where the electron and the hole of the exciton have different wave vectors, direct absorption of FEs like the one shown in Fig. 2.10 is forbidden because momentum is not conserved in such transitions. This is also true for FE recombination only involving the emission of one photon at energy E_{gx}. In PL, what is observed for these materials are FE recombination lines at energies smaller than E_{gx}, assisted by the emission of one or two MC phonons. The FE creation measured in absorption is difficult to follow because it involves the simultaneous creation of a MC phonon, the corresponding absorption taking place

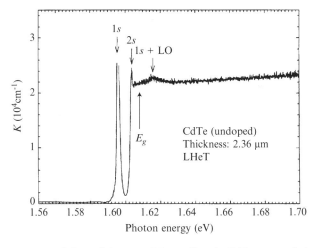

Fig. 2.10 Absorption coefficient of the 1s and 2s FE lines in CdTe at a resolution of 0.1 meV $(0.8\,\text{cm}^{-1})$ between 12502 and 13711 cm^{-1}. The absorption of the 1s line is truncated at 24000 cm^{-1}. The weak 1s + LO(Γ) replica is indicated at about 1.617 eV. (after [133])

at energies above E_g [134]. Energies of MC phonons in Ge, Si, and C_{diam} deduced from the energies of phonon-assisted FE lines are given in Table 2.10.

In indirect-band-gap semiconductors, the measurement of the absorption leading to the creation of FEs has also shown the existence of a splitting[6] of the FE ground state, which can be explained by the departure from cubic symmetry by the introduction of the *CB* ellipsoids [135]. Numerical values of this splitting have been calculated by Lipari and Altarelli [136] for different semiconductors. In ^{nat}Ge, the experimental values of E_{ex} for the indirect FE are 3.14 and 4.15 meV, yielding a ground state splitting of 1 meV [137]. In ^{nat}Ge, the peak of the LA-phonon-assisted recombination of the FE at LHeT has been reported at 713.4 meV with a high-energy shoulder associated with the split component [138]. This peak has often been used as a reference in the study of the electron–hole droplets (EHD) in germanium. In the 1970s, there were many studies of the internal absorption of the indirect FE corresponding to transitions between ground and excited states. These transitions have been measured in germanium in the very far IR ($\sim$1–4 meV) under band-gap excitation (see, for instance, [139–142]). The energies of the observed lines are in good agreement with the energies of the transitions predicted from the calculations taking into account the FE ground-state splitting.

For silicon, a very weak peak at 1.1545 eV, observed at LHeT, is attributed to the no-phonon FE recombination and it corresponds to E_{gx} (a value of E_{gx} of 1.1551 eV has also been given [143]). However, the other FE recombination lines involve the assistance of MC phonons [91,144]; the most intense one-phonon FE recombination band observed at LHeT involves a TO(Δ) phonon, and its energy has been given as 1.09654 eV [91]. The effect of the FE ground state splitting cannot be detected on these peaks, but a much weaker TA-phonon-assisted recombination feature reported near 1.136 eV [145] shows a doublet structure due to this splitting, with a separation of 0.3 meV, in good agreement with previous experimental determinations ([146], and references therein). From the determination of the relevant phonon energies, the above values of E_{gx}, and the accepted value of E_g in silicon at LHeT (1.170 eV), values of the FE binding energy E_{ex} between 14.7 and 15.5 meV are obtained, and an average value of 15 meV is reasonable.

Values of 5.416 and 5.409 eV were measured for E_{gx} in diamond, giving a FE binding energy $\sim$80 meV [147]. The existence of two values of the excitonic gap was explained by the small value of the s–o splitting of the *VB* maximum in diamond ($\sim$6–13 meV), needing to consider two different holes effective masses in the exciton reduced mass. Higher resolution studies using derivative spectroscopy allowed to resolve the MC TA- and TO-assisted FE recombination features into two sets of transitions[7] separated by $\sim$11 meV. It was suggested that this splitting, larger than the one observed at low resolution, was due to some spin coupling between the electron and the hole of the exciton, giving singlet and triplet excitonic states [148].

---

[6]This splitting is often called valley-orbit splitting by analogy with the so-called splitting of the degenerate 1*s* ground state of the effective mass donors in silicon and germanium.

[7]These two sets were also observed in the recombination spectrum of the B-related BE [148].

However, this explanation has been contested ([149], and references therein) and the problem seems to be still open.

The IMC of the excitonic gap E_{gx} of the FE has been measured in qmi Ge samples as $+0.36\,\text{meV}\,\text{u}^{-1}$ [123]. In ^{nat}Ge, ($^{72.59}$Ge), values of E_{gx} between 740.5 and 741.0 meV at LHeT have been given. By adopting a value of 740.8 meV and by adding the FE binding energy taken as 4.2 meV, a value of $\sim$745 meV for the band gap of ^{nat}Ge at LHeT is obtained.

The absorption due to the formation of direct FEs associated with the Γ_7^- CB has also been observed at energies above E_g in very thin silicon and germanium samples [150]. This absorption, which is relatively sharp, is also relatively strong, but resonant with the indirect continuum absorption. For germanium, the energy $E_{ex}(\Gamma_7^-)$ of the direct FE, deduced from Fig. 2.5 of [72], is $\sim$1.5 meV. This value is smaller than the one of the indirect FE ($\sim$4 meV).

Cu_2O is a cubic direct-band-gap semiconductor. There exists a huge amount of experimental work on the spectroscopy of FEs in this material since the first reports [151], and we just give here the main points relevant to their spectroscopy. When considering the coupling of a hole from the $^2\Gamma_7^+$ VB (pseudospin 1/2) of Cu_2O to an electron from the $^2\Gamma_6^+$ CB, excitons with total spin 0 or 1 can be formed, and by analogy with parahelium and orthohelium, the exciton configurations with spin 0 and 1 are called paraexcitons and orthoexcitons, respectively. Transposed to the IRs of O_h, the total symmetry of the 1s exciton is represented by $\Gamma_7^+ \times \Gamma_6^+ \times \Gamma_1^+ = \Gamma_5^+ + \Gamma_2^+$, where Γ_1^+ is the symmetry of the 1s excitonic envelope, and the degeneracy between the triply degenerate Γ_5^+ and nondegenerate Γ_2^+ states is removed by $e-h$ exchange interaction, giving Γ_5^+ for the orthoexciton and Γ_2^+ for the paraexciton [152]. The orthoexcitons are characterized by two main hydrogen-like series, the yellow one related to the 2.174 eV band gap (570 nm), and the green one, to the 2.304 eV band gap (538 nm). The absorption lines of the yellow series with n $\geq$ 2 are forbidden weakly active electric–dipole transitions involving np excitonic states [153], and the 1s transition of the orthoexciton is only quadrupole-allowed [154]. The absorption of the np yellow series is displayed in Fig. 2.11.

Peak absorption coefficients of these lines are given in [156], together with a discussion on their oscillator strengths and asymmetrical shape. For the 2p, 3p, 4p, and 5p lines, peak absorption coefficients at LHeT of the order of 180, 280, 200, and 100 cm^{-1}, respectively, after correction for the continuum, are given in the same reference, and they can be considered as orders of magnitude of the intensities of these excitonic lines.

The energy of the orthoexciton 1s quadrupole line is 2.0337 eV (16403 cm^{-1}) and its intensity is about one order of magnitude smaller than that of the np lines [157]. The 1s paraexciton line can be observed only under external perturbation: it has been observed by PL in a sample submitted to a uniaxial stress of 10 MPa along a $\langle 100 \rangle$ axis at LHeT [158] and by absorption in a polycrystalline sample submitted to a magnetic field larger than $\sim$6 T [159] at an energy of $\sim$2.021 eV (16303 cm^{-1}). Phonon-assisted PL lines involving the 1s paraexciton have also been reported in the absence of external perturbation near LHeT [152]. The admitted splitting of the paraexciton–orthoexciton levels is 11.9 meV or

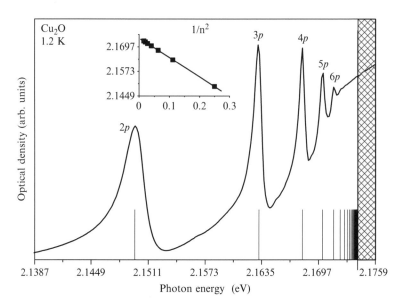

Fig. 2.11 Absorption of the np yellow excitonic series of Cu_2O between 17250 and 17535 cm^{-1}. The spacing between marked abscissa units is 6.2 meV. The hatched region corresponds to the continuum absorption. The *inset* shows the empirical Rydberg fit E_{np} (eV) $= 2.1741 - 0.0975/n^2$ to the positions of the np peaks and the *vertical lines* correspond to the np energies calculated from this expression (after [155])

96 cm^{-1} [152]. The energies of the np exciton lines of the green series have also been fitted to a Rydberg-like expression. The value taken from the review of [160] is E_n green (eV) $= (2.3058 - 0.1514/n^2)$. The spectra of the even-parity ns excitons of the yellow and green series cannot be observed by standard absorption measurements, but they have been detected by two-photon absorption spectroscopy, where a virtual odd-parity state is created [161]. These spectra are characterized by a splitting of the $n = 3$, 4, and 5 states of the yellow series into s and d levels and by the observation of the $1s$ exciton line of the green series at 2.1378 eV, an energy smaller than that of $2p$ of the yellow series (2.1473 eV). Measured energies of the ns and np lines of the yellow and green excitonic series are given in [161]. Energies of the np lines can also be obtained from the above Rydberg fits. The values of the FEs binding energies in Cu_2O deduced from these measurements are ~152 and 168 meV for the yellow and green series, respectively.

Induced absorption spectra for transitions between the $1s$ and np orthoexciton states, where the $1s$ population is produced by TPA, have also been reported in the medium IR region. Their energy positions are in good agreement with the positions expected from the energies measured in the visible region of the spectrum [162].

The FEs can diffuse in the crystal during their lifetime and become weakly trapped by electronic interaction to some impurities and complex centres to become bound excitons (BEs). The binding energy of the electron to the hole of BEs is larger than the binding energy of the FE, the difference being the as already mentioned localization energy of the BE.

Several mechanisms are involved in the lifetime of FEs, usually probed by the time decay of the radiative recombination. The lifetimes of the FEs depend on the nature of the band gap (direct or indirect), on the temperature, and on the chemical nature and concentration of the FAs in the sample with which the FEs can form BEs. As a rule, compared to indirect-band-gap semiconductors, the lifetime of the FEs in direct-band-gap semiconductors is smaller by two or three orders of magnitude because in the latter materials, the FE recombination does not require the assistance of a MC phonon. Several values of the low-temperature lifetime of FEs in silicon ranging from $\sim$0.2 to $\sim$50 μs have been reported ([163], and references therein). In this reference, the FE lifetime has been measured as a function of temperature in high-purity FZ p- and n-type silicon samples. From a temperature of $\sim$12 K, the FE lifetime was shown to decrease nonmonotonously with temperature and to depend also of the concentration and nature of dopant traps, which convert FEs into BEs: in the p-type sample, it varied from $\sim$1 to 30 μs between 2 and 12 K and in the n-type sample, from $\sim$0.6 to 50 μs, in the same temperature range. These variations could be reasonably fitted to a model taking into account the capture, release, and recombination of the FEs at neutral B and P atoms [163]. At temperatures above $\sim$14 K, the FE lifetime decreases because of thermal dissociation [164], and this is correlated to a decrease of the intensity of the FE recombination luminescence (see, for instance, Fig. 14 of [165]). For germanium, a lifetime of the order of 3 μs has been reported without specific indication of the doping level [166].

In ionic crystals, the exciton can be considered as an ion in an excited state. This excitation, called a Frenkel exciton, can also propagate in the crystal through similar ions excitations. The binding energies of the Frenkel excitons are significantly larger than those of the Mott–Wannier excitons. A thorough treatment of the optical properties of excitons in semiconductors and insulators can be found in the books by Yu and Cardona [2] and by Klingshirn [167].

The FE binding energy increases with the ionicity of the semiconductor: it is nearly 2% of the band gap for ZnO ($\sim$60 meV) and about 6% for CuCl ($\sim$190 meV). This rather high value is due to the change from dominant sp^3 hybridization of the orbitals for most of the semiconductors to p–d hybridization for CuCl. The FE binding energies for covalent or mainly covalent crystals are smaller and for Ge, Si, and C_{diam}, the indirect FE binding energies ($\sim$4, 15, and 80 meV, respectively) correspond to 0.56%, 1.32%, and 1.46% of the band gaps.

2.3.3 Electron–Hole Droplets (EHDs), Biexcitons, and Polyexcitons

The analogy of FEs with pseudo-hydrogenic atoms had led to suggest the existence of excitonic molecules (biexcitons), with characteristic binding energies E_{bx}, and of hydrogenic ions [168]. The possibility of electron–hole condensation into a metallic electron–hole liquid (EHL) in germanium was pointed out in 1968 by Keldysh [169] on the basis of electrical measurements made on pure germanium at LHeT by Rogachev [170].

More generally, in a semiconductor, when the exciton concentration becomes very high so that exciton states start to overlap, the exciton gas transforms into an EHL. In some materials with multivalley band structure, the plasma phase can have a lower energy than the exciton gas [171], with an electron–hole pair binding energy $E_l > E_{ex}$. The EHL usually exists in the form of EHD, whose shapes are approximately spherical and whose dimension can reach a few micrometers. The critical temperature T_c below which EHDs are stable [172] and the electron–hole pair concentration n_l in this liquid phase depend on the semiconductor. Like the FEs, the EHDs can diffuse through a crystal, and this diffusion can be modified by the presence of external stresses. At temperatures above T_c, the EHDs dissociate into excitons and polyexcitons.

In 1966, a strong recombination line at ∼1.08 eV observed below 10 K in high-purity silicon was attributed to the recombination of an electron–hole pair in an excitonic molecule made of two electrons and two holes [173]. This line, also phonon-assisted for an indirect band gap, was much more intense than the TO-assisted FE line near 1.10 eV. It was later correctly reattributed to the recombination of an electron–hole pair of EHDs, peaking at 1.0815 eV [174, 175]. A broad recombination-radiation peak, observed at 709.6 meV in intrinsic germanium illuminated at LHeT by the output of filtered Hg arc or of a Xe lamp at levels producing in the sample an average carrier concentration $\geq 1 \times 10^{13}\,cm^{-3}$, was ascribed to electron–hole recombination in a new condensed phase of nonequilibrium carriers, actually EHDs [176]. Since then, many recombination lines or bands due to this EHL have observed in direct- and indirect-band-gap semiconductors and insulators (see, for instance, [177]).

Let us take here the example of germanium to illustrate the main characteristics of the EHDs. The average binding energy of the electron–hole pair in a droplet, deduced from the onset energy of these lines, must be larger than the one of the FE, in order for the droplet to be stable against the exciton gas. In germanium, it is ∼6.0 meV, compared to 4.2 meV for the FE. This leads to a condensation energy of the EHD, ∼2 meV [178]. This condensation energy represents the energy required to evaporate EHDs into an exciton gas.

In the same way as comparing FEs to H atoms, biexcitons can be compared to hydrogen molecules. A very weak TA-phonon replica observed in high-purity silicon at 1.8 K by PL at ∼1.1345 eV, together with the FE TA-phonon replica, has been attributed to the radiative recombination of one exciton in the biexciton complex [179]. A very weak PL line at ∼2.308 eV, about twice the value of the silicon band gap, in the green region of the visible spectrum, has also been observed at 1.6 K in silicon [143, 145]. It has been attributed to the complete no-phonon annihilation of a biexciton. In high-purity silicon, radiative recombination at 20 K of triexcitons, tetraexcitons, and pentaexcitons involving the annihilation of two electron–hole pairs has also been reported in the same spectral region [180].

In silicon, results on the equilibrium between FEs and biexcitons populations confined in strain-induced potential wells have been reported in [181].

In direct-gap semiconductors, the formation of biexcitons has been investigated in cubic ZnS by PL measurements. In this material where the binding energy E_{ex} of the FE is $\sim$36 meV, the binding energy of the biexciton[8] is $\sim$9 meV, its radiative lifetime being 52 ps at 12 K ([182], and references therein).

2.4 Conclusion

The domains of transparency of pure crystals are limited by the intrinsic absorption processes described in this chapter. The presence in these crystals of FAs with shallow-centre characteristics releasing free carriers in the conduction or valence band produces free-carrier absorption (see, for instance, [3], pp. 245–249). The trapping of the free carriers by the ionized shallow centres at temperatures depending on their ionization energies E_i results in neutral hydrogen-like centres which give also a photoionization absorption continuum at energies above E_i. These extrinsic absorption processes limit the transparencies of the crystals, and the absorption of the photoionization continuum can eventually modify their colour, as for the boron acceptor in diamond. For large concentrations of shallow centres, the crystals undergo a metal-to-insulator transition (MIT) due to the transformation of the discrete shallow impurity levels into a impurity band continuum over all the centres, and they become opaque [183]. The critical concentration where this transformation takes place is $\sim 1.9 \times 10^{17}$ cm^{-3} in Ga-doped germanium, but it rises to $\sim 4 \times 10^{20}$ cm^{-3} in B-doped diamond.

References

1. D.L. Greenaway, G. Harbeke, *Optical Properties and Band Structure of Semiconductors* (Pergamon, Oxford, 1968)
2. P.Y. Yu, M. Cardona, *Fundamentals of Semiconductors*, 3rd edn. (Springer, Berlin, 2001)
3. M. Balkanski, R.F. Wallis, *Semiconductor Physics and Applications* (Oxford University Press, Oxford, 2000)
4. R.M. Lambert, R.D. Enoch, Rotable polarizers for the near infra-red. J. Phys. E **2**, 799–801 (1969)
5. R.H. Lyddane, R.G. Sachs, E. Teller, On the polar vibrations of alkali halides. Phys. Rev. **59**, 673–676 (1941)
6. P.H. Borcherds, R.L. Hall, K. Kunc, G.F. Alfrey, The lattice dynamics of gallium phosphide. J. Phys. C **12**, 4699–4706 (1979)
7. J.L. Yarnell, J.L. Warren, R.G. Wenzel, P.J. Dean, in *Neutron Inelastic Scattering Symposium*, Copenhagen, 1968, vol. I (IAEA, Vienna, 1968), p. 301

[8]By comparison, the ionization energy of the H atom is 13.6 eV and the dissociation energy of the H_2 molecule $\sim$4.5 eV.

8. O. Madelung (ed.), *Semiconductors Group IV elements and III–V compounds* (Springer, Berlin, 1991)
9. B. Ulrici, E. Jahne, Phonon frequencies of InP derived from two-phonon absorption. Phys. Stat. Sol. B **74**, 601–607 (1976)
10. P. Giannozzi, S. de Gironcoli, P. Pavone, S. Baroni, *Ab initio* calculation of phonon dispersions in semiconductors. Phys. Rev. B **43**, 7231–7242 (1991)
11. V.G. Plotchinenko, V.O. Nazaryants, E.B. Kryukova, V.V. Koltashev, V.O. Sokolov, A.V. Gusev, V.A. Gavva, M.F. Churbanov, E.M. Dianov, Near to mid-IR refractive index of ^{28}Si, ^{29}Si and ^{30}Si monoisotopic single crystals. Quantum Electron. **40**, 753–755 (2010) (Turpion Ltd)
12. T. Ruf, J. Serrano, M. Cardona, P. Pavone, M. Pabst, M. Krisch, M. D'Astuto, T. Suski, I. Grzegory, M. Leszczynski, Phonon dispersion curves in wurtzite structure determined by inelastic X-ray scattering. Phys. Rev. Lett. **86**, 906–909 (2001)
13. P.W. Sparks, C.A. Swenson, Thermal expansion from 2 to 40 °K of Ge, Si, and four III–V compounds. Phys. Rev. **163**, 779–790 (1967)
14. R. Trommer, H. Müller, M. Cardona, P. Vogl, Dependence of the phonon spectrum of InP on hydrostatic pressure. Phys. Rev. B **21**, 4869–4868 (1980)
15. F. Occelli, P. Loubeyre, R. Letoullec, Properties of diamond under hydrostatic pressures up to 140 GPa. Nature Mater. **2**, 151–154 (2003)
16. K.C. Hass, M.A. Tamor, T.R. Anthony, W.F. Banholzer, Lattice dynamics and Raman spectra of isotopically mixed diamond. Phys. Rev. B **45**, 7171–7182 (1992)
17. R. Vogelgesang, A.D. Alvarenga, H. Kil, A.K. Ramdas, S. Rodriguez, M. Grimditch, T.R. Anthony, Multiphonon Raman and infrared spectra of isotopically controlled diamond. Phys. Rev. B **58**, 5408–5416 (1998)
18. M. Cardona, M.L.W. Thewalt, Isotope effects on the optical spectra of semiconductors. Rev. Mod. Phys. **77**, 1173–1224 (2005)
19. S. Nakashima, H. Harima, Raman investigations of SiC polytypes. Phys. Stat. Sol. A **162**, 39–64 (1997)
20. B. Pajot, C.J. Fall, J.L. Cantin, H.J. von Bardeleben, R. Jones, P.R. Briddon, F. Gendron, Low-frequency vibrational spectroscopy in SiC polytypes. Mater. Sci. Forum **235–236**, 349–352 (Trans Tech, Zurich, 2001)
21. A. Mooradian, G.B. Wright, First order Raman effect in III–V compounds. Solid State Commun. **4**, 431–434 (1966)
22. F. Abelès, J.P. Mathieu, Calcul des constantes optiques des cristaux ioniques dans l'infrarouge à partir du spectre de réflexion. Ann. Phys. (13ème série) **3**, 5–32 (1958)
23. E. Burstein, Interaction of phonons with photons: infrared, Raman and Brillouin spectra, in *Phonons and Phonon Interactions in Solids*, ed. by T.A. Bak (Benjamin, New York, 1964), pp. 276–342
24. D.W. Berreman, Infrared absorption at longitudinal optic frequency in cubic crystal films. Phys. Rev. **130**, 2193–2198 (1963)
25. M.D. Sciacca, A.J. Mayur, E. Oh, A.K. Ramdas, S. Rodriguez, J.K. Furdyna, M.R. Melloch, C.P. Beetz, W.S. Yoo, Infrared observation of transverse and longitudinal polar optical modes of semiconductor films: normal and oblique incidence. Phys. Rev. B **51**, 7744–7752 (1995)
26. C.A. Klein, T.M. Hartnett, C.J. Robinson, Critical-point and phonon frequencies of diamond. Phys. Rev. B **45**, 12854–12863 (1992)
27. T. Ruf, M. Cardona, C.S.J. Pickles, R. Sussmann, Temperature dependence of the refractive index of diamond up to 925 K. Phys. Rev. B **62**, 16578 (2000)
28. J.T. Houghton, S.S. Smith, *Infrared Physics* (Oxford University Press, London, 1966)
29. D. Campi, C. Papuzza, Refractive index dispersion in group IV and binary III–V semiconductors: comparison of calculated and experimental values. J. Appl. Phys. **57**, 1305–1310 (1985)
30. G.A. Samara, Temperature and pressure dependence of the dielectric constant of semiconductors. Phys. Rev. B **27**, 3494–3505 (1983)
31. R.A. Faulkner, Higher donor excited states for prolate-spheroid conduction bands: a reevaluation of silicon and germanium. Phys. Rev. **184**, 713–721 (1969)

32. J.R. Hardy, S.D. Smith, Two-phonon infrared lattice absorption in diamond. Phil. Mag. **6**, 1163–1172 (1961)
33. E.S. Koteles, W.R. Datars, Two-phonon absorption in InSb, InAs, and GaAs. Can. J. Phys. **54**, 1676–1682 (1976)
34. P.J. Gielisse, S.S. Mitra, J.N. Plendl, R.D. Griffis, L.C. Mansur, R. Marshall, E.A. Pascoe, Lattice infrared spectra of boron nitride and boron monophosphide. Phys. Rev. **155**, 1039–1046 (1967)
35. J.L. Birman, Theory of the infrared and Raman processes in crystals: selection rules in diamond and zincblende. Phys. Rev. **131**, 1489–1496 (1963)
36. G.A. Slack, S. Roberts, Zone boundary acoustic phonons in adamantine compounds from far-infrared absorption measurements. Phys. Rev. B **3**, 2613–2618 (1971)
37. E.S. Koteles, W.R. Datars, G. Dolling, Far-infrared absorption in InSb. Phys. Rev. B **9**, 572–582 (1974)
38. P.A. Temple, C.E. Hathaway, Multiphonon Raman spectrum of silicon. Phys. Rev. B **7**, 3685–3697 (1973)
39. M. Ikezawa, M. Ishigame, Far-infrared absorption due to two-phonon difference process in Si. J. Phys. Soc. Jpn. **50**, 3734–3738 (1981)
40. P.G. Sennikov, T.V. Kotoreva, A.G. Kurganov, B.A. Andreev, H. Niemann, D. Schiel, V.V. Emtsev, H.J. Pohl, Spectroscopic parameters of the absorption bands related to the vibrational modes of carbon and oxygen impurities in silicon enriched with ^{28}Si, ^{29}Si, and ^{30}Si isotopes. Semiconductors **39**, 300–307 (2005)
41. M. Balkanski, M. Nusimovici, Interaction du champ de rayonnement avec les vibrations du réseau aux points critiques de la zone de Brillouin du silicium. Phys. Stat. Sol. **5**, 635–647 (1964)
42. F.A. Johnson, R. Loudon, Critical-point analysis of the phonon spectra of diamond, silicon and germanium. Proc. Roy. Soc. Lond. A **281**, 274–290 (1964)
43. S.J. Fray, F.A. Johnson, J.E. Quarrington, N. Williams, Lattice bands in germanium. Proc. Phys. Soc. **85**, 153–158 (1965)
44. M. Ikezawa, T. Nanba, Two-phonon difference absorption spectra in Ge crystals. J. Phys. Soc. Jpn. **45**, 148–152 (1978)
45. R.H. Parmenter, Symmetry properties of the energy bands of the zinc blende structure. Phys. Rev. **100**, 573–579 (1955)
46. L.B. Bouckaert, R. Smoluchowski, E. Wigner, Theory of the Brillouin zones and symmetry properties of wave functions in crystals. Phys. Rev. **50**, 58–67 (1936)
47. M.L. Cohen, T.K. Bergstresser, Band structures and pseudopotential form factors for fourteen semiconductors of the diamond and zinc-blende structures. Phys. Rev. **141**, 789–796 (1966)
48. T.N. Morgan, Symmetry of electron states in GaP. Phys. Rev. Lett. **21**, 819–823 (1968)
49. O. Madelung (ed.), *Semiconductors Other than Group IV Elements and III–V Compounds. Data in Science and Technology* (Springer, Berlin, 1992)
50. W.C. Dash, R. Newman, Intrinsic optical absorption in single-crystal germanium and silicon. Phys. Rev. **99**, 1151–1155 (1955)
51. A. Göbel, T. Ruf, M. Cardona, C.T. Lin, J. Wrzesinski, M. Steube, K. Reimann, J.C. Merle, M. Joucla, Effects of the isotopic composition on the fundamental gap of CuC*l*. Phys. Rev. B **57**, 15183–15190 (1998)
52. L. Kleinman, K. Mednick, Self-consistent energy bands of Cu_2O. Phys. Rev. B **21**, 1549–1553 (1980)
53. F. Biccari, Ph.D thesis, Sapienza Universitá di Roma, Rome (2009)
54. J.B. Grun, M. Sieskind, S. Nikitine, Détermination de l'intensité des forces d'oscillateur de la série verte de Cu_2O aux basses températures. J. Phys. Radium **22**, 176–178 (1961)
55. A. Daunois, J.L. Deiss, B. Meyer, Étude spectrophotométrique de l'absorption bleue et violette de Cu_2O. J. Phys. **27**, 142–146 (1966)
56. W.Y. Ching, Y.N. Xu, K.W. Wong, Ground-state and optical properties of Cu_2O and CuO crystals. Phys. Rev. B **40**, 7684–7695 (1989)
57. E. Ruiz, S. Alvarez, P. Alemany, R. Evarestov, Electronic structure and properties of Cu_2O. Phys. Rev. B **56**, 7189–7196 (1997)

58. G. Dresselhaus, A.F. Kip, C. Kittel, Spin-orbit interaction and the effective masses of holes in germanium. Phys. Rev. **95**, 568–569 (1954)
59. J.M. Luttinger, Quantum theory of cyclotron resonance in semiconductors: general theory. Phys. Rev. **102**, 1030–1041 (1956)
60. M. Cardona, N.E. Christensen, G. Fasol, Terms linear in k in the band structure of zinc-blende semiconductors. Phys. Rev. Lett. **56**, 2831–2833 (1986)
61. C.R. Pidgeon, S.H. Groves, Inversion-asymmetry and warping-induced interband magneto-optical transitions in InSb. Phys. Rev. **186**, 824–833 (1969)
62. G.L. Bir, E.I. Butikov, G.E. Pikus, Spin and combined resonance on acceptor centres in Ge and Si type crystals-I Paramagnetic resonance in strained and unstrained crystals. J. Phys. Chem. Solids **24**, 1467–1474 (1963)
63. P. Lawaetz, Valence-band parameters in cubic semiconductors. Phys. Rev. B **4**, 3460–3467 (1971)
64. N. Casanova, E. Gheeraert, E. Bustarret, S. Koizumi, T. Teraji, H. Kanda, J. Zeman, Effect of magnetic field on phosphorous center in diamond. Phys. Stat. Sol. A **186**, 291–295 (2001)
65. R. Kaplan, R.J. Wagner, H.J. Kim, R.F. Davis, Electron cyclotron resonance in cubic SiC. Solid State Commun. **55**, 67–69 (1985)
66. M. Oshikiri, K. Takehana, T. Asano, G. Kido, Cyclotron resonance of n-GaP in a wide far infrared region. J. Phys. Soc. Jpn. **65**, 2936–2939 (1996)
67. F. Nava, C. Canali, C. Jacoboni, L. Reggiani, F. Kozlov, Electron effective masses and lattice scattering in natural diamond. Solid State Commun. **33**, 475–477 (1980)
68. C.J. Rauch, Millimeter cyclotron resonance experiments in diamond. Phys. Rev. Lett. **7**, 83–84 (1961)
69. M. Willatzen, M. Cardona, N.E. Christensen, Linear muffin-tin-orbital and k. p calculations of effective masses and band structure of semiconducting diamond. Phys. Rev. B **50**, 18054–18059 (1994)
70. M. Willatzen, M. Cardona, N.E. Christensen, Relativistic electronic structure, effective masses, and inversion-asymmetry effects in cubic silicon carbide (3C-SiC). Phys. Rev. B **51**, 13150–13161 (1995)
71. B. Pajot, *Optical Absorption of Impurities and Defects in Semiconducting Crystals – Hydrogen-like Centres* (Springer, Heidelberg, 2010), p. 70
72. G.H. Li, A.R. Goñi, K. Syassen, M. Cardona, Intervalley scattering potentials of Ge from direct exciton absorption under pressure. Phys. Rev. B **49**, 8017–8023 (1994)
73. L. Reggiani, D. Waetcher, S. Zukotynski, Hall-coefficient factor and inverse valence-band parameters of holes in natural diamond. Phys. Rev. B **28**, 3550–3555 (1983)
74. J.C. Hensel, K. Suzuki, Quantum resonances in the valence band of germanium II. Cyclutron resonance in uniaxially stressed crystals. Phys. Rev. B **9**, 4219–4257 (1974)
75. R.E. Lindquist, A.W. Ewald, Optical constants of single-crystal grey tin in the infrared. Phys. Rev. **135**, A191–A194 (1964)
76. J. Weber, M.I. Alonso, Near-band-gap photoluminescence of Si-Ge alloys. Phys. Rev. B **40**, 5683–5693 (1989)
77. R.L. Aggarwal, Stress-modulated magnetoreflectance for the direct transitions $\Gamma_{25}^{\prime 3/2} \rightarrow \Gamma_2^{\prime}$ and $\Gamma_{25}^{\prime 1/2} \rightarrow \Gamma_2^{\prime}$ in germanium. Phys. Rev. B **2**, 446–458 (1970)
78. M. Koshevnikov, B.M. Ashkinadze, E. Cohen, A. Ron, Low temperature electron mobility studied by cyclotron resonance in ultrapure GaAs crystals. Phys. Rev. B **52**, 17165–17171 (1995)
79. M.A. Hopkins, R.J. Nicholas, P. Pfeffer, W. Zawadzki, D. Gauthier, J.C. Portal, M.A. DiForte-Poisson, A study of the conduction band non-parabolicity, anisotropy and spin-splitting in GaAs and InP. Semicond. Sci. Technol. **2**, 568–577 (1987)
80. B. Clerjaud, A. Gélineau, D. Galland, K. Saminadayar, Cyclotron resonance of photoexcited electrons in ZnTe. Phys. Rev. B **19**, 2056–2058 (1979)
81. M.S. Skolnick, A.K. Jain, R.A. Stradling, J. Leotin, J.C. Ousset, An investigation of the anisotropy of the valence band of GaAs by cyclotron resonance. J. Phys. C **9**, 2809–2821 (1976)

82. D. Fröhlich, A. Nöthe, K. Reimann, Determination of valence band parameters in ZnTe. Phys. Stat. Sol. B **125**, 653–657 (1984)
83. G.N. Le-Si-Dang, R. Romestain, Optical detection of cyclotron resonance of electrons and holes in CdTe. Solid State Commun. **44**, 1187–1190 (1982)
84. M.V. Yakunin, Determination of cyclotron masses of heavy holes in InSb from a magnetophonon resonance. Sov. Phys. Semicond. **21**, 859–862 (1987)
85. C.R. Pidgeon, R.N. Brown, Interband magneto-absorption and Faraday rotation in InSb. Phys. Rev. **146**, 575–583 (1966)
86. W.G. Bi, C.W. Tu, Bowing parameter of the band-gap energy of GaN_xAs_{1-x}. Appl. Phys. Lett. **70**, 1608–1610 (1997)
87. L. Bellaiche, S.H. Wei, A. Zunger, Localization and percolation in semiconductor alloys: GaAsN vs GaAsP. Phys. Rev. B **54**, 17568–17576 (1996)
88. J.J. Hopfield, Fine structure of the optical absorption edge of anisotropic crystals. J. Phys. Chem. Solids **15**, 97–107 (1960)
89. G.D. Chen, M. Smith, J.Y. Lin, H.X. Jiang, S.H. Wei, M. Asif Khan, C.J. Sun, Fundamental optical transitions in GaN. Appl. Phys. Lett. **68**, 2784–2786 (1996)
90. D.C. Reynolds, D.C. Look, B. Jogai, C.W. Litton, G. Cantwell, W.C. Harsch, Valence band ordering in ZnO. Phys. Rev. B **60**, 2340–2344 (1999)
91. M.A. Vouk, E.C. Lightowlers, Two-phonon assisted free exciton recombination radiation from intrinsic silicon. J. Phys. C **10**, 3689–3699 (1977)
92. A.T. Collins, M. Kamo, Y. Sato, Intrinsic and extrinsic cathodoluminescence from single-crystal diamonds grown by chemical vapour composition. J. Phys. Condens. Mat. **1**, 4029–4033 (1989)
93. G.G. McFarlane, T.P. McLean, J.E. Quarrington, V. Roberts, Fine structure in the absorption-edge spectrum of Ge. Phys. Rev. **108**, 1377–1383 (1957)
94. N.L. Rowell, Infrared photoluminescence of intrinsic InSb. Infrared Phys. **28**, 37–42 (1988)
95. H.Y. Fan, Temperature dependence of the energy gap in semiconductors. Phys. Rev. **82**, 900–905 (1951)
96. Y.P. Varshni, Temperature dependence of the energy gap in semiconductors. Physica **34**, 149–154 (1967)
97. R. Pässler, Dispersion-related description of temperature dependencies of band gaps in semiconductors. Phys. Rev. B **66**, 085201/1–18 (2002)
98. R. Pässler, Basic model relations for temperature dependencies of fundamental energy gaps in semiconductors. Phys. Stat. Sol. B **200**, 155–172 (1997)
99. L. Viña, S. Logothetidis, M. Cardona, Temperature dependence of the dielectric function of germanium. Phys. Rev. B **30**, 1979–1991 (1984)
100. R. Pässler, Parameter sets due to the fittings of the temperature dependencies of fundamental bandgaps in semiconductors. Phys. Stat. Sol. B **216**, 975–1007 (1999)
101. E. Grilli, M. Guzzi, R. Zamboni, L. Pavesi, High-precision determination of the temperature dependence of the fundamental energy gap in gallium arsenide. Phys. Rev. B **45**, 1638–1644 (1992)
102. Y.W. Tsang, M.L. Cohen, Calculation of the temperature dependence of the energy gaps in PbTe and SnTe. Phys. Rev. B **3**, 1254–1261 (1971)
103. D. Olguín, M. Cardona, A. Cantarero, Electron-phonon effects on the direct band gap in semiconductors: LCAO calculations. Solid State Commun. **122**, 575–589 (2002)
104. S. Zollner, M. Cardona, S. Gopalan, Isotope and temperature shifts of direct and indirect band gaps in diamond-like semiconductors. Phys. Rev. B **45**, 3376–3385 (1992)
105. A.V. Nurmikko, A novel technique for measuring small absorption coefficients in semiconductor infrared laser window materials. Appl. Phys. Lett. **26**, 175–178 (1974)
106. M. Cardona, T.A. Meyer, M.L.W. Thewalt, Temperature dependence of the energy gap of semiconductors in the low-temperature limit. Phys. Rev. Lett. **92**, 196403/1–4 (2004)
107. M.D. Frogley, J.L. Sly, D.J. Dunstan, Pressure dependence of the direct band gap of tetrahedral semiconductors. Phys. Rev. B **58**, 12579–12582 (1998)

108. B. Welber, C.K. Kim, M. Cardona, S. Rodriguez, Dependence of the indirect gap of silicon on hydrostatic pressure. Solid State Commun. **17**, 1021–1024 (1975)
109. A.A. Trojan, M.I. Eremets, P.Yu. Korolik, V.V. Struzhkin, A.N. Utjuh, Fundamental gap of diamond under hydrostatic pressure. Jpn. J. Appl. Phys. **32** (Suppl. 32-1), 282–284 (1993)
110. C.N. Ahmad, A.R. Adams, Electron transport and pressure coefficients associated with L_{1c} and Δ_{1c} minima of germanium. Phys. Rev. B **34**, 2319–2328 (1986)
111. F. Engelbrecht, J. Zeman, G. Wellenhofer, C. Peppermüller, R. Helbig, G. Martinez, U. Rössler, Hydrostatic-pressure coefficient of the indirect gap and fine structure of the valence band of $6H$-SiC. Phys. Rev. B **56**, 7348–7355 (1997)
112. P.Y. Yu, B. Welber, High pressure photoluminescence and resonant study of GaAs. Solid State Commun. **25**, 209–211 (1978)
113. D.J. Wolford, J.A. Bradley, Pressure dependence of shallow bound states in gallium arsenide. Solid State Commun. **53**, 1069–1076 (1985)
114. K. Bulthuis, Effect of uniaxial stress on silicon and germanium. Phys. Lett. **25A**, 512–514 (1967)
115. L.D. Landau, Diamagnetismus der Metalle. Z. Phys. **64**, 629–637 (1930)
116. E. Burstein, G.F. Picus, R.F. Wallis, F. Blatt, Zeeman-type magneto-optical studies of interband transitions in semiconductors. Phys. Rev. **113**, 15–33 (1959)
117. E. Sozontov, L.X. Cao, A. Kazimirov, V. Kohn, N. Konuma, M. Cardona, J. Zegenhagen, X-ray standing wave analysis of the effect of isotopic composition on the lattice constants of Si and Ge. Phys. Rev. Lett. **86**, 5329–5332 (2001)
118. S. Hayama, G. Davies, J. Tan, J. Coutinho, R. Jones, K.M. Itoh, Lattice isotope effects on optical transitions in silicon. Phys. Rev. B **70**, 035202/1–9 (2004)
119. A.T. Collins, S.C. Lawson, G. Davies, H. Kanda, Indirect energy gap of ^{13}C diamond. Phys. Rev. Lett. **65**, 891–894 (1990)
120. H. Holloway, K.C. Hass, M.A. Tamor, T.R. Anthony, W.F. Banholzer, Isotopic dependence of the lattice constant of diamond. Phys. Rev. B **44**, 7123–7126 (1991)
121. D. Karaiskaj, M.L.W. Thewalt, T. Ruf, M. Cardona, M. Konuma, Photoluminescence studies of isotopically enriched silicon: isotopic effects on the indirect electronic band gap and phonon energies. Solid State Commun. **123**, 87–92 (2002)
122. G. Davies, E.C. Lightowlers, K. Itoh, W.L. Hansen, E.E. Haller, V. Ozhogin, Isotope dependence of the indirect energy gap of germanium. Semicond. Sci. Technol. **7**, 1271–1273 (1992), corrigendum, G. Davies, Ibid. **9**, 409 (1994)
123. C. Parks, A.K. Ramdas, S. Rodriguez, K.M. Itoh, E.E. Haller, Electronic band structure of isotopically pure germanium: modulated transmission and reflectivity study. Phys. Rev. B **49**, 14244–14250 (1994)
124. F.J. Manjón, M. Mollar, B. Marí, N. Garro, A.A. Cantarero, R. Lauck, M. Cardona, Effect of isotopic mass on the photoluminescence spectra of β zinc sulfide. Solid State Commun. **133**, 255–258 (2005)
125. A. Tomasulo, M.V. Ramakrishna, Quantum confinement effects in semiconductor clusters. II. J. Chem. Phys. **105**, 3612–3626 (1996)
126. A.D. Yoffe, Semiconductor quantum dots and related systems: electronic, optical, luminescence and related properties of low dimensional systems. Adv. Phys. **50**, 1–208 (2001)
127. K. Sattler, The energy gap of cluster nanoparticles, and quantum dots, in *Handbook of Thin Films Materials, Nanomaterials and Magnetic Thin Films*, vol. 5, ed. by H.S. Nalwa (Academic, New York, 2002), pp. 61–97
128. E. Burstein, Anomalous optical absorption limit in InSb. Phys. Rev. **93**, 632–633 (1954)
129. T.S. Moss, The interpretation of the properties of indium antimonide. Proc. Phys. Soc. Lond. B **67**, 775–782 (1954)
130. E.D. Palik, D.L. Mitchell, J.N. Zemel, Magneto-optical studies of the band structure of PbS. Phys. Rev. **135**, A763–A778 (1964)
131. C. Ghezzi, R. Magnanini, A. Parisini, B. Rotelli, L. Tarricone, A. Bosacchi, S. Franci, Concentration dependence of optical absorption in tellurium-doped GaSb. Semicond. Sci. Technol. **12**, 858–866 (1997)

132. M.A. Abdullaev, S.I. Kokhanovskii, O.S. Koschug, R.P. Seisyan, "Fine" structure of the absorption edge of cadmium telluride crystals. Sov. Phys. Semicond. **23**, 726–728 (1989)
133. P. Horodysky, P. Hlídek, Free-exciton absorption in bulk CdTe: temperature dependence. Phys. Stat. Sol. B **243**, 494–501 (2006)
134. K.L. Shaklee, R.F. Nahory, Valley-orbit splitting of free excitons? The absorption edge of Si. Phys. Rev. Lett. **24**, 942–945 (1970)
135. S. Zwerdling, B. Lax, L.M. Roth, K.J. Button, Exciton and magnetoabsorption of the direct and indirect transitions in germanium. Phys. Rev. **114**, 80–89 (1959)
136. N.O. Lipari, M. Altarelli, Theory of indirect excitons in semiconductors. Phys. Rev. B **15**, 4883–4897 (1977)
137. A. Frova, G.A. Thomas, R.E. Miller, E.O. Kane, Mass reversal effect in the split indirect exciton of Ge. Phys. Rev. Lett. **34**, 1572–1575 (1975)
138. C. Benoit à la Guillaume, M. Voos, Electron-hole drops in pure Ge. Phys. Rev. B **7**, 1723–1727 (1973)
139. V.I. Sidorov, Ya.E. Pokrovskii, Submillimeter exciton photoconductivity in germanium. Sov. Phys. Semicond. **6**, 2015–2016 (1973)
140. N.V. Guzeev, V.A. Zayats, V.L. Kononenko, T.S. Mandelshtam, V.N. Murzin, Submillimeter absorption spectra of the free excitons in ultrapure germanium. Sov. Phys. Semicond. **8**, 1061–1065 (1975)
141. M. Buchanan, T. Timusk, The far-infrared absorption of excitons in germanium, in *Proc. 13th Internat. Conf. Phys. Semicond.*, ed. by F.G. Fumi (Tipografia Marves, Rome, 1976), pp. 821–824
142. D. Labrie, T. Timusk, Far-infrared absorption spectrum of excitons in [111]-stressed germanium: high-stress limit. Phys. Rev. B **27**, 3605–3609 (1983)
143. W. Schmid, Four-particle radiative transitions of biexcitons and multiple bound excitons in Si. Phys. Rev. Lett. **45**, 1726–1729 (1980)
144. P.J. Dean, J.R. Haynes, W.F. Flood, New radiative recombination processes involving neutral donors and acceptors in silicon and germanium. Phys. Rev. **161**, 711–729 (1967)
145. M.L.W. Thewalt, W.G. McMullan, Green and near-infrared luminescence due to the biexcitons in unperturbed silicon. Phys. Rev. B **30**, 6232–6234 (1984)
146. R.R. Parsons, U.O. Ziemelis, J.A. Rostworowski, Resolved splitting of the free exciton luminescence band in silicon. Solid State Commun. **31**, 5–7 (1979)
147. P.J. Dean, E.C. Lightowlers, D.R. Wight, Intrinsic and extrinsic recombination radiation from natural and synthetic aluminum-doped diamond. Phys. Rev. **140**, A352–A368 (1965)
148. R. Sauer, H. Sternschulte, S. Wahl, K. Thonke, Revised fine splitting of excitons in diamond. Phys. Rev. Lett. **84**, 4172–4175 (2000)
149. M. Cardona, T. Ruf, J. Serrano, Comments on revised splitting of excitond in diamond. Phys. Rev. Lett **86**, 3923 (2000)
150. G.G. McFarlane, T.P. McLean, J.E. Quarrington, V. Roberts, Exciton and phonon effects in the absorption spectra of germanium and silicon. J. Phys. Chem. Solids **8**, 388–392 (1959)
151. M. Hayashi, K. Katsuki, Hydrogen-like absorption spectrum of cuprous oxide. J. Phys. Soc. Jpn. **7**, 599–603 (1952)
152. A. Mysyrowicz, D. Hulin, A. Antonetti, Long exciton lifetime in Cu_2O. Phys. Rev. Lett. **43**, 1123–1126; erratum: Phys. Rev. Lett. **43**, 1275–1275 (1979)
153. R.J. Elliot, Intensity of optical absorption by excitons. Phys. Rev. **108**, 1384–1389 (1957)
154. R.J. Elliot, Symmetry of excitons in Cu_2O. Phys. Rev. **124**, 340–345 (1961)
155. D.A. Fishman, PhD thesis, Zernike Institute, The University of Groningen, The Netherlands (2008)
156. S. Nikitine, J.B. Grun, M. Sieskind, Étude spectrophotométrique de la série jaune de Cu_2O aux basses températures. J. Phys. Chem. Solids **17**, 292–300 (1961)
157. P.D. Bloch, C. Schwab, Direct evidence for phonon-assisted transitions to the $1s$ paraexciton level of the yellow exciton series in Cu_2O. Phys. Rev. Lett. **41**, 514–517 (1978)
158. E.F. Gross, F.I. Kreingol'd, V.L. Makarov, Resonant interaction between ortho- and para-excitons with participation of phonons in a Cu_2O crystal. J. Exp. Theor. Phys. Lett. **15**, 269–271 (1972)

159. G. Kuwabara, M. Tanaka, H. Fukutani, Optical absorption due to paraexciton of Cu_2O. Solid State Commun. **21**, 599–601 (1977)

160. V.T. Agekyan, Spectroscopic properties of semiconductor crystals with direct forbidden energy gap. Phys. Stat. Sol. A **48**, 11–42 (1977)

161. Ch. Uihlein, D. Fröhlich, R. Kenklies, Investigation of exciton fine structure in Cu_2O. Phys. Rev. B **23**, 2731–2740 (1981)

162. T. Tayagaki, A. Mysyrowicz, M. Kuwata-Gonokami, The yellow excitonic series of Cu_2O revisited by Lyman spectroscopy. J. Phys. Soc. Jpn. **74**, 1423–1426 (2005)

163. R.B. Hammond, R.N. Silver, Temperature of the exciton lifetime in high-purity silicon. Appl. Phys. Lett. **36**, 68–71 (1980)

164. A. Dargys, S. Žurauskas, A new method for the study of excitons in semiconductors. Semicond. Sci. Technol. **8**, 518–524 (1993)

165. G. Davies, The optical properties of luminescence centres in silicon. Phys. Rep. **176**, 83–188 (1989)

166. C. Benoit à la Guillaume, F. Salvan, M. Voos, Radiative combination in highly excited Ge, in *Proceedings of 10th International Conference on Phys. Semicond.*, S.P. Keller, J.C. Hensel, F. Stern (United States Atomic Energy Commission, Division Technical Information, Germantown, 1970), pp. 516–519

167. C. Klingshirn, *Semiconductor Optics*, 3rd edn. (Springer, Heidelberg, 2005)

168. M.A. Lampert, Mobile and immobile effective-mass particle complexes in nonmetallic solids. Phys. Rev. Lett. **1**, 450–452 (1958)

169. L.V. Keldysh, Concluding remarks, in *Proc. 9th Internat. Conf. Phys. Semicond.*, ed. by M. Ryvkin (Nauka, Leningrad, 1968), pp. 1303–1312

170. A.A. Rogachev, New investigations of excitons in germanium, in *Proc. 9th Internat. Conf. Phys. Semicond.*, ed. by M. Ryvkin (Nauka, Leningrad, 1968), pp. 407–414

171. M. Combescot, P. Nozières, Condensation of excitons in germanium and silicon. J. Phys. C **5**, 2369–2391 (1972)

172. M. Combescot, Estimation of the critical temperature of electron-hole droplets in Ge and Si. Phys. Rev. Lett. **32**, 15–17 (1974)

173. J.R. Haynes, Experimental observation of the excitonic molecule. Phys. Rev. Lett. **17**, 860–862 (1966)

174. A.S. Kaminskii, Ya.E. Pokrovskii, Recombination radiation of the condensed phase of nonequilibrium carriers in silicon. JETP Lett. **11**, 255–257 (1970)

175. Ya.E. Pokrovskii, A. Kaminsky, K. Svistunova, Experimental evidence of the existence of condensed phase of non-equilibrium of charge carriers in Ge and Si, in *Proc. 10th Internat. Conf. Phys. Semicond.*, ed. by S.P. Keller, J.C. Hensel, F. Stern (United States Atomic Energy Commission, Division Technical Information, Germantown, 1970), pp. 504–509

176. Ya.E. Pokrovskii, K.I. Svistunova, Occurrence of a nonequilibrium carrier concentration in germanium. JETP Lett. **9**, 261–262 (1969)

177. S.G. Tikhodeev, The electron-hole liquid in a semiconductor. Sov. Phys. Usp. **28**, 1–30 (1985)

178. T.K. Lo, Spectroscopic determination of condensation energy and density of electron-hole droplets in pure Ge. Solid State Commun. **15**, 1231–1234 (1974)

179. M.L.W. Thewalt, J.A. Rostworowski, Biexcitons in Si. Solid State Commun. **25**, 991–993 (1977)

180. A.G. Steele, W.G. McMullan, M.L.W. Thewalt, Discovery of polyexcitons. Phys. Rev. Lett. **59**, 2899–2902 (1987)

181. P.L. Gourley, J.P. Wolfe, Thermodynamics of excitonic molecules in silicon. Phys. Rev. B **20**, 3319–3327 (1979)

182. N.Q. Liem, V.X. Quang, D.X. Thanh, J.I. Lee, A. Kasi Vismanath, D. Kim, Biexciton photoluminescence in cubic ZnS single crystals. Appl. Phys. Lett. **75**, 3974–3976 (1999)

183. N.F. Mott, *Metal-Insulators Transitions* (Taylor and Francis, London, 1974)

Chapter 3
Instrumental Methods for Absorption Spectroscopy in Solids

3.1 Introduction

The electronic absorption of deep centres is generally located at energies higher than the vibrational absorption of impurity centres, which remains situated in the infrared region of the spectrum. This latter absorption can even go down to the very far infrared when it is associated with pseudo-rotational modes. Absorption spectroscopy of impurity centres in semiconductors has been used for research purposes, where the measurements are usually performed at low temperature, and high resolution is required when the absorption lines are narrow (their true or natural full width at half maximum (FWHM) can be as low as $0.005 \, \text{cm}^{-1}$ or $\sim 0.6 \, \mu\text{eV}$). Absorption spectroscopy has also been used at RT for technological purposes, as for measuring the concentration of interstitial oxygen or substitutional carbon in commercially produced CZ silicon ingots. The observed profile of a spectral line is the convolution of the instrumental function of the spectrometric system, giving the spectral resolution, with the true profile of the line. When the true width of the line is smaller than about three times the spectral resolution, this produces an instrumental broadening of the observed line (and a corresponding decrease of its peak absorption). In this book, unless otherwise specified, the FWHMs indicated are considered to be corrected for the instrumental broadening. There are also experimental situations where the profile of an absorption or PL feature containing several unresolved individual lines cannot be further resolved by increasing the resolution because of the combination of the natural FWHMs of the components and of their separations. It is possible to decrease artificially the FWHMs of the components in order to determine accurately their positions by a method known as self-deconvolution [1].

For isolated lines, the absorption coefficient can be integrated over the whole line to give an integrated absorption A_I:

$$A_I = \int_{\tilde{\nu}_{min}}^{\tilde{\nu}_{max}} K(\tilde{\nu}) d\tilde{\nu}, \qquad (3.1)$$

B. Pajot and B. Clerjaud, *Optical Absorption of Impurities and Defects in Semiconducting Crystals*, Springer Series in Solid-State Sciences 169, DOI 10.1007/978-3-642-18018-7_3, © Springer-Verlag Berlin Heidelberg 2013

where the integration is taken on the spectral extent of the line, here in wavenumber, noted $\tilde{\nu}$. This integrated absorption is independent of the spectral resolution. For absorption in the continuum, K needs only to be specified at the energy at which the measurement is performed.

It is often possible to calibrate the peak absorption coefficient K_{max} of a line associated with an impurity or a centre by measuring independently the concentration N of the impurity or of the centre in the sample. One can then define a more general quantity, the absorption cross-section:

$$\sigma\,(cm^2) = K_{max}/N, \tag{3.2}$$

which is physically significant when instrumental broadening is properly taken into account, or an integrated absorption cross-section $\sigma_{IA}(cm) = A_I/N$, which is independent of the spectral resolution. For a given impurity and a given line of its spectrum, whose true FWHM is much larger than the spectral resolution, the inverse of the cross-section $\sigma(cm^2)$ can be considered as a calibration factor of the absorption. This calibration factor can later be used to determine the impurity concentration from the peak absorption of the line in an unknown sample. For lines with small FWHMs, it is convenient to use for calibration a spectral resolution small enough to keep a reasonable sensitivity. In that case, the calibration factor can be resolution-dependent because of instrumental broadening, and the spectroscopic determination of an impurity concentration must be performed with a spectral resolution equal to (or smaller than) the one used to determine the calibration factor. For a given impurity and a given line of its spectrum, $(\sigma_{IA})^{-1}$ is the impurity concentration for unit A_I of that line, and it constitutes an integrated calibration coefficient of that line.

It must be noted that in a classical absorption measurement with an external IR source, when the sample temperature is higher than the detector temperature, the thermal emissivity of the sample is detected, in addition to the radiation transmitted through the sample. This is for instance the case when the sample is at RT and the detector at LNT. The spectral dependence $\epsilon(\tilde{\nu})$ of the emissivity is related to the absorption coefficient K of a plane parallel sample of thickness d and reflectivity R through:

$$\epsilon(\tilde{\nu}) = (1 - R)(1 - u)/(1 - Ru),$$

where $u = \exp[-Kd]$. This expression is deduced from the article by McMahon [2], and it has indeed been used to obtain values of the absorption coefficients of silicon at different temperatures from emissivity measurements [3]. In classical absorption measurements, for the above case, the contribution of emissivity adds a small extra contribution to the measured absorption coefficient. This contribution can be evaluated by measuring directly the emissivity with the IR source off. When measuring the RT absorption of the vibrational band of interstitial oxygen in silicon at $1107\,cm^{-1}$ in CZ samples with a mercury cadmium telluride detector operated at LNT, the peak absorption coefficient of the band is found to be reduced by $\sim 4\%$ when corrected for the emissivities of the sample and of the reference [4]. Detection of the sample emissivity occurs with spectrometers where the incident and reflected beams are not separated physically, and it is not observed with asymmetrical systems like the modern Fourier transform spectrometer (FTS) where the flat mirrors are replaced by corner cube (roof-top) mirrors, where the beam is incident only on one side of the mirror and reflected back on the reverse side.

For technological purposes, it can be also desirable to obtain a map of the distribution of a defect centre in a semiconductor wafer. One then chooses an absorption or PL line of this centre and scans its intensity at different points of the wafer with a properly focussed beam. This map can be later converted into a concentration map.

In classical optical absorption measurements, one follows the absorption of a sample under different conditions as a function of the energy of the incident electromagnetic radiation. This can be achieved in two ways: one can either take a broadband source and use a spectrometer to disperse the electromagnetic spectrum or use a tunable monochromatic source. There is also a technique known as excitation spectroscopy, which can be used in absorption as well as in PL modes. In this technique, a monochromator is set at the energy of a chosen absorption or PL line of a sample while the sample is illuminated with monochromatic light of varying energy by a second monochromator or a tunable source. The excitation spectrum is a record of the intensity change of the absorption or PL line measured with the first monochromator as a function of the energy of the additional exciting radiation.

A discrete transition between two levels of a centre in a crystal is characterized by its energy and by a FWHM, which is the sum of the widths of the ground and excited states. In the ideal case, for given experimental conditions and for a homogeneously distributed centre, this FWHM is the same throughout the crystal and it can be defined as the homogeneous width of the transition. It turns out that because of local distortions or of inhomogeneities in the local electric field, a small change in the transition energy can occur locally. This can be due for instance to the random distribution of other centres or defects, producing strains in the crystal, or to the random distribution of a centre in a very disproportionate alloy. The observed absorption line is then the superposition of lines with slightly different energies corresponding to sites with different perturbations. The observed line width, which corresponds to the energy distribution of the sum of the different lines, is larger than the homogeneous line width and the line is said to be inhomogeneously broadened. When the lifetime of the excited state of the nonperturbed transition is large, giving a small homogeneous line width, it is possible, by illuminating the sample with laser radiation whose energy is within the inhomogeneous line width of an absorption transition, to excite selectively centres with the same homogeneous width. This produces a dip (a spectral hole) in the inhomogeneous absorption line and the technique is known as hole burning. This possibility was first demonstrated by Szabo [5] to study the effect of a ruby laser illumination on the inhomogeneously broadened R_1 line of Cr in ruby at 693.4 nm. Hole burning informs not only on the homogeneous line width of the transition, but also on resonant excitation transfers [6].

An electronic or vibrational excited state has a finite global lifetime and its de-excitation, when it is not metastable, is very fast compared to the standard measurement time conditions. Lifetime measurement is a part of spectroscopy known as time domain spectroscopy. One method is based on the existence of pulsed lasers that can deliver radiation beams of very short duration and adjustable repetition rates. When the frequency of the radiation pulse of these lasers can be tuned to the frequency of a discrete transition, as with a free-electron laser, this possibility can be used to determine the lifetime of the excited state of the transition in a pump-probe experiment. In this method, a pump energy pulse of moderate intensity produces a transient transmission dip of the sample at the transition frequency. The evolution with time of this dip is probed by a small-intensity pulse at the same frequency, as a function of the delay between the pump and probe pulses.[1] When the decay is exponential, the slope of the decay of the transmission dip as a function of the delay, plotted in a log-linear scale, provides a value of the lifetime of the excited state.

In some experiments, the absorption of a transition is measured at a given energy as a function of the incident power. This is usually performed with a pulsed laser, for which the power dynamics can be adjusted in a broad range and where the repetition rate can be controlled. The transmitted energy can be measured directly with a variable attenuator placed in front of the detector to avoid its saturation for high incident power, or in a pump-probe geometry. What is generally observed as a function of the incident power is first a constant value of the absorption followed, for increasing power by a decrease of the absorption [7], which can reach a point where it goes to zero. Such an effect is known as saturated absorption or optical bleaching, and the kinetics of the absorption decrease, which is observed for electronic as well as vibrational transitions [8, 9] allows also to determine the lifetimes of the excited states.

For sufficiently high power intensities, nonlinear effects can give rise to two-photon absorption (TPA) where simultaneous absorption of two photons with energies $\hbar\omega_1$ and $\hbar\omega_2$ produces a transition at energy $\hbar\omega_1 + \hbar\omega_2$, and the possibility of such an effect was predicted by Göppert-Mayer [10] for $\hbar\omega_1 = \hbar\omega_2$. An experimental verification of TPA was provided by Kaiser and Garett [11], who reported the blue fluorescence at 425.0 nm (2.917 eV) of Eu^{2+} salts in CaF_2 due to absorption in the 3.10–3.72 eV spectral region as a consequence of the illumination with the red light at 694.3 nm (1.786 eV) of a ruby laser. TPA is theoretically explained by the presence of an intermediate virtual state at midpoint between the initial and final states. The possibility to observe TPA of shallow donors in semiconductors has been discussed by Golka and Mostowski [12]. Practically, to investigate TPA of a sample in a spectral domain, it is illuminated with pulses from a high-power monochromatic source at frequency ω_1, and these pulses are superposed

---

[1] In the pump-probe geometry, the two beams are crossed.

to those of a tunable source at frequency ω_2, where the sum $\omega_1 + \omega_2$ falls in the spectral region to study. A detector measures the intensity at frequency $\omega_1 + \omega_2$ when the two pulses are synchronous on the sample, producing eventually the TPA, and a reference is taken by measuring the intensity at frequency $\omega_1 + \omega_2$ when the two pulses are dephased. Experimental techniques for TPA are described in Fröhlich and Sondergeld [13]. TPA has been used in the study of the absorption spectrum of FEs in Cu_2O ([14], and references therein).

Photoconductivity is a peculiar type of absorption measurement where the detector is the sample itself. Classical photoconductivity occurs when the absorption of an electron or of a hole takes place between a discrete state and a continuum, where it can participate to the electrical conductivity. When the final state of a discrete transition is separated from the continuum by an energy comparable to $k_B T$ at the temperature of the measurement, the electron or the hole in this state can be thermally ionized in the continuum and give rise to photoconductivity at the energy of the discrete transition. This temperature-dependent two-step photoconductivity is used in photo-thermal ionization spectroscopy (PTIS). This method concerns the detection of the electronic spectra of the hydrogen-like centres. It presents a very high sensitivity compared to classical absorption spectroscopy, allowing detection limits in the $10^7 \, cm^{-3}$ range, and it is discussed in more details in [15].

Under a directional perturbation as a uniaxial stress or a magnetic field, the absorption of impurities in a crystalline sample can become dichroic with respect to the polarization of the radiation used for the absorption measurement. This means that the features of the spectra are different for a polarization parallel or perpendicular to the direction of the perturbation. This term includes the polarization rules and no mention is generally made of dichroism at this point. In the spectroscopy of paramagnetic centres with related absorption lines, magnetic circular dichroism (MCD) is the difference between the absorption of left- and right-circularly polarized radiation and it can prove to be useful to detect the absorption associated with broad features of paramagnetic centres.

The uncertainty on the measured position of an absorption line depends on the noise in the spectrum, but for negligible noise, as a rule of thumb, it can be considered to be ultimately limited to one-tenth of the FWHM.

At low energies, in the meV energy range, acoustic phonon spectroscopy with superconducting thin film tunnel junctions evaporated onto opposite surfaces of a sample has been used as a technique complementary to far IR optical spectroscopy [16, 17]. In this technique, which has been mainly used in silicon and germanium, phonons are generated and detected by appropriate biasing of the junctions. Biased at voltages $2\Delta_G/e$ above the energy gap $2\Delta_G$ of the superconductor, a phonon line is generated, which can be tuned by the voltage. Biased at voltages below the gap, a junction becomes a detector for phonons with energies sufficient to excite extra quasiparticles (i.e. to break Cooper pairs) in the thin film of the detector junction. With $Al-Al_2O_3-Al$ and $Sn-SnO_x-Sn$ junctions as phonon generators and detectors, respectively, the available phonon spectrum extends from 280 to 3000 GHz ($\sim 9.3-100 \, cm^{-1}$ or $\sim 1.2-12.4 \, meV$) and spectral resolutions of 2 GHz ($\sim 0.07 \, cm^{-1}$ or $\sim 8 \, \mu eV$) can be achieved. The typical sample thickness is 1–2 mm.

The Al critical temperature of 1.2 K determines the operating temperature ($\sim$1 K and below) of this phonon spectrometer (the critical temperature of Sn is 3.2 K). This type of high-resolution acoustic phonon spectroscopy has been developed and used between 1976 and 2000 at the University of Stuttgart to study low energy electronic and vibrational excitations mainly in silicon, germanium, and also in GaAs (Laßmann [18], and references therein, [19]). Phonon spectroscopy has also been developed at the University of Nottingham under the impulsion of Professor Lawrence Challis.

In this chapter, we give basic information on the instrumental methods, and more details can be found in Chap. 4 of [15], specially on the detection methods.

3.2 Spectrometer–Detector Combinations

In the measurement of the optical absorption of a solid sample, some parameters, which can be intercorrelated, must be considered. They are the spectral span of the spectrum, the illumination conditions, the power incident on the sample, and the spectral resolution. There are two main types of optical spectrometers. In the dispersive spectrometers, the radiation of a broad-band source is dispersed spatially into quasimonochromatic radiation by a prism or a diffraction grating. The quasimonochromatic radiation is then analysed sequentially by a detector, and a spectrum of the sample as a function of the energy of the radiation is recorded as a function of time. Another possibility is to separate spatially the beam from a broad-band source by a beam splitter (BS) which redirect the two beams on the two arms of a Michelson interferometer. The mirror of one arm is fixed, but the mirror of the other arm can move in a direction parallel to the incident beam, creating an optical path difference between the two beams when they are reflected back to the BS. The two beams recombine on the BS and the signal from the recombined beam is measured sequentially as a function of the optical path (phase) difference. The resulting signal as a function of time is called an interferogram, and the energy spectrum of the source is obtained by calculating the Fourier transform of the interferogram in the time domain. This is the principle of Fourier transform spectroscopy (FTS). For a review on FTS and of its applications to far IR spectroscopy, see [20].

Most of the first commercial IR spectrometers available from the 1960s to about 1975 were of the dispersive type, based on a prism or grating monochromator with two modulated optical beams (the so-called double-beam spectrometers) which could be used for measuring directly the ratio of the transmission of a sample *vs* that of a reference (air or reference sample). With the best of these spectrometers, the largest spectral ranges for a single scan were $\sim$200–4000 cm^{-1} (25–500 meV) using two or three coupled gratings, and the typical spectral resolution with cooled samples $\sim$1–2 cm^{-1} ($\sim$0.12–0.24 meV). These spectrometers were not intended for the measurement of small samples and vigneting occurred in the absence of additional focussing. Other prism or grating spectrometers existed for the visible and near IR. In most of these machines, the sample was placed between the source and the monochromator, and normally illuminated with the whole spectral output of the radiation source, producing in some cases a local heating of small samples in the absence of optical filtering. The absorption of atmospheric gases (CO_2 and H_2O) was reduced by flushing the spectrometers with dry nitrogen. For the far IR, a vacuum-operated Fourier transform spectrometer (FTS) was commercialized by RIIC, later Beckman-RIIC.

This is the reason why, during this period, most of the low-temperature IR absorption measurements on semiconductors were performed, when possible using prism or grating monochromators like the models 98 or 99 (98G or 99G for the grating version) of the Perkin-Elmer Corporation. These compact monochromators could be used as reasonably monochromatic sources in a spectral range somewhat reduced compared to the one of commercial machines, but they allowed the use of bulky magnetooptical cryostats and accessories including proper focussing of the IR beam on the sample.

The situation changed with the advent of continuous scan FTSs with a spectral range extending from the near IR to the far IR and providing spectral resolutions below $0.1\,cm^{-1}$, like the Bruker IFS 113v model, conceived in the 1970s. The basic interferometer of this particular FTS, known as Genzel interferometer, after Prof. Ludwig Genzel who proposed this geometry, combined interesting particularities like the focussing of the optical beam on the beam splitter (BS), allowing the mounting on a beam-splitter wheel of several small BSs with an effective diameter as small as $\sim 10\,mm$, and the use of a moving double-sided plane mirror replacing the immobile and the moving mirrors of the classical Michelson interferometer. This is best seen in Fig. 3.1 showing the optical layout of this machine.

The use of small diameter BSs in this machine allowed also the use of CsI, with a spectral range down to $\sim 160\,cm^{-1}$ ($\sim 20\,meV$), not possible for larger BSs because of the relative softness of this material, as well as of a metal mesh BS for the far IR. The angle of incidence on the BS was $\sim 15°$ and this limited the polarization effects. Moreover, in these machines, the IR beam was focussed at the centre of the sample compartment and for some of them, the modulated optical beam could be redirected outside from the spectrometer, allowing for bulky experimental set-up.

At the end of the 1970s, a line of FTSs based on a classical Michelson interferometer, but with vertical geometry, was developed by BOMEM, in Canada. The optical layout of these machines, delivered with a series of BSs whose spectral ranges are displayed in Fig. 3.2, is shown in [15]. These FTSs, no longer produced, provided spectral resolutions between 0.013 and $0.0026\,cm^{-1}$ (1.6 and $0.3\,\mu eV$), depending on the model, and they have been widely used in high-resolution spectroscopy of semiconductors.

The FTSs require a better mechanical stability than the dispersive systems because a very good parallelism between the two plane mirrors is required. This is why the plane mirrors of the interferometer are sometimes replaced by corner cube reflectors, as in the very high-resolution Bruker IFS125, or by cat's eyes, which are insensitive to small differences of optical parallelism between the two mirrors, or provided with a dynamic alignment system which controls and maintains the parallelism between the two mirrors. The Bruker IFS 125 FTS

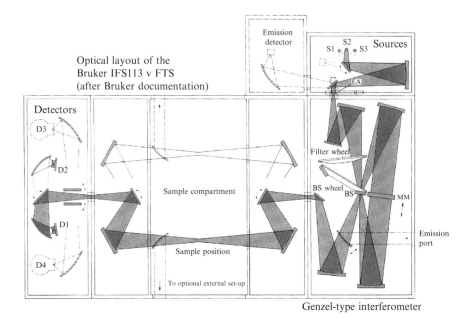

Genzel-type interferometer

Fig. 3.1 Bruker IFS113 v FTS. The source compartment can accommodate three sources, whose output is filtered by one of the filters of the filter wheel. The entrance aperture EA is adjustable between 1.25 and 10 mm. The optical beam from the active source is focussed on one of the BSs of the beam-splitter wheel. The split beams are directed on the two sides of a moving mirror (MM) and the reflected beams redirected on the BS. The modulated beam is directed to the sample compartment, where provision is made for mirrors which can eventually redirect the modulated beam toward external setups. The modulated beam is detected by the most appropriate detector of the detectors compartment. The optional maximum resolution of this FTS was 0.03 cm^{-1}

provides a standard resolution of $0.035\,\mathrm{cm}^{-1}(\sim 0.4\,\mu\mathrm{eV})$ and higher resolutions when necessary: the one at the French SOLEIL synchrotron has a nominal resolution of $0.0007\,\mathrm{cm}^{-1}(\sim 90\,\mathrm{neV})$]. Several FTSs provided with corner cube mirrors, adapted to routine chemical analysis are available on the market and they are not specifically considered here. As already mentioned, another advantage of the corner-cube geometry is that its asymmetric configuration of the optical beams prevents the radiation emitted by the sample in absorption geometry to be detected after modulation when the sample and the detector are at different temperatures.

A notable difference exists between a FTS and a dispersive monochromator. When recording a spectrum made from N spectral elements with the first, each spectral element is measured during the whole recording time of the interferogram, but with the second, each spectral element is recorded only during $1/N$ times the whole recording time. Therefore, for the same recording time of a spectrum of N spectral elements, the gain in the signal over noise (S/N) ratio for the interferometer is $\sqrt{N}$ (Felgett or multiplex advantage). A consequence of this situation is that with a dispersive monochromator, to obtain a large S/N ratio for a spectrum under given spectroscopic conditions, the only way is to increase the gain of the electronic amplifier and its response time, increasing correlatively the recording time, with problems linked to the long-term stabilities of the radiation source, of the detector, and of the electronics. With a FTS, the alternative is the coaddition of interferograms, and the usual method is to coadd first the interferograms of the sample to reach the S/N required, to compute its spectrum from the averaged interferogram, and to do the same for the reference. The differential transmission or absorption spectrum of the sample is then computed directly from the two spectra. There is no miracle and to obtain good S/N ratios, the coaddition times can be very long, leading to the emergence of low-frequency noise and of long-term instabilities, and this is particularly troublesome when trying to observe very weak absorption bands. A procedure where one interferogram of the sample and one interferogram of the reference are recorded alternatively during the coaddition sequence has been described [21]. This method, which requires a special computer-controlled sample changer and the software for fully automated alternate measurements reduces drastically the contributions of the low-frequency noise as well as of long-term instabilities in the reference and sample spectra. Under the same coaddition time as in the classical method, this procedure is claimed to improve the photometric accuracy by a factor of more than 30. An example of the results obtained is shown in Fig. 3.3, where the relative peak transmittance of the nitrogen diinterstitial LVM band at 962.6 cm^{-1}, measured in a one-side polished industrial

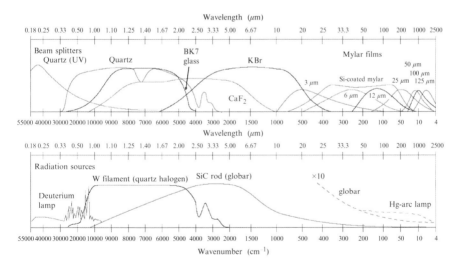

Fig. 3.2 Spectral range of some BSs currently used with FTSs, and of commonly used radiation sources, after an ABB Bomem documentation sheet. The BSs are usually coated to obtain a better reflectivity in the spectral range considered. The quartz BSs for the UV and for the visible–IR are coated with Al and TiO_2, respectively, while the CaF_2 and KBr BSs are coated with Sb_2S_3 and Ge/Sb_2S_3, respectively. The polyethylene terephtalate (Mylar®) BSs are uncoated, except the Hypersplitter™, which is Si-coated to reduce the interference fringes

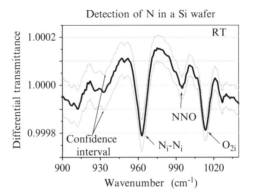

Fig. 3.3 N-related LVMs in a 730-µm-thick one-side polished 200-mm CZ silicon wafer measured in a differential alternate measurement with a total acquisition time of 2 h. O_{2i} is the $1013\,cm^{-1}$ LVM of the interstitial oxygen dimer. The reference sample is a N-lean wafer with a smaller O_i concentration (Akhmetov, private communication). The linear slope of the curves is subtracted (after [21])

200-mm Si wafer, is only ~ 0.9997, corresponding to a variation of $\sim 3 \times 10^{-4}$ of the absolute transmittance.

As a rule, the commercial FTSs are usually provided with different radiation sources and beam splitters adapted to the spectral domain investigated. Figure 3.3 displays the spectral domains of different radiation sources and beam splitters, which cover the spectral domain between 0.5 meV and ~ 6.8 eV (4–54800 cm^{-1} or 2.5–147 nm).

Presently, low-resolution broad-band detection in the 450–5000 cm^{-1} region is made with a RT thermal pyroelectric detector with a KBr window, but high-resolution measurements require the use of a LNT-cooled intrinsic mercury–cadmium telluride ($Hg_{1-x}Cd_xTe$) photoconductive detector with a band gap of ~ 58 meV, corresponding to $x \sim 0.18$–0.20, usually equipped with a ZnSe window. For routine measurements at lower energies, pyroelectric detectors fitted with a polyethylene window are used down to 50 cm^{-1}(~ 6 meV). For experiments requiring a better sensitivity, commercially available germanium or silicon bolometers cooled at LHeT are used below 450 cm^{-1} [for operation below 50 cm^{-1}, liquid helium must be pumped at pressures below the lambda point (~ 2.2 K and 5 kPa or 38 torr)] to increase the sensitivity of the bolometer. These sensitive detectors, which are thermometers using the electron–phonon interaction, have a relatively long time constant. This is not the case of the LHeT-operated free-electron InSb bolometer, also commercially available, based on the heating of hot electrons.

In the near IR, the LNT-cooled intrinsic InGaAs, InAs and InSb photoconductive detectors are used at energies above 6000 and 1800 cm^{-1} (~ 750 and 220 meV). In the UV, visible and near IR regions, Si photodiodes with different load resistors are used, together with photomultiplier tubes with a Cs–Te photocathode, mainly for the UV.

In some experiments, intrinsic photoconductive detectors can display saturation effects above critical incident powers (the electrical output of the detector becomes sublinear) and this leads to use attenuators or to reduce the diameter of the entrance iris of high-resolution FTSs.

A variety of extrinsic photoconductive detectors operated at LHeT have been occasionally used in the medium and far IR in experimental set-ups, for instance Ga-doped germanium, with a low energy cut-off of $\sim$80 cm^{-1}($\sim$10 meV).

3.3 Tunable Sources

Tunable sources are essentially tunable lasers and several kinds of devices of that type are known. Among them, one can sort out the sapphire:Ti laser, the laser diode and the free-electron laser (FEL). The sapphire:Ti laser is based on the fluorescence associated with vibronic transitions of the $3d$ electron of Ti^{3+} replacing Al^{3+} in the α-Al_2O_3 crystal. Under appropriate optical excitation, this fluorescence extends at RT from $\sim$600 to 1000 nm ($\sim$16000 to 10000 cm^{-1} or 2.06 to 1.24 eV), and under pumping with different classical lasers, RT laser emission of Ti^{3+} has been observed in the 662 and 950 nm range ($\sim$15100 to 10500 cm^{-1} or 1.87 to 1.30 eV). Tuning is achieved by inserting in the laser cavity two Brewster-angle silica prisms ([22], and references therein).

The tunable laser diodes are made from direct-gap compound semiconductors, and those whose output extends the farthest in the IR are the $Pb_{1-x}Sn_xTe$ diodes. Most of these particular diodes are operated near LHeT and their peak emission corresponds to the band gap of the alloy. They can be tuned by varying the temperature of the diode in a controlled way or, for a more restricted range, by varying the injection current intensity. Resolutions $\sim$0.001 cm^{-1} (0.12 μeV) near 1000 cm^{-1} (124 meV) have been reported with a temperature-tuned $Pb_{0.86}Sn_{0.14}Te$ laser diode, and such diodes put in use for the study of vibrational modes of ReO_4^- molecules in KI crystals at LHeT [23]. Another possibility is to use magnetic field tuning by changing the Landau levels separation of the semiconductor (the electron effective masses of these lead salts are relatively small).

Free-electron laser (FEL) radiation is the coherent synchrotron radiation of a relativistic electron beam crossing the gaps of a series of magnets arranged to produce zones of alternating magnetic fields. The magnetic fields of this array of magnets, called an undulator (or a wiggler), accelerate sinusoidally the electrons and the coherent radiation emitted depends on the electrons energy. As an example, the CLIO FEL in Orsay, France, could be tuned between 10.3 and 413 meV (120 to 3 μm) with a minimum relative spectral width between 0.2% and 1%. These tunable sources have mainly been used for very high-resolution molecular spectroscopy and also for experiments with semiconductors, like the FEL at Rijnhuisen, in the Netherlands (FELIX).

Tunable sources in the THz range are based on the backward wave oscillator (BWO), known also as carcinotron® (for a review, see [24]). This kind of source has been used in spectrometric configurations in the far IR absorption of semiconductors and other materials in the 0.25–2 mm (5–0.6 meV) spectral region [25, 26].

Impressive results have recently been obtained in the near IR on bound-exciton absorption in ^{28}Si using a tunable Yb-doped fibre laser [27].

3.4   Filtering and Polarization

Optical filters are necessary with grating monochromators to keep only one diffraction order, usually the first, and with FTSs to limit the spectral domain and the radiant power incident on the detector. This can be obtained with low-frequency pass absorption filters with a high-frequency cut-off above which the filter is opaque. From their optical properties, semiconductors are adequate substrates for such filters as they already provide a high-frequency cut-off corresponding to their band-gap energy, but the reflection losses due to the high refractive indices have to be compensated by anti-reflection coatings. Silicon, germanium, indium arsenide, or indium antimonide substrates have been used and the list is not limitative. The low-frequency cut-off of compound crystals due to the onset of the one-phonon absorption can be used when simple high-frequency pass absorption filters are required (note that these compounds become transparent again at frequencies below the one-phonon absorption region). It is also possible to grow on transparent substrates interference filters with different spectral bandwidths and peak transmission energies. Before the advent of FTS machines, filtering for far and very far IR experiments was a very serious problem with dispersive instruments. A decrease of the high-frequency radiation contribution was obtained by using mirror substrates polished with 10 or 20 μm diameter alumina powder grit. The scattering properties of these mirrors for high-frequency radiation made them acceptable reflection filters for the far IR. Similarly, materials transparent in the far IR, but translucent or opaque in the near IR like polyethylene or black polyethylene were and are still used as optical components and filters in the far IR. The selective near-metallic reflection of the alkali halides and alkaline earth halides, due to their strong absorption near from the TO absorption region (see Sect. 2.2.1), has also been used. Practically, in far IR set-ups one metallic mirror was replaced by a "*reststrahlen*" plate made from these compounds, adapted to the spectral range investigated, but they are rarely used presently.

For some absorption experiments on dichroic or anisotropic samples, it can be desirable to use radiation where the orientation of the electric vector with respect to crystal axes is known (linearly polarized radiation). This can be obtained with dispersive monochromators as well as with FTS by inserting a transmission polarizer in the optical path. The most popular ones are the commercially available wire grid polarizers made from a metallic wire grid (Au or A*l*) evaporated on a transparent substrate (ZnSe, AgBr, KRS5, and polyethylene have been used). For wires spacing smaller than the wavelength of the radiations of interest, this array acts as a metallic mirror for electric vector of the radiation parallel to the wires of the grid, but the electric vector component perpendicular to the wires is transmitted with an overall efficiency depending on the metallized area and on the refractive index of the substrate. These polarizers are mounted on a rotating holder so that the orientation of the transmitted electric vector can easily be selected.

3.5 Samples Conditioning

First, there are valuable samples, like cut gemstones, which must be measured as they are, and where conditioning is out of question. Otherwise, the best absorption measurements are made on samples cut from crystals or polycrystals in orthogonal parallelepipedic shapes. The surfaces of the samples intercepting the radiation beam must be reasonably plane and optically polished to prevent scattering of the incident radiation by the surface inhomogeneities with dimensions of the order of the wavelength. This condition becomes less drastic with increasing wavelengths and in the very far IR, samples with ground surfaces are acceptable. However, mechanical cutting and polishing leave disturbed surfaces at the microscopic scale; therefore, as a function of the mechanical properties of the crystals and of the kind of experiment envisaged, it can be necessary to remove the perturbed layer by adequate chemical etching. The surface of cleaved samples has a good optical quality and this is also generally true from the epitaxied samples, with the possible exception of the back surface of the substrate, and these samples do not usually require further mechanical treatment. The absorption measurements on commercial silicon wafers with etched back surfaces are usually performed in the as-received surface state. This surface state reduces the transmission because of the scattering of the back surface and expression (2.2) is no longer valid. A discussion of the methods used to deal with this situation, centred on the absorption of O_i, can be found in the review by Bullis [28].

The spectral transmission of a plane parallel sample of thickness d and refractive index n is modulated by equal-thickness fringes with spacing $\Delta\tilde{\nu}$ in wavenumber, approximately equal to $1/2nd$. When the spectral bandwidth $\delta\tilde{\nu}_d$ is larger than this spacing, the fringes are averaged out, but they become visible at higher resolution, giving what is known as a channelled spectrum.

With thick samples, wedging the sample by $\sim$1–2° can circumvent this drawback, but for very high-resolution spectra in the far IR, the wedging becomes excessive. Moreover, this is not possible when measuring thin wafers. As these fringes come from constructive interferences between successive beams reflecting at the sample interfaces, they disappear when p-polarized radiation is incident at Brewster's angle on two-sided polished samples as the reflection losses are due to s-polarized radiation for that geometry [29]. The results show that the fringes are also suppressed with that configuration for one-sided polished wafers [29, 30].

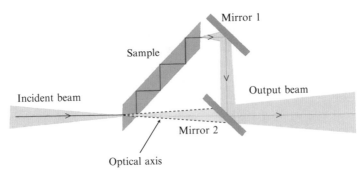

Fig. 3.4 Schematic side view of the positioning in a spectrometer beam of a sample cut with a 45° geometry allowing for multiple internal reflections, and of the two mirrors of the sample holder redirecting the output beam along the optical axis. The beam delimited by dashes is the beam in the absence of this sample holder (C. Naud, unpublished)

When the above methods are not applicable and when the experiments are performed with a FTS, an alternative consists in replacing by zero in the primary interferogram the points of the peaked zone corresponding to the fringes in the FT spectrum. This is not a panacea, however, as, if the channelled spectrum is efficiently removed by this procedure, the wings of sharp absorption lines can show oscillations. At high resolution, a more general, but more time-consuming method, consists in subtracting from the spectral regions of interest a suitable sine function with adequate dispersion and attenuation.

The optical thickness of a sample must be adapted to the peak absorption of the impurity lines to avoid saturation, and this can lead to very thin samples when the impurity concentration is large and cannot be reduced and when the OS is also large. Inversely, the measurement of small impurity concentrations can require thick samples and this puts a limit to the spectroscopic measurements of impurities. In some cases, as an alternative to the increase of the thickness of the sample, it can be cut with a geometry allowing multiple internal reflections, which increases the optical path, as shown schematically in Fig. 3.4.

3.6 Cooling the Samples

Many absorption experiments on impurities and defects are performed at low temperature or as a function of temperature, especially for the observation of discrete spectra. This is a necessity when the population of the ground state level of a transition or of a series of transitions get thermalized at RT. Another reason for using low temperatures is the decreases with temperature of the widths of spectral lines due to the reduced coupling of the levels with lattice phonons. The lowest temperature is obtained using liquid helium as a coolant and several types of liquid-helium optical cryostats have been used for semiconductor spectroscopy.

Presently, the most useful cryostats to cool routinely small samples down to about 5 K are the continuous flow cryostats, through which liquid He is continuously pumped from a container and vaporized in a small exchanger cell. The exchanger cell can eventually be filled with liquid He and pumping on it can allow temperatures near 2 K to be obtained for a short time. For other purposes, the cryostats with a liquid He reservoir are preferable, for instance in experiments where the sample must be processed (implanted or irradiated) at LHeT before optical measurements without breaking the low-temperature conditions, or when measurements between the temperatures of the boiling points of liquid He at atmospheric pressure and at the lambda point are needed with samples mounted in vacuum. When the sample is directly immersed in liquid He, bubbling of the liquid induces a strong scattering of the transmitted radiation. To overcome this, reducing the pressure over the liquid is then necessary to reach temperatures below the lambda point of ^{4}He (50 kPa or 38.3 torr for $T \sim 2.18$ K), where the liquid becomes superfluid with no subsequent bubbling. The cryostats with a liquid He reservoir are thus widely used for transmission and PL experiments between $\sim$2 and 1.2 K.

Below 1.2 K, the cryostats using natural He are replaced by ^{3}He/^{4}He dilution refrigerators. Such refrigerators are commonly used to cool the bolometer/radiation detectors in the mK range (typically $\sim$30–60 mK range). They are used in a limited number of cases for optical studies of impurities in semiconductors [31].

When temperatures $\sim$80–100 K are required routinely, liquid nitrogen, with a boiling point of 77 K, is a convenient cryogenic liquid.

Cooling a sample in vacuum can be obtained by gluing it to a part of the cryostat called a sample holder (cold finger) generally made of copper. This requires for gluing a material with good thermal conductance and mechanical strength. Different types of silicon grease eventually mixed with copper powder have been used for this dual purpose. Accurate temperature measurement also necessitates a temperature sensor glued to the sample. This kind of cooling can be useful for measurements between 2 and $\sim$5 K as bubbling prevents measurements with the sample immersed in liquid He in this temperature range. However, with such mounting, the sample itself must have a good thermal conductance to avoid thermal gradients and it must not be easily cleaved as inhomogeneous mechanical strains are inevitably produced within the sample (mounting of very thin samples is problematic). The presence of these inhomogeneous strains can also lead to inhomogeneous broadening of sharp electronic absorption lines with high piezo-spectroscopic coefficients.

The best way to avoid some of the above problems is to use a cryostat with an extra sample compartment, which can be filled independently with He gas at low pressure from a clean He gas supply with a small oil-free pumping system. Since cooling is insured by gas, the mechanical contact between the sample and the holder can be made loose as long as the sample is immobile during the measurement (for instance, by loosely fitting the sample in an aluminium paper holder fixed to the sample holder by aluminium Scotch® tape, which retains sticking properties down to LHeT). Temperature can be measured by a sensor located close to the sample. The temperature of the sample can be varied easily by controlling the temperature of the gas with an additional heater. Another advantage of gas cooling is that the positions of the samples can be changed or varied with respect to the radiation beam by using sample holders with a thin intermediate tubular section and an extremity at RT. When several samples are mounted on the low-temperature side of such a holder, the use of an appropriate spacer on the room-temperature side of the holder allows to adjust the position of a given sample on the radiation beam. Thin spacers coupled with small cross-sections of the optical beam (down to 0.2 mm^2 with some FTSs) allow also to measure the low-temperature absorption on different points of a sample. Axial rotation and height adjustment of the holder is of course possible through its room-temperature O-ring. The use of spacers allows also to cool the sample in a position above that of the sample beam, avoiding its illumination with room temperature background radiation during cooling-down. This configuration corresponds to a true thermal equilibrium configuration while the usual one (sample in the optical beam) should be called pseudo-thermal equilibrium configuration. The whole holder can even be removed from the sample compartment and replaced by a new one by limiting temporarily the liquid He flow and over-pressurizing the sample compartment with He gas at RT.

The price to pay for these advantages is the necessity of cold optical windows on the exchange gas compartment, in addition to the RT windows of the cryostat. Cold windows are of course mandatory in cryostats with a liquid He reservoir, where the sample is immersed in liquid He. These windows must not be hygroscopic and be resistant to thermal shocks. From the UV to $\sim$0.25 eV, synthetic corundum, improperly called "sapphire" with the c-axis perpendicular to the window surface, to avoid polarization effects, is a good choice. Polycrystalline ZnSe can be used from about 2.7 eV in the visible region of the spectrum down to $\sim$0.03 eV. Such windows, already mounted on metal flanges, are commercially available and the whole units can be mounted on exchange gas as well as liquid He compartments (they are tight to superfluid He) with standard In seals or, for some flanges, with Cu gaskets. KRS5 (thallium bromo-iodide) cold windows are also proposed and this material has the advantage of a relatively extended spectral range (down to $\sim$25 meV or up to $\sim$50 μm) compared to ZnSe, but its reflection losses are higher and its high frequency cut-off is near 2.1 eV. For the far IR, thin polypropylene films ($\sim$30 μm-thick) can be used [32]. These films are slightly permeable to He gas at room temperature, but become He-tight at lower temperatures. Below the one-phonon absorption, the compound-insulating materials become again transparent: at LHeT, corundum and ZnSe windows can again be used below $\sim$23 meV (above $\sim$55 μm). Diamond, which is transparent from the UV region to the far IR, with only a few spectral regions showing absorption presents good optical and mechanical properties, and synthetic diamond windows are commercially available.

Bulk lattice defects or hydrogen-related defects are introduced in semiconductor crystals by high-energy electron or proton irradiation. Even for moderate fluences, these irradiations, made under vacuum, produce a heating of the samples, and for RT irradiations, the samples to be irradiated must be in good thermal contact with a water-cooled metal holder. For the spectroscopic studies of defects which are unstable at RT, like bond-centred isolated hydrogen in silicon, the irradiation must be performed at temperatures near LHeT, and the samples kept at low temperature before the spectroscopic measurements (*in situ* measurements). Thus, the irradiation and the optical measurements must be performed in the same cryostat. To avoid two windows for the irradiation, the sample must be cooled by solid conduction, and it must be positioned externally in front of the irradiation windows, and repositioned in front of the optical windows after irradiation. Figure 3.5 is a schematic drawing of such a custom-made cryostat used for low-temperature irradiation of samples prior to IR absorption measurements (see also [34]).

After the electron irradiation with a Van de Graaff accelerator, this cryostat was transported in an adapted trolley to the IR spectrometer room. Another possibility for the head of this type of cryostat is to dispose the metal window at 90° of the optical windows and to rotate the sample holder by 90° after irradiation (this requires lateral screening of the sample to avoid irradiation of the optical windows by scattered electrons and X-rays during irradiation).

For experiments above RT, an optical oven is used, and the sample is generally heated under vacuum by solid contact. The problem of the emissivity of the oven and of the sample must also be considered to allow for corrections and to avoid saturation of the detector.

3.7 Compressing the Samples

Important information on the atomic properties of impurity centres and defects are obtained by recording the transmission of a sample while subjected to an external pressure. The pressure can be hydrostatic and it can be applied to amorphous as well as monocrystalline samples. This is usually performed by inserting the sample in a diamond anvil cell (DAC). When the samples are monocrystalline, the stress can be applied along one symmetry axis of the crystal. In the following section are described set-ups with which a uniaxial stress can be applied to a sample.

3.7.1 Uniaxial Stresses

Most of the uniaxial stress experiments are performed at LHeT because the mechanical properties of the crystals improve when temperature is lowered, but it can be applied also at higher temperatures for the reorientation experiments. A detailed description of a centrepiece to produce calibrated stresses on samples cooled at LHeT in a cryostat with a liquid helium reservoir is described by Tekippe et al. [35]. Centrepieces adapted to continuous flow cryostats have also been used, allowing a rather large temperature gradient between the room-temperature and LHeT sides with thin stainless steel tubing for the force-transmitting jig. Force is applied to the room-temperature side of the piston by a spring or by pressurized gas, and the sample is located between the piston and a base, as shown in Fig. 3.6. With such a set-up, the heat load is larger than with a classical sample holder so that it is difficult to cool

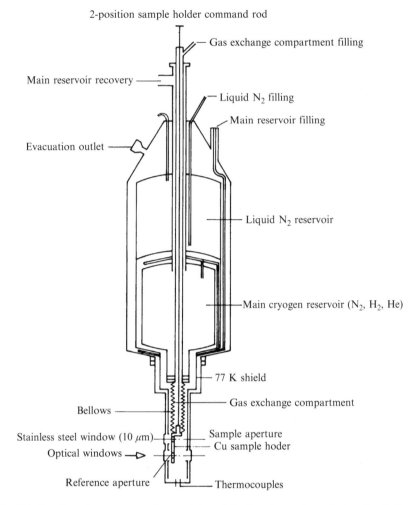

Fig. 3.5 Drawing of a custom-made cryostat used for low-temperature electron irradiation of semiconducting samples through a stainless steel window followed by optical measurements through optical windows (after [33]). The gas exchange compartment could be filled with He and its temperature monitored by a heater (not shown). An optical reference spectrum could be obtained through the reference aperture. The spectrum of the irradiated sample was obtained after pushing-down the sample holder with the command rod to position the sample in front of the optical windows

the samples under stress at temperatures below 8 K. The value of the force applied is measured either by a force transducer or by a manometer reading the gas pressure. The pressure is the ratio of the applied force to the cross-section of the sample.

In the set-up of Fig. 3.6 the pneumatic cylinder can apply a static compressive force on the push-rod, and the magnitude of the force is read on an amplifier connected to the force transducer. In this set-up, the maximum force which can be

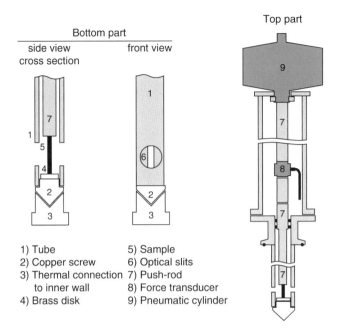

1) Tube   5) Sample
2) Copper screw   6) Optical slits
3) Thermal connection 7) Push-rod
   to inner wall   8) Force transducer
4) Brass disk   9) Pneumatic cylinder

Fig. 3.6 Sketch of a uniaxial stress apparatus of the pneumatic type to be inserted in a continuous-flow optical cryostat for measuring the absorption of a sample under uniaxial stress (after [36])

applied is 5 kN, corresponding to a stress of 1.25 GPa for a sample with a 2×2 mm^2 cross-section. The actual stress that can be applied to the samples depends on their mechanical strength and cleaving properties. Qualitatively, the mechanical strength of crystals increases with covalent bonding and higher pressures can be applied to group IV crystals than to III–V compounds. To apply very high pressures, the pressurized gas set-ups are superiors to those with a spring-loaded piston. Under good experimental conditions, silicon and diamond crystals can withstand uniaxial pressures in the 0.5–1 GPa range.

Beside accurate crystalline orientation, the sample must be cut with a very good parallelism between opposite sides to avoid crushing it when applying stress. A combination of cardboard, Cu or In spacers are placed between the sample and the metallic surfaces to avoid edge effects and minimize the effect of possible misalignment. It is usual to consider that the ratio between the sample length and the largest side of the base section must be ≥ 3 in order to obtain a reasonably uniaxial stress within the central region of the sample.

3.7.2 Hydrostatic Stresses

The optical absorption of small samples subjected to a hydrostatic pressure is usually measured in a diamond anvil cell (DAC), but sapphire has also been used in some set-ups [37]. There are several types of DACs, differing mainly in the way in which pressure is transmitted to the cell [38]. Some of these cells, like the so-called Merril-Basset one, have been modified for absorption spectroscopy at low temperature [39].

The basic part of a DAC is made of two diamonds separated by an indented metal plate, at the centre of which a hole has been drilled to be the sample chamber. The metal plate acts as a seal when pressure is applied. The diamonds used in DACs are brilliant-cut type with the bottom part of the gem (the culet) removed by grinding to obtain another flat surface, or standard cut octagonal diamonds, better suited for very high pressure experiments [40]. These diamonds have a small size and the largest dimension of their tables does not exceed 1 mm, requiring a concentrating optics. Type I diamonds show absorption in the $1100–1400\,cm^{-1}$ spectral region due to nitrogen under different forms, but they are more common (and less expensive!) than the purer IIa ones. They are therefore used for DACs except in situations where access to the above spectral region is needed. Note that the two- and three-phonon intrinsic absorption of diamond between about 1900 and $3900\,cm^{-1}$ limits its transparency in this region. The gasket is made of a metal or alloy (inconel, different varieties of stainless steel, BeCu, Re) adapted to the experiment intended, with a thickness in the 0.1–0.2 mm range. Pressure is exerted on the diamond tables by two metallic plates with apertures for admitting radiation. The largest size of the crystalline samples is of the order of $300\,\mu m$, with thicknesses in the $50\,\mu m$ range. One of the metallic plates is stationary and the other pushed by a movable mechanical device, for instance through a lever arm [41], but for low-temperature measurements, screws mounted directly on the platens exerting the pressure on the metallic tables are preferred [42].

The use of relatively large sapphire anvil cells has also been reported for PL measurements at LHeT in the near IR. This allows a chamber volume for the sample about one order of magnitude larger than the one of DACs at reasonable cost, at the expense of a smaller hydrostatic stress [43].

The sample chamber of a DAC is filled with a medium able to transfer the pressure to the sample as homogeneously as possible, and transparent in the spectral region of interest. At low temperatures, He, Ne and Xe and also homonuclear molecules (H_2, D_2, N_2, or O_2) have been used as pressure-transmitting media. The hydrostatic behaviour of He and H_2 allows experiments at low temperature up to 60 GPa (The kbar unit, traditionally used in many experiments with DACs, is close to 0.1 GPa) and N_2 can be used up to 13 GPa. Loading the sample chamber with the sample and the pressure-transmitting medium is usually performed by the liquid-immersion technique [44]. Hydrostatic pressure measurements in absorption experiments can be obtained from a calibration of the DAC using the pressure-induced shift of the R_1 and R_2 fluorescence lines of Cr^{3+} of a ruby chip near 694 nm, developed by Forman et al. [45]. However, this calibration is performed at room temperature and it must be extrapolated at low temperatures. It has been shown by Hsu [40] that the shift of the vibrational lines of the CO_2 impurities contained in N_2 used for pressure transmission could be used to measure pressure at low temperature.

3.8 Magnetooptical Measurements

The measurement of absorption by impurities and defects in crystalline solids under magnetic fields is mainly intended to observe the Zeeman splitting and shift of their levels or of bound excitons. As for a uniaxial stress, magnetic field is applied along the main symmetry axes of the crystal. When the propagation vector **k** of the radiation is parallel to the magnetic field **B**, one speaks of Faraday configuration, allowing only $\mathbf{E} \perp \mathbf{B}$ polarization, where **E** is the electric vector of the radiation, and when **k** is perpendicular to **B**, of Voigt configuration, allowing both $\mathbf{E} \perp \mathbf{B}$ and $\mathbf{E} \parallel \mathbf{B}$ polarizations.

Minimizing the line widths and preventing thermalization means that most of the optical experiments on impurities under a magnetic field are performed at LHeT. In the first experiments of this kind, the tail of an optical cryostat was inserted in the gap between the poles of a dc electromagnet and samples could be subjected to effective magnetic fields up to near $4\,\mathrm{T}$ ($1\,\mathrm{T} = 10^4\,\mathrm{G}$) [46]. Larger values of the magnetic field ($\sim 10\,\mathrm{T}$) could be obtained in some dedicated places like the Francis Bitter National Magnet Laboratory (Cambridge, MA) using a Bitter solenoid operated at RT. The first commercially available magnetooptical cryostats incorporating superconducting solenoids consisted in a solenoid made from Nb–Ti alloy or Nb_3Sn wire (later cable), whose horizontal bore allowed to insert a sample holder with the sample glued to it. With good thermal contacts, the temperature of the sample was about 8 K. The Faraday configuration was standard, but with diameter sizes of the bore near 20 mm, a sample holder with two parallel mirrors at $45°$ could be used to allow experiments to be performed in the Voigt configuration.

Later on, the solenoids were winded into a split-coil configuration or even replaced by two close solenoids (split pair) with the magnetic field along a vertical axis and a standard geometry in the Voigt configuration. They were also provided with an exchange-gas cryostat in which the sample could be rotated and its temperature adjusted, or with an anti-cryostat for measurements at room temperature. A magnetic field homogeneity of 10^{-3} at the centre of the solenoids combination is generally sufficient for standard Zeeman measurement. For magnetic resonance measurements, an improvement of at least two orders of magnitude in the field homogeneity is necessary and it requires modifications in the design of the overall solenoid structure.

The value of **B** at the centre of a standard solenoid of length L made from N turns of conducting or superconducting material produced by an electric current of intensity I circulating in the solenoid is $\mu_0 N I L^{-1}$. As N and L are generally known, a value of **B** can be deduced from the value of the intensity of the low-voltage dc current. Anyway, whatever the structure of the solenoid, the relationship between the current intensity and the magnetic field is provided by the supplier. The maximum allowable value of I must be kept below a limit corresponding to the transition field B_s above which the material of the solenoid returns into the normal resistive state. The most widely used superconductors in commercial solenoid magnets are the Nb–Ti alloys (40–60% Ti) with transition temperatures

T_s between 10 and 12 K and B_s values near 12 T and Nb_3Sn, with $T_s = 18$ K and $B_s = 22$ T. The setup of large current intensities in the magnetooptical cryostats leads to non-negligible Joule heating of the ohmic metallic leads and electrical contacts with the solenoid, and a corresponding increase of liquid He evaporation. To reduce this evaporation, the manufacturers of magnetooptical cryostats connect the solenoid with a parallel circuit made from the same superconductor. During the set-up of current, a part of this circuit (the so-called superconducting switch) is kept resistive by an external heater so that current flow in the solenoid, but when the desired current intensity is reached, the heater is switched-off and the whole circuit becomes superconducting. The intensity of the current source can then be set down to zero while the solenoid operates in closed loop. This type of operation is called persistent mode.

In magnetooptical experiments, and especially in the ones performed in the very far IR, two methods can be used. In the first one, the transmission or photoconductivity of a sample subjected to a constant magnetic field is analysed in energy with a spectrometer. In the second one, the transmission or photoconductivity of the sample at the energy of a laser line is measured as a function of the magnetic field. This allows a better S/N ratio than the first method because of the low emissivity of the IR sources in the very far IR.

References

1. J.K. Kauppinen, D.J. Moffatt, H.H. Mantsch, D.G. Cameron, Fourier self-deconvolution: a method for resolving intrinsically overlapped bands. Appl. Spectrosc. **35**, 271–276 (1981)
2. H.O. McMahon, Thermal radiation of partially transparent reflecting bodies. J. Opt. Soc. Am. **40**, 376–378 (1950)
3. D.L. Stierwalt, R.F. Potter, Lattice absorption bands observed in silicon by means of spectral emissivity measurements. J. Phys. Chem. Solids **23**, 99–102 (1962)
4. R. Murray, K. Graff, B. Pajot, K. Strijckmans, S. Vandendriessche, B. Griepink, H. Marchandise, Interlaboratory determination of oxygen in silicon for certified reference materials. J. Electrochem. Soc. **139**, 3582–3587 (1992)
5. A. Szabo, Observation of hole burning and cross relaxation effects in ruby. Phys. Rev. B **11**, 4512–4517 (1974)
6. S.P. Love, K. Muro, R.E. Peale, A.J. Sievers, W. Lo, Infrared spectral hole burning of sulfur-hydrogen deep donors in a SiGe crystal. Phys. Rev. B **36**, 2950–2953 (1987)
7. U. Werling, K.F. Renk, Saturation of a resonant phonon-band mode by far-infrared excitation. Phys. Rev. B **39**, 1286–1289 (1989)
8. T. Theiler, H. Navarro, R. Till, F. Keilmann, Saturation of ionization edge absorption by donors in germanium. Appl. Phys. A **56**, 22–28 (1993)
9. M. Budde, G. Lüpke, C. Parks Cheney, N.H. Tolk, L.C. Feldman, Vibrational lifetime of bond-center hydrogen in crystalline silicon. Phys. Rev. Lett. **85**, 1452–1455 (2000)
10. M. Göppert-Mayer, Über elementarakte mit zwei quantensprüngen. Ann. Phys.-Leipzig **9**, 273–294 (1931)
11. W. Kaiser, C. Garett, Two-photon excitation in $CaF_2 : Eu^{2+}$. Phys. Rev. Lett. **7**, 229–231 (1961)
12. J. Golka, J. Mostowski, Two-photon spectroscopy of shallow donor states in semiconductors. Phys. Rev. B **18**, 2755–2760 (1978)

13. D. Fröhlich, M. Sondergeld, Experimental techniques in two-photon absorption. J. Phys. E **10**, 761–766 (1977)
14. Ch. Uihlein, D. Fröhlich, R. Kenklies, Investigation of exciton fine structure in Cu_2O. Phys. Rev. B **23**, 2731–2740 (1981)
15. B. Pajot, *Optical Absorption of Impurities and Defects in Semiconducting Crystals. I. Hydrogen-Like Centres* (Springer, Berlin, 2010)
16. W. Eisenmenger, Superconducting tunneling junctions as phonon generators and detectors, in *Physical Acoustics*, vol. 12, ed. by W.P. Mason, R.N. Thurston (Academic, New York, 1976), pp. 79–153
17. L.J. Challis, Phonon spectroscopy. Contemp. Phys. **24**, 229–250 (1983)
18. K. Laßmann, Acoustic phonon spectroscopy with superconductor tunnel junctions of low-energy defect excitations in semiconductors, in *Advances in Solid State Physics*, vol. 37, ed. by R. Helbig (Springer, Berlin, 1995), pp. 79–98
19. F. Maier, K. Laßmann, Phonon scattering and IR-spectra of oxygen-related defects in gallium arsenide – aspects of quantitative phonon spectroscopy. Physica B **263–264**, 122–125 (1999)
20. L. Genzel, Far-infrared Fourier transform spectroscopy. In *Topics in Applied Physics*, vol. 74, ed. by G. Grüner (Springer, 1998) pp. 169–220
21. V.D. Akhmetov, H. Richter, FTIR spectroscopic system with improved sensitivity. Mat. Sci. Semicon. Proc. **9**, 92–95 (2006)
22. P.F. Moulton, Spectroscopic and laser characteristics of Ti : Al_2O_3. J. Opt. Soc. Am. B **3**, 125–133 (1986)
23. A.R. Chraplyvy, W.E. Moerner, A.J. Sievers, High-resolution spectroscopy of matrix-isolated ReO_4^- molecules. Opt. Lett. **6**, 254–256 (1981)
24. G.V. Kozlov, A.A. Volkov, Coherent source submillimeter wave spectroscopy. In Topics in Applied Physics, vol. 74, ed. by G. Grüner (Springer, 1998) pp. 51–109
25. E.M. Gershenzon, G.N. Goltsman, N.G. Ptisina, Submillimeter spectroscopy of semiconductors. Sov. Phys. JETP **37**, 299–304 (1973)
26. B. Gorshunov, A. Volkov, I. Spektor, A. Prokhorov, A. Mukhin, M. Dressel, S. Uchida, A. Loidl, Terahertz BWO-spectroscopy. Internat. J. Infrared and Millimeter Waves **26**, 1217–1240 (2005)
27. A. Yang, M. Steger, D. Karaiskaj, M.L.W. Thewalt, M. Cardona, K.M. Itoh, H. Riemann, N.V. Abrosimov, M.F. Churbanov, A.V. Gusev, A.D. Bulanov, A.K. Kaliteevskii, O.N. Godisov, P. Becker, H.J. Pohl, J.W. Ager III, E.E. Haller, Optical detection and ionization of donors in specific electronic and nuclear spin states. Phys. Rev. Lett. **97**, 227401 (2006)
28. W.M. Bullis, Oxygen concentration measurements, in *Oxygen in Silicon, Semicond. Semimetals*, vol. 42, ed. by F. Shimura (Academic, New York, 1994), pp. 95–152
29. H. Shirai, Determination of oxygen concentration in single-sided polished Czochralski-grown silicon wafers by p-polarized Brewster angle incidence infrared spectroscopy. J. Electrochem. Soc. **138**, 1784–1787 (1991)
30. H. Saito, H. Shirai, Determination of interstitial oxygen concentration in oxygen-precipitated silicon wafers by low-temperature high-resolution spectroscopy. Jpn. J. Appl. Phys. **34**, L1097–L1099 (1995)
31. L. Podlowski, H. Hoffman, I. Broser, Calorimetric absorption spectroscopy at mK temperatures – an extremely sensitive method to detect non-radiative processes in solids. J. Cryst. Growth **117**, 698–703 (1992)
32. D. Labrie, I.J. Booth, M.L.W. Thewalt, B.P. Clayman, Use of polypropylene for infrared cryogenic windows. Appl. Opt. **25**, 171–172 (1986)
33. A. Brelot, Doctoral thesis, Université Paris VII (1972)
34. P. Vajda, J. Lori, Low-temperature set-up for electron irradiation and subsequent photoconductive studies of semiconductors. Rev. Sci. Instrum. **40**, 690–692 (1969)
35. V.J. Tekippe, P. Fisher, H.R. Chandrasekhar, A.K. Ramdas, Determination of the deformation-potential constant of the conduction band of silicon from the piezospectroscopy of donors. Phys. Rev. B **6**, 2348–2356 (1972)
36. M. Budde, Doctoral thesis, University of Aarhus (1998)

37. K. Furuno, A. Onodera, S. Kume, Sapphire-anvil cell for high pressure research. Jpn. J. Appl. Phys. **25**, L646–L647 (1986)

38. A. Jayaraman, Diamond anvil cell and high-pressure physical investigations. Rev. Mod. Phys. **55**, 65–108 (1983)

39. E.E. Haller, L. Hsu, J.A. Wolk, Far infrared spectroscopy of semiconductors at large hydrostatic pressures. Phys. Stat. Sol. B **198**, 153–165 (1996)

40. L. Hsu, PhD thesis, EO Lawrence Berkeley National Laboratory, University of California, Berkeley (1997)

41. G.J. Piermarini, S. Block, Ultrahigh pressure diamond anvil cell and several semiconductor phase transition processes in relation to the fixed point pressure scale. Rev. Sci. Instrum. **46**, 973–979 (1975)

42. E. Sterer, M.P. Pasternak, R.D. Taylor, A multipurpose miniature diamond anvil cell. Rev. Sci. Instrum. **61**, 1117–1119 (1990)

43. T.W. Steiner, M.K. Nissen, S.M. Wilson, Y. Lacroix, M.L.W. Thewalt, Observation of luminescence from the EL2 metastable state in liquid encapsulated Czochralski-grown GaAs under hydrostatic pressure. Phys. Rev. B **47**, 1265–1269 (1993)

44. D. Schiferl, D.T. Cromer, R.L. Mills, Crystal structure of nitrogen at 25 kbar and 296 K. High Temp. High Press. **10**, 493–496 (1978)

45. R.A. Forman, G.J. Piermarini, J.D. Barnett, S. Block, Pressure measurements made by the utilization of ruby sharp-line luminescence. Science **176**, 284–285 (1972)

46. S. Zwerdling, B. Lax, L.M. Roth, K.J. Button, Exciton and magnetoabsorption of the direct and indirect transitions in germanium. Phys. Rev. **114**, 80–89 (1959)

Chapter 4
Absorption of Deep Centres and Bound Excitons

The deep centres are intrinsic or extrinsic complexes or isolated FAs with ground-state levels deep in the band gap, hence their name. They give rise to relatively high-energy transitions whose excited states cannot be described by donor or acceptor EMT. They can be found in some as-grown crystals, but they are also produced by irradiation with h-e particles or γ-rays, or associated with TMs introduced in the crystals. When their concentration is dominant, the resistivity of the material can reach the intrinsic resistivity and classical resistivity measurements are difficult to perform on such materials. These centres are characterized by the position(s) of their energy level(s) in the band gap, by their point-group symmetries, and by their isotopic distributions. In semiconductors, many high-energy absorption lines are also due to the creation of excitons bound to defects whose electronic properties are only roughly understood, but these lines can bring useful information on the nature of these defects, for instance by their electronic isotope shifts or splitting under a uniaxial stress.

The continuous absorption between the ground state of deep centres and the *VB* or *CB* continuum has also been investigated, and especially its temperature dependence, in connection with the shift between the equilibrium coordinates in the ground and final states, within the configuration coordinates model shown in Fig. 1.9. This temperature dependence, giving a low-energy shift of the absorption with temperature, is considered to be due to the electron–phonon interaction, by thermalization of phonon states represented by harmonic oscillators with energies $\hbar\omega_{ph}$. The relative strength of the electron–phonon interaction is usually measured by the Huang–Rhys factor S, which can be taken as the ratio of the Franck–Condon shift E_{FC} by the phonon energy [1].

In this chapter, with the exception of the divacancy in silicon and of EL2 in GaAs, we focus on lines observed in absorption at low temperature, which can often be observed at the same energy using photoluminescence (PL) methods. These lines, which are often due to the creation or annihilation of excitons bound to the centres, are generally sharp and they are called zero-phonon lines (ZPLs). Usually, one or several localized or resonant vibrational modes with well-defined symmetries are

B. Pajot and B. Clerjaud, *Optical Absorption of Impurities and Defects in Semiconducting Crystals*, Springer Series in Solid-State Sciences 169, DOI 10.1007/978-3-642-18018-7_4, © Springer-Verlag Berlin Heidelberg 2013

associated to these deep centres. The so-called vibronic coupling between the purely electronic transitions and the vibrational modes of these centres can produce mixed transitions involving electronic and vibrational excitations of the centre. These mixed transitions are called vibronic replicas or sidebands of the ZPL. In absorption, the energies of these replicas are the sum of the energies of the ZPL and of the vibrational mode in the electronic excited state, $\hbar\omega_{\mathrm{ve}}$. In the PL experiments, part of the energy of the ZPL electronic recombination is borrowed to excite a vibrational mode in the electronic ground state, and the energy of the photon emitted is the difference between the ZPL energy and the energy $\hbar\omega_{\mathrm{vg}}$ of the vibrational mode in the electronic ground state. Thus, on the frequency scale, the replicas observed in absorption nearly "mirror" those observed by PL with respect to the ZPL. Because of the coupling between the electronic and vibrational transitions, $\hbar\omega_{\mathrm{vg}}$, deduced from the PL measurements, is generally slightly larger than $\hbar\omega_{\mathrm{ve}}$, deduced from the absorption measurements. This produces a small asymmetry in the mirroring of the vibronic transitions in absorption and in PL. It must be noted that the interaction of the bound electrons with lattice phonons can also produce phonon replicas of the ZPLs. One of the interests of the vibronic transitions is that they present isotope shifts due to their vibrational part when the isotopic nature of the host crystal or of the FAs is changed, and they provide therefore valuable information on the nature of the impurity centres and complexes.

Several of the centres discussed in this chapter and in the following ones present structural symmetries lower than the cubic one. When these centres are paramagnetic, this symmetry can be derived from ESR results, but absorption measurements performed under uniaxial stress can also be used for this purpose. The possible symmetries of non-cubic centres in cubic crystals are limited to six, plus triclinic symmetry for centres whose only symmetry element is the identity operation (see Table E.1 of Appendix E). The main symmetry characteristics of these centres and of their response to a uniaxial stress are given in Appendix E. An important property of these centres is their already-mentioned orientational degeneracy, which is the number of different orientations in the crystal preserving their structural symmetry. This degeneracy, noted here R, is shared by the electronic and vibrational lines, in addition to their possible intrinsic degeneracies. For trigonal centres, aligned along the cube diagonals, $R = 4$, and for the orthorhombic-I centres, aligned along the diagonals of the faces of the cube, $R = 6$.

We concentrate here on the electronic absorption of four categories of deep defects in semiconductors and diamond. The first one is the family of radiation defects in silicon, whose optical properties have started being investigated in the 1950s. The second one encompass the centres found in natural and synthetic diamonds, the third one, the family of native defects in III–V semiconductors, whose main representative is EL2 in GaAs, and the last one, the transition metals (TMs) in compound semiconductors.

In silicon and diamond, radiative recombination associated with deep levels is often easier to measure than absorption and many spectroscopic investigations on deep centres in these materials have been made by PL and PLE spectroscopy (see, for instance, [2]), and when relevant, the corresponding results are included. When line positions are expressed in wavenumbers and in meV, unless otherwise specified, the multiplying factor 0.1239842 is used to convert the wavenumbers into meV.

4.1 Radiation Defects in Silicon

Radiation defects have been briefly discussed in Sect. 1.2.4.2, but we remind here some of their properties. They are mainly produced in silicon by irradiation with h-e electrons or fast neutrons with kinetic energies above typically 1 MeV. There is an intrinsic difference between the results obtained by the two methods: in the electron-irradiated samples, where the h-e electrons interact predominantly with the electron cloud of the atoms of the crystal, when the thickness of the sample is adapted to the electrons penetration depth (in some cases, the samples have to be irradiated on both sides), the production of defects is rather homogeneous and one can define a significant value of the Fermi level energy E_F in the sample. In the samples irradiated with fast neutrons, which interact principally with the atomic nuclei, there are regions more disordered than the others, and for moderate neutron fluences, for instance, there can be regions in the irradiated samples where the value of E_F is close to the one in the sample before irradiation [3, 4]. Moreover, neutrons with energies of ~ 10 MeV or more can displace directly or indirectly several atoms of the crystal and produce defects that are more extended than the ones produced by h-e electrons.

The stability of the irradiation defects depends on temperature and on the crystal in which they are produced. One usually distinguishes the intrinsic defects, consisting of one or more lattice vacancies and of self-interstitial atoms, and the extrinsic defects produced by the interaction of intrinsic defects with FAs in the crystal. In silicon, the self-interstitial (I) produced by electron irradiation is not stable in p-type material, even at LHeT, but it is stable up to ~ 140 K in n-type material, while the isolated vacancy (V) is stable only below RT. ESR and DLTS results have, however, shown that irradiation with light ions can extend the domain of stability of the self-interstitial [5]. Most of the irradiation defects are electrically active and they can display several charge states, depending on the initial resistivity of the crystal or on the illumination conditions in the electronic absorption experiments.

An overview of the first absorption results in irradiated silicon can be found in [6]. Depending on the initial resistivity and type of the samples, it showed the existence of several electronic absorption features in silicon samples irradiated with h-e electrons, deuterons, or fast neutrons. From a spectroscopic point of view, the electronic absorption of defects can produce broad bands, but also relatively sharp ZPLs which can often be observed by their PL spectrum, like the C- and P-lines, presented in Sect. 6.7.1 of [7] in relation with the (C,O)-related pseudo-donors. These ZPLs can be accompanied by weaker zero-phonon transitions involving hydrogen-like excited states. Other features related to the ZPLs are vibronic sidebands involving either phonons of the crystal and/or local vibrational modes (LVMs) of the defect. The sharp lines, with typical FWHMs of $\sim 1 \text{cm}^{-1}$ (0.124 meV) or less offer the possibility to observe electronic ISs of atoms of the defect, while, as already mentioned, the vibronic sidebands can also display vibrational ISs of these atoms. Information on the atomic structure and symmetry can be obtained from uniaxial stress experiments on the ZPLs and on the vibronic sidebands, keeping in mind that the electronic stress splittings are significantly larger than the vibrational ones.

When these deep defects are present in two (or more) charge states, one is often paramagnetic, and information on the nature and symmetry of the defect can be obtained from the hyperfine structure of its ESR spectrum and from its ENDOR spectrum. From the theoretical side, when possible atomic structures can be guessed for the defect, *ab initio* calculations can help to make a choice between them and to determine thermodynamic parameters for this structure.

In the first absorption measurements in irradiated silicon samples, absorption bands at 1.8, 3.3, 3.9, 5.5, and 6.0 μm (0.69, 0.38, 0.32, 0.23, and 0.21 eV) were reported [6]. These features were labelled by their wavelength, and among them, the bands at 1.8, 3.3, and 3.6 μm have been widely investigated. On the basis of correlations with ESR measurements in samples where these bands were observed, they were attributed to the absorption of an intrinsic irradiation defect, the divacancy V_2, in different charge states. The absorption features of this centre are presented in the following.

4.1.1   The Divacancy

The divacancy consists in two missing *nn* atoms of the crystal. It is produced in FZ and CZ silicon by low temperature or RT irradiation with h-e electrons, but also by irradiation with fast neutrons or high-energy protons. The electron-energy threshold for the formation of V_2 extrapolated from the paper by Corbett and Watkins [8] seems to be ∼0.6 MeV. This centre dissociates at temperatures between ∼250 and 300°C [9]. The undistorted V_2 centre, with its main axis along a ⟨111⟩ bond direction

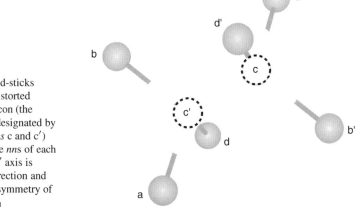

Fig. 4.1 Ball-and-sticks model of an undistorted divacancy in silicon (the vacant sites are designated by the *dashed circles* c and c′) showing the three *nn*s of each vacancy. The $c\,c'$ axis is along a ⟨111⟩ direction and the point-group symmetry of this centre is D_{3d}

should display a D_{3d} symmetry, as shown in Fig. 4.1, but the dangling bonds directed toward the missing atoms are unstable and bond reconstruction occurs. There are four equivalent orientations of the divacancy in the silicon lattice resulting in a fourfold atomic degeneracy of this centre.

In the early 1960s, two ESR spectra (Si-J and Si-C) detected in electron-irradiated silicon were related to the positive and negative charge states of an $S = 1/2$ paramagnetic centre identified as the divacancy [10]. These spectra, later relabelled Si-G6 and Si-G7, were associated with V_2^+ and V_2^-, respectively [11]. These ESR spectra showed evidence of a J-T distortion resulting from bond reconstruction between the atoms in Fig. 4.1. Reconstruction of the bonds between atoms a and d, on the one side, atoms a' and d', on the other side, and the formation of an extended bond between atoms b and b' is shown in Fig. 4.2, and in a static picture, this distortion lowers the symmetry of V_2 from D_{3d} to C_{2h}. As the choice of the bonded atoms of this distorted structure is arbitrary, there are three equivalent V_2 configurations giving an additional threefold degeneneracy, called electronic degeneracy, by opposition to the atomic one. In this configuration, this defect belongs to the category of monoclinic I centres.

In the V_2^+ charge state, there is only one electron in the extended bond b–b', but there are three in the V_2^- state. Electronic reorientation takes place between the three equivalent configurations of the extended bond (from b–b' to a–a' or d–d'), with corresponding changes of the reconstructed bonds. Activation energies for this process have been obtained from lifetime measurements deduced from the

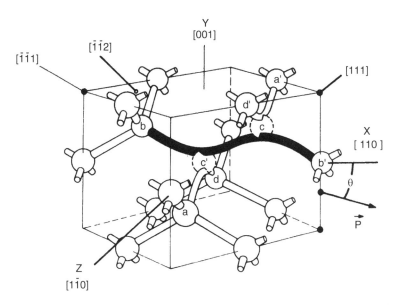

Fig. 4.2 Model of the J-T distorted divacancy. With the notation used, there are two two-electron reconstructed bonds (between atoms a and d and atoms a' and d') and one extended bond, shown *black*, between atoms b and b'. The vacancies are shown as dashed atoms c and c' ([12], after [11]). The point-group symmetry of this centre is C_{2h}. Copyright 1972 by the American Physical Society

temperature dependence of the line width of the ESR spectra. They are 73 and 56 meV for V_2^+ and V_2^-, respectively, (an average value of 0.06 eV is usually considered) and these relatively, small values explain that at about 100 K, the symmetry is averaged to D_{3d} by the frequency of thermal reorientation jumps [11]. The energy necessary to produce atomic reorientation between the four equivalent orientations of V_2 in the silicon lattice (one of them is shown in Fig. 4.1) was first determined from ESR measurements in the following way [11]: a stress of $\sim$180 MPa was first applied for $\sim$1 h along the $[01\bar{1}]$ axis of an irradiated sample at a temperature of 160°C compatible with the thermal stability of V_2. This stress produced an asymmetry in the otherwise equal V_2 populations along the four $\langle 111 \rangle$ axes due to a partial alignment of the divacancies V_2 along the directions with the highest angles with respect to the stress orientation. This asymmetry was maintained down to RT by cooling under stress. Differences in the relative intensities of the components of the low-temperature ESR spectra associated with the different atomic configurations gave full evidence of the atomic reorientation of V_2 along the orientations minimizing the energy with respect to stress. The activation energy for reorientation was obtained by measuring the gradual loss of this difference at low temperature after 15 min annealings at increasing temperatures. Assuming a first-order kinetics for reorientation yielded a pre-exponential term (frequency factor) of 10^{13} s^{-1} and a reorientation energy of 1.3 eV for both charge states [11]. The experimental procedure for reorientation measurements has been described here in detail because the same principle is used, *mutatis mutandis*, for the optical experiments.

The neutral charge state V_2^0 is not in principle paramagnetic. The S = 1 ESR spectrum labelled Si-I1, observed at 35 K in electron-irradiated CZ silicon, attributed to V_2^0 [13] presents the same characteristics as the Si-A14 ESR spectrum [14], and it is attributed in this reference to a vacancy-oxygen complex, presumably $(V_2O)^0$.

The three above-mentioned electronic absorption bands at 1.8, 3.3, and 3.9 μm have been associated with V_2. The 1.8 μm band can be observed from RT in irradiated p- and n-type silicon and it shifts to 1.7 μm at low temperature. The 3.9 μm band can also be observed in irradiated p-type at RT, but the 3.3 μm feature is only observed below $\sim$90 K in n-type silicon, and it is resolved in three components. In the following, these three components are globally labelled the 3.6 μm bands, from the one with the highest intensity. Anticipating the results of measurements of dichroism, it can be stated that the electric dipoles associated with the three above IR band are in the $X-Y$ plane of Fig. 4.2, but the angles θ of these dipoles with the X axis have to be determined experimentally. Besides temperature, the conditions for the observation of these bands depend on several parameters which are: (1) the initial position of the Fermi level in the samples (the type of the samples and their doping level), (2) the nature of the irradiating particles (electrons or neutrons), and (3) the illumination conditions of the samples during the absorption measurements. These conditions suggest that these bands are associated with different charge states of V_2. From an interpretation of the ESR and IR absorption results, it is generally accepted

Table 4.1 Spectroscopic characteristics of the IR bands associated with V_2 in ^{nat}Si at LHeT

| Position | | | FWHM | Charge | ESR |
Wavelength (μm)	Wavenumber (cm^{-1})	Energy (meV)	(meV)	state	spectrum
~1.7	~5880	~730	~90[a]	V_2^0	
	5560 at RT	689 at RT			
3.31	3020	374	~9	V_2^-	Si-G7
3.46	2890	358	8.4	V_2^-	Si-G7
3.614	2767	343.1	2.5[c]	V_2^-	Si-G7
3.6177	2764.2[b]	342.72			
3.9	~2560	~318	~90[d]	V_2^+	Si-G6

The 343 meV band is assumed to be, following [15], a ZPL associated with TA phonon replicas at 358 and 374 meV. In the following, these bands are sometimes denoted by the corresponding charge state of V_2.
[a] Estimated from [16]
[b] [17] in qmi ^{30}Si at LHeT
[c] Estimated from [18]
[d] Estimated at RT from [9]

that the 1.8, 3.6, and 3.9 μm bands are due to the V_2^0, V_2^-, and V_2^+ charge states of V_2, respectively, and their spectroscopic properties are summarized in Table 4.1.

Among the absorption experiments, some were performed on electron-irradiated samples and other ones on samples irradiated with fast neutrons. Besides differences in the irradiation conditions, some of the measurements were performed with the sample illuminated by the unfiltered radiation source while other ones were performed with filtered or monochromatic radiation, and these different experimental conditions can produce different results because of the trapping of photogenerated carriers by the defects. The two above causes can explain some of the differences reported in the published results. An absorption band at 3.07 μm (404 meV) has also been reported at 12 K together with the 3.6 μm bands of Table 4.1 in neutron-irradiated silicon [15]. In this reference, the 404 meV band was attributed to a TO phonon replica of the 343 meV ZPL. A broad band at 2.93 μm (423 meV) has indeed been observed at 15 K in n-type neutron-irradiated silicon diffused with lithium, together with the 3.6 μm bands [19], but illumination conditions which bleached out the 3.6 μm bands did not affect the intensity of the 2.93 μm band so that this latter band cannot be associated with the former ones.

In a paper by Cheng and Vajda [16], the 1.7 and 3.6 μm bands were reported to be observed simultaneously at 15 K with monochromatic radiation in a neutron-irradiated n-type sample, and this was attributed to the inhomogeneity associated with neutron irradiation.

An independent proof of the electronic reorientation of V_2 was provided in this study by illuminating for some time the oriented sample at 1.7 or near 3.6 μm with radiation polarized along the [001] axis: when measuring finally the absorption of the bands with the same polarization, a decrease of the intensities of the bands was observed, and it was due to an optically induced electronic reorientation of the divacancies. It was correlated with an increase of the corresponding intensities of the 1.7 or 3.6 μm bands when the measurement was performed with radiation polarized at 90° along [0$\bar{1}$1] after the same preparation. It can be explained by assuming that electronic reorientation is nearly free in the excited states of the transitions so that de-excitation can take place in any of the three orientations, leading to a depopulation of the configuration which is preferentially excited by polarized radiation, and to an increase of the population of the other configurations. When the same procedure was applied to the 3.9 μm band in an irradiated p-type sample, no reorientation was observed. This can be roughly explained by the fact that the final state of this band is resonant with the *VB* and that the lifetime of the holes in the excited states is reduced, preventing reorientation into the equivalent configurations. The activation energy for the reorientation of the defect associated with the 1.7 μm band is 76 meV, close to the one measured by ESR for V_2^+, but the prefactor for the 1.7 μm band is about one order of magnitude smaller than the one for V_2^+ and this is a justification of the attribution of this band to V_2^0 [16].

A detailed piezospectroscopic study ay LNT of the 1.7, 3.6 μm bands, and of the broad RT band at 3.3 μm (0.38 eV) was performed in different silicon samples irradiated with fast neutrons, and it confirmed the association of all these bands with the divacancy, but it left many questions unanswered [12].

Information on the production of the 3.9 and 1.7 μm bands in electron-irradiated silicon is scarce and the results of [20] are instructive. Figure 4.3 shows the variation of the intensities of these two bands in a silicon sample with [B] $\sim$1.8 × 10^{18} at cm^{-3} irradiated at 90 K with increasing 3 MeV electron fluences.

It is seen that the 3.9 μm band is observed for the measurement at the lowest fluence in strongly p-type samples, but that after a maximum for an electron fluence f_e of 2 × 10^{18} cm^{-2}, it decreases and becomes undetectable for $f_e \sim$1 × 10^{19} cm^{-2}

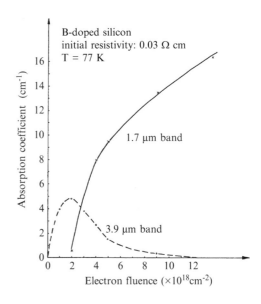

Fig. 4.3 Variation with electron fluence of the absorption coefficients of the 1.7 μm band (sample illuminated with monochromatic radiation) and 3.9 μm band (sample illuminated with white light) of the divacancy in silicon. The value of [B] given for that sample in the original paper is lower than the one expected for the resistivity indicated (after [20]). With permission from the Institute of Physics

because the Fermi level goes above the V_2^0/V_2^+ level. This is correlated with the 1.7 μm band which becomes visible only for $f_e \sim 2 \times 10^{18}$ cm^{-2} and whose intensity increases steadily with increasing fluences. In the samples investigated in this study, which were irradiated at 90 K and not allowed to warm-up after irradiation in order to perform annealing measurements, a decrease by nearly 50% of the 1.7 μm absorption after a 130 K annealing was observed, except in the low-resistivity p-type ones. It was attributed to an onset of migration of self-interstitials in these samples, which recombine with V_2 to form single vacancies. These self-interstitials are certainly not those produced as primary defects in the samples as they are supposed to be already mobile at 90 K. Those recombining with V_2 can possibly originate from the dissociation near 130 K of a secondary defect involving a self-interstitial. What can be better explained is the inverse increase of the 1.7 μm absorption observed for annealing at 180 K. It should correspond to a mobility onset of the single vacancies, which can form V_2. This increase was reduced in a CZ sample and explained by the formation of OV centres [20].

Absorption increases of $\sim$50% at 80 K of the V_2^- bands have been reported in p-type (resistivity $\sim$0.1 Ωcm) and intrinsic samples irradiated to fast neutron fluences of $4-7 \times 10^{16}$ cm^{-2} when monochromatic observation (cryostat placed after the monochromator) was replaced by white light observation (cryostat placed before the monochromator), where photoelectrons and photoholes are produced by band-gap excitation [12]. These samples contained V_2^- and V_2^0, and this increase can be explained by an efficient conversion of V_2^0 into V_2^- by photoelectrons trapping (the conversion of V_2^0 into V_2^+ by trapping photoholes seems to be inexistent). This was correlated with a small decrease of the intensity of the 1.7 μm band (V_2^0) in these samples, but a quantitative comparison is difficult because the intensity of the V_2^0 band is more than one order of magnitude larger than that of V_2^- in these samples under TEC. These authors mentioned that they did not observe the 3.9 μm band of V_2^+ in their irradiated p-type samples and this is simply due to the neutron fluence used. A similar photoconversion of V_2^0 to V_2^- is illustrated more dramatically in Fig. 4.4 for another neutron-irradiated sample for different initial conditions. In this spectrum, the FWHM of the ZPL at 3.61 μm is 3 meV, compared to 2.5 meV in electron-irradiated samples and the increase in the neutron-irradiated sample can be attributed to the lattice perturbation produced by the irradiation.

After suppression of the additional excitation, the return to the initial intensity of the V_2^- bands is very slow at LHeT and it does not fit a simple exponential decrease: there is a fast recombination with a time constant of about 900 s and a slow one with a time constant one order of magnitude larger. The characteristics of this recovery do not seem fundamentally different for initially n-type and p-type samples and the same result is obtained when following continuously the decay curve with time or letting the sample relax in the dark for the same duration [4].

Another kind of experiment was reported by Chen and Corelli [12]: they inserted a long-pass filter with a cutoff at 2 μm ($\sim$0.62 eV) between the white light source and the sample at 90 K and they observed a decrease of $\sim$20% of the V_2^- band in the above-mentioned p-type and intrinsic samples.

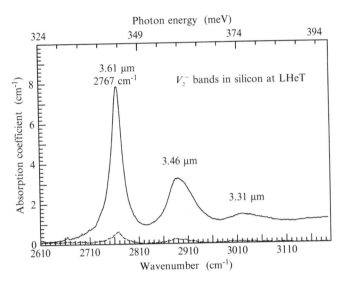

Fig. 4.4 Absorption of the 3.6 μm bands V_2^- bands in FZ n-type ($n = 10^{15}$ cm^{-3}) silicon sample irradiated with a fast neutron fluence of 1.7×10^{17} cm^{-2} at RT (the sample is intrinsic after irradiation). The broken curve is obtained with monochromatic observation and the solid curve with additional pumping by 1 eV photons. The reference sample is a high resistivity FZ sample (after [4]). With permission from the Institute of Physics

Similar absorption experiments at LHeT on p-type and n-type samples irradiated at 21 K with a significantly larger fast neutron fluence (3.5×10^{17} cm^{-2}) have been reported in [19]. The results on the V_2^- bands were different, however: the bands produced with white light illumination could be fully bleached when long-pass filters with cut-off energies between 0.73 and 0.50 eV were inserted between the white-light source and the sample. With a filter with a cut-off at 0.40 eV, only a decrease of the intensity of the V_2^- bands was observed.

Excitation measurements of the 343 meV–2767 cm^{-1} (3.61 μm) component of the V_2^- bands have also been performed at LHeT with the monochromatic beam of a grating monochromator as a probe and the excitation beam from a NaCl prism monochromator [21] and an example of the results is shown in Fig. 4.5.

In the sample of Fig. 4.5, initially 0.8 Ω cm p-type, only V_2^0 is present under equilibrium and the 3.6 μm bands are absent. Starting under equilibrium from 2.2 eV, V_2^0 starts to be ionized in the *VB*, for exciting energies just below E_g, creating V_2^- whose absorption starts to be detected and increases, to take its maximum value for an excitation energy of ∼0.72 eV, close to the ionization energy of V_2^0 in the *VB*. Below this value, V_2^- is no longer photoproduced and the combination of the photoionization of V_2^- in the *CB* (and of the possible production of photoholes from other defects) becomes effective and the absorption of V_2^- starts decreasing, reaching zero for ∼0.43 eV. A second photoionization spectrum was run with a different starting condition: the sample was first illuminated with 0.75 eV photons to saturate the absorption of the V_2^- band. The illumination was then

Fig. 4.5 Excitation spectrum (transmission) at LHeT of the 0.343 eV component of the 3.6 μm band in a 2-mm-thick p-type FZ Si sample irradiated with a fast neutron fluence of 1.8×10^{17} cm^{-2}. A: decreasing energies from 2 eV. B: increasing energies from 0.1 eV after pumping the sample at 0.75 eV ($\sim$1.7 μm). Some critical energies (eV) are indicated. Note the non-linear energy scale (after [21]). With permission from the Institute of Physics

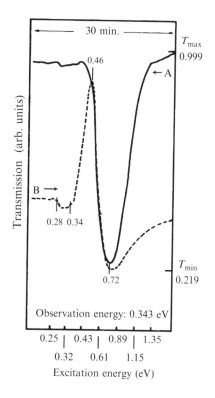

turned off and the sample allowed to reach a new equilibrium without excitation. Then, the excitation beam was scanned toward higher energies. The result is shown in spectrum B of Fig. 4.5. This new spectrum is characterized by the onset at $\sim$0.34 eV for a marked photoinduced decrease of the V_2^- band, up to 0.46 eV. It is followed by a new increase of the absorption of this band, whose maximum is reached for nearly the same excitation energy ($\sim$0.72 eV) as in spectrum A. Above this excitation value, the decrease of the absorption of the band seems to be a purely thermal effect independent of the excitation energy [21]. The interpretation of this spectrum is not straightforward and the decrease of the V_2^- absorption starting at $\sim$0.34 eV could be attributed to an indirect process involving the trapping by V_2^- of electrons photogenerated in the CB by the ionization of a defect separated from the CB by $\sim$0.34 eV to form V_2^{2-}. This mechanism should, however, compete with the photoionization of V_2^{2-} in the CB and it cannot be excluded that this latter effect is the mechanism restoring the V_2^- absorption at the initial level. The extra absorption of the V_2^- band for excitation energies near 0.72 eV is presumably due to the presence of V_2^0 which start again to be ionized in the VB to give V_2^-.

The absorption of the 3.6 μm band as a function of temperature between LHeT and LNT and the results of optical excitations on that band have been reported in 2-MeV-electron-irradiated CZ samples [18]. In this reference, it was pointed out that the 1.7 μm band went unobserved in all the samples investigated under any of the spectroscopic conditions used, so that a doubt was cast on the relation of this

Table 4.2 Spectral characteristics of the electronic lines observed on the low-energy side of the 343 meV (3.61 μm) band of V_2^- at LHe [4]

Position		FWHM (meV)	Relative peak absorption	Label	Annealing (K)
cm^{-1}	meV				
2634.9	326.69	0.4	0.007	Line 7	
2664.9	330.41	0.3	0.038	Line a	∼500
2679.8	332.25	0.3	0.014	Line b	∼400
2688.9	333.38		0.019		
2693.9	334.00	∼0.3	0.037	Line c	∼500
2699.8	334.73		0.016		
2703.9	335.24		0.021		
2708.9	335.86		0.032	Line d	∼500
2715.4	336.67		0.013		
2767	343.1	2.5	1		∼550

There are differences in the lines positions given in [4] and [23]; the ones in Table 4.2 are from [4]. The absorption of the 3.61 μm band of V_2^- is taken as a reference. The annealing temperature above which they are no longer observed is indicated in the last column [23]

band to the divacancy. However, in a more recent reference [22], this was no longer considered.

In neutron-irradiated FZ silicon, the observation at LHeT of about nine weak and relatively sharp lines was reported between ∼327 and 337 meV (∼2635 and 2720 cm^{-1}) on the low-energy side of the 343 meV band of V_2^- [4, 23]. Some of their characteristics are given in Table 4.2.

These new lines, which display the same phototrapping effect as V_2 have clearly an electronic origin, and they are barely detectable at LNT [23]. Origins related to impurity-defect centres have been ruled out as impurity changes over several orders of magnitudes do not change the relative intensities of these lines [23], so that they could be related to intrinsic defect centres. Lines a and c, the most intense of these lines, have been "rediscovered" in 1989, and tentatively attributed to a diinterstitial, which is indeed an intrinsic centre [24].

A calibration factor of 5.5×10^{15} cm^{-2} has been given for the peak absorption at LHeT of the 343.1 meV component of the 3.6 μm band of V_2^-, but it is warned that it may not be universally true [25].

The $Si_{1-x}Ge_x$ alloys have attracted interest for technological applications [26] and the introduction rate of defects in these alloys for small values of [Ge] has been investigated, in connection with radiation hardening. In $Si_{1-x}Ge_x$ alloys with x between 0.02 and 0.15, a low energy shift of the electronic band of V_2^0 at 5560 cm^{-1} (the 1.8 μm band), which can reach ∼400 cm^{-1} for $x \sim 0.12$ has been reported ([27], and references therein). This has been attributed to the contribution to the band of the absorption of a divacancy with a nearby Ge atom (V_2Ge). In the same study, measurements near 90 K of a sample with $x = 0.035$ showed also small (∼20 cm^{-1}) low-energy shifts and broadenings of the peaks at 2767 and 2890 cm^{-1} (3.61 and 3.46 μm) due to[1] V_2^-.

---

[1]In [27], a misprint has inverted the attributions of V_2^+ and V_2^-.

In germanium irradiated with h-e electrons or protons, at LNT and measured *in situ* near LNT, an absorption band at 517 meV (4170 cm^{-1} or 2.40 μm) with an annealing temperature ∼ 200–250 K is attributed to V_2 [28, 29].

4.1.2 The Higher Order Bands (HOBs)

When n- and p-type FZ and CZ silicon samples with resistivities larger than 1 Ωcm (n-type) and 0.1 Ωcm (p-type) are irradiated with fast neutrons and annealed between ∼350 and 500°C, a large number (between 20 and 40) of electronic absorption lines are observed at low temperature between 700 and 1300 cm^{-1} (∼85–160 meV) ([30–32], and references therein). These lines are observed after annealing of samples irradiated with 30-50 MeV electrons [30], fast neutrons [30–32] or with 2.5 MeV protons [33]; they are labelled collectively higher order bands (HOBs). In samples measured under TEC, the HOBs lines are weak or absent, but their intensities grow under above-band-gap-light illumination, as shown in Fig. 4.6.

In the CZ samples, the HOBs are superimposed to the lines of the O-related singly ionized thermal double donors (TDDs$^+$) [32]. Uniaxial stress measurements on 11 of the HOB lines have shown that none of the lines studied exhibited a stress

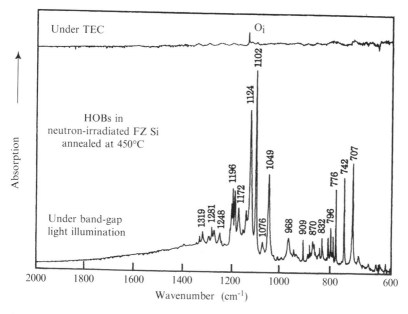

Fig. 4.6 LHeT spectra between 69 and 248 meV after annealing at 450°C of a high-resistivity (>2000 Ω cm) FZ silicon sample irradiated with $f_n = 8 \times 10^{18}$ cm^{-2}. The spectrum taken under thermal equilibrium conditions (TEC), shifted for clarity, shows only the vibrational line of residual O$_i$ at 1136 cm^{-1}. The spectrum taken under illumination shows the HOBs, with their positions (cm^{-1}) indicated for most of them (after [34])

splitting which could fit unambiguously one of the symmetry types established by Kaplyanskii [35]. The best fit assigns the defect associated with line at $1172\,\mathrm{cm}^{-1}$ of Fig. 4.6 to a centre with tetragonal symmetry and those associated with the other lines ($704, 742, 776, 909, 968, 1049, 1076, 1102, 1124$, and $1319\,\mathrm{cm}^{-1}$) to rhombic I-symmetry centres with different orientations of the dipole moment [31].

Several vacancies-related ESR spectra have been reported in FZ silicon irradiated with fast-neutron fluences of $\sim 10^{18}\,\mathrm{cm}^{-2}$ for different annealing temperatures [36]. The dominant one is Si-P1, observed for annealings between ~ 150 and $450°C$, attributed to a nonplanar five-vacancy cluster V_5^- [37]. For higher annealing temperatures, two other spectra, Si-A8 and Si-A11, are observed in the same domain of existence as the HOBs. Si-A8 is expected to arise from a vacancy cluster with more than six vacancies, showing monoclinic I C_{1h} symmetry, whereas Si-A11 is associated with an isotropic structure [36]. This can be taken as an indication that the HOB arise from vacancy clusters.

A ZPL at 1040.0 meV ($8388\,\mathrm{cm}^{-1}$ or $1.192\,\mu\mathrm{m}$) has been reported in absorption and PL measurements in silicon irradiated with neutrons or ions and annealed between ~ 250 and $500°C$, independently from the doping or O and C contents of the samples ([38], and references therein, [39]). From the splitting of this ZPL under uniaxial stress, the related defect can be clearly identified as a centre with tetragonal symmetry, with piezospectroscopic parameters A_1 and A_2 equal to -84 and 32 meV GPa^{-1}, respectively [38], and this symmetry is the same as the one of the HOB at $1172\,\mathrm{cm}^{-1}$. In the paper by Kaminskii et al. [39], this line,[2] noted X_{50}^{11}, is observed by PL at 1039.99 meV at a resolution of 34 $\mu\mathrm{eV}$ ($0.4\,\mathrm{cm}^{-1}$) together with a large number of BE lines. Their splittings under uniaxial stress were investigated and the one found for the X_{50}^{11} line is tetragonal, confirming the results of [38]. Among these ZPLs, the series of lines X_{80}^4, X_{72}^4, X_{66}^4, and X_{60}^4, previously reported by Johnson and Compton as the J_1, J_2, J_3, and J_4 lines [40], were found to display a trigonal symmetry. For this reason, these lines have been associated by Hourahine et al. [41] to the trigonal hexavacancy V_6 defect depicted in Fig. 1.4. These results seem also to indicate that the HOBs are associated with different polyvacancies clusters.

4.1.3 Group V-related Defects and Li-related Defects

4.1.3.1 The Vacancy-Group V Pair

One of the dominant defects produced by RT h-e electron irradiation in n-type silicon is a lattice vacancy trapped as a nn of the group V atom dopant. The identity of this defect, labelled E centre, was first established by Watkins and Corbett [42] for VP using ESR and later by Elkin and Watkins [43] for VAs and VSb. The

---

[2]In [39], the ZPLs are noted X_{yz}^x or B_{yz}^x by specifying the BE energy $xy.z$ with respect to that of the excitonic gap E_{gx} in silicon (1155 meV). The ZPL line at 1040 meV is thus noted X_{50}^{11}, corresponding to $E_{gx} - 115\,\mathrm{meV}$.

stability of this defect increases with the mass of the group V atom and onset of annealing varies from $\sim$100°C for VP to $\sim$130°C for VSb (see Fig. 2 of [44]). The removing of a Si atom to form a vacancy leaves one unpaired electron on each of the three neighbouring Si atoms, and a lone pair on the group V atom, P for definiteness, assumed to be neutral. In this ideal picture, the pair displays C_{3v} symmetry, but a J-T distortion occurs, with the formation of a reconstructed bond between two of the Si atoms. The four atoms and the vacancy lie in the same (110) symmetry plane and the overall symmetry is reduced to C_{1h}. The spin of the unpaired electron on the third Si atom is responsible for the ESR spectrum. When the Fermi level is not too deep in the band gap, a free electron can be trapped by this Si atom to form a lone pair, giving diamagnetic VP$^-$. By analogy with the A^-/A^0 acceptor levels in silicon, the VP$^-/V$P^0 level at $E_c - 0.44$ eV is often referred to as an acceptor level. For VAs and VSb, this acceptor level is located at $E_c - 0.42$ eV and $E_c - 0.39$ eV, respectively ([45], and references therein).

In n-type silicon samples with doping levels $\sim$1–2 $\times$ 10^{17} cm^{-3} irradiated at RT with 1.5 MeV electrons at fluences of 10^{18} cm^{-2}, broad electronic absorption lines were observed at 6150, 6000, and 5500 cm^{-1} (762.5, 743.9, and 681.9 meV or 1.63, 1.67, and 1.82 μm) for P, As, and Sb doping, respectively [45]. They were attributed to the negative charge state of the V-Group-V defect. They were observed under monochromatic spectral conditions in samples where V_2 is in the V_2^- charge state, so that the 1.7 μm absorption band due to V_2^0 could not interfere with these bands. A broad featureless absorption related to the group V–vacancy pair has also been reported in the 8500 cm^{-1} (1050 meV or 1.2 μm) spectral region [45].

This defect, like V_2, is anisotropic and its atomic and electronic reorientations under stress have been investigated from its ESR spectrum, but also from optical dichroism measurements [45]. A description of the J-T distortion for the V-group–V-atom defects in silicon in relation with their optical absorption has been given by Watkins [46]. Positron lifetimes and electron momentum distribution measurements in 2-MeV-electron-irradiated silicon samples with [As] = 10^{20} cm^{-3} have produced evidence of the formation of a centre made of a vacancy surrounded by three As atoms (VAs$_3$) [47].

4.1.3.2   Other Group V-related Defects

The low-temperature absorption of lines in the $\sim$800–1600 cm^{-1} ($\sim$99–200 meV) range has been reported in n-type FZ silicon after RT h-e electron irradiation of P-doped and As-doped FZ silicon [48] or P-doped FZ silicon [49]. With P and As dopings, the dominant features at LNT are located at 1150 and 1260 cm^{-1} (142.6 and 156.2 meV), respectively, with a FWHM estimated to $\sim$20 cm^{-1} ($\sim$2.5 meV) at LNT [48]. The observation of these lines depends on the electron fluence and, for increasing fluences, a maximum of the intensity is observed, followed by a decrease and a vanishing of the lines. They have been observed for doping levels between $\sim$10^{15} and 10^{17} cm^{-3}, but not in Sb-doped silicon with [Sb] $\sim$10^{17} cm^{-3}.

The fluence at the maximum depends on the doping level and the vanishing is related to a Fermi-level dependence. The measurements performed at LHeT by Suezawa et al. [49] enabled to resolve the $1150\,\mathrm{cm}^{-1}$ peak into a doublet at 1150.6 and $1156.5\,\mathrm{cm}^{-1}$ (142.66 and 143.39 meV) with FWHMs estimated to $1.5\,\mathrm{cm}^{-1}$ (0.19 meV). Their photosensitivity and Fermi-level dependence indicate that these lines have an electronic origin and that they are produced by centres having trapped an electron. From the results of uniaxial stress measurements, they have been analysed by Chen et al. [48] as arising from anisotropic donors, but this is unlikely because the lifetime of EM excited states in electron-irradiated silicon is too small to account for the FWHM observed at LHeT. A value of the ground state energy of these transitions in P-doped silicon has been estimated from the temperature dependence of the integrated intensities of some of these lines. It is $\sim E_c - 0.28$ eV [49] and this puts the excited states at $\sim E_c - 0.13$ eV. These lines disappear after annealing near 160°C. From the fluence dependence of the intensity of these bands at low doses, it has been assumed that they involved two vacancies and that they could be produced by the trapping of V by a group V atom, producing VP or VAs, followed by the trapping of a second vacancy giving V_2P or V_2As [49].

4.1.3.3   Li-related Defects

The study of the interaction of lithium with radiation defects started in the 1960s in relation with the interest in the Si:Li radiation detectors. This led *inter alia* to spectroscopic studies of Li in irradiated silicon ([50], and references therein). Li-related electronic absorption bands observed in the near IR are displayed in Fig. 4.7 and the spectrum is dominated by the 2.14 μm line.

The position of the 2.14 μm line at LHeT is $4673\,\mathrm{cm}^{-1}$ (579.4 meV) and its extrapolated FWHM is $\sim 1.6\,\mathrm{cm}^{-1}$ (0.2 meV). The annealing out of the centre giving the 2.14 μm line takes place at about 300°C, a temperature where V_2 dissociates, while the centres giving the 1.36 and 1.6 μm bands are stable up to about 400°C [19, 51].

The positions of the absorption lines and bands corresponding to Fig. 4.7 are given in Table 4.3.

Either in the Voigt or in the Faraday configuration, no Zeeman splitting or broadening of the 2.14 μm ZPL was detected for a magnetic field of 6.4 T indicating the absence of unpaired electron for the corresponding centre [19].

Additionally, in a Li-diffused sample where the 3.6 μm bands of V_2^- were made to appear under band-gap illumination, the observation of a broad band at 2.93 μm (423 meV) and of three sharp lines at 3.15, 3.38, and 3.80 μm (394, 367, and 326 meV), noted 1, 2, 4, and 7, respectively, was also reported [19]. In this experiment, the optical quenching of the 3.6 μm bands was accompanied by those of lines 4 and 7, but the intensity of band 1 did not change while the intensity of line 2 grew by a factor of ~ 2.

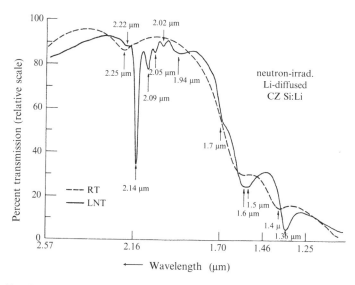

Fig. 4.7   Near IR absorption between ~480 and 1030 meV (~3900 and 8300 cm^{-1}) of Li-diffused CZ silicon irradiated at RT with 5 MeV neutrons at fluence of 8×10^{16} cm^{-2}. The sample was annealed first at 190°C to maximize the intensities of 1.36, 1.6, and 2.14 μm features (after [50]). Copyright 1973, American Institute of Physics

Table 4.3   Positions at LNT of electronic absorption lines and bands in irradiated Li-diffused silicon [50]. The positions in wavenumbers (cm^{-1}) are approximate

μm	1.36	1.50	1.6	1.94	2.02	2.05	2.09	2.14	2.2
meV	912	827	775	639	614	605	593	579	564
cm^{-1}	7350	6667	6250	5155	4950	4880	4785	4670	4545

It has been suggested that these lines and bands are associated with vacancy and polyvacancy defects including one or more Li atoms [50, 51], but a precise knowledge is still missing.

Near band-gap PL and absorption spectra have also been observed in electron-irradiated Li-diffused FZ silicon at LHeT [52]. The so-called Q system consists of three ZPLs Q_L, Q, and Q_H at 1044, 1045, and 1048 meV (8420, 8428, and 8453 cm^{-1} or 1.188, 1.1867, and 1.1832 μm), respectively, and it is accompanied at lower energies by resonant-mode-assisted structures. In silicon samples diffused with ^{nat}Li (^{6}Li: 7.6%, ^{7}Li: 92.4%), the Q ZPL observed in absorption and in PL and the Q_L ZPL observed in PL (this ZPL is too weak in absorption) show a weaker low-energy partially resolved structure. A possible attribution of this structure to a Li isotope effect has been investigated by Canham et al. [53] and by using a sample diffused with ^{7}Li/^{6}Li in a concentration ratio of 42.7/57.3, it has been demonstrated that four Li atoms were involved in the Q centre. The same sample allowed to measure directly the IS (+0.93 meV) between the 4 ^{7}Li and 4 ^{6}Li-atom combinations for the Q and Q_L ZPLs, and to infer that the low-energy structure observed in the samples diffused with ^{nat}Li was due to a doublet originating from the combination of three ^{7}Li atoms with one ^{6}Li atom.

The ZPLs of the Q system are attributed to the recombination or absorption of excitons bound to the isoelectronic $V - Li_4$ centre. In absorption, Q_L is very weak and it is attributed to a forbidden singlet to triplet transition while the Q and Q_H ZPLs are singlet to singlet transitions [53, 54]. The uniaxial stress and Zeeman measurements indicate that the $V - Li_4$ centre has C_{3v} trigonal symmetry [54]. A complete annealing of this centre is obtained for a temperature of 400°C [2]. Calculations indicate that the centre is best described by V^{4-} compensated by four adjacent Li^+ ions [55]. Possible structures of this centre have also been calculated by Myakenkaya et al. [56].

4.1.4 Carbon-related Bands

Spectroscopic measurements of silicon irradiated by h-e electrons have revealed the existence of many carbon-related defects. Two of them, the "C" and the "P" centres, have been discussed in Sect. 6.7.1 of [7] in relation with EM pseudo-donors. The electronic transitions of the carbon-related centres consist usually in a ZPL in the near IR, which can be observed by PL or absorption spectroscopy. These ZPLs, which usually show isotope effects of their own, are often accompanied by vibronic sidebands which can also show a combination of electronic and vibrational isotope effects. When these defects can be produced at higher concentration, the vibrational absorption of their LVMs can be detected at lower energies. In this chapter, to respect the structure of the book, we concentrate on the properties of the ZPLs whereas the vibrational properties of the C-related centres are discussed in Chap. 6.

4.1.4.1 Interstitial Carbon (C_i)

The defect known as interstitial carbon (C_i) is the dominant extrinsic defect produced by h-e electron irradiation at temperatures below RT in C-containing FZ silicon. It is produced when a mobile silicon self-interstitial Si_i (I) is trapped by C_s. C_i is electrically active, with a (0/+) donor level at $E_v + 0.28$ eV and a (−/0) acceptor level at $E_c - 0.10$ eV ([57], and references therein). $(C_i)^0$ has spin zero and the Si-G12 and L6 ESR spectra are associated with $(C_i)^+$ and $(C_i)^-$, respectively [58, 59]. A ball and stick model of C_i deduced from the ESR measurements, where the C atom and inner Si atoms are threefold coordinated, is shown in Fig. 4.8.

This kind of atomic structure is called a split interstitial or interstitialcy and it displays a rhombic-I C_{2v} symmetry. The most recent calculations give for $Si_3 - C_3$ and $C_3 - Si_4$ bond lengths of 0.174 and 0.180 nm, respectively [60]. A line at

Fig. 4.8 Dumbbell or split interstitial model of the C_i centre in silicon showing the dangling bonds on the C and Si atoms. It is represented in the positive charge state where the dangling bond on the Si atom has captured a free hole and is represented *dashed*. The indices represent the coordination numbers of the atoms (after [58]). Copyright 1976 by the American Physical Society

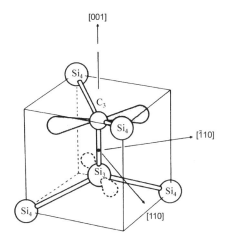

$6903\,\mathrm{cm^{-1}}$ (855.9 meV or 1.4486 μm) observed in absorption at LHeT [61,62] and by PL [61,63] has also been related to C_i. No ^{13}C IS has been reported for this ZPL [61]. The result of the analysis of the stress-induced splitting of the PL line is consistent with the structure shown in Fig. 4.8, with the addition of a small distortion reducing the symmetry of the centre to C_{1h} [63]. The absorption line corresponds to the excitation of an electron from $(C_i)^0$ and the PL line to the recombination of a weakly bound electron with $(C_i)^+$. Alternatively, it could be seen as the recombination or the creation of an exciton bound to C_i. Interstitial carbon starts diffusing near 270 K and C_i anneals out near RT.

In a FZ silicon sample doped with C and Sn (with 85% ^{119}Sn) irradiated with 2 MeV electrons below RT and annealed at RT, a ZPL at $6875\,\mathrm{cm^{-1}}$ (852.4 meV or 1.455 μm) has been reported at LHeT [62]. The observation of this ZPL requires the simultaneous presence of C and Sn in the samples, and it is thus attributed to a defect formed by the trapping of a diffusing C atom by a Sn atom. This ZPL is relatively close to the ZPL of C_i at 856 meV, and it has been therefore related to a metastable centre where C_i is perturbed by a *nnn* Sn_s atom. In this defect, no electrical activity is added to C_i as Sn is isoelectronic to Si, and the ZPL corresponds to the one of C_i perturbed by the Sn atom. This defect dissociates just above RT and more information is given in Sect. 7.2.1, in relation with its vibrational modes.

4.1.4.2   The G Line (969 meV Line) and the C_iSiC_s Defect

In FZ silicon samples electron irradiated at low temperature and annealed to remove C_i, or in samples irradiated at RT with relatively large doses of high-energy electrons or neutrons, a sharp line was first reported in PL measurements [64] at 969 meV, and later in absorption [65]. This ZPL is called the G line and also the 969 meV or 0.97 eV line, and a vast amount of literature has been published on that line. At LHeT, its average energy in FZ ^{nat}Si is 969.43 meV ($7818.9\,\mathrm{cm^{-1}}$ or 1.2789 μm).

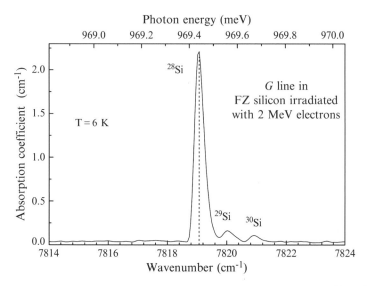

Fig. 4.9 *G* line absorption of a 300 Ω cm p-type FZ silicon sample with [C] ∼10^{16} cm^{-3} irradiated at RT at a fluence of 3×10^{16} cm^{-2}. The resolution is 0.1 cm^{-1} (12.4 μeV). The peak position of the ^{28}Si component is 969.437 meV (7819.04 cm^{-1}) and its FWHM 43 μeV (0.35 cm^{-1}). The ^{29}Si/^{28}Si and ^{30}Si/^{28}Si ISs are 0.120 and 0.229 meV (0.97 and 1.85 cm^{-1}), respectively (Pajot, unpublished)

The observation of C and Si ISs in the absorption and PL spectra of this line has demonstrated the presence of these atoms in the corresponding centre [66, 67].

Its absorption at LHeT in a ^{nat}Si sample is shown in Fig. 4.9. The relative intensities of the Si components (1, ∼0.06, and ∼0.035 for ^{28}Si, ^{29}Si, and ^{30}Si, respectively) practically match the relative natural abundances of the Si (1, 0.05, and 0.034, in the same order), indicating the presence of one Si atom in the related defect, already inferred from previous PL and absorption measurements [70, 71]. The *G* line displayed in Fig. 4.9 shows an asymmetric profile, materialized by the dotted line centred at the peak energy, which could be due to a small gradient of the defect concentration.[3] The temperature dependence of the absorption of the *G* line has been measured between 2 and ∼200 K, where its intensity goes to zero [68]. These results coupled with PL measurements between 2 and 77 K show that the internal structure of the centre is invariant in this temperature range.

The FWHM of the *G* line has been shown to increase as a function of the electron fluence, and the large concentration of oxygen in CZ silicon results also in an increase of the FWHM by a factor of ∼2 for low electron fluences, as shown in Fig. 81 of [2].

---

[3]This profile is observed in the whole series of 2-mm-thick FZ samples irradiated at the same energy.

In absorption, a strong vibronic replica of the G line is observed at 1039.6 meV [66], while in PL, the corresponding replica is observed at 897.55 meV [67]. They correspond to a LVM whose frequency is 580 and 566 cm^{-1} in the electronic ground and excited states, respectively.

In silicon samples enriched with ^{13}C, an extrapolated IS of the G line of $\sim +0.08$ meV (0.65 cm^{-1}) has been reported as the line was too broad to resolve the components of the two C isotopes [69]. The presence of carbon in the related defect was also demonstrated by the observation of a ^{12}C/^{13}C negative IS of about 2 meV (16 cm^{-1}) of the above-mentioned vibronic replica. These results seem to imply that only one C atom is involved in the centre giving the G line. Another weaker vibronic replica has also been reported with an associated local mode frequency of 67.35 meV (543.2 cm^{-1}) for ^{12}C, with a ^{12}C/^{13}C IS of -1.3 meV [69]. These replicas are discussed in Sect. 7.2.2 in connection with the vibrational spectrum of this defect.

In qmi ^{30}Si, the G line is found to shift by $+0.93$ meV ($+7.5$ cm^{-1}), showing the influence of the atoms of the crystal neighbours of the defect, as the value of the ^{30}Si IS of that line in ^{nat}Si is only $+0.23$ meV [70].

The G line shows no detectable Zeeman splitting for magnetic fields up to 5 T ([67], and unpublished results quoted in this reference) indicating zero-spin initial and final states and a global neutral configuration. Inversely, this line splits under uniaxial stresses [67,71,72], and the analysis of its splitting shows that the associated centre has monoclinic I symmetry. The averaged values of the components A_1, A_2, A_3, and A_4 of the piezospectroscopic tensor expressed in Kaplyanskii's notation for this symmetry [see expression (E.1) in Appendix E] measured for that line are $\sim 13.5, -10.4, 4.4$, and ± 9 meV GPa^{-1}, respectively (after [69]).

From the ESR side, an S = 1/2 ESR spectrum, labelled Si-G11, was reported by Watkins [73] and later by Brower [74] near LHeT in p-type FZ silicon with [B] $\sim 3 \times 10^{15}$ cm^{-3} after moderate electron or neutron irradiation. From the use of a C-doped sample enriched with ^{13}C, this Si-G11 spectrum was attributed to a positively charged centre containing two C atoms, with C_{1h} monoclinic 1 symmetry [74]. Now, in a FZ C-doped silicon sample enriched with ^{13}C and irradiated with thermal neutrons fluences between 10^{16} and 10^{17} cm^{-2}, the ODMR spectrum of an S = 1 centre was detected together with the PL of the G line, and the hyperfine (hf) interaction observed in this ODMR spectrum produced evidence of two equivalent C atoms in the centre. Finally, the fact that the amplitudes of the components of the ODMR spectrum, labelled Si-G11*, depended on the polarization of the PL detected at 0.97 eV proved unambiguously that the triplet state responsible for the ODMR spectrum corresponded to an excited state of the centre emitting the G line [75].

The conclusion of these studies is that the atomic structure of the centre associated with the G line includes one Si atom located between two C atoms (see Fig. 7.11). This defect is traditionally referred to as C_iC_s, but we use here C_iSiC_s as it better describes the atomic structure of the centre. The C_iSiC_s centre displays bistability, with two configurations noted here $(C_iSiC_s)_A$ and $(C_iSiC_s)_B$. These configurations are discussed in Sect. 7.2.2, in relation with the associated LVMs. The stable configuration for the neutral charge state of this centre is $(C_iSiC_s)_B$ and it is admitted that the G line is due in absorption to the creation of an exciton bound to $(C_iSiC_s)_B$ and in PL to the recombination of this exciton.

From the annealing temperature for which the G line disappears, C_iSiC_s must transform at $\sim 250°C$ (~ 520 K), and the annealing kinetics of the G line has allowed to determine an annealing activation energy of 1.7 eV [76].

Quantitative measurements have shown that the absorption of the G line in a given silicon sample increases with the electron fluence f_e up to a maximum, followed by a decrease for larger fluences, and that there was a correlation between the value f_e^{max} for which the intensity of the G line is maximum and the initial carbon concentration $[C_s]^0$ of the sample [77]. This allows the detection of C_s in FZ silicon samples at levels $\sim 10^{14}$ cm^{-3}. In FZ silicon, it is found that $f_e^{max} = (2.5\pm 0.5)$ $[C_s]^0$ and that the corresponding integrated absorption of the G line is related to $[C_s]^0$ by a calibration factor of 1.4×10^{16} cm^{-1} [78].

In FZ silicon samples irradiated with high-energy electrons, sharp ZPLs at 951.16, 952.98, 953.96, and 956.91 meV (7671.6, 7686.3, 7694.2, and 7718.0 cm^{-1}) have been observed at 2 K in absorption and by PL together with the G line. These four lines show also uniaxial stress splittings and Si ISs comparable to the ones of the G line, and their vibrational sidebands present C ISs suggesting that these defects may be closely related to each other [69].

In Li-diffused FZ samples doped with carbon, besides the G line, RT irradiation with h-e electrons allows also to observe in absorption and in PL, together with the Q system discussed in Sect. 4.1.3.3, ZPLs near 1083 meV (8735 cm^{-1} or 1.145 μm) ascribed to an S system bearing similarities with the Q system. The ZPLs of the S system are actually attributed to a V-Li$_4$ centre with a C atom neighbour [2].

4.1.4.3   The C and P Lines

The C line, known also as the 0.79 eV line, is observed by absorption or PL spectroscopy at LHeT in CZ silicon samples after irradiation with h-e electrons. This ZPL has already been presented, together with the P line in connection with the pseudo-donors in Sect. 6.7.1 of [7].

At LHeT, from absorption and PL measurements, the average energy of this line in ^{nat}Si is 789.59 meV (6368.5 cm^{-1} or 1.5702 μm). The presence of carbon and oxygen in the related centre is attested for that line by ^{13}C and ^{14}C ISs of $+79$ and $+175$ μeV (0.64 and 1.41 cm^{-1}), respectively [79,80] and by an ^{18}O IS of $+24$ μeV (0.19 cm^{-1}) [80]. The C line observed in a ^{nat}Si sample is represented in Fig. 4.10, and its profile shows a small high-energy dissymmetry.

The energy of the C line decreases with increasing temperature while its FWHM increases, and the values measured at 60 K are ~ 788 and 2 meV, respectively [81].

Presently, this ZPL is related to the neutral charge state of the C_iO_i defect produced by the trapping of C_i by interstitial oxygen (see Sect. 7.2.4.1). More precisely, in absorption, the C line can be considered as being due to the creation in its ground state of an exciton bound to $(C_iO_i)^0$, where the hole part of the exciton is relatively strongly bonded to the defect. The electron part is bonded to the (defect

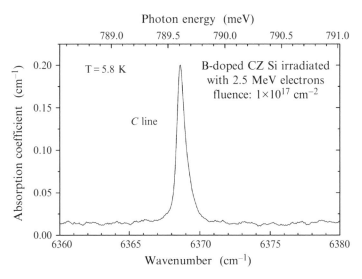

Fig. 4.10 C line absorption under TEC at a resolution of $0.1\,\mathrm{cm}^{-1}$ ($12.4\,\mu\mathrm{eV}$) of a B-doped CZ silicon sample with $p = 1 \times 10^{15}\,\mathrm{cm}^{-3}$ irradiated at RT. The measured peak position is $6368.61\,\mathrm{cm}^{-1}$ ($789.607\,\mathrm{meV}$) and the FWHM $0.65\,\mathrm{cm}^{-1}$ ($81\,\mu\mathrm{eV}$) (Pajot, unpublished)

+ hole) ion in the same way as an effective-mass electron to a shallow ionized donor, in a pseudo-donor or isoelectronic donor configuration. Its observation by PL is due to the recombination of this same BE. This unusual structure gives rise in the PL spectra to vibronic replicas of the C line at lower energies, related to the C_iO_i structure, and in the absorption or PL excitation spectra, to high-energy lines corresponding to the creation of the BE-pseudo-donor in EM excited states. The C line and the corresponding centre disappear after annealings in the 300–350°C range.

Anticipating on the attribution presented in Chap. 7, it is now admitted that this line originates from a C_iO_i centre where the C_i and O_i atoms, both threefold coordinated, are bonded to two shared Si atoms, in a configuration which bears a resemblance with the one of Fig. 1.7 for the N_i–N_i split interstitial pair.

The P line is observed by absorption or PL spectroscopy at LHeT in CZ silicon samples irradiated with h-e electrons after annealing at $\sim$300–400°C, and it anneals out in the 550–600°C range. Its average energy in ^{nat}Si from absorption and PL measurements is 767.17 meV (6187.6 cm^{-1} or 1.6161 μm). A FWHM of $\sim$0.18 meV ($\sim$1.5 cm^{-1}) has been measured for the P line at LHeT. In samples enriched with ^{13}C, it shows an IS of +79 μeV (+0.64 cm^{-1}) and in samples enriched with ^{18}O, an IS < +20 μeV (< + 0.15 cm^{-1}) ([80], and references therein).

PL measurements at LHeT with qmi ^{30}Si samples have shown that the C and P lines experience ISs of +0.84 meV (+6.8 cm^{-1}) and +0.80 meV (+6.1 cm^{-1}), respectively, with respect to their positions in ^{nat}Si [70].

Uniaxial stress measurements indicate a monoclinic I symmetry with point group C_{1h} for the "C" and "P" centres producing the C and P lines [82, 83]. The C and

P lines do not split in magnetic fields up to 5.3 T, and they are therefore attributed to singlet-to-singlet transitions [80].

When observed by PL spectroscopy, these lines present also vibronic sidebands at lower energies due to the coupling of the ZPL with LVMs of the defect [79, 80]. In absorption spectroscopy, the vibronic sidebands should be observed at energies higher than those of the ZPLs but they have apparently not been reported. The analysis of these sidebands is made in Sect. 7.2.4 in relation with the attributions of the LVMs observed by IR absorption (C(3) lines for the *C* line), and compared with *ab initio* calculations to provide finally atomic models of these defects.

In a sample irradiated with an electron fluence of 1×10^{18} cm^{-2}, the FWHM of the *C* line is found to increase by about 35% compared to the value for an electron fluence of 1×10^{17} cm^{-2}, a spectroscopic indication of a larger lattice disorder for a larger fluence. In addition, a new ZPL is observed at 786.95 meV (6347.2 cm^{-1} or 1.5755 μm). This ZPL, noted here X, is shown in Fig. 4.11 for the same initial sample as the one of Fig. 4.10.

The fact that the X ZPL is observed for an electron fluence much larger than the one for which the *C* line is observed alone, but with a reduction of the intensity of the *C* line, could be attributed to the formation of a new centre involving C_iO_i. In CZ silicon, the capture under electron irradiation of a self-interstitial *I* by C_iO_i is known to lead to the formation of IC_iO_i, observed for high fluences ([84], and references

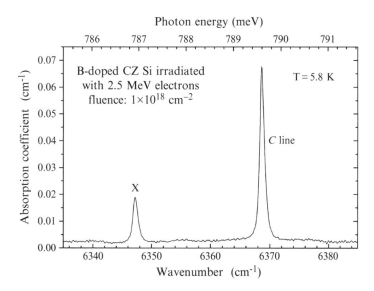

Fig. 4.11 Absorption of the ZPL X on the low-energy side of the *C* line under TEC at a resolution of 0.2 cm^{-1} (25 μeV) in a silicon sample with $p = 1 \times 10^{15}$ cm^{-3} irradiated at RT. The apparent intensity of the *C* line is smaller in this sample than in the one of Fig. 4.10, irradiated at a smaller electron fluence (Pajot, unpublished)

therein), reducing thus $[C_iO_i]$. The (IC_iO_i) centre is electrically active, and the X ZPL could thus be tentatively attributed to one of the charge states of IC_iO_i.

4.1.4.4 The 3942 cm^{-1} Line Defect

In CZ silicon samples irradiated at RT with h-e electrons at high fluences, a line at 3942 cm^{-1} (488.7 meV or 2.537 μm) was observed in absorption [65] and by PL [85]. It was associated with the intensity decrease at high electron fluences of the G and C lines discussed earlier. In samples irradiated with lower fluences, the intensity of the 3942 cm^{-1} line was found to increase after annealing near 150°C, before vanishing at $\sim$225°C in all the samples.

A detailed study of its absorption characteristics has been made by Davies et al. [86, 87]. At LHeT, its FWHM in ^{nat}Si, estimated from Fig. 6a of this reference, is $\sim$1 cm^{-1} ($\sim$0.12 meV). The uniaxial stress measurements show that the line occurs from a centre with monoclinic I (C_{1h}) symmetry with the electric dipole perpendicular to the principal axis of the centre (z dipole). In ^{nat}Si samples enriched with ^{13}C and ^{18}O, the line shows negative ISs of 0.7 cm^{-1} (0.09 meV) and of 0.95 cm^{-1} (0.12 meV) for ^{13}C and ^{18}O, respectively, and the intensities of the isotopic components are in the ratio of the concentration of the isotopes, indicating that there are only one C atom and one O atom per defect. The magnitudes of the shifts are typical of a ZPL, but the signs are opposite to the ones observed for the other ZPLs in silicon [86]. A guess on the nature of another ingredient of the 3942 cm^{-1} line defect has been inferred from the fact that in electron-irradiated silicon, the isoelectronic atom Sn$_s$ acts as a trap for migrating vacancies by forming VSn and that in Sn-doped CZ silicon, the intensity of the 3942 cm^{-1} line is about one order of magnitude smaller than in Sn-free silicon. It has been deduced from this reduction of the defect concentration that the vacancies trapped by Sn were missing to form the 3942 cm^{-1} line defect and that a vacancy was possibly involved in this defect [87]. No LVM has been found to be associated with this ZPL. This line is observed at 3943.6 cm^{-1} in irradiated CZ qmi ^{30}Si sample, shifted by +1.5 cm^{-1} (0.19 meV) with respect to its position (3942.1 cm^{-1}) in ^{nat}Si. This shift has the right sign, but it is significantly smaller than the one observed (from +7 to +14 cm^{-1}) for other ZPLs [17]. The electron-to-phonon coupling for this ZPL has been discussed by Davies et al. [17, 86]. The precise structure of the 3942 cm^{-1} line defect, noted (COV) in the paper by Davies et al. [17], is presently not known.

4.2 Impurity Centres and Defects in Diamond

As already stated, diamonds are roughly classified into two categories, namely, type I diamonds, the most common ones, which contain nitrogen at concentrations which can reach 500 ppma (1 ppma in diamond is 1.76 $\times 10^{17}$ at cm^{-3}), and type II diamonds which contain or not other impurities and eventually a small nitrogen concentration. The majority of the mined diamonds are brown in colour, but some natural diamonds can also be found with other colours, the most common one being yellow, due to the presence of nitrogen (see Fig. 1.1 and [88]). They can also present visible and/or invisible inhomogeneities, the latter being possibly detected spectroscopically or by optical microscopy, and also plastic deformation, which can be detected with polarized light. Defects are produced in some natural diamonds by natural radioactivity, and they are generally located near from the surface of the stone. Synthetic diamond crystals, polycrystals, and polycrystalline films have been and are produced in several laboratories, and also at an industrial scale. This has also allowed the production of diamonds with a controlled proportion of ^{12}C and ^{13}C isotopes, and [89] is an early report of the spectroscopic properties of defects in qmi ^{13}C$_{diam}$. Synthetic diamonds can present in many cases a better homogeneity than the natural stones, but also a larger concentration of unwanted FAs, principally nitrogen, and silicon in the diamond polycrystals obtained by high-temperature CVD on silicon substrates. The optical absorption and PL spectra of many defects and impurity complexes have been reported in diamond and to gain further insight on these defects, natural and artificial diamonds have been electron-, γ-rays-, or neutron-irradiated [90].

As C_{diam} is a metastable phase of carbon at atmospheric pressure, more or less pronounced graphitization of the diamond surfaces occurs during annealing at atmospheric pressure or under vacuum, but this is not very pronounced for short-term annealings at temperatures below 1600°C. Graphitization can be circumvented by annealing under a hydrostatic pressure where C_{diam} is the stable phase (typically above 5 GPa), or in a hydrogen plasma (which introduces hydrogen in diamond). Such so-called low-pressure/high-temperature annealing (LPHT) has been used to improve the optical properties of CVD monocrystalline diamond [91].

At the end of the last century, more than 500 electronic absorption or PL lines or bands were catalogued in diamond [92] and among them only a small number was satisfactorily explained [93]. A partial list of optical features in diamond is given in [94]. An analysis of the absorption lines in diamond is also a difficult task because there are lines with close energies corresponding to different centres. Moreover as experiments and interpretations accumulate, general reviews like the one of Walker [95], while very useful on many points, can become partially outdated. Fundamental information on the structure and chemical nature of defects in diamond has been obtained by ESR experiments. Many centres have been labelled by the name of their ESR signature, and there has been sometimes redundancy with their "spectroscopic" label. Progresses have recently been made in the identification of the ESR and optical features of some defects ([96], and references therein) and they are presented below. For convenience, the energies of the ZPLs with four or more significant digits and the vibrational energies are here expressed in meV. The other ones are expressed in eV.

4.2.1 Intrinsic Defects

4.2.1.1 Self-interstitial-related Centres

Diamond is a group IV crystal where the isolated self-interstitial is stable at RT, and it can be produced in pure diamonds by irradiation, for instance with electrons in the MeV range. Several absorption and PL lines which could be ascribed to self-interstitial-related centres have been reported in type II natural and synthetic diamonds after irradiation with γ-rays, electrons in the MeV range, and fast neutrons.

Two absorption lines at 1685 and 1859 meV (735.8 and 666.9 nm) at LNT, observed in all types of electron-irradiated diamonds, were reported by different authors ([95], and references therein). The line at 1685 meV is a ZPL whose gradual decrease in intensity below 80 K indicates a transition with an initial state $\sim$6 meV above the ground state [97], and the 1859 meV line is a vibronic replica of a dipole-forbidden transition from the ground state to the same final state as that of the 1685 meV line [98]. The frequency of the LVM involved in the vibronic transition is thus $\sim$169 meV ($\sim$1360 cm^{-1}). The annealing behaviour and the symmetry of the $R2(S = 1)$ ESR centre observed at RT [99] were found to correlate with the annealing properties of these two lines, which disappear near 430°C and display a tetragonal symmetry. These features were tentatively attributed to an interstitial complex [95]. As the two absorption lines and the $R2$ spectrum are also observed after irradiation of the purest type IIa diamonds, it has been assumed that the corresponding centre is intrinsic and neutral. At the light of the sum of experimental results and calculations, the current attribution for the 1685 and 1859 meV lines and the $R2$ ESR spectrum is a neutral split self-interstitial I^0 whose ground state is a diamagnetic spin singlet (S $=$ 0), with a spin triplet (S $=$ 1) excited state about 50 meV above the ground state. This excited state becomes populated above $\sim$100 K, allowing the observation of the $R2$ ESR spectrum above this temperature [98, 100].

The above-mentioned absorption lines show a small opposite temperature-dependent shift down to $\sim$20 K: At $\sim$90 K, for synthetic $^{nat}C_{diam}$, the positions of these lines estimated from Fig. 2b of [96] are 1684.67 and 1859.17 meV, shifting to 1684.4 and 1859.45 meV near 20 K.

In qmi $^{13}C_{diam}$, the thermalized ZPL at 1686 meV is shifted by +1.7 meV, but inversely, the vibronic transition is downshifted by -5.35 meV because of the dominant IS contribution of the LVM [97]. Recently, a UV absorption line at 3990 meV (310.7 nm), known for a long time as the R11 line [101], was also related to the split interstitial on the basis of the similarity of its annealing characteristics with those of the 1686 and 1859 meV lines and the $R2$ ESR spectrum [98].

The atomic configuration considered for this single self-interstitial is similar to the one of Fig. 4.8 for C_i in silicon, but with all the atoms identical, yielding a D_{2d} tetragonal symmetry, and this structure has been noted $I_1^{(100)}$ by Goss et al. [100]. It has been assumed that in the ground state, the reconstruction of the two dangling bonds results in an S $=$ 0 structure accompanied by a symmetry lowering perturbing distortion producing a D_2 rhombic-II symmetry of the interstitial [100]. *Ab initio* calculations of the structure of I^0 predict four electronic states when a one-electron doubly degenerate state contains two electrons.

The strong and relatively sharp ZPL at 2462.3 meV (503.53 nm), denoted[4] 3H, has been reported in absorption and PL at LNT in as-irradiated natural and synthetic type II diamond samples, except in a few type IIb samples with a high boron concentration ([95], and references therein, [102]). In $^{13}C_{diam}$, this line shifts by +5.0 meV [89]. This ZPL is accompanied by vibronic transitions corresponding to LVM energies of 169.3, 182.4, 186.7, and 217.8 meV (1365, 1471, 1506, and 1757 cm^{-1}). In a synthetic diamond sample containing 50% ^{13}C, the strongest of the PL vibronic sidebands, at 2244.5 meV (552.38 nm), corresponding to the 217.8 meV LVM, was shown to be split into a triplet corresponding to the $^{12}C-^{12}C$, $^{12}C-^{13}C$, and $^{13}C-^{13}C$ combinations [102]. This indicates that the two C atoms of the strengthened C–C bond of this LVM are equivalent or nearly equivalents. Uniaxial stress measurements indicate a rhombic-I C_{2v} symmetry. These facts have led to ascribe the 3H line to a self-split diinterstitial in the Humble configuration, depicted in Fig. 1.8 ([100], and references therein), and noted I_2^{2NN} in this reference. There are several statements of the annealing of that line, some of them apparently contradictory [95, 102, 103]. It seems that short annealings at temperatures in the 350–380°C range increase the intensity of the line, but that longer annealing in this range or at higher temperature cause the line to disappear, but stability up to $\sim$600°C has also been mentioned. These differences can be possibly related to charge state effects, as illumination with UV radiation can reversibly quench the line [95]. This implies that the 3H centre is paramagnetic, and the calculations seem to show that this centre correspond to the positive charge state of I_2^{2NN} [100]. One could also expect a configuration of the self-diinterstitial corresponding to the one of Fig. 1.7, with C_{2h} symmetry, and the $R1$ ESR signature has indeed been ascribed to that configuration ([104], and references therein, [100]), but no optical signature has been reported up to now for that geometry.

Other absorption systems known as TR12 and 5RL ([95], and references therein) have also been associated with centres involving self-interstitials. The TR12 and 5RL lines are observed at 2638 and 4582 meV (470.0 and 270.6 nm), respectively [105].

4.2.1.2 Vacancy-related Centres

The absorption of a line at 1673 meV (741.1 nm) in diamonds irradiated with h-e electrons or γ-rays was reported by Clark et al. [101] with the GR1 label. Its PL was reported by Mitchell [106] and by Clark and Norris [107], but it was also observed from the surface of natural stones submitted to natural irradiation [108]. This line, observed with samples at LNT or below, shows the highest intensity in type II diamonds, and it is attributed to the neutral vacancy V^0 (see, for instance,

---

[4]Not to be confused with the $H3$ line, due to VN_2^0, nearly at the same energy.

Fig. 4.12 The Dresden
Green Diamond is a natural
type IIa diamond which has
been naturally irradiated, and
its absorption spectrum
contains the GR1 and GR2–8
ZPLs. It is shown here
separated from a hat clasp
including other type IIa
colourless diamonds, which is
the form under which it is
exhibited in the New Green
Vault of the Dresden
Museum [110]

[109]). This centre gives the irradiated diamonds a green or blue/green colour [88]. The Dresden Green Diamond displayed in Fig. 4.12 is a type IIa natural diamond whose green colour is due to the presence of neutral vacancies in the volume of this stone [88].

Another weaker line at 1665 meV (744.6 nm), noted here GR1*, originates from a singlet state 8 meV above the doubly degenerate ground state V^0 ([109, 111], and references therein). At LNT the peak absorption of GR1* relative to that of GR1 is ∼0.05 [112]. Uniaxial stress measurements have shown that the GR1 line occurs at a centre with T_d symmetry; the doublet and singlet states, initial states in absorption of the GR1 and GR1* lines, respectively, are vibronic states resulting from the coupling of the ^1E electronic ground state of V^0 to a doubly degenerate vibrational mode [112]. The absorption of groups of lines, loosely noted GR2 to GR8, observed between 2.88 and ∼3.0 eV (430 and ∼413 nm) is related to that of the GR1 line, but their intensities are notably weaker. These lines are ascribed to transitions from the E state to T_1 excited states of a centre with T_d symmetry, and this is consistent with the results of uniaxial stress measurements [113]. They are observed with the smallest FWHMs in good-quality type IIa natural irradiated diamond where the residual impurity concentration is the smallest compared to N-containing containing diamonds. The absorption of these lines in a natural type IIb semiconducting diamonds irradiated and compensated with h-e electrons is displayed in Fig. 4.13.

The energies and FWHM of the transitions of the GR2–GR8 system and of GR1 at 10 K, corrected for the instrumental resolution, are given in Table 4.4.

At LNT, the GR1 lines are downshifted by ∼0.15 meV with respect to LHeT. The peak absorption of the GR1 line has been measured at LNT in qmi $^{13}C_{diam}$ and found to be shifted by +2.9 meV with respect to $^{12}C_{diam}$ [89].

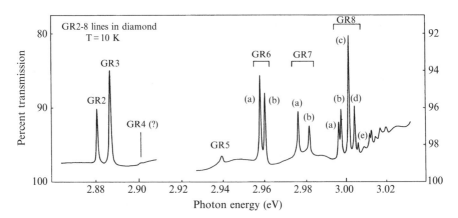

Fig. 4.13 Transmission spectrum of the V^0-related GR2–8 lines in an electron-irradiated IIb natural diamond at a resolution of $\sim$0.2 meV. The transmission scale is 100% to 77.5% for GR2-GR4 and 100% to 91% for GR5–8 so that the inverted transmission spectrum is somewhat similar to an absorption spectrum (after [114])

Table 4.4 Positions and FWHMs (meV) at 10 K of the GR2–8 system in diamond deduced from the spectrum of Fig. 4.13 compared to those of GR1 and GR1*. The corresponding wavelengths (nm) are given in brackets. After [114]

Line	Position	FWHM	Line	Position	FWHM
GR1	1673 (744.6)	0.5	GR1*	1665 (744.6)	1.3
GR2	2880.6 (430.41)	0.73	GR7 (a)	2976.0 (416.61)	0.78
GR3	2886.8 (429.49)	0.92	GR7 (b)	2981.7 (415.82)	0.84
GR4	$\sim$2902 (427.2)	–	GR8 (a)	2996.3 (413.79)	0.69
GR5	$\sim$2938 (422.0)	$\sim$1.4	GR8 (b)	2997.5 (413.63)	0.69
GR6 (a)	2957.7 (419.19)	0.91	GR8 (c)	3001.5 (413.07)	0.68
GR6 (b)	2960.0 (418.87)	0.86	GR8 (d)	3004.4 (412.68)	0.63
			GR8 (e)	3006.3 (412.41)	0.63

Resonant photoconductivity of the GR2-GR8 lines has been observed at 40 K, indicating that the particles in the excited states could be thermalized in the continuum, following the principle of PTIS. Photo-Hall measurements have shown that the charge carriers concerned in this experiment were free holes [115] and this attribution locates the V^0/V^- level at about $E_v + 3$ eV. The most recent determinations of the position of the V^0/V^- level in the diamond band gap from a combination of positron lifetime measurements and optical excitation is $E_v + 2.85$ eV [116].

Ab initio calculations [117] have confirmed that in diamond, the ground state of V^0 has a tetrahedral symmetry while in silicon, it is distorted into a state with D_{2d} (tetragonal) symmetry.

V^0 is diamagnetic, but an isotropic ESR S = 3/2 spectrum (T_d symmetry) has been identified by Isoya et al. [118] as belonging to V^-. This spectrum, first reported by Baldwin [119], had been labelled S1, but not correctly identified at that time. A ZPL at 3150 meV (393.6 nm), only observed in absorption in type-Ia diamonds (the ND1 line), has been ascribed to a transition of V^- ([120], and references therein). The proportionality between its intensity and that of the V^- ESR spectrum as a function of [V^-] indicates that the ground state of the ND1 line is the state responsible for the ESR spectrum [121].

The annealing studies of the GR1 and ND1 ZPLs show that V^0 and V^- display the same annealing curves, starting to migrate near 700°C, with complete annealing near 800°C [122]. The change in the intensity of the absorption of GR1 with temperature has allowed determining an activation energy of 2.3 eV for the migration of V^0 [120].

The divacancy (V_2) has been associated with an absorption band at 2.54 eV (488 nm) labelled TH5 [123], which correlates with the R4/W6 ESR S $=$ 1 spectrum associated to a centre with C_{2h} symmetry. These ESR signatures are produced by annealing of electron-irradiated type IIa diamonds above $\sim$800–850°C ([116], and references therein, [124]). Their observation in diamonds with a low N content has led to ascribe them to V_2^0. The annealing temperature of the TH5 line is $\sim$930°C. In the same samples, absorption lines at 507 (M1), 517 (M2), 595, and 610 nm (2.45, 2.40, 2.08, and 2.03 eV, respectively) have been reported [125], and the centre associated with M1 is stable up to the maximum annealing temperature ($\sim$1030°C).

4.2.1.3   Brown Diamonds

A brown colouration is observed in many natural and synthetic diamonds. The LNT absorption between $\sim$1 and 3.5 eV ($\sim$1240 and 354 nm) of a type I amber-coloured diamond was reported by du Preez [126], quoted in [95], and it showed a steady increase with energy, with structureless broad bands near 2.2 and 3.3 eV (564 and 376 nm, respectively). Putting aside the broad bands, this absorption was very similar to the comparable zone of the RT absorption of a series of brown natural type IIa diamonds shown in Fig. 4.14. The stones for these latter spectra were selected to encompass a colour index varying between 6 (the darkest brown shade) and 1 (the lightest brown shade), and the spectra exhibit a broad featureless absorption which varies monotonously as E^2 or E^3, where E is the photon energy.

Absorption profiles similar to those of Fig. 4.14 have been obtained on CVD brown diamonds [128].

In earlier works, the brown colouration has been linked with dislocations [129]. However, theoretical investigations suggest that the dominant dislocation types found in natural IIa diamonds are optically inactive ([130], and references therein). A confirmation of this point is found with diamonds where the brown colour is concentrated in brown stripes separated by colourless regions (zebra

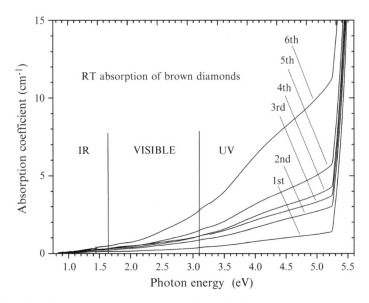

Fig. 4.14 Absorption of natural type IIa brown diamonds showing the variation of absorption when the colour grade varies from the first to the sixth grade in the Diamond Trading Company sorting category. The abrupt absorption rise near 5.3 eV corresponds to the onset of the intrinsic absorption of diamond (after [127])

diamonds): birefringence measurement on these diamonds has shown that the high-strain region associated with dislocations is located in the colourless regions. There is evidence that the brown colouration of natural stones is due to vacancy clusters or vacancy disks, and temperatures above 2200°C are required to remove the absorption continuum. This cannot be the case for CVD diamonds, in which the brown colouration disappears in the 1400–1600°C range.

4.2.2   Nitrogen-related Centres

4.2.2.1   Substitutional and Vacancy-associated Centres

There are many nitrogen-related centres in diamond, some of them being only identified by their ESR signature, and we cannot pretend to be exhaustive. We consider here not only the nitrogen complexes present in type I natural and synthetic diamonds, but also those produced by electron and neutron irradiation or by annealing of the same crystals. There are different labellings for these complexes, based on their atomic composition, ESR spectra, and optical absorption lines or bands (see [94]). There are also refinements in the classification for diamonds containing these complexes. A list of the known substitutional N complexes and of their labellings is given below, together with the data published for other N-related centres.

Substitutional Nitrogen (N_s)

The simplest nitrogen centre is the isolated substitutional N atom (N_s), and the diamonds containing predominantly N_s are classified as type Ib diamonds. These diamonds are relatively rare in nature, but they are the usual type for synthetic diamonds grown by the high-pressure high-temperature (HPHT) method at relatively low temperature (typically $\sim 1400°C$). Early RT photoconductivity and absorption results on N_s were reported by Farrer [131]. When neutral, N_s is a deep donor and the difference between its admitted optical and thermal ionization energies at LNT (~ 2.2 and 1.7 eV, respectively) is attributed to the difference in lattice relaxation between the ground and excited states [95]. Like the other substitutional group V elements in group IV crystals, N_s^0 is paramagnetic and it is characterized by the ESR C-P1 spectrum revealing a trigonal symmetry which indicates that the N atom occupies a substitutional site distorted along one of the C–N bonds [132, 133]. By extension, N_s is sometimes referred to as the C-P1, P1, (or C) centre. More recent ESR results [134] allowed to measure the g factor of N_s and to clearly show its $\langle 111 \rangle$ distortion. Spectroscopically, in type Ib diamond, there is very little electronic absorption below 2.2 eV (above 560 nm). Above 2.2 eV, the photoionization absorption of N_s^0 increases gradually, and at about 2.6 eV (480 nm), the increase becomes more rapid and two broad overlapping bands appear, with maximums at 3.3 and 3.9 eV (~ 380 and 320 nm). At LNT, a ZPL at 4059 meV (305.5 nm) accompanied by a broad vibronic band at 4.6 eV (270 nm) has also been attributed to N_s from the results of absorption measurements under uniaxial stress [135]. The attribution of the 4.6 eV band to N_s has, however, been challenged by Iakoubovskii and Adrianenssens [136] as in some type Ib diamonds, they found that the intensity of the 4.6 eV band was not correlated with the intensity of a N_s-related LVM at 1344 cm^{-1}. This contradiction has been apparently removed by calculations by Jones et al. [137] who ascribe the 4.6 eV band to a transition from the VB to N_s^+, producing N_s^0. In the same reference, the ZPL at 4059 meV is ascribed to a transition from the VB to N_s^0, producing N_s^-.

The temperature dependence of the photoionization spectrum of N_s^0 between 77 and 773 K has been investigated by van Enckevort and Versteegen [138], and the modelling of the results yielded a Huang–Rhys factor S of 16 and a phonon energy of 0.15 eV (~ 1200 cm^{-1}).

RT calibration factors of the absorption coefficients of the photoionization spectrum of N_s^0 at 4.6 eV (270 nm) and 3.1 eV (400 nm) have been deduced from comparisons with the spin-calibrated intensity of the ESR spectrum of N_s^0 ([139, 140], and references therein). From the above-mentioned interference of the N_s^0 continuum at 4.6 eV with a band related to N_s^+ and also from the difficulty to measure the 4.6 eV absorption in type IaA diamond samples, it seems more justified to use the RT calibration factor at 3.1 eV, given as 3.52×10^{17} cm^{-2} or 2.00 ppma cm by De Weerdt and Collins [140], and this absorption provides a detection limit of $\sim$0.3 ppma for [N^0].

The Substitutional N Pair (N$_{2s}$) (A Aggregate or A Centre)

Natural diamonds can also contain *nn* substitutional N atoms pairs (N$_{2s}$), also called *A* aggregates or *A* centres, which are deep donors with an ionization energy of $\sim$4 eV. These pairs are the predominant centres in type IaA diamonds. They can also be formed in HPHT diamonds at temperatures where isolated N becomes mobile (1700–2100°C) ([141], and references therein). The absorption in the UV of three ZPLs at 3757.6, 3901.2, and 3927.8 meV (329.96, 317.81, and 315.66 nm, respectively) has been reported at LNT for this N$_{2s}$ pair, with a FWHM of $\sim$1.2 meV for the lowest energy line [142]. Each of these ZPLs is accompanied by two vibronic replicas involving phonon resonances at 915 and 1280 cm^{-1}. The results of uniaxial stress measurements on the three ZPLs are consistent with their attributions to transitions from a non-degenerate ground state to a non-degenerate (3.758 eV), and doubly degenerate (3.901 and 3.928 eV) excited states of a centre with trigonal symmetry [142] and this had already firmly suggested in [143]. This centre, which is neutral, is diamagnetic, but illumination with UV radiation of natural yellow diamonds containing N$_{2s}$ has led to the observation of an ESR spectrum labelled W24 attributed to an unpaired electron in an orbital on two equivalent *nn* N atoms, and actually due to N$_{2s}^+$ [144]. Detailed ESR and ENDOR results on N$_{2s}^+$ have also been reported [145].

VN$_4$ (B Aggregate or B Centre) and VN$_3$

Complexes of N$_s$ with vacancies are found in natural stones, but some of them can only be produced by irradiation with h-e particles of natural or synthetic diamond. We present first the centres found in natural diamond.

The centre made of four N$_s$ atoms surrounding a vacancy (VN$_4$), known as *B* centre or *B* aggregate, is predominant in type IaB diamonds or in HPHT diamonds after annealing near 2400°C [141]. Three ZPLs at 5254, 5264, and 5279 meV (236.0, 235.5, and 234.9 nm) known as the N9 spectrum are associated with VN$_4$. Generically, the natural diamonds containing both N$_s$ and VN$_4$ are the most common natural diamonds and they are classified as type Ia diamonds. N$_{2s}$ and VN$_4$ are generally found in variable proportions in the same stone and when their concentrations are comparable, one speaks of type IaA/B diamonds.

In the majority of the type IaB diamonds are also found planar defects in (100) planes (platelets), varying in size between a few nanometres and tens of micrometres. It has been suggested that these defects were formed by interstitial C atoms ejected during the formation of VN$_4$ in N$_{2s}$-rich diamonds [146]. Platelets can also contain variable amounts of nitrogen but it was concluded that this nitrogen is present as an impurity and not as an integral part of their structure ([147], and references therein).

The centre made of three N_s first neighbours of a vacancy ($V N_3$) is found in natural types Ia and Ib diamonds. This centre is also known[5] as N3 because of a ZPL so labelled, observed in absorption or PL at 2.985 eV (415.4 nm), with low-energy vibronic replicas when observed by PL ([95], and references therein). The N3 ZPL correlates in intensity with the N2 ZPL observed in absorption at 2.596 eV (477.6 nm). $V N_3$ and N_{2s} concentrations can be increased by annealing type Ib diamonds at 1700–1900°C [149, 150] or by short-term annealing ($\sim$10 min) at 2000°C under an hydrostatic pressure of 8 GPa [151]. The centre $V N_3$ has C_{3v} symmetry, and an ESR (S $= 1/2$) spectrum labelled C-P2, first reported by Smith et al. [132], has been later associated with $V N_3$ [152]. Isotopes effects of the PL spectrum of $V N_3$ have been reported by Davies et al. [153]: the ZPL is shifted by $+0.2 \pm 0.1$ meV in $^{12}C_{diam}$ doped with ^{15}N compared to $^{nat}C_{diam}$. In $^{13}C_{diam}$, the energies of the phonons involved in the vibronic replicas are reduced in the ratio $\sqrt{12/13}$ and the ZPL is shifted by $+4.5 \pm 0.2$ meV compared to $^{nat}C_{diam}$.

The NV Centre

Another N-related centre formed under electron or neutron irradiation followed by annealing above 700°C of N-containing diamonds is $V N$ (also noted N-V), a vacancy trapped by N_s, with two charge states $V N^0$ and $V N^-$, and C_{3v} symmetry. This centre is stable up to at least 1000°C.

A ZPL at 2156.2 meV (575.00 nm or 17391 cm^{-1}) associated with $V N^0$ is observed in absorption and PL,[6] and it has been found to occur from a transition between the E ground state and an A excited state of a centre with trigonal symmetry [155]. The absence of an ESR signature of the S $= 1/2$ ground state has been attributed to a dynamic J-T distortion of this ground state [155,156].

The ZPL of $V N^-$ is observed in absorption or in PL at 1945 meV (637.5 nm or $\sim$15690 cm^{-1}), and it is due in absorption to $^3A_2 \rightarrow {}^3E$ transition of a centre with C_{3v} symmetry [122]. The S $= 1$ ground state gives the C-W15 ESR spectrum ([157], and references therein) with a spin state characterized by a very long coherence time measured by spectral hole burning [158]. The physical properties of $V N^-$ have been much studied (more than 150 papers between 2000 and 2007) since it has been discovered that this centre could be used as a single-photon

---

[5]This centre should not be confused with an ESR S $= 1/2$ centre, also labelled N3 ([148], and references therein), assumed to involve N_s and an O atom.

[6]The energy given for the $V N^0$ ZPL is taken from [154]. In this reference, a ZPL at 575.97 nm (2152.6 meV) *not* related to $V N^0$ has also been reported.

quantum source, with possible applications in solid-state quantum computing and cryptography [159].

VN_2 and V_2N_4

The other N-related defects identified are VN_2, a vacancy trapped by N_{2s}, with C_{2v} symmetry, and V_2N_4, produced by trapping of a vacancy by VN_4 under electronic irradiation of type IaB diamonds.

VN_2 is produced with the highest concentration in type Ia crystals after h-e electron irradiation and annealing between $\sim$500 and 1200°C. In the neutral state, it gives the H3 absorption and PL ZPL at 2463 meV (503.4 nm or $\sim$19870 cm^{-1}), very close to the 3H absorption of the self-split diinterstitial discussed in Sect. 4.2.1.1.[7] This centre has a C_{2v} symmetry with a $\langle 110 \rangle$ symmetry axis ([120, 160], and references therein). The H2 features with a ZPL at 1257 meV (986.4 nm) are attributed to VN_2^-, and uniaxial stress measurements have shown that a vibronic line at 1427 meV (994.3 nm) is also related to VN_2^-. In qmi $^{13}C_{diam}$, the usual positive IS (+0.93 meV) is observed for the pure electronic ZPL, and a negative one for the vibronic replica (-5.28 meV). The values of the optically accessible LVM associated with H2 in $^{nat}C_{diam}$ ($^{12.01}C_{diam}$) and in qmi $^{13}C_{diam}$ are thus 167.08 and 160.87 meV (1347.6 and 1297.5 cm^{-1}), respectively [161]. In a mixed $^{12}C/^{13}C$ diamond sample, the vibronic band is broad and its position is intermediate between those in $^{12.01}C_{diam}$ and qmi $^{13}C_{diam}$, implying the contribution of two C atoms in this LVM, but there is no N-related shift with N isotopic substitution, indicating that the vibrational motion concerns mainly the C atoms [162].

The H4 ZPL is observed at 2498 meV (496.3 nm) by absorption or PL in type IaB diamond samples (containing V_2N_4) after irradiation and annealing at $\sim$650°C or above, and it is ascribed to the V_2N_4 centre ([120], and references therein). It must be noted that some of the N-related centres produced by electron irradiation followed by thermal annealing can also be produced by annealing of brown diamonds [129].

The above description is not exhaustive and electronic absorption features related to other centres can be found in the reviews [95] and [160].

Most of these centres are also responsible for LVMs and for the activation of phonon modes in the infrared, which are discussed in Chap. 5, and the correlation between the two kinds of absorption has been used for the attribution of the electronic lines.

---

[7]To point out again the redundancy in energy between ZPLs associated with different centres, a ZPL at 2463 meV (503.4 nm) almost coinciding with the H3 and 3H ZPLs has been observed in PL together with a ZPL at 2429 meV (510.4 nm) in some natural diamonds, and ascribed to a centre labelled S1 ([95], and references therein).

4.2.2.2 Other N-related centres

There are many unidentified N-related centres in type I diamond. For instance, a line at 2085 meV (594.6 nm) has been observed in electron-irradiated type Ia diamond annealed above $\sim300°C$ [163]. The annealing of the related centre in the 1000–1100°C temperature range is correlated with the observation in the near IR of lines at 612.4 and 640.8 meV (4939 and 5168 cm^{-1}), known as the H1b and H1c lines, respectively. It is stated by Collins et al. [164] that the available evidence suggests that the centres giving rise to the H1b and H1c lines are formed when part or all of the 2085 meV centre is trapped by N_{2s} (for H1b) and VN_4 (for H1c).

A pair of lines at 2153 and 2313 meV (576 and 536 nm) has been reported in the LNT absorption spectrum of an electron-irradiated type IaB diamond annealed at 1600°C for 1 h [154]. The 576 nm line is close to the VN^0 line at 575 nm, but further experiments confirmed that the two lines are distinct. The 536 and 576 nm lines have also been observed in the absorption spectrum of an untreated natural brown diamond [154]. These results have been correlated with those of cathodoluminescence (CL) measurements [154], and references therein), showing the existence of another unidentified N-related defect.

In irradiated type Ib diamonds, the absorption of ZPLs at 2367 and 2535 meV (523.8 and 489.1 nm) has been ascribed to metastable (N_s-I) complexes. Between 300 and 650°C, I becomes mobile and it can be trapped by some (N_s-I) complexes to form (N_i-I) complexes. ZPLs at 2807 and 3188 meV (441.7 and 388.9 nm) have been observed in absorption and more easily in PL [165]. Six vibronic replicas of the 3188 meV ZPL have been observed by PL and their ISs in $^{13}C_{diam}$ and in ^{15}N-enriched $^{12}C_{diam}$ samples indicate the presence of a single N atom and the absence of a vacancy in the corresponding centre [166]. A comparison of these experimental results with the calculation of [167] has led to the attribution of the 3188 meV ZPL to the above-mentioned $(N_i - I)$ complex.

4.2.3 Si-related Absorption

A peak at 738 nm (1.68 eV) was reported by Vavilov et al. [168] in the CL spectrum of CVD diamond. As it was also observed in a Si-implanted diamond sample, this peak was suggested to be related to silicon [169]. This was indirectly confirmed by the CL results of Ruan et al. [170], who observed the 1.68 eV peak only in CVD diamonds grown on silicon substrates. It is shown also in this reference that this Si-related centre is stable up to at least 1350°C. The direct confirmation of the presence of Si in the centre giving the 1.68 eV peak came from a spectroscopic study at a resolution of 0.1 meV of diamond crystals doped with silicon [171]: in this study, PL and absorption measurements showed the resolution of the 1.68 eV peak into twelve components ascribed to four electronic transitions, each with three satellites due to the ^{28}Si, ^{29}Si, and ^{30}Si isotopes. The relative intensities of the isotopic lines matched the relative abundances of the Si isotopes, indicating that only one Si atom

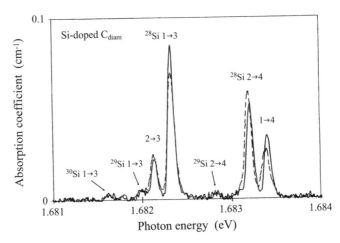

Fig. 4.15 Absorption of the 1.682 eV multiplet structure of a (V, Si) centre in diamond at 6 K (full line spectrum) and 40 K (dashed line spectrum), after [171]. The FWHMs are resolution limited to ~ 0.1 meV. The wavelength range is 737.6–736.2 nm

was involved in the corresponding centre. From the temperature dependence of the relative intensities of the four ^{28}Si transitions between 6 and 40 K, the components could be phenomenologically explained by a splitting of the ground state into two levels 1 and 2 separated by 0.22 meV ($1.8, \text{cm}^{-1}$) and of the excited state into two levels 3 and 4 separated by 0.93 meV ($7.5, \text{cm}^{-1}$), where 1 and 3 correspond to the deepest levels. The energies of the $2 \rightarrow 3$, $1 \rightarrow 3$, $2 \rightarrow 4$, and $1 \rightarrow 4$ transitions[8] deduced from the fit of the spectra for ^{28}Si are 1682.09, 1682.28, 1683.16, and 1683.36 meV, respectively, and the (1–2) and (3–4) separations deduced directly from these energies are 0.20 and 1.07 meV, respectively. These splittings were tentatively attributed to tunnelling between two configurations of the Si atom [171].

The $^{28}\text{Si}/^{30}\text{Si}$ ISs of these four transitions are very comparable (~ 0.7 meV or $\sim 5.6, \text{cm}^{-1}$). The intensity of the 1.682 eV feature in Si-doped diamond irradiated with 2 MeV electrons was also measured by PL at LNT as a function of the annealing temperature. It showed a net increase with respect to the one in as-grown material for annealing at $\sim 700°C$, a temperature where vacancies in diamond become mobile [171], suggesting that the 1682 meV feature could be due to a (V, Si) centre.

The relative intensities of different isotopic transitions of this (V, Si) centre at 6 and 40 K are shown in Fig. 4.15.

The intensity of the 1682 meV feature was also found to decrease to a relative minimum for annealing near 1200°C, followed by a rise to a secondary weaker maximum near 2200°C. At the maximum annealing temperature of 2500°C, the

---

[8] In [171], the ^{28}Si, ^{29}Si, and ^{30}Si components of the $1 \rightarrow 4$ transition were noted A, C, and E, respectively, and those of the $2 \rightarrow 4$ transition B, D, and F, respectively. For the $1 \rightarrow 3$ and $2 \rightarrow 3$ transitions, the same lettering was used, but primed (A′, C′, E′ and B′, D′, F′).

intensity of the peak was much reduced, but it nonetheless showed the wide domain of stability of this (V, Si) centre.

An *ab initio* cluster method has been used to investigate the atomic structure of this centre. It was found that the Si atom at the substitutional site is unstable and spontaneously moves to an interstitial location midway between two *nn* vacancies, yielding a split-vacancy structure with D_{3d} symmetry [172], noted here $V\text{Si}_i V$. In the neutral charge state, the combination of the six dangling bonds of the C atoms *nn*s of the vacancies and of the four dangling bonds of the Si atom gives a centre with $S = 0$ and 1 states, the latter configuration being 0.25 eV more stable than the former, in agreement with first Hund's rule.[9] It can be noted that a similar configuration was proposed by Watkins [173] for the centre responsible of the $S = 1$ Si-G29 ESR spectrum observed at LHeT in electron-irradiated Sn-doped silicon, and attributed to a neutral $V\text{Sn}_i V$ structure. However, in diamond, this configuration, with a $^3A_2$ triplet ground state and ^1E_g and ^1A_g excited states, cannot explain the observed experimental splitting if it is assumed to be due to spin–orbit coupling and J-T interaction. It is only after the trapping of an electron, conferring an acceptor behaviour to this $V\text{Si}_i V^-$ centre, with $S = 1/2$, that the splittings of the doublet states of the internal $^2E_g \rightarrow 2E_u$ transition can explain qualitatively the fine structure observed.

As already mentioned, it has been suggested [171] that the observed fine structure of the 1682 meV feature was due to a tunnelling motion of the Si atom between two equilibrium configurations. The results of molecular orbital (MO) calculations in the restricted-open-shell Hartree–Fock approximation have indeed led to the same conclusion for the $V\text{Si}_i V^0$ configuration, excluding J-T interaction [174]. With the splitting ascribed to tunnelling there is no fundamental objection to ascribe the 1682 meV feature to $V\text{Si}_i V^0$.

A $S = 1$ ESR spectrum noted KUL1 has been detected in CVD polycrystalline diamond and associated to a centre with trigonal symmetry containing only one Si atom, which should correspond to $V\text{Si}_i V^0$ ([175], and references therein, [176]). An $S = 1/2$ centre noted KUL8 has been tentatively ascribed to $V\text{Si}_i V^-$ [177]. The $S = 1/2$ KUL3 centre has monoclinic I symmetry and it has been suggested to be associated to a centre containing $V\text{Si}_i V, H^0$ centre ([176], and references therein).

In electron-irradiated Si-doped diamond, three absorption features at 1679, 1691, and 1711 meV (738.4, 733.2, and 724.6 nm) were reported at LNT after annealing at 800°C, together with the 1682 meV $V\text{Si}_i V$ absorption, but without clear correlation with this latter absorption [178]. The same absorption spectrum was also reported in measurements of neutron-irradiated type Ia natural diamonds annealed at 800°C [103], without explicit mention there of the 1679 meV feature, which was probably mixed with the $V\text{Si}_i V$ absorption because of the relatively large widths of the spectroscopic features in this reference. These latter results seem to show that Si can be also present in natural stones.

---

[9]First Hund's rule states that when two levels with the same electron configuration coexist, the one with the highest spin is the deepest.

There seems to be no vibronic line associated with the 1682 meV ZPL. A weak line near 1.62 eV attributed to a vibronic transition has been reported in PL by Clark et al. [171], but it seems to have no high-energy counterpart in absorption.

Single photon emission at 1.68 eV of the VSi centre has been obtained at RT in Si-implanted type IIa diamonds and this seems to be a promising alternative to the V-N centre [179].

4.3   EL2 in GaAs

4.3.1   EL2 as a Double Donor

A deep donor centre with a level at $E_c - 0.74$ eV and recurrently found in n-type GaAs crystals grown by the HB and LEC methods was reported by Williams [180]. Later on, a native defect with about the same energy level was also found in LEC SI GaAs [181] and correlated with an IR absorption near 1.2 eV [182,183]. This centre, the same as the one reported by Williams and thought to be O-related, had first been called O centre, to be finally coined EL2, after [181].[10] The similarity between the energy of the As_{Ga}^0/As_{Ga}^+ level deduced from photo-ESR experiments and that of the deep donor EL2 led [184] to make a correlation between the two centres. Since As_{Ga} is a double donor, with an As_{Ga}^+/As_{Ga}^{++} level located at $E_v + 0.54$ eV ($E_c - 0.95$ eV), a similar $EL2^+/EL2^{++}$ level is expected for EL2 and the absence of an ESR spectrum associated with $EL2^0$ can be naturally explained. There has been numerous experiments related to the paramagnetism of As_{Ga} and of EL2 and to their correlation, and the results have sometimes been conflicting. Anyway, they confirm the double donor character of EL2 and the position of the $EL2^+/EL2^{++}$ level at about $E_v + (0.52 - 0.54)$ eV. From the atomic side, the conclusion drawn from these experiments was that As_{Ga} is the basic constituent of EL2, with, as an additional ingredient, an interstitial As_i atom [185]. At the light of the discussion by Baraff [186], this latter possibility seemed to be unlikely, but could not be totally excluded. However, from the first-principle calculations of a number of defect structures by Schick et al. [187], it is found that As_i has a very large formation energy, so that its presence in the vicinity of an As_{Ga} antisite can be excluded.

4.3.2   Photoinduced Metastability of EL2

Near-IR photoconductivity at LNT of high-resistivity O-doped GaAs was reported by Lin et al. [188]. This study showed distinct energy-dependent photoresponses when the measurement was performed starting from the low-energy side or from

---

[10]E was for electron and L for Laboratoire d'Électronique Appliquée, Martin et al.'s affiliation.

the high-energy side, and also persistent quenching of the $EL2^0$ photoconductivity by additional illumination with 1.0–1.25 eV photons. This was correlated by near-IR measurements at LHeT, which showed that a broad absorption of EL2 peaking at 1.2 eV could be bleached out up to the band gap of GaAs by additional illumination of the sample with white incandescent light, implying also the existence of a metastable state of EL2 [183]. These observations could also be related with the quenching of the photocapacitance associated with EL2 in GaAs Schottky diodes at LNT, after illumination with photons in the 1.1–1.4 eV range, with a maximum efficiency at 1.13 eV [189, 190].[11] All these results implied that illumination at LNT or below of EL2-containing GaAs converts this centre into an electrically and optically inactive metastable state, noted EL2* or $EL2^m$. A summary of the properties of the EL2 centre as known at the end of 1992 can be found in [186] and [191].

4.3.3   Optical Absorption

The absorption of EL2 is concentrated in the near IR, where features associated with $EL2^0$ and $EL2^+$ are very close and sometimes superimposed, making the task difficult for the spectroscopist. Measurements of the near IR absorption of $EL2^0$ are the superposition of a continuum corresponding to the photoionization $EL2^0$ in the CB, with an onset at ~0.8 eV, and of a broad intracentre band peaking at 1.18 eV at LNT, with a final state resonant with the CB. The absorption of this band was measured by Kaminska et al. [192] by subtracting the contribution of the continuum, evaluated from photocurrent measurements. At LHeT, a ZPL at 1039 meV (1.193 μm or 8380 cm^{-1}) and a series of high-energy replicas separated by 11 meV (~89 cm^{-1}), close to the energy of the TA(X) phonon of GaAs (~81 cm^{-1}), can be observed [192], superimposed or part of the intracentre band. These LNT and LHeT spectra of $EL2^0$ are displayed in Fig. 4.16.

These absorptions occur in the same energy range as the one which turns $EL2^0$ into EL2* and the absorption measurements must be performed with low-intensity monochromatic light, to avoid or at least to reduce the photoconversion of $EL2^0$. This is the usual situation when the sample is placed between the monochromator and the detector.

The above-described near IR absorptions have also been measured at LHeT under hydrostatic pressures up to 1.14 GPa and they show then a mixed behaviour: while the broad intracentre band experiences a pressure-induced red shift of 26.4 meV GPa^{-1}, the ZPL experiences a blue shift of about the same magnitude, with a practically unchanged energy for the phonon replicas [193]. These results indicate that the final states of the broad band and of the ZPL are different. It has been noted by Bardeleben [194] that the above pressure dependence for the ZPL was the same

---

[11]In [189], EL2 is noted "O".

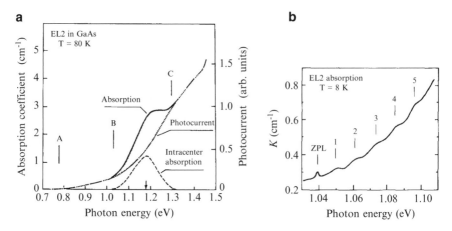

Fig. 4.16 (**a**) Absorption and photocurrent spectra of EL2⁰ in a GaAs sample with n ∼1 × 10¹⁶ cm⁻³ at RT and [EL2] ∼2 × 10¹⁶cm⁻³. The difference (*dashed curve*) between the two spectra defines the intracentre absorption band. A corresponds to the ionization threshold of EL2⁰ in the *CB* and B and C to the limits of the intracentre band. (**b**) Zero-phonon line and TA phonon replicas associated with EL2⁰ (after [192]). Copyright 1983, American Institute of Physics

as that of the L minimum of the GaAs CB^{12} (see Fig. 2.6), for the band structure of a direct-gap semiconductor, implying that the final state of this line could be associated with this minimum.

A fine structure of the ZPL has been reported at LHeT [195] and attributed to non-T_d components, but this structure has not been reproduced in further experiments (see [196], and references therein). The high-resolution measurements of [196], including a hole-burning attempt in the line profile to study the effects of external perturbations on that line, showed that at 15 K, the ZPL, peaking precisely at 1039.04 meV (8380.4 cm⁻¹), with a FWHM of 1.0 meV (8.07 cm⁻¹), has no fine structure, but is asymmetrically broadened by a Fano resonance with the *CB* continuum. The hole-burning experiment, performed at 1039.0 meV with a pulsed narrow-band tunable laser with a line width less than 0.1 cm⁻¹ (12.4 μeV), actually resulted in a quenching of the whole EL2 absorption band. This meant that with this set-up, there was no selective excitation of an EL2 group and that the ZPL was homogeneously broadened.

Piezospectroscopic and Zeeman measurements on the ZPL have indicated that the centre associated with EL2 had tetrahedral symmetry ($A_1 \rightarrow T_2$ transition), supporting the view of an isolated antisite As_{Ga} [197] and the symmetry results were confirmed by Trautman et al. [198].

The metastable EL2* state is produced by illumination with ∼1–1.3 eV photons at low temperature ($T < 130$ K is often quoted for SI GaAs, but a temperature ∼40 K is necessary in n-type GaAs), and the recovery to the normal state can take place under different conditions: thermal recovery has often been measured

---

¹²The band structure of GaAs is qualitatively similar to the one of CdTe, shown in Fig. 2.6.

by monitoring the increase with temperature of the EL2 absorption at 1.25 eV, an energy where its absorption is temperature insensitive. Such annealings show that in the n-type samples, the EL2* → EL2 recovery takes place when the samples are heated above 45 K while in the SI samples, it necessitates temperatures above 125 K (see, for instance, [199]). The activation energies involved for thermal recovery are typically ~0.34–0.38 eV for the SI samples, but only ~0.06–0.1 eV for the n-type samples. Photoinduced recovery or combination of photo and thermal recovery has also been investigated and, as for thermal recovery, these photo effects are sample dependent. For pure photoinduced isothermal recovery at low temperature, partial or total conversions have been claimed for photon energies below and above E_g ([200] and references therein). It has also been demonstrated that the EL2* → EL2 recovery in SI GaAs could be achieved by irradiating the samples with 20 ns pulses of 2 MeV electrons at LNT [201]. In that case, recovery has been attributed to the recombination at the EL2 site of electron–hole pairs produced by irradiation. Photoinduced recovery of EL2* under hydrostatic pressure has also been studied by absorption at LHeT [202], with two salient results: (1) under a pressure of 1 GPa, a SI sample held at LHeT shows a very efficient recovery from EL2* to EL2^0 when illuminated with 1.36 eV photons, and (2) under a pressure of 0.36 GPa, a 1.13 eV illumination, which produces a full EL2→ EL2* conversion in a SI sample, gives in an n-type GaAs sample a broad band centred at 0.85 eV. This band has been associated with EL2*, and its absorption is found to be proportional to the EL2^0 intracentre absorption at 1.2 eV. This new absorption, not observed in the SI sample, is explained by a transition to the continuum from a (EL2*)$^{0/-}$ acceptor level, resonant with the *CB* at zero pressure, but optically accessible when the pressure makes this level to emerge in the band gap. Further DLTS and electrical results on this level are given in [203].

Information on the symmetry of the EL2* centre has been obtained from the study of its thermal recovery under uniaxial stress: by monitoring the increase with temperature of the EL2^0 absorption at 1.24 eV, at the highest applied stress (600 MPa), two components are observed for a [111] stress while no splitting is observed for a [100] stress. These results have been interpreted as an indication that EL2* is distorted into a configuration with trigonal symmetry [204].

In p-type GaAs, EL2 is in the EL2^{++} charge state, where it can be converted into the paramagnetic EL2$^+$ charge state by illumination with photons with energies between 0.52 and 0.74 eV, allowing electrons of the *VB* to be trapped by EL2^{++} (above 0.74 eV, a second electron is captured to form EL2^0). Magnetic circular dichroism (MCD) absorption at LHeT at a magnetic field of 2 T has revealed broad absorption bands at about 1.0 and 1.3 eV, which have been associated with EL2$^+$[205]. Absorption bands at 0.97, 1.07, and 1.32 eV, obtained at LHeT from difference absorption spectra of GaAs samples doped with isoelectronic impurities for different illumination conditions, have also been attributed to the EL2$^+$ charge state by Manasreh et al. [206]. More unexpected, MCD absorption measurements at 140 K have also shown that beside the 1.2 eV band (called diamagnetic as related to the S = 0 state), a weaker diamagnetic band at 0.93 eV is also associated with

EL2^0 [207]. This latter band is close to the one at 0.95–1.0 eV attributed to the photoionization of a hole from EL2$^+$ into the *VB*.

These absorption measurements have been used to try to obtain a calibration of the EL2 concentration in a given charge state, by comparison with photocapacitance measurements. A correlation between the peak absorption of the ZPL at 1.039 eV and the one of the 1.2 eV intracentre band at 2 K and another one between the intensity of the ZPL and the EL2^0 concentration have been made quantitative [208]. From these results, one can deduce for EL2^0 an approximate calibration factor of the 1.2 eV intracentre band of about 3.3×10^{16} cm^{-2}, after subtraction of the background absorption, and this factor should not vary drastically up to LNT. Values of the energy dependence at LNT of the photoionization cross-section $\sigma_n^o(\omega)$ for electrons in the *CB* (including the intracentre absorption) and $\sigma_p^o(\omega)$ for holes in the *VB* have been given by Silverberg et al. [209]. When the paramagnetic and diamagnetic charge states EL2$^+$ and EL2^0 are supposed to be both present under equilibrium in a GaAs sample, the determination of the absorption coefficients at two energies E_1 and E_2 where the photoionization cross-sections are known is supposed to allow an estimation of the [EL2^0] and [EL2$^+$] concentrations [210,211].

4.3.4 Other Optical Properties

The results of PL measurements related to EL2 are mentioned here because they complement absorption results. Two broad PL bands observed at low temperature have been related to EL2. Near band gap excitation ($\sim$1.5 eV) allows the observation at LHeT of a broad PL band at 0.68 eV [212], characterized by an unusually long lifetime, with a sharp structure on its high-energy side, at 702.82 meV [213]. The 0.68 eV band cannot be observed after illumination of the sample at photon energies $\sim$1.24 eV, showing a good correlation with the bleaching of the ZPL absorption at 1039 meV. The 702.82 meV line is a LO(Γ) phonon replica of a weak ZPL at 739.33 meV line, which couples to a variety of GaAs phonons and to LVMs with energies near 11 meV ($\sim$89 cm^{-1}), very close to the energy separating the replicas of the ZPL absorption line at 1039 meV in Fig. 4.16b. Photoexcitation measurements of the PL have shown that the 739.33 meV ZPL is due to the internal recombination on EL2$^+$ of an electron on a shallow excited state with a binding energy of 3.2 meV. This allows determining accurately the separation of the EL2^0/EL2$^+$ level from the *CB* to be 742.5 meV. The 0.68 eV PL band is observed in GaAs samples containing a significant concentration of ionized EL2 and it is best excited at 0.95 eV, where the optical cross section $\sigma_p^o(\omega)$ for the ionization of EL2^0 in the *VB* reaches its maximum [214]. At LHeT, a ZPL at 756 meV on the high-energy side of the 0.68 eV band is attributed to a EL2^0-neutral acceptor (DAP) recombination (the acceptor is carbon); at 40 K, another ZPL is observed at 773 meV, ascribed to the recombination of a hole from the *VB* with EL2^0. The sum of the position of this latter ZPL with the above-determined separation of the EL2^0/EL2$^+$ level from the *CB* is 1516 meV at $\sim$40 K, very close to the band gap of GaAs at this temperature (1515 meV) [213]. Piezospectroscopic

Fig. 4.17 Stress dependence of the energy of the PL LO(Γ) replica at 702.8 meV of the ZPL of EL2 in LEC GaAs samples for stresses σ applied along different crystallographic directions. The PL is excited with a Ti:sapphire laser tuned at 840 nm (1476 meV) and the resolution is 0.12 meV [215]. Copyright 1991 by the American Physical Society

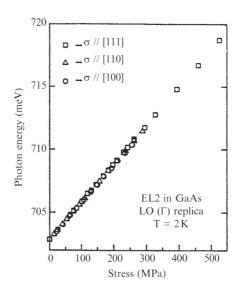

and Zeeman measurements have been made on the strong 702.8 meV LO(Γ) replica of the internal recombination ZPL at 739.33 meV [215]. Figure 4.17 shows the stress dependence of the energy of the LO(Γ) replica of the EL2 ZPL at 739.33 meV.

The measurements for stresses along [100], [110], and [111] crystal axes show no detectable splitting, indicating no orientational degeneracy of the centre, in agreement with the previous piezospectroscopic measurements [197, 198]. This is another proof that the optical measurements consistently indicate a T_d symmetry for EL2.

A PL band associated to EL2* has been reported in slightly n-type and SI LEC GaAs samples excited at 1.17 eV at LHeT under hydrostatic pressure between 0.3 and ∼1 GPa [216]. This band disappears for pressures above ∼1 GPa, as does all the PL excited at 1.17 eV [216].

The most recent spectroscopic studies of EL2 concern the temperature dependence of the recovery from EL2* to EL2^0, and they show that it is characterized by two distinct activation energies [217, 218]. From the latter study, an atomic structure of the EL2 defect has been proposed. For EL2^0, besides As$_{Ga}$, it includes a Ga antisite Ga$_{As}$ and a Ga vacancy V_{Ga} nn and nnn of As$_{Ga}$, respectively. In this configuration, Ga$_{As}$ is third nn of V_{Ga}. Rearrangements of this structure under optical excitation and thermal recovery through intermediate configurations are proposed [218]. The relation between the symmetries implied for these structures and the tetrahedral symmetry deduced from the piezospectroscopic measurements is not clear, even when allowing for a dynamic J-T reorientation [219].

4.4   Transition Metal (TM) Impurities

4.4.1   Introduction

Transition metal impurities in semiconductors have been investigated for decades. The main reason for the interest on this topic is the technological importance of these FAs. They are used to control the minority-carrier lifetime in devices, to obtain semi-insulating substrates, as with Fe in indium phosphide, and in inorganic display technologies. Presently, there is also a great interest for the ferromagnetism of semiconductors heavily doped with TM elements. Books [220, 221] and a large number of reviews (see, for instance, [222–230]) have been devoted to the description of physical properties of TM impurities in semiconductors. Absorption spectroscopy played a role of importance in the investigation of TM impurities in compound semiconductors. Investigations in elemental semiconductors are less developed, and some of them are discussed in [7].

4.4.2   Generalities on TM Impurities

The TMs are characterized by an electronic configuration with an unfilled d shell, and the most investigated ones have been those of the iron group, with an unfilled $3d$ shell (see Appendix D). In compound semiconductors, the TMs usually occupy substitutional site, where they replace the most electropositive atom, but in silicon, interstitial location is the most common. TMs can display different charge states, depending on the location of the Fermi level in the material: for instance, chromium can take four charge states in GaP. This means that TM impurities can introduce electronic levels within the band gap. Most often, these levels are deep within the band gap, and they can be either acceptor or donor levels.

Several spectroscopic properties of TMs in semiconductors mainly depend on the number of electrons in the d shell. For instance, the absorption spectra of "$3d^4$" substitutional chromium with four electrons in the $3d$ shell and none in the $4s$ shell have quite similar features in II–VI and III–V compounds. The free chromium atom has 5 electrons in the $3d$ shell and 1 in the $4s$ shell; therefore, "$3d^4$" chromium, a chromium atom that has spend two electrons in its bonds, is neutral with respect to the lattice in II–VI compounds and negatively charged with respect to the lattice in III–V compounds. Therefore, representing a TM impurity by its charge with respect to the lattice is not the most convenient as it hides the similarities between behaviours in various types of compounds. Therefore, most often, in this section, we use, the ionic notation in which the charge states of a TM is written TM^{n+}; for instance, "$3d^4$" chromium is noted Cr^{2+}, whatever the material in which it is incorporated. The electronic level corresponding to the change of charge state of chromium from "$3d^3$", when the Fermi level is below this level, to "$3d^4$" when it is above, is noted $Cr^{3+}\backslash Cr^{2+}$. In the case of III–V compounds, this level is an acceptor level, but in the case of II–VI compounds, it is a donor level.

When placed in a site of symmetry lower than spherical, T_d symmetry for instance, the metal impurities feel the effect of the crystal field exerted by the other atoms in the crystal. A convenient and widely applied phenomenological approach to describe the effect of the surrounding crystal on the TM impurity is the Tanabe–Sugano approach [231]. In this approach, in the case of T_d symmetry, the energy levels of the system depend on three parameters: two Racah parameters, B and C [232], describing the interactions of d electrons with the core of the TM and the other d electrons, and a "crystal field" parameter describing the interaction of d electrons with the rest of the crystal. The Racah parameters are not the same as in the free ion because of the overlap of the d electron wave functions with the ligands wave functions. This oversimplified approach has several advantages: (1) the intra d shell interactions are easily taken into account, (2) the symmetry of the problem is respected, (3) one can easily draw diagrams, the so-called Tanabe–Sugano diagrams, that allow to visualize the energy levels as a function of the "crystal field" for instance, (4) one can summarize the experimental results in tables of Racah and crystal field parameters. There have been many attempts to improve Tanabe–Sugano approach. For instance, Biernacki [233] realized that e and t_2 orbitals are not affected in the same way by the "ligand field", and therefore that the contribution of Racah parameter A, whose effect is to translate as a whole the energy of the configuration in the Tanabe–Sugano approach, yields there different shifts. Biernacki's approach has, however, the great disadvantage of doubling the number of parameters and therefore of suppressing the simplicity of the Tanabe–Sugano approach. Proper theoretical approaches are described in the review by Zunger [224]. Pragmatically, one uses the Tanabe–Sugano model for describing the spectroscopic experimental results.

Free ion states are split by the "crystal field" felt by the impurity. For instance, in a potential with T_d symmetry, the Cr^{2+} ^5D free ion state is split into a $^5T_2$ state and a ^5E state. These states can further be split by s-o coupling, applied stress, or magnetic field. In the case of electronically degenerated states with dynamic coupling between the electronic states and the lattice vibrations, the vibronic coupling has important effects. These effects are described in the pioneering papers by Ham [234, 235] and in subsequent reviews or books [236–239]. One of the consequences of these effects is that the minima of the potential energy surfaces are not located at tetrahedral sites, and that there are several potential wells corresponding to lower symmetries. Tunnelling among these wells allows keeping the overall tetrahedral symmetry, but it can happen that internal strains stabilize the centres in the wells [240]. In such cases, the apparent symmetry of the centres can look lower than tetrahedral. Another effect of the vibronic interactions is to reduce the effects of perturbations such as s-o coupling, orbital contributions to the gyromagnetic factors, or applied stress.

There is one case where the above statements do not apply. Half filled shells (d^5 configuration) are energetically very stable. Therefore, it is legitimate to question whether, in III–V compounds, with respect to the lattice site, neutral manganese in its ground state has the configuration $3d^4$ or $[3d^5 +$ bound delocalized hole]. In GaAs, the energy of the Mn acceptor level ($E_v + 112$ meV) is rather shallow [241]. ESR measurements [242] have indeed proved that neutral manganese takes a $[3d^5 +$ hole] configuration. Neutral manganese, with ($E_v + 388$ meV), is deeper in GaP [241], and in that case, the ESR measurements [243] have shown that neutral manganese takes a $3d^4$ configuration. The position of the Mn acceptor level in InP is in between these two values, and presently, its electronic configuration in this latter material is not clear. This point is discussed in Sect. 4.4.7.

4.4.3   *Intracentre Transitions*

Intracentre transitions occur between multiplets energy levels belonging to the same charge state of the TM impurity. Their low temperature absorption is often observed as sharp ZPLs followed at higher energies by an unstructured band towards vibronic excited states. At higher temperature (RT, for instance), the ZPLs are no longer visible and only the unstructured band is observed. Figure 4.18 displays in full lines the zero phonon structure of Cr^{2+} in GaAs at LHeT.

This structure corresponds to transitions between the ground $^5T_2$ multiplet and the excited ^5E multiplet. The lines are rather sharp (FWHM ~ 0.3 cm^{-1} or ~ 0.04 meV). The structure of the ground and excited multiplets is depicted in Fig. 4.19.

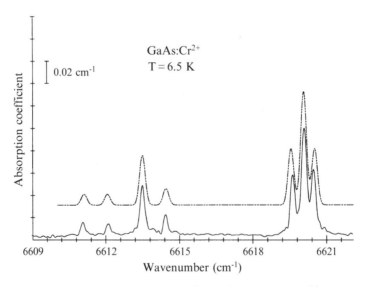

Fig. 4.18 Measured absorption of the ZPLs of the $^5T_2 \rightarrow {}^5E$ transition of Cr^{2+} in GaAs between ~ 819.4 and 821.0 meV (solid line). Simulation of the absorption (dotted line) (after [223])

The structure of the ground state is known from ESR results [244] that have shown that, in this state, Cr^{2+} is apparently tetragonally distorted because of a strain stabilization in the wells of the potential energy surfaces [245]. The dashed line in Fig. 4.18 represents the simulation of the spectrum using on the one hand the values of the spin Hamiltonian parameters for the ground state and on the other hand the value $K = -0.48\,cm^{-1}$ for describing the excited state. The transitions probabilities are proportional to the numbers shown in Fig. 4.19 [245]; the thermal populations of the sublevels of the ground state multiplet have been taken into account for the simulation. The unstructured band associated with these zero-phonon lines has its maximum at $7280\,cm^{-1}$ ($\sim903\,meV$) [246].

The same type of spectra is observed for Cr^{2+} in II–VI compounds [247, 248], InP [249] and GaP [250]. In II–VI compounds, Kaminska et al. [248] observed additional intracentre transitions, in the 500–$2000\,cm^{-1}$ (~60–$250\,meV$) range. They correspond to transitions between the minima of one of the potential energy surfaces and the other potential energy surfaces of the $^5T_2$ ground state. These transitions are vertical in the collective coordinates space and their peak energy is at about three times the Jahn–Teller energy (see Fig. 2 of [246]). The equivalent transition has not been reported in III–V compounds, but the broad band around $2100\,cm^{-1}$ ($260\,meV$) reported for GaAs in the literature [251] is a good candidate. This would correspond to a $700\,cm^{-1}$ ($\sim87\,meV$) Jahn–Teller energy in the $^5T_2$ ground state, a value quite close to the estimations [246, 252].

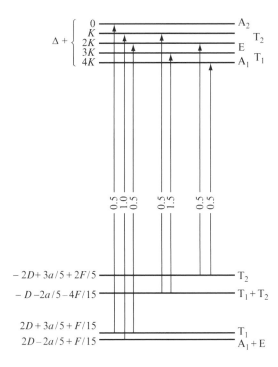

Fig. 4.19 Energy level scheme for Cr^{2+} in GaAs and InP. The numbers are proportional to the transitions probabilities. The energies of the sublevels of the ground states are given on the left side as a function of the spin Hamiltonian parameters [244]. The scheme is not to scale [223]

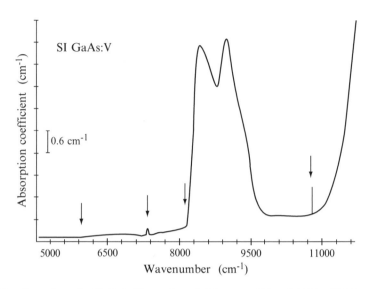

Fig. 4.20 Absorption of a SI GaAs:V sample at 5 K between 620 and 1444 meV. The relative intensities of all the ZPLs marked by arrows are correlated, and they are due to internal transitions of V^{3+} [253]. The absorption increase above $11000\,cm^{-1}$ (1.36 eV) corresponds to the onset of the photoionization of V^{3+} in the *VB*. One abscissa graduation is $300\,cm^{-1}$ ($\sim$37 meV). Copyright 1985, American Institute of Physics

The case of Cr^{2+} in GaAs described earlier can be considered a "simple" one as only one Cr^{2+} excited multiplet is observed. This is not always the case. For instance, Fig. 4.20 [253] shows the global absorption spectrum of V^{3+} in GaAs at 5 K.

The broad band between $\sim$8100 and $9600\,cm^{-1}$ ($\sim$1.0 and 1.2 eV) is due to vibronic coupling associated with the ZPL at $8131\,cm^{-1}$. Its integrated intensity with respect to that of the ZPL is in the ratio e^S, where S is the Huang–Rhys factor, which is $\sim$10 in the present case [254]. Figure 4.21 shows the details of the spectrum near the first arrow on the low-energy side of Fig. 4.20 around $5960\,cm^{-1}$ ($\sim$740 meV). The ZPLs at 5957.85, 5968.05, and $5968.25\,cm^{-1}$ (739.919, 739.944, and 739.969 meV) displayed in Fig. 4.21 are due to transitions between the $^3A_2$ ground state level and the $^3T_2$ multiplet.

The other lines marked by arrows in Fig. 4.20 are located at 7333, 8131, and $10733\,cm^{-1}$ (909.2, 1008.1, and 1330.7 meV), and these ZPLs have been investigated in detail; in particular, the effect of uniaxial stresses has been studied [255]. This allowed their assignments to specific transitions. The Tanabe–Sugano diagram for d^2 ions, to which belongs V^{3+}, is shown in Fig. 4.22.

The line at $7333\,cm^{-1}$ (909.2 meV) is due to the $^3A_2 \rightarrow \,^1E(^1D)$ transition. The line at $8131\,cm^{-1}$ (1008 meV) and the associated broad band are due to the $^3A_2 \rightarrow \,^3T_1(^3F)$ transition, and the line at $10773\,cm^{-1}$ (1330.7 meV) is due to the $^3A_2 \rightarrow \,^1A_1(^1G)$ transition. The integrated intensity of the broad band in the

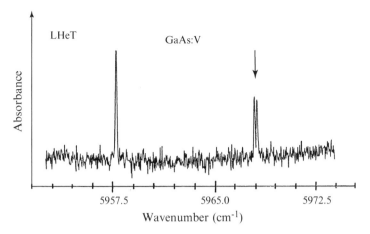

Fig. 4.21 Enlarged detail of the absorption spectrum between 738.1 and 740.9 meV of Fig. 4.20. One abscissa graduation corresponds to 1.5 cm^{-1}. The *lines* marked by an *arrow* are separated by 0.2 cm^{-1} (25 μeV). After [253]. Copyright 1985, American Institute of Physics

Fig. 4.22 Tanabe–Sugano diagram of the spltting of d^2 ions in a crystal field with tetrahedral symmetry. The energy E and crystal field parameter Δ are normalized by the Racah parameter B

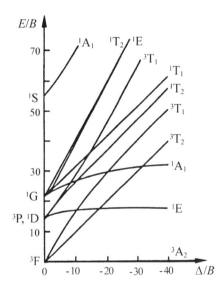

8000–9500 cm^{-1} (~990–1180 meV) range is by far larger than the other transitions because the $^3A_2 \rightarrow {}^3T_1$ transition is the only dipolar electric transition "fully" allowed, the other ones being only allowed by s-o interaction. A large scatter of the FWHMs of the ZPLs is also noticed: it ranges from 0.1 cm^{-1} (~12 μeV) for the three lines of Fig. 4.21 to a few cm^{-1} for the lines at 7333 and 8131 cm^{-1}.

Therefore, the intracentre transitions spectra of a charge state of a TM impurity can be quite "rich". Experiments under uniaxial stress and/or magnetic field allow a precise determination of all the transitions when rather sharp ZPLs are observed.

Fig. 4.23 Absorption
spectrum of Nb^{3+} in GaAs at
LHeT [256]. Reproduced
with permission from Trans
Tech Publications

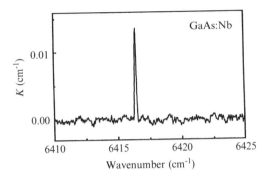

They allow also to obtain quantitative estimations of the vibronic couplings in
each multiplet. Such intracentre transitions have been observed for several $3d$ TM
impurities. In contrast, the absorption results for $4d$ and $5d$ TM impurities are rather
scarce. The very low solubility of these impurities is probably at the source of this
situation. However, a rather sharp absorption transition has been observed for Nb^{3+}
in GaAs [256]. Nb^{3+} is the $4d$ homologue of V^{3+} discussed earlier and therefore,
one can refer to the Tanabe–Sugano diagram of Fig. 4.22. The absorption spectrum
observed for Nb^{3+} is shown in Fig. 4.23.

It consists of a zero phonon line at $6416.4\,cm^{-1}$ (795.53 meV), also observed in
PL, and it has been studied under magnetic field [257]. PL excitation spectroscopy
[256] allows to observe a broad band in the range $8500-9500\,cm^{-1}$ (1.054–
1.178 eV). The sharp absorption line at $6416.4\,cm^{-1}$ is due to the $^3A_2 \rightarrow {}^1A_1({}^1G)$
transition; the broad band observed in PL excitation spectroscopy is most probably
due to the $^3A_2 \rightarrow {}^3T_1({}^3F)$ transition. Experiments under uniaxial stress have been
performed on the line at $6416.4\,cm^{-1}$, and their results are quite surprising [256].
They show a linear splitting of the $^3A_2$ state by a $\langle 001 \rangle$ stress. It is quite unexpected
to split an orbital singlet, but it is not forbidden as the $^3A_2$ state is indeed a Γ_5
state that can be split by a uniaxial stress. This is due to the admixture of $^3T_2$ into
$^3A_2$ by s-o coupling. This means that in fact the splitting of the $^3A_2$ state under
uniaxial stress reflects the one of the $^3T_2$ state. Quantitatively, this leads to an effect
of $\langle 001 \rangle$ stresses in the $^3T_2$ state about 30 times larger for Nb^{3+} than for V^{3+}. The
much larger sensitivity to strain for $4d$ TM than for their $3d$ homologue is indeed
expected as the overlap of $4d$ orbitals with the ligand orbitals is larger than those
of $3d$ orbitals. Another surprising fact is that the shift of the centre of gravity of the
$^3A_2 \rightarrow {}^1A_1$ transition under tetragonal strain is not linear with stress, but quadratic.
This has been explained [256] by a coupling of the ^1E and $^1A_1$ states by tetragonal
strains, and it has been shown that a quite moderate coupling can be responsible for
this quadratic behaviour.

The observation of intracentre transitions is not restricted to impurities in purely
tetrahedral sites. They can be observed also in sites of lower symmetry either
due to lower symmetry crystals or to the complexing of the TM with defects or

other impurities. For instance, V^{3+} and V^{4+} intracentre transition spectra have also been observed in $4H$ and $6H$ polytypes of SiC, where the V atoms replace Si atoms [258]. A good example of TM complexes investigated spectroscopically is the work on Ni^+–donor complexes in GaP and GaAs [259]. The observation of absorption spectra related with intracentre transitions of TM impurities is not restricted either to compound semiconductors. For instance, TMs, mainly nickel and cobalt, that are used as catalysts for obtaining large diamond single crystals by the HPHT technique incorporate in the crystals and have been investigated by optical spectroscopy. Collins [141] and Yelisseyev and Kanda [260] reviewed these spectroscopic investigations of $3d$ TMs in diamond. It is shown that the TMs can occupy three types of sites: substitutional, interstitial, and double semivacancy and form complexes with unintentional nitrogen.

4.4.4 High- or Low-Spin Ground States?

In case of strong "crystal field", the first Hund's rule (maximum spin) can be broken and low-spin ground states can be observed. Whether low-spin ground states can be observed for $3d$ TM impurities in semiconductors have raised a quite large interest. On the one hand, V^{2+} in GaAs [261,262] and GaP [262] have been predicted to have a ^2E low-spin ground state. On the other hand, the calculations in [263] yield a high-spin ground state for V^{2+} in III–V compounds; these authors also predicted Co^{3+} to have a low-spin ground state in GaAs, GaP, and InP. Absorption spectroscopy is not conclusive for V^{2+} in GaAs and GaP, as it shows an unstructured band from which it is not possible to definitively assign the ground state level [253]. The similarity between the spectra in the two materials, however, strongly suggests that the transition occurs between the same levels in the two materials. Other techniques do not provide either a clear-cut view of the ground state nature [228].

The Co^{3+} (neutral) intracentre absorption has been investigated in GaP [264]. The spectra observed at 1.6, 4.2, and 8 K are shown in Fig. 4.24, and they have been interpreted as involving transitions between the $^3T_1$ ground state multiplet and the ^5E excited multiplet [264].

The observation of an increasing number of lines with increasing temperature is due to a thermalization effect within the $^3T_1$ ground state multiplet. From these experiments is deduced the energy level scheme shown in Fig. 4.25.

The excited multiplet consists in five nearly equally spaced levels characteristic of a ^5E multiplet. ^5E would be the ground state in the high-spin configuration; as it is the excited state in this case, this proves that Co^{3+} has a low-spin ground state. The structure of the ground state shown in Fig. 4.25 is fully consistent with a $^3T_1$ multiplet low-spin ground state in which first order s-o effects are nearly fully quenched by a strong vibronic coupling to vibrational modes with E symmetry. Experiments under uniaxial stress show that the ground state of the most intense of the transitions, at 2456.9 cm^{-1} (304.62 meV), splits under $\langle 001 \rangle$ and $\langle 110 \rangle$ stresses and that it only shifts under $\langle 111 \rangle$ stress. Together with the observed polarizations

Fig. 4.24 Absorption of
Co^{3+} (neutral Co) in GaP
between 298.8 and
307.5 meV at increasing
temperatures. The transitions
are labelled by the indices of
the *IR*s of the sublevels of the
ground and excited states. For
instance, 3–4 is for the
$\Gamma_3 \to \Gamma_4$ transition shown in
Fig. 4.25 [264]. Copyright
1996 with permission from
World Scientific Publishing
Co., Singapore

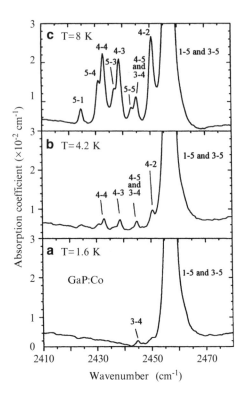

of the lines under stress, this is fully consistent with the levels scheme depicted in
Fig. 4.25. The results of [264] definitely prove that Co^{3+} in GaP has a $^3T_1$ low-spin
ground state, in perfect agreement with the predictions in [263].

4.4.5 Isotope Effects

Isotopic shifts of electronic transitions are not expected as, to the first order, lattice
relaxation energies are independent of the reduced mass of the involved modes of
vibration. However, isotopic shifts of intracentre transitions are sometimes observed
([230, 265–268], and references therein). For the mass of the TM impurity to play
a significant role, relatively localized vibrations around the TM atom have to be
involved; therefore, it is legitimate to reason in terms of clusters. The TM has also
to move during the vibration of the cluster; this implies modes of T_2 symmetry to
be involved.

   As an example, we consider the ZPL absorption of the $^2T_2 \to {}^2E$ transition of
Ni$^+$ in GaAs displayed in Fig. 4.26.

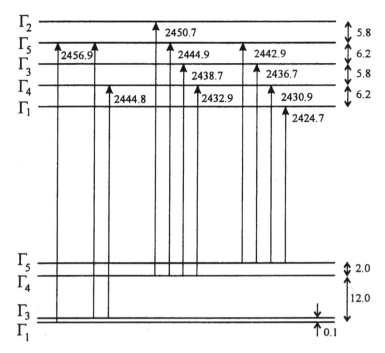

Fig. 4.25 Energy level scheme of the $^3T_1$ and ^5E ground and excited states of Co^{3+} in GaP. The arrows indicate the electric-dipole-allowed transitions. The transition energies and the sublevels spacings are expressed in cm^{-1} [264]. Copyright 1996 with permission from World Scientific Publishing Co., Singapore

The experimental spectrum shows a clear isotopic structure. The reconstructed spectrum takes into account the natural abundances of the five isotopes of ^{nat}Ni (see Appendix E), a positive shift of $0.2\,cm^{-1}\,u^{-1}$ ($\sim25\,\mu eV\,u^{-1}$) and individual FWHMs of $0.20\,cm^{-1}$, and it agrees nicely with the experimental one. The energy of the main peak due to the ^{58}Ni isotope is $\sim4615.1\,cm^{-1}$ (572.20 meV). It is known [269] that the $^2T_2$ ground state is governed by the vibronic coupling to E-symmetry modes of vibration; such a coupling even quenches the coupling to modes of vibration of T_2 symmetry. Therefore, the isotopic shift does not occur within the $^2T_2$ ground state. As, within T_d symmetry, the symmetric product $[E \times E]$ does not contain the T_2 representation, the coupling of a ^2E state to modes of vibrations of T_2 symmetry is not expected; however, this is not fully correct. A ^2E state is in fact a Γ_8 state once spin is included; as the antisymmetric product $\{\Gamma_8 \times \Gamma_8\}$ contains T_2 representation, the vibronic coupling of a Γ_8 state to vibrations of T_2 symmetry is allowed. This means in fact that ^2E and $^2T_2$ states are coupled by both the s-o coupling and vibrations of T_2 symmetry. This is the combined effect of these two interactions that allows the coupling of ^2E states to modes of vibration of T_2 symmetry. This coupling is indeed necessary to explain the behaviour of the ^2E

Fig. 4.26 Measured
absorption between 572.06
and 572.44 meV of the ZPL
of Ni$^+$ in GaAs at LHeT,
showing a Ni isotope effect
(**a**), compared to the
reconstructed profile (**b**). The
FWHMs of the individual
components are ~0.2 cm^{-1}
or 25 μeV. The natural
percent abundances of the
^{58}Ni, ^{60}Ni, ^{61}Ni, ^{62}Ni, and
^{64}Ni isotopes are 68.08,
26.22, 1.14, 3.63, and 0.93,
respectively [268].
Reproduced with permission
from Trans Tech Publications

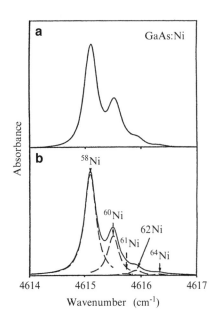

Fig. 4.27 IS of the ZPL
absorption of the Ni$^+$–S pair
in GaAs at LHeT. The
spectral range goes from
548.44 to 548.82 meV [268].
Reproduced with permission
from Trans Tech Publications

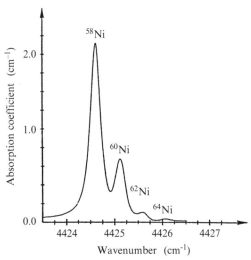

state of Ni$^+$ in GaAs under magnetic field [270]. It has been shown [268,270] that a
quite moderate coupling of ^{2}E and ^{2}T$_2$ states by vibrations of T$_2$ symmetry explains
the isotopic shifts observed and all the properties of the ^{2}E excited state of Ni$^+$ in
GaAs and other materials.

Figure 4.27 displays the ZPL of the Ni$^+$–S pair in GaAs. It shows an isotopic
shift qualitatively comparable to the one of the ZPL of "isolated" Ni$^+$, but larger
(0.24 cm^{-1} u^{-1} or 30 μeV u^{-1}). The energy of the main peak due to ^{58}Ni is
4424.6 cm^{-1} (548.58 meV).

The presence of the nn S atom lowers the symmetry of the centre to C_{3v}. In this case, there is a trigonal component of the "crystal field" felt by the Ni atom. This trigonal component also couples the ^2E and $^2T_2$ states, and it adds a contribution to the isotopic shift, explaining the larger isotopic shifts for the complex than for "isolated" Ni. The important effect of the trigonal component of the "crystal field" on the isotopic shifts seems to be quite general. For instance, Broser et al. [266] did not observe any isotopic effect in cubic ZnS: Ni^{2+} whereas they observed such effects for Ni^{2+} in trigonal polytypic crystals.

4.4.6 Transitions Between TMs Energy Levels and Valence or Conduction Band

Transitions between TMs levels in the band gap and the CB or VB involve a change in the charge state of the TM between the initial and final states (these transitions are sometimes alled charge transfer (CT) bands [230]). Therefore, these transitions are accompanied by a large lattice relaxation that leads to broad absorption bands. The integrated intensities of the transitions between the VB and the TM band gap levels are large, as to the first order they correspond to $p \to d$ transitions that are strongly allowed.

The transitions corresponding to the promotion of an electron from the VB to a TM level are of the type:

$$TM^{n+} + h\nu \to TM^{(n-1)+} + h_{VB}$$

where h_{VB} represents a hole in the VB, and those corresponding to the promotion of an electron from a TM level to the CB:

$$TM^{n+} + h\nu \to TM^{(n+1)+} + e_{CB}$$

where e_{CB} represents an electron in the CB. As mentioned in Sect. 4.4.3, a charge state of a TM can have several electronic levels within the band gap. In such a case, there can be several absorption bands corresponding to transitions towards the various electronic levels. A good example is shown in Fig. 4.28.

This figure shows the absorption spectra of Fe-doped SI GaP, InP, and GaAs materials. In these SI materials, the Fermi level is pinned by the $Fe^{3+}\backslash Fe^{2+}$ acceptor level and at equilibrium both Fe^{3+} and Fe^{2+} charge states are present. The Fe^{2+} free ion has a ^5D ground state that is split into a ^5E ground multiplet and a $^5T_2$ excited multiplet by the tetrahedral "crystal field". The spectra presented in Fig. 4.28 present three regions. Region I, where sharp ZPLs are observed, corresponds to the Fe^{2+} intracentre transitionsbetween the ^5E and $^5T_2$ multiplets. Region II shows a broad absorption band that corresponds to the promotion of an electron from the VB into the ^5E multiplet, described by $Fe^{3+} + h\nu \to Fe^{2+}(^5E) + h_{VB}$. The second broad

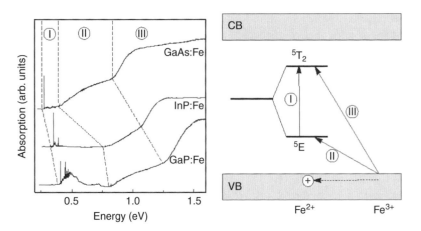

Fig. 4.28 Typical absorption spectra of SI InP, GaAs, and GaP. All these spectra exhibit the internal $^5E \rightarrow {}^5T_2$ transition of Fe^{3+} (I), and two charge transfer bands representing the $Fe^{3+} \rightarrow Fe^{2+}$ charge transfer process resulting in the Fe^{2+} (^5E) ground state (II) and Fe^{2+} ($^5T_2$) excited state (III), with a hole generated in the *VB* [230]. Copyright Wiley-VCH Verlag GmbH & Co. KGa. Reproduced with permission

absorption band, observed in region III of the spectra, corresponds to the promotion of an electron from the *VB* into the $^5T_2$ excited multiplet of Fe^{2+}; the corresponding process is: $Fe^{3+} + h\nu \rightarrow Fe^{2+}(^5T_2) + h_{VB}$.

In direct band gap semiconductors, the transition probability between a TM impurity level and the *CB* is weaker than for a transition between the *VB* and the TM level as, to the first order, the former corresponds to a $d \rightarrow s$ type of forbidden transition. At a difference, in a material like GaP, in which the minima of the conduction band are at X points of the first BZ, the character of the *CB* is p-like at its minima. Therefore, transitions between TM levels and the *CB* are allowed to the first order in this material. One can even have two consecutive transitions [271] such as: $Cr^{3+} + h\nu \rightarrow Cr^{2+} + h_{VB}$ and $Cr^{2+} + h\nu' \rightarrow Cr^{3+} + e_{CB}$. These two consecutive transitions leave a hole in the *VB* and an electron in the *CB*, which can recombine radiatively. The $Cr^{3+} \backslash Cr^{2+}$ level being close to mid-gap in GaP, one can thus obtain an up-conversion with a good efficiency at low temperature in a rather broad infrared energy range [271].

The case of manganese in III–V compounds is specific as, at least in GaAs, Mn neutral with respect to the lattice has a ground state consisting of a hole in its $1S_{3/2}$ state bound to Mn^{2+}. In such a case, the process is:

$$Mn^{2+} + h(1S_{3/2}) + h\nu \rightarrow Mn^{2+} + h_{VB}.$$

The corresponding absorption band has been observed in GaAs [241, 272]. Similar photoionization bands have also been observed in InP [241, 273] and GaP [241]. The photoionization band of Mn in InP is shown in Fig. 4.29.

Fig. 4.29 Overall shape of the photoionization spectrum of manganese in InP between 124 and 620 meV. The D line at 1694 cm^{-1} can only be guessed in this spectrum [273]. The line noted by an asterisk, located at 2272 cm^{-1}, is due to the local mode of vibration of a manganese–hydrogen complex [274]. (After [273])

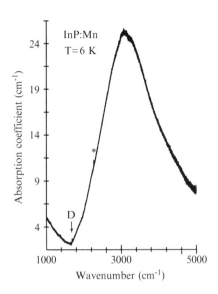

The observation of such photoionization bands does not provide any information on the nature of the ground state of neutral manganese: even though, it is known (see Sect. 4.4.2) that it has a different nature in GaAs and GaP, the photoionization bands look similarly in these two materials. A detailed spectrum of InP:Mn showing effective-mass transitions located at 1650, 1694, and 1719 cm^{-1} (204.6, 210.0, and 213.1 meV) is displayed in Fig. 3 of [273]. These transitions are noted G, D, and C, respectively, in the hydrogen-like spectra of shallow acceptors in silicon (see, for instance, Sect. 7.2.2 in [7]). This introduces appropriately this topic, developed in the next section.

4.4.7   Transitions Involving Hydrogen-Like Hole States

Quite often, photoionization bands present sharp lines at their low-energy onsets. These transitions have been the object of lot of investigations (see, for instance, [229, 230]) and they gave rise to several types of interpretations. For instance, they have been described for a while in terms of exciton bound to neutral (with respect to the lattice) TM. However, the evidence of the implication of gap or resonant vibrational modes specific of the negatively charged (with respect to the lattice) TM impurity [275] contradicted this description that has now been abandoned.

The most clear-cut case is the case of manganese in GaAs [241, 272]. The sharp lines that are observed at the low energy edge of the photoionization band have a nature identical to those observed for shallow acceptors, i.e., they correspond to transitions between energy levels of holes bound to Mn^{2+}. The excited states are reasonably well fitted by effective mass theory [276]. The $J = 1$ ground state resulting from the coupling of the 3/2 hole angular momentum with the $S = 5/2$ spin of Mn^{2+}, observed by ESR, is not in agreement with EMT because of a necessary central cell correction.

A "simple" case is the case of cobalt in III–V compounds; this case is simple because the Co^{2+} ground state is a $^4A_2$ orbital singlet. Figure 4.30 shows the low energy part of the $Co^{3+} + h\nu \rightarrow Co^{2+} + h_{VB}$ photoionization absorption in GaP at a temperature of 6 K [277].

Superimposed on the tail of the broad band, one observes several relatively sharp features: the G and D lines at 3522 and 3572 cm^{-1} (436.7 and 442.9 meV), respectively, as well as vibronic replicas involving Co^{2+}-related vibrational gap mode at 323 cm^{-1} or LO phonon. This indicates that Co^{2+} is involved in the final state of the transition. The energy difference between the D and G lines, 6.2 meV (50 cm^{-1}), is close to the one observed between the $2P_{5/2}(\Gamma_8)$ and $2P_{3/2}(\Gamma_8)$ hole states of shallow acceptors such as Cd: 6.8 meV [278] and with a theoretical values of 6.7 meV [276]. Therefore, the G and D transitions can be considered as involving $2P_{5/2}(\Gamma_8)$ and $2P_{3/2}(\Gamma_8)$ excited states of the hole. The ground state is the same as the one of the photoionization band, i.e., the Co^{3+} ground state. The detailed structure of this state is shown in Fig. 4.25. At 6 K, only the Γ_1 and Γ_3 sublevels of $^3T_1$ are well populated. The G and D lines are therefore part of the $[Co^{3+} + h\nu \rightarrow Co^{2+} +$ bound hole h] spectrum.

At 3352 cm^{-1} (415.6 meV), a weaker line is observed (H line). Depending on the sample, the integrated intensity of this line scales with the absorbance of the photoionization band and with the integrated intensity of the G and D lines: the integrated intensity of the H line is 1/40 of the integrated intensity of the G line at 6 K. The lowest energy hole state is $1S_{3/2}$. Once coupled with the $^4A_2$ Co^{2+}

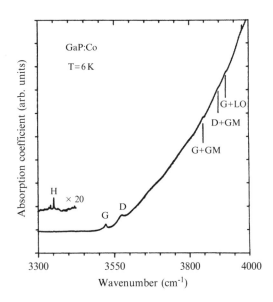

Fig. 4.30 Absorption of cobalt in GaP between 409 and 496 meV, near from the photoionization threshold of Co^{3+} in the *VB*. Lines G and D are transitions to effective-mass-like excited states. Labels G and D are the ones used in the shallow acceptor spectra in germanium. The lines at higher energies are replicas associated to a vibrational gap mode (GM) of Co^{2+} in GaP and to a LO phonon-assisted transition. The weak feature near from the H line is due to the residual absorption of Fe^{2+} [277]

ground state, it gives four levels corresponding to $J = 0, 1, 2, 3$. The $J = 0$ level transforms as Γ_1. The dipolar electric transitions between the lowest Γ_1 and Γ_3 states of Co^{3+} and this $J = 0$ state are forbidden. If this $J = 0$ state is the lowest, the transition between the Co^{3+} ground state and the lowest state of $[Co^{2+} +$ bound hole h] is forbidden at very low temperature. Nevertheless, it is extremely tempting to assign the H line to the $Co^{3+} + h\nu \rightarrow Co^{2+} + h(1S_{3/2})$ transition. Wolf et al. suggested that internal strains that would lower the symmetry of the cobalt site could be responsible for breaking of the selection rule [277]. A more natural proposal is the following: at $6\,K$, the population of the Γ_5 sublevel of $^3T_1$ is 1/30 of the total Co^{3+} amount. The transition between this Γ_5 sublevel and the $J = 0$ sub-state of $[Co^{2+} + h(1S_{3/2})]$ is allowed. Therefore, the H line could be due to the transition: $Co^{3+}\{\Gamma_5(^3T_1)\} + h\nu \rightarrow [Co^{2+} + h(1S_{3/2})]_{J=0}$. Indeed experiments as a function of temperature are necessary for confirming this proposal.

Figure 4.31 shows the bound hole spectra of Mn and Co in GaAs. It clearly evidences the hydrogen-like character of the two spectra, with an identical spacing of the excited acceptor energy levels, producing similar spectra and demonstrating the involvement of holes bound to Mn^{2+} and Co^{2+}. More transitions are observed in the GaAs:Co spectrum than in the GaP:Co one just described above; this confirms indeed the interpretation of the G and D lines in the GaP:Co spectra. According to theory [263], the Co^{3+} ground state should be $^3T_1$ in GaAs as it is in GaP. Therefore,

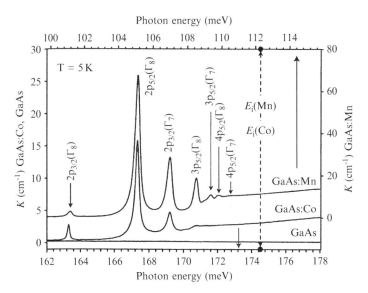

Fig. 4.31 Comparison of the acceptor spectra of Co and Mn in GaAs at a resolution of $0.06\,meV$ ($0.5\,cm^{-1}$) showing the effective-mass character of these TMs in GaAs. The $2p_{3/2}$ lines of these spectra have been brought into coincidence to show the similarity, and the relevant energy scale is indicated by a vertical arrow. The absorption of a pure GaAs sample is shown as a reference. The lower and upper abscissa scale ranges are ~ 1307–1436 and ~ 805–$933\,cm^{-1}$, respectively. After Tarhan et al. [241]. Copyright 2003 by the American Physical Society

the line corresponding to the H line in the GaP:Co spectrum should also exist in the GaAs:Co spectra. To the best of our knowledge, this transition has not been reported.

Mn^{3+} in GaP has a $^5T_2$ ground state that is identical with the one seen in Fig. 4.19 with the spin Hamiltonian parameters determined by Kreissl et al. [243]. Mn^{2+} has a $^6A_1$ orbital singlet ground state. The $[Mn^{2+} + h(nP_j)]$ states are in reasonable agreement with theory [276], and the corresponding transitions have been reported by Tarhan et al. [241]. A central cell correction is necessary for the $[Mn^{2+} + h(1S_{3/2})]$ state. The level associated with this state is the ground state of neutral (with respect to the lattice) manganese in GaAs; in GaP, it is an excited state that should be observable by absorption spectroscopy. It has not been reported yet. It is not clear yet whether the ground state of neutral (with respect to the lattice) manganese in InP is Mn^{3+} or $[Mn^{2+} + h(1S_{3/2})]$; ESR measurements did not provide the answer yet. Absorption measurements could provide the answer: the observation of a transition with a $[Mn^{2+} + h(1S_{3/2})]$ final state would prove that the ground state of neutral manganese in InP is Mn^{3+}.

Transitions involving bound holes are also observed in more complicated cases in which the TM^{2+} state is not an orbital singlet. A good example concerns Fe^{2+} in III–V compounds that has ^5E and $^5T_2$ states. Absorption spectra between the ground state of Fe^{3+} and levels associated with holes bound to Fe^{2+} in its ^5E and $^5T_2$ states have been widely investigated. The results are detailed in the review by Malguth et al. [230]. It must be noted that experiments at variable temperature in the mK range definitively proved that Fe^{3+} is the ground state of the transitions by showing thermalization effects between the energetically close sublevels of the $^6A_1$ ground state of Fe^{3+}. There have also been evidences for transitions towards levels involving $1S_{3/2}$ hole states in GaP, GaAs, and InP.

Copper in GaAs is an interesting case. Figure 4.32 shows the photoionization spectrum of copper in GaAs [241]. In the inset, the transitions involving bound holes are detailed. A remarkable observation is that there are two series of transitions shifted by about $0.84\,meV$ ($6.8\,cm^{-1}$). A more natural explanation for the presence of these two series of transitions than those proposed by Tarhan et al. [241] could be the following. Cu^{2+} has a $^2T_2$ ground state that is split by s-o coupling into a Γ_7 ground sublevel and a Γ_8 upper sublevel. Cu^{2+} in tetrahedral symmetry is known to be rather strongly coupled to E-symmetry modes of vibration [279]. This J-T coupling strongly quenches the splitting between the Γ_8 and Γ_7 sublevels; for instance in ZnS, this splitting is only $14\,cm^{-1}$ [279]. In GaAs, the overlap between the $Cu^{2+}d$-orbitals and the ligands orbitals should be larger than in ZnS and therefore the vibronic coupling should be larger in GaAs than in ZnS and the Γ_8–Γ_7 splitting smaller. The $6.8\,cm^{-1}$ energy difference between the two series of transitions could be the Γ_8–Γ_7 splitting. Therefore, the spectra shown in Fig. 4.32 could reflect the structure of the Cu^{2+} ground state in which case the lowest energy series of lines would correspond to $Cu^{3+} + h\nu \rightarrow [Cu^{2+}\{\Gamma_7(^2T_2)\} + h(nP_j)]$ transitions and the higher energy one to $Cu^{3+} + h\nu \rightarrow [Cu^{2+}\{\Gamma_8(^2T_2)\} + h(nP_j)]$ transitions.

When, in III–V compounds, the holes are bound to TM^{2+} that have a complex or not well-known structure, the hole-involving transitions are extremely hard to

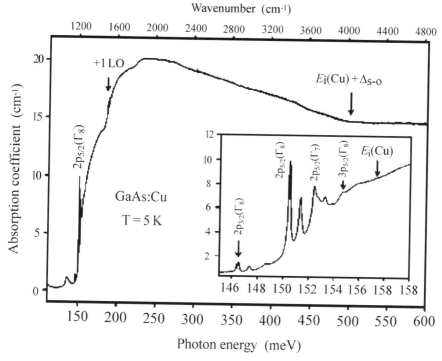

Fig. 4.32 Transitions towards effective mass-like excited states and photoionization continuum of the Cu acceptor spectrum in GaAs. A LO Fano replica superimposed to the photoionization continuum and the s-o split *VB* absorption edge ($E_i(Cu) + \Delta_{s-o}$) are also indicated in the figure. The inset shows a doublet structure for each transition. Resolution: 0.5 cm^{-1} (0.06 meV) [241]. Copyright 2003 by the American Physical Society

interpret. This is, for instance, the case for nickel in GaP where the observed fine structure of the bound hole related lines has not really been interpreted [277].

Transitions involving holes bound to TM impurities are not restricted to compound semiconductors, and in silicon, acceptor spectra of holes bound to gold and platinum have been reported [280]. However, the nature of these acceptors is not precisely known.

References

1. J. Bourgoin, M. Lannoo, *Point Defects in Semiconductors II Experimental Aspects* (Springer, Berlin, 1983)
2. G. Davies, The optical properties of luminescent centres in silicon. Rep. Prog. Phys. **176**, 83–188 (1989)
3. D.F. Daly, H.E. Noffke, An EPR study of fast neutron radiation damage in silicon, in *Radiation Effects in Semiconductors*, ed. by J.W. Corbett, G.D. Watkins (Gordon and Breach, London, 1971), pp. 179–187

4. F. Carton-Merlet, B. Pajot, D.T. Don, C. Porte, B. Clerjaud, P.M. Mooney, Photo-induced changes in the charge state of the divacancy in neutron and electron irradiated silicon. J. Phys. C **15**, 2239–2255 (1982)
5. B.N. Mukashev, Kh.A. Abdullin, Yu.V. Gorelkinskii, S.Zh. Tokmoldin, Self-interstitial related reactions in silicon irradiated with light ions. Mater. Sci. Eng. B **58**, 171–178 (1999)
6. H.Y. Fan, A.K. Ramdas, Infrared absorption and photoconductivity in irradiated silicon. J. Appl. Phys. **30**, 1127–1134 (1959)
7. B. Pajot, *Optical Absorption of Impurities and Defects in Semiconducting Crystals – Hydrogen-Like Centres* (Springer, Berlin, 2010)
8. J.W. Corbett, G.D. Watkins, Production of divacancies and vacancies by electron irradiation in silicon. Phys. Rev. **138**, A555–A560 (1965)
9. L.J. Cheng, J.C. Corelli, J.W. Corbett, G.D. Watkins, 1.8-, 3.3-, and 3.9-μ bands in irradiated silicon: correlation with the divacancy. Phys. Rev. **152**, 761–774 (1966)
10. J.W. Corbett, G.D. Watkins, Silicon divacancy and its direct production by electron irradiation. Phys. Rev. Lett. **7**, 314–316 (1961)
11. G.D. Watkins, J.W. Corbett, Defects in irradiated silicon: electron paramagnetic resonance of the divacancy. Phys. Rev. **138**, A543–A555 (1965)
12. C.S. Chen, J.C. Corelli, Infrared spectroscopy of divacancy-associated radiation-induced absorption bands in silicon. Phys. Rev. B **5**, 1505–1517 (1972)
13. P.R. Brosious, EPR of a spin-1 two-vacancy defect in electron-irradiated silicon, in *Defects and Radiation Effects in Semiconductors 1978. Inst. Phys. Conf. Ser. No 46*, ed. by J.H. Albany (The Institute of Physics, London, 1979), pp. 248–257
14. Y.H. Lee, J.W. Corbett, EPR studies of defects in electron-irradiated silicon: A triplet state of vacancy-oxygen complexes. Phys. Rev. B **13**, 2653–2666 (1976)
15. M.T. Lappo, V.D. Tkachev, Divacancies in silicon irradiated with fast neutrons. Sov. Phys. Semicond. **4**, 1882–1884 (1971)
16. L.J. Cheng, P. Vajda, Effect of polarized light on the 1.8, 3.3, and 3.9-μ radiation induced absorption bands in silicon. Phys. Rev. **186**, 816–823 (1969)
17. G. Davies, S. Hayama, S. Hao, B. Bech Nielsen, J. Coutinho, M. Sanati, S.K. Estreicher, K.M. Itoh, Host isotope effects on midinfrared optical transitions in silicon. Phys. Rev. B **71**, 115212/1–7 (2005)
18. J.H. Svensson, B.G. Svensson, B. Monemar, Infrared absorption studies of the divacancy in silicon: new properties of the singly negative charge state. Phys. Rev. B **38**, 4192–4197 (1988)
19. F. Merlet, B. Pajot, P. Vajda, Near infrared absorption of irradiated lithium-doped silicon and quenching of the 3.3-μm divacancy bands. J. Appl. Phys. **47**, 1729–1731 (1976)
20. B. Massarani, A. Brelot, Evidence for a 130 K annealing stage of divacancy in electron-irradiated silicon, in *Radiation Damage and Defects in Semiconductors. Inst. Phys. Conf. Ser. N⁰ 16*, ed. by J.E. Whitehouse (The Institute of Physics, London, 1973), pp. 269–277
21. F. Carton-Merlet, B. Pajot, P. Vajda, Detection of the photopopulation and photoionisation of intrinsic point defects in irradiated silicon by IR absorption, in *Defects and Radiation Effects in Semiconductors 1978. Inst. Phys. Conf. Series No 46*, ed. by J.H. Albany (The Institute of Physics, Bristol, 1979), pp. 311–316
22. B.G. Svensson, J.L. Lindström, Generation of divacancies in silicon by MeV electrons: Dose rate dependence and influence of Sn and P. J. Appl. Phys. **72**, 5616–5621 (1992)
23. F. Carton-Merlet, B. Pajot, P. Vajda, Extrinsic defects in neutron-irradiated silicon. An infrared study. J. Phys. **39**, L113–L117 (1978)
24. L. Zhong, Z. Wang, S. Wan, L. Lin, Absorption peaks at 2663 and 2692 cm⁻¹ observed in neutron-transmutation-doped silicon. J. Appl. Phys. **66**, 4275–4278 (1989)
25. G. Davies, S. Hayama, L. Murin, R. Krause-Rehberg, V. Bondarenko, A. Sengupta, C. Davia, A. Karpenko, Radiation damage in silicon exposed to high-energy protons. Phys. Rev. B **73**, 165202/1–10 (2006)
26. H.G. Grimmeiss, Silicon-germanium – a promise into the future. Semiconductors **33**, 939–941 (1999)

27. L. Khirunenko, Yu. Pomozov, M. Sosnin, N. Abrosimov, W. Schröder, Interaction of divacancies with Ge atoms in $Si_{1-x}Ge_x$. Physica B **308–310**, 550–553 (2001)
28. H.J. Stein, Divacancy-like absorption in ion-bombarded Ge. *Radiation Damages and Defects in Semiconductors. Inst. Phys. Conf. Ser. N_0 16*, ed. by J.E. Whitehouse (The Institute of Physics, London, 1973) pp 315–321
29. A.B. Gerasimov, N.D. Dolidze, R.M. Donina, B.M. Konovalenko, G.L. Ofengeim, A.A.Tsertsvadze, On the identification and possible space orientation of "light-sensitive" defects in Ge. Phys. Stat. Sol. A **70**, 23–29 (1982)
30. C.S. Chen, R. Vogt-Lowell, J.C. Corelli, Higher order defect infrared absorption bands in Si, in *Radiation Damage and Defects in Semiconductors. Inst. Phys. Conf. Ser. N^0 16*, ed. by J.E. Whitehouse (The Institute of Physics, London, 1973), pp. 210–217
31. J.C. Corelli, D. Mills, R. Gruver, D. Cuddeback, Y.H. Lee, J.W. Corbett, Electronic excitation bands in irradiated silicon, in *Radiation Effects in Semiconductors 1976, Inst. Phys. Ser. No 31*, ed. by N.B. Urli , J.W. Corbett (The Institute of Physics, Bristol, 1977), pp. 251–257
32. Y. Shi, Y.D. Zheng, M. Suezawa, M. Imai, K. Sumino, Investigations on higher order bands in irradiated Czochralski silicon. Appl. Phys. Lett. **64**, 1227–1229 (1994)
33. N.N. Gerasimenko, M. Rollé, L.J. Cheng, Y.H. Lee, J.C. Corelli, J.W. Corbett, Infrared absorption of silicon irradiated by protons. Phys. Stat. Sol. B **90**, 689–695 (1978)
34. Y. Shi, F. Wu, Y. Zheng, M. Suezawa, M. Imai, K. Sumino, Temperature dependent investigation on optically active processes of higher-order bands in irradiated silicon. Phys. Stat. Sol. A **154**, 789–796 (1996)
35. A.A. Kaplyanskii, Noncubic centres in cubic crystals and their spectra in external fields. J. Phys. **28** (C4, suppl. no 8–9), C4-39–C4-48 (1967)
36. Y.H. Lee, P.R. Brosious, J.W. Corbett, New EPR spectra in neutron-irradiated silicon (II). Radiat. Eff. **22**, 169–172 (1974)
37. Y.H. Lee, J.W. Corbett, EPR studies in neutron-irradiated silicon: A negative charge state of a nonplanar five-vacancy cluster (V_5^-). Phys. Rev. B **8**, 2810–2826 (1973)
38. Z. Ciechanowska, G. Davies, E.C. Lightowlers, Uniaxial stress measurements on the 1039.8 meV zero-phonon line in irradiated silicon. Solid State Commun. **49**, 427–431 (1984)
39. A.S. Kaminskii, B.M. Leiferov, A.N. Safonov, Excitons bound to defect complexes in silicon. Sov. Phys. Solid State **29**, 551–556 (1987)
40. E.S. Johnson, W.D. Compton, Recombination luminescence in irradiated silicon – effects of thermal annealing and lithium impurity. Radiat. Eff. **9**, 89–92 (1971)
41. B. Hourahine, R. Jones, A.N. Safonov, S. Öberg, P.R. Briddon, S.K. Estreicher, Identifiction of the hexavacancy in silicon with the B_{80}^4 optical center. Phys. Rev. B **61**, 12584–12597 (2000)
42. G.D. Watkins, J.W. Corbett, Defects in irradiated silicon: electron paramagnetic resonance and electron-nuclear double resonance of the Si-E-center. Phys. Rev. **134**, A1359–A1377 (1964)
43. E.L. Elkin, G.D. Watkins, Defects in irradiated silicon: electron paramagnetic resonance and electron-nuclear double resonance of the arsenic- and antimony-vacancy pairs. Phys. Rev. **174**, 881–897 (1968)
44. G.D. Watkins, Vacancies and interstitials and their interactions with impurities in c-Si, in *Properties of Crystalline Silicon, EMIS Datareviews Series No 20*, ed. by R. Hull (INSPEC, London, 1999), pp. 643–652
45. G.D. Watkins, Optical properties of group-V atom-vacancy pairs in silicon. Radiat. Eff. Defect. Sol. **111–112**, 487–500 (1989)
46. G.D. Watkins, Understanding the Jahn-Teller distortion for the divacancy and the vacancy-group-V-atom pair in silicon. Physica B **376–377**, 50–53 (2006)
47. K. Saarinen, J. Nissilä, H. Kauppinen, M. Hakala, M.J. Puska, P. Hautojärvi, C. Corbel, Identification of vacancy-impurity complexes in highly n-type Si. Phys. Rev. Lett. **82**, 1883–1886 (1999)
48. C.S. Chen, J.C. Corelli, G.D. Watkins, Radiation-produced absorption bands in silicon: piezospectroscopic study of a group-V atom-defect complex. Phys. Rev. B **5**, 510–526 (1972)
49. M. Suezawa, N. Fukata, T. Mchedlidze, A. Kasuya, Many optical absorption peaks observed in electron-irradiated n-type Si. J. Appl. Phys. **92**, 6561–6566 (2002)

50. C.S. Chen, J.C. Corelli, Optical study of lithium defect complexes in irradiated silicon. J. Appl. Phys. **44**, 2483–2489 (1973)
51. F. Carton-Merlet, Doctoral Thesis, University of Paris (1979)
52. E.C. Lightowlers, L.T. Canham, G. Davies, M.L.W. Thewalt, S.P. Watkins, Lithium and lithium-carbon isoelectronic complexes in silicon: luminescence decay-time, absorption, isotope splitting, and Zeeman measurements. Phys. Rev. B **29**, 4517–4523 (1984)
53. L. Canham, G. Davies, E.C. Lightowlers, G.W. Blackmore, Complex isotope splitting of the no-phonon lines associated with exciton decay at a four-lithium-atom isoelectronic centre in silicon. Physica **117B-118B**, 119–121 (1983)
54. G. Davies, L. Canham, E.C. Lightowlers, Magnetic and uniaxial stress perturbations of optical transitions at a four Li atom complex in Si. J. Phys. C **17**, L173–L178 (1984)
55. G.G. DeLeo, W.B. Fowler, G.D. Watkins, Electronic structure of hydrogen- and alkali-metal-vacancy complexes in silicon. Phys. Rev. B **29**, 1819–1823 (1984)
56. G.S. Myakenkaya, G.L. Gutsev, N.P. Afanaseva, V.A. Evseev, R.F. Konopleva, Study of lithium interaction with lattice defects in silicon. Phys. Stat. Sol. B **161**, 91–103 (1990)
57. J.F. Zheng, M. Stavola, G.D. Watkins, Structure of the neutral charge state of interstitial carbon in silicon, in *22nd Internat. Conf. Phys. Semicond.*, ed. by D.J. Lockwood (World Scientific, Singapore, 1995), pp. 2363–2366
58. G.D. Watkins, K.L. Brower, EPR observation of the isolated interstitial carbon atom in silicon. Phys. Rev. Lett. **36**, 1329–1332 (1976)
59. L.W. Song, G.D. Watkins, EPR identification of the single acceptor state of interstitial carbon in silicon. Phys. Rev. B **42**, 5759–5764 (1990)
60. L.I. Khirunenko, V.Yu. Pomozov , M.G. Sosnin, M.O. Trypachko, A. Duvanskii, V.J.B. Torres, J. Coutinho, R. Jones, P.R. Briddon, N.V. Abrosimov, H. Riemann, Local vibrations of interstitial carbon in SiGe alloys. Mater. Sci. Semicond. Process. **9**, 514–519 (2006)
61. R. Woolley, E.C. Lightowlers, A.K. Tipping, M. Claybourn, R.C. Newman, Electronic and vibrational absorption of interstitial carbon in silicon. Mater. Sci. Forum **10–12**, 929–934 (1986)
62. E.V. Lavrov, M. Fanciulli, M. Kaukonen, R. Jones, P.R. Briddon, Carbon-tin defects in silicon. Phys. Rev. B **64**, 125212/1–5 (2001)
63. K. Thonke, A. Teschner, R. Sauer, New photoluminescence defect spectra in silicon irradiated at 100 K: Observation of interstitial carbon? Solid State Commun. **61**, 241–244 (1987)
64. A.V. Yukhnevich, Radiative capture of holes by A-centres in silicon. Sov. Phys. Solid State **7**, 322–323 (1965)
65. A.R. Bean, R.C. Newman, R.S. Smith, Electron irradiation damage in silicon containing carbon and oxygen. J. Phys. Chem. Solids **31**, 739–751 (1970)
66. G. Davies, M.C. do Carmo, Isotope effects on the 969 meV vibronic band in silicon. J. Phys. C **14**, L687–L691 (1981)
67. K. Thonke, H. Klemisch, J. Weber, R. Sauer, New model of the irradiation-induced 0.97-eV (G) line in silicon: a C_s − Si* complex. Phys. Rev. B **24**, 5874–5886 (1981)
68. G. Davies, H. Brian, E.C. Lightowlers, K. Barraclough, M.F. Thomaz, The temperature dependence of the 969 meV "G" optical transition in silicon. Semicond. Sci. Technol. **4**, 200–206 (1989)
69. G. Davies, E.C. Lightowlers, M. do Carmo, Carbon-related vibronic bands in electron-irradiated silicon. J. Phys. C **16**, 5503–5515 (1983)
70. S. Hayama, G. Davies, J. Tan, J. Coutinho, R. Jones, K.M. Itoh, Lattice isotope effects on optical transitions in silicon. Phys. Rev. B **70**, 035202/1–9 (2004)
71. A.V. Yukhnevich, A.V. Mudryi, Deformation splitting of the 0.97 eV luminescence line of irradiated silicon. Sov. Phys. Semicond. **7**, 815–816 (1973)
72. C.P. Foy, M.C. do Carmo, G. Davies, E.C. Lightowlers, Uniaxial stress measurements on the 0.97 eV line in urradiated silicon. J. Phys. C **14**, L7–L12 (1981)
73. G.D. Watkins, A review of the EPR studies in irradiated silicon, in *Radiation Damages in Semiconductors*, ed. by P. Baruch (Dunod, Paris, 1965), pp. 99–113

74. K.L. Brower, EPR of a Jahn-Teller distorted <111> carbon interstitialcy in irradiated silicon. Phys. Rev. B **9**, 2607–2617 (1974). Erratum: Phys. Rev. B **17**, 4130 (1978)
75. K.P. O'Donnell, K.M. Lee, G.D. Watkins, Origin of the 0.97 eV luminescence in irradiated silicon. Physica B **116**, 258–263 (1983)
76. G. Davies, K.T. Kun, Annealing the di-carbon radiation damage centre in silicon. Semicond. Sci. Technol. **4**, 327–330 (1989)
77. G. Davies, E.C. Lightowlers, M.C. do Carmo, J.G. Wilkes, G.R. Wolstenholme, The production and destruction of the C-related 969 meV absorption band in Si. Solid State Commun. **50**, 1057–1061 (1984)
78. G. Davies, R.C. Newman, Carbon in mono-crystalline silicon, in *Handbook on Semiconductors*, vol. 3b, ed. by S. Mahajan (North Holland, Amsterdam, 1994), pp. 1557–1640
79. G. Davies, E.C. Lightowlers, R. Woolley, R.C. Newman, A.S. Oates, Carbon, oxygen and silicon isotope effects in the optical spectra of electron-irradiated Czochralski silicon, in *Proc. 13th Internat. Conf. Defects Semicond.*, ed. by L.C. Kimerling, J.M. Parsey, Jr. (The Metallurgical Society of AIME, New York, 1985), pp. 725–731
80. W. Kürner, R. Sauer, A. Dörnen, K. Thonke, Structure of the 0.767-eV oxygen-carbon luminescence defect in 450°C thermally annealed Czochralski-grown silicon. Phys. Rev. B **39**, 13327–13337 (1989)
81. A.P.C. Hare, G. Davies, A.T. Collins, The temperature dependence of vibronic spectra in irradiated silicon. J. Phys. C **5**, 1265–1276 (1972)
82. C.P. Foy, Uniaxial stress analysis of the 0.79 eV vibronic band in irradiated silicon. J. Phys. C **15**, 2059–2067 (1982)
83. C.P. Foy, Optical studies of vibronic bands in silicon. Physica B **116**, 276–280 (1982)
84. D.J. Backlund, S.K. Estreicher, Theoretical study of the C_iO_i and $I_{Si}C_iO_i$ defects in Si. Physica. B **401–402**, 163–166 (2007)
85. A.V. Yukhnevich, V.D. Tkachev, Optical analog of the Mössbauer effect in silicon. Sov. Phys Sol. State **8**, 1004–1005 (1966)
86. G. Davies, E.C. Lightowlers, M. Stavola, K. Bergman, B. Svensson, The 3942-cm^{-1} optical band in irradiated silicon. Phys. Rev. B **35**, 2755–2766 (1987)
87. G. Davies, E.C. Lightowlers, R. Woolley, R.C. Newman, A.S. Oates, A model for radiation damage effects in carbon-doped silicon. Semicond. Sci. Technol. **2**, 524–532 (1987)
88. A.T. Collins, The detection of colour-enhanced and synthetic gem diamonds by optical spectroscopy. Diam. Relat. Mater. **12**, 1976–1983 (2003)
89. A.T. Collins, G. Davies, H. Kanda, G.S. Woods, Spectroscopic studies of carbon-13 synthetic diamond. J. Phys. C **21**, 1363–1376 (1988)
90. A.T. Collins, Optical centres produced in diamond by radiation damages. New Diam. Frontier Carbon Technol. **17**, 47–61 (2007)
91. Y. Meng, C. Yan, J. Lai, S. Krasnicki, H. Shu, T. Yu, Q. Liang, H. Mao, R. Hemley, Enhanced optical properties of chemical vapor deposited single crystal diamond by low-pressure/high temperature annealing. Proc. Natl. Acad. Sci. U.S.A. **105**, 17620–17625 (2009)
92. A.M. Zaitsev, Optical properties of diamond, in *Industrial Handbook for Diamond and Diamond Films*, ed. by M.A. Prelas, G. Popovici, L.K. Bigelow (Marcel Dekker, New York, 1998), pp. 227–376
93. A.T. Collins, Things we still don't know about optical centres in diamond. Diam. Relat. Mater. **8**, 1455–1462 (1999)
94. F. Bridges, G. Davies, J. Robertson, A.M. Stoneham, The spectroscopy of crystal defects: a compendium of defect nomenclature. J. Phys. Cond. Matter **2**, 2875–2928 (1990)
95. J. Walker, Optical absorption and luminescence in diamond. Rep. Prog. Phys. **72**, 1605–1659 (1979)
96. H.E. Smith, G. Davies, M.E. Newton, H. Kanda, Structure of the self-interstitial in diamond. Phys. Rev. B **69**, 045203/1–9 (2004)
97. G. Davies, H. Smith, H. Kanda, Self-interstitial in diamond. Phys. Rev. B **62**, 1528–1531 (2000)

98. D.C. Hunt, D.J. Twitchen, M.E. Newton, J.M. Baker, T.R. Anthony, W.F. Banholzer, S.S. Vagarali, Identification of the neutral carbon <100>-split interstitial in diamond. Phys. Rev. B **61**, 3863–3876 (2000)

99. E.A. Faulkner, J.N. Lomer, Electron spin resonance in electron-irradiated silicon. Phil. Mag. **7**, 1995–2002 (1962)

100. J.P. Goss, B.J. Coomer, R. Jones, T.D. Shaw, P.R. Briddon, M. Rayson, S. Öberg, Self-interstitial aggregation in diamond. Phys. Rev. **63**, 195208/1–14 (2001)

101. C.D. Clark, R.W. Ditchburn, H.B. Dyer, The absorption spectra of natural and irradiated diamond. Proc. Roy. Soc. Lond. A **234**, 363–381 (1956)

102. J.W. Steeds, T.J. Davis, S.J. Charles, J.M. Hayes, J.E. Butler, 3H luminescence in electron-irradiated diamond samples and its relationship to self-interstitials. Diam. Relat. Mater. **8**, 1847–1852 (1999)

103. K. Iakoubovskii, G.J. Adriaenssens, N.N. Dogadkin, A.A. Shiryaev, Optical characterization of some irradiation-induced centers in diamond. Diam. Relat. Mater. **10**, 18–26 (2001)

104. D.J. Twitchen, M.E. Newton, J.M. Baker, W.F. Banholzer, T. Anthony, Optical spin polarization in the di-<001>-split interstitial (R1) centre in diamond. Diam. Relat. Mater. **8**, 1101–1106 (1999)

105. L. Allers, A.T. Collins, J. Hiscock, The annealing of interstitial-related optical centres in type II natural and CVD diamond. Diam. Relat. Mater. **7**, 228–232 (1998)

106. E.W.J. Mitchell, *Diamond Research 1964* (Industrial Diamond Review 1964), p. 13 (1964)

107. C.D. Clark, C.A. Norris, Photoluminescence associated with the 1.673, 1.944 and 2.468 eV centres in diamond. J. Phys. C **4**, 2223–2229 (1971)

108. S.A. Solin, Photoluminescence of natural type I and type IIb diamonds. Phys. Lett. A **38**, 100–102 (1972)

109. G. Davies, C.P. Foy, Jahn-Teller coupling at the neutral vacancy in diamond. J. Phys. C **13**, 2203–2213 (1980)

110. G. Bosshart, The Dresden Green. J. Gemmology **21**, 351–362 (1989)

111. G. Davies, C.P. Foy, Analysis of uniaxial stress on the GR 2–3 (2.8804, 2.8866 eV) absorption lines in diamond. J. Phys. C **11**, L547–L553 (1978)

112. G. Davies, C.M. Penchina, The effect of uniaxial stress on the GR1 doublet in diamond. Proc. Roy. Soc. Lond. A **338**, 359–374 (1974)

113. C.P. Foy, G. Davies, The GR4 to GR8 absorption lines in diamond. J. Phys. C **13**, L25–L28 (1980)

114. A.T. Collins, High-resolution optical spectra of the GR defect in diamond. J. Phys. C **11**, 1957–1964 (1978)

115. L.A. Vermeulen, C.D. Clark, J. Walker, Optical absorption, photoconductivity and photo-Hall effect in electron-irradiated semiconducting diamond, in *Lattice Defects in Semiconductors 1974, Inst. Phys. Conf. Ser. No 23*, ed. by F.A. Huntley (The Institute of Physics, Bristol, 1975), pp. 294–300

116. A. Pu, V. Avalos, S. Dannefaer, Negative charging of mono- and divacancies in IIa diamonds by monochromatic illumination. Diam. Relat. Mater. **10**, 585–587 (2001)

117. A. Zywietz, J. Furthmüller, F. Bechstedt, Neutral vacancies in group-IV semiconductors. Phys. Stat. Sol. B **210**, 13–29 (1998)

118. J. Isoya, H. Kanda, Y. Uchida, S.C. Lawson, S. Yamasaki, H. Itoh, Y. Morita, EPR identification of the negatively charged vacancy in diamond. Phys. Rev. B **45**, 1436–1439 (1992)

119. J.A. Baldwin, Electron paramagnetic resonance investigation of the vacancy in diamond. Phys. Rev. Lett. **10**, 220–222 (1963)

120. G. Davies, S. Lawson, A.T. Collins, A. Mainwood, S.J. Sharp, Vacancy-related centers in diamond. Phys. Rev. B **46**, 13157–13170 (1992)

121. D.J. Twitchen, D.C. Hunt, V. Smart, M.E. Newton, J.M. Baker, Correlation between ND1 optical absorption and the concentration of negative vacancies determined by electron paramagnetic resonance. Diam. Relat. Mater. **8**, 1572–1575 (1999)

122. G. Davies, M.F. Hamer, Optical studies of the 1.945 eV vibronic band in diamond. Proc. R. Soc. London A **348**, 285–298 (1976)

123. C.D. Clark, R.W. Ditchburn, H.B. Dyer, The absorption spectra of irradiated diamonds after heat treatment. Proc. Roy. Soc. Lond. A **237**, 75–89 (1956)
124. D.J. Twitchen, M.E. Newton, J.M. Baker, T.R. Anthony, W.F. Banholzer, Electron-paramagnetic-resonance measurements on the divacancy defect center R4/W6 in diamond. Phys. Rev. B **59**, 12900–12910 (1999)
125. S. Dannefaer, A. Pu, V. Avalos, D. Kerr, Annealing of monovacancies in electron and γ-irradiated diamond. Physica B **308–310**, 569–572 (2001)
126. L. du Preez, PhD thesis, University of the Witwatersrand, Johannesburg (1965)
127. D. Fisher, D.J.F. Evans, C. Glover, C.J. Kelly, M.J. Sheehy, G.C. Summerton, The vacancy as a probe of the strain in type IIa diamonds. Diam. Relat. Mater. **15**, 1636–1642 (2006)
128. L.S. Hounsome, R. Jones, P.M. Martineau, D. Fisher, M.J. Shaw, P.R. Briddon, S. Öberg, Origin of brown coloration in diamond. Phys. Rev. B **73**, 125203/1–8 (2006)
129. A.T. Collins, H. Kanda, H. Kitakawi, Colour changes produced in natural brown diamonds by high-pressure, high-temperature treatment. Diam. Relat. Mater. **9**, 113–122 (2000)
130. R. Jones, Dislocations, vacancies and the brown colour of CVD and natural diamond. Diam. Relat. Mater. **18**, 820–826 (2009)
131. R.G. Farrer, On the substitutional nitrogen donor in diamond. Solid State Commun. **7**, 685–688 (1969)
132. W.V. Smith, P.P. Sorokin, I.L. Gelles, G.J. Lasher, Electron-spin resonance of nitrogen donors in diamond. Phys. Rev. **115**, 1546–1552 (1959)
133. C.A.J. Ammerlaan, Electron paramagnetic resonance studies of native defects in diamond, *Defects and Radiation Effects in Semiconductors 1980. Inst. Phys. Conf. Ser. N⁰ 59*, ed. by R.R. Hasiguti (The Institute of Physics, Bristol, 1981), pp. 81–94
134. S. Zhang, S.C. Ke, M.E. Zvanut, H.T. Tohver, Y.K. Vohra, g-Tensor for substitutional nitrogen in diamond. Phys. Rev. B **49**, 15392–15395 (1994)
135. M.H. Nazare, A.J. das Neves, Paramagnetic nitrogen in diamond: ultraviolet absorption. J. Phys. C **20**, 2713–2722 (1987)
136. K. Iakoubovskii, G.J. Adriaenssens, Optical transitions at the substitutional nitrogen centre in diamond. J. Phys.: Cond. Matter **12**, L77–L81 (2000)
137. R. Jones, J.P. Goss, P.R. Briddon, Acceptor level of nitrogen in diamond and the 270-nm absorption band. Phys. Rev. B **80**, 033205/1–4 (2009)
138. W.J.P. van Enckevort, E.H. Versteegen, Temperature dependence of the optical absorption by the single-substitutional nitrogen donor in diamond. J. Phys. Cond. Matter **4**, 2361–2373 (1992)
139. H. Sumiya, S. Satoh, High-pressure synthesis of high-purity diamond. Diam. Relat. Mater. **5**, 1359–1365 (1996)
140. F. De Weerdt, A.T. Collins, Determination of the C defect concentration in HPHT annealed type IaA diamonds from UV-VIS absorption spectra. Diam. Relat. Mater. **17**, 171–173 (2008)
141. A.T. Collins, Spectroscopy of defects and transition metals in diamond. Diam. Relat. Mater. **9**, 417–423 (2000)
142. G. Davies, M.H. Nazaré, The ultraviolet absorption by substitutional nitrogen pair in diamond, *Defects and Radiation Effects in Semiconductors 1978. Inst. Phys. Conf. Ser. N⁰ 46*, ed. by J.H. Albany (The Institute of Physics, Bristol, 1979), pp. 334–340
143. G. Davies, The A nitrogen aggregate in diamond – its symmetry and possible structure. J. Phys. C **9**, L537–L542 (1976)
144. J.A. van Wyk, J.H.N. Loubser, Electron spin resonance of a di-nitrogen centre in Cape yellow type Ia diamonds. J. Phys. C **16**, 1501–1506 (1983)
145. G.D. Tucker, M.E. Newton, J.M. Baker, EPR and ¹⁴N electron-nuclear double-resonance of the ionized nearest-neighbor dinitrogen center in diamond. Phys. Rev. B **50**, 15586–15596 (1994)
146. G.S. Woods, Platelets and the infrared absorption of type Ia diamonds. Proc. Roy. Soc. Lond. A **407**, 219–238 (1986)
147. I. Kiflawi, J. Bruley, The nitrogen aggregation sequence and the formation of voidities in diamond. Diam. Relat. Mater. **9**, 87–93 (2000)

148. E.C. Reynhardt, G.L. High, J.A. van Wyk, Temperature dependence of spin-spin and spin-lattice relaxation times of paramagnetic nitrogen defects in diamond. J. Chem. Phys. **109**, 8471–8477 (1998)

149. R.M. Chrenko, R.E. Tuft, H.M. Strong, Transformation of the state of nitrogen in diamond. Nature **270**, 141–144 (1977)

150. T. Evans, Z. Qi, The kinetics of aggregation of nitrogen atoms in diamond. Proc. Roy. Soc. Lond. A **381**, 159–178 (1982)

151. M.R. Brozel, T. Evans, R.F. Stephenson, Partial dissociation of nitrogen aggregates in diamond by high temperature-high pressure treatments. Proc. Roy. Soc. Lond. A **361**, 109–127 (1978)

152. G. Davies, C.M. Welbourn, J.H.N. Loubser, *Diamond Research 1978* (Ascot, De Beers Industrial Diamond, Johannesburg, 1978), pp. 23–30

153. G. Davies, I. Kiflawi, C. Sittas, H. Kanda, The effects of carbon and nitrogen isotopes on the "N3" optical transition in diamond. J. Phys. Cond. Matter **9**, 3871–3879 (1997)

154. A.T. Collins, C.H. Ly, Misidentification of nitrogen-vacancy absorption in diamond. J. Phys. Cond. Matter **14**, L467–L471 (2002)

155. G. Davies, Dynamic Jahn-Teller distortions at trigonal optical centres in diamond. J. Phys. C **12**, 2551–2566 (1979)

156. S. Felton, A.M. Edmonds, M.E. Newton, P.M. Martineau, D. Fisher, D.J. Twitchen, Electron paramagnetic resonance studies of the neutral nitrogen vacancy in diamond. Phys. Rev. B **77**, 081201/1–4 (2008)

157. J.H.N. Loubser, J.A. van Wyk, Electron spin resonance in the study of diamond. Rep. Progr. Phys. **41**, 1201–1248 (1978)

158. N.R.S. Reddy, N.B. Manson, E.R. Krausz, Two-laser spectral hole burning in a colour centre in diamond. J. Lumin. **38**, 46–47 (1987)

159. A. Beveratos, R. Brouri, T. Gacoin, A. Villing, J.P. Poizat, P. Grangier, Single photon quantum cryptography. Phys. Rev. Lett. **89**, 187901/1–4 (2002)

160. G. Davies, The Jahn-Teller effect and vibronic coupling at deep levels in diamond. Rep. Prog. Phys. **44**, 787–830 (1981)

161. S. Lawson, G. Davies, A.T. Collins, A. Mainwood, The "H2" optical transition in diamond: the effects of uniaxial stress perturbation, temperature and isotopic substitution. J. Phys. Cond. Matter **4**, 3439–3452 (1992)

162. A. Mainwood, A.T. Collins, P. Woad, Isotope dependence of the frequency of localised vibrational modes in diamond. Mater. Sci. Forum **143–147**, 29–34 (1994)

163. A.T. Collins, G. Davies, G.S. Wood, Spectroscopic studies of the H1b and H1c absorption lines in irradiated annealed type-Ia diamonds. J. Phys. C **19**, 3933–3944 (1986)

164. A.T. Collins, A. Connor, C.H. Ly, A. Shareef, P.M. Spear, High-temperature annealing of optical centres in type-I diamond. J. Appl. Phys. **97**, 083517/1–10 (2005)

165. G. Davies, The optical properties of diamond, in *Chemistry and Physics of Carbon*, vol. 13, ed. by P.L. Walker, P.A. Thrower (Marcel Dekker, New York, 1977), pp. 1–143

166. A.T. Collins, G.S. Woods, Isotope shifts of nitrogen-related mode vibrations in diamond. J. Phys. C **20**, L797–L801 (1987)

167. J.P. Goss, P.R. Briddon, S. Papagiannidis, R. Jones, Interstitial nitrogen and its complexes in diamond. Phys. Rev. B **70**, 235208/1–15 (2004)

168. V.S. Vavilov, A.A. Gippius, A.M. Zaitsev, B.V. Deryagin, B.V. Spitsyn, A.E. Aleksenko, Investigation of the cathodoluminescence of epitaxial diamond films. Sov. Phys. Semicond. **14**, 1078–1079 (1980)

169. A.M. Zaitsev, V.S. Vavilov, A.A. Gippius, Cathodoluminescence of diamond associated with silicon impurity. Sov. Phys. Lebedev Inst. Rep. **10**, 15–17 (1981)

170. J. Ruan, W.J. Choyke, W.D. Parlow, Si impurity in chemical vapor deposited diamond films. Appl. Phys. Lett. **58**, 295–297 (1991)

171. C.D. Clark, H. Kanda, I. Kiflawi, G. Sittas, Silicon defects in diamond. Phys. Rev. B **51**, 16681–16688 (1995)

172. J.P. Goss, R. Jones, S.J. Breuer, P.R. Briddon, S. Öberg, The twelve-line 1.682 eV luminescence center in diamond and the vacancy-silicon complex. Phys. Rev. Lett. **77**, 3041–3044 (1996)
173. G.D. Watkins, Defects in irradiated silicon: EPR of the tin-vacancy pair. Phys. Rev. B **12**, 4383–4390 (1975)
174. S.S. Moliver, Electronic structure of the neutral silicon-vacancy complex in diamond. Tech. Phys. **48**, 1449–1453 (2003).
175. K. Iakoubovskii, A. Stesmans, M. Nesladek, G. Knuyt, ESR and photo-ESR study of defects in CVD diamond. Phys. Stat. Sol. A **193**, 448–456 (2002)
176. A.M. Edmonds, M.E. Newton, P.M. Martineau, D.J. Twitchen, S.D. Williams, Electron paramagnetic resonance studies of silicon-related defects in diamond. Phys. Rev. B **77**, 245205/1–11 (2008). Erratum: Phys. Rev. B **82**, 249901 (2010)
177. K. Iakoubovskii, A. Stesmans, Characterization of hydrogen and silicon-related defects in CVD diamond by electron spin resonance. Phys. Rev. B **66**, 195207/1–7 (2002)
178. I. Kiflawi, G. Sittas, H. Kanda, D. Fisher, The irradiation and annealing of Si-doped diamond single crystals. Diam. Relat. Mater. **6**, 146–148 (1997)
179. C. Wang, C. Kurtsiefer, H. Weinfurter, B. Burchard, Single photon emission from SiV centres in diamond produced by ion implantation. J. Phys. B **39**, 37–41 (2006)
180. R. Williams, Determination of deep centers in conducting gallium arsenide. J. Appl. Phys. **37**, 3411–3416 (1966)
181. G.M. Martin, A. Mitonneau, A. Mircea, Electron traps in bulk and epitaxial GaAs crystals. Electron. Lett. **13**, 191–192 (1977)
182. D. Bois, A. Chantre, Spectroscopies thermique et optique des niveaux profonds: application à l'étude de leur relaxation de réseau. Rev. Phys. Appl. **15**, 631–646 (1980)
183. G.M. Martin, Optical assessment of the main electron trap in bulk semi-insulating gallium arsenide. Appl. Phys. Lett. **39**, 747–748 (1981)
184. E.R. Weber, H. Ennen, U. Kaufmann, J. Windscheif, J. Schneider, T. Wosinski, Identification of AsGa antisites in plastically deformed GaAs. J. Appl. Phys. **53**, 6140–6143 (1982)
185. B.K. Meyer, D.M. Hofmann, J.M. Spaeth, Energy levels and photo-quenching properties of the arsenic anti-site in GaAs. J. Phys. C **20**, 2445–2451 (1987)
186. G.A. Baraff, The mid-gap donor level EL2 in GaAs: recent developments, in *Deep Centers in Semiconductors,* ed. by S.T. Pantelides (Gordon and Breach, New York, 1992), pp. 547–589
187. J.T. Schick, C.G. Morgan, P. Papoulias, First-principle study of As interstitials in GaAs: convergence, relaxation, and formation energy. Phys. Rev. B **66**, 195302/1–10 (2002)
188. A.L. Lin, E. Omelianowski, R.H. Bube, Photoelectronic properties of high-resistivity GaAs:O. J. Appl. Phys. **47**, 1852–1858 (1976)
189. G. Vincent, D. Bois, Photocapacitance quenching effect for "oxygen" in GaAs. Solid State Commun. **27**, 431–434 (1978)
190. G. Vincent, D. Bois, A. Chantre, Photoelectric memory effect in GaAs. J. Appl. Phys. **53**, 3643–3649 (1982)
191. G.M. Martin, S. Makram-Ebeid, The mid-gap donor level EL2 in GaAs, in *Deep Centers in Semiconductors*, ed. by S.T. Pantelides (Gordon and Breach, New York, 1992), pp. 457–545
192. M. Kaminska, M. Skowronski, J. Lagowski, J.M. Parsey, H.C. Gatos, Intracenter transitions in the dominant deep level (EL2) in GaAs. Appl. Phys. Lett. **43**, 302–304 (1983)
193. M. Baj, P. Dreszer, EL2-intracenter absorption under hydrostatic pressure. Mater. Sci. Forum **38–41**, 101–106 (1989)
194. H.J. von Bardeleben, Metastable state of the EL2 defect in GaAs. Phys. Rev. B **40**, 12546–12549 (1989)
195. M.O. Manasreh, B.C. Covington, Infrared-absorption properties of EL2 in GaAs. Phys. Rev. B **36**, 2730–2734 (1987)
196. C. Hecht, R. Kummer, M. Thoms, A. Winnacker, High-resolution spectroscopy of the zero-phonon line of the deep donor EL2 in GaAs. Phys. Rev. B **55**, 13625–13629 (1997)

197. M. Kaminska, M. Skowronski, W. Kuszko, Identification of the 0.82-eV electron trap, EL2 in GaAs, as an isolated antisite arsenic defect. Phys. Rev. Lett. **55**, 2204–2207 (1985)

198. P. Trautman, J.P. Walczak, J.M. Baranowski, Piezospectroscopic evidence for tetrahedral symmetry of the EL2 defect in GaAs. Phys. Rev. Lett. **41**, 3074–3077 (1990)

199. F. Fuchs, B. Dischler, Infrared studies of the dynamics of transformation between normal and metastable states of the EL2 center in GaAs. Appl. Phys. Lett. **51**, 679–681 (1987)

200. K. Khachaturyan, E.R. Weber, J. Horigan, W. Ford, Interaction of EL2 in semiinsulating GaAs with above bandgap light. Mater. Sci. Forum **83–87**, 881–886 (1992)

201. T. Sugiyama, K. Tanimura, N. Itoh, Recombination-induced metastable to stable transformation of the EL2 center in GaAs. Appl. Phys. Lett. **55**, 639–641 (1989)

202. M. Baj, P. Dreszer, Optical activity of the EL2 metastable state under hydrostatic pressure. Phys. Rev. B **39**, 10470–10472 (1989)

203. M. Baj, P. Dreszer, A. Babinski, Pressure-induced negative charge state of the EL2 defect in its metastable configuration. Phys. Rev. B **43**, 2070–2080 (1991)

204. P. Trautman, J.M. Baranowski, Evidence for trigonal symmetry of the metastable state of the EL2 defect in GaAs. Phys. Rev. Lett. **69**, 664–667 (1992)

205. B.K. Meyer, J.M. Spaeth, M. Scheffler, Optical properties of As-antisite and EL2 defects in GaAs. Phys. Rev. Lett. **52**, 851–854 (1984)

206. M.O. Manasreh, W.C. Mitchel, D.W. Fischer, Observation of the second energy level of the EL2 defect by the infrared absorption technique. Appl. Phys. Lett. **55**, 864–866 (1989)

207. K.H. Wietzke, F.H. Koschnick, K. Krambrock, Correlation of two diamagnetic bands of the magnetic circular dichroism of the optical absorption with EL2^0 in GaAs. Appl. Phys. Lett. **71**, 2133–2135 (1997)

208. M. Skowronski, J. Lagowski, H.C. Gatos, Optical and transient capacitance study of EL2 in the absence or presence of other midgap levels. J. Appl. Phys. **59**, 2451–2456 (1986)

209. P. Silverberg, P. Omling, L. Samuelson, Hole photoionization cross sections of EL2 in GaAs. Appl. Phys. Lett. **52**, 1689–1691 (1988)

210. F.X. Zach, A. Winnacker, Optical mapping of the total EL2-concentration in semi-insulating GaAs-wafers. Jpn. J Appl. Phys. **28**, 957–960 (1989)

211. D.M. Hofmann, K. Krambrock, B.K. Meyer, J.M. Spaeth, Optical and magneto-optical determination of the EL2 concentrations in semi-insulating GaAs. Semicond. Sci. Technol. **6**, 170–174 (1991)

212. P.W. Yu, Deep-center photoluminescence in undoped semi-insulating GaAs: 0.68 eV band due to the main deep donor. Solid State Commun. **43**, 953–956 (1982)

213. M.K. Nissen, T. Steiner, D.J.S. Beckett, M.L.W. Thewalt, Photoluminescence transitions of the deep EL2 level in gallium arsenide. Phys. Rev. Lett. **65**, 2282–2285 (1990)

214. M. Tajima, Radiative recombination mechanism of EL2 levels in GaAs. Jpn. J. Appl. Phys. **26**, L885–L888 (1986)

215. M.K. Nissen, A. Villemaire, M.L.W. Thewalt, Photoluminescence studies of the EL2 defect in gallium arsenide under external pertubations. Phys. Rev. Lett. **67**, 112–115 (1991)

216. T.W. Steiner, M.K. Nissen, S.M. Wilson, Y. Lacroix, M.L.W. Thewalt, Observation of luminescence from the EL2 metastable state in liquid-encapsulated Czochralski-grown GaAs under hydrostatic pressure. Phys. Rev B **47**, 1265–1269 (1993)

217. A. Fukuyama, T. Ikari, Y. Akashi, M. Suemitsu, Interdefect correlation during thermal recovery of EL2 in semi-insulating GaAs: proposal of a three-center model. Phys. Rev. B **67**, 113202/1–4 (2003)

218. P.P. Fávero, J.M.R. Cruz, New EL2 structural model based on the observation of two sequential photoquenching processes. Eur. Phys. J. B **47**, 363–368 (2005)

219. G. Davies, Jahn-Teller coupling of the 1.04-eV *EL2*-related center in GaAs. Phys. Rev. B **41**, 12303–12306 (1990)

220. E.M. Omelyanovskii, V.I. Fistul, *Transition Metal Impurities in Semiconductors* (Adam Hilger, Bristol, 1986)

221. K.A. Kikoin, V.N. Fleurov, *Transition Metal Impurities in Semiconductors* (World Scientific, Singapore, 1994)

222. H.J. Schultz, Optical properties of 3d transition metals in II-VI compounds. J. Cryst. Growth **59**, 65–80 (1982)

223. B. Clerjaud, Transition-metal impurities in III-V compounds. J. Phys. C **18**, 3615–3661 (1985)

224. A. Zunger, Electronic structure of $3d$ transition-atom impurities in semiconductors. Solid State Phys. **39**, 275–464 (1986)

225. B. Clerjaud, Transition-metal impurities in III-V compounds, in *Current Issues in Semiconductor Physics*, ed. by M. Stoneham (Adam Hilger, Bristol, 1986), pp. 117–168

226. H.J. Schultz, Transition metal impurities in semiconductors: experimental situation. Mater. Chem. Phys. **15**, 373–384 (1987)

227. H.J. Schultz, M. Thiede, Optical spectroscopy of $3d^7$ and $3d^8$ impurity configurations in a wide gap semiconductor (ZnO:Co, Ni, Cu). Phys. Rev. B **35**, 18–34 (1987)

228. A.M. Hennel, Transition metals in III/V compounds, Semiconductors and Semimetals **38**, 189–234 (1993)

229. V.I. Sokolov, Hydrogen-like excitations of $3d$ transition-element impurities in semiconductors. Semiconductors **28**, 329–342 (1994)

230. E. Malguth, A. Hoffmann, M.R. Philips, Fe in III-V and II-VI semiconductors. Phys. Stat. Sol. B **245**, 455–480 (2008)

231. S. Sugano, Y. Tanabe, H. Kamimura, *Multiplets of Transition-Metal Ions in Crystals* (Academic, New York, 1970)

232. G. Racah, Theory of complex spectra II. Phys. Rev. **62**, 438–462 (1942)

233. S.W. Biernacki, Splitting of 3d levels of iron group impurities in crystals with zincblende structure. Phys. Stat. Sol B **118**, 525–533 (1983)

234. F.S. Ham, Dynamical Jahn-Teller effect in paramagnetic resonance spectra-orbital reduction factors and partial quenching of spin-orbit interaction. Phys. Rev. **138**, A1727–A1740 (1965)

235. F.S. Ham, Effect of linear Jahn-Teller coupling on paramagnetic resonance in a ^{2}E state. Phys. Rev. **166**, 307–321 (1968)

236. M.D. Sturge, The Jahn-Teller effect in solids. Solid State Phys. **20**, 91–211 (1967)

237. R. Englman, *The Jahn-Teller Effect in Molecules and Crystals* (Wiley, London, 1972)

238. C.A. Bates, Jahn-Teller effects in paramagnetic crystals. Phys. Rep. **35**, 187–304 (1978)

239. I.B. Bersuker, V.Z. Polinger, *Vibronic Interactions in Molecules and Crystals* (Springer, Berlin, 1989)

240. C.A. Bates, K.W.H. Stevens, Localized electron states in semiconductors. Rep. Progr. Phys. **49**, 783–823 (1986)

241. E. Tarhan, I. Miotkowski, S. Rodriguez, A.K. Ramdas, Lyman spectrum of holes bound to substitutional 3d transition metal ions in a III-V host: GaAs(Mn^{2+}, Co^{2+}, or Cu^{2+}), GaP(Mn^{2+}) and InP(Mn^{2+}). Phys. Rev. B **67**, 195202/1–9 (2003)

242. J. Schneider, U. Kaufmann, W. Wilkening, M. Bauemler, Electronic structure of neutral manganese acceptor in gallium arsenide. Phys. Rev. Lett. **59**, 240–243 (1987)

243. J. Kreissl, W. Ulrici, M. El-Metoui, A.M. Vasson, A. Vasson, A. Gavaix, Neutral manganese acceptor in GaP: an electron paramagnetic-resonance study. Phys. Rev. B **54**, 10508–10515 (1996)

244. J.J. Krebs, G.H. Stauss, EPR of Cr^{2+} ($3d^4$) in gallium arsenide: Jahn-Teller distortion and photoinduced charge conversion. Phys. Rev. B **16**, 971–973 (1977)

245. A.S. Abhvani, C.A. Bates, B. Clerjaud, D.R. Pooler, Interpretation of the zero-phonon optical absorption lines associated with substitutional Cr^{2+}:GaAs. J. Phys. C **15**, 1345–1351 (1982)

246. A.M. Hennel, W. Szuszkiewicz, M. Balkanski, M. Martinez, B. Clerjaud, Investigation of the absorption of Cr^{2+} ($3d^4$) in GaAs. Phys. Rev. B **23**, 3933–3942 (1981)

247. J.T. Vallin, G.A. Slack, S. Roberts, A.E. Hughes, Infrared absorption in some II–VI compounds doped with Cr. Phys. Rev. B **2**, 4313–4333 (1970)

248. M. Kaminska, J.M. Baranowski, S.M. Uba, J.T. Vallin, Absorption and luminescence of Cr^{2+} (d^4) in II-VI compounds. J. Phys. C **12**, 2197–2214 (1979)

249. B. Clerjaud, C. Naud, G. Picoli, Y. Toudic, Chromium absorption in InP. J. Phys. C **17**, 6469–6476 (1984)

250. W. Ulrici, J. Kreissl, Optical absorption and electron paramagnetic resonance of the Cr^{2+} impurity in GaP, in *Proc. 23rd Internat. Conf. Phys. Semicond.*, ed. by M. Scheffler, R. Zimmermann (World Scientific, Singapore, 1996), pp. 2833–2836

251. T.F. Deutsch, Absorption coefficient of infrared laser window materials. J. Phys. Chem. Solids **34**, 2091–2104 (1973)

252. B. Deveaud, G. Picoli, B. Lambert, M. Martinez, Luminescence processes at chromium in GaAs. Phys. Rev. B **29**, 5749–5763 (1984)

253. B. Clerjaud, C. Naud, B. Deveaud, B. Lambert, B. Plot, G. Brémond, C. Benjeddou, G. Guillot, A. Nouialhat, The acceptor level of vanadium in III-V compounds. J. Appl. Phys. **58**, 4207–4215 (1985)

254. W. Ulrici, K. Friedland, L. Eaves, D.P. Halliday, Optical and electrical properties of vanadium-doped GaAs. Phys. Stat. Sol. B **131**, 719–728 (1985)

255. D. Côte, Doctoral thesis, Université Pierre et Marie Curie (1988)

256. D. Ammerlahn, B. Clerjaud, D. Côte, L. Köhne, M. Krause, D. Bimberg, Spectroscopic investigation of neutral niobium in GaAs. Mater. Sci. Forum **258–263**, 911–916 (1997)

257. S. Gabillet, V. Thomas, J.P. Peyrade, J. Barrau, The luminescence at 0.795 eV from GaAs:Nb: a Zeeman spectroscopy. Phys. Lett. A **119**, 197–199 (1986)

258. V. Lauer, G. Brémond, A. Souifi, G. Guillot, K. Chourou, M. Anikin, R. Madar, B. Clerjaud, C. Naud, Electrical and optical characterization on vanadium in 4H and 6H-SiC. Mater. Sci. Eng. B **61–62**, 248–252 (1999)

259. H. Ennen, U. Kaufmann, J. Schneider, Donor-acceptor pairs in GaP and GaAs involving the deep nickel acceptor. Appl. Phys. Lett. **38**, 355–357 (1981)

260. A. Yelisseyev, H. Kanda, Optical centers related to $3d$ transition metals in diamond. New Diam. Frontier Carbon Technol. **17**, 127–178 (2007)

261. H. Katayama-Yoshida, A. Zunger, Prediction of a low spin-ground state in the GaAs : V^{2+} impurity system. Phys. Rev. B **33**, 2961–2964 (1986)

262. M.J. Caldas, S.K. Figueiredo, A. Fazzio, Theoretical investigation of the electrical and optical activity of vanadium in GaAs. Phys. Rev. B **33**, 7102–7109 (1986)

263. C. Delerue, M. Lannoo, G. Allan, New theoretical approach of transition-metal impurities in semiconductors. Phys. Rev. B **39**, 1669–1681 (1989)

264. D. Ammerlahn, R. Heitz, D. Bimberg, D. Côte, B. Clerjaud, W. Ulrici, Infrared investigation of neutral cobalt in GaP, in *Proc. 23rd Internat. Conf. Phys. Semicond.*, ed. by M. Scheffler, R. Zimmermann (World Scientific, Singapore, 1996), pp. 2825–2828

265. R.E. Dietz, H. Kamimura, M.D. Sturge, A. Yariv, Electronic structure of copper impurities in ZnO. Phys. Rev. **132**, 1559–1569 (1963)

266. I. Broser, A. Hoffmann, R. Germer, R. Broser, E. Birkicht, High-resolution optical spectroscopy on nickel ions in II-VI semiconductors: isotope shifts of the $^3T_1(F) \leftrightarrows {}^3T_1(P)$ and $^3T_1(F) \leftrightarrows {}^3A_2(F)$, respectively Ni^{2+} transitions in CdS and ZnS crystals. Phys. Rev. B **33**, 8196–8206 (1986)

267. B. Nestler, A. Hoffmann, L.B. Xu, U. Scherz, I. Broser, Investigation of the Jahn-Teller effect through the isotope shift and Zeeman splitting of optical transitions of CdS : Ni^{2+}. J. Phys. C **20**, 4613–4625 (1987)

268. B. Clerjaud, D. Côte, F. Gendron, M. Krause, W. Ulrici, Isotopic effects in GaAs:Ni. Mater. Sci. Forum **38–41**, 775–778 (1989)

269. W. Drozdzewicz, A. Hennel, Z. Wasilewski, B. Clerjaud, F. Gendron, C. Porte, R. Germer, Identification of the double acceptor state of isolated nickel in gallium arsenide. Phys. Rev. B **29**, 2438–2442 (1984)

270. B. Clerjaud, Jahn-Teller effect and vibronic interactions. Acta Phys. Polonica **A73**, 909–923 (1988)

271. B. Clerjaud, F. Gendron, C. Porte, Chromium-induced up conversion in GaP. Appl. Phys. Lett. **38**, 212–214. Erratum: Appl. Phys. Lett. **38**, 952 (1981)

272. R.A. Chapman, G. Hutchinson, Photoexcitation and photoionization of neutral manganese acceptors in gallium arsenide. Phys. Rev. Lett. **18**, 443–445. Erratum: Phys. Rev. Lett. **18**, 822 (1967)

273. B. Lambert, B. Clerjaud, C. Naud, B. Deveaud, G. Picoli, Y. Toudic, Photoionization of Mn acceptor in *Proc. 13th Internat. Conf. Defects in Semiconductors*, ed. by L.C. Kimerling, J.M. Parsey, Jr. (The Metallurgical Soc. of AIME, New York, 1985), pp 1141–1147

274. B. Clerjaud, D. Côte, C. Naud, Evidence for hydrogen-transition metal complexes in as-grown indium phosphide. J. Cryst. Growth **83**, 190–193 (1987)

275. K. Thonke, K. Pressel, Charge-transfer transitions of Fe ions in InP. Phys. Rev. B **44**, 13418–13425 (1991)

276. A. Baldereschi, N.O. Lipari, Cubic contributions to the spherical model of shallow acceptor states. Phys. Rev. B **9**, 1525–1539 (1974)

277. T. Wolf, W. Ulrici, D. Côte, B. Clerjaud, D. Bimberg, New evidence for bound states in the charge transfer spectra of transition-metal doped III-V semiconductors. Mater. Sci. Forum **143–147**, 317–322 (1994)

278. A.A. Kopylov, A.N. Pikhtin, Shallow impurity states and the free exciton binding energy in gallium phosphide. Solid State Commun. **26**, 735–740 (1978)

279. B. Clerjaud, A. Gelineau, Jahn-Teller effect in the $^2T_2$ state of Cu^{2+} in ZnS. Phys. Rev. B **9**, 2832–2837 (1974)

280. G. Armelles, J. Barrau, M. Brousseau, B. Pajot, C. Naud, Effective mass-like states of the deep acceptor level of Au and Pt in silicon. Solid State Commun. **56**, 303–305 (1985)

Chapter 5
Vibrational Absorption of Substitutional Atoms and Related Centres

5.1 Introduction

In this chapter, we try to describe what is known on the vibrational absorption of substitutional FAs atoms and on centres involving substitutional FAs in some semiconductors, but this introduction and the next section on theoretical methods are more general and apply also to centres with interstitial FAs and to centres where the location of the FAs is more difficult to categorize, like the complexes of FAs with crystal defects.

As sketched in Sect. 1.6.2, in a crystal, the presence of FAs usually gives rise to extra vibrational modes, and in some cases to hindered rotational modes. When an atomic structure presents an electric dipole moment, as the one made from two different FAs bonded together, the stretching vibration (stretch mode) along the bond axis is IR-active and its absorption can be observed. Inversely, a similar structure made from two identical atoms presents no dipole moment and its stretching is IR-inactive, but is generally Raman-active and observable by Raman scattering spectroscopy. Some vibrational modes can be both IR- and Raman-active. When a FA is only bonded to one lattice atom, and this situation occur, for instance, with the H atom, it can also vibrate perpendicularly to the bond axis. By analogy with molecular spectroscopy, the corresponding mode is called a bending (or wag) mode. The distinction between IR- and Raman-active modes is true to the first order as, in crystal, electric dipoles can be induced in otherwise IR-inactive structures and make them IR-active. Some impurity or defect modes can also be detected indirectly as vibronic replicas of ZPL PL or sometimes absorption lines. Vibrational modes associated with FAs have been detected in semiconducting crystals and in diamond, which are the crystals considered in this book, but also in ionic crystals like the alkali halides and the alkaline earth fluorides. When their frequencies are above that of the Raman frequency of the crystal, these modes do not propagate in the crystal and they are referred to as local vibrational modes (LVMs). The FAs whose masses are comparable to or larger than those of the atoms of the crystal can give vibrational modes below the Raman frequency of the crystal, known as resonant

B. Pajot and B. Clerjaud, *Optical Absorption of Impurities and Defects in Semiconducting Crystals*, Springer Series in Solid-State Sciences 169,
DOI 10.1007/978-3-642-18018-7_5, © Springer-Verlag Berlin Heidelberg 2013

modes. In compound semiconductors with a phonon gap, and when the energy of an impurity mode falls in this gap, the corresponding mode is called a gap mode.

Isolated interstitial FAs which are not bonded to crystal atoms have very low-frequency vibrational motions which are difficult to detect by optical spectroscopy, but the stretching of interstitial or quasi-interstitial diatomic molecules or radicals can be detected by Raman scattering or infrared absorption at frequencies not too different from the ones measured in the free state in molecular spectroscopy.

The frequencies of the vibrational modes associated with these FAs are related to the masses and binding energies of these atoms in the crystal. They have been actively studied in semiconductors both from the experimental and theoretical sides because they bring useful information on the properties of the centres in which they are involved. These studies mainly concern LVMs because they generally produce well-characterized lines which can be very sharp and easily detected at low temperature. One of the reasons of the usefulness of these studies is that they can be observed for FAs or centres with or without electrical activity (for electrically active centres, the vibration frequency can vary with the charge state). Another interest for the study of these vibrational modes is that when the FAs have stable isotopes, they display isotope frequency shifts rather large compared to the ones for electronic transitions. When the relative concentrations of some stable natural isotopes are too small, like ^{13}C or ^{15}N, synthetic samples can be doped with an isotopically enriched mixture or with an isotope of chosen mass. This allows a convenient chemical identification of some FAs, and this can also bring information on the structure of a vibrating entity. We will also see examples of isotope effects observed at high resolution in compound crystals where the interaction of a FA with its nn can inform unambiguously on its site. However, in the general case where isotope effects are absent, as for FAs with only one isotope, the interpretation of the LVM spectra requires correlations with other experimental results, as those obtained from SIMS, channelling, and PAC measurements, and also with the results of ab $initio$ calculations. A last usefulness of LVM absorption spectroscopy is that at a difference with electronic absorption, the LVMs can generally be observed at room temperature, and for simple technological applications, their observation does not require cooling of the samples. The intensities of their absorption have also been used to determine the concentrations of the related FAs once obtained an adequate calibration factor. One disadvantage of vibrational absorption compared with electronic absorption, however, is that the OSs of the vibrational transitions are smaller than those of the electronic transitions so that, as a rule, for characterization purposes, vibrational absorption is less sensitive than electronic absorption. LVM spectroscopy has also been used to determine segregation coefficients of P and Al in GaAs and InSb [1].

In SI units, the integrated absorption $A_I (\mathrm{m}^{-2})$ of an isotropic oscillator with reduced mass μ at concentration N in a solid with refractive index n is:

$$A_I = N\eta^2 / 4\varepsilon_0 nc^2 \mu, \tag{5.1}$$

where the effective charge η can be considered as a dipole moment per unit of displacement. In cgs units, $A_1(\text{cm}^{-2}) = \pi N \eta^2 / \text{nc}^2 \mu$, and this is the form where it is given in [2]. When μ is expressed in atomic mass units, η, in units of electron charge, is given by:

$$\eta = 4.56 \times 10^7 (\mu \text{n} A_1 / N)^{1/2} \tag{5.2}$$

with N in cm^{-3} and A_1 in cm^{-2}. When the effective charge η is equal to the electronic charge, the integrated absorption is given by:

$$A_1(\text{cm}^{-2}) = 4.81 \times 10^{-16} N(\text{cm}^{-3}) / \text{n} \mu(\text{u}). \tag{5.3}$$

Expression (5.1) is deduced from the one given by [2] and, strictly speaking, it is valid only for a centre with cubic symmetry like a substitutional FA in a cubic crystal. The extension to centres with trigonal symmetry has been treated in [3]. Once the effective charge is determined for a LVM at a given temperature, a calibration factor of the absorption can be obtained from expression (5.3).

To the first order, vibrational transitions of isolated molecules or FAs in crystals are considered as those of harmonic oscillators with a linear restoring force leading to a quadratic potential. One consequence is the prevalence of a selection rule allowing transitions from one state with vibrational quantum number v to another one with $v \pm 1$. The existence of overtones and of combination modes involving different vibrational modes is taken into account by anharmonic potentials including terms of higher order. For the overtones, from a symmetry point of view, when the $v = 1$ state is associated with IR X of a symmetry point group, the levels associated to the $v = 2$ state are obtained from the symmetric direct product X $\times$ X, and those for the $v = 3$ state from the symmetric direct products X $\times$ X $\times$ X. For instance, for a substitutional impurity with T_d symmetry in a cubic crystal, a transition from a $\Gamma_1(A_1$ in Schoenflies notation) ground state to excited states belonging to a triply degenerate $\Gamma_5(T_2$ in Schoenflies notation) IR is symmetry allowed. The symmetric direct product $\Gamma_5 \times \Gamma_5$ is $\Gamma_1 + \Gamma_3 + \Gamma_5$ and the first overtone (or overtone) of the Γ_5 level is split into a triplet, with one IR-allowed transition from the ground state to the Γ_5 level. In the presence of a potential with T_d symmetry, the second overtone of a triply degenerate Γ_5 level is resolved into $\Gamma_1 + \Gamma_4 + 2\Gamma_5$ and two distinct transitions from a Γ_1 ground state to the Γ_5 levels are expected. Transitions to higher overtones are also IR-allowed, but their very weak intensities generally preclude their observation.

To take into account anharmonicity in the expression of the energy levels, cubic and quartic terms can be added to the quadratic expression for the harmonic potential in which an impurity vibrates. For a 1D oscillator, this anharmonic potential can be written:

$$V_a = k x^2 / 2 + \beta x^3 + \gamma x^4. \tag{5.4}$$

The effects of such a potential on the energy levels can be calculated to the first and second order of perturbation theory [4], and the energies ω_{1a} and $\omega_{2a}(\text{cm}^{-1})$ of the fundamental and overtone transitions with respect to the harmonic energy ω_h are given [5] by:

$$\omega_{1a} = \omega_h - B/m, \tag{5.5a}$$

$$\omega_{2a} = 2\omega_h - 3B/m, \tag{5.5b}$$

where m is the mass of the impurity and B an anharmonicity parameter[1]. When the overtone mode at ω_{2a} can be observed, the anharmonicity parameter B is given by $(2\omega_{1a} - \omega_{2a})m$ and the frequency ratio ω_{2a}/ω_{1a} is $2 - B/m\omega_{1a}$. An analysis of the H and D ISs with the Si_{Ga} donor in GaAs has been performed in [4] showing the role of anharmonicity to explain anomalies in the H/D frequency ratio.

The anharmonic potential for a 3D oscillator has been treated by Elliot et al. [6] in their study of the LVMs of negative hydrogen ions on group-VII sites of alkaline earth fluoride.

An alternative to a polynomial potential (5.4) to describe the anharmonicity of LVMs and especially those involving hydrogen is the use of the Morse potential [7, 8], initially intended to describe the vibrational states of diatomic molecules [9]. First, let us define some variables and quantities. For a 1D oscillator with reduced mass μ vibrating at frequency v (Hz) with amplitude x, the harmonic potential energy is $V_h(x) = 2\pi v^2 \mu x^2$. By introducing the dimensionless variable $\xi = x/l$, where $l = (\hbar/2\pi\mu v)^{1/2}$, the Morse potential $V_M(\xi)$ is expressed as:

$$V_M(\xi) = D_e(\exp[-2\alpha_M\xi] - 2\exp[-\alpha_M\xi]), \tag{5.6}$$

where α_M and D_e are parameters. The quasi-exact eigenvalues of the vibrational Hamiltonian with the Morse potential are [9]:

$$\omega_M(v) = \omega_e(v + 1/2) - \omega_e x_e(v + 1/2)^2. \tag{5.7}$$

With the above notations, the anharmonicity parameters ω_e and x_e are related to D_e and α_M by $\omega_e = 4x_e D_e$ and $x_e = \alpha_M^2/2$. It can be checked that ω_e and x_e are related to the measured fundamental ω_1 and first overtone ω_2 by $\omega_e = 3\omega_1 - \omega_2$ and $x_e = (2\omega_1 - \omega_2)/2(3\omega_1 - \omega_2)$. A comparison with expressions (5.5a) and (5.5b) shows that ω_e is the harmonic frequency ω_h. D_e is $(3\omega_1 - \omega_2)^2/2(2\omega_1 - \omega_2)$ and for a diatomic molecule it can be considered as the dissociation energy. In the solid state configuration, however, where the amplitude of vibration of the atoms of the bond is limited by the nearest neighbour (nn) atom, this approximation is questionable.

The ratio $(I_2/I_1)_M$ of the intensities of the first overtone to the fundamental for a Morse potential has been calculated [10] to be:

$$(I_2/I_1)_M = x_e(1 - 5x_e)/(1 - 3x_e)^2 \sim x_e, \tag{5.8}$$

---

[1]In this section the frequency in wavenumber is denoted ω instead of $\tilde{v}$ used in other parts of the book for an easier comparison with review [5].

and this expression is valid as long as the electric dipole moment remains linear as a function of the mode displacement coordinate. When this is no longer the case, one speaks of electrical anharmonicity and this effect can explain the absence of an overtone expected from the value of the mechanical anharmonicity [11].

When two vibrational modes of a centre have close frequencies and display the same symmetry, they can become mixed by anharmonic interaction. This results in frequency shifts and into modifications of the relative intensities. This effect is known as Fermi resonance in molecular spectroscopy [12, 13], and examples of this effect are given in Chap. 8.

Another situation can occur when the absorption of a vibrational mode is superimposed to an electronic absorption continuum. In that case, a resonance effect known as a Fano resonance [14] produces an asymmetric distortion of the profile of the LVM. Such an effect is observed at low temperature in some highly doped p-type GaAs and silicon samples when a LVM is superimposed to the continuous photoionization spectrum of a neutral acceptor [15]. This effect can be cancelled by ionizing the acceptor using some compensation mechanism.

As already mentioned, the vibrational modes are divided into resonant and local modes, plus gap modes in compound crystals with a phonon gap. The frequencies of the resonant modes in the range of the one-phonon modes of the crystal can be studied only in monoatomic crystal because the one-phonon absorption is too strong in compound crystals. The LVMs are observed at frequencies above those of the one-phonon optical modes of the crystal.

It is difficult to make a classification in the vibrational spectroscopy of FAs in semiconductors because if it can concern isolated FAs, it can also involve pairs of identical or different FAs, complexes made of FAs with lattice defects like vacancies or, in the case of compound semiconductors of complexes made of FAs with antisite atoms and eventually defects. A great variety of vibrational modes due to FAs has been reported in semiconductors. A review of the experimental situation in the early 1970s can be found in [16]. Another review, where a large part is devoted to ionic crystals, was published at about the same time [17]. More recent reviews on specific topics, like [2, 5], or [18], are also available. A list of vibrational modes reported in diamond is also given in Sect. 3.2 of [19].

5.2 Overview of Theoretical Methods

Several models have been used to explain, reproduce, or predict the vibrational spectrum of FAs and more complex centres in crystals. The simplest one considers a linear atomic chain that contains a mass defect, under the assumption that the force constant between the atoms of the chain is unchanged. When the equation of motion is applied to such a regular diatomic chain of atoms, with the classical boundary conditions, the model reproduces the resonant, localized, and band modes, as a function of the value of the mass of the FA and of which atom it replaces. An illustrative example of this situation is given for GaP [17]. This linear model shows that the vibration of a light FA with respect to the atoms of the chain remains localized in its vicinity. As pointed out in [5], this means that when extrapolated to the 3D situation, the vibration of such a centre can be viewed as that of a "defect molecule" and that the mathematical tools and symmetry considerations used in molecular spectroscopy (see, for instance, [20]) can be used at first to interpret the experimental data. This has been done indeed to interpret the vibrational spectrum of O_i in silicon (see section 6.1.1.1). A molecular model coupled with a variational method has been shown to reproduce satisfactorily the LVM frequencies of isoelectronic FAs, in III–V and II–VI compounds with sphalerite structure [21].

Green's functions technique allows to calculate the frequencies of LVMs associated with isolated FAs or pairs for atomic displacements of the atoms corresponding to LVMs of definite symmetry [22,23]. With this method, the interactions considered are those of the FAs or of the atoms of the pairs with their nns. It allows to evaluate the frequencies of LVMs of given symmetries using realistic values of the changes in the force constants.

Ab initio or first principles variational calculations using density functional theory (DFT) have been used to predict energetic and structural properties of different defects including their vibrational frequencies (an alternative is Hartree–Fock theory). These calculations are carried out on a periodic supercell [24] or on a cluster of atoms with the defect concerned at its centre and they necessitate to solve a many-body Schrödinger equation. Simplifications can be brought to the procedure by decoupling the electronic motions from the atomic ones by using the Born–Oppenheimer approximation and by using pseudopotentials for the core electrons which do not participate to the atomic bonding instead of the true electronic potentials (for more details, see [25, 26]. For hydrogen, see [27]).

The vibrational absorption of many kinds of centres has been reported in semiconductors. We have tried to categorize them and in the present chapter, we discuss the vibrational modes of substitutional FAs and of their complexes with intrinsic defects of the host lattice.

5.3 Vibrational Spectra of Specific Centres

In this section, the vibrational absorption of different substitutional centres in semiconductors and in diamond is discussed as a function of their symmetries starting from the highest symmetry. It is dominated by the contribution of light atoms and of their complexes as they give rise to local modes which can be used as probes of the properties of the centres.

In a semiconductor with diamond or sphalerite cubic structure, disregarding any asymmetric lattice distortion, a substitutional FA has T_d (tetrahedral) point group symmetry. When its mass is relatively small with respect to the atom it replaces, it produces one triply degenerate local stretch mode ν_s corresponding to the *IR* Γ_5 of T_d. The first overtone is a singlet $2\nu_s$ corresponding to the transition to the Γ_5 sublevel of the $v = 2$ triplet.

In a wurtzite-type crystal, in a similar situation, a substitutional atom has point group symmetry C_{3v}, and it produces two local modes, one, $v_{//s}$, where the FA vibrates along the c-axis of the crystal, with symmetry Γ_1, and another one, $v_{\perp s}$, where it vibrates in the plane perpendicular to this axis, with symmetry Γ_3. The $v = 2$ state is split into a quadruplet and the selection rules allow two transitions, $2v_{1//s}$ and $2v_{2//s}$ for the electric vector $\mathbf{E}$ of the radiation along the c-axis and two other ones, $2v_{1\perp s}$ and $2v_{2\perp s}$, for $\mathbf{E}\perp c$.

A general account of the situation in III–V cubic compounds (mainly GaAs) in 1993 is given in [2].

Isotope shifts of the LVMs of substitutional FAs on a tetrahedral site of a crystal can be easily calculated in the harmonic approximation. In this approximation, the vibrational energy of a FA is $E(v) = \hbar(f/M)^{1/2}(v + 3/2)$, where f is a force constant and M the mass of the FA. We have seen that fundamental vibrational transitions of such a FA with frequencies above the Raman frequency of the crystal are LVMs interacting weakly with the lattice. When the mass M_1 of such a FA is replaced by an isotopic mass M_2, assuming no change in the force constants, the ratio of the frequencies ω_{M1} and ω_{M2} of the LVMs is given, within the harmonic approximation, by:

$$\omega_{M2}/\omega_{M1} = (M_1/M_2)^{1/2}. \tag{5.9}$$

where M_1 is usually taken as the lightest mass, and the isotope shift (IS) $\Delta\omega_{12} = \omega_{M1} - \omega_{M2}$ is given as a function of $\Delta M = M_2 - M_1(M_1 < M_2)$ by:

$$\Delta\omega_{12} \approx \frac{1}{2}\omega_{M1}\frac{\Delta M}{M_2}. \tag{5.10}$$

The measured IS is always less than the one given by expression (5.10), where a harmonic potential is considered for the restoring force, and the difference between the two values can give an idea of the interaction of the FA with the crystal. This is an ideal case where no interaction is assumed between the FA and its *nns*. There are cases where the FWHM of the LVM is small and the interaction with the *nns* strong enough to allow considering the vibrations of a pseudomolecule formed by the FA and its nns. The isotopic distribution of the lattice atoms of this pseudomolecule then produces a host-isotope shift. Examples of this effect are given in the present chapter.

5.3.1 Li, Group-II Atoms, and 3d Transition Metals (TMs)

5.3.1.1 Li

In silicon and germanium, Li is an interstitial donor weakly coupled to the crystal lattice and no LVM or resonant mode is expected from this atom, but, as it is shown later, pairing of interstitial Li with substitutional acceptors produces a series of

LVMs mainly due to the modifications of the acceptors LVMs. They are presented in Sect. 5.3.4.1 in connection with the B acceptor.

In GaAs, substitutional Li is expected to locate on a Ga site and to behave as an acceptor. In GaAs samples from n-type Si-doped GaAs compensated with Li, LVMs observed at 449.64 and 482.31cm^{-1} at LNT have been ascribed to isolated $^{7}Li_{Ga}$ and $^{6}Li_{Ga}$, respectively, and the FWHM of the $^{7}Li_{Ga}$ LVM is 0.45 cm^{-1} [28]. It has been pointed out in [29] that in GaAs, Li, which can also locate on an interstitial site and give Li_i, is a self-compensating impurity (Li_{Ga} is an acceptor and Li_i a donor) whose diffusion in GaAs can result in the passivation of EL2 and, simultaneously, of the deep Cu^{2-} acceptor [30]. For Li_i, there are two possible sites, one is T_{iAs} where the nns are As atoms and the other one T_{iGa} where they are Ga atoms (see Fig. 1.5), and the T_{iAs} site seems to be the most likely. In Li-diffused CdTe, a LVM is ascribed to isolated Li_{Cd} [31], and four LVMs to a (Li_i, Li_{Cd}) close pair with an assumed C_{3v} symmetry [31,32]. The frequencies of these LVMs, observed in ^{6}Li- or ^{7}Li-diffused CdTe samples, are given in Table 5.1, and the frequencies for the (Li_i, Li_{Cd}) pairs are the same for the two references.

In CdTe with large Li concentrations, a vibrational band at 295 (272) cm^{-1} for ^{6}Li (^{7}Li) has been reported and it has been suggested that it could be due to the formation of a second phase, possibly $LiTe_2$ [32]. Li-related LVMs have also been reported in Li-doped ZnSe, and some of them are certainly related to (Li_i, Li_{Zn}) pairs, but their attributions are not obvious ([32], and references therein).

LVMs of Li on a cation site (Li_{cat}) paired with a donor on another nnn cation site (D_{cat}) have been reported in Si-doped GaAs [28], Al- and Ga-doped ZnSe, and in Al-doped CdTe [33]. Such complexes have C_{1h} symmetry, and up to six LVMs are observed for each pair. The LNT frequencies of the LVMs due to the motion of the Li atoms of these pairs are given in Table 5.2, and the importance of the Li IS shows that the coupling between the two atoms of the pair is small. The frequencies of the LVMs of the (Li_{cat}, D_{cat}) pairs which are due to the vibration of the donor atoms do not show Li IS. They are also given in Table 5.2 without indication of the Li isotope.

For comparable masses of the FAs and of the crystal matrices, it is seen that the frequencies of the LVMs are larger in the more covalent GaAs crystal than in ZnSe. In CdTe, the frequencies of the Li LVMs depend weakly on the mass of the donor atom. In GaAs, the LVMs of several Li-related centres involving more than one Li atom have been reported by Levy and Spitzer [34]. In ZnSe, the LVMs of complexes

Table 5.1 Frequencies (cm^{-1}) at LNT of the LVMs of isolated Li_{Cd} and (Li_i, Li_{Cd}) pairs in CdTe. The measured $^{7}Li/^{6}Li$ isotopic frequency ratios in brackets are close to $\sqrt{6/7} = 0.926$ from expression (5.9)

$^{6}Li_{Cd}$[a]	$^{7}Li_{Cd}$[a]	($^{6}Li_i$, $^{6}Li_{Cd}$)[a,b]	($^{7}Li_i$, $^{7}Li_{Cd}$)[a,b]
340	317 (0.932)	271	252 (0.930)
		282	262 (0.929)
		326.5	304 (0.931)
		370	344 (0.930)

[a][31]
[b][32]

Table 5.2 Frequencies (cm^{-1}) at LNT of (Li acceptor, donor nnn) pairs in GaAs, ZnSe, and CdTe. The frequencies of the LVMs of the (Li_{Zn}, Ga_{Zn}) pairs in ZnTe are very close to those of the (Li_{Zn}, Al_{Zn}) pairs and they are not given. The LVM of isolated Li is given for comparison in the last column

GaAs[a]	$(^6Li_{Ga},{}^{28}Si_{Ga})$		470	480	487	$^6Li_{Ga}$	482.31
	$(^7Li_{Ga},{}^{28}Si_{Ga})$		437.75	447.38	454.18	$^7Li_{Ga}$	449.64
	$(Li_{Ga},{}^{28}Si_{Ga})$		374	379	405		
ZnSe[b]	$(^6Li_{Zn},Al_{Zn})$		389	410.5	420.5		
	$(^7Li_{Zn},Al_{Zn})$		359	382	392		
CdTe	$(^6Li_{Cd},Al_{Cd})^{b}$	311	328	349.5	355	$^6Li_{Cd}$	340
	$(^7Li_{Cd},Al_{Cd})^{b}$	299	318	325	332	$^7Li_{Cd}$	317
	$(Li_{Cd},Al_{Cd})^{b}$			292.5	299		
	$(^6Li_{Cd},{}^{28}Si_{Cd})^{c}$	316	331	348	356		
	$(^7Li_{Cd},{}^{28}Si_{Cd})^{c}$	300.5	317	325	332		
	$(Li_{Cd},{}^{28}Si_{Cd})^{c}$			292	299		
	$(^6Li_{Cd},Sn_{Cd})^{c}$		333	350	359		
	$(^7Li_{Cd},Sn_{Cd})^{c}$		312	326	333		

[a][28]
[b][33]
[c][31]

involving Li have also been reported by Ko and Spitzer [32] and in w-CdS in the 350–550 cm^{-1} range by Król and Gołkowska [35].

5.3.1.2 Group-II Atoms

No LVM related to isolated Be has apparently been reported in diamond, silicon, or germanium, but a vibronic band has been related to a (Be_s, Be_i) pair in silicon, corresponding to a frequency of 844 cm^{-1} [36], in reasonable agreement with an ab initio value of 860 cm^{-1} calculated by Tarnow et al. [37].

The LVM of the monoisotopic $^9Be_{III}$ acceptor has been observed in $AlAs$ [38], GaP [16], GaAs [15], InAs and InSb [39]. In $AlAs$, a frequency decrease is observed between the LVM of the neutral charge state $(Be_{Al})^0$ at 481 cm^{-1} and the one of $(Be_{Al})^-$ at 470 cm^{-1} [38]. In GaAs, the ionization energy of the Be acceptor is given as 28 meV and the value should be comparable for $AlAs$, with a relatively large delocalization of the bound hole, and this charge-state effect has not been clearly explained. For GaAs, the frequency of the $(Be_{Ga})^-$ LVM is 482.4 cm^{-1}. In Be-doped samples with a high concentration of Be^0 at low temperature, an asymmetric profile of the $(Be_{Ga})^0$ LVM is observed. It is due to a Fano resonance of this LVM with the photoionization spectrum of Be^0, rising above 27.88 meV or 225 cm^{-1}, the ionization energy of Be^0. This Fano resonance can be suppressed by irradiating the sample with h–e electrons, which compensate the Be acceptor. As a consequence, the profile of the LVM becomes symmetric, but it is then due to $(Be_{Ga})^-$. The two situations are depicted in Fig. 5.1.

Fig. 5.1 Absorption at LHeT of the LVM of the Be_{Ga} acceptor in GaAs in a 2.4 μm epilayer with $p = 3.7 \times 10^{18}$ cm^{-3} (**a**) as grown, (**b**) in a sample irradiated with 2-MeV electrons (after [15]). The asymmetric profile in (**a**) is due to the Fano resonance. With permission from the Institute of Physics

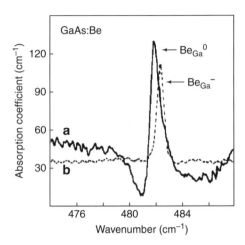

This makes difficult the measurement of a frequency shift between the LVMs in the two charge states, but it seems to be small [15]. The same effect can also be observed *mutatis mutandis* for the Be_{III} LVM in InSb, GaP, and InAs, where the frequencies of the Be LVM are 414, 527, and 434.8 cm^{-1}, respectively. In Mg-doped MOCVD p-type GaN (unless otherwise specified, we refer to the hexagonal form), two modes at 260 and 657 cm^{-1} measured by Raman scattering at RT have been associated with Mg_{Ga} ([40], and references therein). The Mg-doped GaN crystals grown from solution at high pressure are insulating as Mg_{Ga} compensates the residual O_N donor. In these samples, the LHeT absorption of two LVMs at 396 and 782 cm^{-1} has been reported. It is attributed to the two modes of the $^{24}Mg_{Ga}$–O_N pair and the calculated composite spectra including the contribution of ^{25}Mg and ^{26}Mg, with relative isotopic abundances of 0.127 and 0.139 with respect to ^{24}Mg, respectively, reproduce satisfactorily the observed spectra [41]. In the Be-doped spectra of GaN samples obtained by the same method, two LVMs at 1055 and 1063 cm^{-1} are assigned to complexes involving beryllium and oxygen.

The LVM of isoelectronic Be_{II} has also been observed in several II–VI compounds. The absorption of this LVM and of its overtones has been reported in zinc chalcogenides [42] and in cadmium chalcogenides [43,44], and all the fundamental LVMs occur in the 400–500 cm^{-1} spectral region. In CdTe, a resonant mode of Be_{Te} is observed at 61 cm^{-1} [45]. The more recent report of the LVM and first overtone of isoelectronic Mg_{Cd} and Ca_{Cd} in CdTe, and CdSe and of Mg_{Zn} in ZnTe and ZnSe shows that the fundamental frequencies of the LVMs of these group-II atoms are located between about 200 and 300 cm^{-1}, and their ISs are clearly observed[2] [46]. Figure 5.2 shows a low-resolution absorption spectrum of Mg_{Cd} and Ca_{Cd} in CdTe at LHeT.

---

[2]In [46], the atoms of the II–VI compounds are implicitly considered in the ionic form (e.g. $Cd^{2+}Te^{2-}$) and the isoelectronic atoms are atoms are similarly considered (e.g. Ca^{2+} or S^{2-}).

Fig. 5.2 LVM absorption in CdTe of (**a**) Mg_{Cd} and (**b**) Ca_{Cd} (10^{20} cm^{-3}) showing both isotope effect (the relative abundances of the ^{40}Ca, ^{42}Ca, and ^{44}C isotopes are 96.941, 0.657, and 2.086%, respectively, and the ^{40}Ca component is truncated) for a resolution of 0.5 cm^{-1}. The CdTe two-phonon background has been subtracted (after [46]). Copyright 1995 by the American Physical Society

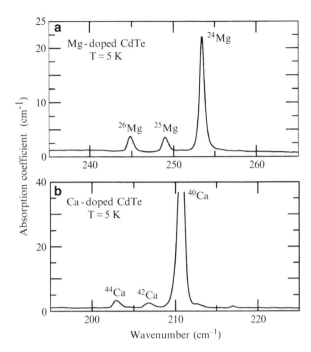

Table 5.3 Frequencies (cm^{-1}) at low-resolution of fundamental (LVM) and overtones associated with substitutional magnesium in some II–VI compounds at LHeT [46]. Expression (5.9) gives 0.980 and 0.961 for the ^{25}Mg/^{24}Mg and ^{26}Mg/^{24}Mg frequency ratios, and the measured ratios are given in brackets below the frequency of the heaviest isotope

	CdTe (c)		ZnTe (c)		ZnSe (w)					
	ν_s	$2\nu_s$	ν_s	$2\nu_s$	$\nu_{//s}$	$2\nu_{1//s}$	$2\nu_{2//s}$	$\nu_{\perp s}$	$2\nu_{1\perp s}$	$2\nu_{2\perp s}$
^{24}Mg	253.3	505.3	272.3	542.9	290.9	557.7	566.8	285.5	566.2	570.8
^{25}Mg	248.9	496.5	267.6	533.6	276.2	549.0	557.7	280.7	557.3	561.6
	(0.983)		(0.983)							
^{26}Mg	244.7	488.2	263.3	524.9	271.9	540.8	549.0	276.4	548.6	552.9
	(0.966)		(0.967)							

The positions of the Mg-related absorptions in some II–VI compounds with sphalerite and wurtzite structures are given in Table 5.3. In ZnSe, with wurtzite structure, two gap modes of Mg_{Zn} have also been reported for $E//c$ and $E\perp c$ at 143.4 and 144.6 cm^{-1}, respectively [46].

High-resolution spectra have revealed the existence of an interaction between the group-II FA and the *nn* Te or Se lattice atoms, which can be observed because the FWHMs of the LVMs are small. The result for ^{40}Ca$_{Te}$ in CdTe is shown in Fig. 5.3. This spectrum is the result of the superposition of the vibrational modes of a ^{40}CaTe$_4$ pseudomolecule taking into account the distribution of the different Te isotopes (see Appendix D). The spectroscopic consequences of this interaction are explained

Fig. 5.3 High-resolution ($0.02\,\mathrm{cm}^{-1}$) absorption spectrum at LHeT of the ^{40}Ca component of the Ca$_{Cd}$ spectrum in CdTe showing the Te isotope splitting, compared with the calculated spectrum for a ^{40}Ca$_{Cd}$Te$_4$ pseudomolecule (after [47])

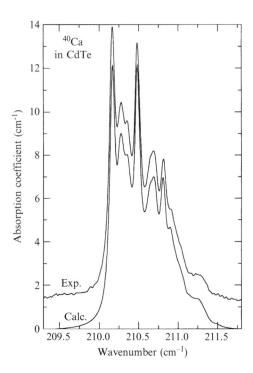

for the simpler case of C$_{As}$ in GaAs in Sect. 5.3.2.2, where the pseudomolecule is ^{12}CGa$_4$. In CdTe, because of the existence of eight Te isotopes instead of two for Ga, the number of combinations is much higher, but a spectrum can still be calculated with the help of a few assumptions.

5.3.1.3 TM Atoms

LVMs associated to $3d$ TMs on cation site in CdTe, ZnTe, and wCdSe have been measured at LHeT by [47, 48]. The high-resolution spectrum of Ti (Ti^{2+}) in CdTe bears a marked resemblance with that of Ca. However, despite a larger atomic mass, the frequencies of the LVM for the different Ti isotopes measured at low resolution (228.5, 226.7, 224.8, 223.0, and 221.2 cm^{-1} for ^{46}Ti, ^{47}Ti, ^{48}Ti, ^{49}Ti, and ^{50}Ti, respectively) are larger than those for the different Ca isotopes (210.6, 206.8, and 202.8 cm^{-1} for ^{40}Ca, ^{42}Ca, and ^{44}Ca, respectively), showing that the presence of $3d$ electrons produces a change in the force constants. Moreover, the frequencies of the LVMs of the TMs display a nonmonotonic trend across the $3d$ series with a minimum for Mn: for ^{51}V, ^{54}Cr, ^{55}Mn, ^{56}Fe, and ^{59}Co in CdTe, the LVM frequencies are 210.2, 201.6, 191.2, 196.1, and 199.0 cm^{-1}, respectively. This softening of the Mn–Te bond is still more marked in ZnTe: for a mass difference of $\sim + 5\,\mathrm{u}$ between Zn and Co, the frequency of the Co LVM in ZnTe is 217.5 cm^{-1} while for a mass difference between Zn and Mn of $+ 9\,\mathrm{u}$, the Mn impurity mode in ZnTe is located at

Table 5.4 Frequencies (cm^{-1}) at LHeT of the MLV and gap modes of Mn, Fe, and Co in wCdSe. The atomic mass considered for the TM atom is given in brackets [48]

TM	LVM		Gap mode	
	$E//c$	$E\perp c$	$E//c$	$E\perp c$
Mn (55 u)	220.0	222.0	137.6	138.7
Fe (56 u)	222.0	224.8	138.2	138.5
Co (59 u)	226.4	229.3	138.5	139.4

$210.3\ cm^{-1}$ and quasi-resonant with the LO phonon modes of this crystal $(210\ cm^{-1}$ at LNT [49]).

Comparable results have been obtained in wCdSe for Mn, Fe, and Co [48] and the relevant frequencies for the LVM and the gap mode are given in Table 5.4.

One can note the small differences in the LVMs of these three TMs in wCdSe, CdTe, and ZnSe.

5.3.2   C

Because the vibrational properties of substitutional C in semiconductors, and especially in GaAs, have been thoroughly studied, and as the results have been used to explain the vibrational properties of other FAs in III–V compounds, we present the results on carbon before those on boron. The results on substitutional Si are presented in Sect. 5.3.5, after those on boron.

5.3.2.1   Silicon and Germanium

Carbon is a residual impurity in many semiconductors, either in bulk material or in epitaxial layers grown by dissociation of metal-organic precursors. In silicon, C_s gives at RT a LVM at $605\ cm^{-1}$ with a FWHM $\sim 6\ cm^{-1}$, whose absorption is partially masked by the rather strong $TO + TA$ two-phonon absorption of the silicon lattice [50]. Taking the ratio of the spectrum with that of a "carbon-free" sample (actually a sample with $[C_s] \sim 10^{14}\ cm^{-3}$) allows to measure the C_s absorption in acceptable conditions, but the use of thick samples $(\sim 1\ cm)$ is forbidden because of the intrinsic absorption. In 400-μm thick industrial wafers, C_s absorption free from interference fringes at a resolution of $2.5\ cm^{-1}$ has been reported [51] by using a Brewster's incidence geometry.

The comparison of the frequencies of the ^{12}C, ^{13}C, and radioactive ^{14}C isotopes has been made a long time ago in ^{nat}Si and the anharmonicity of the C LVM has allowed to observe a weak overtone of this LVM [52]. More recently, the frequency of the fundamental LVM of ^{12}C has also been measured at RT and near LHeT in qmi ^{28}Si, ^{29}Si, and ^{30}Si [53]. The results are summarized in Table 5.5.

In ^{nat}Si, the FWHMs of the ^{12}C and ^{13}C modes are both $\sim 2.0\ cm^{-1}$ at LHeT, but in qmi ^{30}Si, the FWHM increases to $2.7\ cm^{-1}$ [53, 54]. This latter effect can be

Table 5.5 Frequencies (cm^{-1}) of the LVM of C_s for different isotopic compositions of the silicon host crystal at RT and at or near LHeT. The values at LNT are very close to those at LHeT. When measured, the frequencies of the first overtone is given in brackets. The frequencies for ^{13}C and ^{14}C deduced from the IS calculated in the harmonic approximation (HA) using (5.10) are also given on the same line

		^{nat}Si[a]	$qmi^{28}Si$[b]	$qmi^{29}Si$[b]	$qmi^{30}Si$[b]
^{12}C	RT	605.0[b] (1205.8)	605	603.1	600.2
	LHeT	607.5[b] (1211.4)	607.7	605.6	603.8
^{13}C	RT	586.3 (1169.0)			
	LNT	589.1 (1174.5)	$\omega(^{13}C)$ in ^{nat}Si in HA : 584.1		
^{14}C	RT	570.3			
	LNT	572.8	$\omega(^{14}C)$ in ^{nat}Si in HA : 564.1		

[a][52] at RT and LNT
[b][53]

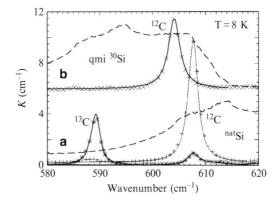

Fig. 5.4 Absorption coefficient at LHeT of the LVMs of C_s: (**a**) in ^{nat}Si samples doped with ^{12}C (+), showing the 607 cm^{-1} LVM, and with ^{13}C (*), showing the 589 cm^{-1} peak and the residual absorption of ^{12}C, after subtraction of the silicon two-phonon absorption of a $FZ^{nat}Si$ sample (dashed spectrum). The lines through the points are fits with Lorentzian line shapes, (**b**) the 604 cm^{-1} LVM of ^{12}C (×) in a $qmi^{30}Si$ sample after subtraction of the two-phonon absorption of a FZ $qmi^{30}Si$ sample (dashed spectrum). The line through the points is a fit with a Lorentzian line shape. The resolution of the C spectra is 0.5 cm^{-1}. For clarity, only representative experimental points are shown and the $qmi^{30}Si$ data have been displaced upward by 6 cm^{-1} (after [54]). Copyright 2005 by the American Physical Society

explained by the increase of the 2-phonon DoS at the frequency of the ^{12}C mode in $qmi^{30}Si$ compared to ^{nat}Si, which decreases the decay time of the C mode [55], and this situation is depicted in Fig. 5.4.

A good agreement between the calculated and experimental frequencies of C_s in silicon has been obtained from models taking into account the lattice distortion induced in the silicon lattice by the small relative size of the C atom [56].

In $Si_{1-x}Ge_x$ alloys, the LVM of C_s LVM is observed to shift to lower frequencies and to broaden with increasing Ge contents. For x above $\sim$0.015, the C_s absorption consists in two unresolved overlapping lines corresponding to different atomic combinations in the vicinity of the C_sSi_4 entity ([57, 58]).

Equilibrium solubility of carbon in crystalline silicon is given as 3.5×10^{17} cm^{-3} near the melting point [59] and it is temperature-dependent, but concentration up to 2×10^{18} cm^{-3} can be found in CZ samples supersaturated with C [60]. There has been several calibrations of the peak absorption coefficient of the 605 cm^{-1} LVM at RT as a function of the concentration of isolated C_s. Based on charged particles activation analysis (CPAA) for the detection of carbon, a calibration factor of 1.00×10^{17} at cm^{-2} is given by [61], which can be extrapolated to $\sim 3.3 \times 10^{16}$ cm^{-2} at LHeT. The calibration factor recommended in [62] is 1.22×10^{17} cm^{-2} or 2.44 ppma/cm^{-1}. These RT calibration factors are valid for a spectral resolution of 2 cm^{-1} or smaller, which should give an observed FWHM of 6 cm^{-1}. They provide a detection limit of $\sim 5 \times 10^{15}$ cm^{-3} or 100 ppba, which can be lowered to ~ 50 ppba with measurements at low temperature, but the method requires a nominally C-free reference sample and a spectral resolution of ~ 0.6 cm^{-1}. This can be avoided in a method described by Alt et al., which is based on the determination of the second derivative of the absorption, which is calculated by numerical interpolation [63]. At LNT, the detection limit with this method is claimed to be better than 10 ppba [63].

In electron-irradiated FZ silicon with a high C content, an ESR spectrum, and LVMs have been attributed to a $(C_s–C_s)$ pair. The properties of this pair are discussed in Sect. 7.2.3, where the vibrational properties of C-related irradiation defects in silicon are presented.

The annealing of silicon supersaturated with C produces the precipitation of SiC, giving at LNT an absorption band peaking at 825 cm^{-1}, within the frequency range of the one-phonon spectrum of SiC [60].

Besides the isolated form, the C atom is part of several complexes in silicon and their vibrational properties are discussed in Sect. 7.2.

Carbon solubility in germanium is very low in the bulk. A value of about 10^8 cm^{-3} in the liquid at the melting point is given by Scace and Slack [64], near 10^{10} times less than in silicon. Concentrations of C-related acceptor complexes measured in quenched Ge samples give a lower limit of the solubility of $\sim 10^{12}$ cm^{-3} [65]. In C-implanted germanium, a LVM measured at 531 (512) cm^{-1} at LHeT has been identified with the T_2 mode of substitutional $^{12}C_s$ ($^{13}C_s$), by comparison with ion channelling measurements. Ab initio calculations predict a frequency of 516 cm^{-1} for this centre well above the Raman frequency of germanium [57]. A Raman scattering line at 530 cm^{-1} has also been observed at RT in a 50-nm-thick Ge$_{0.98}$ C$_{0.02}$ film and ascribed to the LVM of C_s in this material [66].

5.3.2.2 III–V Compounds

In III–V compounds, substitutional C locates on a group-V site where it acts as an acceptor. The results for GaAs are presented first because it is the compound where the properties of C have been the most actively investigated.

A RT, the frequency of the LVM of $^{12}C_{As}$ in GaAs is observed at 579.8 cm^{-1}, and it is accompanied by weaker thermally activated sidebands at 566.2 and 576.6 cm^{-1}

([67], and references therein). The main band at 579.6 cm^{-1} has a FWHM of 1.1 cm^{-1}, whereas the side bands are somewhat broader [67]. At 420 K ($\sim$150°C), the main band shifts to 578 cm^{-1}, with a FWHM $\sim$1.9 cm^{-1}, as deduced from Fig. 1 of [68].

The observation of the $^{12}C_{As}$ LVM was first reported from measurements at LNT, giving a low-resolution position of 582.4 cm^{-1} and a $^{12}C - ^{13}C$ IS of 21.2 cm^{-1} [69]. The same IS has also been reported from low-resolution measurements at RT and LHeT [70]. A clear spectroscopic proof of the site location of C associated with this LVM line was provided by the fine structure observed in high-resolution measurements (0.06 cm^{-1}) at LNT by Theis et al. [71]. This fine structure could be understood by assuming first that the C atom is surrounded by four Ga atoms whose masses reflect the isotopic distribution of this element (^{69}Ga: 0.6, ^{71}Ga: 0.4) and by further assuming that these atoms interact with the C atom, forming the equivalent of a $C_{As}Ga_4$ pseudomolecule. This produces a host-isotope shift which is shown in Fig. 5.5 in the LHeT spectrum of C_{As} in a sample cut from the SI part of a LEC GaAs crystal. In this sample, the Fermi level is in the middle of the band gap and the C acceptor ($E_i = 26.6$ meV) is ionized to give C_{As}^-.

This spectrum is interpreted by assuming that the Ga isotopic combinations can form five different environments for the C_{As} atom. In a pure tetrahedral (T_d) environment ($C^{69}Ga_4$ or $C^{71}Ga_4$) the LVM is triply degenerate and noted $\Gamma_5(3)$.

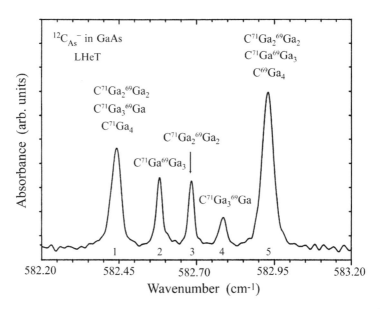

Fig. 5.5 Fine structure at LHeT of the C_{As}^- LVM in SI GaAs at a resolution of 0.013 cm^{-1}. The Ga isotopic contributions are indicated and there are three unresolved transitions in components 1 and 5 (after [52]). The spacing between components 1 and 5 is 0.49 cm^{-1}

Table 5.6 Comparison between the calculated and measured positions (cm^{-1}) of the different $^{12}\text{C}^-\text{Ga}_4$ isotopic LVMs in GaAs (the degeneracy is indicated in brackets in the Symmetry column). The components of Fig. 5.5 are indicated in column C. The frequencies in brackets have been used for the calculation, and the origin is line 3 [72]

Isotopic config.	Symmetry	Position (calc.)	C	Position (meas.)	RI (calc.)	Relative absorption	FWHM (cm^{-1})
C^{71}Ga_4	$\Gamma_5(T_\text{d})$ (3)	582.424					
$\text{C}^{71}\text{Ga}_3^{69}\text{Ga}$	$\Gamma_3(C_{3\text{v}})$ (2)	582.438	1	582.442	0.49	0.41	0.039
$\text{C}^{71}\text{Ga}_2^{69}\text{Ga}_2$	$\Gamma_2(C_{2\text{v}})$ (1)	582.452					
$\text{C}^{71}\text{Ga}^{69}\text{Ga}_3$	$\Gamma_1(C_{3\text{v}})$ (1)	(582.581_5)	2	582.581	0.24	0.26	0.023
$\text{C}^{71}\text{Ga}_2^{69}\text{Ga}_2$	$\Gamma_1(C_{2\text{v}})$ (1)	-	3	582.684	0.24	0.30	0.023
$\text{C}^{71}\text{Ga}_3^{69}\text{Ga}$	$\Gamma_1(C_{3\text{v}})$ (1)	(582.786_5)	4	582.786	0.10	0.11	0.031
$\text{C}^{71}\text{Ga}_2^{69}\text{Ga}_2$	$\Gamma_4(C_{2\text{v}})$ (1)	582.916					
$\text{C}^{71}\text{Ga}^{69}\text{Ga}_3$	$\Gamma_3(C_{3\text{v}})$ (2)	(582.930)	5	582.930	1	1	0.043
C^{69}Ga_4	$\Gamma_5(T_\text{d})$ (3)	582.944					

This degeneracy is lifted[3] by the isotopic distribution, totally for $\text{C}^{69}\text{Ga}_2^{71}\text{Ga}_2(C_{2\text{v}}$ symmetry) giving $\Gamma_1(1)$, $\Gamma_2(1)$, and $\Gamma_4(1)$, and only partially for $\text{C}^{69}\text{Ga}_3^{71}\text{Ga}$ and $\text{C}^{69}\text{Ga}^{71}\text{Ga}_3(C_{3\text{v}}$ symmetry) giving $\Gamma_1(1)$ and $\Gamma_3(2)$ [73]. At temperatures above 200 K, the Ga isotope fine structure is no longer observed because of the line broadening, but in crystals with at high $[\text{C}_{\text{As}}]$, the already mentioned small shoulder located at 576.6 cm^{-1} at RT is observed on the low-energy side of the C_{As} LVM.

The positions and the relative intensities (RI) of the different isotopic transitions of CGa_4 are calculated self-consistently following the method developed in [73]. These results are presented in Table 5.6. The calculated relative intensities for components 1 and 5 correspond to the sums for the three components. LHeT measurements at a resolution of 0.005 cm^{-1} have been reported [74] showing comparable FWHMs and a consistent shift of the frequencies by $\sim +0.008$ cm^{-1} with respect to those of Table 5.6. The calculated relative intensities for components 1 and 5 correspond to the sums for the three components probably due to a small difference of the alignment of the lasers of the two FTS machines.

As the spectrum of Fig 5.5 is due to C_{As}^-, one can wonder if a charge-state effect as the one observed for Be_{Al} in AlAs is also observed for C_{As}^0. In a GaAs sample with $[\text{C}_{\text{As}}^-]/[\text{C}_{\text{As}}^0] \sim 0.25$, the C_{As} LVM was observed at about the same position, but the contrast of the isotopic structure was strongly reduced and a low-energy tail of the structure was apparent. The contrast was restored and the low-energy tail disappeared in the same sample after irradiation with a 2-MeV-electron fluence of 2×10^{16} cm^{-2} which ionized the acceptors, and the effects prior to irradiation were attributed to the interaction of the bound hole with the LVM [75]. Another

---

[3]The components are noted by the IRs of the appropriate point group in Koster's notations and the degeneracy is given in brackets.

test experiment was performed using a different method [76]. Starting from SI, GaAs containing only ionized C_{As}^- and the deep EL2 donor, EL2 was turned into its electrically inactive metastable state by illumination with visible-near IR radiation. After this procedure, the acceptor spectrum of C_{As}^0 was observed in the far IR, but the sharp structure of the C_{As} LVM remained unaltered and no other structure which could have been attributed to C_{As}^0 was observed. It was thus concluded that the charge state of C_{As} had no effect on the vibrational structure and frequency of the LVM. The interpretation of these results has been challenged on the ground that the fraction of acceptors neutralized through hole capture from EL2 was too small for the result to be significant [77].

Evidence for the anharmonicity of the C_{As} LVM has been provided by the observation of an overtone at 1157 cm^{-1} at RT (1164 cm^{-1} at LHeT) [70], with an intensity ~ 0.012 with respect to the fundamental. The $^{12}C - ^{13}C$ IS of this overtone is 43 cm^{-1}, close to two times the IS of the fundamental. At higher resolution and at LHeT, this overtone shows also a fine structure due to the Ga host-isotope shift, which is displayed in Fig. 5.6.

In the description of an anharmonic oscillator with T_d symmetry [6], the C_{As} vibration observed at 579.8 cm^{-1} at RT is ascribed to a $v = 0$ to $v = 1(\Gamma_1 \rightarrow \Gamma_5)$ transition. Within the same description, the overtone at 1156.2 cm^{-1}, shown in the inset of Fig. 5.6, is ascribed to a $v = 0$ to $v = 2(\Gamma_1 \rightarrow \Gamma_5)$ transition. Similarly, the components observed at 566.2 and 576.5 cm^{-1} (~ 70.2 and 71.5 meV) above 240 K are attributed to a $v = 1$ to $v = 2(\Gamma_5 \rightarrow \Gamma_5)$ "hot" transitions of the carbon local oscillator [67].

Several calibrations of the integrated LVM absorption of C_{As} in GaAs have been published and discussed ([67] and references therein). The conclusion is that when considering the product or the peak absorption of the 579.8 cm^{-1} peak at RT by its

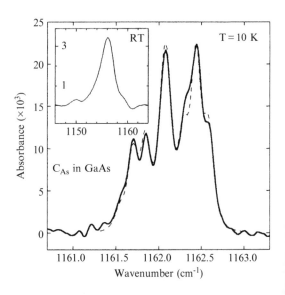

Fig. 5.6 Fine structure at LHeT of the overtone of the LVM of $^{12}C_{As}$ in GaAs (average absorbance ~ 0.01). The inset shows the overtone at RT. The dashed curve is the fit with Leigh and Newman's model (after [68]). Reproduced with permission from Trans Tech Publications

FWHM, a value of $1 \, cm^{-2}$ corresponds to $[C_{As}] \sim 1 \times 10^{16} \, cm^{-3}$. When measuring the true RT-integrated absorption including the $576.6 \, cm^{-1}$ component, the calibration factor becomes $7.4 \times 10^{15} \, cm^{-1}$, and it becomes $7.2 \times 10^{15} \, cm^{-1}$ when measuring the integrated absorption at LNT near $583 \, cm^{-1}$ [67].

The LVM of isolated C has also been observed in GaP [78], AlAs [79], AlSb [80], InP ([81], and references therein), and InAs [82].

In AlAs, absorption measurements at LHeT in samples heavily doped with carbon reveal a LVM at $630 \, cm^{-1}$, attributed to C_{As}, which appears as a Fano profile on the photoionization spectrum of the C_{As} acceptor. Satellites of the $630 \, cm^{-1}$ LVM have also been reported at 615, 633, and $645 \, cm^{-1}$ and tentative assignments have been proposed [83].

The $^{12}C_P$ LVM in InP, located at $546.9 \, cm^{-1}$ at LHeT, lies in a phonon-absorption-free spectral region of InP, and it could be expected to have a small FWHM, comparable to the one of $^{10}B_{In}(0.07 \, cm^{-1})$. However, this C_P LVM is always significantly broadened, with an unusual profile made of a sharp central feature and of two unresolved broader wings, with an overall FWHM $\sim 1 \, cm^{-1}$, displayed in Fig. 5.7.

This profile has been attributed to the perturbation of the ionized C_P^- acceptors by internal random electric fields, produced by the ionized acceptors and compensating donors responsible for the SI character of the samples. This results in an unresolved splitting of the triply degenerate mode into a pseudo-triplet [84]. The high-temperature annealings produce a change in the relative C concentrations on C_P sites and in the dicarbon centres discussed below (at the highest annealing temperature, $[C_P]$ decreases). The net result is a decrease of the random electric fields resulting in a decrease of the contribution of the wings and a narrowing of the central component [81]. A gap mode at $220 \, cm^{-1}$ attributed to C_{In} (donor) has been reported in InP [85].

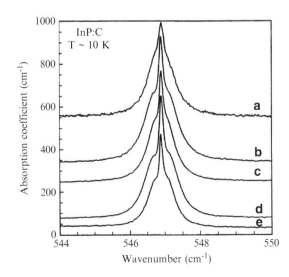

Fig. 5.7 Absorption profile at a resolution of $0.02 \, cm^{-1}$ of the $^{12}C_P$ LVM in a SI InP sample as a function of rapid thermal annealing treatments (5 min) at progressively increasing temperatures. The letters **a, b, c, d, e** refer to the as-grown sample and to annealings at 500, 600, 700, and 800°C, respectively. Total [C] is $4.2 \times 10^{18} \, cm^{-3}$ [81]. Copyright 2001 by the American Physical Society

Table 5.7 Comparison of the low-temperature frequencies (cm^{-1}) measured for the isolated C (C_V) and dicarbon (C–C) LVMs in different III–V compounds. The LVM of ^{12}C in GaP has been reported at $605.7\,cm^{-1}$ at LNT [78]. The relevant references are indicated in the lines noted Ref. C_V and Ref. C–C. The frequencies for ^{13}C deduced from the IS calculated in the harmonic approximation (HA) using (5.10) are given in the fifth line

	AlAs	AlSb	GaAs	InP	InAs
$^{12}C_V$	631.5	591.6	$582.8^b (544)^a$	546.9	530
	$(594)^a$				
$^{13}C_V$	608.5	572.9	561.2^b	526.8	511
	$(575)^a$				
Ref. C_V	[87]	[80]	[88]	[89]	[82]
^{13}C (HA)	607.2	568.8	560.4	525.9	510
$^{12}C-^{12}C$	1752 (T1)		1743 (T1)	1814	1832
	1856 (T2)		1856 (T2)		
$^{12}C-^{13}C$			1708 (T1)		1794
			1824 (T2)		
$^{13}C-^{13}C$			1674 (T1)		1758
			1788 (T2)		
Ref. C−C	[90]		[91]	[81]	[94]

^{a}Calculated (see text)
^{b}Low-resolution values

Ab initio calculations on the C_{As} complexes in GaAs have been published by [86] and in AlAs [83]. They give frequencies of 544 and $594\,cm^{-1}$ for the triply degenerate $^{12}C_{As}$ mode in GaAs and AlAs, respectively, somewhat lower than the experimental values of Table 5.7. A cluster calculation based on the Keating model has also been reported, giving interatomic force constants necessary to reproduce the experimental values [92].

In addition to the absorption of the LVM of the isolated C acceptor and eventually of its overtone, LVMs in the $1800\,cm^{-1}$ spectral region have been observed by Raman scattering in several III–V compounds highly doped with C. In InP and InAs, the observation can be made in as-grown material, but in GaAs and AlAs, it requires annealing at high temperature, and it is correlated with a decrease of the initial free-hole concentration. From the comparison of the results on GaAs samples grown by chemical beam epitaxy and doped with ^{12}C, ^{13}C, and $^{12}C+^{13}C$, these LVMs were attributed to C pairs (dicarbon pairs) formed by migration of C from As sites [91]. The relatively high frequencies of these LVMs, compared to those of isolated C, and their absence in the absorption spectra indicate a direct bonding of two C atoms in a symmetric configuration producing no dipole moment. In the original paper, the low- and high-frequency modes were noted T1 and T2 (for triplet), respectively, and the frequency $(1742\,cm^{-1})$ of the T1 mode for ^{12}C pairs is comparable to the characteristic frequency $(\sim 1650\,cm^{-1})$ of the sp^2 C $=$ C bond in molecules [20]. The LVMs of these dicarbon pairs are displayed in Fig. 5.8, showing two modes for GaAs samples containing either ^{12}C or ^{13}C.

Measurements performed with different laser energies for the Raman excitation showed that T1 and T2 corresponded to different configurations and charge states

Fig. 5.8 Raman scattering spectra at a resolution of $2\,cm^{-1}$ of the T1 and T2 LVMs of dicarbon pairs in GaAs samples annealed at 825°C for 3.5 h with **(a)** $[^{12}C_{As}] = 5 \times 10^{20}\,cm^{-3}$, **(b)** $[^{13}C_{As}] = 5 \times 10^{20}\,cm^{-3}$, **(c)** nearly equal concentrations of $^{12}C_{As}$ and $^{13}C_{As}$ ($1 \times 10^{20}\,cm^{-3}$) before annealing. For **(c)**, the new lines, noted by an asterisk are due to the (^{12}C, ^{13}C) combination [91]. Copyright 1997 by the American Physical Society

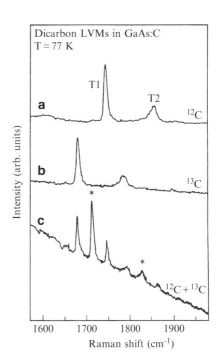

[91]. Similar modes have also been observed in $AlAs$ [90]. A discussion of the atomic structure of the two dicarbon pairs based on the results of *ab initio* calculations favours positively charged split-interstitial symmetric C_2 pairs on an As site, with the C–C axis oriented along [110] or [111] directions, but the uncertainties in the calculations are too large to allow a more precise assignment to T1 or T2 [90]. In InP, two LVMs associated with dicarbon centres are observed, but, at the inverse of $AlAs$ and GaAs, the low-energy one is less intense than the high-energy one [81]. The intensities of these LVMs grow under annealing, showing that dicarbon pairs are formed during the annealing and a third LVM appears at a lower frequency. *Ab initio* calculations predict that this dicarbon centre is a deep donor which is stable only in the positive charge state [83, 93].

The formation of dicarbon centres has also been reported in C-doped InAs ([94], and references therein). A summary of the low-temperature frequencies of the C-related lines in III–V compounds is given in Table 5.7.

In carbon-doped $Al_{0.02}Ga_{0.98}$ As epilayers, beside the C_{As} LVM at $583\,cm^{-1}$, LVMs were observed at 574 and $632\,cm^{-1}$ at LHeT, with relative strength of 2/1. They were attributed to the vibration of C_{As} with one Al and three Ga atoms *nn*, in a structure with C_{3v} symmetry [95]. In heavily carbon-doped epilayers with larger Al concentrations (up to $Al_{0.23}Ga_{0.77}$ As), LVMs at about 566, 574, 583, 608, 632, 638, and $647\,\mathrm{cm}^{-1}$ were observed at LHeT[4] [96]. When adjusted to reproduce the two experimental frequencies of $C_{As}Ga_3Al$, cluster calculations predict the existence of two low-frequency modes at 308.6 and $372.2\,\mathrm{cm}^{-1}$ [92]. The first one is in the one-phonon absorption region of the alloy and cannot be observed while the second one should be in the wing of the strong Al_{Ga} LVM at $362\,\mathrm{cm}^{-1}$ and difficult to observe.

The dependence of the LVMs of C_{As} in GaAs and of C_P in InP on an hydrostatic pressure have been measured up to $\sim$6 GPa and they have been found to be linear in this pressure domain [97, 98]. Under a hydrostatic pressure, the atomic distances between the atoms of a crystal decrease and if the T_d structure of the FA is maintained, an increase of the frequency with pressure is expected, as observed. The pressure coefficients are similar for the two compounds and the pressure-dependent frequencies at LHeT are given by a least-square fit:

$$\tilde{\nu}(\mathrm{GaAs},\ ^{12}C_{As}) = 582.7 + 9.3P,$$

$$\tilde{\nu}(\mathrm{GaAs},\ ^{13}C_{As}) = 561.5 + 9.1P,$$

$$\tilde{\nu}(\mathrm{InP},\ ^{12}C_P) = 546.9 + 9.5P,$$

where the frequencies and pressures are expressed in cm^{-1} and GPa, respectively. This pressure dependence has been found to be very similar to the one for the 2TO mode of InP [98], which can be expressed as

$$\tilde{\nu}(\mathrm{InP}, 2\mathrm{TO}) = 659.0 + 10.7P.$$

Last, but not least, in GaAs layers produced by MBE techniques using beams of trimethyl gallium and As_4 molecules, the electrical activity of the C_{As} acceptor can be passivated by hydrogen by the formation of a hydrogen complex involving C_{As}. The vibrations of this complex involve low-frequency LVMs in the same spectral region as the one of isolated C_{As}, together with high-frequency ones involving the stretching of a hydrogen bond [95, 99]. A similar complex is also produced in MBE or MOVPE Al As layers [87]. The spectroscopy of these centres is described in Sect. 8.5.1.2.

5.3.3   B and Al

Boron is an acceptor in group-IV semiconductors, and it is a dopant atom in synthetic diamond, silicon, and germanium. In III–V compounds, B_{III} is isoelectronic and electrically inactive, but B_V on an antisite is a double acceptor. In II–VI compounds, B_{II} is a single donor. In LEC-grown III–V crystals, B is a residual impurity due to the partial dissociation of the B_2O_3 encapsulant.

---

[4]The frequencies reported are estimated from Fig. 1 of this reference.

5.3.3.1 Group-IV Crystals

In type-IIb diamond, the B atom is substitutional, and one should expect a LVM to be associated with it since B is lighter than C. Up to now, no experimental result related to the vibration of B_s in C_{diam} seems to be available despite the fact that very large concentration of B_s can be introduced in diamond. The results of *ab initio* calculations [100] show that B_s retains tetrahedral symmetry, but at the expense of a small increase of the length of the $B - C$ bonds compared to that of the C–C bonds. This explains why the calculated spectrum is dominated, for ^{11}B, by a resonant mode at $1211 \, cm^{-1}$, due to the weaker bonding of the B atom. For ^{10}B, the calculated peak shifts to $1288 \, cm^{-1}$, but remains resonant.

The solubility of boron in silicon is rather high (in the $10^{20} \, cm^{-3}$ range). Vibrational absorption studies need to be performed on samples with high [B] values because this spectroscopy is not very sensitive. The strong free-carrier absorption or continuous photoionization absorption at low temperature of such samples requires their compensation by electron irradiation, diffusion of interstitial Li donors, or co-doping with group-V donors. This results in an ionized configuration B_s^- of substitutional boron, whatever the temperature, but also in the production of B-related irradiation defects or in pairing of B with compensating atoms, and these situations have led to the observation of many B-related LVMs, sometimes within the same energy range, so that the interpretation of the vibrational spectroscopy of boron in silicon can be a complicated task. We discuss first the results for the isolated atom and the substitutional B pairs (B_s–B_s). Those for B paired with donors are discussed in Sect. 5.3.4, and those for the B-related irradiation defects in section 7.3.1.

The results of studies of the vibrational absorption of boron in silicon by several groups ([101], and references therein, [102]) showed the existence of a LVM of substitutional B_s^- near 642 and $620 \, cm^{-1}$ at RT for the ^{10}B and ^{11}B isotopes, respectively, with an increase in frequency between 2 and $3 \, cm^{-1}$ at LNT and below.[5] The overtone of this mode has also been reported [101]. This LVM has also been observed at RT using Raman scattering in the reflection geometry [117]. The FWHM of the B^- LVM measured at low temperature is ~ 3.5–$4 \, cm^{-1}$, and it increases to 5.1–$5.5 \, cm^{-1}$ at RT [101].

In silicon samples which had been electron-irradiated, local modes attributed to *nn* substitutional B_s–B_s pairs, also called *P* centres (not to be confused with the "*P*" centre discussed in section 4.1.4.3!), have been reported [16, 104], and the origin of these pairs, which are also produced in neutron-irradiated B-doped samples, has remained unclear for some time. These pairs can survive annealings at 300°C, a limit above which absorption measurements become impossible because of free-carrier absorption. The most recent *ab initio* calculations indicate that these pairs are metastable boron clusters with a deep $(-/0)$ level at $\sim E_v + 0.17 \, eV$ [105].

---

[5]In [102], this mode is noted B(2).

Table 5.8 Frequencies (cm^{-1}) of LVMs associated with isolated B^- and nn B pairs (P centre) in silicon at LNT. The RT frequencies of the $^{10}B^-$ and $^{11}B^-$ LVMs are 643.6 and 620.8 cm^{-1}, respectively. The measured $^{11}B_s/^{10}B_s$ frequency ratio is 0.9644 and the mass ratio $\sqrt{10/11} = 0.9535$. The calculated frequencies [105] are in brackets. The IRs associated with the different LVMs are indicated

	$^{10}B_s^-$	$^{11}B_s^-$		$^{10}B_s-^{10}B_s$	$^{11}B_s-^{11}B_s$		$^{10}B_s-^{11}B_s$
$T_2(T_d)$	645.8[a]	622.8[a]	E_u	570.0[b, c]	552.3[b, c]	A_1	560.0[b, c]
	(650)	(626)	(D_{3d})	(568)	(551)	(C_{3v})	(558)
Overtone	1288.5[c]	1242.9					

[a][102]

[b][104]. In this reference, the LVMs at 552, 560, and 570 cm^{-1} are noted P_1, P_2, and P_3, respectively

[c][16]

The observed and calculated frequencies of the corresponding LVMs are given in Table 5.8.

A sharp resonance at 227 cm^{-1} with no B isotope effect has also been attributed to a quasi-local mode of boron in silicon [106, 107].

In germanium, B_s^- produces a LVM at 547 and 571 cm^{-1} at LNT for ^{11}B and ^{10}B, respectively [108].

A resonant mode due to Al_s is expected at 456 cm^{-1} in silicon, but its presence in electron-irradiated Al-doped silicon with $[Al] \sim 10^{19}$ cm^{-3} is not clear [109].

In an Al-doped germanium sample with a hole concentration of 2.5×10^{20} cm^{-3}, a LVM at 365.4 cm^{-1} has been measured by Raman scattering at RT. The value of this frequency, which is certainly softened by the high hole concentration, is none the less comparable to the one (353 cm^{-1}) deduced for a substitutional impurity using a simplified mass-defect secular equation. The above LVM has thus been attributed to Al_{Ge} [110].

5.3.3.2 Compound Crystals

In B-doped 15R-SiC, absorption lines have been reported at 711 (^{11}B) and 718 cm^{-1} (^{10}B) at LHeT, shifting to 709 and 717 cm^{-1} at RT and they have been attributed to a B-related gap mode [111]. In Al-doped 4H-SiC, a LVM at 464 cm^{-1} (LHeT) has been ascribed to Al_{Si} [112].

In III–V compounds, B behaves mostly as an isoelectronic impurity. In GaAs, the LVMs of $^{11}B_{Ga}$ and $^{10}B_{Ga}$ at LHeT are at 517.41 and 540.61 cm^{-1}, respectively [113]. They are close to a strong two-phonon band of GaAs near 530 cm^{-1}, and they show rather different FWHMs (0.85 and 0.34 cm^{-1} for $^{11}B_{Ga}$ and $^{10}B_{Ga}$, respectively). In GaP, the $^{11}B_{Ga}$ and $^{10}B_{Ga}$ LVMs are at 570.00 and 592.75 cm^{-1}, respectively, and their FWHMs (0.9 and 0.6 cm^{-1} for $^{11}B_{Ga}$ and $^{10}B_{Ga}$, respectively) show the same trend as in GaAs. These differences in the FWHMs are explained by the decay of these LVMs into two lattice phonons: the LVMs are superimposed to a two-phonon absorption region, but for both materials, the intensity of the two-phonon DoS is significantly larger at the frequency of $^{11}B_{Ga}$ than at that of $^{10}B_{Ga}$, so that the lifetime of the excited state of the $^{11}B_{Ga}$ LVM is smaller than the one for $^{10}B_{Ga}$. In GaP, the presence of a phonon gap allows the occurrence of a gap mode centred at 283.1 and 284.5 cm^{-1} for $^{11}B_{Ga}$ and $^{10}B_{Ga}$, respectively. This small B IS indicates a small displacement of the B atom for this mode. The shape and FWHM (0.9 cm^{-1}) of this gap mode is explained by the interaction of the B_{Ga} atom with the nnn Ga atoms for this mode. The shapes and frequencies of these B modes have been analysed by a self-consistent model using perfect lattice modes calculated by an *ab initio* density-functional calculation [114]. In InP, the B_{In} LVM is at 522.8 and 543.56 cm^{-1} for ^{11}B and ^{10}B, respectively, and the FWHM of the ^{10}B LVM is only 0.07 cm^{-1} at LHeT [81, 115].

In B-doped GaAs samples irradiated with 2 MeV electrons or with fission neutrons, the observation at LNT of LVMs at 601 and 628 cm^{-1} after annealing near 190°C was reported [116]. In this reference, these LVMs were attributed to the ^{11}B and ^{10}B isotopic components of a centre labelled B(2), and it was suggested that B(2) could be B_{As}. The close similitude of the high-resolution spectrum of B(2) with the one of C_{As}^- presented above led to the firm identification of the B(2) centre with B_{As} [117]. Now, B_{As} is a double acceptor with three possible charge states and one can wonder to what charge state the B(2) vibrational structure is related. A correlation between the disappearance of the electronic absorptions of the two charge states of the so-called 78-meV/203-meV double acceptor related to B_{As} and the growth of the B_{As} LVM at 601 cm^{-1} as a function of 2-MeV-electron fluence indicates that the charge state giving the LVM is B_{As}^{2-} [118]. In unirradiated p-type material, where is was argued that B_{As} was present ([16] and references therein) no LVM is observed because the LVM of B_{As}^0 is assumed to be shifted to energies low enough to escape detection by IR absorption, and the situation for B_{As}^- is not clear.

The high-resolution spectrum of $^{11}B_{As}^{2-}$ in Fig. 5.9 shows a clear similitude with the $^{12}C_{As}^-$ spectrum of Fig. 5.5 due to the Ga host-isotope shift.

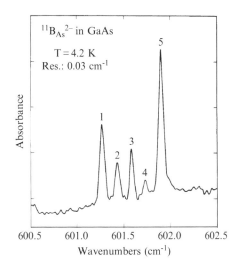

$^{11}B_{As}^{2-}$ in GaAs

T = 4.2 K
Res.: 0.03 cm^{-1}

Fig. 5.9 High-resolution absorption of $^{11}B_{As}^{2-}$ in GaAs. The spacing between the extreme components is 0.64 cm^{-1} (after [2]). Copyright 1993 with permission from Elsevier

Table 5.9 Positions (cm^{-1}) at LNT of the LVMs of B-related centres in LEC GaAs samples [122]

Centre	^{11}B	^{10}B	Isotopic ratio
BX1	1244.0	1285.0	1.033
	1397	1445	1.034
BX2	1325.3	1380.0	1.041
	(1348)		
BX3	1270.5	1309	1.030
	1453	1502	1.034

The attributions of the different components of this spectrum are the same as those of Fig 5.5 replacing C by B. The frequencies of components 1, 2, 3, 4, and 5 measured on this spectrum are 601.26, 601.43, 601.60, 601.73, and 601.90 cm^{-1}, respectively. The smallest FWHMs measured on this spectrum are for the three central components, and they are between 0.05 and 0.06 cm^{-1}, probably close to the true values.

Green's function calculations of the perturbation due to the mass differences of the Ga atoms nns of B$_{As}$ have been performed [119], and the calculated ISs are in good qualitative accord with experiment.

A calibration of the integrated absorption of the 517 cm^{-1} LVM of ^{11}B$_{Ga}$ in GaAs, using SIMS to measure the absolute B concentration, gives [^{11}B$_{Ga}$]~11.0 × 10^{16} at cm^{-3} for an integrated absorption of 1 cm^{-2}, and it is valid between LNT and LHeT ([120], and references therein).

LVMs have been observed at low temperature in O-containing SI GaAs samples in the ~1250–1400 cm^{-1} spectral region [121–123]. Five among these LVMs have been related to three B-related centres, noted BX1, BX2, and BX3, from their B ISs. Their frequencies at LNT are given in Table 5.9.

Aluminium is an isoelectronic impurity (Al$_{III}$) in GaP, GaAs, and GaSb, and the absorption in these materials of one LVM at 444, 362, and 316.7 cm^{-1}, respectively, has been reported [78, 124–126]. In InSb, the Al$_{In}$ LVM shifts to 297.7 cm^{-1}, with an overtone at 590.6 cm^{-1} [127]. This LVM is very near from the frequency of the Al$_{Cd}$ LVM in CdTe (299 cm^{-1}) because of the closeness of the masses of the atoms of the two crystals [128].

5.3.4 B Pairing with Donors

5.3.4.1 Group-IV Crystals

In heavily doped semiconductor crystal, substitutional boron can form pairs with nn FAs. From an energetic aspect, this is facilitated in silicon by the fact that the inward lattice distortion induced by B$_s$ in Si compensates the outward distortion induced by group-V donor atoms. In silicon, pairing of B$_s$ with these atoms produces a centre with C_{3v} symmetry. Such a centre gives two LVMs due to the longitudinal (Γ_1 or A$_1$) and transverse (Γ_3 or E) modes of the pair.

Fig. 5.10 Spectra at LNT of Li-diffused silicon samples from crystals doped with boron enriched with ^{11}B (98%) co-doped with P (**a**), As (**b**), and Sb (**c**). The vertical arrowed peaks indicate LVMs of (B, group-V donor) pairs. For the sample co-doped with As, a third band is observed (after [131]). The strongest line of (**a**) is the LVM of isolated B_s^-. The LVMs ν_1, ν_2, and ν_4 are due to the (^{11}B,Li) pair

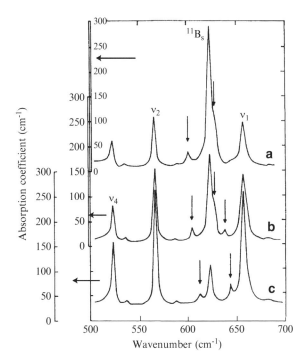

Two independent sets of measurements were performed on samples from silicon crystals co-doped with boron and group-V donors in the 10^{19}–10^{20} cm^{-3} range. These p-type crystals required final compensation which was obtained by electron irradiation [129] and by Li diffusion [130, 131]. The spectra of these samples, performed at LNT, showed weak satellites of the ^{11}B$_s$ LVM at 623 cm^{-1}, whose position depended on the group-V donor, and they are shown in Fig. 5.10.

The experimental positions of the LVMs are given in Table 5.10. They are compared with values obtained from a Green's functions calculation giving the frequencies of the longitudinal A_1 and transverse E modes of the B$-$group-V donor pair as a function of the changes of the force constants. From a thorough analysis, it seems that the LVM with the highest frequency corresponds to the transverse mode [16]. The good agreement between the calculated and experimental frequencies of Table 5.10 is expected as the force constant parameters have been adjusted to give this good agreement.

One notes an unusual increase of the LVMs frequencies with the mass of the group-V atom which indicates that the local perturbations induced by this atom play a more important role than its mass. This is confirmed by the fact that the frequencies of these pairs are close to that of isolated B despite the presence of relatively heavy group-V atoms. In B-doped samples co-doped with P, LVMs at 1201.8, 1231.1, and 1259.3 cm^{-1} have been reported [132]. They are close to the overtones and to the combination mode of the ^{11}B–P pair, and these attributions have been suggested [16].

Table 5.10 Comparison of the LNT frequencies (cm^{-1}) of the B − P, B − As, and B − Sb *nn* pairs in silicon with calculated values assuming that the highest frequency corresponds to the E mode

Pair	Exper.	Symmetry	Calcul.[e]
^{10}B − P	622.9[a]	A$_1$ or $\Gamma_1(C_{3v})$	622.5
	653[c]	E or $\Gamma_3(C_{3v})$	653
^{11}B − P	601.7[d]	A$_1$ or $\Gamma_1(C_{3v})$	601
	628[d] 631[c,*]	E or $\Gamma_3(C_{3v})$	628
^{10}B − As	625*[c]	A$_1$ or $\Gamma_1(C_{3v})$	627
	661.8[c]	E or $\Gamma_3(C_{3v})$	661
^{11}B − As	603.7[b]	A$_1$ or $\Gamma_1(C_{3v})$	604.5
	636.7[b]	E or $\Gamma_3(C_{3v})$	637
	~627[b]		
^{10}B − Sb	635[c]	A$_1$ or $\Gamma_1(C_{3v})$	636.5
	668[c]	E or $\Gamma_3(C_{3v})$	669
^{11}B − Sb	611.9[b]	A$_1$ or $\Gamma_1(C_{3v})$	612
	642.7[b]	E or $\Gamma_3(C_{3v})$	643

[a][130]
[b][131]
[c][129]
[d][132]
[e][133]
* Estimated position

Table 5.11 Measured frequencies (cm^{-1}) at RT of the LVMs of the (B$_s$, Li$_i$) pairs in silicon and of their overtones [134] compared with those in square brackets given in [102]. The values at LNT are about 3 cm^{-1} larger. The frequencies calculated in [22] are given in brackets

	E(ν_1)	A$_1$ (ν_2)	ν_4	$2\nu_1$	$2\nu_2$
(^{10}B, ^{6}Li)	683 [682.4] (678)	584 [584.8] (584)	534 [534.6] (533)	1355	1166
(^{10}B, ^{7}Li)	681 [680.3] (678)	"	"	1354	1167
(^{11}B, ^{6}Li)	657 [656.7] (655)	564 [564.3] (565.5)	522 [522.1] (522)	1312	1126
(^{11}B, ^{7}Li)	655 [653.6] (651)	"	"	1304	1128

It is still easier for the B acceptor to form pairs with Li interstitial donor and the LVMs associated with such pairs have been observed in silicon and germanium. In silicon, three different modes noted ν_1, ν_2, and ν_4 (ν_3 is due to isolated B) by Waldner et al. [134] were reported by these authors and by Balkanski and Nazarewicz [135] to be associated with the (B$_s$, Li$_i$) pairs. These LVMs and their overtones show a B$_s$ IS but a much smaller Li$_i$ IS, and their frequencies are given in Table 5.11.

From its closeness to the Raman frequency of silicon, the ν_4 mode can be considered as a resonant mode. Frequencies assigned to B(3) and B(1) centres, and to a low-frequency mode of the (O, Li$^+$) pair have been reported [102] in Li-compensated B-doped CZ silicon (in this reference, the centre noted B(2) was in fact B$_s$). They actually correspond to the ν_1, ν_2, and ν_4 LVMs of (B, Li), respectively, and they are given as such in Table 5.11. Similar donor–acceptor pairs are produced when lithium is diffused in A*l*- and Ga-doped silicon. The frequency of the equivalent of the v_4 mode is 525 and 520 cm^{-1} for (Al, ^{6}Li) and (Al, ^{7}Li), respectively [109], and 521 and 515 cm^{-1} for (Ga, ^{6}Li) and (Ga, ^{7}Li), respectively [136], and these frequencies correspond to a resonant mode. For the Al and Ga pairs, the equivalents of the v_1 and v_2 modes have apparently not been reported. Because of the interstitial location of the Li atoms, these pairs are expected to reorient along another <111> direction for a moderate energy. This reorientation energy has been determined by anelastic relaxation at a temperature of 150°C and found to be 0.83 eV for the (B, Li) pair [137]. The dissociation energy of the (B, Li) pairs has been measured spectroscopically to be 0.39 eV [138].

In B-doped silicon passivated by a hydrogen plasma, LVMs at 680 and 652 cm^{-1} have been observed by Raman scattering at RT, and attributed to ^{10}B$_s$ and ^{11}B$_s$, respectively. The frequencies of these LVMs remain the same when the H plasma is changed for a D plasma, indicating that in the passivating centre, the B$_s$ acceptor is only perturbed by, but not bonded to the hydrogen atom [139]. One can note the closeness of these frequencies with those for the E mode of (^{10}B, Li) and (^{11}B, Li) in Table 5.11. These LVMs are further discussed in section 8.5.1 in relation with the passivation of the B acceptor by hydrogen.

LVMs associated with (B$_s$, Li$_i$) pairs in germanium have been reported at 518.5 and 610 cm^{-1} for (^{10}B, Li) and 497 and 582.5 cm^{-1} for (^{11}B, Li) by [108].

5.3.4.2 III–V Compounds

In LEC GaAs and GaP co-doped with chalcogens and silicon and in GaP doped with chalcogens, the B$_{As}$ and B$_P$ antisite acceptors can form pairs with substitutional donors. With the ubiquitous Si$_{Ga}$ donor, this is a close B$_{As}$–Si$_{Ga}$ or B$_P$–Si$_{Ga}$ pair with C_{3v} symmetry, but with the group-VI donors on group-V site, the two atoms of the pair are nnn and its symmetry becomes C_{1h} [140]. The frequencies for the LVMs of such pairs are given in Table 5.12. For C_{1h} symmetry, the three modes are singly degenerate, but for C_{3v} symmetry, the v_1 and v_2 modes correspond to the E (doubly degenerate) and A$_1$ IRs, respectively.

5.3.5 Si

Germanium and silicon form a mixed crystal in all proportions. A Si$_{Ge}$ LVM is expected in Ge$_{1-x}$Si$_x$ alloys for small values of x because the mass of the Si atom is much smaller than that of the Ge atom. The first results reported were obtained by Raman scattering at RT, and for $x = 0.01$, a Raman line at 389 cm^{-1} was associated with the Si$_{Ge}$ LVM [141], but it was also observed by IR absorption in Ge$_{1-x}$Si$_x$ alloys with $x < 0.12$ [142]. Other absorption features near 125 and 485 cm^{-1} related to Si were also reported by these authors in these alloys. More recently, the Si isotope effect of the Si$_{Ge}$ LVM has been reported from Raman scattering measurement in a high-quality Ge$_{0.987}$Si$_{0.013}$ crystal, corresponding to

Table 5.12 Frequencies (cm^{-1}) of the LVMs of the (B$_{group-V}$, chalcogen) and B$_{group-V}$ − Si$_{group-V}$ pairs in GaAs and GaP. The experimental value were obtained at LNT. For the B–Si pairs, the average is taken over $2v_1 + v_2$. The frequency v_{av} is the average of v_1, v_2 and v_3, and for B-Si, the average over $2v_1 + v_2$. Note that v_{av} is close to the frequency of isolated B$_V$ of the last column (after [140])

GaAs	v_1	v_2	v_3	v_{av}	$v(B_V)$
(^{11}B$_{As}$, Se$_{As}$)	576.4	609.4	621.7	602.5	601.4
(^{11}B$_{As}$, Te$_{As}$)	580.7	606.5	622.6	603.3	
(^{10}B$_{As}$, Se$_{As}$)	601.8[a]	636.5	649.0	629.1	627.8
(^{10}B$_{As}$, Te$_{As}$)	606.3[a]	633.1	649.6	629.7	
^{11}B$_{As}$ − Si$_{Ga}$	570.9[b]	661.0	-	600.9	601.4
^{10}B$_{As}$ − Si$_{Ga}$	596.0[b]	684.8	-	625.6	627.8
GaP					
(^{11}B$_P$, S$_P$)	595.8	633.3	647.6	625.6	624.5
(^{11}B$_P$, Se$_P$)	599.0	632.2	649.2	626.8	
(^{11}B$_P$, Te$_P$)	600.8	628.7	649.4	626.3	
(^{10}B$_P$, S$_P$)	622.0[a]	661.3	676.1	653.2	652.1
(^{10}B$_P$, Se$_P$)	625.4[a]	660.0	677.8	654.7	
(^{10}B$_P$, Te$_P$)	627.2[a]	656.6	678.2	654.1	
^{11}B$_P$ − Si$_{Ga}$	594.4[a][b]	692.7	-	627.2	624.5
^{10}B$_P$ − Si$_{Ga}$	620.6[a][b]	723.0[a]	-	654.7	652.1

[a] Frequencies calculated assuming a value of 1.044 for the $v(^{10}$B$)/v(^{11}$B$)$ ratio (lines at these positions would be masked by other strong absorption lines)
[b] Assumed doublet

[Si] $= 5.7 \times 10^{20}$ cm^{-3} [143]. The frequency for the ^{28}Si LVM at LNT is 387.3 cm^{-1} and the ^{28}Si–^{29}Si and ^{28}Si–^{30}Si ISs are 6 and 11 cm^{-1}, respectively. This value of the frequency of Si$_{Ge}$ is close to 394 cm^{-1} predicted from the model of Dawber and Eliott [144] for Si$_{Ge}$. For values of $x \geq 0.05$, IR lines at 310 and 370 cm^{-1} and a Raman one at 455 cm^{-1} have been attributed to the LVMs of a *nn* Si–Si pair with D_{3d} symmetry. Raman lines at 439 and 464 cm^{-1} resolved from the 455 cm^{-1} line for $x = 0.098$ have been related to centres with three neighbouring Si atoms. Other IR absorption features of these alloys were observed at frequencies above 500 cm^{-1} for $x < 0.06$, including one at 780 cm^{-1} attributed to the first overtone of the Si$_{Ge}$ LVM near 390 cm^{-1} [143] and references therein. A broad absorption feature near 115 cm^{-1} reported in the latter reference as a quasi-LVM is attributed to a TA mode of germanium activated by the presence of silicon, and it has the same origin as the absorption at 125 cm^{-1} reported in [142].

In III–V compounds, the Si atom displays an amphoteric behaviour, where it goes preferentially on a group-III donor site (Si$_{III}$), and in a much smaller proportion on a group-V acceptor site (Si$_V$). Moreover, for large Si doping values, it can form electrically inactive Si$_{III}$–Si$_V$ pairs, which will be considered later on.

In GaAs, several Si-related LVMs have been reported ([16], and references therein, [28, 145]). The study of these modes has been complicated by the near-superpositions of LVMs belonging to different Si-related centres, leading to different attributions.

In GaAs, the Si_{Ga} donor gives a relatively sharp LVM (FWHM $\sim 0.3\,cm^{-1}$), displaying an isotope effect. For $^{28}Si_{Ga}$, $^{29}Si_{Ga}$, and $^{30}Si_{Ga}$, the LHeT frequencies are 383.9, 378.4, and $373.4\,cm^{-1}$, respectively [145]. The first overtone of the $^{28}Si_{Ga}$ mode, with an intensity of ~ 0.007 with respect to the fundamental, has been reported at $766.5\,cm^{-1}$ at LNT [146].

The situation for the Si_{As} acceptor is also well established and comparable to the above-discussed ones for B_{As} and C_{As}, but for Si_{As}, the nn Ga isotope splitting is only partially resolved into three features at 398.61, 398.08, and $397.62\,cm^{-1}$ for $^{28}Si_{As}$. The $^{28}Si_{As}$ contribution measured at LHeT in a MBE qmi $^{69}GaAs$:Si sample was found to be reduced logically in a single line at $398.9\,cm^{-1}$ due to the only $^{28}Si_{As}^{69}Ga_4$ configuration (the detection of the $^{29}Si_{As}$ components is made difficult by the presence of a LVM of the Si_{Ga}–Si_{As} pair at nearly the same frequency). The measurements on such a sample facilitated the attribution for the centres including an Si_{As} atom by suppressing the nn Ga isotopic contribution [145]. As a consequence of the small differences of the masses of the atoms of the crystals, the frequencies of these LVMs in GaAs are comparable to that of Si_{Ge} in germanium.

The calibration factor of the integrated absorption of the $^{28}Si_{Ga}$ LVM is given as $5 \times 10^{16}\,cm^{-1}$, and a similar factor applies to the $^{28}Si_{As}$ LVM [147].

For high Si doping levels in Bridgman-grown GaAs, two LVMs at 393.05 and $463.42\,cm^{-1}$ were attributed to the transverse and longitudinal modes of Si_{Ga}–Si_{As} nn pairs, and the frequency of the latter mode is larger than those of isolated Si_{Ga} and Si_{As} [28]. In samples doped in equal parts with ^{28}Si and ^{30}Si, the longitudinal mode of this generic pair can be resolved into components corresponding to the different possible isotopic/site combinations, as shown in Fig. 5.11.

The attribution of a LVM at $367\,cm^{-1}$ to an antisymmetric LVM of the Si_{Ga}–Si_{As} pair has raised many questions (see [148]), but the results presented by [145] on Si-doped $^{69}GaAs$ epilayers confirm that, in addition of the LVMs at 393.3 and $464.7\,cm^{-1}$, the LVM observed at LHeT at $366.9\,cm^{-1}$ is due indeed to the Si_{Ga}–Si_{As} pair.

A group of six LVMs near $450\,cm^{-1}$ have been observed at low resolution at temperatures $\sim 230\,K$ in $Ga_{1-x}Al_xAs$ with $x = 0.25$ and 0.43, corresponding to direct and indirect band gap alloys, respectively. They have been attributed to the vibration of the longitudinal mode of the $^{28}Si_{Ga}$–$^{28}Si_{As}$ pair with the $^{28}Si_{As}$ atom interacting with different local environments [149].

In GaAs, besides the LVMs of isolated Si and of the Si–Si pairs, the LVMs of Si-related defects labelled (Si, X) with LVMs at 368.4, 370.0, and $399.6\,cm^{-1}$ and (Si, Y) with LVMs at 366.9, 367.5, and $397.8\,cm^{-1}$ have also been reported[6] ([148], and references therein). The (Si, X) centre was initially identified as a deep acceptor by [28], and tentatively attributed to a (Si_{As}, V_{Ga}) pair while (Si, Y) was

---

[6]There has been some variations in the values given for these LVMs and the values given here are the most recent ones [145].

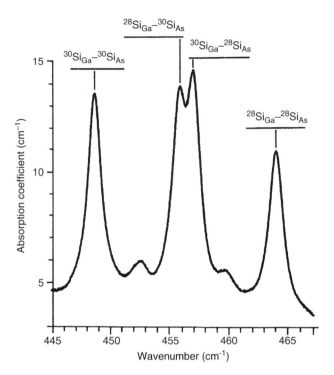

Fig. 5.11 Absorption of the longitudinal mode of the Si_{Ga}–Si_{As} pair at LNT in a GaAs sample doped nominally with equal parts of ^{28}Si and ^{30}Si. The various isotopic combinations of the pairs are indicated [28]. Copyright 1984, American Institute of Physics

attributed to a donor (Si_{As}, As_{Ga}) pair[7] including an As antisite. However, from LVM measurements on plastically deformed and annealed Si-doped GaAs, it was later proposed by [150] that (Si, Y) was Si_{Ga} paired with a nnn V_{Ga} (Si_{Ga}, V_{Ga}). These centres were observed in Si-doped GaAs compensated by electron or neutron irradiation, but also in heavily Si-doped MBE GaAs layers grown at relatively low temperature (250°C) [151]. In these as-grown layers, the LVMs of (Si, Y) were observed, but disappeared under annealing around 300–350°C, while annealing to 550°C produced the (Si, X) centre. When identifying (Si, Y) with (Si_{Ga}, V_{Ga}), (Si, X) could be identified with a (Si, Y) centre with a second adjacent Si atom or another lattice defect perturbing V_{Ga}, and it has also been noted that the two assignments could be interchanged [145]. A STM image of the (Si_{Ga}, V_{Ga}) defect, also identified as an important compensating centre in Si-doped GaAs, is shown in [152].

In GaAs, LEC crystals doped with silicon and germanium, besides the above-discussed LVMs due to isolated Si and Si pairs, two coupled LVMs at 373 and

---

[7]In [28], the A band at $367\,cm^{-1}$ was identified as one of the LVMs of the (Si, X) centre and the C band at $369\,cm^{-1}$ as one of the LVMs of the (Si, Y) centre.

$403\,cm^{-1}$ were reported at LHeT [153]. The LVM at $373\,cm^{-1}$ was superimposed to the $^{30}Si_{Ga}$ LVM and its FWHM was about the same ($0.3\,cm^{-1}$) while the $403\,cm^{-1}$ LVM presented a partially resolved fine structure which could be attributed to a Ge isotope effect.

In GaP the LVM frequency of the Si_{Ga} donor is larger than the one in GaAs because the Si–P bond is stronger than the Si–As bond; thus, in GaP, the $^{28}Si_{Ga}$, $^{29}Si_{Ga}$, and $^{30}Si_{Ga}$ frequencies are reported at 464.9, 461.1, and $456.6\,cm^{-1}$, respectively [78]. The $^{30}Si_{Ga}/^{28}Si_{Ga}$ frequency ratio is smaller in GaAs than in GaP (0.973 and 0.982, respectively), and this is also consistent with a stronger bonding of Si with P than with As ($\sqrt{28/30}$ is 0.966). The LVM of isolated Si_P in GaP has not been identified while the frequencies of Si_{Ga}–Si_P pairs have been observed between $\sim$430 and $455\,cm^{-1}$ [154].

In AlAs, a line at $331\,cm^{-1}$ has been attributed to the $^{28}Si_{Al}$ gap mode, and no charge-state effect of the frequency of this mode is observed [38].

In InP samples doped with Si, including one with Si enriched at 57% with ^{30}Si, absorption features attributed to $^{28}Si_{In}$ and $^{30}Si_{In}$ have been reported at LNT at 430.7 and $422.4\,cm^{-1}$, respectively [155].

5.3.6 N

Nitrogen is considered separately from the other group-V atoms because it is a residual FA in most natural diamonds and in CVD diamonds, and because it is involved in many complexes.

5.3.6.1 Diamond

Isolated substitutional nitrogen, noted here N_s, is the main form of nitrogen present in type-Ib diamond. By analogy with the group-V atoms in group-IV semiconductors, it is expected to be paramagnetic and it is indeed characterized when neutral by the ESR C-P1 spectrum. This spectrum has revealed a trigonal symmetry for N_s, with one of the N_s–C bonds larger than the three other ones (see Sect. 4.2.2.1. A broad resonant mode at $1130\,cm^{-1}$ was correlated to the ESR spectrum of N_s^0 [156]. One LVM, at $1344.3\,cm^{-1}$ at RT, with a FWHM of $\sim$2 cm^{-1} (1345.5 and $0.6\,cm^{-1}$, respectively, at LHeT) was systematically observed in type-Ib diamonds. When natural N (99.6% ^{14}N) is replaced by ^{15}N in synthetic diamonds, this LVM shows no IS, at a difference with the $1130\,cm^{-1}$ resonant mode, which displays a ^{14}N–^{15}N IS of $\sim$5 cm^{-1} [157, 158], and for this reason, the relationship of the $1344\,cm^{-1}$ LVM to N_s^0 was questioned ([158], and references therein). However, the constant ratio observed for many type-Ib samples between the peak absorptions of the $1130\,cm^{-1}$ band and of the $1344\,cm^{-1}$ LVM, and measurements on irradiated synthetic diamond samples enriched with ^{15}N, led finally to the attribution of the $1344\,cm^{-1}$ LVM to N_s^0, or more correctly to the vibration of the

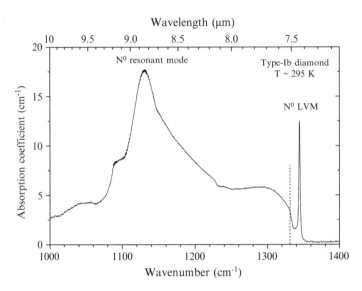

Fig. 5.12 RT absorption of a 0.47 mm-thick synthetic diamond sample showing the $^{14}N_s^0$ LVM at $1344\,\text{cm}^{-1}$ and the associated resonant mode at $1130\,\text{cm}^{-1}$ (resolution: $0.4\,\text{cm}^{-1}$). The position of the Raman frequency of diamond is indicated by the dotted line. $[N^0] \sim 7.7 \times 10^{19}$ at cm^{-3}($\sim$440 ppma). This spectrum is qualitatively very similar to the ones shown in the paper by [162]

nn C atom with the elongated N_s–C bond [159]. This has been confirmed by *ab initio* calculations showing that the LVM involved principally the motion of the C atom while the resonant mode involved the motion of the N atom [160]. These conclusions are substantiated by the large value ($-52\,\text{cm}^{-1}$) of the shift of the N_s^0 LVM in qmi $^{13}C_{\text{diam}}$, which is very close to the ^{13}C–^{12}C IS of the Raman frequency of diamond ($-51\,\text{cm}^{-1}$), and by a more modest value of the corresponding shift ($-32\,\text{cm}^{-1}$) for the resonant mode of N_s^0 [161].

The absorption of these localized and resonant modes of $^{14}N_s^0$, referred to as the *C* absorption by the gemmologists, is displayed in Fig. 5.12.

At RT, the overtone of the $1344\,\text{cm}^{-1}$ LVM is observed at $2688\,\text{cm}^{-1}$, with a FWHM of $3.6\,\text{cm}^{-1}$ and a peak absorption $\sim$2.5% of that of the fundamental (see also [158, 163]). At LHeT, the overtone shifts to $2689.7\,\text{cm}^{-1}$, with the same value of the FWHM as at RT and its absorption is displayed in Fig. 5.13.

The peak absorption coefficient of the $1130\,\text{cm}^{-1}$ resonant mode has been calibrated at RT as a function of the spin concentration of the ESR spectrum of N_s^0 (C-P1), and the calibration factor is $4.4 \times 10^{18}\,\text{cm}^{-2}$ or $25\,\text{ppma cm}$ [162]. Translated to the $1344\,\text{cm}^{-1}$ LVM, it becomes $6.6 \times 10^{18}\,\text{cm}^{-2}$ or $37.5\,\text{ppma cm}$, for a spectral resolution of $1\,\text{cm}^{-1}$ [164]. The RT detection limit of N_s^0 using vibrational spectroscopy depends on the sample thickness, but for a 1-mm-thick sample, judging from the spectrum of Fig. 5.2 of [164], it is $\sim$1 ppma ($1.8 \times 10^{17}\,\text{cm}^{-3}$). At LHeT, using an appropriate resolution ($\leq 0.2\,\text{cm}^{-1}$), an improvement of a factor

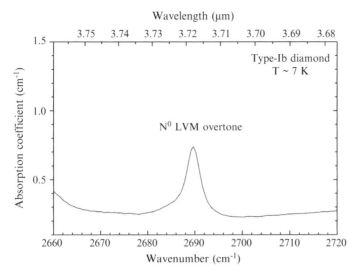

Fig. 5.13 Absorption (resolution: $1.0 \, \mathrm{cm}^{-1}$) of the overtone of the $1345.5 \, \mathrm{cm}^{-1}$ LVM of $^{14}N_{s0}^{0}$ in a diamond sample similar to the one of Fig. 5.12. The peak position is $3.718 \, \mu\mathrm{m}$ (Pajot, unpublished)

of $\sim$3 in the detection limit can be obtained, as the LVM becomes sharper and proportionally more intense at this temperature. A much lower detection limit is obtained using ESR.

In synthetic diamonds containing N_s and irradiated with electrons or neutrons, a relatively sharp absorption at $1332 \, \mathrm{cm}^{-1}$ (the value of the Raman frequency of diamond) has been reported at RT after annealing at high temperatures, together with broader and weaker features at 1115, 1046, and $950 \, \mathrm{cm}^{-1}$, and all these features have been attributed to the positive charge state N_s^+ of substitutional nitrogen[8] [164, 165]. Photoinduced conversion measurements of N^+ into N^0 have been performed and they have allowed to measure the concentration of nitrogen present in the N^+ charge state by determining a calibration factor of $5.5 \, \mathrm{ppma \, cm}$ ($9.9 \times 10^{17} \mathrm{cm}^{-2}$) for the absorption of the $1332 \, \mathrm{cm}^{-1}$ peak at RT [164]. At a difference with N_s^0, N_s^+ is isoelectronic to diamond, and it is expected to display a tetrahedral symmetry.

The electronic absorption near $3.9 \, \mathrm{eV}$ of the substitutional N pairs (N_{2s}) alias A-centre or A aggregate, was presented and discussed in section 4.2.2.1. In the infrared, a broad resonant mode at $1282 \, \mathrm{cm}^{-1}$ with a weaker broad absorption near $1200 \, \mathrm{cm}^{-1}$ (the A absorption of gemmologists), and two other features at $\sim$1090 and $480 \, \mathrm{cm}^{-1}$ are also related to this nitrogen pair [162, 166]. The spacing between the ZPLs of N_{2s} and one vibronic replica is $1280 \, \mathrm{cm}^{-1}$, and this is an

---

[8]Other point defects can also produce absorption peaks at $1332 \, \mathrm{cm}^{-1}$ and the attribution of the peak to N^+ can only be ascertained if the components at 1046 and $950 \, \mathrm{cm}^{-1}$ are also clearly observable [164].

additional proof of the common origin of the UV electronic lines and of the infrared vibrational feature. The RT peak absorption at $1282\,\text{cm}^{-1}$ has been calibrated as a function of the nitrogen concentration in the pairs and the calibration factor is $16.2\,\text{ppma\,cm}$ [162].

The absorption due to the vibrational features of the $V\text{N}_4$ centre (B-centre or B aggregate) found in type-IaB diamonds is characterized by an absorption peak at $1332\,\text{cm}^{-1}$, a flat absorption zone between 1310 and $1230\,\text{cm}^{-1}$ and a broad maximum at $1175\,\text{cm}^{-1}$ with a shoulder at $1100\,\text{cm}^{-1}$ [167]. Besides native samples, type-IaA diamonds can be partially turned into type-IaB diamonds by annealing at $\sim 2600°\text{C}$, giving type-IaA/B diamonds. In "pure" type-IaB diamonds, the nitrogen concentration in the $V\text{N}_4$ centres is currently evaluated by using a calibration factor of $79.4\,\text{ppma\,cm}$ in the plateau region at $1282\,\text{cm}^{-1}$, at the same energy as the peak absorption in the type-IaA diamond samples [167]. Natural diamonds can present mixed types where the absorptions of N_s, N_{2s}, and $V\text{N}_4$ are simultaneously observed ("ABC diamonds" of [163]).

The frequencies of the LVMs of complexes involving N or Si in different configurations have also been calculated with a two-parameter Keating model for a 64-atom supercell [168].

Besides point defects associated with different N_s-related centres, type-Ia natural diamonds often contain extended planar defects in the (100) plane, called platelets, with sizes ranging from nanometres to micrometres, with a relatively small nitrogen content [169]. Their most probable structure is based on a planar C interstitials aggregate, and the perturbed C–C vibration produces an absorption band localized between 1359 and $1375\,\text{cm}^{-1}$ depending on the size of the platelets, known as the B' band [170].

5.3.6.2   Other Group-IV Crystals

In SiC, N locates on a C site. In the $4H$-polytype with two non-equivalent substitutional sites, a gap mode at $625\,\text{cm}^{-1}$ (LHeT) is observed in N-containing samples and attributed to N_C [112], but another N-related LVM at $611\,\text{cm}^{-1}$ (LHeT) is also observed [112,171]. The $625\,\text{cm}^{-1}$ gap mode has also been observed at LHeT in $15R$-SiC [111]. These frequencies are reasonably close to the ones calculated for N_C in $3C$ or $2H$-SiC neglecting any possible J–T distortion [111]. In an N-doped $6H$-polytype, a doublet structure at 630–$635\,\text{cm}^{-1}$ attributed to a gap mode has been observed by Raman scattering at LHeT [172]. The $635\,\text{cm}^{-1}$ peak is twice as strong as the other one and this is in the ratio of the cubic sites to the hexagonal one in $6H$-SiC. In diamond, the frequency is about twice as large, but this is due to the difference in the bond energies.

In silicon, the dominant N-related centre is the N_i–N_i split pair (Fig. 1.7), whose LVMs are discussed in section 6.2.2.1. However, after laser annealing of N-implanted FZ silicon with a 20 ns pulse of a ruby laser (1.783 eV), the observation at LHeT of an ESR spectrum labelled Si-SL5 was reported and attributed to a substitutional N atom with trigonal distortion along a <111> direction; it is noted here N_s [173, 174]. A comparison of the SL5 concentration with the implanted N dose indicated that only $\sim 2.5\%$ of the implanted N atoms were in the paramagnetic N_s form in the laser-annealed samples. Stress-induced electronic reorientation of N_s was measured by cooling the sample under stress between 50 and 25 K and the reorientation energy was found to be 0.11 0 eV [173]. This relatively small value explains why the Si-SL5 spectrum observed at temperatures above 170 K is motionally averaged into the spectrum of a centre with T_d symmetry [174]. Thermal annealing of silicon samples containing N_s above 300°C shows a decrease of the SL5 signal, which disappears around 400°C, and the growth of two ESR spectra, SL6 and SL7. The centres associated with these latter spectra contain both an N atom and the centre associated with SL6 presents also an axial symmetry [174].

In laser-annealed $(^{14}N$-) ^{15}N-implanted FZ silicon, N_s gives also a LVM at $(653\,\mathrm{cm}^{-1})$ 637 cm^{-1} at RT [175], with a frequency shift of $+5\,\mathrm{cm}^{-1}$ at LNT [176]. Another low-frequency LVM at 691 (674) cm^{-1} for ^{14}N (^{15}N) is observed after rapid thermal annealing (RTA) followed by normal thermal annealing of FZ silicon sample [176], and it could be related to the SL6 centre. The ab initio calculations for N_s confirm the distortion deduced from the ESR results [177] and the most recent calculations give frequencies of 637 and 621 cm^{-1} for $^{14}N_s$ and $^{15}N_s$, respectively, for the doubly degenerate mode [178].

The results of ab initio calculations of the structures, physical properties, and LVMs of centres related to N_s in silicon [178] predict the existence of a $N_s - V$ centre, analogous to the donor-vacancy centre discussed in section 4.1.3.1, which is paramagnetic when neutral, and could represent a model for the SL6 or SL7 ESR centre. The frequency of the degenerate LVM calculated for this centre is 663 (646) cm^{-1} for ^{14}N (^{15}N) and the substitution of a ^{29}Si or ^{30}Si atom to one of the three ^{28}Si nns of the N_s atom of this centre splits the LVM by 1.5 or 2.9 cm^{-1}, respectively. The possibility to form, by interaction between (N_i-N_i) and a divacancy, a substitutional pair N_{2s} similar to the one in diamond has been investigated in this study. It is found that the formation is exothermic, with an energy 1.6 eV higher than the one for (N_i-N_i), However, the small value of $[V_2]$ near the silicon melting point makes the detection of N_{2s} in silicon problematic. It could none the less be formed at appreciable concentrations in the presence of an oversaturation of vacancies, as in regions where voids are formed in as-grown silicon crystals or in implantation experiments. For $(^{14}N_s-^{14}N_s)$ and $(^{15}N_s-^{15}N_s)$ with D_{3d} symmetry, the E_u mode is IR active and the calculated frequencies are 668.6 and 651.5 cm^{-1}, respectively. This can be compared with the above-mentioned LVM observed at 691 (674) cm^{-1} in ^{14}N-$(^{15}N$-) implanted FZ silicon [176].

In germanium, a LVM observed at LNT at 577 (559) cm^{-1} for ^{14}N (^{15}N) has been tentatively attributed to N_s [176]. The value of 574.4 cm^{-1} calculated for $^{14}N_s$ in ^{74}Ge and the $^{14}N/^{15}N$ isotopic frequency ratio of 1.032 close to $\sqrt{15/14} = 1.035$ seem to confirm this assignment [179]. A N-related LVM is observed at LNT at 590 and 573 cm^{-1} in ^{14}N- and ^{15}N-implanted laser-annealed germanium, respectively, and it has been related to the 691 (674) cm^{-1} LVM in silicon [176].

5.3.6.3   Other Semiconductors

The vibrational spectroscopy of N in GaAs has been investigated in N-implanted samples [180, 181], in N-containing bulk LEC GaAs [182], and in epilayers grown from a gas mixture containing NH_3 [183]. A broad absorption of a LVM at 480 cm^{-1} at LNT in ^{14}N-implanted GaAs was reported in [181] with a sample-dependent FWHM between 14 and $\sim$40 cm^{-1}, and attributed to isoelectronic N_{As}. This LVM has been reported later at 473 (458) cm^{-1} in ^{14}N- (^{15}N-implanted) GaAs at LNT ([180], and references therein). This ^{15}N/^{14}N frequency ratio of 0.968, remarkably close to $\sqrt{14/15} = 0.966$, confirms this assignment.

At a difference with B_{As} and C_{As}, the N_{As} LVM in N-doped bulk GaAs shows no fine structure related to the nn Ga environment, and the FWHM increases from 2.0 to $\sim$6 cm^{-1} between LHeT-LNT and RT, while the peak position decreases from 471.54 to 469.39 cm^{-1} in the same temperature range [182]. In the N-implanted samples, the FWHMs of the N_{As} LVM at LNT vary from 15 to 18 cm^{-1} for ^{15}N and ^{14}N, respectively [180].

The LVMs observed in GaP (495.8 and 480.3 cm^{-1} at LNT for ^{14}N$_P$ and ^{15}N$_P$, respectively) show a slight frequency increase compared to GaAs, and a ^{15}N/^{14}N frequency ratio of 0.969 [184]. The overtone of the ^{14}N$_P$ LVM has been reported at 988.4 cm^{-1} [185].

These values can be compared to the 553 and 537 cm^{-1} LVMs for the single acceptors ^{14}N$_{Se}$ and ^{15}N$_{Se}$, respectively, observed in ZnSe at RT [186].

A LVM at 638 cm^{-1} has been reported in as-grown nitrogen-rich GaAs crystals [122] and in ^{14}N-implanted GaAs wafers. In N-implanted material, this LVM is broad (see Fig. 5.14b), and it shifts to 619 cm^{-1} in ^{15}N-implanted samples, demonstrating the presence of nitrogen in the related centre. In bulk N-containing crystals, this LVM displays a quadruplet isotopic fine structure with FWHMs of $\sim$0.1 cm^{-1} when observed at high resolution, as shown in Fig. 5.14a ([187], and references therein).

This structure is comparable to the ones of the transverse modes of the C(1) and B(1) centres in electron-irradiated GaAs, presented in sections 7.2.5 and 7.3.2. For the C(1) and B(1) centres, the distortion of the C_{As} and B_{As} atoms from a T_d to a C_{3v} symmetry is attributed to the presence of an As$_i$ atom in a slightly modified nn configuration. In the present case, the distortion lowering the symmetry from T_d to C_{3v} is attributed to the Ga vacancy of a $Ga_3N_{As}V_{Ga}$ or $N_{As} - V_{Ga}$ centre. A self-consistent calculation of the isotopic structure has been performed where the two outermost components are attributed to the doubly degenerate E mode of the ^{14}N^{69}Ga$_3$ and ^{14}N^{71}Ga$_3$ configurations with C_{3v} symmetry (V_{Ga} is omitted). The ^{14}N^{69}Ga$_2$^{71}Ga and ^{14}N^{69}Ga^{71}Ga$_2$ mixed configurations have C_{1h} symmetry and an adjustable parameter fitting the separation between the corresponding LVMs can be chosen [187]. In this model the N atom relaxes along the [111] $V_{Ga}\cdots N_{As}$ direction to stabilize in the plane of its three nn Ga atoms in a natural trivalent configuration [187]. This relaxation and the natural valence of the N atom can be the origin of the small FWHMs observed for the isotopic components.

Fig. 5.14 Comparison
between the structure of the
absorption of the 638 cm^{-1}
LVM in (**a**) a bulk N-doped
GaAs sample showing an a
host-isotope fine structure
and (**b**) a ^{14}N-implanted
GaAs layer. The frequencies
of a host-isotope components
1, 2, 3, and 4 are 637.95,
638.21, 638.41, and
638.67 cm^{-1}, respectively.
The spectral resolution is 0.03
and 0.1 cm^{-1} for (**a**) and (**b**),
respectively (after [187]).
Copyright 2004 by the
American Physical Society

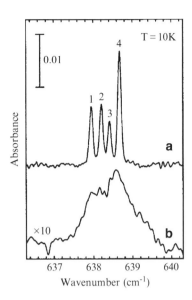

A proportionality coefficient of 7.4×10^{15} cm^{-1} has been established between the integrated absorption at LNT of the 471 cm^{-1} LVM of ^{14}N$_{As}$ and the total N concentration for N-containing bulk GaAs crystals, but it should not be used for N-implanted samples [180], where the presence of the as-grown N$_{As}$–V_{Ga} centres must also be considered. This calibration factor is close to the one of the integrated absorption of the LVM of ^{12}C$_{As}$ at 583 cm^{-1} at LNT (7.2×10^{15} cm^{-1}).

5.3.7 P and As

There seems to be no data on vibrational modes related to Sb and Bi and we only consider P and As. Diamond can be doped with phosphorus, but P solubility in diamond is limited by its atomic radius and P concentrations for the observation of P resonant mode seem impossible to be obtained presently. In silicon samples heavily doped with P and As and compensated by electron irradiation, resonant modes were measured at RT for P at 441 and 491 cm^{-1} and for As at 315 and 336 cm^{-1} [106]. In germanium, a LVM due to P$_{Ge}$ was measured at LNT at 343 cm^{-1} [188].

P forms with GaAs the ternary GaAs$_{1-x}$P$_x$ alloy and at low concentration, P can thus be considered as an isoelectronic impurity. A P$_{As}$ LVM at 355.4 cm^{-1} has been observed at LNT in GaAs samples with [P] $\sim 1 \times 10^{20}$ at cm^{-3}, with a first overtone at 709.7 cm^{-1}, and a pair of lines at 1058 and 1092 cm^{-1} corresponding to the allowed transitions to the second overtone manifold [127]. The LVM at 355 cm^{-1} has also been reported in P-implanted GaAs [126]. In GaSb and InSb, the P$_{Sb}$ LVM frequency is located at 324.0 and 292.7 cm^{-1}, respectively, at LNT, and the first overtone of the P_{Sb} mode at $585.9\,cm^{-1}$ in InSb [124, 126]. In CdTe, one among several LVMs, observed at $322\,cm^{-1}$ at LNT, could be related to P_{Te} [189].

In the linear chain model of GaP, it is found that a heavy atom on a P site can produce a gap mode, while an atom heavier than Ga on a Ga site cannot [17]. In GaP, the phonon gap extends from 255 to $326\,cm^{-1}$, and a gap mode centred at $269\,cm^{-1}$ has indeed been observed at LHeT. This mode displays the characteristic Ga isotope multiplet structure with five sharp components (FWHMs $\sim 0.15\,cm^{-1}$) similar to the ones observed for the LVM of B_{As} and C_{As} in GaAs. Calculations based on mass changes converge toward the attribution of this gap mode to As_P, an isoelectronic electrically inactive impurity, and this is confirmed by the absence of a main isotope effect for this gap mode since As has only one natural isotope. The measured frequencies of the components of the isotopic structure of As_P corresponding to components 1, 2, 3, 4, and 5 of Fig 5.5 are 367.95, 268.34, 268.68, 269.02, and $269.37\,cm^{-1}$, respectively [190].

The Raman scattering of vibrational modes related to As have been reported in GaN at 97, 178, and $235\,cm^{-1}$ [191].

5.3.8   O, S, and Se

There is no evidence of truly substitutional oxygen in silicon or germanium because of the strongly divalent character of oxygen in these crystals, but the chalcogen atoms (S, Se, and Te) are well-known substitutional double donors. However, solubilities of chalcogens in silicon and germanium are in the 10^{16}–$10^{17}\,cm^{-3}$ range and this makes the observation of vibrational features related to these atoms difficult. Moreover, in silicon, one can only expect resonant modes associated with S because of the mass differences. In compound crystals, group-VI FAs usually locate at the most electronegative (anion) site.

5.3.8.1   III-V Compounds

In partially ionic GaP, an O atom can locate on a P site to give a deep O^0/O^+ donor level at E_c–$0.898\,eV$ ($E_v + 1.452\,eV$) [192]. A ZPL at 841 meV ($6783\,cm^{-1}$) is observed by PL in O-containing GaP and it has been attributed to radiative recombination from the 1s(E) excited state to the 1s(A$_1$) ground state of O_P^0, and this ZPL displays a ^{16}O–^{18}O IS of $+0.7\,meV^9$ ($+5.6\,cm^{-1}$) [192, 193]. No vibrational absorption of O_P has been directly observed, but vibronic replicas of the 841 meV ZPL are observed by PL at lower energy, corresponding to local modes at 24.7 and 28.4 meV (199 and $229\,cm^{-1}$), with ^{16}O–^{18}O ISs of $+1.6$ and $+0.5\,meV$ ($+13$ and $+4\,cm^{-1}$), respectively, taking into account the positive ^{16}O–^{18}O IS of the ZPL

---

[9]This means that the energy of the no-phonon ^{18}O-related electronic transition is *smaller* than that of the ^{16}O-related one.

[193]. The trapping of an electron by O_P^0 to give O_P^- is accompanied by a strong relaxation of the O atom. The localized and resonant modes of O_P in GaP have also been calculated using a Green's function approach and the deformable bond approximation [194].

In GaN, it is assumed that the strong residual n-type conductivity in the material grown from solution under high pressure is due to O_N acting as a shallow donor, which can be present at concentrations in the $1–5 \times 10^{19}\,cm^{-3}$ range ([195], and references therein). No absorption measurement can be made in these conditions, but a so-called Q mode at $544\,cm^{-1}$ is observed by Raman scattering and it has been ascribed to a gap mode of O_N. Under hydrostatic pressure $\sim 20\,GPa$, the shallow donor undergoes a transition to a deep gap state, which has been attributed to a DX-like state of oxygen. Raman vibrational modes labelled Q_1, Q_2, and Q_3 associated with O are induced by pressure and they are attributed to different charge states of the DX-like centre [195].

In GaP, the absorption of a broad gap mode of sulphur has been observed at LHeT at 275.2 and $266.2\,cm^{-1}$ for ^{32}S and ^{34}S, respectively [196]. S_P is a single donor in GaP and the study has also been performed as a function of compensation, introduced by electron irradiation or by in-diffusion of copper. Interestingly, these experiments show that the gap mode is due to S_P^+ and that the contribution of S_P^0 is undetectable in the uncompensated samples. This fact and the broad line width compared to that of the gap mode of As_P are discussed and explained in terms of quenching of the dipole moment for S_P^0 and of anharmonic coupling with the lattice modes [196].

5.3.8.2   II–VI Compounds

The foreign group-VI atoms replacing a group-VI atom of the crystal matrix are isoelectronic and normally electrically inactive. However, the non-Coulombic potential near from the FA allows the bonding of an exciton giving an isoelectronic bound exciton (IBE). PL and absorption spectroscopies of these IBEs have been a way to gain information on these centres and to measure the energies of particular phonons of the matrix crystal [197, 198]. Let us start with the S and Se chalcogen atoms. The doping of CdTe and ZnTe with sulphur produces a LVM clearly identified as a ^{32}S component, because of the observation of down-shifted ^{34}S and ^{33}S components, and the overtone of the ^{32}S component is also observed for CdTe and ZnTe. Besides this LVM, a gap mode is also observed for CdTe and ZnTe [46]. The positions at LHeT of the S-related lines measured in CdTe and ZnSe are given in Table 5.13.

In the above results, it is tacitly assumed that the S atom replaces a group-VI atom of the crystal. There is no LVM of Se_S in ZnS because the Se mass is about two times larger than that of S, but a gap mode of Se_S has been reported in ZnS at LHeT near $230\,cm^{-1}$ [47]. The spectrum (Fig. 5.15) shows a structure which is the superposition of a normal Se isotope effect (six isotopes) confirming the chemical nature of the FA, and of the nn Zn host-isotope shift (five isotopes) indicating its site.

Table 5.13 Frequencies (cm^{-1}) of the fundamental LVM ν_S, 1st overtone ($2\nu_S$), and gap modes associated with substitutional sulphur in CdTe and ZnSe at LHeT [46]. Expression (5.9) gives 0.985 and 0.970 for the $^{33}S/^{32}S$ and $^{34}S/^{32}S$ frequency ratios, respectively, and the measured ratios are given in brackets to the right of the frequency for the heaviest isotope

	CdTe			ZnTe		
	ν_S	$2\nu_S$	Gap mode	ν_S	$2\nu_S$	Gap mode
^{32}S	254.1	507.6	106.1	272.7	543.8	144.6
^{33}S	250.7 (0.987)					
^{34}S	247.5 (0.974)	494.1		265.7 (0.974)		

Fig. 5.15 Observed fine structure of the gap mode of Se$_S$ in cubic ZnS at LHeT showing the normal isotope effect of the five most abundant Se isotopes and the Zn *nn* host-isotope shifts, compared with the calculated one. The strongest peak for each Se isotope corresponds to Se^{64}Zn$_4$. The resolution is 0.1 cm^{-1} and the whole structure is only partially resolved (after [47]). Copyright 1996 by the American Physical Society

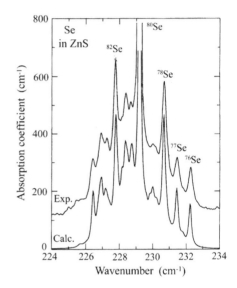

In CdTe doped with oxygen by the addition of CdO, with excess Cd added to prevent the creation of V_{Cd}, a LVM at 349.79 cm^{-1}, with a FWHM of 0.24 cm^{-1} has been reported at LHeT [199, 200]. The intensity of this LVM increases with the initial amount of CdO, and the order of magnitude of its frequency is in agreement with the one expected for isoelectronic O$_{Te}$. The ^{18}O mode, with a natural relative intensity $\sim 2 \times 10^{-3}$ with respect to ^{16}O was not observed in these experiments. Another weaker LVM is also reported at 695.72 cm^{-1} and ascribed to the first overtone of O$_{Te}$.

In samples from CdSe crystals (wurtzite structure) grown with the addition of CdO and excess Se to suppress the occurrence of V_{Se}, two LVMs noted μ_1 and μ_2, centred at low resolution at 1991.77 and 2001.3 cm^{-1}, respectively, with intensities proportional to the CdO added to the starting material, were reported [201]. Actually, two modes with symmetries Γ_1 and Γ_3 are expected from an atom at a substitutional site (C_{3v} symmetry) in the wurtzite structure. Under high-resolution (0.03 cm^{-1}) and polarized radiation, μ_1 and μ_2 show a Se isotope fine structure and polarization effects similar to the ones observed for the two local modes Γ_1 and Γ_3

of Mg_{Cd} in CdSe [47]. Now, if O replaces Se, two LVMs of isoelectronic O_{Se} are expected in the 350–450 cm^{-1} spectral range, but no evidence for O_{Se} was found in these samples. The conclusion drawn from these results is that the μ_1 and μ_2 LVMs are due to the unexpected double donor O_{Cd}. The large frequency of the LVMs of O_{Cd} is attributed to the four extra electrons which increase the force constants responsible for μ_1 and μ_2. [201, 202]. In CdSe samples grown with a larger Se amount, intended not only to suppress the formation of V_{Se}, but also to promote that of V_{Cd}, the μ_1 and μ_2 LVMs are no longer observed, but three new LVMs, noted γ_1, γ_2, and γ_3, were reported at 1094.1, 1107.45, and 1126.33 cm^{-1}, respectively [201]. It has been proposed that the μ_1 and μ_2 LVMs were due to Se–H modes with Se atoms nns of V_{Cd} [203], but the presence of hydrogen in the samples used has been ruled out [202].

In CdTe crystals grown with the addition of CdO, but also with excess Te, where the presence of V_{Cd} was expected, instead of the previously mentioned low-frequency LVM of O_{Te}, two LVMs noted ν_1 and ν_2 were reported at LHeT at 1096.78 and 1108.35 cm^{-1}, respectively, with FWHMs $\sim$0.15 cm^{-1} and a ν_2/ν_1 peack intensity ratio $\sim$1.7. LVMs noted ν_4 and ν_5, observed at 2198.66 and 2210.5 cm^{-1}, respectively, have also been attributed to overtones of ν_1 and ν_2 [199, 200]. The results of thermoelectric effect spectroscopy measurements on undoped CdTe crystals grown by the Bridgman method had been explained by the existence of an electrically active (O_{Te}, V_{Cd}) centre, where one of the Cd nns of O_{Te} is replaced by V_{Cd} [204]. The correlation of the intensities of ν_1 and ν_2 with the CdO amount added led to ascribe these two modes to the vibration of the O_{Te} atom of the (O_{Te}, V_{Cd}) centre, with point group symmetry C_{3v} [200]. The ν_1 and ν_2 LVMs were ascribed to the vibrations Γ_1 and Γ_3, respectively, of the O atom along and perpendicularly to the immaterial axis between O and V_{Cd}, noted $\hat{c}$ in this reference. While the corresponding absorptions are polarized parallel and perpendicular to this axis, no net polarization is observed because of the fourfold orientational degeneracy of the (O_{Te}, V_{Cd}) centre. For increasing temperatures, an upward shift of the frequency of ν_1 and a downward shift of the frequency of ν_2 are observed, and near RT, these two modes coalesce into a single absorption peak at 1102 cm^{-1} with a FWHM of 15 cm^{-1}. While the downward shift with increasing temperature is usual, the inverse effect observed for ν_1 is less common [199, 200].

As for the γ modes in CdSe, the large difference between the frequency of isolated O_{Te} in CdTe and those of the ν_1 and ν_2 LVMs attributed to the (O_{Te}, V_{Cd}) centre led to test the latter attribution by first principle calculations and to show its inconsistency [205]. From such calculations emerged also models of centres with high vibrational frequencies involving O in CdTe, like the stretch mode of an O_2 molecule complexed with V_{Cd} [206].

Recent experiments propose an alternative answer: in CdTe samples annealed at 850°C in sealed tubes with SO_4Cd, the ν_1 and ν_2 LVMs and the ν_4 and ν_5 LVMs have been observed together with weaker lines in the 1100 cm^{-1} spectral region [207]. The lines in this region are divided into group I (1096.8 (ν_1), 1089.5, 1082.8, and 1081.2 cm^{-1}), and group II (1108.3 (ν_2), 1101.1, and 1094.3 cm^{-1}). The observation that the relative intensities of the three highest energy lines of each group match the natural relative isotopic abundance of sulphur (see Appendix D) has been considered as a proof that the centre related to these two groups of lines actually contains one S atom (the weaker lines of group II can also be seen in the spectra in [200]).

As oxygen must also be present in this centre, a SO_n^* complex has been proposed instead of (O_{Te}, V_{Cd}) [207]. In group I, the LVM at $1081.2\,cm^{-1}$ has been attributed to the counterpart of ν_1, but involving ^{18}O. The intensity of the $^{32}S^{16}O_{n-1}{}^{18}O^*$ isotopic configuration should occur with an intensity close to $n \times 0.2\%$ of the intensity of $^{32}S^{16}O_n$. The intensity of the $1081.2\,cm^{-1}$ LVM fulfils approximately this ratio for $n = 2$ (see Fig. 1 in [207]), and the centre giving ν_1 and ν_2 should thus be SO_2^* (the stretch frequency of the S–O mode in the gas phase is $1138\,cm^{-1}$) [207]. A self-consistent fit of the isotopic frequencies of groups I and II attributed in decreasing order to the asymmetric mode of $^{32}S^{16}O_2^*$, $^{33}S^{16}O_2^*$, $^{34}S^{16}O_2^*$ and of $^{32}S^{16}O^{18}O^*(I)$ has been made for harmonic frequencies, within the valence force approximation. This fit predicts a frequency of $1192.5\,cm^{-1}$ for $^{32}S^{16}O^{18}O^*$ (II) and the broadened feature observed at this frequency is tentatively attributed to a Fano resonance effect (see Fig. 1 in [207]). In addition, the molecular parameters deduced from this fit for SO_2^* are comparable to the ones for the free SO_2 molecule. Considering the above-mentioned temperature dependence of the frequencies of ν_1 and ν_2, it is implicitly assumed these LVMs are due to two slightly different locations of SO_2^* in the CdTe lattice. Lattice vacancies or tetrahedral interstitial sites have been suggested as possible locations [207].

Modes ν_4 and ν_5 at 2198.7 and $2210.5\,cm^{-1}$ were attributed to overtones of modes ν_1 and ν_2, respectively [200]. Now, for a centre with mass m showing vibrational anharmonicity, the ratio of the vibrational frequencies of the first overtone ω_{2a} to the fundamental transition ω_{1a} is $2 - B/m\omega_{1a}$, where B is a positive anharmonicity parameter (see expressions (5.5a) and (5.5b)). As the frequency of mode ν_4 is larger than 2 times the one of mode ν_1, the overtone attribution must certainly be revised for this mode.

Before the introduction of the SO_2^* model, it had already been suggested that the closeness of the frequencies of the γ_1, γ_2, and γ_3 modes in CdSe and of the ν_1 and ν_2 modes in CdTe could be due to the presence of the same centre in the two materials, and, as mentioned before, the possibility of an O–H wag mode had been proposed [203], but ruled out in [202]. Very weak lines at 1093.5, 1099.7, and $1112.5\,cm^{-1}$ can be spotted in Fig. 3 in [201], and the first two ones fit a S isotope effect of γ_2 while the one at $1112.5\,cm^{-1}$ fits a ^{34}S isotope shift of γ_3. In CdSe treated with $CdSO_4$, the same γ modes as the one reported in [201] were observed [207] and this seems to prove that the same kind of SO_2^* centre as in CdTe is also produced in CdSe grown or annealed in a O-Contained atmosphere. In this latter crystal, the presence of a third line can be explained by the hexagonal symmetry of this crystal [207].

References

1. L.H. Skolnik, W.P. Allred, W.G. Spitzer, An infrared localized mode technique for measuring segregation coefficients. J. Phys. Chem. Solids **32**, 1–14 (1971)
2. R.C. Newman, Local vibrational mode spectroscopy of defects in III/V compounds, in *Imperfections in III-V compounds, Semiconductors and Semimetals*, vol. 38, ed. by E.R. Weber (Academic Press, 1993), pp. 117–187
3. B. Clerjaud, D. Côte, On the determination of "effective charge" associated with local vibrational modes. J. Phys.: Cond. Matt. **4**, 9919–9926 (1992)
4. R.C. Newman, The effect of anharmonicity on the vibration of hydrogen impurity pairs in gallium arsenide. Physica B **170**, 409–412 (1991)
5. M. Stavola, Vibrational spectroscopy of light element impurities in semiconductors, in *Semiconductors and Semimetals vol. 51B "Identification of Defects in Semiconductors"*, ed. by M. Stavola (Academic Press, San Diego, 1999), pp. 153–224
6. R.J. Elliot, W. Hayes, C.D. Jones, H.F. MacDonald, C.T. Sennett, Localized vibrations of H^- and D^- ions in the alkaline earth fluoride. Proc. Roy. Soc. London A **289**, 1–33 (1965)
7. R. Darwiche, B. Pajot, B. Rose, D. Robein, B. Theys, R. Rahbi, C. Porte, F. Gendron, Experimental study of the hydrogen complexes in indium phosphide. Phys. Rev. B **48**, 17 776–17 790 (1993)
8. W.B. Fowler, R. Capeletti, E. Colombi, XH defects in non-metallic solids: Isotope effects and anharmonicity as probes of the defect environment. Phys. Rev. B **44**, 2961–2968 (1991)
9. P.M. Morse, Diatomic molecules according to the wave mechanics. II Vibrational levels. Phys. Rev. **34**, 57–64 (1929)
10. J.E. Rosenthal, Intensities of vibration rotation bands. Proc. Natl. Acad. Sci. **21**, 281–285 (1935)
11. R. Jones, J. Goss, C. Ewels, S. Öberg, *Ab initio* calculations of anharmonicity of the C-H stretch mode in HCN and GaAs. Phys. Rev. B **50**, 8378–8388 (1994)
12. E. Fermi, *Über den Ramaneffekt des Kohlendyoxids*. Z. Phys. **71**, 250–259 (1931)
13. J.T. Houghton, S.D. Smith, *Infra-Red Physics* (Oxford, at the Clarendon Press, 1966), p. 66
14. U. Fano, Effect of configuration interaction on intensities and phase shifts. Phys. Rev. **124**, 1866–1878 (1961)
15. R. Murray, R.C. Newman, R.S. Leigh, R.B. Beall, J.J. Harris, M.R. Brozel, A. Mohades-Kassai, M. Goulding, Fano profiles of absorption lines from the localised vibrational modes of Be acceptors in GaAs and B acceptors in silicon. Semicond. Sci. Technol. **4**, 423 426 (1989)
16. R.C. Newman, *Infra-red Studies of Crystal Defects* (Taylor and Francis, London, 1973)
17. A.S. Barker, A.J. Sievers, Optical studies of the vibrational properties of disordered solids. Rev. Mod. Phys. **47**, S1-S179 Suppl No 2, 180.(1975)
18. M.D. McCluskey, Local vibrational modes of impurities in semiconductors. J. Appl. Phys. **87**, 3593–3617 (2000)
19. A.M. Zaitsev, *Optical Properties of Diamond. A Data Handbook* (Springer, Berlin, Heidelberg, 2001)
20. G. Herzberg, *Molecular Spectra and Molecular Structure*, I and II (Van Nostrand, New York, 1964)
21. A. Grimm, Extension of the molecular model for isoelectronic impurities. Solid State Commun. **10**, 1305–1308 (1972)
22. R.J. Elliot, P. Pfeuty, Theory of vibrations of pairs of defects in silicon. J. Phys. Chem. Solids **28**, 1789–1809 (1967)
23. M. Vandevyver, P. Plumelle, Local force variations due to substitution impurities in nine compounds with the zinc-blende structure. Phys. Rev. B **17**, 675–685 (1977)
24. R.M. Nieminen, Supercell methods for defects in semiconductors, in *Theory of Defects in Semiconductors*, ed. by D.A. Drabold, S.K. Estreicher *Theory of Defects in Semiconductors*, vol. 104 of Topics in Applied Physics (Springer, Heidelberg, 2007), pp. 27–62

25. C. Ewels, Doctoral thesis, the University of Exeter (1997)
26. R. Jones, P.R. Briddon, The *ab initio* cluster method and the dynamics of defects in semiconductors. In: Stavola M (ed) Semiconductors and Semimetals vol. 51A "Identification of Defects in Semiconductors" (Academic Press, 1998), pp. 287–349
27. S.K. Estreicher, Hydrogen-related defects in crystalline semiconductors. Mat. Sci. Engin. Rep. **14**, 319–412 (1995)
28. W.M. Theis, W.G. Spitzer, High-resolution measurements of localized vibrational mode infrared absorption of Si-doped GaAs. J. Appl. Phys. **56**, 890–898 (1984)
29. R.T. Chen, W.G. Spitzer, Infrared absorption and microstructure of Li-saturated Si-doped GaAs. J. Electrochem. Soc. **127**, 1607–1617 (1980)
30. T. Egilsson, B. Yang, H.P. Gislason, Passivation of shallow and deep levels by lithium in GaAs. Physica Scripta **T54**, 28–33 (1994)
31. R. Zielińska-Purgal, J. Piwowarczyk, W. Nazarewicz, Point defects in CdTe containing simultaneously lithium and a group IV impurity. Phys. Stat. Sol. (b) **186**, 355–365 (1994)
32. J.S. Ko, W.G. Spitzer, Localised vibrational modes and defects of Li-doped CdTe and ZnTe. *Ibid.* **15**, 5593–5604 (1982)
33. J.S. Ko, W.G. Spitzer, Infrared absorption of A*l*-Li defect pairs in ZnSe and CdTe. J. Phys. C **14**, 4891–4906 (1981)
34. M.E. Levy, W.G. Spitzer, Localized vibrational modes of Li defect complexes in GaAs. J. Phys. C **6**, 3223–3244 (1973)
35. A. Król, A. Gołkowska, Lithium-related point defects in CdS. J. Phys. C **21**, 259–269 (1988)
36. M.O. Henry, K.G. McGuigan, M.C. Do Carmo, M.H. Nazare, E.C. Lightowlers, A photoluminescence investigation of the local mode vibrations of the beryllium pair centre in silicon. J. Phys.: Cond. Matt. **2**, 9697–9700 (1990)
37. E. Tarnow, S.B. Zhang, K.J. Chang, D.J. Chadi, Theory of Be-induced defects in Si. Phys. Rev. B **42**, 11 252–11 260 (1990)
38. H. Ono, T. Baba, Charge effect on vibrational modes of impurities in A*l*As. Phys. Rev. B **47**, 16 628–16 630 (1993)
39. R. Addinall, R. Murray, R.C. Newman, J. Wagner, S.D. Parker, R.L. Williams, R. Droopad, A.G. DeOliveira, I. Ferguson, R.A. Stradling, Local vibrational mode spectroscopy of Si donors and Be acceptors in MBE InAs and InSb studied by infrared absorption and Raman scattering. Semicond. Sci. Technol. **6**, 147–154 (1991)
40. H. Harima, T. Inoue, S. Nakashima, M. Ishida, M. Taneya, Local vibrational modes as a probe of activation process in *p*-type GaN. Appl. Phys. Lett. **75**, 1383–1385 (1999)
41. B. Clerjaud, D. Côte, C. Naud, R. Bouanani-Rahbi, D. Wasik, K. Pakula, J.M. Baranowski, T. Suski, E. Litwin-Staszewska, M. Bockowski, I. Grzegory, The role of oxygen and hydrogen in GaN. Physica **308–310**, 117–121 (2001)
42. A. Manabe, Y. Ikuta, A. Mitsuishi, H. Komyia, S. Ibuki, Infrared absorption and Raman scattering due to localised vibrational mode of Be in zinc-chalcogenides. Solid State Commun. **9**, 1499–1502 (1971)
43. A. Manabe, A. Mitsuishi, H. Komyia, S. Ibuki, Infrared absorption due to localized vibrational mode of beryllium in cadmium sulfide and selenide. Solid State Commun. **12**, 337–340 (1973)
44. W. Hayes, A.R.L. Spray, Infra-red absorption of beryllium in cadmium telluride: I. J. Phys. C 1969, **2**, 1129–1136 (1969)
45. C.T. Sennet, D.R. Bosomworth, W. Hayes, A.R.L. Spray, Infrared absorption of beryllium in cadmium telluride: II. J. Phys. C **2**, 1137–1145 (1971)
46. M.D. Sciacca, A.J. Mayur, N. Shin, I. Miotkowski, A.K. Ramdas, S. Rodriguez, Local vibrational modes of substitutional Mg^{2+}, Ca^{2+}, and S^{2-} in zinc blende and wurtzite II-VI semiconductors. Phys. Rev. B **51**, 6971–6978 (1995)
47. M.D. Sciacca, A.J. Mayur, H. Kim, I. Miotkowski, A.K. Ramdas, S. Rodriguez, Host-isotope fine structure of local and gap modes of substitutional impurities in zinc-blende and wurtzite II-VI semiconductors. Phys. Rev. B **53**, 12 878–12 883 (1996)

48. A.J. Mayur, M.D. Sciacca, H. Kim, I. Miotkowski, A.K. Ramdas, S. Rodriguez, Local and gap modes of substitutional $3d$ transition-metal ions in zinc-blende and wurtzite II-VI semiconductors. Phys. Rev. B **53**, 12 884–12 888 (1996)

49. D.L. Peterson, A. Petrou, W. Giriat, A.K. Ramdas, S. Rodriguez, Raman scattering from the vibrational modes in $Zn_{1-x}Mn_xTe$. Phys. Rev. B **33**, 1160–1165 (1986)

50. R.C. Newman, J.B. Willis, Vibrational absorption of carbon in silicon. J. Phys. Chem. Solids **26**, 373–379 (1965)

51. J. Leroueille, Carbon measurements in thin silicon wafers ($\sim$400 μm) by infrared absorption spectroscopy. Appl. Spectroscopy **36**, 153–155 (1982)

52. R.C. Newman, R.S. Smith, Vibrational absorption of carbon and carbon-oxygen complexes in silicon. J. Phys. Chem. Solids **30**, 1493–1505 (1969)

53. P.G. Sennikov, T.V. Kotoreva, A.G. Kurganov, B.A. Andreev, H. Niemman, D. Schiel, V.V. Emtsev, H.J. Pohl, Spectroscopic parameters of the absorption bands related to the vibrational modes of carbon and oxygen impurities in silicon enriched with ^{28}Si, ^{29}Si, and ^{30}Si isotopes. Semiconductors **39**, 300–307 (2005)

54. G. Davies, S. Hayama, S. Hao, B. Bech Nielsen, J. Coutinho, M. Sanati, S.K. Estreicher, K.M. Itoh, Host isotope effects on midinfrared optical transitions in silicon. Phys. Rev. B **71**, 115212/1–7 (2005)

55. G. Davies, S. Hayama, S. Hao, J. Coutinho, S.K. Estreicher, M. Sanati, K.M. Itoh, Lattice isotope effects on the widths of optical transitions in silicon. J. Phys.: Cond. Matt. **17**, S2211–S2217 (2005)

56. C.S. Chen, D.K. Schroder, Lattice distortion and vibrational modes of substitutional impurities in silicon. Phys. Rev. B **35**, 713–717 (1987)

57. L. Hoffmann, J.C. Bach, J. Lundsgaard Hansen, A. Nylandsted Larsen, B. Bech Nielsen, P. Leary, R. Jones, S. Öberg, Substitutional carbon in Ge and $Si_{1-x}Ge_x$. Mater. Sci. Forum **258–263**, 97 (1997)

58. L. Khirunenko, Yu. Pomozov, M. Sosnin, V.J.B. Torres, J. Coutinho, R. Jones, N.V. Abrosimov, H. Riemann, P.R. Briddon, Local vibrations of substitutional carbon in SiGe alloys. Mater. Sci. Forum **514–516**, 364–368 (2006)

59. T. Nozaki, Y. Yatsurugi, N. Akiyama, Y. Endo, Y. Makide, Behaviour of light impurity elements in the production of semiconductor silicon. J. Radioanal. Chem. **19**, 109–128 (1974)

60. A.R. Bean, R.C. Newman, The solubility of carbon in pulled silicon crystals. J. Phys. Chem. Solids **32**, 1211–1219 (1971)

61. J.L. Regolini, J.P. Stoquert, G. Ganter, P. Siffert, Determination of the conversion factor for infrared measurement of carbon in silicon. J. Electrochem. Soc. **133**, 2165–2168 (1987)

62. Annual Book of ASTM Standards (1996) (ASTM, Philadelphia), p. 515

63. H.Ch. Alt, Y. Gomeniuk, B. Wiedemann, H. Riemann, A new method for the determination of carbon in silicon by infrared absorption spectroscopy. J Electrochem. Soc. **150**, G498–G501 (2003)

64. R.I. Scace, G.A. Slack, Solubility of carbon in silicon and germanium. J. Chem. Phys. **30**, 1551–1555 (1959)

65. E.E. Haller, W.L. Hansen, P. Luke, R. McMurray, B. Jarrett, Carbon in high-purity germanium. IEEE Transac. Nucl. Sci. **29**, 745–750 (1982)

66. W.H. Weber, B.K. Yang, M. Krishnamurthy, The Ge − C local mode in epitaxial GeC and Ge-rich GeSiC alloys. Appl. Phys. Lett. **73**, 626–628 (1998)

67. H.Ch. Alt, B. Wiedemann, J.D. Meyer, R.W. Michelmann, K. Bethge, Analysis of electrically active carbon in semi-insulating gallium arsenide. Jpn J. Appl. Phys, Part 1 **38**, 6611–6616 (1999)

68. H.Ch. Alt, Anharmonicity of the C_{As} local oscillator in gallium arsenide. Mater. Sci. Forum **196–201** (Trans Tech Publications, Switzerland), pp. 1577–1782 (1995)

69. R.C. Newman, F. Thompson, M. Hyliands, R.F. Peart, Boron and carbon impurities in gallium arsenide. Solid State Commun. **10**, 505–507 (1972)

70. L.Z. Zhang, Y.C. Du, Y.H. Wang, B.C. Ma, G.G. Gin, The localised vibrational mode two-phonon absorption of carbon in gallium arsenide. J. Phys.: Condens. Matter **1**, 4025–4028 (1989)
71. W.M. Theis, K.K. Bajaj, C.W. Litton, W.G. Spitzer, Direct evidence for the site of substitutional carbon impurity in GaAs. Appl. Phys. Lett. **41**, 70–72 (1982)
72. C. Song, (1992) Doctoral thesis, Université Paris 7
73. R.S. Leigh, R.C. Newman, Host isotope fine structure of local modes in GaAs. J Phys. C: **15**, L1045–L1051 (1982)
74. N. Nagai, High-resolution measurements of the carbon localized mode in gallium arsenide. J. Appl. Phys. **89**, 8345–8347 (2001)
75. B.V. Shanabrook, W.J. Moore, T.A. Kennedy, P.P. Ruden, Line-shape anomaly in the local vibrational mode of a shallow acceptor in GaAs. Phys. Rev. B **30**, 3563–3565 (1984)
76. D.W. Fischer, M.O. Manasreh, The effect of charge state on the local vibrational mode absorption of the carbon acceptor in semi-insulating GaAs. J. Appl. Phys. **68**, 2504–2506 (1990)
77. W.J. Moore, S.V. Shanabrook, Comment on: "The effect of charge state on the local vibrational mode absorption of the carbon acceptor in semi-insulating GaAs". Appl. Phys. Lett. **69**, 6731–6732 (1991)
78. F. Thompson, R.C. Newman, Localized vibrational modes of light impurities in gallium phosphide. J. Phys. C **4**, 3249–3257 (1971)
79. B.R. Davidson, R.C. Newman, D.A. Robbie, M.J.L. Sangster, J. Wagner, A. Fischer, K. Ploog, The lattice site of carbon in highly doped AlAs:C grown by molecular beam epitaxy. Semicond. Sci. Technol. **8**, 611–614 (1993)
80. M.D. McCluskey, E.E. Haller, P. Becla, Carbon acceptors and carbon-hydrogen complexes in AlSb. Phys. Rev. B **65**, 045201/1–4 (2001)
81. R.C. Newman, B.R. Davidson, J. Wagner, M.J.L. Sangster, R.S. Leigh, Thermal stability of InP epilayers: the role of dicarbon and carbon-hydrogen centers. Phys. Rev. B **63**, 205307/1–8 (2001)
82. S. Najmi, X.K. Chen, A. Yang, M. Steger, M.L.W. Thewalt, S.P. Watkins, Local vibrational mode study of carbon-doped InAs. Phys. Rev. B **74**, 113202 (2006)
83. R. Jones, S. Öberg, Theory of carbon complexes in aluminum arsenide. Phys. Rev. B **49**, 5306–5312 (1994)
84. R.S. Leigh, M.J.L. Sangster, R.C. Newman, Anomalous shape of a triplet vibrational mode line: Random electric-field broadening. Phys. Rev. B **60**, 10 845–10 851 (1999)
85. M. Ramsteiner, P. Kleinen, K.H. Ploog, J. Oh, M. Konagai, Y. Takahashi, Raman scattering from vibrational modes in metalorganic molecular beam epitaxy grown carbon doped InP: Spectroscopic search for the carbon donor. Appl. Phys. Lett. **67**, 647–649 (1995)
86. R. Jones, S. Öberg, Theory of the structure and dynamics of the C impurity and C–H complexes in GaAs. Phys. Rev. B **44**, 3673–3677 (1991)
87. R.H. Pritchard, B.R. Davidson, R.C. Newman, T.J. Bullough, T.B. Joyce, R. Jones, S. Öberg, The structure and vibrational modes of H-C$_{As}$ pairs in passivated AlAs grown by chemical beam epitaxy. Semicond. Sci. Technol. **9**, 140–149 (1994)
88. B.R. Davidson, R.C. Newman, T.J. Bullough, T.B. Joyce, Hydrogen wag modes and transverse carbon modes of H-C$_{As}$ complexes in GaAs doped with ^{12}C and ^{13}C. Semicond Sci. Technol. **8**, 1783–1785 (1993)
89. B.R. Davidson, R.C. Newman, C.C. Button, Vibrational modes of carbon acceptors and hydrogen-carbon pairs in semi-insulating InP doped using CCl_4. Phys. Rev. B **58**, 15609–15613 (1998)
90. B.R. Davidson, R.C. Newman, C.D. Latham, R. Jones, J. Wagner, C.C. Button, P.R. Briddon, Raman scattering observations and *ab initio* models of dicarbon complexes in AlAs. Phys. Rev. B **60**, 5447–5455 (1999)
91. J. Wagner, R.C. Newman, B.R. Davidson, S.P. Westwater, T.J. Bullough, T.B. Joyce, C.D. Latham, R. Jones, S. Öberg, Di-carbon defects in annealed highly carbon doped GaAs. Phys. Rev. Lett. **78**, 74–77 (1997)

92. M.J.L. Sangster, R.C. Newman, G.A. Gledhill, S.B. Upadhyay, Cluster calculations of local vibrational mode frequencies of impurities in III-V semiconductors: application to detect complexes involving C_{As} in GaAs. Semicond. Sci. Technol. **7**, 1295–1305 (1992)

93. B-H Cheong, K.J. Chang, Compensation and diffusion mechanisms of carbon dopants in GaAs. Phys. Rev. B **49**, 17436–17439 (1994)

94. S. Najmi, X.K. Chen, M.L.W. Thewalt, S.P. Watkins, Dicarbon defects in as-grown and annealed carbon-doped InAs. J. Appl. Phys. **102**, 083528/1–5 (2007)

95. K. Woodhouse, R.C. Newman, R. Nicklin, R.R. Bradley, M.J.L. Sangster, The lattice sites of carbon and hydrogen incorporation in GaAs grown by MOVPE revealed by infrared spectroscopy. J. Cryst. Growth **120**, 323–327 (1992)

96. H. Ono, N. Furuhata, Vibrational modes of carbon-aluminum complexes in $Al_x Ga_{1-x}$ As grown by metalorganic molecular beam epitaxy. Appl. Phys. Lett. **59**, 1881–1883 (1991)

97. M.D. McCluskey, E.E. Haller, J. Walker, N.M. Johnson, J. Wetterhöffer, J. Weber, T.B. Joyce, R.C. Newman, Local vibrational modes in GaAs under hydrostatic pressure. Phys. Rev. B **56**, 6404–6407 (1997)

98. M.D. McCluskey, K.K. Zhuravlev, B.R. Davidson, R.C. Newman, Pressure dependence of local vibrational modes in InP. Phys. Rev. B **63**, 125202/1–4 (2001)

99. B. Clerjaud, F. Gendron, F. Krause, W. Ulrici, Electronic level of interstitial hydrogen in GaAs. Phys. Rev. Lett. **65**, 1800–1803 (1990)

100. S.J. Breuer, P.R. Briddon, Ab initio study of substitutional boron and the boron-hydrogen complex in silicon. Phys. Rev. B **49**, 10332–10336 (1994)

101. J.F. Angress, A.R. Goodwin, S.D. Smith, A study of the vibration of boron and phosphorus in silicon. Proc. Roy. Soc. London A **287**, 64–88 (1965)

102. R.M. Chrenko, R.S. McDonald, E.M. Pell, Vibrational spectra of lithium-oxygen and lithium-boron complexes in silicon. Phys. Rev. **138**, A1775–A1784 (1965)

103. W. Nazarewicz, M. Balkanski, J.F. Morhange, C. Sébenne, Raman scattering from localized vibrational modes of boron impurities in silicon. Solid State Commun. **9**, 1719–1721 (1971)

104. R.C. Newman, R.S. Smith, Local mode absorption from boron pairs in silicon. Phys. Lett. **24A**, 671–672 (1967)

105. J. Adey, J.P. Goss, R. Jones, P.R. Briddon, Identification of boron clusters and boron-interstitial clusters in silicon. Phys. Rev. B **67**, 245325/1–5 (2003)

106. J.F. Angress, A.R. Goodwin, S.D. Smith, One-phonon band-mode absorption by impurity resonances in diamond and silicon. Proc. Roy. Soc. London A **308**, 111–124 (1968)

107. R.S. Leigh, M.J.L. Sangster, Infrared absorption spectrum of B-doped Si. Phys. Rev. B **27**, 6331–6345 (1983)

108. W. Nazarewicz, J. Jurkowski, Infrared absorption by localized vibrational modes of boron impurity in germanium. Phys. Stat. Solidi (b) **31**, 237–243 (1969)

109. S.B. Devine, R.C. Newman, One phonon absorption from aluminium complexes in silicon compensated by lithium or electron irradiation. J. Phys. Chem. Solids **31**, 685–700 (1970)

110. G. Contreras, M. Cardona, A. Compaan, Vibrational local mode of Al-implanted and laser annealed germanium. Solid State Commun. **53**, 857–859 (1985)

111. B. Pajot, C.J. Fall, J.L. Cantin, H.J. von Bardeleben, R. Jones, P.R. Briddon, F. Gendron, Low-frequency vibrational spectroscopy in SiC polytypes. Mater. Sci. Forum vol. **235–236**, (Trans Tech Publications, Switzerland, 2001), pp. 349–352

112. W. Götz, A. Schöner, W. Suttrop, G. Pensl, W.J. Choyke, R.A. Stein, S. Leibenzeder, Nitrogen donors, aluminum acceptors and strong impurity vibrational modes in 4H-silicon carbide (4H-SiC). Mater. Sci. Forum **143–147**, 69–74 (1994)

113. H.Ch. Alt, M. Maier, Assessment of the boron impurity in semi-insulating gallium arsenide by localized vibrational modes spectroscopy. Semicond. Sci. Technol. **6**, 343–347 (1991)

114. D.A. Robbie, M.J.L. Sangster, E.G. Grosche, R.C. Newman, T. Pletl, P. Pavone, D. Strauch, Measurement and analysis of the gap and local modes of B_{Ga} impurities in GaP. Phys. Rev. B **53**, 9863–9868 (1996)

115. R.C. Newman, F. Thompson, J.B. Mullin, B.W. Straughan, The localized vibrations of boron impurities in indium phosphide. Phys. Lett. **33A**, 113–114 (1970)

116. F. Thompson, S.R. Morrison, R.C. Newman, Infrared local mode absorption in irradiated GaP and GaAs. Proc. Internat. Conf. Rad. Damage and Defects Semicond. Reading, 1972, Inst. Phys. Conf. Ser. **16**, 371–376 (1973)

117. G.A. Gledhill, R.C. Newman, J. Woodhead, Direct evidence for the existence of B_{As} antisite centres in GaAs. J. Phys. C **17**, L301–L304 (1984)

118. R. Addinall, R.C. Newman, Ga_{As} and Be_{As} antisite defects in gallium arsenide. Semicond. Sci. Technol. **7**, 1005–1007 (1992)

119. D.N. Talwar, M. Vandevyver, K.K. Bajaj, W.M. Theis, Gallium-isotope fine structure of impurity modes due to defect complexes in GaAs. Phys. Rev. B **33**, 8525–8539 (1986)

120. R. Addinall, R.C. Newman, Y. Okada, F. Orito, A calibration of the localized vibrational mode absorption line due to isovalent boron impurities in gallium arsenide. Semicond. Sci. Technol. **7**, 1306–1309 (1992)

121. B. Ulrici, R. Stedman, W. Ulrici. Local vibrational mode absorption of hydrogen and oxygen centres in LEC-grown GaP and GaAs, Phys. Stat. Sol. B **143**, K135–K139 (1987)

122. G. Gärtner, T. Flade, M. Jurisch, A. Köhler, J. Korb, U. Kretzer, B. Weinert, Oxygen incorporation in undoped LEC-GaAs. J. Cryst. Growth **198–199**, 355–360 (1999)

123. C. Song, B. Pajot, F. Gendron, Local mode spectroscopy and photo-induced effects of oxygen-related centers in semi-insulating gallium arsenide. J. Appl. Phys. **67**, 7307–7312 (1990)

124. W. Hayes, Localized vibrations of phosphorus and aluminum impurities in GaSb. Phys. Rev. Lett. **13**, 275–277 (1964)

125. O.G. Lorimor, W.G. Spitzer, M. Waldner, Local mode absorption of A*l* and P in GaAs. J. Appl. Phys. **37**, 2509 (1966)

126. L.H. Skolnik, W.G. Spitzer, R.G. Kahan A Hunsperger, Infrared localized-vibrational mode absorption of ion-implanted aluminum and phosphorus in gallium arsenide. J. Appl. Phys. **42**, 5223–5229 (1971)

127. S.D. Smith, R.E.V. Chaddock, A.R. Goodwin, Localized modes of substitutional impurities in intermetallic compounds. J. Phys. Soc. Jpn., vol. **21**, suppl.: 68–71 (1966)

128. B.V. Dutt, M. Al-Delaimi, W.G. Spitzer, Point defects, localized vibrational modes and free-carrier absorption of A*l*-doped CdTe. J. Appl. Phys. **47**, 565–572 (1975)

129. R.C. Newman, R.S. Smith, (1968) Local mode absorption from boron complexes in silicon, in *Localized Excitations in Solids*, ed. by R.F. Wallis (Plenum, New York, 1968), pp. 177–184

130. V. Tsvetov, W. Allred, W.G. Spitzer, Localized vibrational modes in silicon: B-P pair bands. Appl. Phys. Lett. **10**, 326–329 (1967)

131. V. Tsvetov, W. Allred, W.G. Spitzer, (1968) Localized vibrational modes of defect pairs in silicon, in *Localized Excitations in Solids*, ed. by R.F. Wallis (Plenum, New York, 1968), pp. 185–192

132. A.R. Bean, S.R. Morrison, R.C. Newman, R.S. Smith, Electron irradiation damage in silicon containing high concentrations of boron. J. Phys. C **5**, 379–400 (1972)

133. P.M. Pfeuty, (1968) Theory of vibrations of pairs of defects in silicon, in *Localized Excitations in Solids*, ed. by R.F. Wallis (Plenum Press, New York, 1968), pp. 193–202

134. M. Waldner, M.A. Hiller, W.G. Spitzer, Infrared combination mode absorption in lithium-boron-doped silicon. Phys. Rev. **140**, A172–A176 (1965)

135. M. Balkanski, W. Nazarewicz, Infrared study of localized vibrations in silicon due to boron and lithium. J. Phys. Chem. Solids **27**, 671–684 (1966)

136. W.G. Spitzer, M. Waldner, Localized mode measurements of boron- and lithium-doped silicon. J. Appl. Phys. **36**, 2450–2453 (1965)

137. B.S. Berry, An elastic relaxation in silicon doped with lithium and boron. J. Phys. Chem. Solids **31**, 1827–1834 (1970)

138. W.G. Spitzer, M. Waldner, Pairing energy of lithium and boron in silicon. Phys. Rev. Lett. **14**, 223–224 (1965)

139. C.P. Herrero, M. Stutzmann, Microscopic structure of boron-hydrogen complexes in crystalline silicon. Phys. Rev B **38**, 12668–12671 (1988)

140. S.R. Morrison, R.C. Newman, F. Thompson, The behaviour of boron impurities in n-type gallium arsenide and gallium phosphide. J. Phys. C **7**, 633–644 (1974)

141. D.W. Feldman, M. Ashkin, J.H. Parker, Raman scattering by local modes in germanium-rich silicon-germanium alloys. Phys. Rev. Lett. **17**, 1209–1212 (1966)
142. A.E. Cosand, W.G. Spitzer, Infrared absorption of lattice modes and the silicon local mode in Ge_xSi_{1-x} alloys. J. Appl. Phys. **42**, 5241–5249 (1971)
143. M. Franz, K.F. Dombrowski, H. Rücker, B. Dietrich, K. Pressel, A. Barz, U. Kerat, P. Dold, K.W. Benz, Phonons in $Ge_{1-x}Si_x$ bulk crystals. Phys. Rev. B **59**, 10 614–10 621 (1999)
144. P.G. Dawber, R.J. Elliot, The vibrations of an atom of different mass in a cubic crystal. Proc. R. Soc. London Ser. A **273**, 222–236 (1963)
145. M.J. Ashwin, R.C. Newman, K. Muraki, The infrared vibrational absorption spectrum of the Si-X defect present in heavily Si doped GaAs. J. Appl. Phys. **82**, 137–141 (1997)
146. F. Thompson, R.C. Newman, Localized vibrational modes in gallium arsenide containing silicon and boron. J. Phys. C **5**, 1999–2010 (1972)
147. R. Murray, R.C. Newman, M.J.L. Sangster, R.B. Beall, J.J. Harris, P.J. Wright, J. Wagner, M. Ramsteiner, The calibration of the strength of the localized vibrational modes of silicon impurities in epitaxial GaAs revealed by infrared absorption and Raman scattering. J. Appl. Phys. **66**, 2589–2596 (1989)
148. M.R. Brozel, R.C. Newman, B. Özbay, Silicon donor-acceptor pair defects in gallium arsenide. J. Phys. C **12**, L785–L788 (1979)
149. P. Kaczor, Z.R. Żytkiewicz, L. Dobaczewski, New local vibrational modes related to silicon in bulk A*l*GaAs. Act. Phys. Polon. A **88**, 759–762 (1995)
150. H. Ono, R.C. Newman, The complexing of silicon impurities with point defects in plastically deformed and annealed GaAs. J. Appl. Phys. **66**, 141–145 (1989)
151. S.A. McQuaid, R.C. Newman, M. Missous, S. O'Hagan, Heavily Si or Be doped MBE GaAs grown at low temperatures. J. Cryst. Growth **127**, 515–518 (1993)
152. J. Gebauer, R. Krause-Rehberg, C. Domke, Ph. Ebert, K. Urban, Identification and quantification of defects in highly Si-doped GaAs by positron annihilation and scanning tunneling spectroscopy. Phys. Rev. Lett. **78**, 3334–3337 (1997)
153. G.A. Gledhill, R.C. Newman, J. Sellors, Isotopic fine structure of the local mode absorption from [$Si_{Ga} - Ge_{As}$] pairs in gallium arsenide. Semicond. Sci. Technol. **1**, 298–301 (1986)
154. A.H. Kachare, W.G. Spitzer, O.G. Lorimor, F.K. Euler, R.N. Brown, Infrared absorption of silicon isotopes in gallium phosphide. J. Appl. Phys. **45**, 5475–5477 (1974)
155. M.R. Brozel, R.C. Newman, M.G. Astles, Vibrational modes of silicon in indium phosphide. J. Phys. C **11**, L377–L380 (1978)
156. R.M. Chrenko, H.M. Strong, R.E. Tuft, Dispersed paramagnetic nitrogen content of large laboratory diamonds. Phil. Mag. **23**, 313–318 (1971)
157. M.I. Samoilovich, G.N. Bezrukhov, V.P. Butuzov, L.D. Podol'skikh, Peculiarities in electron paramagnetic resonance and infrared spectra of diamonds alloyed with the isotope N^{15}. Sov. Phys. Dokl. **19**, 409–410 (1975)
158. A.T. Collins, G.S. Woods, An anomaly in the infrared absorption spectrum of synthetic diamonds. Philos. Mag. A **46**, 77 (1982)
159. A.T. Collins, M. Stanley, G.S. Woods, Nitrogen isotope effects in synthetic diamonds. J. Phys. D **20**, 969–974 (1987)
160. P.R. Briddon, R. Jones, Theory of impurities in diamond. Physica B **185**, 179–189 (1993)
161. A.T. Collins, G. Davies, H. Kanda, G.S. Woods, Spectroscopic studies of carbon-13 synthetic diamond. J. Phys. C **21**, 1363–1376 (1988)
162. I. Kiflawi, A.E. Mayer, P.M. Spear, J.A. van Wyk, G.S. Woods, Infrared absorption of the single nitrogen and A defect centres in diamond. Phil. Mag. B **69**, 1141–1147 (1994)
163. T. Hainschwang, F. Notari, E. Fritsch, L. Massi, Natural, untreated diamonds showing the A, B and C infrared absorption ("ABC" diamonds) and the H2 absorption. Diam. Relat. Mater. **15**, 1555–1564 (2006)
164. S.C. Lawson, D. Fisher, D.C. Hunt, M.E. Newton, On the existence of positively charged single-substitutional nitrogen in diamond. J. Phys.: Cond. Matt. **10**, 6171–6180 (1998)
165. Y. Mita, H. Kanehara, Y. Adachi, Y. Nisida, M. Okada, M. Kobayashi, Optical modulation of bulk one-phonon state in diamond. Appl. Phys. Lett. **73**, 1358–1360 (1998)

166. G.B.B.M. Sutherland, D.E. Blackwell, W.G. Simeral, The problem of the two types of diamond. Nature **174**, 901–904 (1954)
167. S.R. Boyd, I. Kiflawi, G.S. Woods, Infrared absorption by the B nitrogen aggregate in diamond. Phil. Mag. **72**, 351–361 (1995)
168. P.J. Lin-Chung, Local vibrational modes of impurities in diamond. Phys. Rev. B **50**, 16 905–16 913 (1994)
169. G.S. Wood, Platelets and the infrared absorption of type Ia diamonds. Proc. R. Soc. Lond. A **407**, 219–238 (1986)
170. J.P. Goss, B.J. Coomer, R. Jones, C.J. Fall, P.R. Briddon, S. Öberg, Extended defects in diamond: The interstitial platelet. Phys. Rev. B **67**, 165208/1–15 (2003)
171. W.E. Carlos, W.J. Moore, G.C.B. Braga, J.A. Freitas, Jr, E.R. Glaser, B.V. Shanabrook, Contactless studies of semi-insulating 4H-SiC. Physica B **308–310**, 691–694 (2001)
172. P.J. Colwell, W.D. Compton, M.V. Klein, L.B. Schein, (1970) Localized phonon gap mode in N-doped SiC. In: Keller S P, Hensel J C, Stern F (eds) Proc. 10th Internat. Conf. Phys. Semicond. Cambridge (Mass), Washington DC: USAEC Div. Technical Information, pp. 484–485
173. K.L. Brower, Jahn-Teller distorted nitrogen donor in laser-annealed silicon. Phys. Rev Lett. **44**, 1627–1629 (1980)
174. K.L. Brower, Deep-level nitrogen centers in laser-annealed ion-implanted silicon. Phys. Rev. B **26**, 6040–6052 (1982)
175. H.J. Stein, Infrared absorption band for substitutional nitrogen in silicon. Appl. Phys. Lett. **47**, 1339–1341 (1985)
176. H.J. Stein, Implanted nitrogen in germanium. Appl. Phys. Lett. **52**, 153–154 (1988)
177. R. Jones, S. Öberg, F. Berg Rasmussen, B. Bech Nielsen, Identification of the dominant nitrogen defect in silicon. Phys. Rev. Lett. **72**, 1882–1885 (1994)
178. J.P. Goss, I. Hahn, R. Jones, P.R. Briddon, S. Öberg, Vibrational modes and electronic properties of nitrogen defects in silicon. Phys. Rev. B **67**, 045206/1–11 (2003)
179. F. Berg Rasmussen, R. Jones, S. Öberg, Nitrogen in germanium: Identification of the pair defect. Phys. Rev. B **50**, 4378–4384 (1994)
180. H.Ch. Alt, A.Yu. Egorov, H. Riechert, B. Wiedemann, J.D. Meyer, R.W. Michelmann, K. Bethge, Infrared absorption study of nitrogen in N-implanted GaAs and epitaxially grown GaAs$_{1-x}$N$_x$ layers. Appl. Phys. Lett. **77**, 3331–3333 (2000)
181. A.H. Kachare, W.G. Spitzer, A. Kahan, F.K. Euler, T.A. Whatley, Ion-implanted nitrogen in gallium arsenide. J. Appl. Phys. **44**, 4393–4399 (1973)
182. H.Ch. Alt, B. Wiedemann, K. Bethge, Spectroscopy of nitrogen-related centers in gallium arsenide. Mater. Sci. Forum **258–263**, 867–872 (1997)
183. V. Riede, H. Neumann, H. Sobotta, R. Schwabe, W. Seifert, S. Schwetlick, The localized vibration mode of nitrogen in GaAs. Phys. Stat. Sol. (a) **93**, K151–K155 (1986)
184. F. Thompson, R. Nicklin, Localized vibrational modes due to isotopes of nitrogen in gallium phosphide. J. Phys. C **5**, L223–L225 (1972)
185. S.R. Morrison, R.C. Newman, Interstitial nitrogen defects in gallium phosphide. J. Phys. C **6**, L223–L225 (1973)
186. H.J. Stein, Vibrational mode for nitrogen in zinc selenide. Appl. Phys. Lett. **64**, 1520–1522 (1994)
187. H.Ch. Alt, Y. Gomeniuk, B. Wiedemann, Spectroscopic evidence for a N-Ga vacancy defect in GaAs. Phys. Rev. B **69**, 125214/1–4 (2004)
188. A.E. Cosand, W.G. Spitzer, Localized vibrational modes of Li and P impurities in germanium. Appl. Phys. Lett. **11**, 279–281 (1967)
189. B.V. Dutt, W.G. Spitzer, A localized vibrational mode study of infrared absorption of CdTe:P. J. Appl. Phys. **48**, 954–960 (1977)
190. E.G. Grosche, M.J. Ashwin, R.C. Newman, D.A. Robbie, M.J.L. Sangster, T. Pletl, P. Pavone, D. Strauch, Nearest-neighbor isotopic fine structure of the As$_P$ gap mode in GaP. Phys. Rev. B **51**, 14758–14761 (1995)

191. H. Siegle, A. Kaschner, A. Hoffmann, I. Broser, C. Thomsen, S. Einfeldt, D. Hommel, Raman scattering from defects in GaN: the question of vibrational or electronic scattering mechanism. Phys. Rev. B **58**, 13 619–13 626 (1998)

192. P.J. Dean, (1986) Oxygen in gallium phosphide. In: Pantelides S T (ed) *Deep Centers in Semiconductors A State-of-the-Art Approach* (Gordon and Breach, Switzerland, 1992), pp. 215–377

193. P.J. Dean, C.H. Henry, Electron-capture ("internal") luminescence from the oxygen donor in gallium phosphide. Phys. Rev. **176**, 928–937 (1968)

194. R.J. Hauenstein, T.C. McGill, R.M. Feenstra, Vibrational modes of oxygen in GaP including second-nearest-neighbor interaction. Phys. Rev. B **29**, 1858–1969 (1984)

195. C. Wetzel, H. Amano, I. Akasaki, J.W. Ager III, A. Grzegory, B.K. Meyer, DX-like behavior of oxygen in GaN. Physica B **302–303**, 23–38 (2001)

196. E.G. Grosche, R.C. Newman, D.A. Robbie, R.S. Leigh, M.J.L. Sangster, S_P gap mode in GaP: Force constant reductions and the loss of nearest neighbor isotopic structure. Phys. Rev. B **56**, 15701–15711 (1997)

197. J.J. Hopfield, D.G. Thomas, R.T. Lynch, Isoelectronic donors and acceptors. Phys. Rev. Lett. **17**, 312–315 (1966)

198. M.J. Seong, I. Miotkowski, A.K. Ramdas, Oxygen isoelectronic impurities in ZnTe: Photoluminescence and absorption spectroscopy. Phys. Rev. B **58**, 7734–7739 (1998)

199. G. Chen, I. Miotkowski, S. Rodriguez, A.K. Ramdas, Stoichiometry driven impurity configuration in compound semiconductors. Phys. Rev. Lett. **96**, 035508/1–4 (2006)

200. G. Chen, I. Miotkowski, S. Rodriguez, A.K. Ramdas, Control of defect structure in compound semiconductors with stoichiometry: oxygen in CdTe. Phys. Rev. B **75**, 125204/1–10 (2007)

201. G. Chen, J.S. Bhosale, A.K. Ramdas, I. Miotkowski, Spectroscopic signature of novel oxygen-defect complexes in stoichiometrically controlled CdSe. Phys. Rev. Lett. **101**, 195502/1–4 (2008)

202. G. Chen, J.S. Bhosale, A.K. Ramdas, I. Miotkowski, Reply to: Comment on "Spectroscopic signature of novel oxygen-defect complexes in stoichiometrically controlled CdSe". Phys. Rev. Lett. **102**, 209602/1–1 (2009)

203. L. Zhang, J.T. Thienprasert, M.H. Du, D.J. Shing, S. Limpijumnong, Comment on "Spectroscopic signature of novel oxygen-defect complexes in stoichiometrically controlled CdSe. Phys. Rev. Lett. **102**, 209601/1–1 (2009)

204. S.A. Awadalla, A.W. Hunt, K.G. Lynn, H. Glass, C. Szeles, S.H. Wei, Isoelectronic oxygen-related defect in CdTe crystals investigated using thermoelectric effect spectroscopy. Phys. Rev. B **69**, 075210/1–4 (2004)

205. J.T-Thienprasert, S. Limpijumnong, A. Janotti, C.G. Van de Walle, L. Zhang, M.-H. Du, D.J. Singh, Vibrational signatures of O_{Te} and $O_{Te} - V_{Cd}$ in CdTe: A first-principe study. Comput. Mater. Sci. **49**, 5242–5245 (2010)

206. W. Cheng, L. Liu, P.Y. Yu, Z.X. Ma, S.S. Mao, A tale of two vacancies, Ann. Phys. (Berlin) **523**, 129–136 (2011)

207. E.V. Lavrov, D. Bastin, J. Weber, J. Schneider, A. Fauler, M. Fiederle, Reassignment of the $O_{Te} - V_{Cd}$ complex in CdTe. Phys. Rev. B **84**, 233201/1-4 (2011)

Chapter 6
Vibrational Absorption of O and N Interstitial Atoms and Related Centres

Instead of replacing an atom of the crystal, to become substitutional, with a possible distortion along one bond or in a lattice plane, a FA can also be inserted in the crystal by doping or contamination without loss of a crystal atom. Potential locations for such FAs are represented in Fig. 1.5 by the small grey spheres, and they define what is called broadly "interstitial" sites in the title of the chapter, even if more accurate terms as "bond centred" or "antibonded" can later be used. Interstitial locations can be occupied by small or relatively small atoms like H, Li or Mg, O, and N, and also by relatively heavy metal atoms like Cu. However, to produce vibrational modes in the mid-IR, relatively strong bonds must exist between the FA and the atoms of the crystal and such a situation seems to be only met for H, N, or O. However, it has been found that in compound semiconductors, two light atoms with strong binding energies, like C and N, can form a practically free $C \equiv N$ interstitial pair with small interaction with the lattice atoms, which vibrates at frequencies close to those in vacuum, as in alkali halides. Such a pair can eventually display hindered rotation.

Another way to produce interstitial atoms is to eject crystal atoms or FAs from their normal sites by irradiation with h–e particles like electrons or neutrons. In that case, some of the self-interstitials may recombine with the lattice vacancies produced by the irradiation or diffuse to the surface of the crystal, but they can also be stabilized and form bonds with the neighbouring crystal atoms. In C-containing silicon, self-interstitials produced by h–e electron irradiation under $100\,K$ can be trapped by substitutional carbon to produce a defect shown schematically in Fig. 4.8, where the C atom is displaced from its substitutional site, to form what is called a C interstitialcy, also noted C_i. The vibrational properties of this centre and of derived centres are discussed in Sect. 7.2.1.

In the present chapter, we limit ourselves to the vibrational spectra of interstitial O and N, and of their complexes. The vibrational absorption of H and D is discussed in Chap. 8.

B. Pajot and B. Clerjaud, *Optical Absorption of Impurities and Defects in Semiconducting Crystals*, Springer Series in Solid-State Sciences 169, DOI 10.1007/978-3-642-18018-7_6, © Springer-Verlag Berlin Heidelberg 2013

6.1   Bond-Centred O in Semiconductors

The configuration of a bond-centred (*BC*) atom is shown in Fig. 1.5. The *BC* atoms have usually small atomic radii with respect to the atoms of the crystal and a reduced valence. They can be symmetrically bonded to two *nn* atoms of the crystal lattice, like O in Fig. 1.6, or asymmetrically bonded like H in some H complexes. When the sum of the two bond lengths between the *BC* atom and its *nn*s exceeds the unperturbed *nn* distance, the extra forces between the three atoms can be such that their relaxation must be found in an equilibrium puckered configuration of the *BC* atom in a plane perpendicular to the axis between the *nn* atoms (a <111> axis in diamond and sphalerite crystals) and it is characterized by an apex angle 2α less than 180°. A classical *BC* atom is O in silicon, germanium, GaAs and GaP; another one is H in several semiconductors where it can passivate the electrical activity of acceptors. These atoms are often called interstitial only by opposition to the substitutional location and this has been coined for the O configuration. The vibrational spectroscopy of interstitial oxygen in silicon, germanium, GaAs and GaP is discussed in the next sections.

6.1.1   Interstitial Oxygen in Silicon

6.1.1.1   Isolated Oxygen (O_i)

From its divalent bonding in the crystal, isolated interstitial oxygen in silicon or germanium is electrically inactive, and vibrational methods of investigation are the only ones providing detailed information on this centre. It is the dominant residual impurity in CZ silicon, grown from a melt contained in a silica crucible, with concentrations $\sim 10^{18}$ cm^{-3}. This has consequences on its mechanical properties [1] and on the formation kinetics of the O-related thermal donors [2]. The vibrational absorption of interstitial oxygen (O_i) in silicon has thus been actively investigated in the last 50 years, and some aspects of this spectroscopy have been discussed in several reviews (see, for instance [3–5]). We follow here an historical presentation based on the low-temperature results. There have been several notations for the IR lines related to changes in the spectroscopic attributions and to the atomic models, and we have ourselves added new notations for some transitions, but we hope the presentation remains intelligible.

In the first IR absorption measurements on silicon samples from crystals contaminated with oxygen, a vibrational absorption was systematically observed at RT at a wavelength of 9 μm or ~ 1110 cm^{-1} [6], and an experimental correlation was established between oxygen concentration and the intensity of this band [7]. The so-called 9 μm band was reported to be rather symmetric, with a FWHM ~ 30 cm^{-1}, and when observed at lower temperatures, it shifted to higher energies and displayed a temperature-dependent structure [8]. The firm attribution of this band to O_i was established by the observation at LHeT in a sample enriched with ^{18}O of a low-frequency satellite of the 9 μm band [8]. The existence of other oxygen-related bands, one at $515\,cm^{-1}$, and another one at $1205\,cm^{-1}$ only observed at low temperatures, was also reported in [8]. They deduced from their results that these bands were due to the vibration of an interstitial O atom bonded to two *nn* Si atoms in a non-linear Si–O–Si configuration, which has become since then the basic model for interstitial oxygen in semiconductors.

An interstitial location of oxygen was also deduced from X-ray diffraction studies showing a lattice expansion of silicon containing high concentrations of oxygen [9], and this was made later quantitative [10] (see Sect. 1.3.2). This is explained locally by the fact that the *nn* distance in the silicon crystal is 235 pm and that the average value of the Si-O bond length for α-quartz and different chemical compounds with a Si-O bond is 160 pm. This implies an outward distortion of the Si atoms of the Si-O-Si structure to accommodate the two Si-O bonds, as the puckering of the O atom alone cannot compensate the increase in the length of the chemical bonds.

The infrared work was further refined in a subsequent paper [11] where the temperature dependence of the structure of the 9 μm band at low temperature was explained by a 2D low-frequency motion (2DLFM) of the O atom in a plane perpendicular to the Si$\cdots$Si axis, in a puckered configuration reproduced in Fig. 6.1.

In molecular spectroscopy, a pseudomolecule like Si_2O can be compared, to a non-linear symmetrical XYX molecule with C_{2v} symmetry, with three distinct normal modes (see, for instance, ref. [20] of chapter 5) an antisymmetric one, ν_3, with an electric dipole parallel to the $Si_e \cdots Si_c$ axis (π or z dipole), corresponding to the B_1 *IR* of C_{2v}, and two symmetric ones, ν_1 and ν_2, with electric dipoles perpendicular to the $Si_e \cdots Si_c$ axis (σ or (x,y) dipoles), both corresponding to the A_1 *IR*. In this model of O_i in silicon, the 9 μm band was attributed to the B_1 antisymmetric mode ν_3. In [11], the 1205 and $515\,cm^{-1}$ absorption bands were ascribed to the ν_1 and ν_2 modes, respectively. This attribution of the $1205\,cm^{-1}$ band

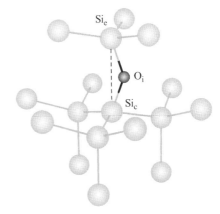

Fig. 6.1 Ball and stick model of the static location of *BC* O_i in a silicon crystal (the off-axis location of O_i has been exaggerated). In addition to the four equivalent locations of O_i around the central Si_c atom, there are six equivalent locations of O_i in a plane perpendicular to the $Si_e \cdots Si_c$ axis. The configuration is identical for O_i in germanium

was questioned in [12] where it was suggested that this band was the combination of the ν_3 mode with the low-frequency motion of the O atom. The fine structure of the 9 μm band observed at higher resolution[1] at LHeT was later explained by a Si isotope effect of the ν_3 mode [13].

An important step forward was the observation of four far-IR absorption lines associated with O_i in silicon [14]. The attribution of these lines to oxygen was made clear from an ^{18}O IS of the lowest energy one. Their temperature dependence was a direct spectroscopic proof of the 2DLFM of O_i postulated by Hrowstowski and Alder [11] to explain the temperature dependence of the components of the 9 μm band. This study included uniaxial-stress measurements of these far-IR lines and of the already-known mid-IR lines [14]. The conclusions were a reattribution of the ν_2 mode to the far-IR lines observed in this study while the $1205\,\text{cm}^{-1}$ absorption was reattributed to a combination of this new ν_2 mode with ν_3, in agreement with a previous suggestion [12]. Because of its proximity with the Raman frequency of silicon, the $515\,\text{cm}^{-1}$ line was described as an impurity-induced lattice absorption [14]. The far IR absorption lines observed in the original reference are shown in Fig. 6.2.

A quantitative investigation of the temperature dependence of the intensities and FWHMs of these far-IR lines has been published by Yamada-Kaneta [15]. Evidence of these low-energy excitations of O_i in silicon was also obtained by phonon spectroscopy near 1 K [16, 17], and references therein). Two dips are observed in the phonon transmission spectrum of FZ silicon samples, the strongest one, at 878 GHz ($29.3\,\text{cm}^{-1}$), and a weaker one at 818 GHz ($27.3\,\text{cm}^{-1}$), corresponding to the IR lines at 29.3 and $27.2\,\text{cm}^{-1}$ associated to ^{16}O and ^{18}O, respectively.

The analysis of the results of [14] was based on a motion of O_i in a plane perpendicular to the Si$\cdots$Si axis, which could be described either by a perturbed 2D harmonic oscillator with states defined by quantum numbers v and l or, alternatively, by a central force oscillator model with a principal quantum number n ($= 0, 1, 2,\dots$) for the radial dependence, and l ($= 0, \pm1, \pm2, \dots$) for the angular dependence of this oscillator. The central forces oscillator model is described here in some detail, following closely the presentation of [14]: in the limit of a large potential barrier preventing reorientation of O_i by tunnelling through the Si$\cdots$Si axis, the expression for the energy of the oscillator, expanded about the bent equilibrium position of O_i is given by:

$$E(n,l) = \hbar\omega_r \left(n + \frac{1}{2} \right) + \left[B - \beta \left(n + \frac{1}{2} \right) \right] l^2 - Dl^4 + \dots, \qquad (6.1)$$

where the rotational constant B is $\hbar^2/2I_B$, $\beta = -6B^2/\hbar\omega_r$ and $D = 4B^3/(\hbar\omega_r)^2$ is the centrifugal distortion coefficient. Here, ω_r is the radial pulsation (angular

---

[1]A silicon IS of the ν_3 mode of O_i in silicon can be observed in Fig. 1 of [12], but it is not commented.

Fig. 6.2 Far IR absorption of
O_i in silicon due to the
2DLFM of the O atom in a
plane perpendicular to the
$Si_c \cdots Si_c$ axis of Fig. 6.2. (**a**)
At the lowest temperature,
only one line of ^{16}O is
observed at 29.3 cm^{-1}. (**b**) At
35 K, other lines due to the
thermalization of upper levels
appear. The positions (cm^{-1})
of the lines are indicated. (**c**)
Same spectrum as in (**a**) in a
sample enriched with ^{18}O.
The spectral resolutions are
indicated graphically [14].
Reproduced with permission
from the Royal Society of
London

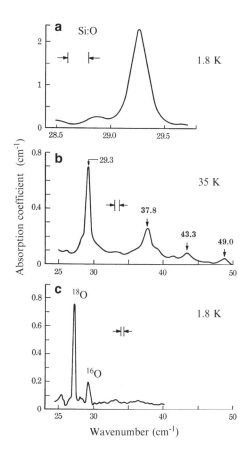

frequency) of the vibration around the equilibrium bent configuration, and the
moment of inertia I_B corresponds to the position of O_i in the bent configuration.
The two O_i lines of lowest energy observed at 29.3 and 37.8 cm^{-1} were attributed to
the $|0,0\rangle \rightarrow |0,1\rangle$ and $|0,1\rangle \rightarrow |0,2\rangle$ transitions, respectively, in the above non-rigid
bent molecule scheme where the states are labelled $|n,l\rangle$. From this attribution were
derived values of 19.8 cm^{-1} for B and of 87 cm^{-1} for the energy $\hbar\omega_r$ of the radial
vibration [14]. From the value of B, a value of 162° was obtained for the apex angle
2α, very close to the one obtained with the 2D oscillator model. One of the merits
of this interpretation is that, by combining the far IR and near IR results with and
without uniaxial stress, an experimental level scheme for the far and near-IR lines
of O_i was provided [14]. This included the already signalled new attribution of the
1205 cm^{-1} line to a combination of the transition giving the 1136 cm^{-1} line with
the low-frequency excitation (LFE) of the radial vibration ($|1,0\rangle$) state).

A self-consistent determination of the spectroscopic parameters associated with
the ISs of the ν_3 mode of O_i in silicon was also undertaken in [18] based on
a molecular model, and as it will be used later for other centres with the same
symmetry, it is presented here. The method is based on the calculation of the IS

of a non-linear symmetrical XYX molecule with apex angle 2α when X and Y are replaced by isotopes X′ and Y′ [19]. It is applied to a pseudomolecule in a crystal, and its bonding with crystal atoms is taken into account by adding to the masses M_X and $M_{X'}$ of atoms X and X′ an interaction mass m which represents, in a simplistic way, the interaction of these atoms with the surrounding crystal [18]. When the frequency of the antisymmetric mode of this pseudomolecule is $\tilde{\nu}_{3(XYX)}$, changing X by X′ and Y by Y′ gives the general expression:

$$\left(\frac{\tilde{\nu}_3(X'Y'X')}{\tilde{\nu}_3(XYX)}\right)^2 = \frac{M_Y(M_X + m)(M_{Y'} + 2(M_{X'} + m)\sin^2\alpha)}{M_{Y'}(M_{X'} + m)(M_Y + 2(M_X + m)\sin^2\alpha)}. \tag{6.2}$$

For a substitution from XYX to XY′X, it gives:

$$\left(\frac{\tilde{\nu}_3(XY'X)}{\tilde{\nu}_3(XYX)}\right)^2 = u = \frac{M_Y(M_{Y'} + 2(M_X + m)\sin^2\alpha)}{M_{Y'}(M_Y + 2(M_X + m)\sin^2\alpha)} \tag{6.3}$$

and similarly for a substitution from XYX to X′YX′:

$$\left(\frac{\tilde{\nu}_3(X'YX')}{\tilde{\nu}_3(XYX)}\right)^2 = v = \frac{(M_X + m)(M_Y + 2(M_{X'} + m)\sin^2\alpha)}{(M_{X'} + m)(M_Y + 2(M_X + m)\sin^2\alpha)}. \tag{6.4}$$

When u and v have been determined experimentally, relations between $\sin\alpha$ and the experimental ratios u and v can be obtained, and equating the expressions for $\sin^2\alpha$, the following expression for m is derived:

$$m = \frac{M_X(u - 1 + \Lambda) + M_{X'}(1 - u - \Lambda v)}{\Lambda(v - 1)}, \tag{6.5}$$

where $\Lambda = (1 - M_Y/M_{Y'})$. When u and v can be obtained experimentally, a value of m can be derived from expression (6.5), from which, in turn a value of 2α can be deduced. Using for instance expression (6.3), one obtains:

$$\sin^2\alpha = \frac{M_Y M_{Y'}(1 - u)}{2(M_X + m)(uM_{Y'} - M_Y)}. \tag{6.6}$$

One can wonder about the physical meaning of such a determination for a pseudomolecule strongly bonded to the crystal. The tacit assumption leading to the relatively simple expression (6.2) is that the forces acting on a given atom of the molecule are the resultant of the attractions or repulsions of the other atoms and this is equivalent to assuming that the potential energy is a purely quadratic form of the central force coordinates. This assumption is strong and it is not obvious that self-consistency of the results corresponds necessarily to strict physical reality.

A more sophisticated phenomenological model of the vibration of O_i in silicon, also aimed toward a self-consistent calculation of the transition energies was worked out by Yamada-Kaneta et al. [20]. It was based on a $Si_3 \equiv Si - O - Si \equiv Si_3$ structure where the $Si - O - Si$ part was considered as linear ($2\alpha = 180°$), giving a D_{3d} symmetry point-group. With this symmetry, the three vibrational modes associated with the $Si - O - Si$ part correspond to *IR*s A_{2u} (the antisymmetric mode corresponding to B_1 (ν_3) for a non-linear XYX molecule with C_{2v} symmetry), E_u (a doubly degenerate one), and A_{1g}, a fully symmetric IR-inactive mode corresponding to ν_1 in the molecular description (see Table B.5 of Appendix B). These authors considered the Hamiltonians for the motion of the O atom in the above radial oscillator model, and the antisymmetric O_i mode (A_{2u} *IR* of D_{3d}) coupled by an interaction term. In the interaction scheme, a state is described by $|k, l, N\rangle$, where N is the number of excitations of the A_{2u} mode[2] and k corresponds to n in expression (6.1). In this description, the 29, 1136, and 1206 cm^{-1} lines are attributed to transitions from the $|0,0,0\rangle$ ground state to the $|0, \pm 1,0\rangle$, $|0,0,1\rangle$, and $|1,0,1\rangle$ states, respectively. The calculations of the eigenvalues of the global Hamiltonian including the interaction term were performed self-consistently as a function of four adjustable parameters, two being the same for ^{16}O and ^{18}O [20]. At about the same time, the results of *ab initio* calculations of the antisymmetric and symmetric modes of O_i were also published, yielding a value of $172°$ for the Si–O–Si apex angle [21].

Frequencies

The overall agreement between the experimental values of frequencies of some interstitial oxygen transitions in silicon and the self-consistent calculation of [20] can be appreciated from Table 6.1. In this table and later, in the case of O_i silicon, the $|0,0,0\rangle \rightarrow |0, \pm 1,0\rangle$, $|0, \pm 1,0\rangle \rightarrow |0, \pm 2,0\rangle$, $|0, \pm 2,0\rangle \rightarrow |0, \pm 3,0\rangle$, and $|0, \pm 1,0\rangle \rightarrow |1,0,0\rangle$ transitions are noted 2DLFM1, 2DLFM2, 2DLFM3, and 2DLFM4, respectively[3] (for 2DLFM). They correspond to the mode labelled ν_2 for the initial puckered $Si - O - Si$ configuration [14]. The $|0,0,0\rangle \rightarrow |0,0,1\rangle$, $|0, \pm 1,0\rangle \rightarrow |0, \pm 1,1\rangle$, and $|0, \pm 2,0\rangle \rightarrow |0, \pm 2,1\rangle$ transitions, corresponding globally to the ν_3 mode, are noted $A_{2u}(0)$, $A_{2u}(1)$, and $A_{2u}(2)$, respectively.

  The reason for the rather good agreement between the experimental and calculated values is that in the calculations, the only interaction considered is the one between the A_{2u} mode and the 2DLFM. The experimental energies of the low-frequency levels $|0, \pm 2,0\rangle$, $|0, \pm 3,0\rangle$, and $|1,0,0\rangle$ (the $|0,1\rangle$, $|0,2\rangle$, $|0,3\rangle$, and $|1,0\rangle$

---

[2]Bosomworth et al. [14] used the notation $|N; n, l\rangle$ where n is identical to k of $|k, l, N\rangle$.
[3]The low-frequency motion was noted generically 2D LEAE (2D low-energy anharmonic excitation) by Yamada-Kaneta et al. [20].

Table 6.1 Frequencies (cm^{-1}) of the O_i transitions involving the v_2 (2DLFM) and v_3 modes in silicon, measured at LHeT and in the low-temperature region. The frequencies in square brackets are values calculated by Yamada-Kaneta et al. [20] from the experimental data. The states are noted $|k, l, N\rangle$. The $^{16}O/^{18}O$ ISs of $A_{2u}(0)$, $A_{2u}(1)$, and $A_{2u}(1)$ are comparable to the one (48.6 cm^{-1}) for the 9 μm band at RT [22]. For the transitions where a Si isotope effect is observed, only the $^{28}Si_2O$ frequency is given here

Initial state	Final state	^{16}O	^{17}O	^{18}O	Observation (see text)
$|0,0,0\rangle$	$|0,\pm1,0\rangle$	29.25[a] [29.4]	28.2	27.2[d] [27.5]	2DLFM1 From LHeT
$|0,\pm1,0\rangle$	$|0,\pm2,0\rangle$	37.7[a] [37.2]		35.3[d] [34.7]	2DLFM2 Above ~10 K
$|0,\pm2,0\rangle$	$|0,\pm3,0\rangle$	43.3[a] [43.0]		[40.0]	2DLFM3 Above ~25 K
$|0,\pm1,0\rangle$	$|1,0,0\rangle$	48.6[a] [48.4]		[44.8]	2DLFM4 Above ~10 K
$|0,\pm2,0\rangle$	$|0,\pm2,1\rangle$	1121.9[c] [1121.2]	1095.1[b]	1071.8[b] [1072.6]	$A_{2u}(2)$ Above ~25 K
$|0,\pm1,0\rangle$	$|0,\pm1,1\rangle$	1128.3[c] [1127.7]	1101.3[b]	1077.6[b] [1077.7]	$A_{2u}(1)$ Above ~8 K
$|0,0,0\rangle$	$|0,0,1\rangle$	1136.4 [1135.7]	1109.5[b]	1085.0 [1084.1]	$A_{2u}(0)$ From LHeT
$|0,0,0\rangle$	$|1,0,1\rangle$	1205.7[c] [1203.8]	1176.7[b]	1151.1[b] [1148.2]	$A_{2u}(0)$+LFE1 From LHeT
$|0,\pm1,0\rangle$	$|1,\pm1,1\rangle$	1216.7[c] [1216.5]	ND	1162.4[e] [1161.3]	Above 20 K

[a][15]
[b][3]
[c][18]
[d][14]
[e][22]

levels in the $N = 0$ vibrational state) for ^{16}O can be deduced from Table 6.1. In the $N = 1$ vibrational state, the energies of the $|0,1\rangle$ and $|0,2\rangle$ states are smaller than in the $N = 0$ state. The $|0,0,0\rangle \rightarrow |1,0,1\rangle$ transition gives the 1206 cm^{-1} line. It is a combination of the $A_{2u}(0)$ mode with the $|1,0\rangle$ low-frequency excitation (LFE) in the $N = 1$ vibrational state. It is noted here $A_{2u}(0)$ + LFE1.

The static host-lattice potential for O_i as a function of the distance from the true BC location ($r = 0$, where r is the distance of O_i from the Si$\cdots$Si axis) was also calculated to show a small minimum at $r \sim 0.03$ nm from the BC location, the potential maximum at $r = 0$ (1 meV) remaining smaller than the ground state energy [20]. This justifies the initial assumption to consider the equilibrium position of O_i at the true BC location (see Fig. 3 of [23]).

An O IS is plain evidence for the lines of Table 6.1. When the FWHM allows it and when the mode implies an appreciable motion of the Si atoms, a Si host-isotope shift is also observed [13]. This is the case for $A_{2u}(0)$ and $A_{2u}(1)$ of Table 6.1. These combined O and Si ISs are shown for $A_{2u}(0)$ in Fig. 6.3a, which displays the spectrum of a FZ sample diffused with ^{17}O and ^{18}O, but containing also "residual" ^{16}O ($[^{16}O_i]$, $[^{17}O_i]$, and $[^{18}O_i]$ are approximately 7×10^{16}, 1×10^{17}, and 7.5×10^{15} cm^{-3}, respectively). Figure 6.3b shows a spectrum of the same sample where the spectral range and ordinate expansion allow the observation of the $A_{2u}(0)$ + LFE1 combination band near 1 200 cm^{-1} for the three O isotopes. This latter band is too broad to give an observable Si IS.

In the spectrum of Fig. 6.3a, for each O isotope, the highest energy component is due to ^{28}Si$_2$O, followed by ^{28}SiO^{29}Si and ^{28}SiO^{30}Si. The relative intensities of the LVMs associated with combinations of isotopes X of a symmetrical XYX molecule are proportional to the product of the natural abundances of isotopes X when they are the same and to two times this product when they are different. As the percent natural abundances are 92.23, 4.67, and 3.10 for ^{28}Si, ^{29}Si, and ^{30}Si, respectively, the ^{30}Si$_2$O and ^{29}SiO^{30}Si isotopic components are too weak to be observed in this sample (whatever the sample and $[O_i]$, the ^{29}Si$_2$O component is undetectable in ^{nat}Si because its frequency nearly coincides with that of the much stronger ^{28}SiO^{30}Si component). The FWHMs of the Si isotopic components of $A_{2u}(0)$ present a particularity related to the O isotope: while their values for ^{16}O and ^{18}O at LHeT are $\sim$0.6 cm^{-1}, those for ^{17}O are $\sim$1.4 cm^{-1}, and this point is discussed in the section on linewidths. In Fig. 6.3a, the integrated absorption (*IA*) of $A_{2u}(^{16}O_i)$ measured between 1124 and 1146 cm^{-1} is 8.68 cm^{-2}, compared to a peak absorption of 8.51 cm^{-1} for $A_{2u}(0)(^{28}$Si$_2^{16}$O) and the ratio of the peak absorption to the *IA* is $\sim$1 at LHeT when the spectral resolution is ≤ 0.2 cm^{-1}. For larger values of the spectral resolution, this ratio decreases because of instrumental broadening.

When temperature is increased above LHeT, thermalization of the population of the $|0, \pm1, 0\rangle$ and $|0, \pm2, 0\rangle$ levels of O_i occurs and the $A_{2u}(1)$ and $A_{2u}(2)$ thermalized spectra are observed, as in Fig. 6.4.

The FWHMs of $A_{2u}(0)(^{28}$Si$_2^{16}$O) and of $A_{2u}(0)(^{28}$Si$_2^{18}$O) measured at 30 K on the spectrum of Fig. 6.4 are $\sim$0.9 and $\sim$0.8 cm^{-1}, respectively, and they experience a small instrumental broadening, while the FWHMs of $A_{2u}(0)(^{28}$Si$_2^{17}$O) and $A_{2u}(1)(^{28}$Si$_2^{16}$O) with a FWHM $\sim$1.4 cm^{-1} are unaffected. Near from and at LNT, the $A_{2u}(0)$, $A_{2u}(1)$, and $A_{2u}(2)$ components of $^{16}O_i$ merge into a structure showing only two maximums at 1135.9 and 1127.5 cm^{-1} with a shoulder at 1121.4 cm^{-1} [13, 24, 25]. At 140 K, only one asymmetric broad feature due to $A_{2u}(^{16}O_i)$ is observed with a maximum near 1125 cm^{-1} (see, for instance, figure 1 of [25]). The profile of the $A_{2u}(^{16}O)$ mode observed at different temperatures has been nicely fitted using an anharmonic potential taking into account the interaction of A_{2u} with the symmetric A_{1g} mode and the 2DLFM (ν_2 mode) [25].

The absorption of $A_{2u}(^{16}O_i)$ has been measured above RT in CZ silicon [26], and for a temperature of 500°C (775 K), a broad absorption maximum is observed at 1080 cm^{-1} (9.26 μm). A crude extrapolation to $^{18}O_i$ at this temperature gives an absorption maximum at 1030 cm^{-1} (9.7 μm).

The frequency of the allowed $|1, 0, 0\rangle \rightarrow |1, 0, 1\rangle$ thermalized transition deduced from Table 6.1 is 1127.9 cm^{-1}. This frequency is very close to that of the

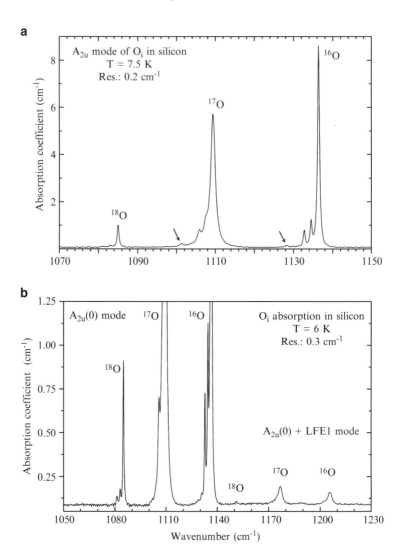

Fig. 6.3 (a) Absorption of $A_{2u}(0)$ in a ^{nat}Si sample enriched with ^{17}O and ^{18}O. For each O isotope, the three components are due to ^{28}Si$_2$O, ^{28}SiO^{29}Si, and ^{28}SiO^{30}Si. The arrows indicate the ^{28}Si$_2$O component of $A_{2u}(1)$, very weak at 7.5 K. Note the difference between the FWHMs of ^{17}O$_i$ and those of ^{16}O$_i$ and ^{18}O$_i$. (b) Absorption of the same sample on an extended ordinate scale, allowing the additional observation of the $A_{2u}(0) +$ LFE1 combination band. The peak absorption of $A_{2u}(0)$ in this spectrum is $\sim$5 cm^{-1}, due to some instrumental broadening and to some fluctuation in $[^{16}$O$_i]$. The relative natural abundances of the O isotopes can be appreciated in Fig. 6.13 (Pajot, unpublished).

$|0, \pm 1, 0\rangle \rightarrow |0, \pm 1, 1\rangle$ transition of $A_{2u}(1)$ (1128.3 cm^{-1}). For this reason and because of the FWHM of the latter line near 30 K, the 1127.9 cm^{-1} line is practically impossible to detect [20].

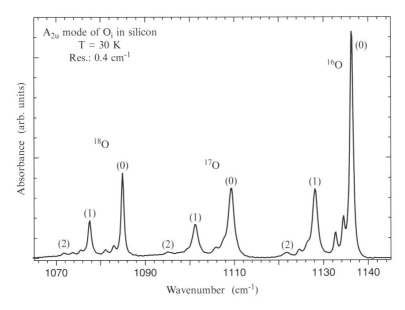

Fig. 6.4 Absorption at 30 K of the A_{2u} modes (noted (0), (1), and (2) for $A_{2u}(0)$, $A_{2u}(1)$, and $A_{2u}(2)$, respectively) of O_i in an In-doped FZ ^{nat}Si sample enriched by diffusion with ^{17}O and ^{18}O

The measured frequencies of the O and of some Si isotopic combinations of A_{2u} between LHeT and ~30 K are given in Table 6.2.

Values of 1118.3 and 1114.7 cm^{-1} have been measured for $A_{2u}(2)$ of ^{29}Si$_2^{16}$O and ^{30}Si$_2^{16}$O in qmi ^{29}Si and qmi ^{30}Si samples [27].

Values of the frequencies of $A_{2u}(0)$ for Si^{16}OSi and Si^{18}OSi have been calculated *ab initio*, with ISs in good agreement with experiment [28]. Calculations of frequencies for the symmetric Si isotopic configurations taking into account anharmonicity and lattice coupling give a very good agreement with the measured ones (see Table II of [29]).

When the measured frequencies of Table 6.2 are used in expressions (6.2), (6.3), and (6.4) for the ISs of a Si$_2$O quasimolecule with C_{2v} symmetry embedded in a crystal, a value of the interaction mass m of 2.47u is obtained, slightly smaller than the value of 3u estimated in [18]. For this "exact" value of m, the apex angle is practically equal to 180°. One can note that the frequencies of $A_{2u}(0)$ for ^{30}Si$_2^{16}$O measured in ^{nat}Si(1129.16 cm^{-1}) and in qmi ^{30}Si(1129.1 cm^{-1}) are practically the same, showing that the interaction of O_i with the *nnn* Si atoms is weak, and this is another a posteriori other justification of the use of the molecular model. In qmi samples, the contribution of the symmetric configuration of the monoisotopic species is predominant, and it is the only one for an ideal 100% monoisotopic crystal.

Table 6.2 Measured frequencies (cm^{-1}) of the Si isotopic components of fundamental and first thermalized transitions associated with the A_{2u} mode of O_i in ^{nat}Si for the different O isotopes. The relative peak intensities (*RI*) of these components are indicated in second column for $A_{2u}(0)$. The $A_{2u}(0)$ values are given for LHeT. At 30 K, they are ~0.1 cm^{-1} smaller. The values in italics are measured in qmi ^{28}Si, qmi ^{29}Si, and ^{30}Si samples [27]. The accuracy of the $A_{2u}(0)$ frequencies for ^{16}O and ^{18}O is 0.02 cm^{-1}

$A_{2u}(0)$	RI	^{16}O	^{17}O	^{18}O
^{28}Si$_2$O	1	1136.36 *1136.5*[a]	1109.4	1085.01 *1084.4*[a]
^{28}SiO^{29}Si	0.102	1134.46 *1134.4*[b]	1107.7	1083.00
^{28}SiO^{30}Si	0.067	1132.72 *1132.0*	1105.9	1081.20
^{29}Si$_2$O	0.0026	1132.56[d] *1132.5*[b]	1105.7[d]	1080.99[d] *1081.0*[b]
^{29}SiO^{30}Si	0.0034	1130.82 *1130.8*[b]		
^{30}Si$_2$O	0.0011	1129.16 *1129.1*[c]	[1102.09]	1077.4[d]
$A_{2u}(1)$		^{16}O	^{17}O	^{18}O
^{28}Si$_2$O		1128.3	1101.3	1077.5
^{28}SiO^{29}Si		1126.6 *1126.4*[b]	1099.6	1075.7
^{28}SiO^{30}Si		1124.8		1073.8
^{29}Si$_2$O		*1124.5*[b]		
^{30}Si$_2$O		*1121.1*[c]		
$A_{2u}(2)$		^{16}O	^{17}O	^{18}O
^{28}Si$_2$O		1121.8	1095	1071.7

[a] qmi ^{28}Si [b] qmi ^{29}Si [c] qmi ^{30}Si [d] Interpolated/extrapolated experimental values obtained by taking the position of the asymmetric combination as the average of the ones of the two symmetric combinations (see text)

The $|0,0,0\rangle \rightarrow |0,0,2\rangle$ overtone transition is symmetry-forbidden.[4] A very weak structureless resonance with an inflection point at 2270 cm^{-1} (2×1135 cm^{-1}) has been observed at LHeT in CZ silicon at a resolution of 0.75 cm^{-1} (Pajot, unpublished).

Some O and Si ISs of $A_{2u}(0)$ transitions have also been calculated using first-principle calculations [28, 30, 31] and they are in reasonable agreement with the experimental ones.

As already mentioned, the band at 515 cm^{-1} reported in the first observations by Hrostowski and Kaiser [8] is below the Raman frequency of silicon at 524 cm^{-1} and must be considered as a resonant mode. It shows neither O nor Si IS, and it shifts to 518 cm^{-1} at LHeT. Piezospectroscopic measurements [32] have shown that the components of its dipole moment transform like coordinates x and y in a plane perpendicular to the $<111>$ Si$\cdots$Si axis (the z axis), and it is ascribed to the doubly degenerate E_u mode. A rather weak and broad absorption band has been reported at 560 cm^{-1} at RT, which disappears at low temperature. Careful checks have shown that it is associated with O_i and its ground state has been tentatively

---

[4]Lines at exactly twice the energies of the ^{28}Si$_2^{16}$O, ^{28}Si^{16}O^{29}Si, and ^{28}Si^{16}O^{30}Si components of the $|0,0,0\rangle \rightarrow |0,0,1\rangle$ transition have been reported [27]. They arise probably from aliasing effects due to insufficient optical filtering in the FTS machine used.

Fig. 6.5 RT absorption of the E_u mode of O_i and of a related excited state at $560 \, \text{cm}^{-1}$ in CZ silicon. The exact position of the E_u mode in this spectrum is $514.1 \, \text{cm}^{-1}$. The two-phonon background has been subtracted (Pajot, unpublished)

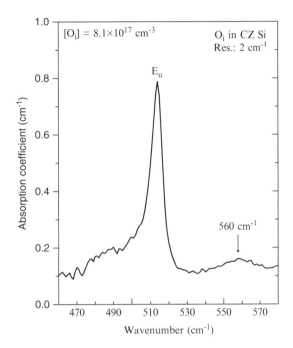

related to an excited state of the E_u mode [22]. The absorption of the E_u mode and of the $560 \, \text{cm}^{-1}$ band at RT is displayed in Fig. 6.5.

The frequency calculated *ab initio* for this E_u mode is $519 \, \text{cm}^{-1}$, with no O IS and an IS of $1 \, \text{cm}^{-1}$ between $^{28}\text{Si}_2\text{O}$ and $^{28}\text{SiO}^{30}\text{Si}$ [33].

A weak band observed at $1013 \, \text{cm}^{-1}$ at RT in as-grown CZ Si [34, 35] was ascribed by Pajot and Cales [18] to an overtone of the E_u mode. It is actually a fundamental mode of the interstitial oxygen dimer O_{2i} discussed in Sect. 6.1.1.2. The overtone of the E_u mode has, however, been invoked to explain the effect of an hydrostatic pressure on the A_{2u} mode of O_i [36].

A weak line at $5.81 \mu\text{m}$ ($\sim 1720 \, \text{cm}^{-1}$) attributed to a combination band involving O_i was reported by Lappo and Tkachev [37] in the RT spectrum of O_i-containing silicon. At LHeT, this line shifts to $1749 \, \text{cm}^{-1}$, (more precisely $1748.6 \, \text{cm}^{-1}$), with a FWHM $\sim 5 \, \text{cm}^{-1}$ [34, 38]. On the one hand, this line displays O ISs ($28.2 \, \text{cm}^{-1}$ for $^{16}\text{O} - ^{17}\text{O}$ and $52.6 \, \text{cm}^{-1}$ for $^{16}\text{O} - ^{18}\text{O}$), practically identical to those of the A_{2u} mode. On the other hand, the $^{28}\text{Si}_2 - ^{28}\text{Si}^{29}\text{Si}$ and $^{28}\text{Si}_2 - ^{28}\text{Si}^{30}\text{Si}$ ISs of the Si atoms bonded to O_i for the $1749 \, \text{cm}^{-1}$ line are 6.4 and $13.2 \, \text{cm}^{-1}$, compared to 1.90 and $3.64 \, \text{cm}^{-1}$ for A_{2u} [39]. This difference in the Si ISs can be appreciated in Fig. 6.6.

The frequencies of the $^{29}\text{Si}_2^{16}\text{O}$ and $^{30}\text{Si}_2^{16}\text{O}$ isotopic combinations for this line measured in qmi samples are 1734.5 and $1721.2 \, \text{cm}^{-1}$, respectively [27].

This line was first attributed to a combination of the A_{2u} mode with a $TA + TO(X)$ phonon combination of the silicon lattice, because the energy separation matched the energy of the two-phonon combination [18]. The weak point

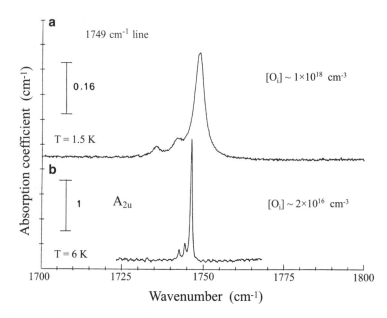

Fig. 6.6 (a) The line at $1748.6\,\mathrm{cm}^{-1}$ in CZ ^{nat}Si Si compared to (b) the A_{2u} mode at $1136\,\mathrm{cm}^{-1}$ in FZ ^{nat}Si displayed with the same energy scale. The wavenumbers refer to spectrum (a). The spectral range of spectrum (b) is 1095–1171 cm^{-1}. The large components are due to $^{28}\mathrm{Si}_2^{16}\mathrm{O}$ and the small ones to $^{28}\mathrm{Si}^{16}\mathrm{O}^{29}\mathrm{Si}$ and $^{28}\mathrm{Si}^{16}\mathrm{O}^{30}\mathrm{Si}$ [39]

of this interpretation was the value of the Si ISs of the $1749\,\mathrm{cm}^{-1}$ line, which implied an abnormally large Si IS of the phonon band. This discrepancy was removed by Yamada-Kaneta et al. [20], who attributed rightly that line to the combination $A_{2u} + A_{1g}$ of the A_{2u} mode with the IR-inactive A_{1g} mode (the ν_1 mode in the molecular description), yielding a frequency $\sim 612\,\mathrm{cm}^{-1}$ for this A_{1g} mode. In this symmetric stretch mode, the two Si atoms bonded to O_i vibrate in antiphase and for Si atoms of equal isotopic masses, the O atom is static and no O isotope shift is expected. At temperatures above LHeT, absorptions at ~ 1742 and $1736\,\mathrm{cm}^{-1}$ are also observed [18,40] and they are due to the combination of the hot bands of A_{2u} and A_{1g}.

Weak lines related to O_i have also been reported later [22,41]. One is located near $648\,\mathrm{cm}^{-1}$ (LHeT) and it displays $^{16}\mathrm{O} - ^{18}\mathrm{O}$ IS of $3.1\,\mathrm{cm}^{-1}$. As all the fundamental modes of O_i in silicon have been identified, new lines must be related to combination modes. The $648\,\mathrm{cm}^{-1}$ line has indeed been attributed to the coupling of the IR-inactive A_{1g} mode with the 2DLFM1 in the A_{1g} vibrational state [22,42] and its absorption spectrum, showing a Si isotope effect is displayed in Fig. 6.7.

As 2DLFM1 is an O_i motion presenting practically no Si isotope effect, the Si ISs of Fig. 6.7 correspond to those of the A_{1g} mode. For this mode, the isotope shifts from the $^{28}\mathrm{Si}_2^{16}\mathrm{O}$ component for $^{28}\mathrm{Si}^{16}\mathrm{O}^{29}\mathrm{Si}$ and $^{28}\mathrm{Si}^{16}\mathrm{O}^{30}\mathrm{Si}$ are 4.9 and $9.6\,\mathrm{cm}^{-1}$, respectively. These values compare with 4.5 and $9.5\,\mathrm{cm}^{-1}$ obtained by subtracting from the Si ISs of the $(A_{2u} + A_{1g})$ combination at $1749\,\mathrm{cm}^{-1}$ those of the A_{2u}

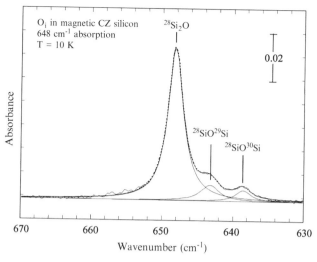

Fig. 6.7 Absorbance spectrum at LHeT of the A_{1g} + 2DLFM1 combination of O_i modes in silicon showing Si ISs. The spectrum is fitted with Lorentzian profiles centred at 648.1, 643.2, and 638.5 cm^{-1}. The resolution is 0.5 cm^{-1} (after [41])

mode [39]. O and Si ISs of the E_u and A_{1g} modes have also been calculated by Artacho et al. [30]. The difference of the order of 7 cm^{-1} between the values of the frequency of A_{1g} deduced from the attributions of the 1749 cm^{-1} and 648 cm^{-1} lines can be attributed to anharmonicity. When the frequency of $A_{1g}(^{28}Si_2O)$ is taken as 612 cm^{-1}, those of $A_{1g}(^{28}SiO^{29}Si)$ and $A_{1g}(^{28}Si_2O^{30}Si)$ are ~608 and 602 cm^{-1}, respectively, and the extrapolation to $A_{1g}(^{30}Si_2O)$ gives ~594 cm^{-1} for that mode. The frequencies of $A_{1g}(^{28}Si^{16}O^{30}Si)$ and of $A_{2u}(0)$ + A_{1g} calculated *ab initio* are 602 and 1789 cm^{-1}, respectively [33].

The most recently reported line of O_i in silicon [43] is one at 1819.5 cm^{-1} (LHeT), with a thermalized "hot" band at 1831.3 cm^{-1} ([44], and references therein). These two bands are attributed to a 3-mode combination of the $|0,0,0\rangle \rightarrow |1,0,1\rangle$ and $|0,\pm1,0\rangle \rightarrow |1,\pm1,1\rangle$ transitions of Table 6.1 with the A_{1g} mode.

To take into account the presence of the new lines, Yamada-Kaneta [42] has proposed a model of the O_i absorption in silicon expanded from the one of [20]. In the new notation, a state is described as $|k,l,N,M\rangle$ by merely adding the number of excitations M of the A_{1g} mode, compared to $|k,l,N\rangle$ for the states of Table 6.1. This model reproduces self-consistently the positions of all the different lines observed for the $^{28}Si_2^{16}O$ combination, at the exception of the ones at 518 and 560 cm^{-1}, associated with the E_u mode.

Several lines arising from the combination of the A_{1g} mode of O_i with other fundamental vibrations from O_i have been observed and some of them have already been discussed. Table 6.3 supplements Table 6.1, giving the observed positions of lines involving the IR-inactive A_{1g} mode of O_i.

Table 6.3 Measured frequencies (cm^{-1}) at low temperature of the ^{28}Si$_2$O transitions involving the A$_{1g}$ mode of O$_i$ in ^{nat}Si. The frequency of the $|0,0,0,0\rangle \rightarrow |0,0,1,1\rangle$ transition for ^{17}O is 1720.4 cm^{-1}. The states are noted $|k, l, N, M\rangle$, where M is the number of excitations of the A$_{1g}$ mode (after [42])

Initial state	Final state	^{16}O	^{18}O	Remarks
$|1,0,0,0\rangle$	$|0,\pm1,0,1\rangle$	571		
$|0,\pm2,0,0\rangle$	$|0,\pm1,0,1\rangle$	580.2		
$|0,\pm1,0,0\rangle$	$|0,0,0,1\rangle$	588.4		
$|0,0,0,0\rangle$	$|0,\pm1,0,1\rangle$	648.2[a]	645.1[a]	A$_{1g}$ + 2DLFM1 LHeT
$|0,\pm1,0,0\rangle$	$|0,\pm2,0,1\rangle$	657.4		
$|0,\pm1,0,0\rangle$	$|1,0,0,1\rangle$	668.0		
$|0,\pm2,0,1\rangle$	$|1,0,1,1\rangle$	1734.1	1683[a]	Above 30 K
$|0,\pm1,0,0\rangle$	$|0,\pm1,1,1\rangle$	1740.9	1689.3[a]	"
		1727.6[d] 1714.2[e]		
$|0,0,0,0\rangle$	$|0,0,1,1\rangle$	1748.6[b]	1696.0[b]	A$_{2u}$(0) + A$_{1g}$LHeT
		1734.5[d] 1721.5[e]		
$|0,0,0,0\rangle$	$|1,0,1,1\rangle$	1819.5[c]	1764[c]	A$_{2u}$(0) + A$_{1g}$ + LFE1"
$|0,\pm1,0,0\rangle$	$|1,\pm1,1,1\rangle$	1831.3		

[a] [22]
[b] [39]
[c] [43]
[d] in qmi ^{29}Si
[e] in qmi ^{30}Si

In [14], the interpretation of the far IR results led to a value of 162° for the Si − O − Si apex angle while the fit of the ISs presented earlier indicates a value ~180°. There have been many calculations of the atomic parameters of O$_i$ in silicon and comparisons of the results for the apex angle 2α are given in [45] and [33], showing values between 140° and 180°. The conclusion of the most recent *ab initio* calculations confirms the highly anharmonic character of the O$_i$ motion, which can be considered as a vibration of the O atom along the Si$\cdots$Si axis coupled to its motion in a plane perpendicular to this axis. The calculations agree with the conclusions of Artacho et al. [23] and the earlier assumptions of Yamada-Kaneta et al. [20] that this structure should be treated as linear rather than puckered, in spite of the fact that the linear configuration does not correspond to the global energy minimum of the system [28]. In this reference and in [31] are presented results of *ab initio* calculation of Si and O ISs for the A$_{2u}$, A$_{1g}$, and E$_u$ modes of O$_i$ in good agreement with the experimental values.

A comparison of the energies (cm^{-1}) of all the ^{16}O$_i$ levels related to the 2DLFM, A$_{2u}$ and A$_{1g}$ modes deduced from the experimental results with the most recent ones obtained from self-consistent variational calculations trying to reproduce the relatives intensities of the O$_i$ lines and also from *ab initio* calculations is presented in Table 6.4.

The energy of the $|1,0,1,0\rangle$ state (A$_{2u}$ + LFE1) of ^{29}Si$_2$^{16}O and ^{30}Si$_2$^{16}O in qmi ^{29}Si and ^{30}Si has been measured to be 1201.4 and 1197.1 cm^{-1}, respectively, compared to 1205.7 cm^{-1} for ^{nat}Si$_2$^{16}O [27]. A comparison with the values of the energies of the $|0,0,1,0\rangle$ state (A$_{2u}$(0)) of Table 6.2 for these isotopic combinations

Table 6.4 Energies (cm^{-1}) of the vibrational levels of ^{28}Si$_2$O involving the 2DLFM, A$_{2u}$ and A$_{1g}$ modes of ^{16}O$_i$ in ^{nat}Si. The values deduced from the IR absorption spectra at low temperature (exp) are compared with the calculated (calc) values [44]. The levels are noted $|k, l, N, M\rangle$, where M is the number of excitations of the A$_{1g}$ mode, and the ground state is $|0,0,0,0\rangle$. The identifications of some of these levels with vibrational modes of O$_i$ are indicated. After [44]

Level	^{16}O exp	^{16}O calc.[a]	Remarks
$|0, \pm1, 0, 0\rangle$	29.25	32.8	2DLFM1, $N = M = 0$
$|0, \pm2, 0, 0\rangle$	67.0	68.8	
$|1, 0, 0, 0\rangle$	77.9	82.4	LFE0 ($N = M = 0$)
$|0, \pm3, 0, 0\rangle$	110.3	105.0	
$|1, \pm1, 0, 0\rangle$		127.6	
$|0, 0, 0, 1\rangle$	617.7	616.6	A$_{1g}$
		615.5[b]	
$|0, \pm1, 0, 1\rangle$	648.1	646.7	A$_{1g}$ +2DLFM1, $N = 0, M = 1$
$|0, \pm2, 0, 1\rangle$	686.7	683.5	
$|1, 0, 0, 1\rangle$	697.3	697.1	
$|0, \pm3, 0, 1\rangle$	~731	719.9	
$|0, \pm1, 0, 1\rangle$		744	
$|0, 0, 1, 0\rangle$	1136.4	1131.9	A$_{2u}$(0)
		1160.6[b]	
$|0, \pm1, 1, 0\rangle$	1157.6	1157.1	
$|0, \pm2, 1, 0\rangle$	1188.9	1187.6	
$|1, 0, 1, 0\rangle$	1205.7	1201.9	A$_{2u}$(0) +LFE1 ($N = 1, M = 0$)
$|0, \pm3, 1, 0\rangle$		1224.4	
$|1, \pm1, 1, 0\rangle$	1245.1	1251.6	
$|0, 0, 1, 1\rangle$	1748.6	1746.3	A$_{2u}$(0) +A$_{1g}$
		1776.0[b]	
$|0, \pm1, 1, 1\rangle$	1770.2	1772.0	
$|0, \pm2, 1, 1\rangle$	1801.1	1802.4	
$|1, 0, 1, 1\rangle$	1819.5	1816.6	A$_{2u}$(0) +A$_{1g}$ +LFE2 ($N = 1, M = 1$)
$|0, \pm3, 1, 1\rangle$		1840.8	
$|1, \pm1, 1, 1\rangle$	1860.6	1866.6	

[a] [44]
[b] [31] *ab initio*

shows that the $|1, 0, 1, 0\rangle$ - $|0, 0, 1, 0\rangle$ spacing decreases from qmi ^{29}Si (68.9 cm^{-1}) to qmi ^{30}Si (68.0 cm^{-1}) state. Such an energy decrease is also expected for the $|1, 0, 0, 0\rangle$ level (LFE0), and this could also be the case for the $|0, \pm1, 0, 0\rangle$ level (2DLFM1 line). It turns out that the frequency of the 2DLFM1 line has recently been measured at LHeT in a qmi ^{30}Si sample using a tunable source made of a BWO operating between 30 GHz and about 1.5 THz (1 to 50 cm^{-1}). At LHeT, the position of the 2DLFM1 line in the qmi ^{30}Si sample decreases by 0.4 cm^{-1} compared to the one in ^{nat}Si (Lassmann et al., to be published in Phys. Rev. B), in qualitative agreement with the above observations. Using the ratio of the frequency shifts between ^{nat}Si and qmi ^{30}Si for A$_{2u}$(0) given in Table 6.2 and for the 2DLFM1 line, one obtains for the 2DLFM1 transition approximate frequency shifts with respect to ^{28}Si$_2^{16}$O of only -0.10 and -0.20 cm^{-1} for the ^{28}Si^{16}O^{29}Si and ^{28}Si^{16}O^{30}Si components, respectively.

Linewidths and Lifetimes

At LHeT, the value of the linewidth (FWHM) of $A_{2u}(0)(Si_2^{18}O)$ measured in ^{nat}Si CZ samples with a resolution $\leq 0.2\,cm^{-1}$ is $0.53\,cm^{-1}$, a value slightly smaller than the one (0.58–$0.62\,cm^{-1}$) for $^{28}Si_2^{16}O$ in the same material or in FZ silicon, and these values are not expected to change significantly up to $\sim 10\,K$. Variation of the FWHMs of the 1136 ($A_{2u}(0)$) and $1128\,cm^{-1}$ ($A_{2u}(1)$) transitions of $^{28}Si_2O$ between 5 and 100 K has been calculated [24] using a model proposed by Yamada-Kaneta et al. [20] and values of 0.598 and $0.958\,cm^{-1}$ for $A_{2u}(0)$ and $A_{2u}(1)$, respectively, at LHeT. The FWHMs at 30 K obtained with this calculation are comparable to the ones deduced from Fig. 6.4. At 100 K, the FWHMs obtained by deconvolution are ~ 3 and $6\,cm^{-1}$ for $A_{2u}(0)$ and $A_{2u}(1)$, respectively. Besides the already mentioned results of De Gryse and Clauws [24], the temperature dependence of the FWHMs of several $^{16}O_i$-related LVMs has been measured between 10 and $\sim 60\,K$ in ^{nat}Si [46], and this dependence is represented in Fig. 6.8.

This figure shows that at 70 K, the FWHM of the $1136\,cm^{-1}$ line increases to about $1.8\,cm^{-1}$, and that the FWHM of the 2DLFM1 line (the $|0,0,0\rangle \rightarrow |0,0,1\rangle$) transition) experiences a spectacular increase from $0.09\,cm^{-1}$ at LHeT to about $1.6\,cm^{-1}$ near 60 K.

The difference between the LHeT value ($\sim 1.4\,cm^{-1}$) of the FWHM of $A_{2u}(0)(^{28}Si_2^{17}O)$ and those for the ^{16}O and ^{18}O isotopes has already been signalled and it is related below to lifetime measurements.

The homogeneous linewidth Γ_0 (cm^{-1}) of a lifetime-broadened transition is determined by the natural lifetime T_1 of the excited state, and it is given by:

$$\Gamma_0 = (2\pi c T_1)^{-1} \text{ or } \Gamma_0(cm^{-1}) = 5.31/T_1(ps). \tag{6.7}$$

Fig. 6.8 Temperature dependence of the FWHMs of the mid-IR $^{28}Si_2^{16}O$ components of $A_{2u}(0)$ and $A_{2u}(1)$ at 1136.4 and $1128.2\,cm^{-1}$, respectively, and of the far-IR DLFM1, 2DLFM2, and 2DLFM4 transitions of ^{16}O at 29.3, 37.7, and $48.6\,cm^{-1}$, respectively, in ^{nat}Si. Calculated curves are superimposed to the discrete experimental points (after [46])

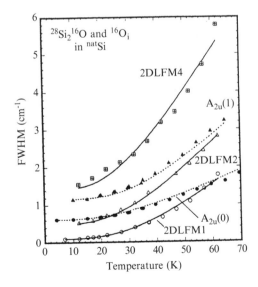

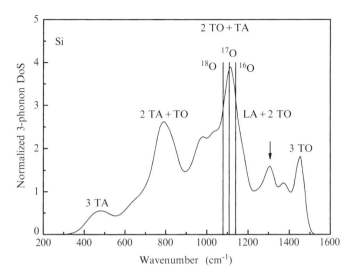

Fig. 6.9 Normalized three-phonon density of state of silicon at 0 K. The vertical lines indicate the positions of the A_{2u} lines of the different O isotopes (after [47]). Copyright 2004 by the American Physical Society

The differences on the FWHMs should transpose on the lifetimes if the IR lines are homogeneously broadened. The lifetimes at LHeT of the excited states of the $A_{2u}(0)$ transition for ^{16}O and ^{17}O in a ^{nat}Si sample from the same crystal as the one of Fig. 6.3 have indeed been measured in a pump-and-probe experiment (see [47] for the details) using a spectral band width of $\sim 15\,cm^{-1}$ for the optical excitation, and these lifetimes were found to be 11.5 ± 0.5 and 4.5 ± 0.4 ps for $A_{2u}(0)(^{28}Si_2^{16}O)$ and $A_{2u}(0)(^{28}Si_2^{17}O)$, respectively. The natural linewidths deduced from these lifetimes are 0.43 and $1.18\,cm^{-1}$ for ^{16}O and ^{17}O, respectively, in qualitative agreement with the spectroscopic values, and this indicates that homogeneous broadening is the main contribution to the FWHMs of these lines. In an attempt to explain these lifetimes by a thermal decay involving three lattice phonons [47], the normalized three-phonon DoS in silicon at 0 K is plotted in Fig. 6.9, where the vertical lines indicate the LHeT positions of the ^{16}O, ^{17}O, and ^{18}O A_{2u} modes of Fig. 6.3. It is seen that the ^{17}O mode coincides practically with the 2TO + TA density peak whereas the ^{16}O and ^{18}O modes lie on both sides of this peak. The corresponding values of the FWHMs seem to indicate that in ^{nat}Si, the lifetimes of the LVMs excited states are controlled by the decay of the excitation through coupling with phonons of the silicon lattice, and that the smaller the phonon DoS contribution at the LVM frequency, the longer the lifetime. This model predicts for the lifetime of $A_{2u}(0)$ of $Si_2^{18}O$ a value slightly smaller than the one for $Si_2^{16}O$, still in agreement with the small decrease of the FWHM observed for $Si_2^{18}O$ compared to $Si_2^{16}O$.

The positions and FWHMs of $^{16}O_i$ lines at LHeT in qmi ^{28}Si, ^{29}Si, and ^{30}Si samples have also been reported [27]. Very small differences with respect to the positions in ^{nat}Si are observed (see Table 6.2), but these qmi samples allow also the measurement of the positions of lines not possible with ^{nat}Si, like $A_{2u}(0)$ of $^{29}Si_2^{16}O$. A significant decrease of the FWHMs of $A_{2u}(0)$ of $Si_2^{16}O$ is observed in the qmi samples compared with ^{nat}Si: When it is $\sim 0.6\,cm^{-1}$ for $Si_2^{16}O$ in ^{nat}Si, they become 0.40, 0.23, and 0.30 cm^{-1} for $^{28}Si_2^{16}O$, $^{29}Si_2^{16}O$, and $^{30}Si_2^{16}O$ in qmi ^{28}Si, ^{29}Si, and ^{30}Si, respectively. The three-phonon DoS in qmi ^{28}Si is not expected to differ sensibly from the one in ^{nat}Si, and the decrease of the FWHM of $A_{2u}(0)(^{28}Si_2^{16}O)$ in the former material could then be attributed to the absence of isotope disorder or randomness (this absence has also been invoked [48] to explain the decrease of the electronic linewidth of shallow donors and acceptors in qmi Si). In qmi ^{29}Si and ^{30}Si, the effect of the low-frequency shift of the three-phonon DoS with respect to ^{nat}Si or qmi ^{28}Si should be partially attenuated by the lower frequencies of the $^{29}Si_2^{16}O$ and $^{30}Si_2^{16}O$ A_{2u} modes, and there again, the absence of isotopic disorder should play a role in the reduction of the FWHM in the qmi materials.

Besides ^{16}O and ^{17}O dependences, silicon isotope dependence of the lifetime of the $A_{2u}(0)$ mode of oxygen in silicon at LHeT has also been reported using a spectral band width of 1 cm^{-1} [49]. The lifetimes measured in this investigation for the $^{28}Si_2^{16}O$ and $^{28}Si_2^{17}O$ components in ^{nat}Si (11 and 4 ps, respectively) are practically the same as those found by Sun et al. [49], and they correlate with the values of the FWHMs of the corresponding IR components. On the other hand, the lifetimes measured for $^{28}Si^{16}O^{29}Si$ and $^{28}Si^{16}O^{30}Si$ are larger by a factor of ~ 2 and ~ 3, respectively, than the one measured for $^{28}Si_2^{16}O$. This difference is intriguing as the FWHM of the $^{28}Si^{16}O^{29}Si$ and $^{28}Si^{16}O^{30}Si$ measured on the IR spectra is nearly the same (0.56 cm^{-1}) as that of $^{28}Si_2^{16}O$.

The actual decay of an excited state can take place through more than one channel. The explanation proposed for the decay is that it includes invariably the production of a A_{1g} excitation of O_i. The remaining energy is $E(A_{2u}(0)) - E(A_{1g})$ and this difference corresponds to a spectral region where the one- and the two-phonon densities of state of silicon coexist, but while the two-phonon DoS is relatively monotonous, the one-phonon DoS shows a relatively important contribution below the silicon Raman frequency down to $\sim 475\,cm^{-1}$. It turns out that for ^{17}O, the above energy difference falls within the one-phonon DoS, while for the ^{16}O isotopic combinations, it falls at higher energies where the deexcitation involves two lattice phonons. This is an explanation for the relatively high values of the lifetime measured for the excited states of the $A_{2u}(0)(^{28}Si^{16}O^{29}Si)$ and $A_{2u}(0)(^{28}Si^{16}O^{30}Si)$ transitions [49].

Recent measurements of the temperature dependence of the lifetime of the excited state of the $A_{2u}(0)$ transition of $^{28}Si_2^{16}O$ at 1136.4 cm^{-1} have allowed to propose to identify the three phonons involved in the three-phonon decay channel of this excited state as the A_{1g} mode of $^{28}Si_2^{16}O$ at 612 cm^{-1}, a lattice phonon at 489 cm^{-1} and a low-energy phonon at 35 cm^{-1} [50]. The combination of A_{1g} with such a low-energy phonon is very close to the combination $A_{1g} + 2DLFM1$ giving the weak IR band at 648 cm^{-1} listed in Table 6.4.

Stress Effects and Reorientation

The dynamic configuration of O_i in silicon depicted in Fig. 1.6 results in four equally populated configurations along the <111>-oriented bonds from the central silicon atom, and it is characteristic of a trigonal centre in a cubic crystal (see Appendix E). The jump from one orientation to an equivalent one is energy consuming as at least one Si–O and one Si–Si bonds must be broken. The activation energy E_r for this reorientation has been measured in stress-induced dichroism experiments aimed toward the determination of the diffusion coefficients of O_i in silicon and germanium [51]. The method used is based first on the creation of a dissymmetry between the populations of the four configurations. It consists in applying to a crystallographically oriented CZ sample a uniaxial stress (values of $\sim$300 MPa have been used) at a temperature $\sim$400°C along a preferential crystal axis during a time $\sim$30 min and to cool down the sample under stress to RT, where the stress is released.[5] When the stress direction is properly chosen, this produces a difference between the populations $N_{//}$ and $N_\perp$ of centres whose <111> axes make angles $\theta_{//}<90°$ and $\theta_\perp \geq 90°$ with respect to the stress direction. This difference can be quantified by the alignment ratio $A = (N_\perp - N_{//})/(N_\perp + N_{//})$, which is positive. The stress-induced dichroism of a vibrational transition of the <111>-oriented O_i centre is then determined by measuring the absorption coefficients $K_{//}$ and $K_\perp$ for electric vectors of the radiation parallel and perpendicular, respectively, to the initial aligning stress. The dichroic ratio DR is $(K_\perp - K_{//})/(K_\perp + K_{//})$ and its sign depends on the orientation of the dipole moment of the transition with respect to the <111> orientation of the main axis of the centre: when the dipole moment is parallel to this axis (π or z dipole), as $N_\perp$ is larger than $N_{//}$, $K_\perp$ is also larger than $K_{//}$, but when the dipole moment is perpendicular to this axis (σ or (x, y) dipole), $K_{//}$ becomes larger than $K_\perp$. DRs for the O_i modes have been measured at RT [32, 35] and at LHeT [52]. The results obtained are summarized in Table 6.5 for aligning stresses along <110>.

These results confirm previous measurements for A_{2u} and they show that the dipole moment for the 515 cm^{-1} mode is in a plane perpendicular to the <111> axis.

In the seminal experiment of Corbett et al. [51], an aligning stress of 270 MPa was applied at 400°C along a <111> direction for 30 min, and the dichroic ratio of the 9 μm band at RT[6] for a viewing axis along <110> was 0.13 (as expected, no dichroism was observed for an aligning stress along <100>). From isothermal annealing of dichroism at 337°C, a value of the reorientation energy E_r of 2.56 eV was measured for O_i in silicon, from which the diffusion coefficient D given by:

$$D(\text{cm}^2\text{s}^{-1}) = 0.23\exp[-2.56/k_B T]$$

---

[5]The concentration of O-related thermal donors produced by this annealing can be neglected with respect to $[O_i]$.

[6]In this reference, the quantity measured was $K_\perp/K_{//}$ and its value was 1.30, from which a DR of 0.13 is deduced.

Table 6.5 Dichroic ratios (*DR*s) and dipole orientations of the main IR modes of O_i in silicon obtained by stress-induced dichroism. The alignment conditions are the stress magnitude, the temperature, and the duration. The lines positions are the ones given in the references. In all these measurements, the aligning stress is along <110> and the propagation vector of the polarized radiation (viewing axis) parallel to <100>. Dipole orientations have also been obtained from uniaxial stress measurements (see Table 6.6)

O_i mode	T	Position (cm^{-1})	Alignment conditions	*DR*	dipole
E_u	RT	515[a]	280 MPa at 420°C 30 min	−0.06	(x, y)
A_{1g} + 2DLFM1	LHeT	648.2[b]	280 MPa at 410°C 30 min	−0.079	(x, y)
A_{2u}	RT	1106[a]	280 MPa at 420°C 30 min.	0.12	z
$A_{2u}(0)$ +LFE1	LHeT	1206 (1203)[c]	250 MPa at 450°C 30 min.	0.18	z
"	"	"b	280 MPa at 410°C 30 min.	0.164	z
	RT	1227[d]	280 MPa at 430°C 30 min	0.23	z
$A_{2u}(0)$ + A_{1g}	RT	1720[d]	"	0.20	z
"	LHeT	1749[b]	280 MPa at 410°C 30 min	0.165	z
$A_{2u}(0)$ + A_{1g} + LFE1	"	1819.5[b]	"	0.0021	z

[a][32]
[b][52]
[c][14]
[d][35]

Table 6.6 Values of the piezospectroscopic parameters A_1 and A_2 and of the stress splitting coefficients $|8A_2/3|$ and $|2A_2|$ (cm^{-1} GPa^{-1}) of the O_i lines in silicon. They are deduced from measurements with σ // <111> and σ // <110>, respectively, at LHeT for 2DLFM1 and at 20 K for the other lines [14]. The lines positions are the ones given in the reference. The (x, y) and z notations correspond to dipole moments perpendicular and parallel to the Si⋯Si axis, respectively

Line position (cm^{-1})		A_1	A_2	$|8A_2/3|$	$|2A_2|$	Dipole moment
29.3	2DLFM1	−3.8 ± 0.3	−3.8 ± 0.3	10.1	7.6	(x, y)
517	E_u	∼1.1	∼0.3	∼0.8	∼0.6	(x,y)
1136	$A_{2u}(0)$	0.2 ± 0.1	−0.2 ± 0.1	0.5	0.4	z
1203	$A_{2u}(0)$ + LFE1	−0.9 ± 0.2	−2.3 ± 0.3	6.1	4.6	z

was derived [51]. This value is comparable to the one given in Table 1.2 and it shows that in this particular case the use of absorption measurements coupled with the application of stress can provide dynamic parameters associated with impurities.

In their work on O_i in silicon, Bosomworth et al. [14] also performed uniaxial stress measurements on all the O_i lines known at that time, from which they deduced values of the piezospectroscopic parameters A_1 and A_2 associated with a trigonal centre (see Table E.2 of Appendix E). These parameters are given in Table 6.6.

It is seen that the value of the piezospectroscopic parameters of the A_{2u} mode is small, and that this mode is relatively insensitive to stress.

The absorption of the A_{2u} mode of $^{16}O_i$ has been measured under hydrostatic pressures up to 7 GPa at temperatures from 4 to 20 K [53]. It shows a reduction with pressure of the separation between the $A_{2u}(0)$ ground and $A_{2u}(1)$ thermalized components of Table 6.2, which can no longer be resolved for pressures above at 5.5 GPa, and an overall decrease of their energies is observed. The reduction of the spacing between the two components is correlated with an increase of the intensity of the thermalized one with respect to the ground state one. These effects can be understood by assuming a reduction of the lattice spacing of silicon with pressure, which produces an outward buckling of the $^{16}O_i$ atom from the dynamic linear configuration, which transforms progressively the 2DLFM into a rotational motion and decreases the anharmonic coupling between this motion and the A_{2u} mode. The potential calculated from the experimental results at 6 GPa as a function of the radial distance r from the $Si_c \cdots Si_c$ axis of Fig. 6.6 shows a substantial increase of its value at the true BC configuration ($r = 0$) and a minimum at a value of r about twice the one at zero pressure, corresponding to the new equilibrium configuration of $^{16}O_i$. The energy of the $^{18}O_i$ A_{2u} mode follows the same negative shift with hydrostatic pressure as the $^{16}O_i$ mode. Inversely, the E_u resonant mode at $518\,cm^{-1}$ is expected to follow the positive shift with pressure of the Raman mode of silicon. The $^{18}O_i$ A_{2u} mode and the overtone of the E_u mode have been brought into resonance by hydrostatic pressure tuning at LHeT [36]. Below $\sim$3 GPa, the $^{18}O_i$ absorption follows the same trend as the one reported above for $^{16}O_i$, namely, a merging of the $A_{2u}(0)$ and $A_{2u}(1)$ transitions, but around 4.2 GPa, a new absorption peak appears on the low-energy shoulder of the merged peaks, which grows in size with increasing pressure, whereas the A_{2u} contribution disappears for pressures above 4.4 GPa. This has been attributed to the pressure-induced anticrossing of the A_{2u} mode with the components of the E_u overtone, where the new peak borrows some characteristics of the overtone. A quantitative model of this anticrossing has been worked out [36], which provides an excellent agreement with the observed energies and FWHM of the spectroscopic absorption features.

The profiles of the A_{2u} mode measured between LHeT and RT have been fitted with an anharmonic potential including the coupling between the A_{2u}, the A_{1g}, and the 2DLFM (ν_2 mode), with parameters derived from the components of the modes that are resolved at low temperature [25]. This vibrational potential has been shown to be consistent with the above hydrostatic stress results if the $Si - O - Si$ structure is stiffened by a factor of $\sim$4.5 relative to the nearest $Si - Si$ bonds. The coupling of the A_{1g} mode with the 2DLFM has been found to explain quantitatively the off-axis displacement of O_i under hydrostatic pressure [25].

O_i *Perturbation by Ge and Sn Atoms*

Silicon and germanium form mixed $Si_{1-x}Ge_x$ crystal alloys in all proportions, as the covalent atomic radii are not too different (0.11 nm for silicon and 0.12 nm for germanium). Tables relating the lattice parameters and the densities of these alloys to the values of x can be found in [54]. LVMs of O_i have been observed in these alloys, with x-dependent FWHMs, and in all cases these FWHMs are broader than in pure silicon.

In CZ silicon samples with [Ge] $= 3 \times 10^{20}$ cm^{-3} ($x = 0.006$ or $\sim$6000 ppma), in addition to the $A_{2u}(0)$ mode at 1136 cm^{-1}, absorption bands at 1129.8 and 1118.5 cm^{-1} were observed [55] at LHeT and attributed to the perturbation of the $A_{2u}(0)$ mode of O_i by a *nnn* Ge atom. A negative shift of the A_{2u} mode and positive shift of the 2DLFM1 mode of O_i with increasing [Ge] were also reported [56]. Experimental investigations for values of x between 1.3×10^{-4} and 0.0134 have globally confirmed these results and brought new insight on their understanding [57]. In the sample with $x = 0.0134$, at LHeT, besides the strong $A_{2u}(0)$ mode noted $O_i - I$, corresponding to the $A_{2u}(0)$ mode weakly perturbed by distant Ge atoms, two other modes are observed at lower frequencies, noted $O_i - II$ and $O_i - III$, (see Table 6.7). The relative intensities of these components can be appreciated from the absorption spectra of the $A_{2u}(0)$ mode displayed in Fig. 6.10 for different values of x of $Si_{1-x}Ge_x$ alloys.[7]

A later study of the perturbation of the absorption of the $A_{2u}(0)$ mode by Ge atoms has shown that $O_i - II$ could be resolved into the relatively strong component at 1130.1 cm^{-1} and a weaker one at 1127.2 cm^{-1} [58], noted here $O_i - II'$.

In CZ silicon samples doped with Sn at a concentration $\sim 2 \times 10^{19}$ cm^{-3} ($x = 0.0004$), LVMs corresponding to $O_i - II$, $O_i - II'$, and $O_i - III$ were also reported at frequencies slightly lower than those for Ge (see Table 6.7) [58]. The small shift of the $O_i - I$ LVM in the $Si_{1-x}Ge_x$ alloys can be due to the average increase of the silicon lattice parameter and to the average perturbation of the more distant Ge atoms.

In the $Si_{1-x}Ge_x$ alloys, positive shifts have been reported for the 2DLFM1 and $|0,0,0\rangle \rightarrow |1,0,1\rangle$ modes and their frequencies are given in Table 6.7. The reduction of the intensity of the $|0,0,0\rangle \rightarrow |1,0,1\rangle$, also observed with increasing [Ge] is attributed to a reduction of the anharmonic coupling between the A_{2u} mode and the low-frequency motion.

These LVMs, whose frequencies rule out the direct bonding of a Ge atom with O_i, which would produce a much larger negative shift, are ascribed by order of increasing energy to the $A_{2u}(0)$ mode perturbed by the presence of a Ge or Sn atom as a *nnn*, 3rd *nn*, and 4th *nn* of O_i, and the larger frequency shift for tin must be due to its atomic covalent radius, which is larger (0.14 nm) than the one of germanium (0.12 nm).

---

[7]In this figure the ordinate scale meter, initially 0.5, has been modified to take into account the O_i concentration re-evaluated with the O_i calibration factor presently admitted.

Fig. 6.10 Ge concentration dependence of the O_i-II and O_i-III components of the $A_{2u}(0)$ mode of $Si_{1-x}Ge_x$ alloys. For the lowest values of x, the absorption of O_i-I is truncated. Decomposition of the spectrum is shown for $x = 0.0134$, where the deconvoluted individual bands are represented by dotted curves (after [57])

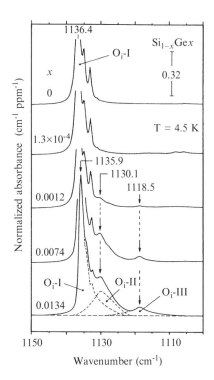

Table 6.7 Frequencies (cm^{-1}) at LHeT of the $O_i - I$, $O_i - II$, $O_i - II'$, and $O_i - III$ components of different O_i modes in a $Si_{1-x}Ge_x$ alloy compared with the position in pure silicon [57,58]. The frequencies in brackets correspond to the modes reported for $Si_{1-x}Sn_x$ with $x = 0.0004$ ([Sn] $= 2 \times 10^{19} cm^{-3}$)

x	Mode:	2DLFM1	$A_{2u}(0)$	$A_{2u}(1)$	$A_{2u}(2)$	$A_{2u}(0) + LFE1$
0		29.3	1136.4	1128.2	1121.9	1205.7
0.0134	$O_i - I$	30.1	1135.9	1128.1	1121.2	1206.3
0.0134	$O_i - II$	~33	1130.1 (1129.7)[a]	1126.2	1123	1213
0.004	$O_i - II'$		1127.2[a] (1124.8)[a]			
0.0134	$O_i - III$	~36	1118.5 (1109.2)[a]	1115.6		

[a][58]

Complementary spectroscopic measurements in $Si_{1-x}Ge_x$ alloys at LHeT for $0.024 < x < 0.066$ have been reported [59]. The higher Ge concentration required a deconvolution of the raw absorption spectra, but it still gave the three components $O_i - I$, $O_i - II$, and $O_i - III$. The deconvolution allows to determine the shift of $O_i - I$ of A_{2u} with x, which seems to be linear up to $x = 0.066$, with a value of $\sim -2\, cm^{-1}$ for this value of x. The shift of O_i-I of the $|0,0,0\rangle \rightarrow |1,0,1\rangle$ transition is somewhat larger, but sublinear with x [59].

These conclusions are qualitatively supported by the *ab initio* calculations for $Si_{1-x}Ge_x$ [28].

RT measurements performed on the whole Ge concentration range have shown that the 9 μm band can be followed up to $x \sim 0.6$ [60]. They show also that the shift of the peak frequency of this band is still linear with x, but that the shift rate is somewhat smaller than the one at LHeT reported earlier.

The use of the anharmonic potential has been extended to calculate the profiles of $A_{2u}(^{16}O)$ at different temperatures to O_i in $Si_{1-x}Ge_x$ [25]. This has allowed to investigate the profiles at 16 K of $O_i - I$, $O_i - II$, and $O_i - III$ components for x up to 0.066 and to give a better picture of the location of the Ge atoms in the vicinity of O_i.

Relative Intensities and Calibration

In order of increasing energies, the RT spectrum of O_i in silicon consists in the E_u mode shown in Fig. 6.5, the 9 μm band, a weak band at $1227 \, cm^{-1}$ and another one at $1720 \, cm^{-1}$ [34, 35]. The 9 μm band at $1107 \, cm^{-1}$ corresponds to the A_{2u} mode whose main component $A_{2u}(0)$ is observed at $1136.4 \, cm^{-1}$ at LHeT, the low-frequency shift at higher temperatures coming mainly from the thermalization of the population of the lower 2D levels (at 300 K, $k_B T$ is $\sim 210 \, cm^{-1}$). The $1720 \, cm^{-1}$ band is due to the combination of the A_{2u} and A_{1g} modes giving the $1748 \, cm^{-1}$ of Fig. 6.6a at LHeT. The band at $1227 \, cm^{-1}$ observed at RT is attributed to the combination of the A_{2u} mode with low-energy 2D excitations, but one must note that its frequency is larger than that of the $|0,0,0\rangle \rightarrow |1,0,1\rangle$ transition giving the line at $1206 \, cm^{-1}$ at LHeT, involving also 2D excitations (see Table 6.1). Assuming a value of $1106 \, cm^{-1}$ for A_{2u} at RT leads to a value of $121 \, cm^{-1}$ for the RT low-energy excitation. One notes that the frequency of the $1227 \, cm^{-1}$ band is close to two times that of the A_{1g} IR-inactive mode, but the existence of an IR-active overtone of A_{1g} seems very unlikely.

Table 6.8 gives a correlation between the bands observed at RT and the lines and bands observed at LHeT. The peak intensities of the 9 μm band and of the $A_{2u}(0)$ mode of $^{28}Si_2^{16}O$ are taken as the references at RT and at LHeT, respectively. Except for the $518 \, cm^{-1}$ band, the increase of the FWHM from LHeT to RT is not truly representative because the bands at RT are the envelopes of several thermalized components.

From the result of Murin et al. [43], the RI (peak) at LHeT of the combination mode at $1819.5 \, cm^{-1}$ (see Table 6.3) is $\sim 4 \times 10^{-5}$. Relative integrated intensities of the 1205.7, 1748.6, and $1819.5 \, cm^{-1}$ line with respect to that at $1136.4 \, cm^{-1}$ have been calculated [44]. They are 3.78×10^{-2}, 1.84×10^{-2}, and 2.6×10^{-4}, respectively, and it must be noted that the first two ones scale well with the experimental values of Table 6.8.

When the global absorption coefficient of the 9 μm band of a CZ silicon sample is measured at RT with an air or vacuum reference, a two-phonon contribution of the absorption of the silicon lattice at this frequency, amounting to $\sim 0.85 \, cm^{-1}$, must be subtracted. Correction factors for the transmission measurement arising from the back-surface conditions on commercial wafers polished on one side and etched on the other or from the free-carrier absorption of the samples are discussed in the review by Bullis ([61], and references therein). One notes in Table 6.8 an inversion in the relative intensities of the bands near 1200 and $1700\,\text{cm}^{-1}$ between RT and LHeT. A value of $35.2\,\text{cm}^{-1}$ has been measured for the ratio of the integrated to peak absorptions of the $9\,\mu\text{m}$ band at RT (Pajot, unpublished). At LNT, the peak absorption of the O_i band at $1128\,\text{cm}^{-1}$ is about 2.6 times larger than the one of the $9\,\mu\text{m}$ band at RT [62].

The ratio of the peak absorptions of the $A_{2u}(0)(^{28}\text{Si}_2^{16}\text{O})$ at $1136\,\text{cm}^{-1}$ at LHeT (<6 K) and of the $9\,\mu\text{m}$ band at RT cannot be measured on the same sample because the peak absorption is much larger at LHeT than at RT. A value of ~ 33 can be deduced from Fig. 5 of [24], obtained with thinned samples for the LHeT measurement. From the value of the peak absorption of $A_{2u}(0)(^{28}\text{Si}_2^{18}\text{O})$ at $1085\,\text{cm}^{-1}$, knowing the relative abundances of the ^{18}O and ^{16}O isotopes, a value of ~ 38 is found for this ratio (Pajot, unpublished).

The knowledge of the O_i concentration $[O_i]$ in CZ silicon is of technological importance because it determines the value of some parameters of thermal treatments related to the production of the O-related thermal double donors or of oxygen precipitation. As O_i is electrically inactive, the use of vibrational absorption is the only practical choice for a non-destructive determination of $[O_i]$. Many values of the calibration factor between the peak absorption coefficient at RT of the $9\,\mu\text{m}$ band and $[O_i]$ measured on plane-parallel polished samples have been proposed (Table VI of [61]). They are based on the determination of the total O concentration using mainly charged particle activation analysis and/or gas fusion analysis to determine the absolute oxygen concentration (for a review, see [61]). The most recent values of the RT peak calibration factor are similar and close to $3.0 \times 10^{17}\,\text{cm}^{-2}$ or $6.0\,\text{ppma/cm}^{-1}$ [63–66]. The calibration factor of [66] ($3.096 \times 10^{17}\,\text{cm}^{-2}$ or $6.204\,\text{ppma/cm}^{-1}$), rounded to $3.1 \times 10^{17}\,\text{cm}^{-2}$ or $6.2\,\text{ppma/cm}^{-1}$ is the one used here. A calibration factor of $4.8 \times 10^{17}\,\text{cm}^{-2}$ or $9.6\,\text{ppma/cm}^{-1}$ (ASTM F-121 1979), based on [67], was still used in some papers in the 2000s and it is mentioned here for comparison.

Table 6.8 Characteristics at LHeT and at RT of the peak absorptions of $^{16}O_i$ bands or lines in natural silicon. The relative peak and integrated intensities (*RI*) are normalized to that of A_{2u} ($A_{2u}(0)$ at LHeT). The positions of the $^{18}O_i$ bands at RT are indicated in brackets. At RT, the A_{2u} absorption is known as the $9\,\mu\text{m}$ band

Attribution	2DLFM1	E_u	A_{1g}+2DLFM1	A_{2u}	A_{2u}+LFE1	A_{2u} +A_{1g}
			LHeT			
Position (cm^{-1})	29.25	517.8	648.2	1136.36	1205.7	1748.6
FWHM (cm^{-1})	0.09[a]	~ 4.8	~ 3	0.58	3.0	3.7
RI (peak)	4×10^{-2}	1.5×10^{-2}	9×10^{-4}	1	9×10^{-3}	2.8×10^{-3}
RI (integrated)	5×10^{-3}	0.25		1	3.4×10^{-2}	1.6×10^{-2}
			RT			
Position (cm^{-1})		514.1		1107.1	1226.7	1720.1
		$(513.5)^{\text{b}}$		$(1058.5)^{\text{b}}$	$(1172.6)^{\text{b}}$	$(1670)^{\text{b}}$
FWHM (cm^{-1})	-	7.6		33	22	31
RI (peak)		0.26		1	1.1×10^{-2}	1.6×10^{-2}

[a][15]
[b][22]

From the value of the ratio (38) of the peak absorptions of the A_{2u} 1136 cm^{-1} mode of $^{16}O_i$ at LHeT to the RT 9 μm band at 1107 cm^{-1}, the extrapolated calibration factor of the peak of the 1136 cm^{-1} line below $\sim$7 K for a resolution $\leq$0.2 cm^{-1} is 8.2 × 10^{15} cm^{-2} or 0.164 ppma/cm^{-1}. Calibration factors for larger spectral resolutions and an expression allowing to determine [O_i] by absorption measurements up to 100 K have also been given [24]. The LHeT peak calibration factor proposed in this reference for a resolution of 1 cm^{-1} is 2.2 × 10^{16} cm^{-2}. The peak calibration factor of the absorption of the 1085 cm^{-1} line of $A_{2u}(0)(^{18}O_i)$ at LHeT for a resolution $\leq$0.2 cm^{-1} to determine [$^{18}O_i$] is 7.5 × 10^{15} cm^{-2}, due to the slightly lower value of the FWHM for $^{18}O_i$ compared to $^{16}O_i$. From this, under the same experimental conditions, a value of [$^{16}O_i$] can be eventually estimated from the peak absorption of the 1085 cm^{-1} line using a calibration factor of 3.6 × 10^{18} cm^{-2} (Pajot, unpublished).

An integrated calibration factor of 9.58 × 10^{15} cm^{-1} (0.192 ppma/cm^{-2}) for the whole isotopic structure of the $A_{2u}(0)$ mode of ^{16}O at LHeT has been estimated (Pajot, unpublished) giving [$^{16}O_i$] from the integrated absorption of the whole $A_{2u}(^{16}O_i)$ mode at LHeT. Such a calibration factor should be independent from the spectral resolution.

In fact, LHeT absorption measurements of [O_i] in CZ Si are necessary only with thin samples ($d \sim$0.2 mm or below), and such measurements are useful in heavily doped CZ silicon which display a strong photoionization background absorption [68] and strong free-carrier absorption at RT. The LHeT measurements with thicker samples are restricted to FZ samples, and the absorption of the $A_{2u}(0)$ mode of O_i in such a sample is shown in Fig. 6.11. Evidence of O_i absorption in a FZ sample at LHeT has also been given by Sassella [69].

The estimated limit of detection of O_i in silicon by measuring the 1136 cm^{-1} absorption at LHeT in $\sim$ 5-mm-thick FZ samples at a resolution of 0.2 cm^{-1}

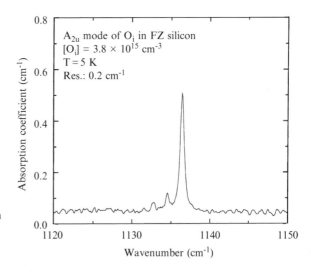

Fig. 6.11 Absorption of the $A_{2u}(0)$ mode of $^{16}O_i$ at 1136 cm^{-1} in a 2-mm-thick FZ silicon sample. The peak absorption is 0.46 cm^{-1}. The interference fringes have been locally suppressed by subtracting a sine function Pajot, unpublished

A_{2u} mode of O_i in FZ silicon
[O_i] = 3.8 × 10^{15} cm^{-3}
T = 5 K
Res.: 0.2 cm^{-1}

Absorption coefficient (cm^{-1})

Wavenumber (cm^{-1})

is $\sim 5 \times 10^{13}\,\mathrm{cm}^{-3}$, and it can be lowered to $\sim 10^{13}\,\mathrm{cm}^{-3}$ using a multi-reflection geometry like the one of Fig. 3.4. This limit is comparable to the one obtained with acoustical phonon absorption at 1 K of the 2DLFM1 at 878 GHz [17].

The IR calibration factors are determined with CZ Si samples where all the oxygen present in thermal donors or silica precipitates has been converted into O_i by annealing near 1350°C followed by quenching. Thus, the measurement of the oxygen content of an unknown silicon sample from the IR absorption of the $1108\,\mathrm{cm}^{-1}$ band informs only on the O_i concentration. In CZ samples containing silica precipitates, the IR absorption of these precipitates interferes with the 9 μm band of O_i at $1108\,\mathrm{cm}^{-1}$. It has been suggested [70] to use then the weak absorption of the $A_{2u}+A_{1g}$ combination band at $1720\,\mathrm{cm}^{-1}$ at RT with a proper calibration for the determination of $[O_i]$, at the expense of a loss of sensitivity (see Table 6.8). In the presence of silica precipitates, the use of the peak absorption of the $1205\,\mathrm{cm}^{-1}$ band at LHeT has also been proposed, with a calibration factor not too different from the one for $A_{2u}(0)$ scaled by the relative peak intensity of the $1205\,\mathrm{cm}^{-1}$ band given in Table 6.8 [24].

The absorption measurements of O_i at RT have been used not only to measure the average oxygen concentration of the crystals, but also, at a smaller scale, to probe the radial O_i distribution in CZ silicon wafers. It has been shown that in CZ silicon grown in the absence of a magnetic field, $[O_i]$ decreases near from the surface of the CZ ingot [62, 71]. Radial fluctuations of the oxygen distribution in CZ silicon crystals have also been measured using a PbTe–PbSnTe laser tuned at $1106\,\mathrm{cm}^{-1}$, and correlated with swirls patterns [72], or by using a micro-FTIR mapping system [73].

6.1.1.2 The Dimer (O_{2i}) and the TDDs

The formation in silicon with a dispersed O_i distribution of O-related thermal double donors (TDDs) containing several O atoms implies the existence at the beginning of O_i atoms close from each other. The simplest entity is the O_i dimer, usually noted O_{2i} (called also dioxygen), and the next one is the O_i trimer, O_{3i}. Possible structures for these centres are shown in Fig. 6.12.

These complexes should display vibrational modes close to the ones of O_i. The stabilities of different configurations of O_{2i} have been calculated [74, 75]. The most stable O_{2i} configuration calculated by Öberg et al. [74] is the skewed one, but the energy of the staggered configuration is only 0.14 eV higher, while the inverse result, is found in [75] with the energy of the skewed configuration 0.1 eV higher. In another O_{2i} geometry more symmetric than the skewed one, the two O atoms of the $O_i - Si - Si - O_i$ configuration would lie in the same $\{110\}$ plane, but this configuration is found to be less stable than the other ones by $\sim$1 eV.

Two relatively broad absorptions at 1012 and $1060\,\mathrm{cm}^{-1}$, observed at LHeT in as-grown CZ silicon near from the A_{2u} absorption of O_i, are attributed to O_{2i} [76], and they can be seen in Fig. 6.13.

Fig. 6.12 Calculated structures of interstitial oxygen complexes in silicon: (**a**) isolated O_i in the C_{1h} (puckered) configuration, (**b**) staggered oxygen dimer, (**c**) skewed $O_i - Si - Si - O_i$ configuration of the dimer (the two O_i atoms are not in the same {110} plane), (**d**) staggered oxygen trimer, (**e**) ring-like oxygen trimer. Dark spheres represent Si atoms and light grey spheres O atoms [45]. Copyright 1999 by the American Physical Society

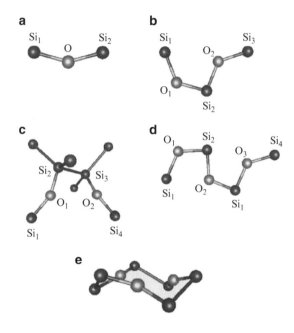

Their positions do not vary much between RT and LHeT, but a small uncommon negative energy shift between RT and LHeT [34, 35] is observed.[8] Two lower frequency LVMs at 690.1 and 555.8 cm^{-1} (LHeT) are also associated with the above LVMs, with a normal positive energy shift between RT and LHeT (note that the (C_s, O_i) centre, discussed at the end of Sect. 6.1.1.4, gives also a LVM at 690 cm^{-1}). When ^{16}O is replaced by ^{18}O in the silicon samples, the three LVMs with the highest frequencies shift from 1059.8, 1012.4, and 690.1 cm^{-1} to 1011.8, 969.2, and 679.7 cm^{-1}, respectively, (LHeT), but the one at 555.8 cm^{-1} shows practically no IS [77]. In silicon samples containing the same proportions of ^{16}O and ^{18}O, new LVMs are observed, in addition to the pure ^{16}O- and ^{18}O-related LVMs, representing coupled modes of mixed $^{16}O-^{18}O$ dimers [74]. It has been noted that the intensities of these LVMs, and hence the O_{2i} concentration, increase in samples electron irradiated above 250°C [78]. Another sharper LVM at 1105.4 cm^{-1} only observed at LHeT, also shown in Fig. 6.13, has been attributed to another O_{2i} configuration, with a $^{16}O - ^{18}O$ IS of 48.2 cm^{-1}, but this configuration does not seem to be produced in notable proportions by electron irradiation [76].

The LVMs of O_{2i} shown in Fig. 6.13 have also been observed in qmi ^{29}Si and qmi ^{30}Si [79]. In these samples, they invariably show a low-frequency shift with respect to ^{nat}Si of $\sim$2–5 cm^{-1} for qmi ^{29}Si and of $\sim$6–8 cm^{-1} for qmi ^{30}Si.

---

[8]The line at 1012.3 cm^{-1} was reported at 1013 cm^{-1} at RT [38, 39], but not attributed to the O_i dimer at that time.

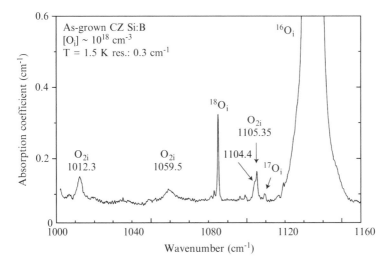

Fig. 6.13 Peak absorption (cm^{-1}) of vibrational modes of O_{2i} in the vicinity of the $A_{2u}(0)$ mode of O_i in $CZ^{nat}Si$ (see text). The mode at $1104.4\,cm^{-1}$ is due to LVM A of the (C_s, O_i) centre. The extrapolated peak absorption of $A_{2u}(0)(^{28}Si_2^{16}O)$ at $1136.4\,cm^{-1}$ is $\sim123\,cm^{-1}$ and the one of $A_{2u}(0)(^{30}Si_2^{16}O)$ at $1129.2\,cm^{-1}\sim0.14\,cm^{-1}$. There is a small instrumental broadening of the $A_{2u}(0)$ mode of $^{18}O_i$ (and of $^{16}O_i$!) (after [18])

On the basis of first-principles calculations [75], the four LVMs at 556, 690, 1012, and $1060\,cm^{-1}$ are presently attributed to the staggered configuration, and the $1105\,cm^{-1}$ LVM to the skewed configuration of O_{2i} shown in Fig. 6.12. With these attributions, the relatively strong interaction between the two O atoms of the staggered geometry can be the origin of the relatively large widths of the 1012 and $1060\,cm^{-1}$ LVMs. Inversely, the LVM attributed to skewed geometry is much sharper and this could reflect the smaller interaction between the two O atoms in this configuration, but while other LVMs have been calculated to be associated with the skewed configuration, no other mode than the one observed at $1105\,cm^{-1}$ has apparently been reported for that geometry. One LVM at $1006.8\,cm^{-1}$ observed in CZ silicon samples after heat treatment in the temperature region where TDDs start to be produced has been ascribed to an oxygen trimer O_{3i} ([80], and references therein). The next steps in the aggregation lead to the formation of TDDs.

Since O_{2i} and O_{3i} are the precursors of the TDDs and as LVMs have also been found to be associated with TDDs [81], it seems logical to try present here an atomic model of the TDDs [82], which will help for a discussion of the LVMs associated with these centres. The starting point is a staggered O_{ni} structure along a $<110>$ direction, an extension of the staggered trimer of Fig. 6.12, where the O_i atoms are bonded to nearest neighbour Si atoms. In the notation of the Exeter group [82], who proposed this TDDs model, this structure is labelled O-1NN, where 1NN is for nearest neighbour. The distortion induced by such a structure produces a transition to a structure where the O atoms become bonded to second nearest neighbours

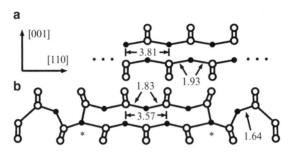

Fig. 6.14 Most stable calculated structures in silicon for (**a**) O-2*NN* infinite chain of alternated Si and O atoms and (**b**) the TDD model O$_9$–2*NN*. All the lengths are in Å. O and Si atoms are shown as black and white circles, respectively [83]. The asterisks indicate the tri-coordinated O atoms. Because of the 2D rendering, there are two Si atoms superposed in the upper and lower circles of each chain, hence the apparent double bonds. Copyright 2001 by the American Physical Society

(2*NN*). To preserve the normal valence of the O and Si atoms, this transition induces a splitting of the initial O-1*NN* structure into two parallel chains where bonding takes place only within one chain. This new structure, labelled O-2*NN*, is shown schematically in Fig. 6.14a. These two new chains, which have approximately the same length, must rebond to the linear Si structure along <110>. This can take place through the overcoordination of the two terminal O atoms of the O-2*NN* structure. Such a structure becomes then electrically active as it leaves an unpaired electron on each of the two overcoordinated O atoms at the ends of the chain, and this is an explanation of the double-donor character of similar structures.

It is found that in real silicon crystal, the most stable structure is obtained by adding one O$_i$ atom at both ends of the O-2*NN* structure, as shown in Fig. 6.14b and this configuration should correspond to the general TDD structure in silicon and in germanium. If the O-2*NN* structure contains N (here, N is an integer) O atoms, the new TDD structure contains n $= N + 2$ atoms, and it is labelled O$_n$ − 2*NN*. The symmetry of this structure depends on the parity of N: for N odd, the symmetry is C_{2v}, and it is C_{2h} for N even. The differentiation between the various TDDs comes from the different values of N. Calculated stresses for different O$_n$ − 2*NN* centres show that O$_5$–2*NN*, O$_6$–2*NN*, and O$_7$–2*NN* could, for instance, be related to TDD2, TDD3, and TDD4 in silicon, respectively [83].

A comparable model for TDDs has also been proposed by the group of the Helsinki University of Technology ([84], and references therein).

In silicon, LVMs related to the TDDs are observed in the 945–1000, 700–730, and 575–580 cm^{-1} spectral regions and a comparison of the observed and calculated values in the high-frequencies region are given in Table 6.9.

These high-frequency LVMs are attributed to the vibration of the O$_i$ atoms at the very ends of the O$_n$ − 2*NN* structures, as the one in Fig. 6.14b. When using samples containing a mixture of ^{16}O and ^{18}O, no other LVM than the high-frequency ones already observed in samples containing only ^{16}O or ^{18}O is observed, showing that the O atom giving this LVM is decoupled from the other O atoms. This differs from

Table 6.9 Comparison of the high-frequency LVMs (cm^{-1}) of TDDi observed at RT with the ones calculated for the $O_{i+4}-2NN$ as a function of i [83]

i		1	2	3	4
Calc.	^{16}O	940	951	963	969
Obs.	^{16}O	975	988	999	1006
ʺ	^{18}O		945	955	

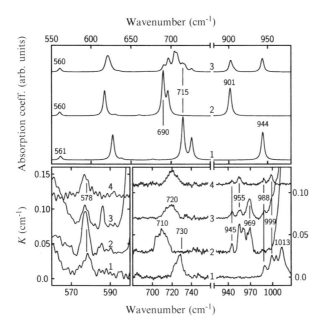

Fig. 6.15 Calculated LVMs absorption spectra of O_7-2NN (*top*) compared to RT spectra observed after annealing at 470°C in silicon (*bottom*). Labels 1, 2, and 3 correspond to ^{16}O-rich, ^{18}O-rich, and $^{16}O/^{18}O$ mixed samples, respectively. In spectrum 4, the spectrum of an unannealed sample was subtracted from 3. The lines at 1013 and 969 cm^{-1} in the observed spectra 1 and 2, respectively, are those of O_{2i} [83]. *K* is the absorption coefficient. Copyright 2001 by the American Physical Society

the LVMs in the 700–730 cm^{-1} region, where mixed modes are observed with the $^{16}O-^{18}O$ mixture, proving that the O atoms giving these latter LVMs are coupled together and in a close spatial proximity. The calculated IR spectra for O_7-2NN are compared in Fig. 6.15 with LVM spectra of TDD2 and TDD3 observed at RT.

The electronic spectra of the first TDDs display metastability [85] discussed in [86]. This is observed in n-type silicon as a decrease of the intensities of the spectra of these TDDs in samples cooled from RT under TEC down to ∼100 K and below, compared to the ones in the samples cooled under band-gap light illumination. Metastability can be complete (no spectrum) or partial (a reduction of the intensity). For these samples, the stable state at RT and after cooling under TEC is electrically inactive, and it is generally noted X, while the metastable state is the usual TDD.

When focusing on LVMs, the effect of metastability has been observed at RT by comparing the LVM spectrum obtained in samples illuminated by the unfiltered IR source with the one obtained with a filter blocking band-gap radiation [87]. For TDD1 and TDD2, a new LVM at $1020\,\text{cm}^{-1}$ is attributed to the stable states X1 and X2 of these two centres.

6.1.1.3   Oxygen Precipitation

CZ silicon is oversaturated with O_i at RT and TDDs are already present in as-grown material. When annealing CZ silicon at temperatures above $\sim600°C$, absorption bands attributed to SiO_x precipitates are observed in the $800–1300\,\text{cm}^{-1}$ region.[9] These precipitates have been shown to be very effective sinks for polluting TM impurities introduced during technological steps, and for this reason, the kinetics of their formation has been actively investigated. Different annealing sequences are used to produce these precipitates, including usually a pre-treatment intended to generate nuclei for further precipitation, followed by another heat treatment at a higher temperature where further precipitation takes place. This can end by a normal return to RT or better by quenching to prevent a modification during the cooling down. A broad SiO_x absorption band at $1225\,\text{cm}^{-1}$ was reported in CZ silicon samples annealed at 1000°C for 64 h after a pre-anneal of 15 min at 1300°C, while a pre-anneal of 24 h at 1200°C produced another band near $1100\,\text{cm}^{-1}$, indicating the existence of different kinds of precipitates [88]. The results of IR absorption measurements on CZ silicon samples with different values of [C] have been correlated with those of transmission electron microscopy (TEM) by Gaworzewski et al. [89]. With reference to Fig. 6.16, they can be summarized as follows: prolonged annealings (~100–250 h) at 600°C produce a very broad band at $1080\,\text{cm}^{-1}$, correlated with the observation in TEM of rod-like defects (a). Annealings between 850 and 1000°C in samples with high [C] produce a band near $1100\,\text{cm}^{-1}$, less symmetrical than the preceding one, and correlated with a high density of small defects (b). For smaller [C], annealing between ~900 and 1000°C produces a pair of bands at about 1120 and $1225\,\text{cm}^{-1}$, correlated with irregularly shaped precipitates of relatively large size (~200 nm) and platelets (c). In a sample with undetectable [C], annealing at 1275°C for 2 h produces two bands at 1105 and $\sim470\,\text{cm}^{-1}$ comparable to the ones observed in bulk silica, with octahedral-faceted precipitates and long extended stacking faults (d).

The pair of bands near 1100 and $1225\,\text{cm}^{-1}$ has also been reported in CZ samples with a ring-shaped distribution of precipitates in the ring area, where the density of stacking faults is high [90]. The LHeT absorption of O precipitates in samples submitted to prolonged nucleation and thermal treatments can show structures associated with different kinds of precipitates [91].

---

[9]It is likely that a small concentration of precipitates is already present in as-grown crystals, but the related bands are too broad to be observed when they are weak.

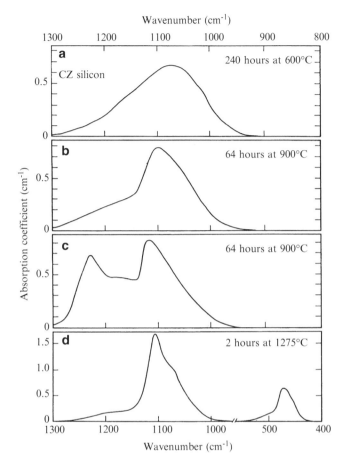

Fig. 6.16 Absorption bands at RT of different types of (Si,O) precipitates produced in CZ silicon with initial concentrations $[O_i]_0$ by different annealings (**a**): $[O_i]_0 = 1.4 \times 10^{18}$ cm^{-3}. (**b**): $[O_i]_0 = 1.5 \times 10^{18}$ cm^{-3}, $[C] \sim 10^{17}$ cm^{-3}. (**c**): $[O_i]_0 = 1.4 \times 10^{18}$ cm^{-3}, $[C] \sim 6 \times 10^{16}$ cm^{-3}. (**d**): $[O_i]_0 = 1.6 \times 10^{18}$ cm^{-3} (after [89])

The nature and stoichiometry of these precipitates has been studied from the interpretation of IR absorption measurements and from the results of TEM measurements, and it is now admitted that the band near 1100 cm^{-1} in Fig. 6.16c is due to polyhedral precipitates while the one near 1225 cm^{-1} to 2D defects. These 2D defects are expected to show a {100} preferential location and this has been proven from absorption measurements with polarized radiation on samples cut from wafers of CZ crystals grown along a <100> axis [92]. In these samples, the intensity of the 1230 cm^{-1} band was observed to be maximum when the electric vector of the radiation was parallel to the <100> growth axis, but zero for radiation polarized in the (100) plane of the wafer perpendicular to the growth axis. This effect showed that the platelets are formed preferentially in this plane and not in the (010) and

(001) planes (in that case, the $1230\,\text{cm}^{-1}$ band would have been also observed for radiation polarized in the (100) plane). Such a selective formation is in relation with the thermal gradient determined by the pulling conditions [92].

A phenomenological model of the SiO_x precipitates based on effective medium theory has been used by Borghesi et al. [93], yielding $x \sim 1.8$, close to a SiO_2 precipitate. This result has been challenged by introducing additional constraints for the solution of the model equations and relying on the basic assumption that the SiO_x precipitates consist in a mixture of amorphous silicon and amorphous SiO_2, yielding $x \sim 1.2$ [94].

In O-doped germanium crystals grown in silica crucibles, a SiO_x band is observed at $1100\,\text{cm}^{-1}$ [95]. For ^{18}O doping of germanium, this band has also been observed, shifted to $1057\,\text{cm}^{-1}$ [96].

6.1.1.4 O_i Interaction with Other Foreign Atoms

The interaction of O_i with Ge (and Sn) has been discussed separately in Sect. 6.1.1.1 because Ge forms a mixed crystal with silicon, but the atoms discussed in the present section must be rather considered as impurity atoms. Oxygen has an electronegativity of 3.44 in Pauling's scale and it can interact with atoms with small electronegativity. Interstitial oxygen produces also a local outward distortion of the silicon lattice, so that the elastic energy of the lattice can be reduced by interaction with substitutional atoms with small radii, but O_i can also interact with interstitial atoms. The interaction of O_i with the interstitial N split pair is discussed in Sect. 6.2.2.2.

Carbon

In CZ silicon, the interaction of O_i with residual carbon produces several complexes which have been investigated since the end of the 1960s. In as-grown CZ silicon, LVMs observed at 1104 and $1052\,\text{cm}^{-1}$ at LHeT, denoted the A and B lines, respectively, were reported ([97], and references therein). The A LVM at $1104\,\text{cm}^{-1}$ is shown blended with an O_{2i} mode in Fig. 6.13. The A LVM displays no C IS but it is correlated to LVMs at 589, 640, and $690\,\text{cm}^{-1}$, noted X, Y, and Z, respectively, which display a C IS (see Table 3 of [98]). The B LVM is correlated to three C_s-related LVMs at 716.1, 724.9, and $744.1\,\text{cm}^{-1}$, noted P, Q, and R, respectively [97]. In addition to A and B, two satellites of A at 1100 and $1108\,\text{cm}^{-1}$ and another LVM at $1129.1\,\text{cm}^{-1}$ were also observed in CZ silicon samples with $[C_s]$ $\approx 8 \times 10^{17}\,\text{cm}^{-3}$ [58]. The atomic configurations and vibrational frequencies of (C,O) complexes in silicon have been calculated by Kaneta et al. [99]. In this reference, the $1104\,\text{cm}^{-1}$ LVM is attributed to the $A_{2u}(0)$ mode of O_i perturbed by a C_s *nnn* atom, and the low-frequency modes X, Y, and Z (very weak in as-grown material) to the vibration of this C_s atom, in a centre noted CO-2, while the LVM at $1052\,\text{cm}^{-1}$ is tentatively attributed to a centre involving two O atoms. A structure noted CO-4 involving possibly C_s atom 3rd *nn* of O_i, for which the perturbation of the O_i mode would be smaller than the one for the $1104\,\text{cm}^{-1}$ was left unattributed, but it could be attributed now to the LVM at $1129\,\text{cm}^{-1}$. In [58], in correlation with the spectra observed in the $Si_{1-x}Ge_x$ and $Si_{1-x}Sn_x$ alloys discussed in Sect. 6.1.1.1, there are some differences with the above attributions: the B, A and $1129\,\text{cm}^{-1}$ LVMs, are assumed there to be the vibration of the $A_{2u}(0)$ mode of O_i perturbed by C_s atoms nnn, 3rd nn, and 4th nn of O_i, respectively, and the satellites of A are explained by non-equivalent locations of the 3rd nn C_s atom. One can note that in C-doped CZ silicon, the lattice expansion induced by BC O_i is reduced when C_s is located close to it due to the smaller covalent atomic radius of C ($\sim 70\%$ the one of Si) while the presence of Ge_s or Sn_s increases this expansion.

In CZ samples annealed at $500°C$, LVMs associated with other (C_s, O_i) complexes were also reported at 1027, 1055, 1100, $1108\,\text{cm}^{-1}$ and at 588.3, 640.1, and $666.9\,\text{cm}^{-1}$ [97].

Different (C,O) complexes are produced in CZ silicon irradiated with h–e electrons and they are discussed later in Sect. 7.2.4.

Hydrogen

The interaction of O_i with hydrogen has been studied in CZ samples where hydrogen was introduced either by proton or deuteron implantation at low temperature [100], or by high-temperature anneal in a hydrogen atmosphere followed by quenching [101, 102]. The atomic configuration of hydrogen interacting with O_i has been found to be different in the two methods of introduction.

In unannealed samples implanted at $20\,\text{K}$ with protons, in addition to the LVM of isolated H_{BC}^+ at $1998\,\text{cm}^{-1}$ (see Sect. 8.2.1), the absorption measurements performed at LHeT with a resolution of $1\,\text{cm}^{-1}$ revealed an O-related LVM at $1077\,\text{cm}^{-1}$ and an H-related one at $1879\,\text{cm}^{-1}$ [100]. Additional absorption measurements at LHeT on ^{16}O and ^{18}O-containing samples implanted at $\sim 20\,\text{K}$ with protons and/or deuterons, either as-implanted or as a function of annealing between 80 and $240\,\text{K}$ were explained by assuming the existence of two centres involving oxygen and hydrogen. The LVMs of the one noted OH_I are observed in the as-implanted samples, and they are no longer observed in samples annealed at $140\,\text{K}$. The LVMs of the centre noted OH_{II} start to be observed after annealing at $\sim 120\,\text{K}$ and they disappear after annealing at $\sim 240\,\text{K}$. The spectra of samples co-implanted with protons and deuterons show no new LVM, indicating that these centres contain only one hydrogen atom. The probability of trapping of hydrogen by as-grown O_{2i} is considered as negligible so that a single O atom (O_i) is assumed to be involved in OH_I and OH_{II}. The frequencies of the LVMs associated with these centres for different isotopic combinations are given in Table 6.10.

One notes that in OH_I, the O-related mode is practically unchanged when H is replaced by D, indicating a weak interaction between O and hydrogen in this centre, but in OH_{II}, an unusual positive shift is observed in that case ($+\sim 4$ and $+0.6\,\text{cm}^{-1}$ for ^{16}O and ^{18}O, respectively). The intensities of the hydrogen-related

Table 6.10 Frequencies (cm^{-1}) at LHeT of the O- and hydrogen-related local modes of the OH_I and OH_{II} centres in silicon for different isotopic combinations [100]

Isotopic combination	OH_I		OH_{II}	
	O-relat.	Hydrogen-relat.	O-relat.	Hydrogen-relat.
(^{16}O, H)	1076.9	1879.2	1028.5	1830.4
(^{16}O, D)	1076.7	1358.0	1032.3	1329.1
(^{18}O, H)	1028.3	1879.1	987.1	1830.1
(^{18}O, D)	1028.0	1358.0	987.7	1330.6

LVMs of Table 6.10 are larger than those of the corresponding O-related LVMs, with intensity ratios of 6 ± 1 and 3 ± 0.5 for OH_I and OH_{II}, respectively [100].

Ab initio calculations considering an ionized (O$_i$, H$_{BC}^+$) centre with H at first, second, or third BC sites from O$_i$ (O$_i$H$_{BC1}^+$, O$_i$H$_{BC2}^+$ or O$_i$H$_{BC3}^+$, respectively) have been performed. From their results, O$_i$H$_{BC1}^+$ has been eliminated as a possible candidate because of the rather large ISs associated with this structure. It has been suggested that OH_{II} may be represented as H$^+$ located at the second- or third-nearest BC site from O$_i$, and that in OH_I, H$^+$ occupies a third- or fourth-nearest BC site [100].

In CZ silicon samples annealed at ~1300°C in a H$_2$ or D$_2$ atmosphere and quenched to RT, the observation at LHeT of the absorption of an (O$_i$, hydrogen) LVM peak (P$_H$ or P$_D$) LVM at 1075.1 or 1076.6 cm^{-1} has been reported, with an unresolved shoulder (S$_H$ ir S$_D$), as shown in Fig. 6.17 [102]. These frequencies are close to those of Table 6.10 for the O-related mode of (^{16}O, H) and (^{16}O, D) of OH_I.

The deconvolution of the (P$_H$, S$_H$) and (P$_D$, S$_D$) structures of Fig. 6.17a, b, assumed to be associated with the A$_{2u}$(^{16}O$_i$) perturbed by the presence of hydrogen has allowed to evaluate ^{28}Si$_2^{16}$O$-^{28}$Si^{16}O^{30}Si ISs of 3.6 and 3.7 cm^{-1} for the P$_H$ and P$_D$ modes, respectively, comparable to the value of 3.6 cm^{-1} for isolated A$_{2u}$(^{16}O$_i$). A deconvolution of the spectra obtained with samples annealed under gas mixtures with different H$_2$/D$_2$ proportions showed that two hydrogen atoms were involved in the hydrogen perturbing centre, and that this centre could be a hydrogen molecule. This latter point and the Si isotope shift can be explained by the perturbation of isolated O$_i$ by a hydrogen molecule. The existence of two distinct structures P$_H$ and S$_H$ of the O$_i$ mode could be explained by distinct locations of the hydrogen molecule near from O$_i$ or to a splitting associated with a single hydrogen molecule present in the ortho ($I = 1$) or para ($I = 0$) state.

The absorption of sharp (FWHMs ~0.1–0.2 cm^{-1}) stretch modes of a perturbed hydrogen molecule H$_2$ in the ~3600–3800 cm^{-1} spectral range, with isotope effects for HD and D$_2$ at lower frequencies was also reported in the same samples [102]. For each isotopic combination, two different hydrogen stretch modes, noted ν_1, ν_2, could be related to each isotopic composition. Correlation measurements performed between the integrated absorptions of modes ν_1 and ν_2 and that of the P$_H$–S$_H$ structure showed that they were related to the same centres [102]. On the basis of a hydrogen molecule partially polarized by its location in the silicon lattice, ν_1 and ν_2 were finally ascribed to distinct configurations of this hydrogen molecule perturbing

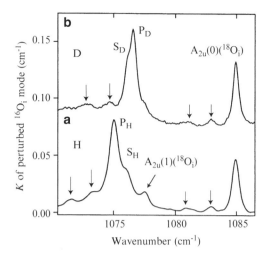

Fig. 6.17 Absorption spectra at $\sim 10\,K$ of CZ silicon samples annealed in (**a**) hydrogen and (**b**) deuterium. They show the absorption of unperturbed A_{2u} ($^{18}O_i$) mode, with Si isotopic components marked by arrows (see Table 6.2), and the absorption of A_{2u} (0) ($^{16}O_i$) perturbed by two configurations of nearby hydrogen molecules, giving the unresolved P and S modes observed near $1075\,cm^{-1}$ for H and for D. The Si isotopic pattern of these modes is also marked by arrows. The resolution is $0.1\,cm^{-1}$ (after [102])

Table 6.11 Frequencies (cm^{-1}) at LHeT of the perturbed modes P_H and S_H of $^{16}O_i$ and of the ν_1 and ν_2, modes of the polarized perturbing hydrogen molecule in hydrogenated CZ silicon for different substitutions of the hydrogen isotopes. The unperturbed frequency of the interstitial hydrogen molecule in silicon and in the gas form is given for reference (after [102])

	P_H	S_H	ν_1	ν_2	Unperturbed	Hydrogen gas
H_2	1075.1	1075.8	3788.9	3780.8	3618.4	4161.1
HD	1076.3		3304.3	3285.3	3265.0	3632.1
D_2	1076.6	1076.0	2775.4	2715.0	2642.6	2993.6
				2716.0		

O_i [102]. The frequencies of the perturbed O_i modes and of the ν_1 and ν_2 modes are given in Table 6.11.

Further investigations of the absorption of the hydrogen molecule perturbing O_i in silicon have been reported ([103] and Sect. 8.3.6). They show that splittings related to the ortho and para states of the hydrogen molecule are indeed at the origin of modes ν_1 and ν_2, and that only one lattice configuration is concerned. This point is discussed in Sect. 8.3 devoted to the vibrational spectroscopy of molecular hydrogen in semiconductors.

These absorptions associated with O_i perturbed by molecular hydrogen are observed to disappear for annealings between ~ 110 and $125°C$, but they can be retrieved by subsequent annealing at $50°C$. However, annealing above $250°C$ produces an irreversible bleaching of the absorption. An activation energy of $0.78\,eV$ has been found for the migration of H_2 from the vicinity of O_i [104].

Lithium

Electronic EM spectra associated of Li_i donors perturbed by O_i have been reported by several groups [105–107]. In CZ B-doped silicon with high [B] values and diffused with Li to reduce the free carrier absorption, interaction between interstitial lithium and O_i has been spectroscopically demonstrated [12]. At RT, a LVM at $1006.5\,cm^{-1}$ is related to the perturbation of ν_3 (A_{2u}) by Li^+. At LHeT, the corresponding LVM is observed at $1016.6\,cm^{-1}$, with a FWHM of $\sim 3\,cm^{-1}$ and no Li IS is observed, confirming that this LVM is mainly related to the O_i motion. Compared to the FWHM of isolated O_i, the width can be related to the presence of several kinds of (Li,O) complexes. No counterpart of the $1207\,cm^{-1}$ band is observed in the Li-diffused samples. LVMs observed at 524.8 and $537.3\,cm^{-1}$ at LNT in samples diffused with ^7Li_i and ^6Li_i, respectively, were attributed by Chrenko et al. [12] to the perturbation by Li of the E_u mode of O_i, but these modes are actually due to a resonant mode of the Li^+ ion of the B^-Li^+ pair (see Table 5.11).

To end this presentation, it is worth mentioning that in CZ silicon, the ESR spectra A17 and A18 observed after electron irradiation at 100 K and below, and related to LVM modes at $935\,cm^{-1}$ and at 945 and $956\,cm^{-1}$, respectively, were attributed to complexes of O_i with unknown entities [108]. In this reference, the centres associated with A17 and A18 were shown to anneal out at $\sim 400\,K$ ($\sim 120°C$) and $\sim 280\,K$ ($\sim$ RT), respectively. From IR studies of the LVMs produced in Ge-doped silicon electron-irradiated at 90 K, it has been suggested that the entity associated with O_i in the centre giving the $935\,cm^{-1}$ LVM was the self-interstitial I [109, 110]. At the light of more recent results [111], it seems that the 945 and $956\,cm^{-1}$ LVMs observed at LNT in CZ silicon electron-irradiated at LNT can be ascribed to (IO_i) and that the $935\,cm^{-1}$ LVM, whose intensity increases after RT annealing is due to the complexing of two interstitials with O_i (I_2O_i). Combined ESR and DLTS measurements on CZ silicon samples irradiated at LNT with light ions (protons and α-particles) have shown that the A18 ESR spectrum was related to a monoclinic-I configuration of a metastable centre identified as an (IO_i) complex, and characterized by two other ESR spectra, AA13 and AA14, with trigonal and triclinic symmetries, respectively. The DLTS measurements have shown that an energy level at $E_v + 0.13\,eV$ was associated with this centre ([112], and references therein).

6.1.2   Interstitial Oxygen in Germanium

6.1.2.1   Isolated Oxygen (O_i)

In CZ germanium, oxygen is not a residual impurity because graphite crucibles are generally used to contain the melt, but germanium can be doped with oxygen at a maximum concentration $\sim 7 \times 10^{17}$ at cm^{-3}, for instance by adding water vapour to the growth atmosphere. The interest of a comparison between the behaviour of oxygen in silicon and germanium led to the preparation and to the study of oxygen-doped germanium, including its infrared absorption [7, 113]. At RT, the main absorption, due to the vibration of O_i in germanium, is an asymmetric band at 855.6 cm^{-1} ($\sim$11.7 μm) with a FWHM of 5.3 cm^{-1}, showing a low-frequency shoulder at about 843 cm^{-1} [96,113–115]. This band is similar, *mutatis mutandis*, to the 9 μm band in silicon, and in germanium, it is due to the antisymmetric vibration of an interstitial O atom bonded to two *nn* Ge atoms, with symmetry B_1. In the following, it is referred to as the ν_3 band or mode. A weak band at 1260 cm^{-1}, with a FWHM of $\sim$20 cm^{-1}, was also reported to grow in intensity proportionally to the 856 cm^{-1} band in O-containing germanium [114]. It is ascribed to the combination ($\nu_3 + \nu_1$) of the ν_3 band with the symmetric IR inactive ν_1 bending mode of Ge_2O, and it is equivalent to the 1720 cm^{-1} band observed at RT in silicon. Another weak O_i-related band is also observed at 927 cm^{-1} and it been attributed to a vibration-rotation combination band of ν_3 with the rotational motion of O_i [96]. The shoulder of the 856 cm^{-1} band and the 927 cm^{-1} band appear in the spectra measured above $\sim$100 K and their intensities increase at the expense of the 856 cm^{-1} peak. The spectroscopic and isotopic characteristics of these three bands at RT are given in Table 6.12.

When the temperature is lowered, the 856 cm^{-1} band shifts to higher frequencies. Qualitative results were reported by Kaiser [114], and a shift to 862 cm^{-1} at 80 K is shown in [116], together with a $^{16}O - ^{18}O$ IS of +44 cm^{-1} similar to the one at RT. Further discussion on the low-temperature spectrum of O_i in germanium needs a mention of the isotopic composition of germanium. In silicon, it has been shown that because of Si natural isotopic abundance, three relatively sharp components due to the most intense Si isotopic combinations of the ν_3 or $A_{2u}(0)$ mode of Si_2O are observed in ^{nat}Si at LHeT. Now, ^{nat}Ge has five isotopes, ^{70}Ge, ^{72}Ge, ^{73}Ge, ^{74}Ge, and ^{76}Ge with percent abundances of 20.38, 27.31, 7.76, 37.62, and 7.83, respectively. Fifteen different vibrational Ge $-$ O $-$ Ge combinations are thus expected, from which only ten have different masses or average masses (AMs) M of the Ge atoms for a given combination M_1-M_2 of the Ge masses (among them, 70 u for 70 – 70, 72.5 u for 72 – 73, and 74 u for 74 – 74 and 72 – 76).

A high-resolution absorption measurement of the ν_3 band of O_i in ^{nat}Ge at LHeT near 862 cm^{-1} showed in a spectral interval of $\sim$3.3 cm^{-1}, 25 resolved or partially resolved sharp isotopic components with FWHMs $\sim$0.03 cm^{-1} [117]. From the study of the well-isolated highest-energy $^{70}Ge_2O$ combination, it was found that at LHeT, three lines were associated with each isotopic combination. They were attributed to the coupling of the antisymmetric vibration of Ge_2O_i with a

Table 6.12 Spectroscopic and isotopic characteristics of the O_i-related bands in ^{nat}Ge at RT. The value for the IR inactive ν_1 mode is the difference between $\nu_3 + \nu_1$ and ν_3 (after [96])

		ν_3	$\nu_3 +$ rotation	$\nu_3 + \nu_1$	ν_1
RT position (cm^{-1})	$^{16}O_i$	855.6	927	1263.7	408
	$^{18}O_i$	811.7	882	1209	397
FWHM (cm^{-1})		5.3	$\sim$18	18	
Relative intensity		1	0.003	0.012	

low-frequency rotational motion of the O_i atom about the Ge$\cdots$Ge axis, the highest energy one, noted $A_{2u}(0)$ in silicon, corresponding to the rotational ground state, and the two other ones, $A_{2u}(1)$ and $A_{2u}(2)$ in silicon, to the coupling with thermalized rotational levels. These thermalized transitions are already observed at LHeT in germanium because of the smaller values of the rotational energies.

It was not possible to observe the modes of all the isotopic combinations in ^{nat}Ge because of interferences between components with comparable energies. Measurements on qmi germanium samples [118, 119] allowed to observe in better conditions the three above mentioned lines, noted I, II, and III by order of decreasing energy in the latter reference, and to detect a fourth one (IV), as shown in Fig. 6.18. Near LHeT, the relative intensities of these four lines are temperature dependent.

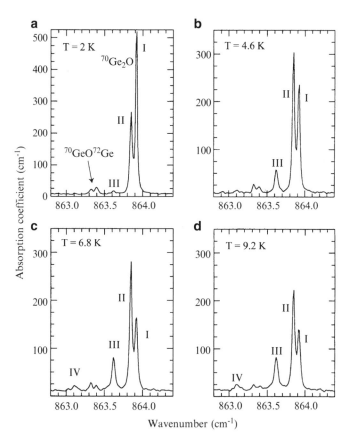

Fig. 6.18 Absorption of the vibration-rotation lines I, II, III, and IV of the ν_3 mode of ^{70}Ge$_2^{16}$O in qmi ^{70}Ge in spectra (**a**) to (**d**) for increasing temperatures. They show the increasing population of the higher rotational levels with temperature. The unapodized resolution is $0.015\,\text{cm}^{-1}$ (after [119]). In [117], lines I, II, and III were noted H, C, and L, respectively. Copyright 1994 by the American Physical Society

Above ~ 10 K, the broadening with temperature of these lines is very rapid and the Ge isotopic structure disappears near 15 K [118].

In the rigid rotator model, the rotational energy E_l of the O atom is given by:

$$E_l = \frac{\hbar^2}{2I_B}l^2 - Dl^4, \tag{6.8}$$

where $l = 0, \pm 1, \pm 2, \ldots$ The moment of inertia I_B is μr_0^2 is the product of the reduced mass μ of the Ge_2O quasi-molecule by the square of r_0, the distance of the O atom from the Ge$\cdots$Ge axis, and D is the centrifugal distortion coefficient of expression (6.1). In describing the states $|n, l, N\rangle$ (N is the number of excitations of the ν_3 mode), a notation borrowed from [20], the four lines correspond to transitions from the $|0, l, 0\rangle$ ground state to the $|0, l, 1\rangle$ state.

At a difference with O_i in silicon, the hindered rotation of O_i in germanium has been measured directly only by phonon spectroscopy, as measurements by optical spectroscopy require to work at or below 1 K in the $\sim$5–20 cm^{-1} (0.6–2.5 meV or 150–600 GHz) energy domain.

Phonon spectroscopy experiments in the 0.4–1 K temperature domain have allowed to determine the energies of the first $|0, l, 0\rangle$ states and to show that the $l = \pm 3$ state is split into $l = -3$ and $+3$ states by the sixfold angular perturbation of the potential due to the six nnn of the O atom hindering free rotation [120]. In ^{nat}Ge, rotational energies for the $|0, \pm 1, 0\rangle$, $|0, \pm 2, 0\rangle$, $|0, -3, 0\rangle$, $|0, +3, 0\rangle$, $|0, \pm 4, 0\rangle$, and $|0, \pm 5, 0\rangle$ states averaged for ^{nat}Ge isotopic composition were given as 0.18, 0.67, 1.37, 1.65, 2.66, and 4.08 meV (1.45, 5.4, 11.0, 13.3, 21.5, and 32.9 cm^{-1}), respectively, from the results of phonon resonance spectroscopy under uniaxial stress [120]. Later, slightly lower values were reported [121].

Lines I, II, III, and IV of Fig. 6.18 correspond to the $|0,0,0\rangle \rightarrow |0,0,1\rangle$, $|0, \pm 1, 0\rangle \rightarrow |0, \pm 1, 1\rangle$, $|0, \pm 2, 0\rangle \rightarrow |0, \pm 2, 1\rangle$, and $|0, -3, 0\rangle \rightarrow |0, -3, 1\rangle$ transitions, respectively, with rotational energies slightly different in the vibrational ground and excited states. The transitions labelled I, II, and III in germanium correspond to those we have labelled $A_{2u}(0)$, $A_{2u}(1)$, and $A_{2u}(2)$, respectively, in silicon. A line attributed to the $|0, +3, 0\rangle \rightarrow |0, +3, 1\rangle$ transition has been observed for $^{76}Ge_2^{16}O$ in ^{nat}Ge (Pajot, unpublished results) and it has been labelled line V.

The rotation of the O atom about the Ge$\cdots$Ge axis, for a configuration similar to the one of Fig. 6.1, is not easy to figure out as it requires considerable flexibility of the Ge$-$O bonds. This weak coupling with the Ge atoms leads to the possibility of a contribution of the Ge atoms to the rotation resulting in a rotational isotope effect of germanium, and this effect will be discussed later. In [117], a self-consistent value of the apex angle of $140°$ for the Ge $-$ O $-$ Ge puckered structure was obtained from the interpretation of this spectrum, indicating that the motion of the O atom in the plane perpendicular to the Ge$\cdots$Ge axis is a hindered rotation rather than a 2D vibration, as in silicon. Another consequence of the reduced value of the apex angle is that the symmetry of the $Ge_3 \equiv Ge - O - Ge \equiv Ge_3$ structure is no longer D_{3d}, but C_{1h}, so that the *IR* associated with the antisymmetric vibration is A", but this mode is usually noted ν_3, following the molecular use.

Frequencies

The overall infrared spectrum of O_i in germanium has been calculated with a dynamical matrix model based on first-principles total-energy calculations describing the potential energy for the atomic motions ([26], and references therein). In this calculated spectrum appear four distinct features at 877, 416, 230, and 0 cm^{-1}. The latter frequency corresponds to the free rotation of oxygen, as the model used does not describe the potential that hinders O rotation, and there is therefore a free O rotation that results in a zero-frequency mode in the harmonic approximation used for these calculations. The other three frequencies correspond, in order of decreasing energy to the ν_3 antisymmetric mode, the ν_1 bending mode, and ν_2 to the rocking mode where the O atom vibrates in a plane perpendicular to the Ge$\cdots$Ge axis. The observation of an IR ($\nu_3 + \nu_1$) combination band at 1264 cm^{-1} has been mentioned before, but the ν_2 mode has not been observed.

An overall spectrum of the absorption of $\nu_3({}^{16}O_i)$ in ^{nat}Ge at 1.6 K is shown in Fig. 6.19.

One notes that the resolution of the I–II doublet is less for $M = 75$u than for $M = 70$u, and this is correlated with a measured FWHM of 0.030 cm^{-1} for component I of ^{70}Ge$_2$O compared to 0.045 cm^{-1} for ^{76}GeO^{74}Ge. Without further details, typical FWHMs $\sim$0.02 cm^{-1} for lines I and II and $\sim$0.03 cm^{-1} for lines III and IV have been reported from measurements at a resolution of 0.009 cm^{-1} [119].

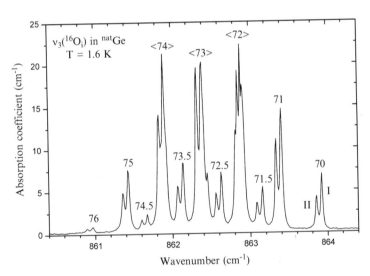

Fig. 6.19 Absorption spectrum of the ν_3 mode of Ge$_2^{16}$O$_i$ in ^{nat}Ge. At 1.6 K, only one doublet (lines I and II) is clearly seen for eight combinations involving only one isotopic pair, with the AMs of the Ge atoms indicated. The three largest features, showing also saturation, are due to two (for <72> and <74>) and three (for <73>) different isotopic pairs with the same value of the AM. The average I–II spacing is 0.07 cm^{-1} and the unapodized resolution 0.013 cm^{-1} [23]. Copyright 1997 by the American Physical Society

In the spectrum of Fig. 6.19, because of the reduced temperature, only lines I and II are observed for the eight distinct Ge mass combinations 70–70, 70–72, 70–73, 72–73, 73–74, 73–76, 74–76, and 76–76. For <72>, <73>, and <74>, involving two or three isotopic combinations with the same AMs, more than two unresolved or partially resolved lines are observed. They are explained as follows: In a linear approximation, the frequency for an asymmetric isotopic substitution XYX' of a symmetric X_2Y molecule is taken as the mean of the frequencies of the XYX and $X'YX'$ molecules. For the cases where the frequency of XYX' could also be measured (see Table 6.2 for $^{28}Si^{16}O^{30}Si$ in silicon), it has indeed been found that this approximation gave a very small deviation from the experimental value. Now, in germanium, where different Ge isotopic combinations for Ge_2O_i can also correspond to the same AM, due to the small linewidths and the resolution used, different combinations with the same AM are observed to vibrate at slightly different frequencies. Fig. 6.20 shows a partial view of another spectrum of the ν_3 vibration rotation of $^{16}O_i$ in ^{nat}Ge at 1.6 K illustrating this point.

In Fig. 6.20, the two weak doublets noted 72.5 and 73.5 correspond to the 72–73 and 73–74 Ge pairs, respectively (see Table 6.14). The vertical lines in this figure represent relative intensities of lines I and II of different Ge_2O atomic combinations calculated from the Ge isotopic abundances. The line labels in this figure correspond to the bars with the highest and lowest energies for each AM.

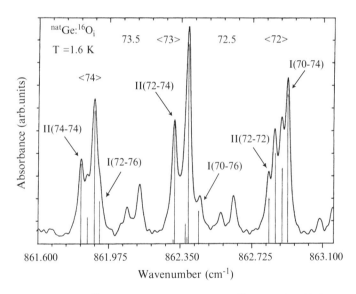

Fig. 6.20 Portion of the $\nu_3(^{16}O_i)$ absorption spectrum in a ^{nat}Ge sample with less oxygen than the one used for Fig. 6.19. It shows only lines I and II of different Ge isotopic combinations (the I/II intensity ratio is ~ 1.7 in this spectrum) and frequency differences of the vibrations of Ge_2O atomic combinations with the same Ge AM can be observed (see text for details). The unapodized resolution is $0.013\,cm^{-1}$ and one abscissa graduation is $0.075\,cm^{-1}$ (after [23]). Copyright 1997 by the American Physical Society

For $<73>$, the intensities of the lines of the $73 - 73$ combination are too weak for their observations. The components resolved in the spectrum for AM $= 72$u compare reasonably with (1) the relative intensities predicted for $^{74}\text{GeO}^{70}\text{Ge}$ and $^{72}\text{Ge}_2\text{O}$ and (2) the relative intensities of I and II at 1.6 K; the observed splittings for AM $= 73$ and 74 u can be similarly explained. These results show that for a Ge $-$ O $-$ Ge atomic structure with the same Ge AM, the one with the largest difference of the Ge masses has the highest frequency. The experimental shifts observed for the ν_3 mode of $\text{Ge}_2^{16}\text{O}_i$ are compared in Table 6.13 with the ones derived from the spectra calculated following the method used to calculate O and Si ISs of Si_2O of O_i in silicon [30].

In an *ad hoc* modelling, this behaviour can be rationalized by fitting the mass deviation $\Delta M_{\text{Ge}} = (M_{\text{Ge1}} - M_{\text{Ge2}})/2$ to a parameter C through:

$$\tilde{\nu}_3(\Delta M_{\text{Ge}}) = \tilde{\nu}_3(0) - C(\Delta M_{\text{Ge}})^2, \qquad (6.9)$$

where $\tilde{\nu}_3(0)$ is the frequency of the mode for the symmetric isotopic configuration with the Ge mass $M = (M_{\text{Ge1}} + M_{\text{Ge2}})/2$. It is assumed that a linear term cannot exist since the frequency of the mode cannot depend on which one of the Ge atoms is the heavier. From the experimental results, a value of $C \sim 0.06$–0.07 cm^{-1} is obtained, to be compared with a calculated value of 0.010 cm^{-1} [23]. This effect is not particular to germanium as for the A_{2u} mode of $^{16}\text{O}_i$ in $^{\text{nat}}\text{Si}$, the frequency of the $^{28}\text{SiO}^{30}\text{Si}$ asymmetric combination is about 0.1–0.2 cm^{-1} larger than that of the $^{29}\text{Si}_2\text{O}$ symmetric one (see Table 6.2).

Measured positions of the $\nu_3(^{16}\text{O}_i)$ lines are listed in Tables 6.14 and 6.15. In $^{\text{nat}}\text{Ge}$, except for the first and last isotopic combinations, the interferences between the positions of the lines allow to observe only lines I and II. The average values of the I–II and I–III spacings are 0.07 and 0.30 cm^{-1}, respectively.

Lines IV and V of $^{76}\text{Ge}_2\text{O}$ in $^{\text{nat}}\text{Ge}$ have been measured at 860.18 ± 0.01 and 859.97 ± 0.01 cm^{-1}, respectively.

The positions of the lines in qmi samples differ slightly from those in $^{\text{nat}}\text{Ge}$, and they can also differ with the isotopic purities of the isotopic samples. They are given for some combinations in Table 6.15.

The peak frequency of $\nu_3(^{16}\text{O}_i)$ at RT has also been measured in different qmi samples (Pajot, unpublished results). For qmi ^{70}Ge, ^{73}Ge, ^{74}Ge, and ^{76}Ge samples, the values (accuracy: ± 0.1 cm^{-1}) are $857.1, 855.4, 855.0$, and 854.1 cm^{-1},

Table 6.13 Ge ISs (cm^{-1}) measured at LHeT of the frequencies of line I of the ν_3 mode for Ge_2^{16}O combinations with the same AM (u), but a different isotopic configuration compared with the calculated ones. For the values of $C(\Delta M_{\text{Ge}})^2$, see text (after [23])

	AM	Experim.	Calc.	$C(\Delta M_{\text{Ge}})^2$
I(70 – 74)– I(72 – 72)	72	$+0.031$	$+0.044$	$4C$
I(70 – 76)– I(72 – 74)	73	$+0.063$	$+0.084$	$8C$
I(72 – 74)–I(73 – 73)	73	$\sim + 0.006$	$+0.010$	C
I(72 – 76)–I(74 – 74)	74	$\sim + 0.02$	$+0.040$	$4C$

Table 6.14 LHeT positions (cm^{-1}) of the isotopic combinations (Combin.) of lines I, II, and III lines of $v_3(^{16}O_i)$ in ^{nat}Ge ($^{72.64}$Ge). The relative intensities (RI) refer to that of the (70 - 72) combination. The values in brackets are calculated from expression (6.4) with $m = 11.86$ u, $\alpha = 69.18°$ and the frequency of ^{70}Ge$_2$O as a reference. The shifts of the last column refer to line I. The estimated accuracy is ± 0.003 cm^{-1} unless otherwise indicated

Combin.	AM	RI	I	II	III	Shift/^{70}Ge$_2$O
				Line:		
70 – 70	70	0.373	863.923	863.853	863.628	–
70 – 72	71	1.000	863.401	863.332		0.522
			863.401^a	863.334^a		
70 – 73	71.5	0.284	863.152	863.083		0.771
70 – 74	72	1.378	862.912	862.844		1.011
72 – 72	//	0.670	862.881	862.812		1.042
72 – 73	72.5	0.381	862.630	862.561		1.293
			862.629^a	862.564^a		
70 – 76	73	0.287	862.452	862.38 ± 0.01		1.471
72 – 74	//	1.846	862.389	862.318		1.534
73 – 73	//	0.054	(862.385)	(862.315)	(862.091)	(1.538))
73 – 74	73.5	0.525	862.141	862.069		1.782
72 – 76	74	0.384	861.93 ± 0.01	861.85 ± 0.01		1.99
74 – 74	//	1.271	861.896 ± 0.006	861.826 ± 0.006		2.027
73 – 76	74.5	0.109	861.674	861.604		2.249
74 – 76	75	0.529	861.430	861.361	861.133	2.493
			861.428^a	861.365^a	861.137^a	
76 – 76	76	0.055	860.962	860.894	860.659	2.961
			(860.950)	(860.880)	(860.656)	(2.973)

^{a}[119]

respectively, compared to 855.6 cm^{-1} in ^{nat}Ge ($^{72.64}$Ge). The difference between the frequencies for qmi ^{70}Ge and ^{76}Ge is 3.0 cm^{-1} at RT, close to the one (2.94 cm^{-1}) between the frequencies of I(^{70}Ge$_2$O) in qmi ^{70}Ge and of I(^{76}Ge$_2$O) in qmi ^{76}Ge at LHeT. The FWHM of v_3 in the qmi samples at RT is ~ 5 cm^{-1}, not very different from the one in ^{nat}Ge, and this shows that the FWHM of the mode at RT is principally related to the rotation of O$_i$ and to the interaction with the lattice.

A comparison of Tables 6.14 and 6.15 shows that the LHeT positions of the lines of the same v_3 isotopic combinations of Ge$_2^{16}$O$_i$ in ^{nat}Ge and in qmi Ge are very close, but that they consistently reveal small differences, which can be detected because of the small FWHMs of the lines: when measured in a qmi sample enriched with $\sim 90\%$ ^{76}Ge, the positions of lines I, II, and III of ^{76}Ge$_2^{16}$O are 0.019, 0.017, and 0.031 cm^{-1} higher in energy, respectively, than in ^{nat}Ge. This is illustrated for lines I and II of ^{76}Ge^{16}O^{74}Ge in a ^{nat}Ge and in a qmi ^{76}Ge sample in Fig. 6.21.

Inversely, the comparison of the frequencies of the components of ^{70}Ge$_2^{16}$O in ^{nat}Ge and in qmi ^{70}Ge in Tables 6.14 and 6.15 shows that the frequencies of the lines in the qmi sample are slightly smaller than in ^{nat}Ge. This is better seen in the values of the I(^{70}Ge$_2$O) $-$ I(^{76}Ge$_2$O) spacings in ^{nat}Ge and qmi ^{70}Ge and ^{76}Ge in

Table 6.15 LHeT Positions (cm^{-1}) of some of the I, II, III, and IV lines of $v_3(^{16}O_i)$ in some qmi Ge samples (Pajot, unpublished), [119]. The estimated accuracy is ± 0.005 cm^{-1} except when otherwise specified. The shift refers to line I. The qmi ^{73}Ge used was too thick to measure lines I and II of 73 – 73

Combin.	AM		I	II	III	IV	Shift/^{70}Ge$_2$O
70 – 70	70	a	863.919	863.848	863.621	863.10 ± 0.01	-
		e,a	863.918	863.846	863.620	863.105	
70 – 72	71	a	863.396	863.324	863.095		0.523
		e			863.095		
70 – 74	72	b	862.921	862.849	862.628		0.998
72 – 73	72.5	c	862.632	862.555			1.287
72 – 74	73	b	862.399	862.325			1.520
73 – 73	73	e,c	862.384	862.315	862.091	861.600	1.535
		c			862.082	861.586	
73 –74	73.5		862.143^b	862.071^b	861.840^c		1.776
72 –76	74	d	861.905	861.835	861.610		2.014
74 – 74	74	b	861.899	861.828	861.606	861.11 ± 0.01	2.020
		e				861.118	
74 – 76	75	d	861.442	861.371	861.153		2.477
		b	861.434				2.485
76 – 76	76	d	860.975	860.905	860.684	860.20 ± 0.01	2.944
		e,d	860.976	860.910	860.689	860.199	

^{a}In a qmi ^{70}Ge sample ^{b}In a qmi ^{74}Ge sample
^{c}In a qmi ^{73}Ge sample ^{d}In a qmi ^{76}Ge sample
^{e}[119]

^{nat}Ge, it is 2.961 cm^{-1} while in qmi Ge, it is 2.944 cm^{-1}. The difference is small, but significant.

The v_3 mode of Ge$_2^{16}$O$_i$ is superimposed to a region of the three-phonon spectrum of germanium whose absorption decreases with increasing energy. In qmi ^{76}Ge and ^{70}Ge, this three-phonon absorption is shifted toward lower and higher energies, respectively. Thus, the positive frequency shift of the ^{76}Ge$_2$O lines could be explained by a smaller phonon interaction, and the negative frequency shift of the ^{70}Ge$_2$O lines by a larger phonon interaction [122]. However a flaw in this explanation is that, by comparison with what is observed for ^{17}O$_i$ and ^{16}O$_i$ in silicon, shown in Fig. 6.3, no appreciable change is observed in the FWHMs of the lines in the qmi and natural Ge samples.

One must also notice that the change of the mass of the atom of an elemental crystal produces a change of the lattice parameter because of the change in the phonon interaction, including the zero-point vibration. For qmi crystals made of different isotopes, the lattice spacing is the smallest for the heaviest isotope, and this has been measured for qmi Ge crystals [123]. In the present case, the reduction of the lattice spacing of qmi ^{76}Ge should produce a very small decrease of the apex angle of ^{76}Ge–^{16}O–^{76}Ge, while for qmi ^{70}Ge, the increase of the lattice spacing

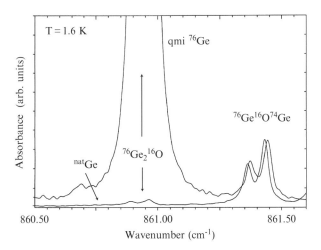

Fig. 6.21 Low-energy part of the ν_3 absorption of $Ge_2^{16}O_i$ in ^{nat}Ge superimposed with the same energy scale on the absorption of a qmi ^{76}Ge sample containing $\sim$89% of ^{76}Ge and $\sim$3.4% of ^{74}Ge. The unapodized resolution is $0.013\,cm^{-1}$. It shows a small positive shift ($\sim$0.010 cm^{-1}) of the $^{76}GeO^{74}Ge$ doublet in qmi ^{76}Ge compared to ^{nat}Ge [122]. Reproduced with permission from Trans Tech Publications

should produce a very small increase of the apex angle of $^{70}Ge-^{16}O-^{70}Ge$. For line I of these combinations, associated to a pure vibrational mode, corresponding frequency increase and decrease are expected in qmi germanium with respect to ^{nat}Ge. The values for these differences are $+0.019\,cm^{-1}$ and $-0.004\,cm^{-1}$ for $^{76}Ge_2O$ and $^{70}Ge_2O$, respectively. This change of the apex angle with the mass of the Ge atom of the qmi crystal should also produce a change of the equilibrium distance r_0 involved in the moment of inertia I_B of $^{16}O_i$, slightly changing the rotational frequency of $^{16}O_i$.

We have mentioned earlier the measurement of O_i rotational energies in ^{nat}Ge by phonon spectroscopy, and a $^{16}O/^{18}O$ IS of these energies has been reported [121, 124, 125]. These energies have also been measured in different qmi Ge samples and in these samples, the rotational energies are found to decrease when the Ge isotopic mass increases [121, 124, 125]. The experimental results showing this rotational isotope effect of germanium are summarized in Table 6.16. The energy threshold of the Sn detector does not allow measurements below $\sim$1.18 meV ($9.5\,cm^{-1}$ or $\sim$1 mm) and the energies of the $|0, \pm 1,0\rangle$ and $|0, \pm 2,0\rangle$ levels ($\sim$0.18 and 0.67 meV, respectively) have to be determined indirectly.

The energy of the $|0, \pm 1,0\rangle$ rotational level deduced from the energy difference between the $|0,0,0\rangle \rightarrow |0, \pm 5,0\rangle$ and $|0, \pm 1,0\rangle \rightarrow |0, \pm 5,0\rangle$ transitions for $^{73}Ge_2^{16}O$ is 0.168 meV ($1.36\,cm^{-1}$). The Ge IS is in agreement with a non-rigid rotation of O_i involving a small displacement of the two *nn* Ge atoms [23]. This explanation is also comforted by an average value of $\sim$1.07 the $^{16}O/^{18}O$ IS ratio, smaller than the one expected (1.125) in the rigid rotator approximation. The Ge ISs

Table 6.16 Rotational transition energies (meV (cm^{-1} in brackets)) of Ge$_2$O measured in ^{nat}Ge and in qmi Ge samples by phonon spectroscopy. The values for Ge$_2^{18}$O were obtained on ^{18}O-implanted samples [121]

	^{70}Ge$_2^{16}$O	^{nat}Ge$_2^{16}$O	^{nat}Ge$_2^{18}$O	^{73}Ge$_2^{16}$O	^{74}Ge$_2^{16}$O	^{76}Ge$_2^{16}$O
$|0,0,0\rangle \rightarrow |0,-3,0\rangle$	1.378 (11.11)	1.352 (10.90)	1.26 (10.2)	1.353 (10.91)	1.336 (10.78)	1.323 (10.67)
$|0,\pm1,0\rangle \rightarrow |0,+3,0\rangle$	1.485	1.459	1.38	1.454	1.445	1.430
$|0,0,0\rangle \rightarrow |0,+3,0\rangle$	1.652	1.625	1.53	1.632	1.611	1.597
$|0,\pm1,0\rangle \rightarrow |0,\pm4,0\rangle$	2.494	-		2.452	2.434	2.408
$|0,0,0\rangle \rightarrow |0,\pm4,0\rangle$	2.666	2.618	2.45	2.620	2.597	2.572
$|0,\pm1,0\rangle \rightarrow |0,\pm5,0\rangle$	3.923		3.78	3.865	3.832	3.789
$|0,0,0\rangle \rightarrow |0,\pm5,0\rangle$	4.100	4.037		4.033	3.998	3.955
$|0,\pm1,0\rangle \rightarrow |0,\pm6,0\rangle$				5.552	5.510	
$|0,0,0\rangle \rightarrow |0,\pm6,0\rangle$				5.720	5.676	

of the rotational transitions depends linearly on the AM of the Ge atom and their amplitudes depend also on the rotational transition considered: The ^{70}Ge–^{76}Ge IS varies from 0.055 to 0.145 meV (0.44 to 1.17 cm^{-1}) for the $|0,0,0\rangle \rightarrow |0,+3,0\rangle$ and $|0,0,0\rangle \rightarrow |0,\pm5,0\rangle$ transitions, respectively. The Ge isotope effect of the rotational transitions is also present in ^{nat}Ge, but it cannot be observed because the FWHMs of the phonon scattering resonances are $\sim$0.04 meV or $\sim$0.3 cm^{-1} [125]. The changes in the equilibrium distance r_0 in qmi samples due to zero-point vibrational motion is expected to produce a very small Ge IS contribution inverse to the one reported. The energy spacing between components I and III is the difference between the $|0,\pm2,0\rangle$ rotational energies in the $N = 0$ and $N = 1$ vibrational states of $\nu_3(O_i)$. From Tables 6.14 and 6.15, this spacing decreases for ^{76}Ge$_2$O by 0.012 cm^{-1} (1.5 μeV) between ^{nat}Ge and qmi ^{76}Ge, but for ^{70}Ge$_2$O, it increases by 0.003 cm^{-1} (0.4 μeV) between ^{nat}Ge and qmi ^{70}Ge. It can be assumed that the decrease of the I–III spacing of ^{76}Ge$_2$O for qmi ^{76}Ge is related to the decrease of the rotational energy for that combination in this material, due to the decrease of the lattice spacing. The inverse conclusion should be valid for the increase of the I–III spacing of the ^{70}Ge$_2$O combination in qmi ^{70}Ge.

^{17}O isotope effects of $\nu_3(O_i)$ in ^{nat}Ge have been reported [126], and the fine structure of $\nu_3(^{17}O_i)$ observed at LHeT between $\sim$837 and 841 cm^{-1} as well as the FWHMs of the lines are comparable to the ones for $\nu_3(^{16}O_i)$. ^{18}O isotope effects have been also reported [119, 126]. The fine structure of $\nu_3(^{18}O_i)$ is observed at LHeT between 816 and 820 cm^{-1}, but the FWHMs of the components observed at high resolution show an increase from ^{76}Ge$_2^{18}$O ($\sim$0.05 cm^{-1}) to ^{70}Ge$_2^{18}$O ($\sim$0.06 cm^{-1}), where the splitting between lines I and II, which can be appreciated in Fig. 6.22, is not resolved as well as for ^{70}Ge$_2^{16}$O.

The positions of the $\nu_3(^{17}O_i)$ and $\nu_3(^{18}O_i)$ lines that can be identified with a minimum of interferences with other lines are listed in Table 6.17.

The ^{16}O/^{18}O and ^{16}O/^{17}O frequency ratios deduced from Tables 6.14 and 6.17 are 1.054 and 1.028, respectively, whatever the line.

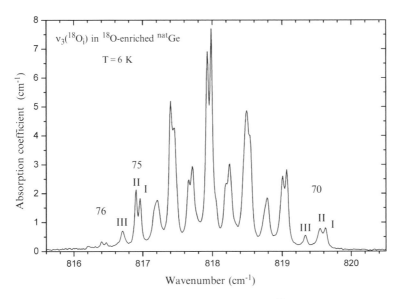

Fig. 6.22 Absorption of $v_3({}^{18}O_i)$ in a ^{nat}Ge sample enriched with ^{18}O. Some of the Ge masses are indicated. At 6 K, line III of a combination where the Ge AM is M interferes with lines I and II of the combination where it is $M - 0.5$ u. Note the change in the separation of the components for $M = 70$ u. The apodized resolution is 0.02 cm^{-1} [126]. Copyright 2000 by the American Physical Society

The differences between the frequencies of the 70–70 and 76–76 isotopic combinations of $v_3(O_i)$ in ^{nat}Ge are the same for components I and II, but they display a small increase with the O isotope mass: they are 2.96, 3.06, and 3.15 cm^{-1} for ^{16}O, ^{17}O, and ^{18}O, respectively. In the same way as for O_i in silicon, the combination of the O and Ge ISs can be been used to calculate a value of the interaction mass m using expressions (6.3), (6.4), and (6.5). Using three experimental input for ^{16}O and ^{18}O (the values for I(^{70}Ge$_2^{16}$O), I(^{70}Ge$_2^{18}$O), and I(^{76}Ge$_2^{16}$O)), a value of (11.86 ± 0.05)u is obtained for the interaction mass m, with an apex angle of 138°4, and comparable values are obtained with components II and III. These values are close to the ones ($m = 11.65$ u and $2\alpha \sim 140°$) determined empirically in [117], and the apex angle is close to the one (140°) obtained from first-principles total-energy calculations [23]. As mentioned earlier, there are some small variations of the FWHMs of the Ge$_2$O vibrational transitions in germanium at LHeT, but they remain in the 0.03–0.06 cm^{-1} region, and they are discussed later.

The observation at LHeT of very weak sideband spectra of the $v_3(^{16}O_i)$ spectrum has been reported in a ^{nat}Ge sample cut in a multireflection geometry to provide an optical path of 22 mm [126]. The FWHMs of these sidebands are $\sim$0.08 cm^{-1} and their intensities $\sim$10^{-4} to 5 × 10^{-4} of the one of the absorption maximum of the $v_3(^{16}O_i)$ spectrum They consist in series of distinct sharp line spectra related to O_i, with well-defined Ge ISs, which have been attributed to different $|0, l, 0\rangle$ to $|0, l', 1\rangle$

Table 6.17 Observed positions (cm^{-1}) at LHeT of $\nu_3(^{17}O_i)$ and $\nu_3(^{18}O_i)$ lines of different Ge_2O isotopic combinations in ^{nat}Ge. The IS shift refers to line I unless otherwise indicated. The uncertainty is $\pm 0.01\ cm^{-1}$ (after [126]). The values in bracket are calculated *ab initio* [33]

AM	O isotope	I	II	III	Shift/I($^{70}Ge_2^{16}O$)
70	^{17}O	840.72	840.65		23.20
"	^{18}O	819.62	819.55	819.33	44.30 (44)
71	^{17}O	840.18	840.11	839.90	23.74
"	^{18}O	819.07	819.01	818.79	44.85
72	^{17}O	839.68			24.24
"	^{18}O	818.54	818.48		45.38
72.5	^{17}O		839.32		24.53*
"	^{18}O	818.26[a]	818.19		45.66
73	^{17}O	839.13	839.07	838.86	24.79
"	^{18}O	817.98	817.93		45.94
73.5	^{18}O	817.72[a]	817.67[a]		46.20
74	^{17}O	838.62	838.55		25.30
"	^{18}O	817.45	817.40		46.47 (46)
75	^{17}O	838.14	838.07	837.86	25.78
"	^{18}O	816.96	816.90	816.71	46.96
76	^{17}O		837.59		26.26*
"	^{18}O	816.46	816.40	816.2	47.46

[a][119]
*Measured for line II

Table 6.18 Frequencies (cm^{-1}) at LHeT of the series of H and L sidebands of $\nu_3(^{16}O_i)$ in ^{nat}Ge, for different values of the average mass M of the Ge atoms. The rounded position of lines I of Table 6.14 due to the $|0,0,0\rangle \rightarrow |0,0,1\rangle$ transition are given as references. The last line of the table gives the frequency differences between $M = 71$ and 75u

L_1	L_2	L_3	L_4	I	M	H_1	H_2	H_3	H_4
847.43				863.92	70	868.97			
846.92	952.3		858.07	863.40	71	868.46	872.19	874.84	878.33
			857.83	863.15	71.5				
846.41		853.24	857.56	862.90	72	867.95	871.69	874.32	877.84
	851.6			862.63	72.5	867.67	871.41		
845.94	851.33	852.8	857.06	862.39	73	867.43	871.17	873.81	877.32
	851.10		856.80	862.14	73.5	867.18	870.93		
845.46	850.86	852.3	856.58	861.90	74	866.95	870.69	873.32	876.84
				861.67	74.5	866.7			
845.00			856.12	861.43	75	866.47	870.21		876.36
1.92			1.95	1.97	71–75	1.99	1.98		1.97

transitions of the ν_3 mode with $l' \neq l$. The series on the high- and low-energy sidebands are labelled H_i and L_i, respectively, and their positions for different Ge AMs are given in Table 6.18.

These Ge isotopic series are clearly related to the central band and the series on the high-energy side look slightly more intense than those on the low-energy side. The spectrum of series H_1 is the most intense at LHeT, and it is shown in Fig. 6.23.

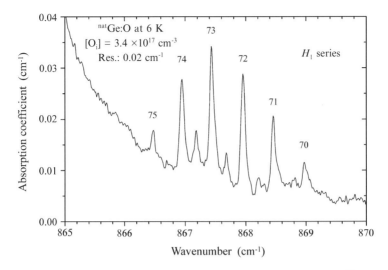

Fig. 6.23 Absorption spectrum of series H_1 of the vibrational sidebands of $\nu_3(^{16}O_i)$ in ^{nat}Ge. The components are referenced by the average mass M of the Ge atoms. The peak absorption for $M = 73$ is about 3×10^{-4} times the peak absorption of component I for this mass at $862.39\,\text{cm}^{-1}$. This series is ascribed to the $|0,0,0\rangle \rightarrow |0,\pm 2,1\rangle$ transition (after [126]). Copyright 2000 by the American Physical Society

In a spectrum obtained at 15 K, the high-energy series display a small relative decrease with respect to the low-energy ones. It is difficult to obtain significant data above ~ 15 K because of the broadening of the lines.

Based on the rotational energies of O_i in the $N = 0$ ground state obtained from phonon spectroscopy and on the first rotational energies in $N = 1$ excited state deduced from the optical spectroscopy results, it is possible to make for the L_i and H_i series plausible spectroscopic attributions which fit reasonably the observed positions of Table 6.16. The best agreement is obtained by attributing lines H_1, H_2, H_3, and H_4 to $|0,0,0\rangle \rightarrow |0,\pm 2,1\rangle$, $|0,\pm 1,0\rangle \rightarrow |0,-3,1\rangle$, $|0,\pm 1,0\rangle \rightarrow |0,+3,1\rangle$, and $|0,\pm 2,0\rangle \rightarrow |0,\pm 4,1\rangle$ transitions, respectively, and lines L_1, L_2, and L_3 to $|0,\pm 4,0\rangle \rightarrow |0,\pm 2,1\rangle$, $|0,-3,0\rangle \rightarrow |0,0,1\rangle$, and $|0,-3,0\rangle \rightarrow |0,\pm 1,1\rangle$ transitions, respectively.

By combining the energies of the I, II, III, and IV transitions of the $\nu_3(^{16}O_i)$ spectrum, corresponding to the selection rule $\Delta l = 0$, with those of the corresponding lines of these sidebands once attributed to specific transitions, it is possible to derive values of the first $|0,l,0\rangle$ and $|0,l,1\rangle$ rotational levels of O_i in germanium. These values are given in Table 6.19, and one notes that, as usual, the rotational energies are higher in the vibrational ground state than in the vibrational excited state.

The weak band observed at RT at $1264\,\text{cm}^{-1}$ in O-doped germanium shifts to $1269.4\,\text{cm}^{-1}$ at LHeT, with a reduction by a factor of two of its FWHM ($\sim 10\,\text{cm}^{-1}$ at LHeT) and a $^{16}O - ^{18}O$ IS of $+53.6\,\text{cm}^{-1}$. Its Ge IS can only be measured from differences in qmi Ge samples because of its large FWHM [23]. As already

Table 6.19 Energies (cm^{-1}) of the first hindered-rotational levels of $^{16}O_i$ in ^{nat}Ge in the ground ($N = 0$) and excited ($N = 1$) vibrational states deduced from the IR values of the $|0, l, 0\rangle$ to $|0, l', 1\rangle$ transitions. The values in meV are given in brackets. The IR values are determined from the experimental isotopic frequencies for $M_{Ge} = 73$ u (after [126]). The Phonons values are obtained by phonon spectroscopy

$|0,0,0\rangle$ to	$|0, \pm1, 0\rangle$	$|0, \pm2, 0\rangle$	$|0, -3, 0\rangle$	$|0, +3, 0\rangle$	$|0, \pm4, 0\rangle$
IR	1.50 (0.186)	5.34 (0.662)	11.07 (1.373)		21.49 (2.664)
Phonons	[a](0.18)	(0.67)	(1.37)	(1.65)	(2.66)
	[b](0.168)		(1.352)	(1.625)	(2.618)

$|0,0,1\rangle$ to	$|0, \pm1, 1\rangle$	$|0, \pm2, 1\rangle$	$|0, -3, 1\rangle$	$|0, +3, 1\rangle$	$|0, \pm4, 1\rangle$
IR	1.43 (0.177)	5.03 (0.624)	10.27 (1.273)	12.95 (1.606)	20.26 (2.512)

[a][120]
[b][121]

Table 6.20 Experimental positions (cm^{-1}) of the ν_3 and $\nu_1 + \nu_3$ modes for some isotopic combinations in qmi Ge at LHeT, from which are interpolated the corresponding ν_1 positions (in square brackets). The calculated values of the $\nu_1 + \nu_3$ mode are extrapolated as the sum of the calculated values of the ν_3 and ν_3 modes, neglecting anharmonicity (after [23])

	ν_3 (line I)		$\nu_1 + \nu_3$		ν_1	
	Exper.	Calc.	Exper.	Calc.	Exper.	Calc.
$^{70}Ge_2^{16}O$	863.9	877.48 847[a]	1275 ± 1	[1293.30]	[411]	415.82
				[1248][a]		401[a]
$^{73}Ge_2^{16}O$	862.4	875.45	1269.1	[1287.12]	[407]	411.67
$^{74}Ge_2^{16}O$	861.9	874.80 845[a]	1266.7	[1285.15]	[405]	410.35
				[1240][a]		395[a]
$^{76}Ge_2^{16}O$	861.0	873.56	1263.1	[1281.42]	[402]	407.86
$^{70}Ge/^{74}Ge$ IS	2.0	2[a]	8	[8][a]	[6]	6[a]
$^{70}Ge/^{76}Ge$ IS	2.9	3.92	11.9	[11.88]	[9]	7.96

[a][33]

mentioned, this band, attributed to the $\nu_1 + \nu_3$ combination of the $\nu_3(O_i)$ mode with the IR-inactive symmetric bending mode $\nu_1(O_i)$. The frequencies of the $\nu_1 + \nu_3$ combination and of the extrapolated values for ν_1 are given in Table 6.20. In this table, the choice of the frequency of I(ν_3) instead of an average value of I, II, and III makes only a small difference.

The frequency of the ν_1 mode of $^{16}O_i$ in ^{nat}Ge is 408 cm^{-1} at RT (see Table 6.12) and it is practically temperature independent. The extrapolated $^{70}Ge - ^{76}Ge$ and $^{16}O - ^{18}O$ ISs of the ν_1 mode are $\sim+9$ and $\sim+10$ cm^{-1}, respectively [23,96,126]. The existence of an O IS of the symmetric ν_1 mode is due to the puckered structure of Ge $-$ O $-$ Ge.

The absorption of the ν_3 mode has been measured recently in ^{nat}Ge samples with Sn concentrations of 2 and 4×10^{19} cm^{-3} [127]. Tin is isovalent to germanium, but its covalent radius is 0.14 nm compared to 0.12 nm for germanium. In these samples, at LHeT, the components of the ν_3 mode of O_i centred at $\sim$862.4 cm^{-1} broaden considerably and it is no longer possible to follow the individual isotopic

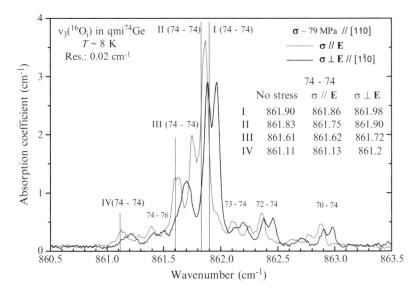

Fig. 6.24 Splittings under a uniaxial stress of 79 MPa parallel to the [110] crystal axis of lines I, II, III, and IV of the ν_3 mode of $^{74}Ge_2^{16}O$ (74– 74) in a qmi ^{74}Ge sample for $\sigma//E$ (in red) and $\sigma\perp E//[1\bar{1}0]$. The vertical lines represent the positions of the $^{74}Ge_2O$ lines at zero stress and their frequencies (cm^{-1}) are indicated for comparison with the values under stress. Weaker features due to combinations of ^{74}Ge with the other residual Ge isotopes are globally identified by the Ge masses involved (Pajot, unpublished results). The zero-stress spectrum, not shown, is comparable, *mutatis mutandis*, to the one of Fig. 6.18d for qmi ^{70}Ge

components, and two new smaller sets of lines are observed, centred at 842.2 and 854.5 cm^{-1}. In these two sets, a Ge isotopic structure can be observed, but without evidence for rotation of the O atom. The low-frequency tail of a third set, very close in frequencies with the unperturbed set centred at $\sim$862.4 cm^{-1}, is observed near 860 cm^{-1}. The existence of this latter set explains very probably the broadening of the components of the ν_3 mode centred at $\sim$862.4 cm^{-1} by an overlap of the isotopic components of the two sets. By order of increasing energies, the three new sets of lines are attributed to the vibration of the ν_3 mode perturbed by a Sn atom *nnn*, 3^{rd} *nn*, and 4^{th} *nn* of O$_i$ [127].

The effect of a uniaxial stress σ // [110] of the ν_3 mode of O$_i$ in germanium has been investigated for values of the stress between 10 and $\sim$300 MPa on an O-doped qmi ^{74}Ge sample with [^{74}Ge] $\sim$87.7% (Pajot, unpublished). Excluding line IV, the overall amplitude of the splitting for the set of lines I, II, and III at 302 MPa is $\sim$0.6 cm^{-1} compared to a zero stress overall separation of 0.3 cm^{-1} for theses three lines. The splitting pattern for a stress of 79 MPa is shown in Fig. 6.24. For higher values of the stress, it is difficult to resolve the observed spectra into well-defined patterns because of the overlap of the split or shifted components and the interpretation is far from obvious.

Lines III and IV, due to the $|0, \pm 2,0\rangle \rightarrow |0, \pm 2,1\rangle$ and $|0, -3,0\rangle \rightarrow |0, -3,1\rangle$ transitions, respectively, show practically no stress dependence[10] for $\mathbf{E} \, // \, [110]$, but for $\mathbf{E} \, // \, [1\bar{1}0]$, an upward shift estimated to $1.3 \, \mathrm{cm}^{-1}\mathrm{GPa}^{-1}$ is observed. There is a strong feature appearing progressively at low energy for $\sigma//\mathbf{E}$ for increasing stresses. A crude estimation indicates that this feature, with a piezospectroscopic coefficient of $\sim - 1.2 \, \mathrm{cm}^{-1}\mathrm{GPa}^{-1}$, originates from line II due to the $|0, \pm 1,0\rangle \rightarrow |0, \pm 1,1\rangle$ transition, while components III and IV show no response to stress for that configuration.

For the measurements performed up to 200 MPa, lines I and II seem to show for $\mathbf{E} \, // \, [1\bar{1}0]$ ($\mathbf{E}\perp\sigma$) a high-energy component with a shift of $\sim 0.7 \, \mathrm{cm}^{-1}\mathrm{GPa}^{-1}$, but the value for 300 MPa gives a shift about two times less. For $\mathbf{E} \, // \, \sigma$, line I shows a component with a downward energy shift of $\sim - 0.6 \, \mathrm{cm}^{-1}\mathrm{GPa}^{-1}$.

At a difference with lines II, III, and IV, which involve the vibrational mode and the difference between the energies of the rotational state in the ground and excited vibrational states, line I is only associated to a π dipole oriented along a $<111>$ axis. For the stress orientation used, only one stress component is expected for each electric field polarization, with shifts of $A_1 + A_2$ and $A_1 - A_2$ for $\mathbf{E}//[110]$ and $\mathbf{E} \, // \, [1\bar{1}0]$, respectively. The experiment shows that $A_1 - A_2$ is positive. From the above measurements, a value of A_2 between 0.5 and $0.7 \, \mathrm{cm}^{-1}\mathrm{GPa}^{-1}$ is deduced, comparable in absolute value to the one reported for $A_{2u}(0)$ of O_i in silicon in Table 6.6.

Piezospectroscopic measurements of the rotational transitions have been performed by phonon spectroscopy near 1 K in ^{nat}Ge for values of the stress up to ~ 200 MPa [120]. The results can be fitted by assuming a splitting of the $|0, \pm 1,0\rangle$ rotational level which amounts to $\sim 1.1 \, \mathrm{meV\,GPa}^{-1}$ ($8.9 \, \mathrm{cm}^{-1}\mathrm{GPa}^{-1}$) for $\sigma // [110]$, while the upper levels are practically unaffected by stress. This is qualitatively similar to what is observed in the mid-IR, where what is probed is the difference of the stress splitting of the rotational state in the ground and excited vibrational states.

Linewidths and Lifetimes

Because of the small values of their FWHMs, the components of the ν_3 mode of O_i in germanium have attracted interest in lifetime measurements. A comparison of these values measured for the ^{70}Ge$_2$O and ^{76}Ge$_2$O components at LHeT for the different O isotopes in ^{nat}Ge is given in Table 6.21.

Table 6.21 FWHMs (cm^{-1}) measured at LHeT for component I of the ^{70}Ge$_2$O and ^{76}Ge$_2$O combinations of ν_3 in ^{nat}Ge for the different O isotopes

	^{16}O	^{17}O	^{18}O
^{70}Ge$_2$O	0.030	$\sim$0.05	$\sim$0.06
^{76}Ge$_2$O	0.045	$\sim$0.05	$\sim$0.05

---

[10]Component III(^{74}GeO^{76}Ge) is very close to component IV(^{74}Ge$_2$O), but it is much weaker.

In germanium, the $\nu_3(^{16}O_i)$ mode lies on the high-energy side of the 3TO DoS, with negative slope, while the $\nu_3(^{17}O_i)$ modes lies on the low-energy side of this 3TO DoS, with positive slope, and the $\nu_3(^{18}O_i)$ mode lies on the negative slope of the 2TO + LO DoS. The IR activity of these three-phonon processes is smaller than the 2TO + TA process in silicon and this can be a qualitative explanation of the smaller FWHMs of the O_i modes in germanium compared to silicon [47]. With the same assumptions as the ones made about the FWHMs of the $A_{2u}(0)$ modes of O_i in silicon, a much larger lifetime ($\sim$100 ps) is expected for the excited state of the $\nu_3(O_i)$ transitions in germanium at LHeT. Assuming that, at a local scale, the FWHMs of the LVMs of the Ge isotopic combinations depend on their interaction with the three-phonon DoS, one could expect for $\nu_3(^{16}O_i)$ smaller FWHMs for the $^{70}Ge_2O$ combinations than for the $^{76}Ge_2O$ combinations, as observed, because the former are higher in energy. The inverse is expected for $\nu_3(^{17}O_i)$, but there, referring to Table 6.21, no difference is observed. For $\nu_3(^{18}O_i)$, the same trend as for $\nu_3(^{16}O_i)$ is expected, but there, the inverse is observed, showing that the initial assumption should be reconsidered.

An average lifetime of 125 ps has been measured at 7.5 K using an excitation at 862 cm^{-1} for the $\nu_3(Ge_2^{16}O)$ excited states in ^{nat}Ge, corresponding to an average FWHM of 0.04 cm^{-1} for the different $Ge_2^{16}O$ isotopic combinations [47]. In this study, a decay of the excited state by a process involving three lattice phonons is assumed. A comparable value (115 ps) has been reported by Davies et al. [128] using an excitation band width of 1 cm^{-1}, compared to 15–20 cm^{-1} in the former study, with an additional value of 110 ps for $\nu_3(Ge_2^{18}O_i)$.

It is assumed in [128] that, as for $A_{2u}(O)$ in silicon, $\nu_3(Ge_2^{16}O)$ decays predominantly with the emission of a local excitation, which is the symmetric $\nu_1(Ge_2^{16}O)$ mode. For specific values, the $^{70}Ge_2^{16}O$ isotopic combination is considered first. The frequencies of the ν_3 and ν_1 modes of this combination are 863.9 and 411.2 cm^{-1}, respectively, leaving a difference of 452.7 cm^{-1}, well above the Raman frequency of the germanium lattice (304.5 cm^{-1} at 80 K). The decay of ν_3 requires thus the emission of at least two lattice phonons and values for these phonons have been obtained from a fit of the experimental values of the temperature dependence of the decay time of the excited state of $\nu_3(Ge_2^{16}O)$ [128]. Within this two-phonon-decay scheme, the important point is that the ν_3–ν_1 differences for the $^{70}Ge_2^{16}O$ and $^{76}Ge_2^{16}O$ combinations lie on the positive slope of a two-phonon feature of the two-phonon DoS of germanium, but as the ν_3–ν_1 difference is larger for $^{76}Ge_2^{16}O$ than for $^{70}Ge_2^{16}O$ (459 and 453 cm^{-1}, respectively), the two-phonon interaction is expected to be stronger for $^{76}Ge_2^{16}O$. As a consequence, a larger FWHM is expected for the corresponding LVMs than for the $^{70}Ge_2^{16}O$ ones, and this is what is observed. The differences $\nu_3 - \nu_1$ are 424 and 419 cm^{-1} for $^{76}Ge_2^{18}$ and $^{70}Ge_2^{18}O$ respectively, using the ^{16}O–^{18}O IS value of +10 cm^{-1} given by Litvinov et al. [96]. In that case, the ν_3–ν_1 differences lie on the negative slope of another germanium two-phonon feature of lower energy, but as the difference for $^{76}Ge_2^{18}O$ is larger, the interaction for this isotopic combination is expected to be smaller than the one for $^{70}Ge_2^{18}O$, producing a smaller FWHM for $^{76}Ge_2^{18}O$. This is what is indeed observed. The above analysis shows that a qualitative agreement explaining the Ge isotopic dependence of the FWHMs of the v_3 LVMs is obtained by assuming that the v_3 mode of Ge_2O decays into the local v_1 mode and two phonon modes of germanium.

Reorientation and Calibration

The method based on stress-induced dichroism used to determine the reorientation energy of O_i in silicon was also used for germanium [51]. Annealing a Ge sample at 350°C for 30 min under a stress of 115 MPa along a <111> axis produced a dichroic ratio of 0.074. A reorientation energy of O_i of 2.08 eV in germanium was deduced from the annealing of the dichroism. This energy is smaller than the one in silicon (2.56 eV) because of the smaller bond energies in germanium. In a similar way as for silicon, this energy was considered as the activation energy for O_i diffusion in germanium [51].

Several calibrations of the RT absorption coefficient of the $v_3(^{16}O_i)$ mode as a function of $[O_i]$ have been proposed ([96], and references therein); the RT calibration factor recommended in this reference is $\sim 1 \times 10^{17}\,cm^{-2}$, about twice the value of the old calibration factor of [129]. The ratio of the intensities of the absorptions of the v_3 and $v_1 + v_3$ modes at RT is 81 so that the $(v_1 + v_3)$ combination mode can eventually be used for the determination of $[O_i]$ in germanium. The relative intensities of the $v_3(^{16}O_i)$ mode at RT and at LHeT can be appreciated from Fig. 6.25. The resolution used for the LHeT spectrum is such that there is no instrumental broadening.

In Fig. 6.25, the peak absorption at LHeT is at $862.39\,cm^{-1}$ and it corresponds mainly to component I ($^{72}Ge O^{74}Ge$) with the largest weight (see Table 6.14). The ratio of this peak absorption to the one at RT is ~ 31, and it is slightly less than the corresponding one in silicon (~ 36). Such a large value is due to the absence of instrumental broadening at LHeT, and in the early times, when the spectrometers did not allow high resolutions, much lower values were observed (see, for instance, Fig. 1 of [114]). The RT measurement of the v_3 band is the easiest way to measure the O_i concentration in germanium, and a detection limit of ~ 2–$3 \times 10^{15}\,cm^{-3}$ can be estimated. It can of course be lowered by more than one order of magnitude by measurements at LHeT.

6.1.2.2   The Dimer (O_{2i}) and the TDDs

LVMs attributed to the dimer O_{2i} have also been observed in O-doped germanium. After quenching from 900°C, only a LVM at $818\,cm^{-1}$ is observed (besides the O_i absorption at $862\,cm^{-1}$). After h–e electron irradiation and annealings at increasing temperatures, other LVMs due to O_{2i} are observed together with those of the TDDs [130]. This situation is summarized in Fig. 6.26.

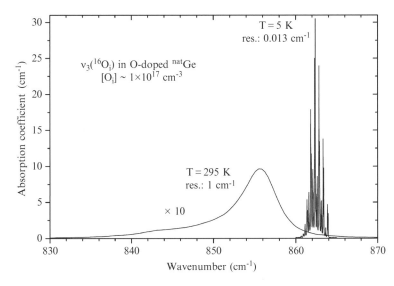

Fig. 6.25   Comparison of the absorption coefficients at RT and at LHeT of the ν_3 mode of O_i in a ^{nat}Ge sample. For a better appreciation, the ordinates of the RT spectrum, with a peak absorption of $0.97\,\mathrm{cm}^{-1}$ and a FWHM of $5.3\,\mathrm{cm}^{-1}$, have been multiplied by a factor of 10 (Pajot, unpublished). With the abscissa scale used, the Ge_2O components at LHeT look like Dirac peaks

Concerning O_{2i}, as in silicon, only one mode is ascribed to the skewed geometry while four modes are attributed to the staggered geometry shown in Fig. 6.12. The fair agreement between the experimental frequencies of four of these modes and the ones obtained from density-functional calculations can be appreciated from Table 6.22.

As seen in Fig. 6.27 for the spectra of the samples annealed at 280°C and above, three broad vibrational features near 600, 740, and 780 cm^{-1} are associated with TDDs. The intensities of these bands grow up and their peak positions are shifted towards higher energies with increasing annealing time. Now it is known that for increasing annealing times, in germanium as well as in silicon, higher order TDDs containing more O atoms are formed at the expense of the lower order ones. As the broad bands are the superposition of the LVMs of the individual TDDs species, the above changes in the spectra can be explained by assuming that the LVM frequencies depend on the order of the TDDs and that the highest the order, the highest the frequencies. The correctness of this assumption was checked by measurements at low temperature where the absorption of the LVMs associated with different TDDs could partially or totally resolved, providing a way to identify them [131, 133].

Metastability effects, which are observed on the electronic spectra of the TDDs in germanium (see [86]), are also observed on their LVM spectra [131, 133]. Before going further, we remember that the TDDs in germanium are noted TDi (I = 1, 2, etc.). In the LVM studies, the electrically active form is noted generically DD

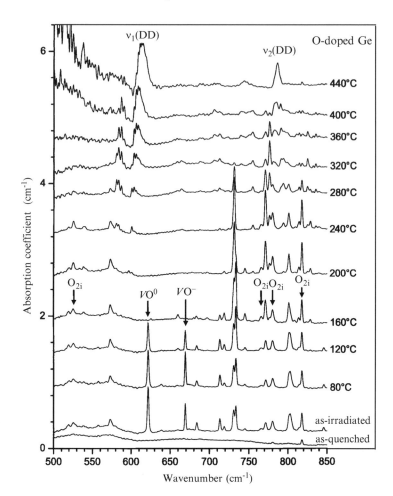

Fig. 6.26 LVM absorption at LHeT of O-doped germanium irradiated with 2 MeV electrons and annealed for 20 min at increasing temperatures [130]. The LVMs of the dimer are noted O_{2i} and the two main LVM bands of the electrically active thermal double donors ν_1(DD) and ν_2(DD) according to [131]. Coyright 2001 by the American Physical Society

Table 6.22 Comparison between the measured (LHeT) and calculated frequencies (cm^{-1}) of the O_{2i} dimer on germanium for the different configurations. The values in brackets are measured at RT

O_{2i}		Skewed		Staggered		
Calc.[a]	^{16}O	843–849	784	749		517
Exp.[b,c]	^{16}O	856.8	817.9	780.3	766.0	525.6
Exp.[c]	^{18}O	811	776 (772)	741 (737)		

[a][33]
[b][130]
[c][132]

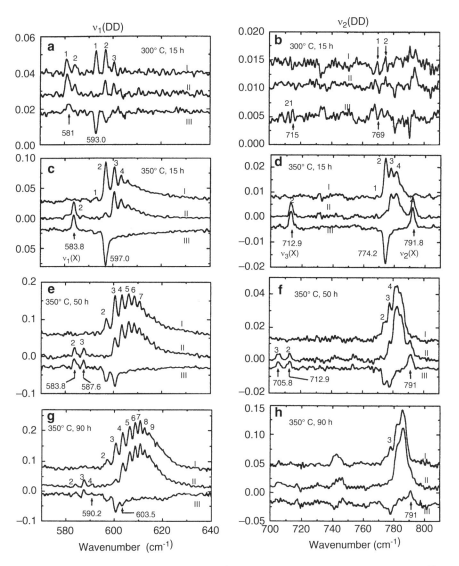

Fig. 6.27 Absorption coefficient spectra (cm^{-1}) at LHeT of the LVMs of the DDs in Ge:^{16}O samples annealed at 300 and 350°C for different durations. Spectra I and II are obtained for samples cooled under band-gap illumination and under TEC, respectively. Spectra III are the difference between spectra II and I [131]. Letters **a** to **h** correspond to the annealing conditions. Copyright 1999 with permission from Elsevier

(double donor) and the inert metastable form (X). The measurements are performed by comparing the spectra obtained with samples cooled under band-gap illumination to those obtained with samples cooled under TEC. Following [131], the LVMs of the DD and X forms with the lowest and highest frequencies are noted ν_l

and v_2, respectively. For the X form, there is a third category of LVMs, $v_3(X)$, with frequencies in the $700\,cm^{-1}$ region. The situation for increasing annealing temperatures and times is shown in Fig. 6.27.

In this figure, the spectra I(a), I(c), I(e), and I(g) show the $v_1(DD)$ mode for an increasing variety of TDi (noted simply i), which can be identified up to i = 9 in I(g), and the spectra I(b), I(d), I(f), and I(g) the $v_2(DD)$ mode, with a smaller separation between the modes of the different TDi. The spectra II(a), II(c), II(e), and II(g) show the $v_1(X)$ mode for i = 1, 2, 3, and 4, due to the total or partial conversion of the corresponding DD centres when the sample is cooled under TEC, and the correlated absence or decrease of the corresponding $v_1(DD)$ modes, better seen in the spectra III(a), III(c), III(e), and III(g). The spectra II(b), II(d), II(f), and II(h) show the $v_3(X)$ and $v_2(X)$ modes, the latter one being close to $v_2(DD)$. In spectrum I(h), the LVM at $740\,cm^{-1}$ is clearly visible, but without any structure.

The corresponding frequencies of the LVMs associated are given in Table 6.23.

Spectroscopic measurements on the metastability of TDDs in germanium have also been presented in [134].

One can note that the frequency ratio $v_2(DD)^{16}O/v_2(DD)^{18}O$ for TD2 and TD3 is close to the one ($\sim$1.05) for $v_3(^{16}O_i)/v_3(^{18}O_i)$ and this can be taken as an indication that the $v_2(DD)$ LVM is due to the two O_i end atoms of structure b) of Fig. 6.14. From comparisons with the calculation of the frequency of the tricoordinated O atom, it has also been suggested that $v_1(DD)$ could be due to the vibration of these tricoordinated O atoms [131].

An average number of O atoms per TDD in germanium can be evaluated from the correlation between the decrease of the intensity of $v_3(O_i)$ and the increase of the TDD concentration. For annealing times between 20 and 60 hours at 350°C, this number goes from 7 to 8, with some uncertainty on the absolute value due to the spectroscopic calibration factor. Average numbers of vibrating O atoms associated with $v_1(DD)$ and $v_2(DD)$ under the same conditions have also been estimated by assuming that the same calibration coefficient as for $v_3(O_i)$ apply for the increase of the corresponding bands. For $v_2(DD)$, a value of 2 atoms is extrapolated ([135], and

Table 6.23 Positions (cm^{-1}) at LHeT of the LVMs associated with the electrically active (DD) and inert (X) forms of the TDDs in germanium. For ^{18}O, the position of $v_3(X)$ of TD2 is $672.5\,cm^{-1}$. Compiled from [130, 131, 134]

		TD1	TD2	TD3	TD4	TD5	TD6	TD7	TD8	TD9
^{16}O	$v_1(DD)$	593.0	597.0	600.6	603.5	606.1	608.6	610.6	612.5	614.5
	$v_2(DD)$	769	774.5	777.6	780.4	782.7	785.1	789.4		
^{18}O	$v_1(DD)$	569	573.9	575.6	578.4					
	$v_2(DD)$		735.2	739						
^{16}O	$v_1(X)$	581.0	583.6	587.4	590.6					
	$v_2(X)$	794.1	791.9	790.5	790.3					
	$v_3(X)$	715	712.9	706.0	706.0					
^{18}O	$v_1(X)$	555.8	559.4	562.5						

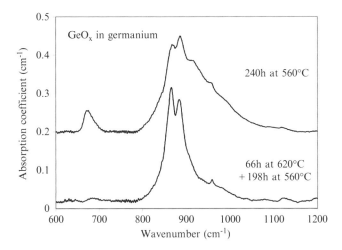

Fig. 6.28 RT absorption of GeO_x precipitates in heat-treated O-doped germanium samples ($[O_i]_0 = 5 \times 10^{17}\,cm^{-3}$). The absorptions from remaining O_i and SiO_x inclusions have been subtracted [95]. Copyright 2006 with permission from Elsevier

references therein), comforting the above attribution of this mode to the two O_i end atoms of the thermal donor structure.

6.1.2.3 Oxygen Precipitation

A broad absorption band centred at $11.5\,\mu m$ ($870\,cm^{-1}$) observed in O-doped germanium annealed above $600°$ C was attributed by Kaiser and Thurmond [129] to a precipitated GeO_2 phase. More recent spectra of the absorption of GeO_x precipitates are displayed in Fig. 6.28.

This figure shows the existence of at least two different kinds of precipitates. The values of $[O_i]$ measured after annealing show that in the sample without pre-annealing, 83% of O_i initially present has precipitated, but only 67% in the sample with pre-annealing at $620°C$. The difference between the spectrum of the pre-annealed sample compared to the one of the sample without pre-anneal is the absence in the first one of the broad band extending from about 800 to $1100\,cm^{-1}$ and of the smaller band near $650\,cm^{-1}$. The absence of the broad band allows a better appreciation of the relatively narrower doublet peaking at 865 and $884\,cm^{-1}$. A small narrow peak at $958\,cm^{-1}$ emerges also in the sample with a pre-anneal. The same model as the one for the SiO_x precipitates in silicon [94] has been used for a fitting of the profiles of the main precipitate bands of Fig. 6.28. In this fitting, the doublet in the pre-annealed sample is attributed to spherical precipitates and the broad band to the contribution of platelet precipitates.

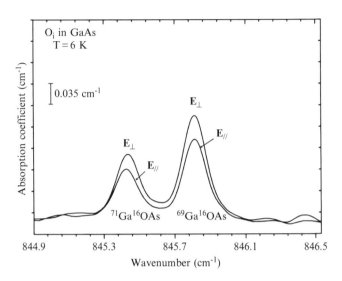

Fig. 6.29 Stress-induced dichroism of the isotopic doublet of the $Ga - O - As$ stretch mode after application of a 140 MPa aligning stress at RT along a $<110>$ axis (resolution: $0.1\,cm^{-1}$). The polarizations are indicated with respect to the aligning stress. The residual stress is ~ 1 MPa (after [3, 138]). Copyright 1994 with permission from Elsevier

6.1.3   Interstitial Oxygen in GaAs and GaP

In as-grown GaAs, an O atom can participate to the VO centre discussed in the Sect. 7.1.3 on quasi-substitutional oxygen. It can also locate interstitially in the lattice, where it binds to one Ga and one As nn atoms, in a way similar to the situation described above in silicon or germanium. A LVM observed at RT at $836\,cm^{-1}$, with a FWHM of $10\,cm^{-1}$ and a $^{16}O - ^{18}O$ IS of $46\,cm^{-1}$ was attributed to the antisymmetric stretch mode of this Ga–O–As structure by Akkerman et al. [136]. At LHeT, this mode shifts to $\sim 845\,cm^{-1}$ at LHeT and it splits into a doublet corresponding to the ^{69}Ga and ^{71}Ga isotopes as there is only one stable As isotope [137]. The frequencies are 845.82 and $845.44\,cm^{-1}$ for ^{69}Ga and ^{71}Ga, respectively, and the relative intensities match the ratio (1.51) of the abundances of the two Ga isotopes. In samples with low internal strain, a FWHMs of $\approx 0.07\,cm^{-1}$ of these components has been reported [138], but the FWHM at LHeT is usually closer to $0.13\,cm^{-1}$ [139]. The line broadening and the spacing of these components is such that at LNT, they have already merged into a single feature at $845.1\,cm^{-1}$. The $^{16}O - ^{18}O$ IS of $46\,cm^{-1}$ measured by Schneider et al. [137] is not very different from the one measured in germanium ($44\,cm^{-1}$).

While the piezospectroscopic measurements of O_i in germanium are difficult to interpret because of the number of isotopic components, such measurements have been made in GaAs [140]. The apex angle of the $Ga - O - As$ pseudomolecule is estimated experimentally to $\sim 150°$ and the centre should be considered as having C_{1h} symmetry rather than trigonal symmetry. Piezospectroscopy at LHeT for a stress along a <110> direction gives a fully polarized doublet, with component I, with positive slope $C_1 \sim 0.9\,\mathrm{cm^{-1}GPa^{-1}}$, polarized perpendicular to the stress, and component II, with negative slope $C_2 \sim -0.5\,\mathrm{cm^{-1}GPa^{-1}}$, polarized perpendicular to the stress. Such characteristics are consistent with a dipole moment oriented along a <111> axis and this was not a priori obvious considering the fact that the Ga $-$ O and As $-$ O bond lengths are expected to differ (first-principle total-energy calculations of O_i in GaAs [141] predict an apex angle of $131°$ and Ga$-$O and As$-$O bond lengths of 0.183 and 0.176 nm, respectively). Within the trigonal symmetry approximation, the piezospectroscopic parameters deduced from experiment are $A_1 = 0.2\,\mathrm{cm^{-1}GPa^{-1}}$ and $A_2 = -0.7\,\mathrm{cm^{-1}GPa^{-1}}$. Stress-induced dichroism at LHeT of O_i in GaAs has been measured following the method already described, and a result is shown in Fig. 6.29.

This dichroism shows the existence of atomic reorientation of the O atom, and as for O_i in silicon, it is seen that in GaAs, it reorients in directions making the largest possible angle with the direction of stress. There is a large uncertainty on the reorientation energy determined from the annealing of dichroism and it is estimated to be $1.0 \pm 0.2\,\mathrm{eV}$ (after [3, 138]), about one half the one measured for O_i in germanium. Considering the dissociation energies of diatomic molecules (285 kJ mole^{-1} for Ga–O and 481 kJ mole^{-1} for As–O), one would expect reorientation to take place about the As atom.

A calibration factor of the integrated absorption of the O_i LVM at 845 cm^{-1} has been proposed [142]. It is based on the fact that in samples where this mode and the only A LVM of off-centre oxygen[11] at 730 cm^{-1} are observed, there is an inverse linear correlation between the intensities of the two LVMs under annealing between 670 and 700°C, showing that the ratio of the oscillator strengths of the two LVMs is equal to unity. The calibration factor proposed for the A LVM (8×10^{16} cm^{-1}) is therefore suggested for the O_i LVM. However, other sources of dispersed O in GaAs crystals must also be considered when the total O concentration is sought for.

Phonon spectroscopy performed in O-containing GaAs samples at 0.6 and 0.95 K has shown, among several resonances, two sharp peaks at 1.86 and 1.98 meV (15.0 and 16.0 cm^{-1}) labelled S1 and S2, respectively, which have been related to O_i, and they seem to provide evidence for a low-frequency motion of O_i in GaAs [143], but there is no IR evidence of a rotational motion of the O_i atom comparable to the one observed in germanium. In the same reference, two broad and weak lines at 2.1 and 2.7 meV (17 and 22 cm^{-1}) labelled B1 and B2, respectively, have been attributed to B – O complexes as they are observed only in B-contaminated samples, but no LVM which could be attributed to such centres has been reported at higher energies.

In O-doped GaP, a line at 1007 cm^{-1} has been related to interstitial oxygen [144, 145]. The mode is observed at 1006.8 cm^{-1} at LHeT and it is attributed to the stretch mode of the Ga–O–P structure, but its FWHM of 1.2 cm^{-1} at LHeT precludes the observation of a Ga isotope effect. Its absorption is shown in Fig. 6.30

---

[11]The absorption features of the A LVM in GaAs are discussed in Sect. 7.1.3.

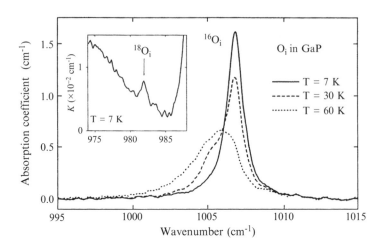

Fig. 6.30 Vibrational absorption of $^{16}O_i$ in GaP at different temperatures. The RT position given by Barker et al. [144] is $1002\,cm^{-1}$. The *inset* shows the absorption of the 0.2%-abundant ^{18}O isotope [145]. Copyright 1999 with permission from Elsevier

and the presence of oxygen in the centre is attested by the observation of a LVM due to ^{18}O at $981.9\,cm^{-1}$ [145]. The first overtone of the ^{16}O LVM has been reported at $1998.8\,cm^{-1}$, with an intensity of 3×10^{-3} of the fundamental one [145].

The measured $^{16}O - ^{18}O$ IS is $+24.9\,cm^{-1}$, a value significantly smaller than the one for O_i in GaAs ($45\,cm^{-1}$). The relatively high frequency of the stretch mode, not too different from the one of the $P - O$ bond is an indication that the O atom is more strongly bonded to P than to Ga. The shift with temperature of this mode between LHeT and LNT is $1.8\,cm^{-1}$ and it is significantly larger than the same shift in GaAs ($\sim 0.4\,cm^{-1}$).

Combined uniaxial stress and dichroism measurements give evidence of a puckered configuration of the O atom. The direction of the dipole moment of the O_i mode deduced from the piezospectroscopic measurements has been interpreted to be not too different from the $P - O$ bond direction. In this description, the $P - O - Ga$ structure lies in a (110) symmetry plane including also another Ga atom *nn* of P, resulting in a threefold degeneracy of the O atom for a given $<111>$ P$\cdots$Ga direction [145]. In silicon and germanium, the O atom laid in a plane bisecting two (110) symmetry planes and the orientational degeneracy of O for a given $<111>$ direction was sixfold.

6.2   Interstitial-N-related Centres

6.2.1   Diamond

A LVM at $0.181\,eV$ (near $1450\,cm^{-1}$) observed after electron irradiation and annealing at $900°C$ in different kinds of natural diamonds was first reported by Clark et al. [146] under the H1a label, and it was related to nitrogen by Runciman and Carter [147]. In synthetic type Ib diamonds, annealing at 650°C after irradiation was found to be necessary to produce H1a [148]. This centre is stable up to $\sim$1400°C.

The observation of only one LVM associated with each N isotope (1450 and 1426 cm^{-1} for ^{14}N and ^{15}N, respectively), and the absence of a mixed (^{14}N–^{15}N) mode led Woods and Collins [149] to attribute this LVM to a centre involving only one N atom bonded to two C atoms. A detailed investigation of that LVM in electron-irradiated synthetic diamonds with different ^{12}C/^{13}C ratios containing mainly ^{14}N showed the existence of a symmetrical bonding of the C atoms to nitrogen, and this can appreciated from Fig. 6.31 when the C isotopic ratio varies between nearly pure ^{12}C and ^{13}C diamond [150].

The frequencies of these LVMs show a dependence on the mass of the C isotope: For qmi ^{12}C$_{diam}$, the one for (^{12}C,^{12}C) is 1450.6 cm^{-1}, for qmi ^{13}C$_{diam}$, that for (^{13}C,^{13}C) is 1424.1 cm^{-1}. For intermediate ratios of ^{12}C to ^{13}C, the (^{12}C,^{13}C) peak is at 1438.7 cm^{-1}. These are RT values. At LNT, the peak attributed to (^{12}C,^{12}C)

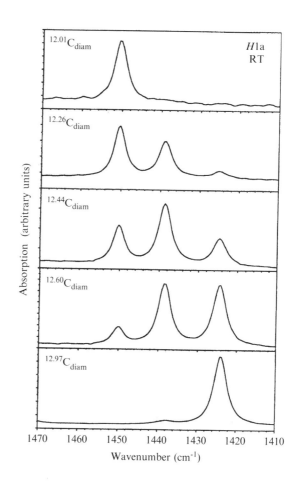

Fig. 6.31 RT absorption of the H1a centre in electron-irradiated HPHT-grown synthetic diamonds with different average ^{12}C–^{13}C isotopic compositions indicated by the average C mass after some annealing at a temperature between 500 and 1300°C (after [150]). The annealing temperature for which the spectra presented here are obtained is not indicated in this reference. These lines, first attributed to ^{12}C^{14}N^{12}C, ^{12}C^{14}N^{13}C, and ^{13}C^{14}N^{13}C by order of decreasing energy, are attributed now to ^{12}C^{14}N$_i^{14}$N$_i^{12}$C, ^{12}C^{14}N$_i^{14}$N$_i^{13}$C, and ^{13}C^{14}N$_i^{14}$N$_i^{13}$C, with no interaction between the two N$_i$ atoms oriented along a <100> axis [151]. Copyright 1996 by the American Physical Society

shifts to $1452.2\,\text{cm}^{-1}$. These ISs were initially explained by the vibration of an N atom symmetrically bonded to two C atoms, and an interstitial BC puckered N_i configuration comparable to the one for O_i in silicon or germanium was then proposed in [150]. It was later shown that the calculated activation energy for diffusion of this bond-centred N_i configuration was too small to account for the relatively high annealing temperature [151]. Moreover, it was also shown in this reference that, contrary to intuition, the $H1a$ centre actually included two decoupled N atoms in a $<100>$ interstitial configuration ($N_{2i}[100]$) sharing a substitutional site, and that this configuration should better account for the stability of this centre. This configuration displays D_{2d} symmetry and it is similar to the one for C_i in silicon shown in Fig. 4.8, but with atoms C_3 and Si_3 of this figure replaced by two tricoordinated N atoms. The existence of other LVMs at 1502, 1706, and $1856\,\text{cm}^{-1}$ observed under different annealing conditions and involving more than one interstitial N atom, together with ZPLs at 2.807, 3.188, 2.535, and 2.367 eV mentioned in the last paragraph of Sect. 4.2.2.2 was also considered [151]. Extensive calculations on different configurations were also performed and atomic structures responsible for these spectroscopic features were proposed (see Table VIII of [151]). Properties of LVMs at 1424, 1564, and $1612\,\text{cm}^{-1}$ related to a family of self-interstitial-nitrogen centres produced by electron irradiation in type 1b diamonds are discussed in [152].

6.2.2 Silicon and Germanium

6.2.2.1 The N_i-N_i Split Pair

The interest for nitrogen in silicon started for technological reasons in the early 1980s, and the observation in N-doped FZ silicon of the absorption of two LVMs of comparable intensities at 763 and $964\,\text{cm}^{-1}$ was reported by Abe et al. [153]. Measurements in FZ silicon implanted with ^{14}N, ^{15}N or ^{14}N and ^{15}N showed that these LVMs could be attributed to a close pair of N atoms, which was the dominant centre introduced by nitrogen in silicon [154]. This was confirmed by channelling measurements and by $ab\ initio$ LDF calculations coupled with further IR measurements [155]. The calculations showed that the local structure with the lowest energy is the interstitial split N_i-N_i structure with C_{2h} symmetry depicted in Fig. 1.7. In this configuration, the two N atoms of the split pair are located at equivalent distorted interstitial sites. Each of the two Si atoms separating the N atoms (atoms 4 and 5 in Fig. 1.7) is bonded to the two N atoms in order to realize trivalent bonding of the N atoms, making this pair electrically inactive. The distance between the Si atoms 3 or 6 of this figure and their three nn Si atoms, like Si atom 7, predicted from these calculations is $\sim 4\%$ larger than the normal Si $-$ Si bond length [156].

From the calculations, four modes of the split (N_i-N_i) structure are predicted above the Raman frequency of silicon, and the two ones with even (*gerade*) symmetry with respect to inversion (A_g modes) are expected to be IR inactive (but Raman active), whereas the two ones with odd (*ungerade*) symmetry with respect to inversion (B_u modes) are expected to be only IR active, in agreement with observation. In a formal view, for the mixed ($^{14}N_i$-$^{15}N_i$) pair, the symmetry is reduced to C_{1h} and the four modes are both IR and Raman active.

The formation energy of the isolated interstitial N (N_i) is very close to that of substitutional N discussed in Sect. 5.3.6.2. The configuration with the lowest energy for N_i is comparable to the one of C_i in Fig. 4.8, but with a tilt of the central N-Si$_3$ bond of 13° with respect to the <100> axis, giving a C_{1h} symmetry. The frequencies calculated for the (^{14}N, ^{28}Si) isotopic combination of this centre are 550, 773, and 885 cm^{-1} for the A", A', and A' modes, respectively [157].

Similar studies have also been performed for nitrogen in germanium and the same conclusions as for silicon have been reached [158]. A comparison between the measured and calculated frequencies of the N_i–N_i pairs in these two materials can be found in Table 6.24. For silicon, one can note a satisfactory agreement between the experimental values and those from the most recent calculations.

In N-doped FZ silicon, practically no change is observed in the intensity of the 964 cm^{-1} LVM of the split N_i–N_i pair after a 6-h annealing at 1100°C, but in N-doped CZ silicon, the annealing at the same temperature is very rapid ($\sim$30 min) and it is suggested that the N pairs aggregate on grown-in voids and/or oxide precipitates [160]. In the N-implanted germanium samples, the modes attributed to N_i-N_i are observed after annealing at $\sim$400°C and they anneal out near 600°C. When annealing the implanted germanium samples above 450°C, new N-related modes appear including, in ^{14}N-implanted germanium, one mode at the same frequency as the low-frequency mode of N_i-N_i (659 cm^{-1}). It can be differentiated from the N_i–N_i mode because it anneals out at $\sim$700°C. The frequencies at LNT of the other N-related LVMs observed in N-implanted germanium are for ^{14}N (^{15}N): 620 (608), 644 (626), 659 (637), 740 (717), 749 (726), and 799 (774) cm^{-1} and the associated centres all anneal out at $\sim$700°C [158].

Table 6.24 Comparison of the frequencies (cm^{-1}) of the LVMs of the split N_i-N_i diinterstitial measured in silicon and in germanium with the calculated ones for the B_u modes. The FWHMs are given in brackets

	Silicon				Germanium			
	Exp.[a]	Calc.[b]	Exp.[a]	Calc.[b]	Exp.[d]	Calc.[c]	Exp.[d]	Calc.[c]
$^{14}N^{14}N$	765.9,	772.9	962.6,	967.8	658.6	632.1	825.3	753.8
	771[c](8.0)		967[c](9.6)		(3.0)		(4.1)	
$^{14}N^{15}N$	758.7	766.5	947.2	952.5	653.7	620.6	809.7	739.1
	(7.0)		(8.2)		(7.5)		(4.1)	
$^{15}N^{15}N$	748.5,	755.4	936.3,	941.4	640.0	611.8	799.4	730.3
	753[r](6.9)		942[c](8.7)		(6.3)		(7.6)	

[a][158] at RT [b][157] [c][158]
[d][158] at LNT [e][159] at LNT

Fig. 6.32 Ball and stick
model of the NNO centre in
silicon. The black atoms
correspond to nitrogen
whereas the dark grey atom
represents oxygen. The
structure is asymmetric with
respect to the N atoms, noted
N1 and N2 (after [165]).
Copyright 2007 with
permission from Elsevier

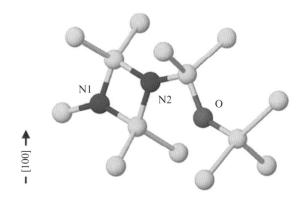

6.2.2.2   Complexes with Oxygen

In N-doped CZ silicon, nitrogen interacts with O_i, and it was suggested that additional LVMs observed at 802, 996, and 1027 cm^{-1} (RT values) with approximate relative intensities of 2, 1, and 1, respectively, were due to complexes of O with the N_i-N_i split pair [161]. Under annealing in the 400–800°C range, the intensities of the 802, 996, and 1027 cm^{-1} show a variation complementary to those of the LVMS of N_i-N_i [162, 163], which can be interpreted by the trapping of O_i by a N_i–N_i pair in the 400–600°C range, followed by a detrapping of O_i for higher annealing temperatures. At a difference with the situation in FZ material, in N-doped CZ silicon, the intensity of the 964 cm^{-1} LVM of the split N_i–N_i pair starts to decrease for annealings above 800°C and it is no longer observed for a 30-min annealing at 1100°C [160]. The results of absorption measurements at LNT on FZ silicon samples implanted with ^{16}O, ^{17}O, and either ^{14}N, ^{15}N, or $^{14}N+^{15}N$ have been reported [164]. Besides the (N_i–N_i, O_i) modes at 805, 999, and 1030 cm^{-1} (we give the LNT frequencies of [164]), a new mode at 739 cm^{-1}, with a small intensity and the same annealing characteristics as the three other ones was observed in the (^{14}N,^{16}O)-implanted sample.

The precise assignments of these LVMs have required the help of *ab initio* LDF calculations. They are based on the N_i–N_i structure of Fig. 1.7 with an O_i atom located between Si atoms 6 and 7. This location is chosen because it is slightly larger than the normal Si–Si bond length while the other Si atoms bordering the N pair along the [100] axis are compressed, and the resultant structure with point-group symmetry C_{1h} is represented in Fig. 6.32.

This structure introduces in principle no donor or acceptor level in the band gap (an equally electrically inactive alternative structure, where O_i is located in the *nnn BC* site, is found to be 0.7 eV higher in energy and it is not considered in the calculations). The calculations have first been performed with the pseudo-molecule embedded in a 134-atom cluster [164], and more recently in a 216-atom cubic silicon supercell [165]. We consider here the frequencies calculated in the latter reference for the $^{14}N^{14}N^{16}O$ structure and for complete or partial substitution

Table 6.25 Comparison between the frequencies (cm^{-1}) of the LVMs of the (NNO) centre in silicon for different isotopic combinations measured at LNT [164] with the calculated ones [165]. The first and second columns show the frequencies for $^{14}N^{14}N^{16}O$ and the atoms of Fig. 6.32 with the main atomic contribution indicated in brackets. The next columns give the experimental and calculated (in brackets) downward ISs for the isotopic combinations indicated

$^{14}N^{14}N^{16}O$		$^{15}N^{15}N^{16}O$	$^{14}N^{15}N^{16}O$	$^{15}N^{14}N^{16}O$	$^{14}N^{14}N^{17}O$	
Exp.	Calc.	Exp. (calc.)	Exp. (calc.)	Exp. (calc.)	Exp.	Calc.
	1132 (N2,N1)	(34)	(23)	(8)		(0)
1030	1016 (O)	2 (10)	2 (4)	(9)	17	(6)
999	992 (O,N1)	25 (18)	25 (5)	0 (13)	8	(18)
805	807 (N2)	19 (19)	19 (15)	0 (2)	1	(0)
739	749 (N1)	18 (18)	(4)	(15)	0	(0)
	662 (O)	(1)	(1)	(0)		(5)

with ^{15}N and also for $^{14}N^{14}N^{17}O$. The frequencies of six modes with A' symmetry are obtained between 662 and $1132\,cm^{-1}$ for $^{14}N^{14}N^{16}O$, limited to five in the 700–$1150\,cm^{-1}$ region. The correspondences of these five calculated frequencies with the measured ones given in Table 6.25 are made on the following arguments. The high-frequency mode at $1132\,cm^{-1}$ shows no ^{17}O IS, indicating that it involves the motions of atoms N1 and N2. The two low-frequency modes at 807 and $749\,cm^{-1}$ must be nitrogen related as their ^{17}O IS is predicted to be zero. The mode at 1016 and $992\,cm^{-1}$ must correspond to those measured at 1030 and $999\,cm^{-1}$. However, the corresponding ISs differ markedly: for the mode of highest energy, the calculated $^{16}O - ^{17}O$ IS is $+6\,cm^{-1}$ compared to $+17\,cm^{-1}$ for the measured one, and inversely, the calculated and measured $^{14}N - ^{15}N$ ISs are $+10\,cm^{-1}$ and $+2\,cm^{-1}$, respectively. Similar anomalies also occur for the mode at $992\,cm^{-1}$. This has been explained by the proximity of the two modes, which enhances their coupling: when ^{15}N replaces ^{14}N, the high-frequency mode becomes strongly localized on O, producing a marked O IS, while it becomes strongly localized on N when ^{17}O replaces ^{16}O. The opposite occurs for the low-frequency mode and this anticrossing is attributed to the relatively small separation of the calculated modes $(24\,cm^{-1})$ while the measured separation is $31\,cm^{-1}$ [165]. In Table 6.25, the atoms of the (NNO) centre with the most important vibrational amplitude are indicated in brackets for the $^{14}N^{14}N^{16}O$ combination.

The weak LVMs reported at 810 and $1018\,cm^{-1}$ at RT [161, 166] are suggested to arise from another centre labelled NNO2, and two possible models have been proposed for it: one, noted NNOO, where an additional O_i is added on a *BC* site *nn* of the O_i of NNO, and another one, noted ONNO, where the additional O_i is more or less symmetrical of the first one with respect to N_{2i} [165]. The ONNO structure is found to be marginally more stable than NNOO (0.1 eV) and the vibrational modes of this structure have been calculated for $^{16}O^{14}N_2^{16}O$ and for full ^{15}N and ^{17}O substitutions. For $^{16}O^{14}N_2^{16}O$, the two highest frequency modes are at 1167 and $1065\,cm^{-1}$ and their ISs show that they are N related. From their ISs, the next two modes at 1003 and $993\,cm^{-1}$ are O-related. Other N-related modes are also found at 813 and $780\,cm^{-1}$. The LVMs reported at 810 and $1018\,cm^{-1}$ have been tentatively

Table 6.26   Calculated frequencies (cm^{-1}) of the local modes of the (NO) and ($NO_{(3)}O_i$) centres in silicon. The atoms in parentheses refer to the localization of the modes on these atoms [168]

$^{14}N^{16}O$	1001	801	722		
$^{14}N^{18}O$	999	800	690		
$^{15}N^{16}O$	974	783	721		
$^{14}N^{16}O^{16}O$	1022 (N)	977 (O_i)	812 (N, $O_{(3)}$)	794 (N, $O_{(3)}$)	670 (O_i)
$^{14}N^{16}O^{18}O$	1022 (N)	934 (O_i)	812 (N, $O_{(3)}$)	793 (N, $O_{(3)}$)	661 (O_i)
$^{14}N^{18}O^{16}O$	1018 (N)	976 (O_i)	805 (N)	766 ($O_{(3)}$)	667 (O_i)
$^{14}N^{18}O^{18}O$	1018 (N)	933 (O_i)	805 (N)	765 ($O_{(3)}$)	657 (O_i)
$^{15}N^{16}O^{16}O$	996 (N)	976 (O_i)	805 ($O_{(3)}$, N,)	781 (N)	670 (O_i)

attributed to the N-related and O-related modes at 813 and 1003 cm^{-1}, respectively [165].

The RT absorption of other weak LVMs at 856, 973, 984, and 1002 cm^{-1} has been reported in N-doped CZ silicon [166]. A comparison between the annealing of these new LVMs and that of the $2p_{\pm 1}$ electronic lines at 240 and 250 cm^{-1} of N-related STDs corresponding[12] to STDD or N $-$ O-3 and STDG or N $-$ O-5 of Table 6.29 of [86] shows that the LVM at 1002 cm^{-1} could be related to theses STDs. These N-related STDs have been found to contain only one N atom [167] and they can be compared with models of (NO) and (NOO) centres in silicon proposed in [168], and references therein. These models are similar to the N pair and to the (NNO) centre, respectively, with one of the N atoms of the N pair replaced by a tri-coordinated O atom for (NO), and atom N1 of (NNO) in Fig. 6.32 replaced by a tri-coordinated $O_{(3)}$ atom for (NOO). Both centres display a C_{1h} symmetry and the (NOO) structure differs from the (ONO) symmetrical one proposed previously by the same group, where one tri-coordinated interstitial N was separated from two O_i atoms by common Si atoms. The calculated frequencies of the LVMs of (NO) and (NOO) for different isotopic combinations are given in Table 6.26, where the tri-coordinated and interstitial O atoms are noted $O_{(3)}$ and O_i, respectively.

The corresponding frequencies for the (ONO) structure have also been calculated [168] and it is found that the highest N frequency is larger (1084 cm^{-1} for $^{16}O^{14}N^{16}O$) than the one calculated for (NOO).

Weak LVMs at 860 cm^{-1} (probably the LVM reported at 856 cm^{-1} at RT by Inoue et al. [166] and 1070 cm^{-1} have recently been observed at LNT in N-containing CZ silicon samples. These LVMs disappear for annealings between 600 and 700°C. Such an effect is also observed for the $2p_{\pm 1}$ of the N $-$ O-3 SD at 240 cm^{-1}, and it has been suggested that these LVMs were due to N $-$ O-3 [169].

The (NO) centre is expected to be a deep donor and its LVMs cannot be compared with those associated with the STDs. The LVM at 1022 cm^{-1} can possibly be related to the one observed at 1002 cm^{-1}.

---

[12]There are several labels for the N-related SDs.

The nitrogen concentration in FZ and CZ silicon depends on the doping conditions and out-of-equilibrium solubilities can be met in implanted materials. A solid solubility of about $5 \times 10^{15}\,\mathrm{cm}^{-3}$ [170] is often quoted, but concentration values up to $\sim 2 \times 10^{16}\,\mathrm{cm}^{-3}$ have been reported [171]. To determine [N] in silicon, an infrared calibration factor of $(1.83 \pm 0.24) \times 10^{17}\,\mathrm{cm}^{-2}$ is generally used for the absorption coefficient of the $964\,\mathrm{cm}^{-1}$ LVM of N_i–N_i at RT [172], and it is valid in FZ silicon, where the concentration of (N,O) complexes can be neglected. In CZ silicon, N_i–N_i coexists with NNO and NNO2, and for the differential absorption coefficient at RT of the $964\,\mathrm{cm}^{-1}$ LVM of N_i–N_i after annealing sequences of 2 h at 600 and at 750°C, a calibration factor of $9.4 \times 10^{17}\,\mathrm{cm}^{-2}$ has been proposed [163].

6.3 Some Interstitial Centres in III–V Compounds

In neutron-irradiated N-doped GaP, LVMs at 946.7 and $925.2\,\mathrm{cm}^{-1}$ have been related to a ^{14}N- and ^{15}N-related defect, respectively [173]. The relatively high frequency of this mode, noted N(1), compared to the one of substitutional $^{14}N_P$ LVM in GaP, at $495.8\,\mathrm{cm}^{-1}$, is unusual, and it has been tentatively ascribed to an interstitial centre involving only one N atom paired with a substitutional atom. The annealing temperature of this LVM is the same (~ 170°C) as the one of the GaP C(1) LVM mentioned at the end of Sect. 7.2.5.

In as-grown GaP with C and N concentrations larger than $\sim 10^{16}$ and $10^{17}\,\mathrm{cm}^{-3}$, respectively, LVMs observed at 2087.12 and $2030.66\,\mathrm{cm}^{-1}$ at LHeT have been related to C as the intensity ratio of these two LVMs corresponds exactly to the ratio of the natural abundances of ^{12}C and ^{13}C [174]. In the samples with the highest values of [C], a weak line is also observed at $2048.45\,\mathrm{cm}^{-1}$, and its intensity ratio with the $2087.12\,\mathrm{cm}^{-1}$ line corresponds to the ratio of the natural abundances of ^{15}N and ^{14}N. From the value of the stretching frequency of the $C \equiv N^-$ cyanide ion in alkali halide crystals ([175], and references therein) as well as from the comparison of the observed ISs with that of a CN free radical in the harmonic approximation, these LVMs have been ascribed to the stretching mode of an interstitial CN centre. Uniaxial-stress measurements have shown that the most likely location for $C \equiv N$ is a <100> orientation at an interstitial T_{iP} site (see Fig. 1.5). A LVM at $2088.5\,\mathrm{cm}^{-1}$ has been observed at LHeT in C- and N-containing GaAs. Its shift and broadening with temperature are the same as those of the $2087.1\,\mathrm{cm}^{-1}$ LVM in GaP, and it has been attributed to the vibration of $C \equiv N$ in a <100> orientation at an interstitial T_{iAs} site [174].

After long-time annealing (typically $10 - 40\,\mathrm{h}$) at ~ 700°C, within a temperature range of 700 ± 100°C, of LEC GaAs samples containing C_{As}, O_i, and off-centre oxygen (OV_{As}), an absorption line is observed at $2059.61\,\mathrm{cm}^{-1}$ at LHeT [176]. This line, which is relatively sharp, with a FWHM of $0.07\,\mathrm{cm}^{-1}$ at LHeT, is also observed at RT at $2054.0\,\mathrm{cm}^{-1}$ with a FWHM of $1.8\,\mathrm{cm}^{-1}$, indicating that it is associated with a LVM. The presence of C in the associated centre is deduced from the observation at LHeT of a satellite at $2003.76\,\mathrm{cm}^{-1}$, whose intensity matches the ratio of the natural $^{13}C/^{12}C$ isotopic abundances. Because that LVM was observed only in GaAs samples containing a sufficiently large oxygen concentration, the centre has been attributed to the stretching of a $^{12}C - ^{16}O$ bond. The frequency of this LVM is comparable to the one of carbon monoxide in the gas phase ($2143\,cm^{-1}$) and this indicates that this CO entity is decoupled from the GaAs lattice. Under the same annealing conditions, two additional weak LVMs not correlated with the former one were also observed in the same O-containing samples at 2083.68 and $2099.18\,cm^{-1}$.

In the same O-containing GaAs samples, a pair of lines observed at LHeT at 1828.18 and $1891.41\,cm^{-1}$ with FWHMs of $\sim 0.05\,cm^{-1}$ were tentatively attributed to (B, O) complexes as the ratio IA^{1828}/IA^{1891} of their integrated absorptions (4.0) matched exactly the one of the natural abundances of ^{11}B and ^{10}B and as the measured $^{10}B - ^{11}B$ IS ($63.2\,cm^{-1}$) was very close to the one calculated for the B$-$O bond in the harmonic approximation ($53.7\,cm^{-1}$). These LVMs were reported to disappear after annealing above 900°C, but they reappeared after subsequent annealing at 700°C [176].

6.4   Self-Interstitial-related Centres in Diamond

In diamond, the self-interstitial C atom is stable at RT and it can be produced by electron or neutron irradiation. In irradiated type Ib and IIa diamonds, a LVM at $1530\,cm^{-1}$ has been reported as a primary defect annealing in the 400–450°C range. This annealing leads to the observation of a new LVM at $1570\,cm^{-1}$ [177, 178], and references therein). Measurements with qmi $^{13}C_{diam}$ samples have given $^{12}C - ^{13}C$ ISs of 59 and $61\,cm^{-1}$ for the 1530 and $1570\,cm^{-1}$ LVMs, respectively, and the ratios of the ^{12}C and ^{13}C isotopic frequencies are very close to the ratio $\sqrt{12/13}$ for these two LVMs, indicating that they involve only C atoms [177]. In mixed $^{12}C/^{13}C$ diamond crystals, the $1570\,cm^{-1}$ LVM splits into three components, showing that this mode actually involves a pair of C atoms [179]. These LVMs have been ascribed to self-interstitial-related centres [177].

References

1. K. Sumino, I. Yonenaga, Oxygen effect on mechanical properties, in *Oxygen in Silicon, Semicond. Semimet.*, vol. 42, ed. by F. Shimura (Academic Press, San Diego, 1994), pp. 449–511
2. P. Wagner, J. Hage, Thermal double donors in silicon. Appl. Phys. A **49**, 123–138 (1989)
3. B. Pajot, Some atomic configurations of oxygen, in *Oxygen in Silicon, Semicond. Semimet.*, vol. 42, ed. by F. Shimura (Academic Press, San Diego, 1994), pp. 191–249
4. M. Stavola, Vibrational spectroscopy of light element impurities in semiconductors, in *Semiconductors and Semimetals vol. 51B "Identification of Defects in Semiconductors"*, ed. by M. Stavola (Academic Press, 1999), pp. 153–224
5. M.D. McCluskey, Local vibrational modes of impurities in semiconductors. J. Appl. Phys. **87**, 3593–3617 (2000)

6. R.J. Collins, H.Y. Fan, Infrared lattice absorption bands in germanium, silicon and diamond. Phys. Rev. **93**, 674–678 (1954)

7. W. Kaiser, P.H. Keck, C.F. Lange, Infrared absorption and oxygen content in silicon and germanium. Phys. Rev. **101**, 1264–1268 (1956)

8. H.J. Hrostowski, R.H. Kaiser, Infrared absorption of oxygen in silicon. Phys. Rev. **107**, 966–972 (1957)

9. W.L. Bond, W. Kaiser, Interstitial versus substitutional oxygen in silicon. J. Phys. Chem. Solids **16**, 44–45 (1960)

10. Y. Takano, M. Maki, Diffusion of oxygen in silicon, in *Semiconductor Silicon*, ed. By H.R. Huff, R.R. Burgers (Electrochemical Soc., Pennington, NJ, 1973), pp. 469–481

11. H.J. Hrowstowski, B.J. Alder, Evidence for internal rotation in the fine structure of the infrared absorption of oxygen in silicon. J. Chem. Phys. **33**, 980–990 (1960)

12. R.M. Chrenko, R.S. McDonald, E.M. Pell, Vibrational spectra of lithium-oxygen and lithium-boron complexes in silicon. Phys. Rev. **138**, A1775–A1784 (1965)

13. B. Pajot, J.P. Deltour, États vibrationnels associés au groupement Si_2O dans le silicium. Infrared Phys. **7**, 195–200 (1967)

14. D.R. Bosomworth, W. Hayes, A.R.L. Spray, G.D. Watkins, Analysis of oxygen in silicon in the near and far infrared. Proc. Roy. Soc. Lond. A **317**, 133–152 (1970)

15. H. Yamada-Kaneta, Far infrared absorption by interstitial oxygen impurities in silicon crystals. Phys. Rev. B **58**, 7002–7006 (1998)

16. W. Forkel, M. Welte, W. Eisenmenger, Evidence for 870-GHz phonon emission from superconducting A*l* tunnel diodes through resonant scattering by oxygen in silicon. Phys. Rev. Letters **31**, 215–216 (1973)

17. C. Würster, E. Dittrich, W. Scheitler, K. Laßmann, W. Eisenmenger, W. Zulehner, Quantitative phonon spectroscopy of interstitial oxygen in silicon. Physica B **219–220**, 763–765 (1996)

18. B. Pajot, B. Cales, Infrared spectroscopy of oxygen in silicon. Mater. Res. Soc. Symp. Proc. **59**, 39–44 (1986)

19. R.E. DeWames, T. Wolfram, Vibrational analysis of substituted and perturbed molecules. I. The exact isotope rules for molecules. J. Chem. Phys. **40**, 853–859 (1964)

20. H. Yamada-Kaneta, C. Kaneta, T. Ogawa, Theory of local-phonon-coupled low-energy anharmonic excitation of the interstitial oxygen in silicon. Phys. Rev. B **42**, 9650–9656 (1990)

21. R. Jones, A. Umerski, S. Öberg, *Ab initio* calculation of the local vibratory modes of interstitial oxygen in silicon. Phys. Rev. B **45**, 11321–11323 (1992)

22. T. Hallberg, L.I. Murin, J.L. Lindström, V.P. Markevich, New infrared absorption bands related to interstitial oxygen in silicon. J. Appl. Phys. **84**, 2466–2470 (1998)

23. E. Artacho, F. Yndurain, B. Pajot, R. Ramirez, C. Herrero, L.I. Khirunenko, K.M. Itoh, E.E. Haller, Interstitial oxygen in germanium and silicon. Phys. Rev. B **56**, 3820–3833 (1997)

24. O. De Gryse, P. Clauws, Quantification of the low temperature infrared vibrational modes from interstitial oxygen in silicon. J. Appl. Phys. **87**, 3294–3300 (2000)

25. G. Davies, The spectral bandshape of the v_3 ($1136\,cm^{-1}$) vibration of oxygen in silicon and dilute silicon-germanium alloys. J. Phys. Cond. Matt. **22**, 505801/1–9 (2010)

26. C.S. Chen, F.Q. Zeng, Y.X. Huang, H.J. Ye, C.M. Hu, D.K. Schroder, Thermal configurations of oxygen in silicon. Appl. Phys. A **55**, 317–323 (1992)

27. J. Kato, K.M. Itoh, H. Yamada-Kaneta, H.J. Pohl, Host isotope effect on the localized vibrational modes of oxygen in isotopically enriched ^{28}Si, ^{29}Si, and ^{30}Si single crystals. Phys. Rev. B **68**, 035205/1–6 (2003)

28. S. Hao, L. Kantorovich, G. Davies, Interstitial oxygen in Si and $Si_{1-x}Ge_x$. Phys. Rev. B **69**, 155204/1–9 (2004)

29. R.N. Pereira, B. Bech Nielsen, J. Coutinho, V.J.B. Torres, R. Jones, T. Ohya, K.M. Itoh, P.R. Briddon, Anharmonicity and lattice coupling of bond-centered hydrogen and interstitial oxygen defects in monoisotopic silicon crystals. Phys. Rev. B **72**, 115212/1–13 (2005)

30. E. Artacho, A. Lizón-Nordström, F. Yndurain, Geometry and quantum delocalization of interstitial oxygen in silicon. Phys. Rev. B **51**, 7862–7865 (1995)

31. A. Carvalho, R. Jones, J. Coutinho, P.R. Briddon, *Ab initio* calculation of the local vibrational modes of the interstitial boron-interstitial oxygen defect in silicon. J. Phys.: Condens. Matter **17**, L155–L159 (2005)
32. M. Stavola, Infrared spectrum of interstitial oxygen in silicon. Appl. Phys. Lett. **44**, 514–516 (1984)
33. J. Coutinho, R. Jones, P.R. Briddon, S. Öberg, Oxygen and dioxygen in Si and Ge: Density functional calculations. Phys. Rev. B **62**, 10824–10840 (2000)
34. B. Pajot, H.J. Stein, B. Cales, C. Naud, Quantitative spectroscopy of interstitial oxygen in silicon. J. Electrochem. Soc. **132**, 3034–3037 (1985)
35. A.S. Oates, M. Stavola, Infrared spectrum of oxygen in silicon. J. Appl. Phys. **61**, 3114–3116 (1987)
36. L. Hsu, M.D. McCluskey, J.L. Lindström, Resonant interaction between localized and extended vibrational modes in Si :^{18}O under pressure. Phys. Rev. Lett. **90**, 095505/1–4 (2003)
37. M.T. Lappo, V.D. Tkachev, Vibration spectra of lattice defect-oxygen complexes in silicon irradiated with fast neutrons. Sov. Phys. Semicond. **4**, 418–422 (1970)
38. K. Krishnan, S.L. Hill (1981) Detailed Fourier transform infrared (FTIR) study of the temperature dependence of oxygen impurity in silicon, in *Proc. SPIE*, vol. 289 ed. by H. Sakai (SPIE, 1981), pp. 27–29
39. B. Pajot, E. Artacho, C.A.J. Ammerlaan, J.M. Spaeth, Interstitial O isotope effects in silicon. J. Phys. Cond. Matter **7**, 7077–7085 (1995)
40. H. Yamada-Kaneta, Vibrational energy levels of oxygen in silicon up to one-A_{2u}-phonon plus one-A_{1g}-phonon states. Physica B **303–303**, 172–179 (2001)
41. H. Yamada-Kaneta, Silicon isotope shifts of the 648 cm^{-1} infrared absorption line of oxygen in silicon. Solid State Phenomena **82–84**, 87–92 (2002)
42. H. Yamada-Kaneta, Expanded model for anharmonic vibrational excitation of oxygen in silicon. Physica B **308–310**, 309–312 (2001)
43. L.I. Murin, V.P. Markevich, T. Hallberg, J.L. Lindström, New infrared vibrational bands related to interstitial and substitutional oxygen in silicon. Solid State Phenomena **69–70**, 309–314 (1999)
44. H. Yamada-Kaneta, Extended theory for local-phonon-coupled low-energy anharmonic excitation of oxygen in silicon: Calculation of line-intensity ratio. Phys. Stat. Sol. (c) **0**, 673–679 (2003)
45. M. Pesola, J. von Boehm, T. Mattila, R.M. Nieminen, Computational study of interstitial oxygen and vacancy-oxygen complexes in silicon. Phys. Rev. B **60**, 11449–11463 (1999)
46. H. Yamada-Kaneta, Temperature-dependent widths of infrared and far-infrared absorption lines of oxygen in silicon. Mater. Sci. Forum **258–263** (Trans Tech), pp. 355–360 (1997)
47. B. Sun, Q. Yang, R.C. Newman, B. Pajot, N.H. Tolk, L.C. Feldman, G. Lüpke, Vibrational lifetimes and isotope effects of interstitial oxygen in silicon and germanium. Phys. Rev. Lett. **92**, 185503/1–4 (2004)
48. M. Steger, A. Yang, D. Karaiskaj, M.L.W. Thewalt, E.E. Haller, J.W. Ager III, M. Cardona, H. Riemann, N.V. Abrosimov, A.V. Gusev, A.K. Kaliteevskii, O.N. Godisov, P. Becker, H.J. Pohl, Shallow impurity absorption spectroscopy in isotopically enriched silicon. Phys. Rev. B **79**, 205210 (2009)
49. K.K. Kohli, G. Davies, N.Q. Vinh, D. West, S.K. Estreicher, T. Gregorkiewcz, I. Izeddin, K.M. Itoh, Isotope dependence of the lifetime of the 1136-cm^{-1} vibration of oxygen in silicon. Phys. Rev. Lett. **96**, 225503/1–4 (2006)
50. G. Davies, G. Liaugaudas, N.Q. Vinh, K. Litvinenko, Three phonon decay mode of the 1136 cm^{-1} ν_3 vibration of oxygen in silicon. Phys. Rev. B **81**, 033201/1–3 (2010)
51. J.W. Corbett, R.S. McDonald, G.D. Watkins, The configuration and diffusion of isolated oxygen in silicon and germanium. J. Phys. Chem. Solids **25**, 873–879 (1964)
52. H. Takahashi, H. Yamada-Kaneta, M. Suezawa, Stress-induced-dichroism of the 642.2- and 1819.5 cm^{-1} infrared absorption lines of oxygen in silicon. Physica B **340–342**, 592–595 (2003)

53. M.D. McCluskey, E.E. Haller, Interstitial oxygen in silicon under hydrostatic pressure. Phys. Rev. B **56**, 9520–9523 (1997)
54. J.P. Dismukes, L. Ekstrom, R.J. Paff, Lattice parameter and density in germanium-silicon alloys. J. Phys. Chem **68**, 3021–3027 (1964)
55. B. Pajot, Doctoral Thesis, Faculté des Sciences de l'Université de Paris (1969)
56. L.I. Khirunenko, V.I. Shakhovtsov, V.K. Shinkarenko, Investigation of vibrational absorption spectra of oxygen in Si:Ge solid solutions. Sov. Phys. Semicond. **20**, 1388–1389 (1986)
57. H. Yamada-Kaneta, C. Kaneta, T. Ogawa, Infrared absorption by interstitial oxygen in germanium-doped silicon crystals. Phys. Rev. B **47**, 9338–9345 (1993)
58. L.I. Khirunenko, Yu.V. Pomozov, M.G. Sosnin, V.K. Shinkarenko, Oxygen in silicon doped with isovalent impurities. Physica B **273–274**, 317–321 (1999)
59. D. Wauters, P. Clauws, Ge content dependence of the infrared spectrum of interstitial oxygen in crystalline Si-Ge. Mater. Sci. Forum **258–263**, 103–108 (1997)
60. I. Yonenaga, M. Nonaka, N. Fukata, Interstitial oxygen in GeSi alloys. Physica B **308–310**, 539–541 (2001)
61. W.M. Bullis, Oxygen concentration measurements, in *Oxygen in Silicon*, Semicond. Semimet., vol. 42, ed. by F. Shimura (Academic Press, San Diego, 1994), pp. 95–152
62. K. Graff, E. Grallath, S. Ades, G. Goldbach, G. Tölg, Bestimmung von parts per billion Sauerstoff in Silizium durch Eichung der IR-Absorption bei 77°K. Solid State Electron. **16**, 887–893 (1973)
63. T. Iizuka, S. Takasu, M. Tajima, T. Arai, T. Nozaki, N. Inoue, M. Watanabe, Determination of conversion factor for infrared measurement of oxygen in silicon. J. Electrochem. Soc.**132**, 1707–1713 (1985)
64. J.L. Regolini, J.P. Stoquert, G. Ganter, P. Siffert, Determination of the conversion factor for infrared measurement of carbon in silicon. J. Electrochem. Soc. **133**, 2165–2168 (1986)
65. R. Murray, K. Graff, B. Pajot, K. Strijckmans, S. Vandendriessche, B. Griepink, H. Marchandise, Interlaboratory determination of oxygen in silicon for certified reference materials. J. Electrochem. Soc. **139**, 3582–3587 (1992)
66. B.G. Rennex, J.R. Ehrstein , R.I. Scace, Methodology for the certification of reference specimens for determination of oxygen concentration in semiconductor silicon by infrared spectrophotometry. J. Electrochem. Soc. **143**, 258–263 (1996)
67. J.A. Baker, Determination of parts per billion of oxygen in silicon. Solid State Electron. **13**, 1431–1434 (1970)
68. Q.Y. Wang, T.H. Cai, Y.H. Yu, L.Y. Lin, Low-temperature (10 K) infrared measurement of interstitial oxygen in heavily-doped antimony-doped silicon via wafer thinning. Semicond. Sci. Technol. **12**, 464–466 (1997)
69. A. Sassella, Measurement of interstitial oxygen concentration in silicon in silicon below 10^{15} atoms/cm^3. Appl. Phys. Lett. **79**, 4339–4341 (2001)
70. Y. Kitagawara, H. Kubota, M. Tamatsuka, T. Takenaka, K. Takamizawa, Method and apparatus for determination of interstitial oxygen concentration in silicon crystal, US Patent no. 5386118 (1995)
71. K. Hoshikawa, X. Huang, Oxygen transportation during Czochralski crystal growth. Mater. Sci. Engin. B **72**, 73–79 (2000)
72. A. Ohsawa, K. Honda, S. Ohkawa, R. Ueda, Determination of oxygen concentration profiles in silicon crystals observed by scanning IR absorption using semiconductor laser. Appl. Phys Lett. **36**, 147–148 (1980)
73. I. Fusegawa, H. Yamagishi, Evaluation of interstitial oxygen along striations in CZ silicon single crystals with a micro-FTIR mapping system. Semicond. Sci. Technol. **7**, A304–A310 (1999)
74. S. Öberg, C.P. Ewels, R. Jones, T. Hallberg, J.L. Lindström, L.I. Murin, P.R. Briddon, First stage of the oxygen aggregation in silicon: The oxygen dimer. Phys. Rev. Lett. **81**, 2930–2933 (1998)
75. M. Pesola, J. von Boehm, R.M. Nieminen, Vibration of the interstitial oxygen pairs in silicon. Phys. Rev. Lett. **82**, 4022–4025 (1999)

76. L.I. Murin, T. Hallberg, V.P. Markevich, J.L. Lindström, Experimental evidence of the oxygen dimer in silicon. Phys. Rev. Lett. **80**, 93–96 (1998)
77. T. Hallberg, L.I. Murin, V.P. Markevich, The oxygen dimer in silicon: Some experimental observations. Mater. Sci. Forum **258–263**, 361–366(1997)
78. J.L. Lindström, T. Hallberg, D. Åberg, B.G. Svensson, L.I. Murin, V.P. Markevich, Formation of oxygen dimers in silicon during electron-irradiation above 250°C. Mater. Sci. Forum **258–263**, 367–372 (1997)
79. D. Tsurumi, K.M. Itoh, H. Yamada-Kaneta, Host-isotope effecy on the localized vibrational modes of oxygen dimer in isotopically enriched silicon. Physica B **376–377**, 959–962 (2006)
80. L.I. Murin, V.P. Markevich, M. Suezawa, J.L. Lindström, M. Kleverman, T. Hallberg, Early stages of oxygen clustering in hydrogenated Cz-Si: IR absorption studies. Physica B **302–303**, 180–187 (2001)
81. J.L. Lindström, T. Hallberg, Clustering of oxygen atoms in silicon at 450°C: A new approach to thermal donors. Phys. Rev. Lett. **72**, 2729–2732 (1994)
82. R. Jones, J. Coutinho, S. Öberg, P.R. Briddon, Thermal double donors in Si and Ge. Physica B **308–310**, 8–12 (2001)
83. J. Coutinho, R. Jones, L.I. Murin, V.P. Markevich, J.L. Lindström, S. Öberg, P.R. Briddon, Thermal double donors and quantum dots. Phys. Rev. Lett. **87**, 235501/1–3 (2001)
84. Y.J. Lee, M. Pesola, J. von Boehm, R.M. Nieminen, Local vibrations of thermal double donors in silicon. Phys. Rev. B **66**, 075219/1–4 (2002)
85. V.D. Tkachev, L.F. Makarenko, V.P. Markevich, L.I. Murin, Modifiable thermal donors in silicon. Sov. Phys. Semicond. **18**, 324–328 (1984)
86. B. Pajot, *Optical Absorption of Impurities and Defects in Semiconducting Crystals. I Hydrogen-like Centres* (Springer, Berlin Heidelberg, 2010)
87. T. Hallberg, J.L. Lindström, The bistability of thermal donors in silicon. Appl. Phys. Lett. **68**, 3458–3460 (1996)
88. W.J. Patrick, The precipitation of oxygen in silicon and its effect on surface perfection. Silicon Device Processing, NBS Special Publication 337 (National Bureau of Standard, Wasington, DC, 1970), pp. 442–449
89. P. Gaworzewski, E. Hild, F.G. Kirscht, L. Vecsernyés, Infrared spectroscopical and TEM investigations of oxygen precipitation in silicon crystals with medium and high oxygen concentrations. Phys. Stat. Sol. (a) **85**, 133–147 (1984)
90. H. Ono, T. Ikarashi, S. Kimura, A. Tanikawa, Anomalous ring-shaped distribution of oxygen precipitates in a Czochralski–grown silicon crystal. J. Appl. Phys. **78**, 4395–4400 (1995)
91. A. Sassella, A. Borghesi, P. Geranzani, G. Borionetti, Infrared response of oxygen precipitates in silicon: Experimental and simulated spectra. Appl. Phys. Lett. **75**, 1131–1133 (1999)
92. A. Borghesi, B. Pivac, A. Sassella, Polarization effect on infrared absorption of oxygen precipitates in silicon. Appl. Phys. Lett. **60**, 871–873 (1992)
93. A. Borghesi, A. Piaggi, A. Sassella, A. Stella, B. Pivac, Infrared study of oxygen precipitate composition in silicon. Phys. Rev. B **46**, 4123–4127 (1992)
94. O. De Gryse, P. Clauws, J. Van Landuyt, O. Lebedev, C. Claeyes, E. Simoen, J. Vanhellemont, Oxide phase determination in silicon using infrared spectroscopy and transmission electron microscopy techniques. J. Appl. Phys. **91**, 2493–2498 (2002)
95. O. De Gryse, P. Vanmeerbeek, J. Vanhellemont, P. Clauws, Infrared analysis of the precipitated oxide phase in silicon and germanium. Physica B **376–377**, 113–116 (2006)
96. V.V. Litvinov, B.G. Svensson, L.I. Murin, J.L. Lindström, V.P. Markevich, A.R. Peaker, Determination of interstitial oxygen concentration in germanium by infrared absorption. J. Appl. Phys. **100**, 033525/1–5 (2006)
97. A.R. Bean, R.C. Newman, The effect of carbon on thermal donor formation in heat treated pulled silicon crystals. J. Phys. Chem. Sol. **33**, 255–268 (1972)
98. R.C. Newman, R.S. Smith, Vibrational absorption of carbon and carbon-oxygen complexes in silicon. J. Phys. Chem. Solids **30**, 1493–1505 (1969)

99. C. Kaneta, T. Sasaki, H. Katayama-Yoshida, Atomic configuration, stabilizing mechanism, and impurity vibrations of carbon-oxygen complexes in crystalline silicon. Phys. Rev. B **76**, 13179–13185 (1992)

100. B. Bech Nielsen, K. Tanderup, M. Budde, K. Bonde Nielsen, J.L. Lindström, R. Jones, S. Öberg, B. Hourahine, P. Briddon, Local vibrational modes of weakly bound O-H complexes in Si. Mater. Sci. Forum **258–263**, 391–398 (1997)

101. V.P. Markevich, M. Suezawa, K. Sumino, Optical absorption due to vibration of hydrogen-oxygen pairs in silicon. Mater. Sci. Forum **196–201**, 915–920 (1995)

102. R.E. Pritchard, M.J. Ashwin, J.H. Tucker, R.C. Newman, E.C. Lightowlers, M.J. Binns, S.A. McQuaid, R. Falster, Interaction of hydrogen molecules with bond centered interstitial oxygen and another defect center in silicon. Phys. Rev. B **56**, 13118–13125 (1997)

103. E.E. Chen, M. Stavola, W.B. Fowler, Ortho and para $O - H_2$ complexes in silicon. Phys. Rev. B **65**, 245208 (2002)

104. V.P. Markevich, M. Suezawa, Hydrogen-oxygen interaction in silicon at around 50°C. J. Appl. Phys. **83**, 2988–2993 (1998)

105. T.E. Gilmer, R.K. Franks, R.J. Bell, An optical study of lithium and lithium-oxygen complexes as donor impurities in silicon. J. Phys. Chem. Solids **26**, 1195–1204 (1965)

106. C. Jagannath, Z.W. Grabowski, A.K. Ramdas, Linewidths of the excitation spectra of donors in silicon. Phys. Rev. B **23**, 2082–2098 (1981)

107. Z. Yu, Y.X. Huang, S.C. Shen, New shallow donors in high-purity silicon single crystal. Appl. Phys. Lett. **55**, 2084–2086 (1989)

108. Y.H. Lee, J.C. Corelli, Corbett, Oxygen vibrational bands in irradiated silicon. Phys. Lett. **60A**, 55–57 (1977)

109. A. Brelot, J. Charlemagne, Infrared studies of low temperature electron irradiated silicon containing germanium, oxygen and carbon, in *Radiation Effects in Semiconductors*, ed. by J.W. Corbett, G.D. Watkins (Gordon and Breach, London, 1971), pp. 161–169

110. A. Brelot, Selective trapping of vacancies, in *Radiation Damages and Defects in Semiconductors*. Conf. Ser. No 16, ed. by E.W.J. Mitchell (The Institute of Physics, London, 1973), pp. 191–201

111. J. Hermansson, L.I. Murin, T. Hallberg, V.P. Markevich, J.L. Lindström, M. Kleverman, B.G. Svensson, Complexes of the self-interstitial with oxygen in irradiated silicon: a new assignment of the 936 cm^{-1} band. Physica B **303–303**, 188–192 (2001)

112. B.N. Mukashev, A. Abdullin Kh, V. Gorelkinskii Yu, S.Z.h. Tokmoldin, Self-interstitial related reactions in silicon irradiated with light ions. Mater. Sci. Engin. B **58**, 171–178 (1999)

113. J. Bloem, C. Haas, P. Penning, Properties of oxygen in germanium. J. Phys. Chem. Solids **12**, 22–27 (1959)

114. W. Kaiser, Electrical and optical investigations of the donor formation in oxygen-doped germanium. J. Phys. Chem. Solids **23**, 255–260 (1962)

115. H.J. Stein, Localized vibrational modes for implanted oxygen in germanium. J. Appl. Phys. **44**, 2889–2890 (1973)

116. R.E. Whan, Investigations of oxygen-defect interactions between 25 and 700 K in irradiated germanium. Phys. Rev. **140**, A690–A698 (1965)

117. B. Pajot, P. Clauws, High resolution local mode spectroscopy of oxygen in germanium, in *Proc. 18th Internat. Conf. Phys. Semicond.*, ed. by O. Engström (World Scientific, Singapore, 1988), pp. 911–914

118. L.I. Khirunenko, V.I. Shakhovtsov, V.K. Shinkarenko, F.M. Vorobkalo, Structure of infrared absorption by oxygen in germanium. Sov. Phys. Semicond. **24**, 663–665 (1990)

119. A.J. Mayur, M.D. Sciacca, M.K. Udo, A.K. Ramdas, K. Itoh, J. Wolk, E. Haller, Fine structure of the asymmetric stretching vibration of dispersed oxygen in monoisotopic germanium. Phys. Rev. B **49**, 16293–16299 (1994)

120. M. Gienger, M. Glaser, K. Laßmann, Phonon spectroscopy of the low energy vibrations of interstitial oxygen in germanium. Solid State Commun. **86**, 285–289 (1993)

121. C. Linsenmaier, Diplomarbeit, University of Stuttgart (1998)

122. B. Pajot, E. Artacho, L.I. Khirunenko, K. Itoh, E.E. Haller, Matrix-induced isotope shift of a vibrational mode of interstitial oxygen in silicon. *Defects in Semiconductors 19*. Mater. Sci. Forum **258–263**, 41–46 (1997)
123. M.Y. Hu, H. Sinn, A. Alatas, W. Sturhahn, E.E. Alp, H.C. Wille, Yu.V. Shvyd'ko, J.P. Sutter, J. Bandaru, E.E. Haller, V.I. Ozhogin, S. Rodriguez, R. Colella, E. Kartheuser, M.A. Villerest, Effect of isotopic composition on the lattice parameter of germanium measured by x-ray backscattering. Phys. Rev. B **67**, 113306/1–4 (2003)
124. N. Aichele, U. Gommel, K. Laßmann, F. Maier, F. Zeller, E.E. Haller, K.I. Itoh, L.I. Khirunenko, V. Shakovtsov, B. Pajot, E. Fogarassy, H. Müssig, Isotopic shifts of the rotational states of interstitial oxygen in germanium. *Defects in Semiconductor 19*, Mater. Sci. Forum **258–263**, 47–52 (1997)
125. K. Lassmann, C. Linsenmaier, F. Maier, F. Zeller, E.E. Haller, K.M. Itoh, L.I. Khirunenko, B. Pajot, H. Müssig, Isotopic shifts of the low-energy excitations of interstitial oxygen in germanium. Physica B **263–264**, 384–387 (1999)
126. B. Pajot, P. Clauws, J.L. Lindström, E. Artacho, O isotope shifts and vibration-rotation lines of interstitial oxygen in germanium. Phys. Rev. B **62**, 10165–10172 (2000)
127. L.I. Khirunenko, V. Pomozov Yu, M.G. Sosnin, A.V. Duvanskii, N.V. Abrosimov, H. Riemann, Oxygen in Ge:Sn. Semiconductors **44**, 1253–1257 (2010)
128. G. Davies, K.K. Kohli, P. Clauws, N.Q. Vinh, Decay mechanism of the v_3 865 cm^{-1} vibration of oxygen in crystalline germanium. Phys. Rev. B **80**, 113202/1–3 (2009)
129. W. Kaiser, C.D. Thurmond, Solubility of oxygen in germanium. J. Appl. Phys. **32**, 115–118 (1961)
130. P. Vanmeerbeek, P. Clauws, Local vibrational mode spectroscopy of dimer and other oxygen-related defects in irradiated germanium. Phys. Rev. B **64**, 245201/1–6 (2001)
131. V.P. Markevich, L.L. Murin, V.V. Litvinov, A.A. Kletchko, J.L. Lindström, Local vibrational mode spectroscopy of thermal donors in germanium. Physica B **273–274**, 570–574 (1999)
132. V.V. Litvinov, L.I. Murin, J.L. Lindström, V.P. Markevich, A.A. Klechko, Vibration modes of oxygen in germanium. Semiconductors **35**, 864–869 (2001)
133. P. Clauws, P. Vanmeerbeek, Infrared vibrational mode absorption from thermal donors in germanium. Physica B **273–274**, 557–560 (1999)
134. P. Vanmeerbeek, Doctoral thesis, Ghent University (2004)
135. P. Clauws, in *Germanium-Based Technologies, From Materials to Devices*, ed. by C. Clayes, E. Simoen (Elsevier, Amsterdam, 2007), Chap. 4
136. Z.L. Akkerman, L.A. Borisova, A.F. Kravchenko, Infrared absorption spectra of oxygen-doped gallium arsenide. Sov. Phys. Semicond. **10**, 590–591 (1976)
137. J. Schneider, B. Dischler, H. Seelewind, P.M. Mooney, J. Lagowski, M. Matsui, D.R. Beard, R.C. Newman, Assessment of oxygen in gallium arsenide by infrared local vibrational mode spectroscopy. Appl. Phys. Lett. **54**, 1442–1444 (1989)
138. C. Song, Doctoral thesis, Université Paris 7 (1992)
139. C. Song, B. Pajot, F. Gendron, Local mode spectroscopy and photo-induced effects of oxygen-related centers in semi-insulating gallium arsenide. J. Appl. Phys. **67**, 7307–7312 (1990)
140. C. Song, B. Pajot, C. Porte, Piezospectroscopy of interstitial oxygen in gallium arsenide. Phys. Rev. B **41**, 12330–12333 (1990)
141. W. Orellana, A.C. Ferraz, Structural properties and energetics of oxygen impurities in GaAs. Phys. Rev. B **62**, 5326–5331 (2000)
142. M. Skowronski, S.T. Neild, R.E. Kremer, Calibration of the isolated oxygen interstitial localized vibrational mode absorption line in GaAs. Appl. Phys. Lett. **58**, 1545–1547 (1991)
143. F. Maier, K. Lassmann, Phonon scattering and IR-spectra of oxygen-related defects in gallium arsenide- aspects of quantitative phonon spectroscopy. Physica B **263–264**, 122–125 (1999)
144. A.S. Barker, R. Berman, H.W. Verleur, Localized vibrational modes of interstitial oxygen and oxygen complexes in GaP. J. Phys. Chem. Solids **34**, 123–132 (1973)
145. W. Ulrici, B. Clerjaud, D. Côte, Local-vibrational-mode absorption of interstitial oxygen in GaP. Physica B **273–274**, 807–810 (1999)

146. C.D. Clark, R.W. Ditchburn, H.B. Dyer, The absorption spectra of irradiated diamonds after heat treatment. Proc. R. Soc. London A **237**, 78–89 (1956)
147. W.A. Runciman, T. Carter, High resolution infra-red spectra of diamond. Solid State Commun. **9**, 315–317 (1971)
148. G.S.Woods, Infrared absorption studies of the annealing of irradiated diamonds. Phil. Mag. B **50**, 673–688 (1984)
149. G.S. Woods, A.T. Collins, The $1450\,cm^{-1}$ infrared absorption in annealed electron-irradiated type I diamonds. J. Phys. C **15**, L949–L952 (1982)
150. I. Kiflawi, A. Mainwood, H. Kanda, D. Fisher, Nitrogen interstitial in diamond. Phys. Rev. B **54**, 16719–16726 (1996)
151. J.P. Goss, P.R. Briddon, S. Papagiannidis, R. Jones, Interstitial nitrogen and its complexes in diamond. Phys. Rev. B **70**, 235208/1–15 (2004)
152. A.T. Collins, A. Dahwich, The annealing of interstitial-related optical centres in type Ib diamond. Diam. Relat. Mater. **13**, 1959–1962 (2004)
153. T. Abe, K. Kikuchi, S. Shirai, S. Muraoka, Impurities in silicon single crystals, in *Semiconductor Silicon*, ed. by H.R. Huff, R.J. Kriegler, Y. Takeishi (The Electrochem. Soc., Pennington, 1981), pp. 54–71
154. H.J. Stein, Evidence for pairing of implanted nitrogen in silicon, in *Proc. Internat. Conf. Defects Semicond.*, Coronado, Calif., ed. by L.C. Kimerling, J.M. Parsey (The Metallurgical Society of AIME, Warrendale PA, 1985), pp. 839–845
155. R. Jones, C. Ewels, J. Goss, J. Miro, P. Deák, S. Öberg, F. Berg Rasmussen, Theoretical and isotopic infrared absorption investigations of nitrogen-oxygen defects in silicon. Semicond. Sci. Technol. **9**, 2145–2148 (1994)
156. R. Jones, S. Öberg, F. Berg Rasmussen, B. Bech Nielsen, Identification of the dominant nitrogen defect in silicon. Phys. Rev. Lett. **72**, 1882–1885 (1994)
157. J.P. Goss, I. Hahn, R. Jones, P.R. Briddon, S. Öberg, Vibrational modes and electronic properties of nitrogen defects in silicon. Phys. Rev. B **67**, 045206/1–11 (2003)
158. F. Berg Rasmussen, R. Jones, S. Öberg, Nitrogen in germanium: Identification of the pair defect. Phys. Rev. B **50**, 4378–4384 (1994)
159. H.J. Stein, Implanted nitrogen in germanium. Appl. Phys. Lett. **52**, 153–154 (1988)
160. K. Tanahashi, H. Yamada-Kaneta, N. Inoue, Annealing behavior of interstitial nitrogen pair in Czochralski silicon observed by infrared absorption method. Jpn. J. Appl. Phys. **42**, L436–L438 (2004)
161. P. Wagner, R. Oeder, W. Zulehner, Nitrogen-oxygen complexes in Czochralski silicon. Appl. Phys. A **46**, 73–76 (1988)
162. M.W. Qi, S.S. Tan, B. Zhu, P.X. Cai, W.F. Gu, X.M. Xu, T.S. Shi, The evidence for interaction of the NN pair with oxygen in Czochralski silicon. J. Appl. Phys. **69**, 3775–3777 (1991)
163. K. Tanahashi, H. Yamada-Kaneta, Technique for determination of nitrogen concentration in Czochralski silicon by infrared absorption measurement. Jpn. J. Appl. Phys. **42**, L223–L225 (2003)
164. F. Berg Rasmussen, S. Öberg, R. Jones, C. Ewels, J. Goss, J. Miro, P. Deák, The nitrogen-oxygen pair defect in silicon. Mater. Sci. Engin. B **36**, 91–95 (1996)
165. N. Fujita, R. Jones, S. Öberg, P.R. Briddon, Local vibrational modes of $N_2 - O_n$ defects in Cz-silicon. J. Mater. Sci: Mater. Electron. **18**, 683–687 (2007)
166. N. Inoue, M. Nakatsu, Y. Inoue, Infrared absorption peaks in nitrogen doped CZ silicon. Mater. Sci. Engin. B **134**, 202–206 (2006)
167. H.Ch. Alt, H.E. Wagner, W. von Hammon, F. Bittersberger, A. Huber, L. Koester, Chemical composition of nitrogen-oxygen shallow donor complexes in silicon. Physica B **401–402**, 130–133 (2007)
168. N. Fujita, R. Jones, S. Öberg, P.R. Briddon, First-principles study on the local vibrational modes of nitrogen-oxygen defects in silicon. Physica B **401–402**, 159–162 (2007)
169. H.Ch. Alt, H.E. Wagner, Comparative mid- and far-infrared spectroscopy of nitrogen-oxygen complexes in silicon. Physica B **404**, 4549–4551 (2009)

170. Y. Yatsurugi, N. Akiyama, Y. Endo, T. Nozaki, Concentration, solubility and equilibrium distribution coefficient of nitrogen and oxygen in semiconductor silicon. J. Electrochem. Soc. **120**, 975–979 (1973)
171. M.W. Qi, T.S. Shi, S.S. Tan, B. Zhu, P.X. Cai, L.Q. Liu, D.L. Que, L.B. Li, On the determination of nitrogen in Czochralski silicon. Mater. Sci. Forum **83–87**, 263–268 (1992)
172. Y. Itoh, T. Nozaki, T. Masui, T. Abe, Calibration curve for the infrared spectrophotometry of nitrogen in silicon. Appl. Phys. Lett. **47**, 488–489 (1985)
173. S.R. Morrison, R.C. Newman, Interstitial nitrogen defects in gallium phosphide. J. Phys. C **6**, L223–L225 (1973)
174. W. Ulrici, B. Clerjaud, Nitrogen-related defects in GaP and GaAs. Physica B **376–377**, 560–563 (2006)
175. C.E. Mungan, R.C. Spitzer, J.P. Sethna, A.J. Sievers, Infrared spectrum study of the dressed rotation of the CN^- isotopes in alkali halide crystals. Phys. Rev B **43**, 43–61 (1991)
176. W. Ulrici, M. Jurisch, Thermally-induced carbon-oxygen complexes in GaAs. Phys. Stat. Sol. (B) **242**, 2433–2439 (2005)
177. A.T. Collins, G. Davies, H. Kanda, G.S. Woods, Spectroscopic studies of carbon-13 synthetic diamond. J. Phys. C **21**, 1363–1376 (1988)
178. Y. Mita, Y. Yamada, Y. Nisida, M. Okada, T. Nakashima, Infrared absorption studies of neutron-irradiated type Ib diamond. Physica B **376–377**, 288–291 (2006)
179. Y. Nisida, Y. Yamada, H. Kanda, T. Nakashima, S. Satoh, S. Kobayashi, M. Okada, in *Defects in Insulating Materials*, ed. by O. Kanert, J.M. Spaeth (World Scientific, Singapore, 1993), pp. 496–498

Chapter 7
Vibrational Absorption of Quasi-substitutional Atoms and Other Centres

This chapter deals almost only with the absorption of centres produced in semiconductors during irradiation with γ-rays and fast electrons or neutrons or during subsequent annealing treatments. Together with PL studies, this domain has been actively investigated because electrical measurements are usually difficult to perform after irradiation treatments due to the high resistivity of the samples. Technically, the observed spectra generally depend on the irradiation temperature, and as for the ESR and electronic spectra in these materials, an integrated set-up can be required allowing to perform the optical measurement at the temperature of irradiation when it is below room temperature. A large number of LVMs associated with radiation defects has been reported, especially in silicon. The interpretation of these LVMs can be far from simple because of the possible overlap of some bands and also because of some metastability effects. Potential modelling of the related centres can also be made hard because of the diversity of the possible atomic structures. In this respect, the use of quasi-monoisotopic crystals and the doping with selected isotopes has been of a great help.

7.1 Quasi-substitutional O (VO)

The vibrational spectroscopy of centres associated with interstitial oxygen in different semiconductors has been discussed in Sect. 6.1. In as-grown silicon and O-doped germanium, isolated O is normally found in the electrically inactive interstitial BC location bonded to two nn atoms of the crystal as shown in Fig. 1.6. In GaAs and GaP, it can also be found in an interstitial site (see Sect. 6.1.3), but the substitutional O_P location in GaP is also known for a long time, with three possible charge states, where the negative one shows a strong relaxation. In as-grown LEC GaAs, a defect associated with oxygen has been associated to an OV_{As} centre, which can be seen as an off-centre quasi-substitutional O atom (O_{oc}) bonded to only two of its nn Ga atoms.

B. Pajot and B. Clerjaud, *Optical Absorption of Impurities and Defects in Semiconducting Crystals*, Springer Series in Solid-State Sciences 169, DOI 10.1007/978-3-642-18018-7_7, © Springer-Verlag Berlin Heidelberg 2013

In irradiated silicon and germanium, isolated O is also found in a quasi-substitutional off-centre location, which has been modelled by considering an oxygen-vacancy centre (VO). The resultant atomic configuration in silicon and germanium is shown in Fig. 1.3 with the off-centre oxygen atom distorted along a $\langle 100 \rangle$ axis. This centre is electrically active and its vibrational properties are discussed here in relation with its other properties in the three semiconductors. The vibrational absorptions of several complexes involving lattice vacancies and O atoms, deriving mainly from V and V_2, have been reported and some of them are also discussed in this chapter. It has not been possible to report all the published values of the frequencies and the tables generally give the most recent ones.

7.1.1 O-Vacancy Centres in Silicon

7.1.1.1 Isolated VO

VO is produced by h-e electron or neutron irradiation at RT in CZ silicon by the trapping of a vacancy produced by irradiation by an interstitial O atom and a rebonding of the O atom to two *nnn* Si atoms, which are nearest neighbours of the vacancy. In silicon, the presence of a vacancy produces an outward distortion of the nearest neighbours of the missing atom. Their unperturbed normal separation is 384 pm and the comparison with an unperturbed Si–O bond length of ~ 160 pm implies an inward distortion of the two Si atoms bonded to O. VO is stable up to about 300°C (~ 570 K) and beside O_i, it is the simplest O-related defect in silicon (LVMs associated with other O-related defects have also been observed in silicon and they are discussed later on). In silicon, VO can take two charge states, and a clear correlation was established by Corbett et al. [1] between a $S = 1/2$ ESR spectrum, noted Si–B1 [2] and also Si–A [3, 4], and a LVM observed at 836 cm^{-1} at low temperature in electron-irradiated CZ silicon [1]. This showed that the 836 cm^{-1} LVM and the ESR spectrum were associated with the same VO centre, noted at that time Si–A or A centre. The ESR spectrum, observed at low temperature, is actually associated with the negative charge $(VO)^-$, but the 836 cm^{-1} LVM is due to $(VO)^0$. This comes from the fact that in the above experiments, the initial electron concentration in the samples was about one order of magnitude smaller than $[VO]$, so that $[(VO)^0]$ was much larger than $[(VO)^-]$. As LVM spectroscopy requires a relatively large concentration of centres, only the IR absorption of $(VO)^0$ was observed. In irradiated n-type silicon samples with a larger free-electron concentration, a LVM at 884 cm^{-1} at LNT was later associated with $(VO)^-$ [5]. The existence of these two charge states is related to a deep acceptor level at $E_c - 0.17$ eV, whose presence was derived from electrical measurements [6]. A definite spectroscopic proof of the presence of an O atom in the Si–A centre was brought by the observation, in electron-irradiated silicon samples enriched with ^{18}O, of an ^{18}O isotope replica of the $(V^{16}O)^0$ with a frequency of 798 cm^{-1} at LNT [1, 7]. Based on the VO atomic model with C_{2v} symmetry of Fig. 1.3, the LVM modes at 836 and 885 cm^{-1} can be associated with the antisymmetric B_1 or Γ_2 mode of a Si-O-Si pseudomolecule with C_{2v} symmetry where the two Si atoms are *nnn* along a $\langle 110 \rangle$ direction in the silicon lattice. This $\langle 110 \rangle$ orientation defines a crystallographic centre with an orthorhombic-I (or rhombic-I) symmetry (see appendix E). The $\langle 110 \rangle$ orientation of the electric dipole producing the 836 cm^{-1} mode has been confirmed by modulated stress measurements [8]. Such a centre should also give rise to an IR inactive symmetric A_1 (Γ_1) stretching mode of the two Si atoms with a lower frequency [9]. Besides the B_1 modes, weaker vibrational lines have indeed been observed in the 1360 cm^{-1} region in electron-irradiated CZ Si samples containing ^{16}O and ^{18}O, and attributed to $A_1 + B_1$ combination modes [10, 11]. These two sets of lines are shown in Fig. 7.1. In this figure, the line at 865.9 cm^{-1} (LHeT) is associated with the neutral charge state of the $C_i O_i$ centre, and this particular LVM has the particularity to show a very small ^{16}O $-$ ^{18}O IS (see Table 7.10).

One can note that, in Fig. 7.1, the ^{16}O $-$ ^{18}O ISs of the B_1 modes of VO and of the two $A_1 + B_1$ lines in the 1360 cm^{-1} region are practically the same, and a linear relationship between the integrated intensities of the corresponding lines of the two sets has also been found. These two observations have led to the attribution of the

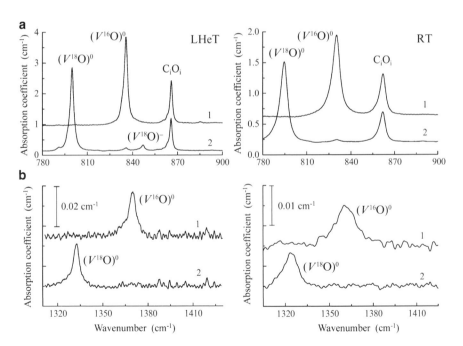

Fig. 7.1 Absorption of (**a**) the B_1 mode and (**b**) the $A_1 + B_1$ combination mode of $(VO)^0$ at LHeT and at RT of CZ Si samples irradiated with an electron fluence of 1×10^{18} cm^{-2}. Spectra noted 1: [^{16}O]~1×10^{18} cm^{-3}, 2: [^{16}O] $= 6.5 \times 10^{16}$ cm^{-3} and [^{18}O] $= 1.7 \times 10^{18}$ cm^{-3}. [C] is 3×10^{17} cm^{-3} for both samples (after [10]). The abscissa scales are the same for all the spectra

Table 7.1 Measured frequencies (cm^{-1}) of the vibrational modes of the VO centre in ^{nat}Si [11] compared with different calculated values

Line attribution	$(VO)^0$				$(VO)^-$			
	^{16}O		^{18}O		^{16}O		^{18}O	
	LHeT	RT	LHeT	RT	LHeT	RT	LHeT	RT
B_1(exp.)	835.8	830.4	799.9	794.6	885.2	877.1	847.0	839.0
	835[a]	828[b]	798[a]	791[b]	884[c]			
B_1(calc.)	843[d], 839[e], 832[f]		806[d], 802[e]		850[d], 872[e]		834[e]	
$B_1 + A_1$(exp.)	1370.0	1361.4	1332.5	1323.9	1430.1	–	1391.0	–
A_1(exp.)	~534	~531	~533	~529	~545	–	~544	–
A_1(calc.)	540[d], 548[e], 565[f]		537[d], 547[e]		539[d], 532[e]		531[e]	

The experimental values for the A_1 LVM are the differences between those of the $B_1 + A_1$ and B_1 modes. Other reported experimental and calculated values are also given
[a][7] at LNT
[b][1]
[c][5] at LNT
[d][9]
[e][12]
[f][13]

lines in the $1360\,cm^{-1}$ region to the $B_1 + A_1$ combination mode of $(VO)^0$. The frequencies of the different transitions associated at RT and at LHeT with $(VO)^0$ and $(VO)^-$ are given in Table 7.1.

In qmi ^{30}Si, the frequency of the B_1 mode of $(V^{16}O)^0$ at LHeT is shifted by $-5.9\,cm^{-1}$ with respect to the frequency in ^{nat}Si, and this is close to the IS calculated for the $^{30}Si - {}^{16}O - {}^{30}Si$ combination in ^{nat}Si. This is in agreement with the fact that the FWHMs of this mode remain the same ($2.45\,cm^{-1}$ at LHeT) in both ^{nat}Si and qmi ^{30}Si [14].

The assumption of a small O IS of the A_1 mode is comforted by the calculations, and this is due to a small displacement of the O atom for this mode. No feature which could be related to the A_1 mode of VO has been reported near $530-540\,cm^{-1}$ because the absorption coefficient of this symmetric mode is expected to be small and superimposed to the two-phonon absorption of silicon. The "experimental" values of Table 7.1 for the A_1 mode are thus taken as the differences between the experimental positions of the combination and B_1 modes. No IR vibrational measurement has been reported on electron-irradiated samples enriched with ^{17}O, as this requires relatively large concentrations of the isotope, but in the absence of resonance with phonons, the frequencies of the ^{17}O-related LVMs can be safely interpolated from those of the ^{16}O- and ^{18}O-related LVMs, giving a frequency of $\sim817\,cm^{-1}$ for $(V^{17}O)^0$ at LNT. ENDOR transitions of $(V^{17}O)^-$ have been detected in FZ silicon enriched with ^{17}O [15].

As can be judged from Fig. 7.1, the FWHMs at LHeT of the VO LVMs in silicon are rather broad ($\sim1.5-2.5\,cm^{-1}$) compared to those of the A_{2u} mode of $^{16}O_i$ and $^{18}O_i$, and no Si isotope effect can be observed. The band shape of the B_1 mode of $(V^{16}O)^0$, which is slightly asymmetric, has however been analyzed in ^{nat}Si in order to be able to measure the positions of LVM of (V,O) complexes blended with this mode and the deconvolution of the Si_2O isotopic components undertaken from the profiles observed at 20 K and at RT. The 20 K frequencies for the $^{28}Si^{16}O^{28}Si$, $^{28}Si^{16}O^{29}Si$, and $^{28}Si^{16}O^{30}Si$ combinations of $(V^{16}O)^0$ deduced from this fitting are 835.8, 834.2, and 832.7 cm^{-1}, respectively, and the RT frequencies 830.2, 828.7, and 827.25 cm^{-1}, respectively. The fitting FWHM at 20 K was the same (2.34 cm^{-1}) for the three isotopic components [16, 17].

Looking at the atomic structure of VO shown in Fig. 1.3, two parts can be considered: (a) the Si-O-Si structure whose vibrational modes produce the IR absorption, and (b) the reconstructed bond between the two Si atoms. In $(VO)^0$, two antiparallel electrons are involved in this bond (a bonding state in a molecular-orbital description). Under band gap optical excitation, one of these electrons can be promoted into an antibonding excited state, producing a S = 1 paramagnetic state $(VO^*)^0$, giving the Si–S1 ESR spectrum reported by Brower [18], later changed to Si–SL1 (in order to comply with Sandia Laboratories, Brower's affiliation). As far as we know, no LVM distinct from that of $(VO)^0$ has been associated to $(VO^*)^0$. One reason can be that the appropriate observation has not been done, but if the relaxation of the two bonded Si atoms due to the change in their bonding energy does not modify the Si–O–Si structure, no sizeable frequency difference of the O mode is expected. Information on $(VO)^+$ is scarce: One ESR spectrum with orthorhombic-I symmetry has been related to this charge state by Brosious [19], but there has been no confirmation of this result. Electron spin echo experiments locate the donor state of VO at $E_c - 0.76$ eV and calculations predict for $(VO)^+$ a configuration distorted along a $\langle 111 \rangle$ axis, with trivalent bonded O atom [20], in contradiction with Brosious' result.

The IR absorption of the VO mode has been used to determine thermodynamic properties of this centre [1]. The method is developed in the following by identifying the orientation of a given VO centre with that of the dipole moment of the corresponding Si-O-Si LVM. There are six equally populated orientations for a centre like VO in the diamond lattice, corresponding to the six distinct $\langle 110 \rangle$ directions, noted I, II, III, IV, V, and VI in Fig. 7.2.

The aim is to determine the energy required for the jump from one orientation to another. Such a jump is thermally activated, and it can be defined at temperature T by an attempt frequency ν given by:

$$\nu = \nu_0 \exp(-E_r/k_B T) \qquad (7.1)$$

where E_r is the atomic reorientation energy of the centre. This energy can be evaluated by creating first a difference of population between the six equally populated VO orientations. This is achieved by submitting an oriented sample containing VO to a stress along a given orientation at a temperature T_{reo} where effective reorientation can take place at a reasonable rate without annealing the centre, and a value of T_{reo} near room temperature seems to be adequate. In these conditions, the VO centres reorient along directions where their energy is minimized with respect to the stress. Considering the atomic configuration of VO, the energy of the stretched Si-O-Si structure is reduced when the two Si atoms come closer from one another. This means that the structure lowers its energy by aligning the Si-O-Si bond axes along the stress direction, and the two Si atoms in the reconstructed bond structure in a direction perpendicular to the stress [4]. Thus, for a stress along orientation [1] of Fig. 7.2, the centres along directions I and II will partially reorient along directions III, IV, V, and VI and the populations $N_{I,II}$ become smaller than N_{III} (the three other orientations are omitted). After annealing under such a stress at T_{reo} for a given time, the sample is cooled down under stress to the temperature of measurement (LNT or LHeT) and the stress is released. This results in a frozen population difference between otherwise equivalent configurations. The absorption coefficients $K_{//}$ and $K_{\perp}$ of the VO LVMs are then measured with radiation polarized parallel (index $//$) and perpendicular (index $\perp$) to the aligning stress, and from the premise, the initial dichroic ratio $DR_0 = (K_{//} - K_{\perp})/(K_{//} + K_{\perp})$ must be positive. The sample is then annealed for a given time t at a temperature $T_{an} < T_{reo}$, cooled down again to LNT or LHeT and a new dichroic ratio DR measured. DR/DR_0 can be expressed as $\exp(-k_r t)$ where the term k_r is $\nu_0 \exp(-E_r/k_B T_a)$. The annealing procedure is repeated (isochronal annealing) for increasing values of T_{an} and from a plot of k_r as a function of T_a^{-1}, values of ν_0 and E_r can be obtained [1, 4]. The initial value at LHeT of $K_{//} - K_{\perp}$ of VO in silicon after alignment under stress is shown in Fig. 7.3a for an n-type sample irradiated at RT with 2 MeV electrons at a fluence of 2×10^{17} cm^{-2}. It shows the return to a near isotropic VO distribution after annealing in Fig. 7.3b.

The atomic reorientation energy of VO in silicon was assumed to be independent of the charge state and the average values of 3×10^{12} Hz and 0.38 eV have been obtained for ν_0 and E_r, respectively, by combining ESR results on $(VO)^-$ and IR

Fig. 7.2 Equivalent orientations of a $\langle 110 \rangle$-oriented centre, like VO, in a cubic crystal. I, II, III, IV, V, and VI correspond to [110], [1$\bar{1}$0], [$\bar{1}$01], [101], and [0$\bar{1}$1] orientations, respectively

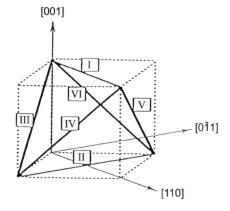

Fig. 7.3 Dichroism of the
absorption of the $(V^{16}O)^0$
and $(V^{16}O)^-B_1$ LVMs after
(**a**) alignment by a 240 MPa
stress along [101] for 15 min
at 240 K and cooling down
under stress at LHeT (see
text); (**b**) followed by
annealing for 30 min at
140 K. Spectrum (**a**)
represents approximately the
relative concentrations of the
two VO charge states. The
residual stress is ∼7 MPa
[21]. Reproduced with
permission from Trans Tech
Publications

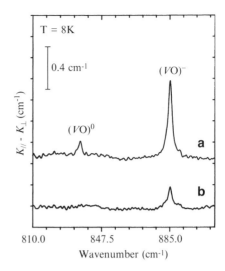

Table 7.2 Values of the piezospectroscopic parameters ($cm^{-1}GPa^{-1}$) of the $(VO)^0$ and $(VO)^-$ centres in silicon. The last line gives the hydrostatic coefficient

	$(VO)^0$		$(VO)^-$
	[21]	[8]	[21]
A_1	-2.8 ± 0.2	1.7 ± 0.3	~ 0
A_2	2.8 ± 0.2	2.5 ± 0.3	3.13 ± 0.2
A_3	4.6 ± 0.3	3.5 ± 0.3	4.0 ± 0.5
$A_1 + 2A_2$	2.8	6.7	6.16

results on $(VO)^0$ [1]. For $(VO)^-$, IR values of E_r between 0.4 and 0.5 eV have been obtained by this procedure [21].[1]

As already mentioned, the VO LVMs in silicon are rather broad and the splittings observed at LHeT for the $(VO)^0$ mode for stresses parallel to [110] and [111] are only partially resolved, leading to some uncertainties on the values of the piezospectroscopic parameters, but the results are in global qualitative agreement with the expectations from Fig. 7.2 for Si-O-Si dipoles oriented along $\langle 110 \rangle$ [21]. The piezospectroscopic parameters of $(VO)^0$ have also been obtained from modulated stress measurement of the 836 cm^{-1} mode [8] and there are some differences between the piezospectroscopic parameter A_i so obtained and the one measured from the stress splittings [21]. These values are compared in Table 7.2, where one can note the large difference between the values of parameter A_1 for the two charge states.

Formation energies and atomic parameters of VO in silicon have been calculated by the density-functional theory in the local density and local mass approximation and compared with previous calculations [9, 12]. They give for $(VO)^0$ and for

---

[1] In [21], the contents of pages 970 and 971 are inverted.

$(VO)^-$ comparable values for the Si-O bond length ($\sim$0.17 nm) combined with a Si-O-Si apex angle of $\sim$150° and a location of the O atom $\sim$0.1 nm from the substitutional site. The calculated frequencies of the A_1 and B_1 modes obtained in these references are given in Table 7.1.

A RT calibration factor of the peak absorption coefficient of the 830 cm^{-1} LVM of $(VO)^0$ of 6.1×10^{16} cm^{-2} has been obtained [22], while for the $(VO)^0$ and $(VO)^-$ peaks, the LHeT calibration factors are 1.8×10^{16} and 3.5×10^{16} cm^{-2}, respectively [23]. In this reference, the ratio of the RT/LHeT calibration coefficients has been determined to be 2.3, which is the same value as the one deduced from Fig. 7.1. An average value of $\sim$5 $\times 10^{16}$ cm^{-2} for the RT calibration factor of the 830 cm^{-1} LVM of $(VO)^0$ seems thus to be reasonable.

7.1.1.2 V_nO_m Centres

The LVMs of the VO defect start to disappear in the spectra after annealing of the samples at $\sim$300°C, and near 400°C, a new absorption band at 889 cm^{-1} at RT emerges clearly[2] [24]. It is attributed to a vacancy-dioxygen centre VO_2 produced by the trapping of VO by O_i. The structure generally admitted for this centre results from the breaking of the reconstructed bond of Fig. 1.3, allowing the insertion of a second O atom in a symmetrical configuration with symmetry D_{2d}. The two Si-O-Si structures of VO_2 are identical, incorporating all the electrons available, and the near impossibility to add or remove an electron from a Si-O bond makes thus VO_2 electrically inactive. VO_2 is characterized by a decoupling of the vibrations of the two O atoms so that only one antisymmetric O-related LVM is observed. As a consequence, in samples containing both ^{16}O and ^{18}O, only the unshifted lines observed at 889 or 850 cm^{-1} in samples containing ^{16}O or ^{18}O, respectively, are observed, showing no interaction between the two O atoms, and this has made first the VO_2 attribution questioned[3] [25]. In fact, in the vibrational modes of VO_2, there are (1) two essentially degenerate (E) LVMs, consisting in the asymmetric stretching of one O atom, the other O atom being immobile, (2) a mode corresponding to the out-of-phase stretching of the O atoms along the O$\cdots$O immaterial axis (A_1 mode), which is IR inactive, and (3) the in-phase stretching of the O atom along the O$\cdots$O axis (B_2 mode), which is IR active, but with a small vibrational amplitude of the O atoms [9]. The only VO_2-related LVM observed corresponds to the E modes and their frequencies calculated using LDF theory are in agreement with the observed ones (see Table 7.3). The frequencies calculated for the A_1 and B_2 modes for $V^{16}O_2$ and $V^{18}O_2$ are about 600 and 550 cm^{-1}, respectively, and they display rather small O IS [12], and references therein.

The bistability of one of the O atoms of VO_2 has recently been demonstrated, with the simultaneous observation near LHeT of two LVMs at 928.4 and

---

[2]In the original paper, the frequency was given as 887 cm^{-1}.

[3]In [25], the VO_2 centre is simply noted A', with reference to VO, often noted A at that time.

Table 7.3 Comparison of the frequencies (cm^{-1}) of the observed O-related LVMs of V_nO_m defects in ^{nat}Si at RT and at low temperature (LT), mainly near LHeT, with calculated values

	$V^{16}O_2$	$V^{18}O_2$	$V^{16}O_2{}^*$	$V^{16}O_3$	$V^{16}O_4$	$V_2^{16}O$	$V_3^{16}O$
RT	888[a]	850[b]		904[a] 968[a] 1000[a]	985[a] 1010[a]	826[c]	839[c]
LT	896.6[d]		928.4[d] 1003.7[d]			833.4[c] 837.0[c]	842.4[c] 848.7[c]
Calc.	912[e] 893[f]	871[e] 852[f]				829[e]	

The frequencies for $(VO)^0$, close to those of V_2O and V_3O, are given in Table 7.1

[a][26]
[b][25]
[c][17]
[d][27]
[e][9]
[f][12]

1003.7 cm^{-1} attributed to a VO_2^* bistable [26] configuration, together with the 896.6 cm^{-1} LVM of the stable form. The relaxation process from VO_2^* to VO_2 was found to be characterized by an energy barrier of $\sim$2 eV, a relatively high value [27].

The decrease of the intensity of the VO_2 LVM in the RT spectrum of electron-irradiated CZ silicon and Ge-doped CZ silicon samples after annealing near 500°C is correlated with the observation of three new LVMs at 904, 968, and 1000 cm^{-1}. These LVMs, which display the same relative intensities and annealing characteristics, have been ascribed to VO_3 [24,26]. In this centre the three Si–O–Si structures have different configurations: in the model proposed by Corbett et al. [24] for VO_3, two of them correspond to those of VO_2, and the third one to an O_i atom bonded to Si atoms nn and nnn from V, explaining the observation of O-related LVMs at different frequencies for this centre. Along the same line, the LVMs at 985 and 1010 cm^{-1} appearing after further annealing of these samples have been ascribed to VO_4 (a LVM at 984 cm^{-1} had already been reported in electron-irradiated CZ silicon after a relatively long-time annealing near 500°C [23,24]). These latter LVMs anneal out near 650°C [26]. Very recently, the 985 cm^{-1} LVM has been observed in very sensitive RT measurements of as-grown CZ silicon p-type wafers subjected to RTA of 30s from 1250°C in an oxygen-containing Ar atmosphere [28,29]. This has been attributed to the presence of thermally produced VO_4 defects produced by the RTA. The absence of the other LVM of VO_4 at 1010 cm^{-1} in this latter experiment was explained by its near coincidence with the LVM of O$_{3i}$ at 1005 and the 1013 cm^{-1} LVM of O$_{2i}$ discussed in Sect. 6.1.1.2.

In silicon, the divacancy V_2 is a well-known irradiation defect (see Sect. 4.1.1). In heavily electron-irradiated or neutron-irradiated CZ silicon, the ESR spectrum Si–A14 [30] was attributed to a V_2O centre. No LVM could be related to this centre besides one at 836 cm^{-1}, at the same frequency as the one due to $(VO)^0$.

However, considering for V_2O a Si-O-Si configuration similar to the one for VO, it was suggested that the Si-O-Si frequencies in both centres should be close [31], and this was confirmed by *ab initio* calculations [32]. Recent IR measurements as a function of annealing of CZ samples irradiated with fast neutrons have indeed shown that a LVM observed at 833.4 cm^{-1} near LHeT could be related to V_2O. This LVM was associated with another one at 837.0 cm^{-1}, possibly related to another charge state of V_2O [17]. Another LVM at 842.4 cm^{-1} observed in the same study was tentatively attributed to V_3O.

As the V_nO_m defects (n and m different from 1) present some local structural similarities, some of the frequencies of their LVMs lie in the same spectral range, and their attribution is distinguished from their annealing properties. The experimental frequencies reported by different groups are summarized in Table 7.3. For VO_2, the frequencies correspond to the E modes.

7.1.1.3 V O and Foreign Atoms

It has been found that depending on the doping-alloying or hydrogenation of silicon, the immediate atomic environment of the isolated VO centre of Fig. 1.3 could be modified, with correlated changes in the experimental frequencies of the stretch modes of Table 7.1.

To illustrate this point, CZ silicon samples doped in the 10^{17} cm^{-3} range with P, As, and Sb were irradiated near RT with doses of 2 MeV electron in the 1–5 $\times$ 10^{18} cm^{-2} range. In these samples, the annealing of the LVMs of the VO bands was anomalously rapid and the growth of the VO_2 LVMs was strongly reduced; for each dopant, two new LVMs associated with different charge states of the same defect were observed [33]. In a preliminary study with P-doping, where these LVMs were observed at 859 and 910 cm^{-1} at RT, the ODMR spectrum of a VO-like centre was also observed, involving two P atoms, and this led to propose for this centre a model where the two Si atoms of VO bonded to O were replaced by two P atoms [34]. This model was tentatively extrapolated to the other n-type dopants [33], but the measured frequencies[4] for the three dopants are too close (within one or two wavenumbers) to account for such a structure while the frequencies expected with the above model should differ significantly. A possibility which would take into account the ODMR results should be the substitution of the *nnn* Si atoms of the reconstructed bond by the group-V atoms.

Another example concerns the $Si_{1-x}Ge_x$ alloys, already mentioned at the end of Sect. 4.1.1 in relation with the electronic spectra of V_2. In an electron-irradiated alloy with $x \sim 0.04$ $(2 \times 10^{21}$ cm$^{-3})$, the presence of two LVMs at 834.6 and 839.2 cm^{-1} was deduced from the deconvolution of the RT profile near 830 cm^{-1} involving the $(VO)^0$ LVM [35]. This observation indicates a perturbation of VO by a Ge atom,

---

[4]The frequency of the lowest energy mode is close to one of the LVMs associated with (C_iO_i) shown in Fig. 7.1, associated with the (VO) modes.

Fig. 7.4 Ball and stick
model of the VO centre in
silicon where the six distinct
Si atoms *nn* and *nnn* of V are
identified following Fig. 9 of
[13]. The Si $\cdots$ Si axes of the
Si$-$O$-$Si structure and of the
reconstructed bond are taken
as the [110] and [1$\bar{1}$0] axes,
respectively, while V and O
are along a [1] axis

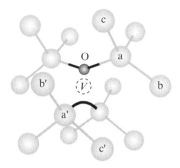

but the frequencies are too close to those of $(VO)^0$ to account for the direct bonding
of Ge to the O atom. Now, with the notations of Fig. 7.4, there are six distinct Si
neighbours of O: the a and a$'$ *nn* and the b, c, b$'$, and c$'$ *nnn* of V.

The simplest assumption is a perturbation of VO by the nearest possible Ge atom.
Calculations of the stabilities of the structures resulting from the substitution of a Ge
atom to one of the six atoms identified in Fig. 7.4 have been performed [13]. They
show that there are only three stable structures, namely VO–Ge$_b$, VO–Ge$_{a'}$, and
VO–Ge$_{b'}$ (the index indicates the substituted Si atom). The frequencies of the Si-O-
Si stretch mode calculated in this reference for VO–Ge$_{a'}$, VO–Ge$_{b'}$, and VO–Ge$_b$
are 854, 852, and 852 cm^{-1}, respectively, and the comparison with the experimental
values [35] leads to the attribution of the 839.2 cm^{-1} LVM to VO–Ge$_{a'}$ while the
834.6 cm^{-1} LVM is attributed to the VO–Ge$_b$ and/or VO–Ge$_{b'}$ structures.

The last example is related to hydrogen. Many impurities or defect centres
can form complexes with hydrogen, and some of these complexes can passivate
the electrical activity of the impurities or centres [36, 37]. IR measurements on
irradiated CZ silicon samples containing VO and subjected to hydrogen exposure
revealed LVMs at 870 and 891 cm^{-1} attributed to the Si-O-Si stretch mode of VOH
and VOH$_2$ centres [38]. In another study [39], P-doped CZ silicon samples with
$n \sim 5 \times 10^{15}$ cm^{-3} were annealed for 1 h with H$_2$ or D$_2$ at 1200°C, quenched, and
then irradiated with 3-MeV electrons at about 100°C. In the as-irradiated samples,
the LVM of $(VO)^-$ at 885 cm^{-1} was observed, but after short-term annealing at
200°C, IR measurements at LHeT showed an increase in the intensity of a LVM
at 943.5, 943.2, and $\sim$943.5 cm^{-1} in samples annealed in H$_2$, D$_2$, and H$_2$ + D$_2$,
respectively. This LVM was correlated with the absorption at higher energy of
two LVMs at (2151.5 and 2126.4), (1567.4 and 1549.1), and (2140.6 and 1557.3)
cm^{-1} in samples annealed in H$_2$, D$_2$, and H$_2$ + D$_2$, respectively (see the end of
Sect. 8.2.3.1). This clearly showed that the LVM near 943, cm^{-1} was due to a
centre containing two hydrogen atoms, and it was attributed to the stretching of
the Si $-^{16}$O $-$ Si structure of the VOH$_2$ defect. These experimental results were
confirmed by the results of *ab initio* calculations [39].

The atomic structure of the defect results from the passivation of VO by two
hydrogen atoms breaking the reconstructed bond between the two Si atoms of
Fig. 7.4, and giving the VOH$_2$ centre shown in Fig. 7.5. When the two hydrogen

Fig. 7.5 Ball and stick model of the electrically inactive VOH_2 centre in silicon. The *grey balls* are Si atoms. The configuration where the H atoms are outside of the vacancy region, in an antibonding configuration, has been calculated by Markevich et al. [39] to be 1.25-eV higher in energy

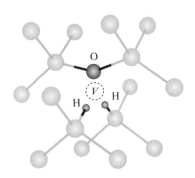

isotopes are the same, the point group symmetry of the centre is C_{2v}, reducing to C_{1h} when they differ.

A calculated value of 1076 cm^{-1} is obtained for a B_2 mode of the $V^{16}OH_2$ defect, corresponding to the B_1 mode of VO. It is independent of the mass of the hydrogen isotope, and it is in fair agreement with the experimental value (943 cm^{-1}). These values are larger than the observed and calculated ones for the B_1 mode of $(V^{16}O)^0$ of Table 7.1. This shows that the saturation by hydrogen of the Si-Si reconstructed bond leads to a strengthening of the bonds of the Si-O-Si structure. A bending mode of the Si-O-Si structure of $V^{16}OH_2$ with B_1 symmetry is also predicted at 774 cm^{-1}, and it has the particularity to have a H-D IS of 9 cm^{-1}, but it has not been observed. The calculations predict a $^{16}O - ^{18}O$ IS of 55 cm^{-1} for the B_2 mode of VOH_2, compared to 37 cm^{-1} for the B_1 mode of $(VO)^0$ (Table 7.1).

In these hydrogenated samples, annealing in the 300–400°C temperature range leads to the disappearance of the 943.5 cm^{-1} LVM, and the growth of a LVM at 891.5 cm^{-1} which has been tentatively attributed to a $V^{16}O_2H_2$ centre, consisting of a VO_2 centre with a H_2 molecule at the nearest tetrahedral interstitial site, producing a small perturbation of the isolated $V^{16}O_2$ LVM at 895 cm^{-1} [39]. In the same study, a LVM at 872.2 cm^{-1} is observed after annealing at 300°C, very probably the same as the one at 870 cm^{-1} reported in [38].

Besides the stretch modes of the Si-H bonds of VOH_2, the calculations predict a Si-H wag mode with A_1 symmetry at 771 cm^{-1} (698 cm^{-1} for Si-D), which is not observed [39].

A stepwise passivation of VO by hydrogen should result in a VOH structure, with the passivation of a single dangling bond, preceding the formation of VOH_2. VOH should be electrically active because of the presence of an unsaturated dangling bond. DFT calculations indeed predict a neutral $(VOH)^0$ and a negative $(VOH)^-$ charge state for this defect, whose calculated local and resonant modes are given in Table XIV of [12]. The frequency of the Si-O-Si structure calculated for this defect lies between the ones for VO and VOH_2. No LVM which could be related to VOH has been detected, but the ESR (S = 1/2) spectra of a monoclinic-1 defect (point group C_{1h}) has been observed in the temperature ranges 180–240 and 230–290 K in H$^+$- and D$^+$-implanted CZ silicon samples, respectively [40]. The ESR spectrum in H$^+$-implanted sample shows a proton hyperfine splitting implying the presence of one H atom in this defect. This point, linked to a large hyperfine ^{29}Si splitting typical of a vacancy-type defect has led to the attribution of these ESR spectra to $(VOH)^-$ and $(VOD)^-$. The structure of these ESR spectra starts changing above 240 K (below RT) because of the dynamic reorientation of the hydrogen atom between the two dangling bonds in the (110) symmetry plane of the centre. This motional averaging transforms the centre with a C_{1h} static symmetry into a centre with a dynamic C_{2v} symmetry, the same as the one of VOH_2. The attempt frequency ν_0 and reorientation energy E_r of expression (7.1) are (1.1 × 10^{12} Hz and 0.18 eV) and (1.9 × 10^{12} Hz and 0.26 eV) for Si-H and Si-D, respectively [40]. Further information on VOH had been obtained from Laplace-transform DLTS measurements under uniaxial stress [41]. They provide for the $(VOH)^{-/0}$ acceptor level and $(VOH)^{0/+}$ donor level experimental energies of $E_c - 0.31$ eV and $E_v + 0.35$ eV, respectively, in agreement with previous determinations or estimations. They show also that the annealing of the DLTS acceptor level and of the ESR spectrum attributed to $(VOH)^-$ coincides, confirming that the centres involved are the same. Finally, reorientation effects involving both the H and O atoms explain the absence of stress sensitivity of the $(VOH)^+$ donor state by a dynamic T_d symmetry [41].

7.1.2 VO in Germanium

The earlier results on O-related LVMs in germanium doped with a mixture of ^{16}O and ^{18}O isotopes, after electron irradiation at 80 K and different annealing, were reported by Whan [42]. From their annealing characteristics, a correlation was established between a ^{16}O LVM at 620 cm^{-1}, first observed at LNT after annealing above about 120 K, and an ESR spectrum reported by Baldwin [43], both attributed to the VO centre. The VO centre is also produced in O-containing germanium by electron irradiation at RT. As in silicon, the ESR spectrum is due to $(VO)^-$ and the 620 cm^{-1} LVM to $(VO)^0$. The LVM associated with the antisymmetric B_1 mode of $(VO)^-$ is located at 667 cm^{-1}, and the same qualitative upward shift of the frequency of the LVM with respect to that for $(VO)^0$ as in silicon is observed in germanium. In a recent spectroscopic study, a LVM at 716 cm^{-1} (LHeT), already observed by Whan [42] at 715 cm^{-1} at LNT, has been ascribed to a third $(VO)^{2-}$ charge, corresponding to a double acceptor level [44]. There has been some confusion in the presentations of the values of the single and double acceptor levels of VO in germanium. We take here the values given in [44]: $E_v + 0.27$ eV for the single acceptor level $(VO)^-/(VO)^0$, and $E_c - 0.21$ eV or $E_v + 0.49$ eV for the double acceptor level $(VO)^{2-}/(VO)^-$, with E_g(Ge) taken as ~0.70 eV near LNT.

The ^{18}O counterparts of the 620 cm^{-1} LVM of $(V^{16}O)^0$ has been observed at 589 cm^{-1} [42]. More generally, the VO LVMs exhibit O ISs with $[\tilde{\nu}(^{16}O)/\tilde{\nu}(^{18}O)]^2$ ~ 1.11 [44]. At a difference with VO in silicon, some of the LVMs of VO in germanium are relatively sharp and their LHeT absorption shows Ge isotope splitting at high resolution. This is the case for $(VO)^-$ and $(VO)^{2-}$ [45,46]. The Ge

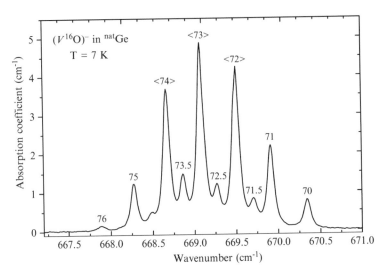

Fig. 7.6 Ge isotopic structure of the $(V^{16}O)^-$ LVM in ^{nat}Ge irradiated at RT with 2-MeV electrons due to different isotopic combinations of ^{nat}Ge (the average mass of the two Ge neighbours of the O atom are indicated). The components with values in *triangular brackets* are the sums of two or three isotopic combinations. The apodized resolution is $0.02\,\mathrm{cm}^{-1}$ (after [46])

isotopic structure of the $(V^{16}O)^-$ LVM centred near $669\,\mathrm{cm}^{-1}$ at LHeT is shown in Fig. 7.6.

The isotopic structure of this spectrum bears some resemblance with the one of interstitial oxygen (see for instance Fig. 6.22), but the rotational fine structure is absent because the structure of the defect prevents rotation of the O atom. The featureless appearance of the $(V^{16}O)^0$ LVM mode near $620\,\mathrm{cm}^{-1}$ preventing the observation of a Ge fine structure [45] is attributed to a relatively small reorientation energy of the $(VO)^0$ charge state allowing reorientation at low temperature [45,47].

The measured frequencies of the isotopic components of the $(VO)^-$, $(VO)^{2-}$ LVMs, and of a LVM at $731\,\mathrm{cm}^{-1}$, probably due to VO_2 [46, 47], are given in Table 7.4.

The existence of this isotope effect has been used to obtain information on the geometrical parameters of the centre. For VO in germanium, using the experimental ISs $^{16}O/^{18}O$ for $^{70}Ge - O - ^{70}Ge$ and $^{70}Ge/^{76}Ge$ for $Ge - ^{16}O - Ge$ for $(VO)^-$, values of 25.9 u and 100° are estimated for the interaction mass m and apex angle 2α using expressions (6.2) to (6.5) for the antisymmetric mode of a puckered Ge-O-Ge structure. They compare with the values of 23 u and 102° obtained from a fit of the experimental data [46]. The value of the apex angle obtained spectroscopically is slightly less than the value of 107° between two sp^3 bonds in the diamond structure; this implies a reduction of the distance between the two *nnn* Ge atoms and a location of the O atom relatively close to the substitutional site. It must be clear that the above results are based on self-consistent fits (SCFs) of the experimental frequencies based on harmonic approximation. The results of

Table 7.4 Measured frequencies (cm^{-1}) of isotopic components of VO and VO_2 centres in ^{nat}Ge at LHeT as a function of the average masses M_{av} of the Ge atoms (after [46])

M_{av}	$(VO)^-$		$(VO)^{2-}$	VO_2
	^{16}O	^{18}O	^{16}O	^{16}O
70	670.34	636.88	717.6	732.94
71	669.91	636.41	717.1	732.50
71.5	669.71	636.19		732.28
72	669.50	635.96	716.7	732.06
72.5	669.27	635.68		731.85
73	669.07	635.50	716.3	731.62
73.5	668.86	635.25		731.40
74	668.67	635.05	715.8	731.19
74.5	668.49	634.78		730.96
75	668.27	634.62	715.4	730.78
76	667.89			
70 − 75	2.07	2.26	2.2	2.16

By comparison, the 70 − 75 spacing for $^{16}O_i$ in germanium is 2.49 cm^{-1}

Table 7.5 Frequencies (cm^{-1}) of the B$_1$ mode (ν_3 in the molecular description) of VO in ^{nat}Ge for different isotopic combinations

	$(VO)^0$		$(VO)^-$			$(VO)^{2-}$		
	Calc.[a]	Exp.	Calc.[a]	Exp.	SCF	Calc.[a]	Exp.	SCF
$^{74}Ge_2^{16}O$	603	621.4[b]	684	668.67	668.64	694	715.80	715.83
$^{72}Ge_2^{16}O$				669.50	669.48	693.2	716.72	716.68
$^{74}Ge^{16}O^{72}Ge$				669.08	669.05	683.6	716.26	716.25
$^{74}Ge_2^{18}O$				635.05	635.06	657.7		
$^{nat}Ge_2^{18}O$	571.6	589.6[b]	648.2	635.4[b]		657.7	680.4[b]	

The ISs are calculated from $^{74}Ge_2^{16}O$ [47, 48]. They are compared with the experimental values at LHeT and with the self-consistent fit (SCF) using expressions (6.3) and (6.4) with $m = 23$u, $2\alpha = 102°$ and $^{71}Ge_2^{16}O$ as the reference [46]
[a][47]. The average atomic weight for ^{nat}Ge is taken as 72.64 u
[b]No or unresolved Ge fine structure [45]

local-density functional *ab initio* calculations without initial assumptions of the VO centre in germanium have been published [47]. They predict for $(VO)^-$ a value (154°) of 2α substantially larger than the spectroscopic one and a more off-centre position of the O atom; for $(VO)^{2-}$, the value of the calculated angle is comparable (156°), but it is slightly reduced (140°) for VO^0. These results may be physically more significant than those obtained from SCFs. Within the quasimolecular scheme, inverting expression (6.6) shows that a value of 154° of the Ge-O-Ge apex angle 2α for $(VO)^-$ corresponds to an interaction mass m of ∼60 u. A comparison of the experimental frequencies of some isotopic combinations for $(VO)^0$, $(VO)^-$, and $(VO)^{2-}$ with the ones derived from the local-density functional calculations is given in Table 7.5.

Contrary to the situation in silicon, there seems to be no experimental determination of the reorientation energy E_r of VO in germanium, but the calculated values for $(VO)^0$, $(VO)^-$ and $(VO)^{2-}$ are 0.11, 0.23, and 0.40 eV, respectively [47]. The correlation between the decrease of a DLTS peak associated with $(VO)^{2-}/^-$ with the increase of new DLTS peaks and *ab initio* calculations has led to the attribution of these new peaks to VO_2 [48]. They could be also correlated with the LVMs at 731 and 771 cm^{-1} [49].

7.1.3 Off-centre Oxygen (O_{oc}) in GaAs

In GaAs samples from as-grown LEC SI crystals or from crystals grown by the HB method, two LVMs centred at 729 and 714 cm^{-1}, noted A and B, respectively, were observed at low temperature [50]. In some samples, only mode A was observed under TEC while in other samples, under the same conditions, A and B were observed simultaneously. These LVMs, which are photosensitive and show a triplet fine structure at LHeT, were ascribed to the vibration of the Ga-O-Ga structure of an OV_{As} centre in combination of the charge states, similar to the VO centre in silicon and germanium. Oxygen was expected to be present in these samples (in the LEC crystals, its presence results from the partial dissociation of the B_2O_3 encapsulant), and the triplet structure could be explained by the combination of the ^{69}Ga and ^{71}Ga isotopes [51]. The presence of oxygen in the centre was confirmed by the observation of an ^{18}O IS at 679 cm^{-1} of the B mode, confirming the Ga-O-Ga structure [52]. No ESR spectrum has been correlated with these two modes.

In samples where only mode A is observed under TEC, it is possible to observe the decrease of the absorption of mode A and the increase of that of B by illumination at low temperature (usually the temperature of the absorption measurement) with photons with energies ≥ 0.8 eV [50,53]. During this conversion, a transient LVM, noted B', shifted from B by only ~ -0.7 cm^{-1}, is observed[5] [54, 55]. This photoconversion is explained (a) by the presence in the samples of neutral EL2 centres which are photoionized in the CB into EL2$^+$ (the EL2^0/EL2$^+$ level is located in the band gap of GaAs at $E_c - 0.75$ eV), and (b) by the trapping of the photoelectrons by OV_{As}. The presence of the transient LVM B' is attributed to a negative-U behaviour of OV_{As}, leading to a transient unstable charge state where only one electron is trapped, before the trapping of a second electron, giving a final stable state associated with the B LVM [56, 57]. The charge state associated with B' is metastable at low temperature. When the illumination of a sample held at LHeT is stopped at an intermediate stage, the sample is in a metastable state where the relative intensities of the modes do not change appreciably with time. Thermal recovery from the metastable state B' is achieved by annealing at temperatures above 90 K, and it can be fitted with an activation energy of 0.14 eV [57]. Spectra showing the relative intensities of A, B, and B' at such an intermediate stage are displayed in Fig. 7.7, where indexes H and L stand for heavy (^{71}Ga) and L for light (^{69}Ga) Ga isotopes, respectively.

---

[5]In [55,58–61] this mode is noted C.

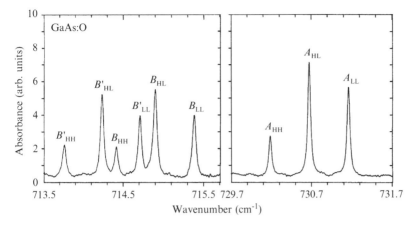

Fig. 7.7 Absorption at LHeT of modes B', B, and A of $^{16}OV_{As}$ in GaAs in metastable state (resolution: 0.013 cm^{-1}). Indexes HH, HL, and LL are for $^{71}Ga_2O$, $^{71}GaO^{69}Ga$, and $^{69}Ga_2O$, respectively (after [58]). Copyright 1992 with permission from World Scientific Publishing, Singapore

Table 7.6 Positions and FWHM (cm^{-1}) at LHeT of isotopic components of the A, B, and B' modes of OV_{As} in GaAs. Rounded values of the A, B and B' modes are given in [55]

	A	B		$B'(C)$	604 cm^{-1}LVM[a]
	^{16}O[b]	^{16}O[b]	^{18}O[c]	^{16}O[b]	^{16}O
$^{69}Ga_2O$(LL)	731.161	715.384	679.38	714.711	605.14
$^{69}GaO^{71}Ga$(HL)	730.675	714.901	678.86	714.236	604.77
$^{71}Ga_2O$(HH)	730.186	714.415	678.35	713.759	604.37
FWHM	0.045	0.055		0.056	0.14[b]

The spectral characteristics of the 604 cm^{-1} LVM observed in O-rich GaAs are also given for comparison
[a][55]
[b][59]
[c]Song, unpublished results

The measured positions of isotopic components of modes A, B, and B' are given in Table 7.6.

It has been pointed out above that in EL2-containing GaAs samples where only the A LVM is observed under TEC, illumination with photons with energies >0.8 eV resulted in the photoconversion of this LVM into LVM B via the intermediate LVM B'. This is only part of the story: under subsequent illumination of the sample with photons of the same energy, the inverse photoeffect is observed, leading to a final state where only LVM A is again observed. The state of the sample after this second photoconversion is, however, not the same as the one in the initial state as further illumination does not modify the situation [54]. The A to B conversion is a relatively fast process, while the inverse effect is much slower. This is illustrated in Fig. 7.8, showing the time dependence of this double interconversion for illumination with 1.25 eV photons, using as a probe the intensities of the LL

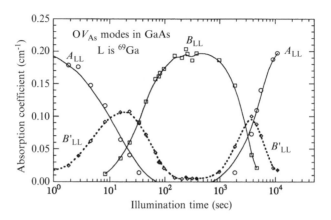

Fig. 7.8 Change with illumination time at 1.25 eV of the absorption coefficients at LHeT of the A_{LL}, B_{LL}, and B'_{LL} components of the OV_{As} modes in a SI GaAs sample with [EL2] $\sim 1 \times 10^{16}$ cm^{-3} and [C_{As}] $\sim 1 \times 10^{15}$ cm^{-3}. The photon flux is normalized to 2.5×10^{15} cm^{-2}s^{-1} (see text) (after [60]). Copyright 1995 by the American Physical Society

component (^{69}Ga$_2$O) for the three modes. In this figure, the time scale for the photoconversion from A to B (up to 10^2 s) has been normalized to the photon flux used for the photoconversion from B to A (2.5×10^{15} cm^{-2}s^{-1}). The reason was that the actual photon flux for the photoconversion from A to B was limited to 0.4×10^{15} cm^{-2}s^{-1}, to ensure a good accuracy, giving for the same effect an illumination time $\sim$5.4 times larger than the one shown in the figure [59].

In the first step of the above process, the photoconversion from A to B is attributed to the fast capture by OV_{As} in the more positive A configuration of electrons photoionized from EL2^0 in the CB (a small fraction of EL2^0 can be also converted into EL2*), leaving EL2$^+$. Under further illumination, the most negative charge state of OV_{As} can de-excite at a slower rate by electron emission or more probably hole capture, leaving at the end the most positive charge state giving mode A, while EL2$^+$ is slowly re-converted into EL2^0 by photoionization in the VB. It is assumed that the story ends there because the ultimate conversion of EL2^0 into EL2* under photon illumination is more efficient than photoionization, which predominated in the first step.

The results of a detailed piezospectroscopic study of modes A, B, and B' have been reported [58, 59]. For a uniaxial stress along $\langle 110 \rangle$ and $\langle 111 \rangle$ directions, these modes split into three and two components, respectively, as expected from Fig. 7.2 for Ga–O–Ga dipoles oriented along $\langle 110 \rangle$ (see [61] for mode A). The splittings of the three modes for σ // [111] and different polarizations are displayed in Fig. 7.9.

For a centre with rhombic-I symmetry, the slopes of the split components 1, 2, and 3 of Fig. 7.9 expressed as a function of the piezospectroscopic coefficients A_i are $A_2 - A_3$, $(A_1 + A_2)/2$, and $A_2 + A_3$, respectively [61]. The values of the slopes and of the relative intensities of the components allow to determine the piezospectroscopic

Fig. 7.9 Splittings at LHeT of the isotopic components of modes A, B, and B' of OV_{As} in GaAs as a function of the stress applied along the [110] direction. LL, HL, and HH have the same meaning as in Fig. 7.7. The split components noted 1 are observed for $\mathbf{E}//[1\bar{1}0]$, those noted 3 for $\mathbf{E}//[111]$ and those noted 2 for both polarizations [58]. Copyright 1992 with permission from World Scientific Publishing, Singapore

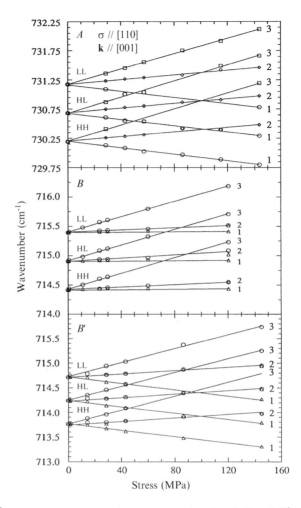

Table 7.7 Values ($cm^{-1}GPa^{-1}$) of the piezospectroscopic parameters of modes A, B, and B' attributed to the vibration of the orthorhombic I OV_{As} centre in GaAs in different charge states [59]

Piezospectrocopic parameter	A_1	A_2	A_3
Mode A	2.08 ± 0.22	2.05 ± 0.12	4.77 ± 0.12
Mode B	-1.44 ± 0.19	3.53 ± 0.08	3.36 ± 0.08
Mode B'	1.21 ± 0.25	2.10 ± 0.15	5.28 ± 0.15

parameters A_1, A_2, and A_3 of the three centres. The values of these parameters given in Table 7.7, are deduced from the data for $\sigma//$ [111].

Attempts to observe atomic reorientation of the OV_{As} centres by applying a stress along a [111] axis at room temperature and cooling under stress at LHeT, in the way described for VO in silicon have given results which were sample dependent: in a sample where only mode A was observed under TEC, no reorientation was observed. However, in another sample where modes A and B were observed under TEC, modes A, B, and B' were observed after the above stress application procedure and dichroism of modes A and B' were observed, indicating reorientation, but it was not clear if the effect was due to atomic or electronic effects [59, 61].

Up to now, the absolute charge states of OV_{As} related to the different IR modes have not been specified because they could only be guessed. No ESR spectrum is related to the A and B LVMs, but optically detected EPR associated with the B' LVM has been reported [62], and an extended investigation of optically detected EPR has led to the conclusion that the B' state corresponded to the neutral charge state of the centre ([63], and references therein). This contrasted with the results of LDF ab $initio$ calculations on the OV_{As} centre in GaAs, where the A, B', and B LVMs are associated to the -1, -2, and -3 charge states of a structure comparable with VO in silicon (off-centre oxygen) [64]. In this study, the frequencies of the symmetric and antisymmetric modes for $(OV_{As})^-$ and $(OV_{As})^{3-}$ have also been calculated for some isotopic configurations. These calculations predict the existence of positive and neutral charge states where the O atom is quasi-substitutional. Similar conclusions have been reached from calculations based on the plane-wave pseudopotential method [65]. Within the LDF approximation, Taguchi and Kageshima [66] have considered not only the above OV_{As} structure, but also another one, where the O atom, bonded to two Ga atoms, is close to the T_{iIII} interstitial site of Fig. 1.5. In the latter structure, noted O_I, the A, B', and B modes are associated with the neutral, $(-)$, and $(2-)$ charge states of the centre, and the $(2-)$ charge state is found to be more stable than the singly negative one, reproducing the experimentally observed negative-U character of the centre.

These views have finally been challenged by Pesola et al. [67], who compared the classical OV_{As} configuration, noted O_{As} by these authors, and the above O_I structure with a new one, in which the two Ga atoms nn of V_{As} not bonded to the O atom are replaced by As_{Ga} antisites. This new structure, noted $(As_{Ga})_2 - O_{As}$, reproduces qualitatively the charge-induced frequency shifts of the IR modes, and it exhibits strong negative-U characteristics for the neutral and $(-)$ charge states. Within Pesola et al.'s model, the $(+)$ and $(-)$ stable charge states correspond to the A and B LVMs and the metastable neutral one-electron state to B'. This new structure has indeed some appealing properties, and it explains naturally the absence of atomic reorientation mentioned above for the OV_{As} centre. When keeping the $(As_{Ga})_2 - O_{As}$ attribution for the A, B, B' LVMs, one can wonder if isolated quasi-substitutional O exists in GaAs. The calculations show that for the $(+)$ and neutral charge state, it should be nearly substitutional, but would become off-centre for more negative charge state, with a spectroscopic signature similar to the A-like LVMs. Such a triplet is consistently observed in O-rich GaAs at $604\,\mathrm{cm}^{-1}$ (see Table 7.4), but no O isotope effect or charge state dependence of this LVM has still been reported. The attribution of the $(+)$ and $(-)$ charge states to the A and B LVMs has been recently confirmed experimentally by Alt et al. [68], and it favours the $(As_{Ga})_2 - O_{As}$ microscopic model of [67].

The positions in the band gap of the electrical levels associated with the A, B, and B' LVMs have been deduced from thermal decay studies of the B and B' LVMs [56, 57] and from optical cross-section threshold studies [56]. With the attributions of [67], the 0/+ level is located at $E_c - 0.14$ eV and the $-/0$ level at E_c–0.58 eV, locating the virtual $-/+$ level of the negative-U centre at about E_c–0.4 eV. Incidentally, the comparison between absorption spectroscopy of the A and B LVMs and DLTS studies of NTD GaAs samples have revealed that the $-/0$ level at $E_c - 0.58$ eV is very probably the one identified by Martin et al. [69] as EL3 [70].

The shifts with temperature of the individual components of the A and B LVMs from the positions at LHeT given in Table 7.6 can be followed up to 150–160 K. The monotonous shifts are small up to ~30 K (~ − 0.1 cm^{-1}). At 150 K, they are -2.1 and -1.2 cm^{-1} for A and B, respectively. At RT, the position of the A LVM is 725 cm^{-1}, with a FWHM ~6 cm^{-1}. Because of its metastability, the B' LVM can be followed only up to 90 K. This LVM has the peculiarity to show first a very small positive shift with temperature (~ + 0.02 cm^{-1}) up to 30–40 K, followed by a small negative shift which depends on the components (-0.15 and -0.3 cm^{-1} for HH and LL, respectively) at 90 K [59].

During the above-described photoconversion measurements of modes A, B, and B', observations in contradiction with what is predicted for the relative intensities of the isotopic components of the Ga-O-Ga pseudomolecule have been made [58]. The natural abundances of ^{69}Ga and ^{71}Ga are 0.601 and 0.399, respectively, and three isotopic components HH (^{71}Ga$_2$O), HL (^{71}GaO^{69}Ga) and LL (^{69}Ga$_2$O) are normally observed with peak absorptions, normalized to unity for HH, of 3.01 for HL and 2.27 for LL (the FWHMs are assumed to be the same). This situation is approximately represented in Fig. 7.7.

Two distributions of abnormal intensity ratios, noted Γ and Λ, can be observed in the pseudo-equilibrium states and they are shown for mode B in Fig. 7.10, where they are compared with the normal (N) situation.

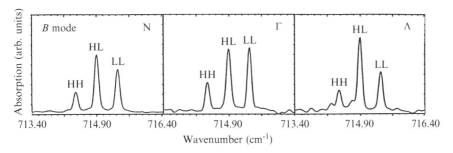

Fig. 7.10 Relative intensities at LHeT of the components of the triplet of the B mode of OV_{As} in GaAs due to the two Ga isotopes for the normal (N) and abnormal (Γ and Λ) distributions at a resolution of 0.1 cm^{-1} (after [58]). Copyright 1992 with permission from World Scientific Publishing, Singapore

Within experimental accuracy, the LL/HH intensity ratios for the Γ and Λ cases are not too different from the one in the normal case. The most visible difference comes from the HL intensity, which is comparable to that of LL in the Γ case, and is about two times that of LL in the Λ case.

When such an anomaly occurs in a pseudo-equilibrium situation in a sample where A and B are both observed, it is found that the anomalies of the two modes are opposite while in a sample where A and B' are both observed, the anomaly is the same for both modes (the anomaly is of course opposite for B and B'). A possible explanation could be that atomic reorientation takes place when the charge state associated with B' traps a second electron or when the charge state associated with B traps a hole. The anomaly has also been followed as a function of temperature in a sample showing B with anomaly Λ at LHeT. This anomaly persists up to 60 K, but at 80 K it has turned into a strong Γ anomaly and at 120 K as far as it can be judged because of the broadening of the lines, the anomaly has disappeared [58,59].

Isochronal annealing measurements on the A LVM of the positive charge state of $(As_{Ga})_2 - O_{As}$ at 730 cm^{-1} show a decrease of its the intensity near 650°C, with a correlated increase of the intensity of the O_i LVM near 846 cm^{-1} [71]. It can be attributed to a thermal dissociation of the $(As_{Ga})_2 - O_{As}$ centre, but the situation is not simple as further annealing at higher temperature shows an increase of the intensities of the A and O_i LVMs, suggesting the existence of another source of oxygen in the crystal. Calibration factors of 3.6 and 4.2×10^{16} cm^{-1} for O involved in the positive and negative charge states of $(As_{Ga})_2 - O_{As}$ have been proposed from the integrated absorptions of the A and B LVMs, respectively [68]. They are based on an anticorrelation with the C_{As} concentration measured from the calibration factor of 7.2×10^{15} cm^{-1}. These results are in between previous estimates from a comparison with DLTS measurements [72] and from the result of optically induced charge transfer [53], giving calibration factors of 8 and 2×10^{16} cm^{-1}, respectively.

In GaP, self-consistent semi-empirical calculations of the geometry of O_P^- indicate that the equilibrium configuration for this charge state corresponds to a relaxation of the O atom along a $\langle 100 \rangle$ direction, to an inward relaxation of the two nearest Ga atoms which bind to the O, and to an outward relaxation of the two other Ga atoms [73]. This leads to a puckered Ga-O-Ga structure similar to the OV structure in silicon and germanium.

7.2   C-Related Irradiation Defects

The spectroscopic study of irradiation defects in semiconductors started early. One of the first reports on the vibrational absorption spectra of defects in silicon is the paper by Fan and Ramdas [74] and the domain is still active [75]. The vibrational properties of VO centre in silicon and germanium, an irradiation defect in these materials, have been presented above. Partly due to the presence of carbon as a residual impurity in silicon, there have been many studies on carbon-related radiation defects, and we try to show here how the results obtained by vibrational spectroscopy have contributed, in connection with other methods, to a better understanding of these centres. Electronic absorption associated with some of these defects has been discussed in Sect. 6.7.1 of [76]. The optical signatures of these centres consist in LVMs observed by IR absorption or in ZPLs due to the creation or recombination of excitons bound to these defects. The vibronic sidebands associated with the ZPLs provide useful information on the vibrational modes which can be compared with that obtained from IR absorption by LVMs and both results are presented. There have been many labellings for these centres and some of them are indicated.

7.2.1   Interstitial Carbon (C_i), (IC_i) and (C_i, Sn_s) in Silicon

The defect known as interstitial carbon (C_i) is the dominant extrinsic defect produced by electron irradiation at temperatures typically below 100 K in C-containing FZ silicon. It is produced when a mobile silicon self-interstitial Si_i or I is trapped by C_s. The ball and sticks model of C_i (Fig. 4.8) deduced from the ESR measurements shows what is also called a split interstitial or interstitialcy configuration where the C and inner Si atoms are threefold coordinated. Beside the ZPL at 856 meV (see Sect. 4.1.4.1), two LVMs (the C(1) lines) with intensities in a $\sim 2/1$ ratio are associated with this defect [77, 78]. Their frequencies for ^{12}C (^{13}C) at LHeT are 922.1 (893.1) and 932.1 (905.9) cm^{-1}, corresponding to longitudinal and transverse modes of the C and inner Si (Si_3 in Fig. 4.8) atoms, respectively [79], and they can be seen on the high-energy side of spectra (a) and (c) of Fig. 7.13. In ^{14}C-enriched silicon, these LVMs have been reported at 867 and 881 cm^{-1} at LNT [81]. The most recent calculated *ab initio* values of these longitudinal and transverse modes for ^{12}C (^{13}C) are 930 and 945 cm^{-1} (904 and 915) cm^{-1}, respectively [79]. In $Si_{1-x}Ge_x$ alloys with $x \leq 0.044$, a broadening of the LVMs of C_i has been reported, together with small opposite shifts of the transverse and longitudinal modes (positive and negative, respectively). Each band can be resolved by deconvolution into two bands, none of them implying direct Ge-C bonds in the corresponding structures [82].

The measurement of the stress-induced dichroism at LHeT of the ^{13}C LVMs at 892 and 904 cm^{-1} has confirmed the C_{2v} symmetry deduced from the ESR measurements, showing that the low-frequency LVM is polarized along the C_2 [100] axis (A_1 mode) and the high-frequency one along the perpendicular [110] axis (B_1 mode). In addition, the Fermi level position of the samples used for these measurements showed that the LVMs are associated with the neutral charge state C_i^0 [83]. The *ab initio* LDF calculations of the LVMs frequencies of C_i are in fair agreement with the experimental values [84]. The electronic transition at 856 meV associated with C_i^0 has been discussed in Sect. 4.1.4.1. The annealing of the C_i centre near RT correlates with the observation of a new and very weak LVM at 858 cm^{-1} (LNT), but not much else has been reported on this LVM, attributed to a centre labelled C(2) [81, 85].

In FZ and CZ silicon samples electron-irradiated at RT, a pair of LVMs at 960 and 966 cm^{-1}, the C(5) lines, have been reported by Chappell and Newman [86] and attributed to the trapping of a self-interstitial by C$_i$, producing the IC$_i$ centre, containing actually two Si$_i$. This centre is stable up to $\sim$200°C and in qmi ^{30}Si samples, the C(5) LVMs show ^{30}Si ISs of -5 and -8 cm^{-1} for the high- and low-frequency modes, respectively [14].

The electronic absorption of a ZPL at 852.4 meV has been reported at the end of Sect. 4.1.4.1. This line, observed in FZ Sn-doped silicon samples electron-irradiated below RT and annealed at RT, is associated with the C$_i$, Sn$_s$ defect. It is accompanied in the infrared by two LVMs at 872.5 and 1025 cm^{-1} at LHeT, whose intensities increase when leaving the sample at RT for increasing time. This is correlated with the intensity decreases of the C(1) LVMs of C$_i$ [87]. The "interstitial" C$_3$ atom of Fig. 4.8 has two kinds of nnn atoms: one is a Si$_4$ atom bonded to atom Si$_3$ and another is a regular Si lattice atom not shown in the figure. The calculations show that in the (C$_i$, Sn$_s$) defect, the Sn atom replaces the regular lattice atom nnn of C$_3$. This configuration is noted (C$_i$−Sn$_s$)^{2N} in [87]. The point group symmetry of this defect is C_{1h}, with the C$_i$ and Sn$_s$ atoms lying in a (110) plane. The calculated LVMs of this defect are 905 and 1036 (878 and 1003) cm^{-1} for ^{12}C (^{13}C) [87]. For the lowest frequency LVM, the vibration of the C$_i$ atom corresponds to the IR A$'$ of C_{1h}, and the atom vibrates in the (110) mirror plane. For the other mode, the vibration corresponds to the IR A$''$ and the C$_i$ atom moves out of this plane.

7.2.2   The C$_i$SiC$_s$ Defect in Silicon

The C$_i$SiC$_s$ (or C$_i$$IC_s$) defect, often labelled C$_i$C$_s$ defect,[6] contains two C atoms separated by a Si atom, and it can be formed in electron-irradiated FZ silicon under diffusion of the above C$_i$ centres followed by their trapping by C$_s$. In that case, IR measurements as a function of the annealing temperature have shown that C$_i$$IC_s$ is produced via a metastable centre, noted here C$_i$$IC_s^*$, giving LVMs at 960 and 966 cm^{-1} at LNT. The ^{12}C−^{13}C IS of the 966 cm^{-1} LVM is $+38$ cm^{-1} and the ^{12}C −^{14}C IS of the 960 cm^{-1} LVM $+48$ cm^{-1} [88]. C$_i$SiC$_s$ is also produced directly in silicon irradiated at RT, and its concentration usually increases with annealing up to 100–150°C. The G-line whose absorption has been discussed in Sect. 4.1.4.2 is due to this defect. It is electrically active and bistable, with two configurations labelled A and B whose respective stabilities depend on their charge states [89]. The energy of the A form represents a minimum for the C$_i$SiC$_s^+$ and C$_i$SiC$_s^-$ charge states. The atomic structure of (C$_i$SiC$_s$)$_A$ is similar to that of C$_i$, with one of the Si$_4$ atoms bonded to the Si$_3$ atom of Fig. 4.8 replaced by a fourfold coordinated C atom (C$_4$) as shown in Fig. 7.11.

---

[6]There exists a defect made of two contiguous C atoms, whose hydrogenated form HC$_i$C$_s$ gives the PL T-line at 935.1 meV (see Sect. 8.4.2.1).

The energy of the $(C_iSiC_s)_B$ form represents a minimum for the neutral charge state. In this latter configuration, the Si_3 atom of Fig. 7.11 becomes twofold coordinated and the C_3 atom of the A form is changed into a C_4 atom (the index denotes the coordination number of the atoms), in a nearly symmetric structure which bears a resemblance with that of interstitial oxygen O_i [84]. The symmetry of $(C_iSiC_s)_A$ is monoclinic-I (C_{1h}), but $(C_iSiC_s)_B$ represented in Fig. 7.12 is found to display a C_{3v} symmetry for $T > 15$ K because of thermal averaging and C_{1h} symmetry at lower temperatures, and its structure can be described approximately as $C_sSi_iC_s$ [89].

The $0/+$ energy level of the A form is located at $E_v + 0.09$ eV and $(C_iSiC_s)_A^+$ gives a $S = 1/2$ ESR spectrum known as Si-G11 ([91], and references therein). The $-/0$ energy level for this form is located at $E_c - 0.17$ eV and the S = 1/2 ESR spectrum associated with $(C_iSiC_s)_A^-$ is known as Si-G17. The $-/0$ energy level of the B form is located at $E_c - 0.11$ eV and the $S = 1/2$ ESR spectrum of a centre with trigonal symmetry, known as Si-L7, is associated with $(C_iSiC_s)_B^-$ ([89], and references therein).

The first optical signature of C_iSiC_s was the already-discussed electronic G-line, corresponding to the neutral charge state $(C_iSiC_s)_B^0$. Two vibronic sidebands of this line have been observed by PL in silicon samples, and they are attributed to LVMs of $(C_iSiC_s)_B^0$. Six LVMs are expected, however, from a C-Si-C defect with C_{1h} symmetry, with three modes in the C-Si-C plane, corresponding to the IR A' of C_{1h},

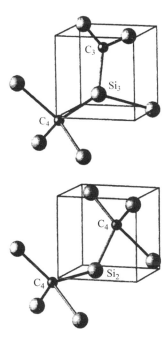

Fig. 7.11 Structure of the $(C_iSiC_s)_A$ form of the C_iSiC_s centre in electron-irradiated silicon resulting from *ab initio* calculations, showing a marked difference between the C_i atom (C_3) and the C_s atom (C_4) [84]. Copyright 1997 by the American Physical Society

Fig. 7.12 Structure of $(C_iSiC_s)_B$ in electron-irradiated silicon resulting from *ab initio* calculations, showing a nearly BC Si atom between two nearly C_s atoms [84]. Copyright 1997 by the American Physical Society

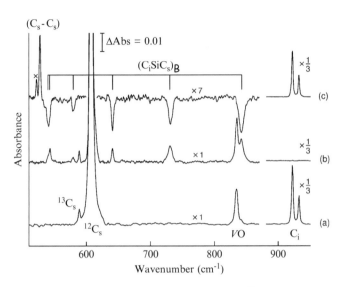

Fig. 7.13 Sections of the spectra at 10 K of electron-irradiated Si :^{12}C showing LVMs of different C-related centres (**a**): the C_i doublet after irradiation below RT ($f_e = 8 \times 10^{17}$ cm^{-2}). (**b**) the B form of C_iSiC_s after RT annealing for 210 min. (**c**) the $(C_s–C_s)^0$ LVM at 527 cm^{-1} in a spectrum obtained by subtraction of (**b**) from a spectrum obtained after a second irradiation ($f_e = 6 \times 10^{17}$ cm^{-2}) below RT. The *line* noted × on the low-energy side of the $C_s–C_s$ line is not related to this centre (after [80]). The dominant line near 600 cm^{-1} is due to substitutional carbon. Copyright 2000 by the American Physical Society

and three modes out-of-plane with symmetry[7] A''. Three additional LVMs have indeed been observed for the B form in absorption measurements [92] and they correspond to the out-of-plane modes of $C_iSiC_s^0$. Five of these LVMs can be seen in Fig. 7.13b, c. By converting first the B form of $C_iSiC_s^0$, stable at LHeT in samples with Fermi level at midgap, into charged states by band-gap light illumination at a temperature where the B → A conversion is efficient (above 50 K) and by cooling down the sample to LHeT quickly enough to limit the inverse conversion, LVMs associated with the A form have also been observed [92].

The atomic parameters of the centres in the B and A forms and the frequencies of the LVMs have been calculated by *ab initio* LDF methods [84, 93]. For the nearly symmetric B form, the two highest energy LVMs with symmetry of the *IR* A' of the C_{1h} group correspond to the antisymmetric (ν_3) and breathing mode of an isolated puckered C-Si-C molecule. Si ISs are also expected for these LVMs, but their intensities are too small for detection. Lower frequency LVMs with symmetry A'' correspond to C modes out of the C-Si-C plane. Between them is a A' mode reminiscent of the symmetric stretch mode of the isolated molecule, but with a stronger component for one of the two C atoms due to some asymmetry of the

---

[7]In the papers quoted in this section, the *IR* A' and A'' of C_{1h} are noted A and B, respectively.

B configuration. The lowest energy mode, also with symmetry A', is the same as the preceding one for the second C atom. The frequency calculations have been performed for the $^{12}C_iSi^{12}C_s$, $^{12}C_iSi^{13}C_s$, and $^{13}C_iSi^{13}C_s$ combinations [93].

For the A form, the two highest frequency modes correspond to the in-plane (A') and out-of-plane (A'') vibrations of the C_3 atom. The two highest modes of the C_4 atom are in-plane and the lowest mode is out-of-plane.

The measured frequencies of the LVMs of the A and B forms of $C_iSiC_s^0$ for samples containing ^{12}C or ^{13}C are given in Table 7.8.

Because the two C atoms of C_iSiC_s are separated by a Si atom, the frequencies of the C-related modes in samples containing both ^{12}C and ^{13}C are expected to differ marginally from those observed in the $^{12}C-$ and ^{13}C-containing samples: for the $(^{12}C_iSi^{12}C_s)_B$ A' modes at 579.5 and 543.1 cm^{-1}, $(^{12}C_iSi^{13}C_s)_B$ ISs of -0.6 and -0.1 cm^{-1}, respectively, have been reported [95]. In the A form, a marked disymmetry exists between the C_i (C_3) and C_s (C_4) atoms, and this reflects on the ISs of the modes for the mixed isotopic combinations. The values of these shifts depend on the amplitude of the C atoms involved in the mode and this can be appreciated by considering the calculated values of the LVMs for the different isotopic combinations given in Table 7.9.

From this table, it is seen that the C atom with the largest vibrational amplitude in LVM1 and LVM2 is the C_i atom as a change in the mass of the C_s atom does not alter the frequencies of the modes substantially. Inversely, for the other modes, the amplitude of the C_s atom is seen to be predominant. This table indicates only trends as these are calculated values. In this context, when comparisons are possible, there is a general good agreement between the calculated and measured frequencies. The

Table 7.8 Symmetries and measured frequencies (cm^{-1}) at LHeT of the LVMs of the $C_iSiC_s^0$ defect under the B and A forms in silicon for different isotopic combinations [92]

Symmetry	A'	A'	A''	A'	A'	A''
$(^{12}C_iSi^{12}C_s)_B$	842.4	730.4	640.6	579.8	543.3	540.4
		731.1[a]		579.5[a]	543.1[a]	
$(^{13}C_iSi^{13}C_s)_B$	818.5	707.7	621.8	564.6	532.9	529.3
		709.4[a]		564.4[a]	532.2[a]	
$(^{12}C_iSi^{12}C_s)_A$	953	722.4	872.6	596.9		594.6
$(^{12}C_iSi^{13}C_s)_A$		700.6	845.4	579.9		577.7

A' and A'' are used for the IRs of C_{1h} instead of A and B
[a][94], PL

Table 7.9 Calculated frequencies (cm^{-1}) of the $(C_iSiC_s)_A$ LVMs in silicon for the different ^{12}C and ^{13}C combinations (after [93])

	Symm.	$^{12}C_iSi^{12}C_s$	$^{12}C_iSi^{13}C_s$	$^{13}C_iSi^{12}C_s$	$^{13}C_iSi^{13}C_s$
LVM1	A'	889.9	889.6	863.1	862.5
LVM2	A''	874.1	874.1	847.2	847.2
LVM3	A'	721.8	701.6	720.3	700.3
LVM4	A'	567.5	550.6	567.3	550.4
LVM5	A''	557.1	540.9	557.1	540.9

frequencies calculated by the cluster method in [84] deviate at most by $16\,\text{cm}^{-1}$ from the ones measured for the B configuration, while those calculated *ab initio* in a local-density approximation [93] deviate a bit more for the two LVMs with the lowest frequencies.

The intensities of the LVMs of the C_iSiC_s centre are discussed in [92]. The evolution of the intensities of these LVMs with the annealing temperature shows that dissociation of this centre is complete for an annealing temperature $\sim 250°C$ (520 K) (see also Sect. 4.1.4.2).

7.2.3   The Dicarbon nn Pair ($C_s - C_s$) in Silicon

In III–V compounds, the existence of Raman-active interstitial C pairs on a cation site is mentioned at the end of Sect. 5.3.2.2. In C-containing silicon electron-irradiated at RT, the presence of the ESR GGA-2 S $= 1/2$ spectrum, first reported without an attribution [96], was attributed to a substitutional pair of *nn* C atoms with a trigonal symmetry. The presence of C is attested by weak replicas of the central line of GGA-2 due to hyperfine interaction with ^{13}C with a nuclear spin $\mathbf{I} = 1/2$ [97].

An ESR signal means that this dicarbon pair is electrically active, a property which is not obvious at first sight as C is isoelectronic to Si. The calculated electronic spectrum for a neutral silicon cluster containing a $(C_s - C_s)^0$ unit shows indeed an unoccupied level in the upper half of the band gap, closely similar to an antibonding state. For a normal $C - C$ bond, the bonding/antibonding energy difference greatly exceeds the silicon band gap, preserving the electrically inactive character of the dicarbon pair. However, when embedded in the Si cluster, the two C atoms are pushed apart by the six stretched $Si - C$ bonds (the $C - C$ bond length is 155 pm compared to 235 pm for $Si - Si$). This reduces the energy difference between the bonding and antibonding states, and the antibonding state locates then below the conduction-band edge of silicon. When an electron is added to the cluster, it enters the antibonding $C-C$ orbital, and the $(C_s-C_s)^-$ pair becomes paramagnetic [97].

When cooling an electron-irradiated FZ silicon sample with $[^{12}C_s] = 4 \times 10^{17}\,\text{cm}^{-3}$ from RT in the dark (pseudo-TEC), the spectrum observed at LHeT with a filter blocking frequencies above $3000\,\text{cm}^{-1}$ showed a new LVM at $748.7\,\text{cm}^{-1}$ and electronic lines at 2554 and $2695\,\text{cm}^{-1}$ (316.7 and 334.1 meV) [79]. In the spectrum obtained after removing the filter, these three lines disappeared and a new LVM at $527.4\,\text{cm}^{-1}$ was observed. The observation of lines at $726.3\,\text{cm}^{-1}$ and $2551\,\text{cm}^{-1}$ (316.3 meV) when ^{12}C was replaced by ^{13}C proved the implication of C in the related centre. The equivalent of the $527\,\text{cm}^{-1}$ LVM could not be detected with ^{13}C as its frequency is expected to be less than the Raman frequency of silicon ($524\,\text{cm}^{-1}$ at LHeT) so that the coupling with lattice phonons prevents its observation. The correlation between these lines and $C_s - C_s$ was established by comparing the intensities of the $2554\,\mathrm{cm}^{-1}$ line and of the GGA-2 ESR spectrum as a function of the electron fluence. $(C_s - C_s)^0$ is associated with the $527\,\mathrm{cm}^{-1}$ LVM (see Fig. 7.13c) and $(C_s - C_s)^-$ to the other lines [80]. The ESR spectrum, the electronic lines and the two LVMs have an identical annealing behaviour and they disappear at $\sim 500°C$ ($\sim 730\,\mathrm{K}$).

It has been proposed that this dicarbon centre was formed by the capture of a mobile vacancy by C_iSiC_s [80]. *Ab initio* calculations have been performed on a $Si_3 \equiv C - C \equiv Si_3$ entity with D_{3d} symmetry embedded in a cluster of 86 atoms. Two starting structures were envisaged with (1) a short $C - C$ bond ($\sim 160\,\mathrm{pm}$) and (2) a long $C - C$ bond ($\sim 290\,\mathrm{pm}$). It was found that for $(C_s - C_s)^0$ the short-bond structure was the most stable, while for $(C_s - C_s)^-$ it was the long-bond structure. It is also found that for both charge states, the frequency of the E_u mode is larger than that of the A_{2u} mode, so that the LVMs observed should correspond to E_u symmetry, and that, in agreement with experiment, the frequency is the largest for $(C_s - C_s)^-$. The frequencies in cm^{-1} calculated (measured) for the E_u mode are 522 (527) and 706 (749) for $(^{12}C_s - {}^{12}C_s)^0$ and $(^{12}C_s - {}^{12}C_s)^-$, respectively, and 469 and 684 (726) for $(^{13}C_s - {}^{13}C_s)^0$ and $(^{13}C_s - {}^{13}C_s)^-$, respectively [80].

Figure 7.13 shows LHeT vibrational spectra, where the LVMs of the three above-discussed C-related irradiation defects in FZ silicon can be seen.

7.2.4   The Carbon–Oxygen Centres in Silicon

7.2.4.1   The C_iO_i Centre ["C" or C(3) Centre]

In C-doped CZ silicon electron-irradiated at RT, a LVM at $865.2\,\mathrm{cm}^{-1}$ has been reported a long time ago ([98], and references therein). It can be seen in Fig. 7.1 and its FWHM is $\sim 2\,\mathrm{cm}^{-1}$ at LHeT. It was shown later to be associated with another LVM at $1115.5\,\mathrm{cm}^{-1}$, which can only be observed at LHeT because it is hidden by the $A_{2u}(1)$ and $A_{2u}(2)$ modes of $^{16}O_i$ at higher temperatures [85]. The FWHM of this latter LVM is $\sim 2\,\mathrm{cm}^{-1}$ and its peak intensity is about half of the $865\,\mathrm{cm}^{-1}$ LVM. These two LVMs are often called the C(3) lines, and they are ascribed to a C_iO_i centre [81]. Besides these LVMs, other ones given in Table 7.10, are also related to this centre [14, 99].

This centre, also responsible for the electronic C-line at $790\,\mathrm{meV}$, is electrically active with a donor level at $E_v + 345\,\mathrm{meV}$, and its paramagnetic positive charge state gives the Si-G15 ESR spectrum [100]. In the earlier literature, the C_iO_i centre can also be found labelled as the 0.79 eV centre, the "C" centre or the C(3) centre. The ZPL C-line presented in Sect. 4.1.4.3 shows vibronic replicas [101–103] whose energy differences from the ZPL are close to the energies of the LVMs of C_iO_i measured by IR absorption, as can be judged from Table 7.10. These LVMs have been observed in samples where the Fermi level was above $E_v + 0.35\,\mathrm{eV}$ so that they are associated with $(C_iO_i)^0$.

Table 7.10 Frequencies (cm^{-1}) at LHeT of the local modes of the C_iO_i or C(3) centre measured by IR absorption (first column) and deduced from the vibronic PL sidebands of the C-line (second column). The *IR* associated with the LVMs are given in the last column.

Absorp.[a,b] $(^{12}C_i^{16}O_i)^0$	PL[c]	Negative ISs with respect to $^{12}C,^{16}O$					*IR*
		$^{13}C^a$	$^{14}C^d$	$^{18}O^a$	$^{13}C,^{18}O^{a-}$	$^{12}C,^{16}O$ in qmi$^{30}Si^b$	
529.6 (544.2)	530	0.1	–	5.2	5.5	14.2	A$'$
549.8 (555.8)	550	0.2	–	0.3	0.5	17.7	A$''$
$\sim$588	588	–	–	$\sim$3	3.3	19	A$'$
742.8 (759.6)	743	0.5	–	33.4	35.8		A$'$
865.9 (861)	866	23.9	46.0	0.15	24.6	8.1	A$'$
1116.3 (1114)	1116	36.4	68.5	1.0	38.8	4.8	A$'$

The ISs for the indicated substitutions are taken from the absorption measurements. The calculated frequencies for $(^{12}C_i^{16}O_i)^0$ are given in brackets [79, 99]
[a][99]
[b][14]
[c][101]
[d][85]

On the basis of the experimental results including uniaxial stress measurements [104, 105], a structure of the C_iO_i centre was initially proposed [103]. It derived from the structure of the C_i split-interstitial of Fig. 4.8, with one O_i atom bonded to one of the Si_4 atoms *nn*s of the C_3 atom, and to a Si atom *nnn* of C_3 of this figure. This structure presents a (110) symmetry plane containing the C_i − Si and Si–O_i pairs, and displays a monoclinic-I symmetry (C_{1h} symmetry point group). Theoretical modelling of this centre [106, 107] showed, however, that the O atom was rather bonded to three Si atoms, in a way similar to the O atoms of the TDD structure noted with an asterisk in Fig. 6.14. This latter model bears a resemblance with the one for the $N_i–N_i$ pair in silicon (Fig. 1.7), with one N atom replaced by a tri-coordinated C atom and the other one by a tri-coordinated O atom. The main evidence to support this model is the absence in Table 7.10 of a LVM in the vicinity of the isolated O_i mode with a $^{16}O - ^{18}O$ IS comparable to the one of the A_{2u} mode of isolated O_i (see Table 6.4).

In the above series of LVMs, the only one at 742.8 cm^{-1}, with an unusually broad FWHM of $\sim$10 cm^{-1}, presents a notable O IS showing that this specific mode mainly involves the motion of the O atom. The LVM at 549.8 cm^{-1}, which has also been reported at 545.5 cm^{-1} at RT [108], shows small C and O ISs, but a notable Si-related shift in the qmi ^{30}Si sample, and it must be related to the vibration of the Si atoms separating the C and O atoms. All the modes associated with C_iO_i anneal out between 300 and 400°C [99] in qualitative agreement with the annealing conditions of the C-line.

In silicon samples irradiated near LNT and measured without heating, the C_iO_i lines are not observed, but the spectra measured for increasing annealing temperatures between $\sim$250 and 340 K have allowed to identify three sets of LVMs, which can be associated with metastable centres precursors of C_iO_i produced during the diffusion of interstitial carbon in the vicinity of O_i ([79], and references therein).

Table 7.11 Frequencies (cm^{-1}) of the ^{12}C$_i^{16}$O$_i$ precursors and of some of their ISs measured at LHeT under appropriate annealing of silicon electron-irradiated at LNT

(^{12}C$_i^{16}$O$_i$)$_{p1}$	^{13}C	^{18}O	(^{12}C$_i^{16}$O$_i$)$_{p2}$	^{13}C	^{18}O	(^{12}C$_i^{16}$O$_i$)$_{p3}$	^{13}C	^{18}O
910.1	28.4	0	885	25	0	812.2 (811)	20 (23)	0 (0)
942.7	26.4	0	1013.6	31.1	0	967.4 (934)	0 (1)	42.3 (42)
1097.3	0	49.2	1059.6	0	51.8	1086.2 (1087)	34.9 (35)	ND (0)

Calculated values are given in brackets [78]. The ISs are indicated for the isotope considered in the first line

The frequencies and ISs of the LVMs of these precursors, noted here (C$_i$O$_i$)$_{p1}$, (C$_i$O$_i$)$_{p2}$, and (C$_i$O$_i$)$_{p3}$ by order of increasing annealing temperature and intensities of the LVMs, are given in Table 7.11.

These precursors have a domain of existence of $\sim$30 K and the most stable is (C$_i$O$_i$)$_{p3}$, whose LVMs can be observed after annealing between $\sim$275 and 330 K. The ISs of Table 7.11 indicate that for (C$_i$O$_i$)$_{p1}$ and (C$_i$O$_i$)$_{p2}$, the LVM of highest energy is due to O$_i$ slightly perturbed by the C atom. For (C$_i$O$_i$)$_{p3}$, the structure calculated to give an O mode at 967 cm^{-1} with a twofold bonding [79] bears a resemblance with the Humble model of the split interstitial shown in Fig. 1.8, where the C and O atoms are the I atoms, but in this (C$_i$O$_i$)$_{p3}$ structure, if the C atom is tri-coordinated, atom 5 of Fig. 1.8 is bonded to atom 6, producing its overcoordination and leaving the O atom twofold coordinated as a classical O$_i$ atom.

7.2.4.2   The (IC$_i$O$_i$) Centre

LVMs at 937 cm^{-1} (LHeT) and 1021 cm^{-1} (LNT) were reported in electron- and neutron-irradiated CZ silicon samples [98]. In a correlation between ESR spectra and IR absorption bands of oxygen in electron irradiated silicon, it was also noted that a LVM at 936 cm^{-1} and the ESR signature of a centre with monoclinic-I symmetry (C$_{1h}$ point group) labelled A17 had the same annealing temperature [31].[8] The two C(4) LVMs appeared first as such in Fig. 4.4 of [109], showing, in a sample with an equal concentration of ^{12}C and ^{13}C, a clear ^{13}C IS, about 2.3 times larger for the low-frequency mode than for the high-frequency one. The frequencies of these two C(4) LVMs modes are 939.8 and 1024.2 cm^{-1} in ^{nat}Si at LHeT, with ^{13}C ISs of -28.6 and -11.8 cm^{-1}, respectively and ^{18}O ISs of -0.6 and -41.7 cm^{-1}, respectively, and this shows that the high-frequency mode corresponds to a motion of the O atom [75]. The ^{30}Si ISs of these LVMs have been measured in qmi ^{30}Si to be -6.6 and -6.7 cm^{-1} for the low- and high-frequency modes, respectively. The corresponding centre, already produced by electron irradiation at RT, is labelled (IC$_i$O$_i$) and it is assumed to result from the trapping of I by C$_i$O$_i$ [14].

The evolution upon annealing of the LVMs of C-O related defects produced by RT electron irradiation in CZ silicon has been reported ([75] and references therein).

---

[8]This is apparently the only reference where the Si-A17 spectrum is mentioned.

It shows that $(I C_i O_i)$ anneals at $\sim 175°C$ giving first a centre noted $(I C_i O_i)^*$, with three LVMs at 724.5, 951.8, and 973.1 cm^{-1}. Near $\sim 250°C$, $(I C_i O_i)^*$ transforms in turn into $(I C_i O_i)^{**}$, with three LVMs at 951.3, 977.8, and 969.3 cm^{-1}. This series ends between ~ 300 and 325° C, where $(I C_i O_i)^{**}$ disappears. The ^{18}O ISs of these modes have also been measured. For $(I C_i O_i)^*$, they show that the 724.5 cm^{-1} LVM mainly corresponds to a motion of the O atom, and that for $(I C_i O_i)^{**}$, this role is played by the 969.3 cm^{-1} LVM (the other LVMs show practically no ^{18}O IS). In this study, a pair of rather weak lines at 991 and 998 cm^{-1} observed in the as-irradiated samples, is apparently unaffected by the annealing sequence up to the maximum temperature of 325°C. These LVMs are ascribed to $(I_3 C_i O_i)$ and $(I_2 C_i O_i)$ centres, respectively [75]. The main annealing results of this study are summarized in Fig. 7.14, showing the amplitudes of the LVMs related to the $(I C_i O_i)$, $(I C_i O_i)^*$, and $(I C_i O_i)^{**}$ centres as a function of the annealing treatment.

The P-line at 767.15 meV, observed in CZ silicon samples annealed at 450 C for several hours and electron-irradiated or not displays a ^{13}C IS of $+79$ μeV, but no clear ^{18}O IS. Vibronic sidebands of the P-line have been reported by Kürner et al. [103] corresponding to LVM frequencies of 504, 581, 826, 899, 1107, and 1163 cm^{-1}, not too different from the ones of Table 7.10 for $(C_i O_i)$. The analysis of the pseudodonor behaviour of the "P" centre and its similarity with the "C" centre indicates for the former the existence of a donor level (a hole trap) at $E_v + 368$ meV. The only structure proposed for the "P" centre is a defect resulting from the combination of the C_i intersticialcy of Fig. 4.8 with a vacancy-dioxygen structure ($V O_2$) where the two O atoms are aligned along the same $\langle 100 \rangle$ orientation as C_i. In this geometry, the (110) symmetry plane would contain C_i and the two O atoms of this (C_i, $V O_2$) defect [103].

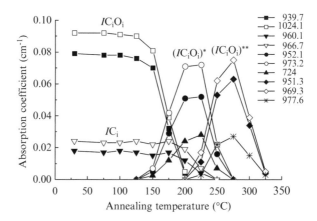

Fig. 7.14 Amplitudes of the absorption bands related to the $I C_i O_i$, $(I C_i O_i)^*$ and $(I C_i O_i)^{**}$ defects in silicon as a function of the ischronal (30 min) annealing temperature [75]. The LVMs are identified by their frequencies (cm^{-1}) on the right side of the figure. Copyright 2006 with permission from Elsevier

7.2.5   The C(1) Centre in GaAs and GaP

In GaAs samples where the LVM of $^{12}C_{As}$ is observed near $580\,cm^{-1}$ (see Sect. 5.3.2.2), two C-related lines, referred to as the C(1) lines and located at 577 and $606\,cm^{-1}$ at LNT, are observed after electron or neutron irradiation, together with a reduction of the intensity of the $^{12}C_{As}$ LVM [110]. The annealing temperature of the C(1) lines is $\sim 190°C$ and this annealing produces a recovery of the intensity of the isolated $^{12}C_{As}$ LVM. The integrated intensity ratio of the 577–606 cm^{-1} modes is 2/1 and the C(1) centre[9] has been attributed to a trigonal defect involving one C_{As} atom, with a lattice defect on one of the four $\langle 111 \rangle$ axes. This defect induces a $\langle 111 \rangle$ anisotropy axis and the 606 and 577 cm^{-1} modes are defined as longitudinal and transverse modes, respectively, with respect to this axis in ref. [111]. High-resolution measurements show that the longitudinal mode of C(1) is a doublet separated by $0.54\,cm^{-1}$, and the relative intensities of this doublet lines match the $^{69}Ga/^{71}Ga$ relative isotopic abundances. This doublet, shown in Fig. 7.15a, has thus been attributed to the perturbation of the C_{As} mode by the Ga nn atom along the anisotropy axis. Similarly, the transverse mode is a quadruplet (Fig. 7.15b) with an overall splitting of $\sim 0.5\,cm^{-1}$ due to the Ga isotope effect associated with the three off-axis Ga atoms in ref. [112].

A possible structure for the C(1) centre, where the lattice defect is an interstitial As_i atom in an antibonding position with respect to C_{As}, has been proposed from a cluster calculation based on a Keating model [112, 113]. However, *ab initio* calculations have shown that this (C_{As}, As_i) structure relaxes into a structure of lower energy, which results from the fact that when the As_i atom comes closer to a Ga atom on a lattice site, it switches it initial position with that of the Ga atom, to become what can be considered as As_{Ga} while the Ga atom stabilizes in a BC location between C_{As} and what can be considered as As_{Ga} [114]. This structure is slightly nonlinear, with a $C_{As}-Ga-As_{Ga}$ apex angle of 178°, but the C_{3v} symmetry can be considered to be preserved. The corresponding atomic structure is represented in Fig. 7.16.

The averaged value of the highest frequency calculated for this latter structure, due the antisymmetric $^{12}C\,-^{69}Ga_{BC}$ stretch, is $649\,cm^{-1}$, while the one of a transverse mode involving the C atom is $510\,cm^{-1}$, significantly lower than the ones calculated for the ($^{12}C_{As}$, As_i) structure (881 and 703 cm^{-1}) [114].

In electron-irradiated GaP, comparable C(1) LVMs have been reported at LNT at 599 and 642 cm^{-1}, not far from the LVM of isolated C_P (ref. [110]). The annealing temperature of these LVMs is $\sim 170°C$. The structure of the related centre should be the same, *mutatis mutandis*, as the one in GaAs.

---

[9]In silicon, LVMs denoted as C(1) are related to the C_i split interstitial configuration (see Sect. 7.2.1).

Fig. 7.15 Absorption of the C(1) centre in LEC GaAs irradiated with 2 MeV electrons (res.: 0.03 cm^{-1}). (**a**) Longitudinal LVM showing isotope splitting of the Ga atom on the anisotropy axis compared with a calculated profile showing the contribution of the ISs of the other Ga atoms. (**b**) Transverse LVM showing isotope splitting of the three nonaxial Ga atoms compared with a calculated profile (after [112]). Copyright 1991 with permission from Elsevier

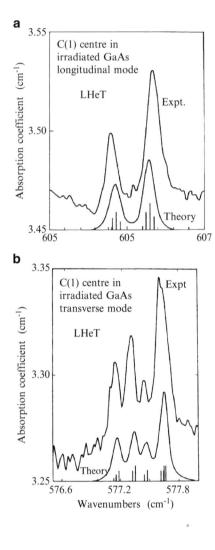

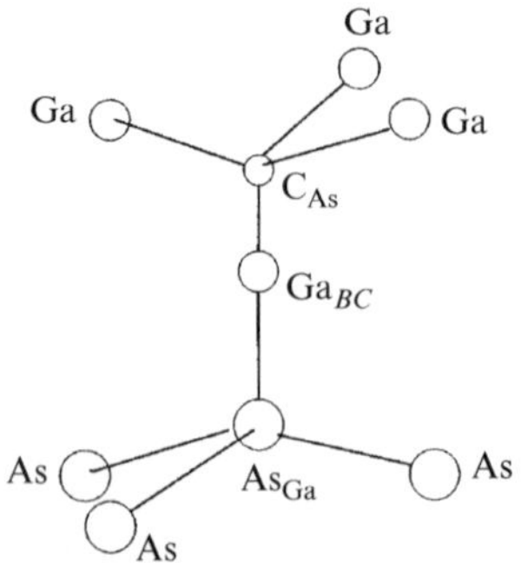

Fig. 7.16 Ball-and-stick model of the C(1) or C_{As}–Ga_{BC}–As_{Ga} centre in GaAs. The bare atoms are at their normal (more or less relaxed) lattice sites (after [114]). With permission from the Institute of Physics

7.3   B-Related Irradiation Defects

7.3.1   Silicon

The vibrational absorption of the metastable B_s–B_s pair, or P centre, has been discussed in Sect. 5.3.3.1. There have been several reports of the absorption of B-related defects in electron- and neutron-irradiated silicon ([115], and references therein), and a detailed study of the absorption of the LVMs related to interstitial B in electron-irradiated silicon under successive annealing treatments is presented in this reference. A firm conclusion is that in B-doped material irradiated with h-e electrons below typically 100 K, a self-interstitial I can be trapped by B_s to form an electrically active centre known as the boron interstitial (B_i) or (BI) centre, which anneals at $\sim$230 K. LVMs observed at LNT at 757 and 730 cm^{-1} for ^{10}B and ^{11}B, respectively (the R lines) were attributed to this centre [115].

A combination of DLTS and ESR experiments on B_i has given evidence for negative-U properties of this centre with a (0/+) donor level at $E_c - 0.13$ eV and a ($-$/0) acceptor level at $E_c - 0.37$ eV ([116], and references therein). The neutral metastable state B_i^0 is paramagnetic and it must be produced by neutralizing B_i^+ by continuous illumination with near-band-gap radiation in order to observe the $S = 1/2$ Si–G28 ESR spectrum. A monoclinic-I symmetry (C_{1h}) of B_i^0 was deduced from this ESR spectrum and atomic models were suggested for B_i [116]. First-principles calculations of the structure of B_i confirmed the metastability of this centre, with a configuration with C_{3v} symmetry including a B_s atom and a self-interstitial I, competing with another configuration with C_{1h} symmetry, where the B atom is also interstitial [117–119]. From now this centre is noted BI. The *ab initio* calculations of BI [119] have been of a great help in elucidating their structure and in allowing to attribute observed LVMs to these structures.

The ground-state structure of $(BI)^+$ and $(BI)^0$ is calculated to be the C_{3v} configuration, but for $(BI)^0$, the C_{1h} configuration is only 0.18 eV less stable, and it is the one actually deduced from the Si-G28 ESR spectrum. The frequencies of LVMs of $(BI)^+$ calculated *ab initio* are within 30 cm^{-1} from those of the R lines, and the experimental B IS is reproduced within 2 cm^{-1} (see Table 7.12). This strengthens the attribution of the R lines to $(BI)^+$. A breathing mode at 522.2 cm^{-1} for $(^{11}BI)^+$ is also predicted from the calculations, but the experimental equivalent should lie below the Raman frequency of silicon, in which case it may escape detection ([120], and references therein).

After the annealing of the (BI) centre, two series of LVMs, the S lines, near $\sim$910 and 600 cm^{-1}, appear after annealing in a rather narrow domain of temperature near 250 K. The annealing out of the S lines leads to the observation of the Q lines, with frequencies close to those of the R lines (see Table 7.12), but which are stable up to above RT [115, 121]. From their relative intensities and from the positions of the triplet near 910 cm^{-1}, the S lines were first ascribed to an interstitial (B_iB_i) pair with axial symmetry, and the Q lines to clusters of three or more B_i atoms. The *ab initio* calculations of structures involving close B atoms

Table 7.12 Measured frequencies (cm^{-1}) at LNT of the B-related centres observed in electron-irradiated silicon (after [115]) compared to the frequencies calculated for the corresponding assignments [119]

LVM label	Frequencies		Assignment	T_{max} (K)
	Measured	Calcul.		
^{10}R	757	726	^{10}BI	~230
^{11}R	730	697	^{11}BI	
S_1	903	919	$^{11}B_s-I-^{11}B_s$	270
S_2	917	934	$^{10}B_s-I-^{11}B_s$	
S_3	928	946	$^{10}B_s-I-^{10}B_s$	
S_4	599	611	$^{11}B_s-I-^{11}B_s$	
S_5	603	628	$^{10}B_s-I-^{11}B_s$	
^{10}Q	760	785	$^{10}B_i^{10}B_i[101]$	>300
^{11}Q	733	756	$^{11}B_i^{11}B_i[101]$	
P_1	552	551	$^{11}B_s-^{11}B_s$	>573
P_2	560	558	$^{10}B_s-^{11}B_s$	
P_3	570	568	$^{10}B_s-^{10}B_s$	

The last column indicates the maximum temperature T_{max} for stability

led the stability of different structures to be investigated. Besides the *nn* metastable B_s-B_s pair discussed in Sect. 5.3.3.1, the lowest energy structure is a B_iB_i pair sharing a substitutional site and oriented along a $\langle 100 \rangle$ axis. This pair is similar to the C_i centre of Fig. 4.8, but with atoms C_3 and Si_3 of this figure replaced by two tri-coordinated B atoms. Such a structure, with D_{2d} symmetry, is comparable to $N_{2i}[100]$ (the H1a centre) in diamond (see Sect. 6.2.1), and it can be labelled $B_{2i}[100]$. The Q lines have been re-assigned to the LVMs of this structure. The other structure is a metastable linear B_s-I-B_s pair along a $\langle 111 \rangle$ axis, which is 0.4 eV less stable than $B_{2i}[100]$. The point group symmetry of this $B_s - I - B_s$ centre is D_{3d} when the B isotopes are the same and C_{3v} otherwise, and the S lines have been re-assigned to the LVMs of this centre (S_1 and S_3 to the A_{2u} mode and S_4 to E_u) [119]. Table 7.12 allows to compare the measured and calculated LVMs and the agreement seems reasonable. For the sake of completeness, the frequencies of the irradiation-produced B_s-B_s pairs already given in Table 5.8 have been added to Table 7.12.

For the $B_{2i}[100]$ structure, the calculation shows that the interaction between the two B_i atoms is weak and changing the mass of one of them has a negligible effect on the frequency of the second one, hence the observation of only two Q lines for the $^{10}B_i^{11}B_i$ isotopic combination. It is interesting to note that the same effect is also observed for $N_{2i}[100]$ (the H1a centre), produced in diamond after electron irradiation and some annealing, discussed in Sect. 6.2.1.

In CZ silicon, simultaneously to the annealing of BI, DLTS measurements show the growth of an electrical level at $E_c - 0.23$ eV. The related centre, which is also produced by electron irradiation of CZ silicon at RT, has been attributed to a (B_iO_i) centre [122, 123], stable up to ~150–200°C. In a density functional modelling study of this centre, it was found that its most stable configuration was the one with C_{1h} symmetry, where the B_i and O_i atoms were tri-coordinated, with two bonds of each atom to the two same Si atoms. This configuration is the same as the one for the split nitrogen pair shown in Fig. 1.7, with the interstitial N atoms replaced by the B_i and O_i atoms [120]. Such a resemblance had already been noted in Sect. 7.2.4.1 for the (C_iO_i) centre. Six LVMs with their ^{10}B and ^{18}O ISs have been calculated for $(^{11}B_i^{16}O_i)$, together with the host crystal IS of $(^{11}B_i^{16}O_i)$ for qmi ^{30}Si [120]. From the values of the ISs, A′ LVMs at 736.2 and 716.2 cm^{-1} are predominantly localized on the O and B atoms, respectively, while low-frequency LVMs at 539.7 and 545.9 cm^{-1} correspond to motions of the Si atoms. One can note that the O mode at 736.2 cm^{-1} is close to the one of (C_iO_i) observed at 742.8 cm^{-1} (Table 7.10), involving also the O atom. However, up to now no electronic or vibrational absorption or PL feature which could be related to (B_iO_i) has been reported.

7.3.2   III–V Compounds

After h-e electron irradiation of undoped or p-type GaAs, the intensity of the LVM mode of B_{Ga}, discussed in Sect. 5.3.3.2, is observed to decrease and three new LVMs associated with boron appear ([124], and references therein). These LVMs, observed irrespective of the presence of other dopants in the crystal, are known as the B(1) lines. These lines disappear after annealing at ∼190–200°C. For GaP, only one LVM associated to B(1) is observed, and this line (and the B(1) centre) anneals in the range 250–300°C (ref. [110]). The frequencies of the LVMs related to the B(1) centre in these two materials are given in Table 7.13. The FWHMs quoted for GaAs are obtained with a resolution of 0.03 cm^{-1}.

The ratio of the integrated absorptions of the 763 and 641 cm^{-1} LVMs in GaAs is ∼2 and the integrated absorption of the 372 cm^{-1} LVM is ∼0.7 times that of the 641 cm^{-1} LVM [126]. When considering the attribution of B(1) to a (B, defect) pair, the relative sharpness of these LVMs indicates a B atom on a Ga site (no Ga isotope

Table 7.13 Frequencies and FWHMs (cm^{-1}) at LHeT of the LVMs of the B(1) centre in electron-irradiated GaAs and GaP ^{10}B and ^{11}B. The *ab initio* cluster calculations have been made using either a ^{69}GaAs or a ^{71}GaAs cluster. This is indicated by ^{69}Ga or ^{71}Ga in columns 5, 6, or 7

	^{10}B (exp.)[a]	^{11}B (exp.)[a]	FWHM	^{10}B(calc.) ^{69}Ga	^{11}B(calc.) ^{69}Ga	^{11}B(calc.) ^{71}Ga
GaAs	796	763	0.35	841.94	808.86	805.85
	669	641	0.56	651.56	623.44	623.44
	387	372	0.3	405.27	387.86	387.80
	−	−		303.38	303.37	300.94
GaP	882[b]	849[b]				

The LVMs frequencies are calculated with four ^{69}Ga or ^{71}Ga atoms surrounding the B(1) entity [125].
[a][124]
[b][110]

splitting similar to the one of B_{As} shown in Fig. 5.9). The relatively high frequencies of the LVMs indicates for the defect an interstitial atom rather than a vacancy, and the fact that the LVMs show no doublet splitting, which would occur for Ga_i, leaves only As_i. The B(1) LVMs have thus been attributed to a $(B_{Ga}-As_i)$ pair made from a B_{Ga} atom bonded to an interstitial As atom in a $\langle 100 \rangle$ split interstitial configuration comparable to the one of Fig. 4.8 [126]. *Ab initio* calculations of the structure of the $(B_{Ga}-As_i)^+$ and $(B_{Ga}-As_i)^-$ pairs show that the former pair is stable, with a mid-gap occupied level whereas the latter has an occupied level close to the *CB* [125]. As the Fermi level is pinned to mid-gap in irradiated materials, this negative charge state should not be stable, and it is considered that the B(1) centre results from the capture of As_i^+ by B_{Ga} [125]. Four LVMs are calculated for this structure with C_{2v} symmetry. The highest mode is the [100]-oriented $B-As_i$ stretch mode and the next one involves the motion of B and its As_s *nn* in the $\langle 110 \rangle$ plane, while the third mode involves a motion of B out of this plane. A small IS of the stretch mode is found when the four surrounding ^{69}Ga atoms are replaced by ^{71}Ga. The above situation is summarized in Table 7.13. Note that the calculated $^{10}B-^{11}B$ ISs deduced from this table are very close to the experimental ones.

References

1. J.W. Corbett, G.D. Watkins, R.M. Chrenko, R.S. McDonald, Defects in irradiated silicon. II Infrared absorption of the Si-A center. Phys. Rev. **121**, 1015–1022 (1961)
2. G. Bemski, Paramagnetic resonance in electron-irradiated silicon. J. Appl. Phys. **30**, 1195–1198 (1959)
3. G.D. Watkins, J.W. Corbett, R.M. Walker, Spin resonance in electron-irradiated silicon. J. Appl. Phys. **30**, 1198–1203 (1959)
4. G.D. Watkins, J.W. Corbett, Defects in irradiated silicon. I. Electron spin resonance of the Si − A center. Phys. Rev. **121**, 1001–1014 (1961)
5. A.R. Bean, R.C. Newman, An infra-red study of defects produced in n-type silicon by electron irradiation at low temperatures. Solid State Commun. **9**, 271–274 (1971)
6. G.K. Wertheim, D.N.E. Buchanan, Electron-bombardment damage in oxygen-free silicon. J. Appl. Phys. **30**, 1232–1234 (1959)
7. F.A. Abou-el-Fotouh, R.C. Newman, Electron irradiation damage in silicon containing carbon and diffused ^{18}O. Solid State Commun. **15**, 1409–1411 (1974)
8. D.R. Bosomworth, W. Hayes, A.R.L. Spray, G.D. Watkins, Analysis of oxygen in silicon in the near and far infrared. Proc. Roy. Soc. Lond. A **317**, 133–152 (1970)
9. M. Pesola, J. von Boehm, T. Mattila, R.M. Nieminen, Computational study of interstitial oxygen and vacancy-oxygen complexes in silicon. Phys. Rev. B **60**, 11449–11463 (1999)
10. L.I. Murin, V.P. Markevich, T. Hallberg, J.L. Lindström, New infrared vibrational bands related to interstitial and substitutional oxygen in silicon. Solid State Phenom. **69–70**, 309–314 (1999)
11. J.L. Lindström, L.I. Murin, V.P. Markevich, T. Hallberg, R.G. Svensson, Vibrational absorption from vacancy-oxygen-related complexes (VO, V_2O, VO_2) in irradiated silicon. Physica B **273–274**, 291–295 (1999)
12. J. Coutinho, R. Jones, P.R. Briddon, S. berg, Oxygen and dioxygen centres in Si and Ge: Density functional calculations. Phys. Rev. B **62**, 10824–10840 (2000)
13. V.P. Markevich, A.R. Peaker, J. Coutinho, R. Jones, V.J.B. Torres, S. Öberg, P.R. Briddon, L.I. Murin, L. Dobaczewski, N.V. Abrosimov, Structure and properties of vacancy-oxygen complexes in $Si_{1-x}Ge_x$ alloys. Phys. Rev. B **69**, 125218/1–11 (2004)

14. G. Davies, S. Hayama, S. Hao, B. Bech Nielsen, J. Coutinho, M. Sanati, S.K. Estreicher, K.M. Itoh, Host isotope effects on midinfrared optical transitions in silicon. Phys. Rev. B **71**, 115212/1–7 (2005)

15. R. Van Kemp, M. Sprenger, E.G. Sieverts, C.A.J. Ammerlaan, Oxygen-vacancy complex in silicon. II. ^{17}O electron-nuclear double resonance. Phys. Rev. B **40**, 4054–4061 (1989)

16. L.I. Murin, B.G. Svensson, J.L. Lindström, V.P. Markevich, C.A. Londos, Trivacancy-oxygen complex in silicon: Local vibrational mode characterization. Physica B **404**, 4568–4571 (2009)

17. L.I. Murin, B.G. Svensson, J.L. Lindström, V.P. Markevich, C.A. Londos, Divacancy-oxygen and trivacancy-oxygen complexes in silicon: local vibrational modes studies. Solid State Phenom. **156–158**, 129–134 (2010)

18. K.L. Brower, Electron paramagnetic resonance of the neutral (S = 1) one-vacancy-oxygen center in irradiated silicon. Phys. Rev. B **4**, 1968–1982 (1971)

19. P.R. Brosious, EPR evidence for a positively charged vacancy-oxygen defect in silicon. Appl. Phys. Lett. **29**, 265–267 (1976)

20. A.B. Van Oosten, A.M. Frens, J. Schmidt, Metastable triplet state of the vacancy-oxygen center in silicon: an *ab initio* study. Phys. Rev. B **50**, 5239–5246 (1994)

21. B. Pajot, S. McQuaid, R.C. Newman, C. Song, R. Rahbi, A piezo-spectroscopic study of oxygen-vacancy centres in silicon. Mater. Sci. Forum **143–147**, 969–974 (1994)

22. A.S. Oates, R.C. Newman, Involvement of oxygen-vacancy defects in enhancing oxygen diffusion in silicon. Appl. Phys. Lett. **49**, 262–264 (1986)

23. A. Nakanishi, N. Fukata, M. Suezawa, Vacancy-oxygen pairs and vacancy-oxygen-hydrogen complexes in electron-irradiated n-type Cz-Si pre-doped with hydrogen. Jpn. J. Appl. Phys. **42**, 6737–6741 (2003)

24. J.W. Corbett, G.D. Watkins, R.S. McDonald, New oxygen infrared bands in annealed irradiated silicon. Phys. Rev. **135**, A1381–A1385 (1964)

25. H.J. Stein, Oxygen isotope effect on the $889 \, cm^{-1}$ band in silicon. Appl. Phys. Lett. **48**, 1540–1541 (1986)

26. C.A. Londos, A. Andrianakis, A. Aliprantis, H. Ohyama, V.V. Emtsev, G.A. Oganesyan, IR studies of oxygen-vacancy defects in electron-irradiated Ge-doped Si. Physica B **401–402**, 487–490 (2007)

27. J.L. Lindström, T. Hallberg, L.I. Murin, B.G. Svensson, V.P. Markevich, T. Hallberg, The VO_2^* defect in silicon. Physica B **340–342**, 509–513 (2003)

28. V. Akhmetov, G. Kissinger, W. von Ammon, Interaction of oxygen with thermally induced vacancies in Czochralski silicon. Appl. Phys. Lett. **94**, 092105/1–3 (2009)

29. V. Akhmetov, G. Kissinger, W. von Ammon, Formation of vacancy and oxygen containing complexes in Cz-Si by rapid thermal annealing. Physica B **404**, 4572–4575 (2009)

30. Y.H. Lee, J.W. Corbett, EPR studies of defects in electron-irradiated silicon: A triplet state of a vacancy-oxygen defect. Phys. Rev. B **13**, 2653–2665 (1976)

31. Y.H. Lee, J.C. Corelli, J.W. Corbett, Oxygen vibrational bands in irradiated silicon. Phys. Lett. **60A**, 55–57 (1977)

32. C. Ewels, R. Jones, S. Öberg, A first principles investigation of vacancy-oxygen defects in Si. Mater. Sci. Forum **196–201**, 1297–1301 (1995)

33. T. Hallberg, J.L. Lindström, B.G. Svensson, K. Swiatek, Annealing of electron-irradiated P-, As-, Sb- and Bi-doped Czochralski silicon. Mater. Sci. Forum **143–147**, 1239–1244 (1994)

34. J.L. Lindström, B.G. Svensson, W.M. Chen, A new defect observed in annealed highly phosphorus-doped electron-irradiated silicon. Mater. Sci. Forum **83-87**, 333–338 (1992)

35. Yu.V. Pomozov, M.G. Sosnin, L.I. Khirunenko, V.I. Yashnik, N.V. Abrosimov, W. Schröder, M. Höhne, Oxygen-containing radiation defects in $Si_{1-x}Ge_x$. Semiconductors **34**, 989–993 (2000)

36. S.J. Pearton, J.W. Corbett, M. Stavola, *Hydrogen in Crystalline Semiconductors* (Springer, Berlin, 1992)

37. S.K. Estreicher, Hydrogen-related defects in crystalline semiconductors. Mat. Sci. Eng. Rep. **14**, 319–412 (1995)

38. B.N. Mukashev, S.Zh. Tokmoldin, M.F. Tamendarov, V.V. Frolov, Hydrogen passivation of vacancy-related centres in silicon. Physica B **170**, 545–549 (1991)
39. V.P. Markevich, L.I. Murin, M. Suezawa, J.L. Lindström, J. Coutinho, R. Jones, P.R. Briddon, S. Öberg, Observation and theory of the $V - O - H_2$ complex in silicon. Phys. Rev. B **61**, 12964–12969 (2000)
40. P. Johannesen, B. Bech-Nielsen, J.R. Byberg, Identification of the oxygen-vacancy defect containing a single hydrogen atom in crystalline silicon. Phys. Rev. B **61**, 4659–4666 (2000)
41. J. Coutinho, O. Andersen, L. Dobaczewski, K. Bonde Nielsen, A.R. Peaker, R. Jones, S. Öberg, P.R. Briddon, Effect of stress on the energy levels of the vacancy-oxygen-hydrogen complex in Si. Phys. Rev. B **68**, 184106/1–11 (2003)
42. R.E. Whan, Investigations of oxygen-defect interactions between 25 and 700 K in irradiated germanium. Phys. Rev. **140**, A690–A698 (1965)
43. J.A. Baldwin, Electron paramagnetic resonance in irradiated oxygen doped germanium. J. Appl. Phys. **36**, 793–795 (1965)
44. V.P. Markevich, V.V. Litvinov, L. Dobaczewski, J.L. Lindström, L.I. Murin, A.R. Peaker, Radiation-induced defects and their transformations in oxygen-rich germanium crystals. Phys. Stat. Sol. C **0**, 702–706 (2003)
45. V.P. Markevich, V.V. Litvinov, L. Dobaczewski, J.L. Lindström, L.I. Murin, S.V. Vetrov, I.D. Hawkins, A.R. Peaker, Vacancy-oxygen complex in Ge crystals. Physica B **340–342**, 844–848 (2003)
46. P. Vanmeerbeck, P. Clauws, H. Vrielinck, B. Pajot, L. Van Hoorebeke, A. Nylandsted Larsen, High-resolution local vibrational mode spectroscopy and electron paramagnetic resonance study of the oxygen-vacancy complex in irradiated germanium. Phys. Rev. B **70**, 035203/1–8 (2004)
47. A. Carvalho, R. Jones, J. Coutinho, V.J.B. Torres, S. Öberg, J.M. Campanera Alsina, M. Shaw, P.R. Briddon, Local-density calculations of the vacancy-oxygen center in germanium. Phys. Rev. B **75**, 115206/1–8 (2007)
48. A. Carvalho, V.J.B. Torres, V.P. Markevich, J. Coutinho, V.V. Litvinov, A.R. Peaker, R. Jones, P.R. Briddon, Identification of stable and metastable forms of VO_2 centers in germanium. Physica B **401–402**, 192–195 (2007)
49. P. Vanmeerbeck, P. Clauws, Local vibrational mode spectroscopy of dimer and other oxygen-related defects in irradiated germanium. Phys. Rev. B **64**, 245201/1–6 (2001)
50. C. Song, W. Ge, D. Jiang, C. Hsu, Pair of local mode absorption bands related to EL2 defects in semi-insulating GaAs. Appl. Phys. Lett. **50**, 1666–1668 (1987)
51. X. Zhong, D. Jiang, W. Ge, C. Song, Model study of the local vibration center related to EL2 levels in GaAs. Appl. Phys. Lett. **52**, 628–630 (1988)
52. J. Schneider, B. Dischler, H. Seelewind, P.M. Mooney, J. Lagowski, M. Matsui, D.R. Beard, R.C. Newman, Assessment of oxygen in gallium arsenide by infrared local vibrational mode spectroscopy. Appl. Phys. Lett. **54**, 1442–1444 (1989)
53. H.Ch. Alt, Photosensitivity of the 714 and 730 cm^{-1} absorption bands in semi-insulating GaAs: Evidence for a deep donor involving oxygen. Appl. Phys. Lett. **54**, 1445–1447 (1989)
54. H.Ch. Alt, Fine structure of the oxygen-related local mode at 714 cm^{-1} in GaAs. Appl. Phys. Lett. **55**, 2736–2738 (1989)
55. C. Song, B. Pajot, F. Gendron, Local mode spectroscopy and photo-induced effects of oxygen-related centers in semi-insulating gallium arsenide. J. Appl. Phys. **67**, 7307–7312 (1990)
56. H.Ch. Alt, Experimental evidence for a negative-U center in gallium arsenide related to oxygen. Phys. Rev. Lett. **65**, 3421–3424 (1990)
57. M. Skowronski, S.T. Neild, R.E. Kremer, Location of an energy level of oxygen-vacancy complex in GaAs. Appl. Phys. Lett. **57**, 902–904 (1990)
58. C. Song, B. Pajot, C. Porte, Some properties of the oxygen-vacancy center in gallium arsenide, in *Proc. 21st Intern. Conf. Phys. Semicond.*, Beijing, 11–14 August 1992, ed. by P. Jiang, H Zheng (World Scientific, Singapore, 1992), pp. 1629–1633
59. C. Song, Doctoral thesis, Université Paris 7 (1992)

60. C. Song, B. Pajot, W.K. Ge, D.S. Jiang, Relation between the metastability of EL2 and the photosensitivity of local vibrational modes of semi-insulating GaAs. Phys. Rev. B **52**, 4864–4869 (1995)
61. C. Song, B. Pajot, C. Porte, Piezospectroscopy of interstitial oxygen in gallium arsenide. Phys. Rev. B **41**, 12330–12333 (1990)
62. M. Linde, J.M. Spaeth, H.Ch. Alt, The paramagnetic charge state of substitutional oxygen in GaAs. Appl. Phys. Lett. **67**, 662–664 (1995)
63. F.K. Koschnick, M. Linde, M.V.B. Pinheiro, J.M. Spaeth, Optically detected electron-paramagnetic resonance investigations of the substitutional oxygen defect in gallium arsenide. Phys. Rev. B **56**, 10221–10227 (1997)
64. R. Jones, S. Öberg, Multiple charge states of substitutional oxygen in gallium arsenide. Phys. Rev. Lett. **69**, 136–139 (1992)
65. T. Mattila, R.M. Nieminen, *Ab initio* study of oxygen point defects in GaAs, GaN, and A*l*N. Phys. Rev. B **54**, 16676–16682 (1996)
66. A. Taguchi, H. Kageshima, Diffusion and stability of oxygen in GaAs and A*l*As. Phys. Rev. B **60**, 5383–5391 (1998)
67. M. Pesola, J. von Boehm, V. Sammalkorpi, T. Mattila, R.M. Nieminen, Microscopic structure of oxygen defect in gallium arsenide. Phys. Rev. B **60**, R16267–R16270 (1999)
68. H.Ch. Alt, Y.V. Gomeniuk, U. Kretzer, Charge state and quantitative infrared spectroscopy of electrically oxygen centers in gallium arsenide. J. Appl. Phys. **101**, 073516/1-6 (2007)
69. G.M. Martin, A. Mitonneau, A. Mircea, Electron traps in bulk and epitaxial GaAs crystals. Electron. Lett. **13**, 191–192 (1977)
70. U. Kaufmann, E. Klausmann, J. Schneider, H.Ch. Alt, Negative-U, off-center O_{As} in GaAs and its relation with the EL3 defect. Phys. Rev. B **43**, 12106–12109 (1991)
71. M. Skowronski, S.T. Neild, R.E. Kremer, Calibration of the isolated oxygen interstitial localized vibrational mode absorption line in GaAs. Appl. Phys. Lett. **58**, 1545–1547 (1991)
72. S.T. Neild, M. Skowronski, J. Lagowski, Signature of the gallium-oxygen-gallium defect in GaAs by deep level transient spectroscopy measurements. Appl. Phys. Lett. **58**, 859–861 (1991)
73. G.S. Khoo, C.K. Ong, O^- in GaP: a negative-U centre. J. Phys. Condens. Matt. **5**, 3917–3924 (1993)
74. H.Y. Fan, A.K. Ramdas, Infrared absorption and photoconductivity in irradiated silicon. J. Appl. Phys. **30**, 1127–1134 (1959)
75. L.I. Murin, J.L. Lindström, G. Davies, V.P. Markevich, Evolution of radiation-induced carbon-oxygen related defects in silicon upon annealing: LVM studies. Nucl. Instrum. Meth. B **253**, 210–213 (2006)
76. B. Pajot, *Optical Absorption of Impurities and Defects in Semiconducting Crystals – Hydrogen-Like Centres* (Springer, Berlin, 2010)
77. R.E. Whan, Oxygen-defect complexes in neutron-irradiated silicon. J. Appl. Phys. **37**, 3378–3382 (1966)
78. A.R. Bean, R.C. Newman, Low temperature electron irradiation of silicon containing carbon. Solid State Commun. **8**, 175–177 (1970)
79. L.I. Khirunenko, M.G. Sosnin, V.Yu. Pomozov, L.I. Murin, V.P. Markevich, A.R. Peaker, L.M. Almeida, J. Coutinho, V.J.B. Torres, Formation of interstitial carbon-interstitial oxygen complexes: Local vibrational mode spectroscopy and density functional theory. Phys. Rev. B **78**, 155203/1-8 (2008)
80. E.V. Lavrov, B. Bech Nielsen, J.R. Byberg, B. Hourahine, B. Jones, S. Öberg, P.R. Briddon, Local vibrational modes of two neighboring substitutional carbon atoms in silicon. Phys. Rev. B **62**, 158–165 (2000)
81. R.C. Newman, *Infra-red Studies of Crystal Defects* (Taylor and Francis, London, 1973)
82. L.I. Khirunenko, Yu.V. Pomozov, M.G. Sosnin, M.O. Trypachko, A. Duvanskii, V.J.B. Torres, J. Coutinho, R. Jones, P.R. Briddon, N.V. Abrosimov, H. Riemann, Local vibrations of interstitial carbon in SiGe alloys. Mater. Sci. Semicond. Process. **9**, 514–519 (2006)

83. J.F. Zheng, M. Stavola, G.D. Watkins, Structure of the neutral charge state of interstitial carbon in silicon, in *22nd Internat. Conf. Phys. Semicond.*, ed. by D.J. Lockwood (World Scientific, Singapore, 1995), pp. 2363–2366

84. P. Leary, R. Jones, S. Öberg, V.J.B. Torres, Dynamic properties of interstitial carbon and carbon-carbon pair defects in silicon. Phys. Rev. B **55**, 2188–2194 (1997)

85. R.C. Newman, A.R. Bean, Irradiation damage in carbon-doped silicon irradiated at low temperatures by 2 MeV electrons. Rad. Eff. **8**, 189–193 (1971)

86. S.P. Chappell, R.C. Newman, The selective trapping of self-interstitials by interstitial carbon impurities in electron irradiated silicon. Semicond. Sci. Technol. **2**, 691–694 (1987)

87. E.V. Lavrov, M. Fanciulli, M. Kaukonen, R. Jones, P.R. Briddon, Carbon-tin defects in silicon. Phys. Rev. B **64**, 125212/1–5 (2001)

88. S.P. Chappell, G. Davies, E.C. Lightowlers, R.C. Newman, A metastable precursor to the di-carbon centre in crystalline silicon. Mater. Sci. Forum **38–41**, 481–485 (1989)

89. L.W. Song, X.D. Zhan, B.W. Benson, G.D. Watkins, Bistable interstitial-carbon-substitutional-carbon pair in silicon. Phys. Rev. B **42**, 5765–5783 (1990)

90. L.W. Song, X.D. Zhan, B.W. Benson, G.D. Watkins, Bistable defect in silicon: the interstitial-carbon-substitutional-carbon pair. Phys. Rev. Lett. **60**, 460–463 (1988)

91. K.P. O'Donnell, K.M. Lee, G.D. Watkins, Origin of the 0.97 eV luminescence in irradiated silicon. Physica B **116**, 258–263 (1983)

92. E.V. Lavrov, L. Hoffmann, B. Bech Nielsen, Local vibrational modes of the metastable dicarbon center ($C_s - C_i$) in silicon. Phys. Rev. B **60**, 8081–8086 (1999)

93. R.B. Capaz, A. Dal Pino, J.D. Joannopoulos, Theory of carbon-carbon pairs in silicon. Phys. Rev. B **58**, 9845–9850 (1998)

94. E.C. Lightowlers, A.N. Safonov, Photoluminescence vibrational spectroscopy of defects containing the light impurities carbon and oxygen in silicon. Mater. Sci. Forum **258–263**, 617–622 (1997)

95. G. Davies, R.C. Newman, Carbon in mono-crystalline silicon, in *Handbook on Semiconductors*, vol. 3b, ed. by S. Mahajan (North Holland, Amsterdam, 1994), pp. 1557–1640

96. H. Horiye, E.G. Wikner, Three new electron spin resonance centers in electron-irradiated silicon. J. Appl. Phys. **40**, 3879–3880 (1969)

97. J.H. Byberg, B. Bech Nielsen, M. Fanciulli, S.K. Estreicher, P.A. Fedders, Dimer of substitutional carbon in silicon studied by EPR and *ab-initio* methods. Phys. Rev. B **61**, 12939–12945 (2000)

98. A.K. Ramdas, M.G. Rao, Infrared absorption spectra of oxygen-defect complexes in irradiated silicon. Phys. Rev. **142**, 451–456 (1966)

99. J. Coutinho, L.I. Murin, V.P. Markevich, J.L. Lindström, Interstitial carbon-oxygen center and hydrogen related shallow thermal donors in Si. Phys. Rev. B **65**, 014109/1–11 (2001)

100. J.M. Trombetta, G.D. Watkins, Identification of an interstitial carbon-interstitial oxygen complex in silicon. Appl. Phys. Lett. **51**, 1103–1105 (1987)

101. S. Hayama, G. Davies, J. Tan, J. Coutinho, R. Jones, K.M. Itoh, Lattice isotope effects on optical transitions in silicon. Phys. Rev. B **70**, 035202/1–9 (2004)

102. K. Thonke, G.D. Watkins, R. Sauer, Carbon and oxygen isotope effects in the 0.79 eV defect photoluminescence spectrum in irradiated silicon. Solid State Commun. **51**, 127–130 (1984)

103. W. Kürner, R. Sauer, A. Dörnen, K. Thonke, Structure of the 0.767-eV oxygen-carbon luminescence defect in 450°C thermally annealed Czochralski-grown silicon. Phys. Rev. B **39**, 13327–13337 (1989)

104. C. Foy, Uniaxial stress analysis of the 0.78 eV vibronic band in irradiated silicon. J. Phys. C **15**, 2059–2067 (1982)

105. K. Thonke, A. Hangleiter, J. Wagner, R. Sauer, 0.79 (C line) defect in irradiated oxygen-rich silicon: excited state structure, internal strain and luminescence decay time. J. Phys. C **18**, L795–L801 (1985)

106. R. Jones, S. Öberg, Oxygen frustration and the interstitial carbon-oxygen complex in Si. Phys. Rev. Lett. **68**, 86–89 (1992)

107. D.J. Backlund, S.K. Estreicher, Theoretical studies of the C_iO_i and $I_{Si}C_iO_i$ defects in silicon. Physica B **401–402**, 163–166 (2007)
108. C.A. Londos, M.S. Potsidi, G.D. Antonaras, A. Andrianakis, Isochronal annealing studies of the carbon-related defects in irradiated silicon. Physica B **376–377**, 165–168 (2006)
109. G. Davies, A.S. Oates, R.C. Newman, R. Wooley, E.C. Lightowlers, M.J. Binns, J.G. Wilkes, Carbon-related radiation damage centres in Czochralski silicon. J. Phys. C **19**, 841–855 (1986)
110. F. Thompson, S.R. Morrison, R.C. Newman, Infrared local mode absorption in irradiated GaP and GaAs. Proc. Internat. Conf. Rad. Damage and Defects Semicond. Reading, 1972, Inst. Phys. Conf. Ser. **16**, 371–376 (1973)
111. R.C. Newman, Local vibrational mode spectroscopy of defects in III/V compounds, in *Imperfections in III-V Compounds, Semiconductors and Semimetals*, vol. 38, ed. by E.R. Weber (Academic, New York, 1993), pp. 117–187
112. G.A. Gledhill, S.B. Upadhyay, M.J.L. Sangster, R.C. Newman, Fine structure of the LVM-lines from $(C_{As} - As_i)$ complexes in irradiated GaAs. J. Mol. Struct. **247**, 313–319 (1991)
113. M.J.L. Sangster, R.C. Newman, G.A. Gledhill, S.B. Upadhyay, Cluster calculations of local vibrational mode frequencies of impurities in III-V semiconductors: application to detect complexes involving C_{As} in GaAs. Semicond. Sci. Technol. **7**, 1295–1305 (1992)
114. R. Jones, S. Öberg, Instabilities of simple models of $C - As_i$ complexes in gallium arsenide. Semicond. Sci. Technol. **7**, 855–857 (1992)
115. A.K. Tipping, R.C. Newman, An infrared study of the production, diffusion and complexing of interstitial boron in electron-irradiated silicon. Semicond. Sci. Technol. **2**, 389–398 (1987)
116. R.D. Harris, J.I. Newton, G.D. Watkins, Negative-U defect: interstitial boron in silicon. Phys. Rev. B **36**, 1094–1104 (1987)
117. E. Tarnow, Theory of the B interstitial related defect in Si. Europhys. Lett. **16**, 449–454 (1991)
118. M. Hakala, M.J. Puska, R.M. Nieminen, First-principles calculations of interstitial boron in silicon. Phys. Rev. B **61**, 8155–8161 (2000)
119. J. Adey, R. Jones, P.R. Briddon, J.P. Goss, Optical and electrical activity of boron interstitial in silicon. J. Phys. Condens. Matter **15**, S2851–S2858 (2003)
120. A. Carvalho, R. Jones, J. Coutinho, P.R. Briddon, *Ab initio* calculation of the local vibrational modes of the interstitial boron-interstitial oxygen defect in silicon. J. Phys. Condens. Matter **17**, L155–L159 (2005)
121. R.C. Newman, R.S. Smith, Local mode absorption from boron pairs in silicon. Phys. Lett. **24A**, 671–672 (1967)
122. P.M. Mooney, L.J. Cheng, M. Süli, J.D. Gerson, J.W. Corbett, Defects energy level in boron-doped silicon irradiated with 1-MeV electrons. Phys. Rev. B **15**, 3836–3843 (1977)
123. L.C. Kimerling, M.T. Asom, J.L. Benton, P.J. Drevinsky, C.E. Caefer, Interstitial defect reactions in silicon. Mater. Sci. Forum **38–41**, 141–150 (1989)
124. J.D. Collins, G.A. Gledhill, R. Murray, P.S. Nandhra, R.C. Newman, The selective trapping of arsenic interstitial atoms by impurities in gallium arsenide. Phys. Stat. Sol. B **151**, 469–477 (1989)
125. R. Jones, S. Öberg, Theory of $B - As_i$ complexes in gallium arsenide. Semicon. Sci. Technol. **7**, 429–431 (1992)
126. M.R. Brozel, R.C. Newman, A low-symmetry interstitial boron centre in irradiated gallium arsenide. J. Phys. C **11**, 3135–3146 (1978)

Chapter 8
Vibrational Absorption Associated with Hydrogen

8.1 Introduction

The role and properties of hydrogen in semiconductors have been widely investigated during the last decades, and several books and reviews have been dedicated to this topic [1–9]. One of the reasons for this interest arose from the discovery, in 1983, that the electrical activity of boron acceptor in silicon could be greatly reduced, by exposure of the silicon sample to atomic hydrogen, in the volume of the sample where hydrogen could penetrate [10, 11]. Another one comes from the fact that, because of growth methods where hydrogen is involved, it is a residual impurity in several semiconductors.

Hydrogen being a very light and reactive atom, LVM spectroscopy has been a major tool for its investigation and review papers focused on this approach have been published [12, 13]. With reference to Fig. 1.5, two vibrational modes of hydrogen are of importance in the spectroscopy of this element in semiconductors. One is due to the stretching of a bond-centred (BC) hydrogen atom along the axis of its bond with one of the atoms it separates.[1] The other one is due to the stretching of a hydrogen atom antibonded (AB) to one of the atoms of a broken crystal bond. In the AB location, it is easier for the H atom to vibrate perpendicularly to the axis of the chemical bond and this stretch mode is often accompanied by a bending mode (or wag mode) at a lower frequency.

It is out of the scope of this volume to make an exhaustive review of the spectroscopic investigations of hydrogen in semiconductors and only the present understanding of the roles of hydrogen will be summarized.

In the gas form, hydrogen is present as a diatomic molecule and the atomic form is unstable. Hydrogen is usually introduced in semiconductors under the atomic form, and as it is a very small species, it diffuses easily in the materials until it

---

[1] In the BC location, the bonding of the H atom may be more or less equally shared with the two atoms, but generally, one bond remains predominant.

B. Pajot and B. Clerjaud, *Optical Absorption of Impurities and Defects in Semiconducting Crystals*, Springer Series in Solid-State Sciences 169, DOI 10.1007/978-3-642-18018-7_8, © Springer-Verlag Berlin Heidelberg 2013

is trapped by some impurity or defect. Under heavy hydrogenation, for instance by H plasma exposure or reactive ion etching, hydrogen-induced platelets can be formed. They are planar extended defects typically a few tens of nanometres in size that are aligned predominantly along (111) crystallographic planes, and in silicon, for instance, they involve the formation of 2D ordered arrays of Si − H bonds. These bonds and the presence of trapped H_2 molecules have been revealed by IR absorption and Raman scattering measurements [14–16]. Platelets do not fall explicitly in the categories of defects studied in this volume, but the vibration of H_2 molecules in platelets is still discussed in Sect. 8.3.8. A brief description of their properties can be found in [8]. See also [17, 18].

We consider in the next three sections the various types of intrinsic defects formed once hydrogen has been trapped in a semiconductor.

8.2   Intrinsic Hydrogen Centres

8.2.1   "Isolated" Hydrogen

We call "isolated" hydrogen a hydrogen atom stable at a lattice location where it has not been trapped by an impurity or a lattice defect. This state of hydrogen is relatively not well documented, except in silicon. The usual method for its observation is the irradiation of the samples with protons or deuterons in the MeV range at low temperature while keeping the sample at low temperature before performing optical, ESR, or electrical experiments (*in situ* measurements).

8.2.1.1   Diamond

Despite the presence of hydrogen in some natural diamonds and in diamonds grown by CVD [19], there is no spectroscopic evidence for isolated interstitial hydrogen in diamond. Calculations show that in diamond, the bond centred (*BC*) interstitial configuration is the lowest one in energy for isolated hydrogen [20], but that it produces an increase of the C·· ··C distance by ∼40% with respect to the normal bond length [9]. This results in a puckered configuration for H_{BC}^+, corresponding to *IR* A' of the C_{1h} symmetry point group, with, while H_{BC}^0 and H_{BC}^- keep a D_{3d} symmetry (*IR* A$_{2u}$). In the harmonic approximation, the frequencies calculated for the A' LVMs of H_{BC}^+ are 2456 and 2086 cm^{-1} and the one for the A$_{2u}$ LVM of H_{BC}^0 is 2919 cm^{-1} [21]. Theory predicts a very small effective charge for H_{BC} in diamond. As the intensity of a vibrational transition is proportional to the square of the effective charge of the vibrational mode (see expression (5.1)), this could explain why large concentrations of H_{BC} escape spectroscopic detection in diamond [21]. For the negative charge state, the effective charge is larger and the frequency calculated for A$_{2u}$ is 2730 (1952) cm^{-1} for H_{BC}^- (D_{BC}^-), respectively.

8.2.1.2 Silicon, Germanium, and SiGe Alloys

In high-resistivity FZ silicon samples irradiated with protons and/or deuterons at ~ 30 K, *in situ* absorption measurements have revealed a strong LVM at $1998\,\mathrm{cm}^{-1}$. The strength of this absorption has been quantified by an effective charge of $3.0e$ [22]. This LVM was first reported by Stein at $1990\,\mathrm{cm}^{-1}$ in a measurement at 80 K on a CZ silicon sample implanted at the same temperature with protons [23]. In Budde's study [22], another LVM at $1449\,\mathrm{cm}^{-1}$ was observed in a deuteron-implanted sample, and the spectra of samples co-implanted with protons and deuterons showed only the 1998 and $1449\,\mathrm{cm}^{-1}$ LVMs, and no extra one, indicating that the centre responsible for these LVMs contains only one hydrogen atom. The intensity of the $1998\,\mathrm{cm}^{-1}$ LVM was observed to decrease under band gap light illumination with respect to the intensity under TEC, but the effect was found to be reversible and the intensity returned to the initial value when the illumination was turned off [22]. A comparison can be made with the Si-AA9 $(S = 1/2)$ ESR spectrum,[2] observed at low temperature under band-gap illumination in FZ or CZ silicon implanted with protons near LNT [24]. The centre giving this ESR spectrum contains two equivalent Si atoms and a H atom aligned along $<111>$ trigonal axes [25].

The isochronal annealing curves of the AA9 ESR spectrum and of the $1998\,\mathrm{cm}^{-1}$ LVM were the same, with an annealing temperature around 200 K [22]. This fact and the above-mentioned effect of band gap light illumination on the $1998\,\mathrm{cm}^{-1}$ LVM have led to attribute this LVM to H^+ in a bond-centred location (H_{BC}^+) and the one giving the AA9 ESR spectrum to H_{BC}^0 [25]. The decrease of the intensity of the $1998\,\mathrm{cm}^{-1}$ under band gap light illumination is interpreted as the reversible H_{BC}^+ to H_{BC}^0 conversion. The 1998 and $1449\,\mathrm{cm}^{-1}$ LVMs, with a r(H/D) frequency ratio of 1.3789, are thus ascribed to the asymmetric mode (A_{2u}) of the H^+ and D^+ equilibrium states, respectively, of interstitial hydrogen when considering a D_{3d} symmetry with the H atom bonded symmerically to the two Si atoms. An unsymmetrical bonding reduces the symmetry to C_{3v} and the *IR* of the asymmetric mode becomes A_1. On the spectroscopic side, the trigonal symmetry of H_{BC}^+ has been confirmed from uniaxial stress measurements on the $1998\,\mathrm{cm}^{-1}$ LVM, giving for this LVM piezospectroscopic parameters A_1 and A_2 of 9 and $11\,\mathrm{cm}^{-1}\mathrm{GPa}^{-1}$, respectively [26].

The H_{BC}^0/H_{BC}^+ donor level has been identified by *in situ* DLTS as the E'_3 level located at $E_c - 0.2$ eV (the quantity measured by DLTS is the ionization enthalpy, and it is 0.175 eV) [27]. No LVM associated with H_{BC}^0 has been reported so far and this must be due to its small effective charge.

It is admitted that in silicon, interstitial hydrogen is a negative-U centre with stable H^+ and H^- charge states and a metastable paramagnetic neutral charge state. The positive and neutral charge states are H_{BC}^+ and H_{BC}^0 while the negative one is located near from the tetrahedral interstitial site. The calculated location of the H_{BC}^-/H_{BC}^0 acceptor level is $E_c - 0.6$ eV ([28], and references therein).

---

[2]AA stands for Alma-Ata (now Almaty), the capital of Kazakstan till 1997.

The stretch frequencies of the H_{BC}^+ (D_{BC}^+) calculated by an *ab initio* supercell method are 2 125 (1 512) cm^{-1}, and the difference with the experimental values 1 998 (1449) cm^{-1} can be accounted for by the neglect of anharmonicity [28].

The FWHM at LHeT of the LVM of H_{BC}^+ and D_{BC}^+ in Budde's experiments were ~ 1 and 1.5 cm^{-1}, respectively, and no Si isotope effect could be observed [22]. The absorption of this LVM and the dependence of its FWHM on the hydrogen concentration are shown in Fig. 8.1.

Measurements on qmi ^{28}Si, ^{29}Si, and ^{30}Si crystals revealed, however, very small shifts of the LVMs with the mass of the silicon isotope [30,31], opposite for the H^+ and D^+ modes: the frequency of the H^+ mode increased with the Si mass while the one of the D^+ mode decreased. Compared to the frequency of the H^+ LVM in a qmi ^{28}Si sample (1997.73 cm^{-1}), the ^{28}Si–^{29}Si and ^{28}Si–^{30}Si ISs measured by comparison with qmi ^{29}Si and ^{30}Si samples are −0.05 and −0.18 cm^{-1}, respectively. For the D^+ LVM, observed at 1448.43 cm^{-1} in a qmi ^{28}Si sample, the corresponding shifts are +0.19 and +0.35 cm^{-1}, respectively. The sign of the shifts measured for H^+ is rather puzzling as, in vibrational spectroscopy, one expects a positive IS for $M_1 - M_2$ with $M_1 < M_2$. However, considering the smallness of the effect, one must correct the frequencies measured in qmi silicon by a volumetric term due to the fact that the $Si \cdots Si$ distances in qmi ^{29}Si and ^{30}Si are slightly smaller than the one in qmi ^{28}Si. This volumetric contribution has been estimated for qmi ^{29}Si and ^{30}Si, and the ^{28}Si–^{29}Si and ^{28}Si–^{30}Si ISs corrected for this contribution become positive for H_{BC}^+ (+0.19 and +0.30 cm^{-1}, respectively), and they increase for D_{BC}^+ (+0.36 and +0.69 cm^{-1}, respectively). Self-consistent fits of the H_{BC}^+ and D_{BC}^+ frequencies to a

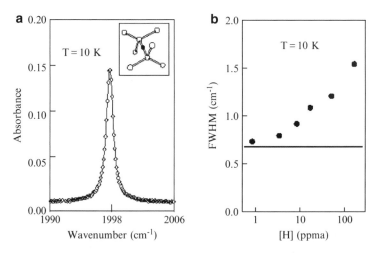

Fig. 8.1 (a) Absorbance spectrum of the LVM at 1998 cm^{-1} of the H_{BC}^+ structure shown in the inset (H atom in black) of a ^{nat}Si FZ sample implanted with protons at 80 K. The solid line is a Lorentzian fit to the data. [H] is 17 ppma or 8.5 × 10^{17} cm^{-3}. (b) FWHM of the 1998 cm^{-1} LVM vs. [H]. The horizontal line represents the natural FWHM deduced from the lifetime (7.8 ps) of this mode using expression (6.7). Note the logarithmic abscissa scale (after [29]). Copyright 2000 by the American Physical Society

Si$-$H$-$Si linear structure taking into account anharmonicity and the coupling of the A_{2u} mode with the symmetric A_{1g} mode give a Si$-$H force constant of $8.43\,\mathrm{eV\mathring{A}}^{-2}$, significantly smaller than those of the Si$-$H bonds ($15.38\,\mathrm{eV\mathring{A}}^{-2}$ and $12.60\,\mathrm{eV\mathring{A}}^{-2}$ for Si$-$H$_{BC}$ and Si $-$ H$_{AB}$, respectively) of the H$_2^*$ centre discussed in Sect. 8.2.2 [31]. This difference is assumed to be caused by the fact that in H$_2^*$, the Si$-$H bonds share two electrons while only one is involved in the Si $-$ H bonds of H$_{BC}^+$.

The vibrational lifetime at LHeT of the A_{2u} mode of H$_{BC}^+$ in ^{nat}Si has been measured by time-resolved transient bleaching spectroscopy and compared with the one deduced from the FWHM of the A_{2u} LVM [29]. The value measured by transient bleaching spectroscopy is 7.8 ps at $\sim$20 K, corresponding to a FWHM of $0.62\,\mathrm{cm}^{-1}$ for a homogeneously broadened line using expression (6.7). The value of the FWHM of the $1998\,\mathrm{cm}^{-1}$ LVM measured by optical absorption depends on [H], as shown in Fig. 8.1, but the value measured at the lowest concentration at LHeT is $0.73\,\mathrm{cm}^{-1}$ comparable to the one measured by transient bleaching spectroscopy [29].

The above-mentioned annealing at 200 K of the $1998\,\mathrm{cm}^{-1}$ LVM of H$_{BC}^+$ is achieved under TEC, but under band gap light illumination, the annealing temperature is reduced to 160 K. This effect has been interpreted as due to the photo-induced diffusion of vacancies produced by the irradiation, leading to the formation of vacancy-related H defects with H$_{BC}$ [22].

Near complete annealing of the $1998\,\mathrm{cm}^{-1}$ LVM has also been observed by irradiation with h–e ^{3}He at 80 K of a silicon sample implanted with protons at the same temperature [32]. At a difference with thermal annealing, no LVM associated to H$_2^*$ is observed as a consequence of this annealing. The presence of very weak LVMs associated with VH, VH$_2$, and IH$_2$ centres discussed later in this chapter cannot account for hydrogen lost in the annealing of the $1998\,\mathrm{cm}^{-1}$ LVM. It has therefore been suggested that the loss of H$_{BC}^+$ could be related to the formation of molecular hydrogen by radiation-induced diffusion of interstitial hydrogen [32].

It must be pointed out that photo-induced annealing to $\sim$75% of the $1998\,\mathrm{cm}^{-1}$ LVM has also been reported under above-band-gap illumination at 80 K [23]. It has been tentatively interpreted as the result of light-induced motion of vacancies from VH clusters [23], but as in the preceding case, the formation of H$_2$ is a possible channel.

In intrinsic germanium samples implanted at about 20 K with protons and/or deuterons in the MeV range, $in\ situ$ IR absorption measurements at about 10 K allowed to observe LVMs at 1794 and $1293\,\mathrm{cm}^{-1}$ in proton- and deuteron-implanted samples, respectively [26], and no new LVM in samples implanted with both protons and deuterons. Uniaxial stress measurements showed that the $1794\,\mathrm{cm}^{-1}$ LVM is due to the stretch mode of centres aligned along trigonal axes of the crystal, with piezospectroscopic parameters relatively close to those of the $1998\,\mathrm{cm}^{-1}$ mode in silicon ([26], Budde and Bech Nielsen, unpublished). This led to the attribution of the $1794/1293\,\mathrm{cm}^{-1}$ LVMs, with a r(H/D) frequency ratio of 1.3875, to the A_{2u} modes of H$_{BC}^+$/D$_{BC}^+$ in germanium [22, 26]. The centre anneals at about 210 K, and it is assumed to be the same as the one giving in silicon the $1998/1449\,\mathrm{cm}^{-1}$ LVMs.

In the same experiments, Budde et al. [26] also observed LVMs at $745/535\,\mathrm{cm}^{-1}$ in proton/deuteron-implanted germanium samples, and in proton-implanted germanium samples, two weak LVMs at 1480 and $1488\,\mathrm{cm}^{-1}$. All these LVMs anneal at the same temperature ($\sim 150\,\mathrm{K}$), significantly lower than the annealing temperature of the $1794\,\mathrm{cm}^{-1}$ LVM of $\mathrm{A_{2u}(H_{BC})}$ ($\sim 210\,\mathrm{K}$), and the intensity of the $1794\,\mathrm{cm}^{-1}$ LVM slightly increases after this annealing. These lower frequency LVMs cannot be thus associated with H^+_{BC}. Experiments under uniaxial stress reveal that the $745\,\mathrm{cm}^{-1}$ LVM is due to a doubly degenerate bending mode of a centre aligned along trigonal axes of the lattice, and the weak lines at 1480 and $1488\,\mathrm{cm}^{-1}$ are attributed to overtones of this mode. Comparison with theory [33] has led to attribute the $745\,\mathrm{cm}^{-1}$ LVM to the bending mode of H^- in an AB location [26]. A very surprising fact is that no stretching mode due to this centre could be detected.

In Si:Ge (Si-rich) alloys proton implanted at low temperature, *in situ* measurements at LHeT showed the absorption of a vibrational structure near $1998\,\mathrm{cm}^{-1}$, allowing to identify four LVMs at 1996.5, 1980.4, 2003.5, and $1991.2\,\mathrm{cm}^{-1}$, noted S_∞, S_1, S_2, and S_3, respectively (see figures 2 of [34]). In Ge:Si (Ge-rich) alloys, a corresponding vibrational structure is observed around $1800\,\mathrm{cm}^{-1}$ with four components noted G_∞, G_1, G_2 and G_3 at 1796.1, 1832.8, 1787.9, and $1800.0\,\mathrm{cm}^{-1}$, respectively, whose positions are a mirror image of the components in the Si:Ge alloy (see Fig. 5 of [34]). On the one hand, components S_∞ and G_∞ correspond to the $\mathrm{A_{2u}}$ mode of H^+_{BC} between two Si or Ge atoms, respectively, perturbed by a remote atom of the other type. S_1 and G_1 (at $\sim 1838\,\mathrm{cm}^{-1}$), on the other hand, correspond to the vibration of H^+_{BC} in a $\mathrm{Si - H - Ge}$ structure, and the relative positions of these components with respect to S_∞ and G_∞ stem from the fact that in the Ge-rich alloy, the $\mathrm{Si - Ge}$ bond is shorter than a $\mathrm{Ge - Ge}$ bond while in the Si-rich alloy, it is longer than a $\mathrm{Si - Si}$ bond. The S_2 and G_2 LVMs are associated to a configuration where H^+_{BC} is located between two atoms of the same type perturbed by an *nnn* atom of the other type. For S_3 and G_3, the perturbation arises from a third *nn* atom of the other type [34]. Similar sets of lines were observed in deuteron-implanted samples with frequencies reduced by a factor near from $\sqrt{2}$ [34].

In addition, in the Ge:Si samples as implanted with protons, LVMs at 816 and $1430\,\mathrm{cm}^{-1}$, together with the overtone at $1631\,\mathrm{cm}^{-1}$ of the $816\,\mathrm{cm}^{-1}$ LVM, were also observed at LHeT, with counterparts at 587 and $1065\,\mathrm{cm}^{-1}$ in the deuteron-implanted samples. These LVMs have the same annealing temperature of $200\,\mathrm{K}$, and they are not observed in Si-free germanium; they have been attributed to a negatively charged hydrogen atom antibonded to a Si atom [35].

As already mentioned, the vibrational dynamics of the centres in silicon has been investigated with time-resolved transient spectroscopy and by analysing the temperature dependence of the width and shape of the H_{BC} line [29, 36]. At $10\,\mathrm{K}$, the vibrational lifetime of H^+_{BC} in silicon is in the $10\,\mathrm{ps}$ range, and the measurements in germanium give a comparable value [37]. These lifetimes are governed by anharmonic couplings to low frequency pseudo-localized modes.

8.2.1.3 Compound Semiconductors

Isolated hydrogen is much less documented in compound semiconductors. Proton implantation at 20 K of HVPE-grown w-GaN has been performed followed by an *in situ* spectroscopic investigation at 8 K [38]. This allowed observing[3] a LVM at 1456 cm^{-1}, whose intensity is constant up to 175 K, where it starts to decline, reaching half of its maximum intensity at 225 K. The line completely disappears after annealing at 325 K while two lines previously identified as due to N − H stretch modes of (V_{Ga}, H) defects appear at 3027 and 3139 cm^{-1}, an indirect proof of the implication of hydrogen in the 1456 cm^{-1} LVM. Pereira et al. [38] performed *ab initio* calculations for identifying the centre responsible of the 1456 cm^{-1} line. A reasonable theory/experiment agreement is reached for the transverse mode of interstitial H$^-$ at the centre of the c-axis channel of the wurtzite GaN structure. It has to be noted that such a centre has also a longitudinal mode of vibration, but it cannot be observed with the experimental configuration used [38].

It is known for long that hydrogen gives rise to n-type conductivity in zinc oxide [39]. Lavrov et al. [40] investigated ZnO samples in which hydrogen and/or deuterium was introduced either by plasma diffusion or via annealing in sealed quartz ampoules. They observed both by IR absorption and Raman scattering LVMs at 3611 and 2668 cm^{-1} in H- and D-containing samples, respectively, with a r(H/D) factor of 1.3534. From the frequencies and from uniaxial-stress measurements, they attribute these transitions to the O–H and O–D stretch mode of H$^+$ and D$^+$ bond centred on a Zn − O bond aligned along the c-axis of the ZnO lattice [40]. These authors observed also an electronic transition at 330 cm^{-1} (40.9 meV), correlated with these LVMs. This transition is due to the $1s \rightarrow 2p$ transition of the "isolated" hydrogen shallow donor. This allowed deducing for this hydrogen donor an ionization energy of 53 meV. The intensity of the O–H stretch mode can be appreciated from an effective charge η of 0.28e for the 3611 cm^{-1} LVM. A calibration factor of the integrated absorption of this mode with respect to hydrogen is $(4.6 \pm 0.4) \times 10^{16}$ cm^{-1} [40].

The thermal stability of this centre has been investigated [40]. For hydrogenated materials, the intensities of the 3611 and 330 cm^{-1} transitions remain constant up to 110°C and anneal out at 190°C, whereas for deuterated materials, the 2668 cm^{-1} and 330 cm^{-1} lines retain constant intensities up to 150°C and anneal out at about 230°C. Therefore, in ZnO, positively charged "isolated" hydrogen seems to be stable at room temperature, and BC deuterium is more stable than BC hydrogen. The high apparent stability of isolated H$^+$ in ZnO is surprising: on the one hand, it contrasts with other materials and on the other hand it does not agree with experimental and theoretical investigations of hydrogen/deuterium diffusion in ZnO; for the diffusion of deuterium in ZnO, a low activation energy (0.17 ± 0.12) eV was found experimentally [41] while calculations gave a value around 0.4–0.5 eV

---

[3]With the experimental arrangement used, vibrations along the wurtzite structure c-axis could not be detected.

for the migration barrier for hydrogen [42], and that should lead to mobile hydrogen already at or below 200 K.

The differences in apparent thermal stability of H^+ in ZnO and silicon or germanium can be understood in terms of centres formed by mobile H^+. In silicon, once it migrates, H^+ is trapped by negatively charged deep acceptors, eventually due to defects created by the proton implantation, and it forms complexes with them; such complexes have been detected through their local vibrational modes [22]. The ZnO samples investigated by Lavrov et al. [40] are not proton implanted; in this respect, Ip et al. [41] point out a difference in the diffusivity of deuterium in plasma-diffused and deuteron-implanted materials due to a trapping of deuterium by residual damages produced in the proton-implanted samples by the nuclear stopping process. Therefore, negatively charged trapping centres for hydrogen are not present with a concentration comparable to the hydrogen concentration in the samples investigated by Lavrov et al. [43], who suggest that after annealing, two released hydrogen atoms form a molecule. As the released entities are in fact positively charged, the Coulomb repulsion between neighbouring H^+ slows down the formation rate of H_2 that explains the relative stability of BC hydrogen in ZnO. Actually, Lavrov et al. [43] observed by Raman scattering that the decrease of the H^+ LVM upon annealing is accompanied by the appearance of the modes of interstitial H_2 molecules that are discussed later in this chapter.

8.2.2 The Hydrogen Dimer H_2^*

The relative stabilities of different configurations of a pair of H atoms in a diamond crystal have been determined by *ab initio* calculations [20,44]. The result is that the lowest energy configuration is realized by a hydrogen dimer when a H_{BC} atom in a $C - H - C$ structure is associated with a H_{AB} atom bonded to one of the C atoms of the previous structure in a linear configuration with trigonal symmetry, as shown in Fig. 8.2.

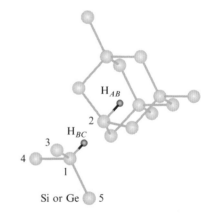

Fig. 8.2 Ball and stick model of the $(Si - H_{BC}Si - H_{AB})$ dimer structure in a diamond-type lattice. In the absence of hydrogen, atom 1 is tetrahedrally bonded to atoms 2, 3, 4, and 5. This structure has been labelled H_2^* [45]

In diamond, the stability of this configuration, later labelled H_2^* [21] by analogy with the situation in silicon [45] has been confirmed by Goss et al. [21]. It is electrically inactive and diamagnetic. The calculated frequencies of the LVMs associated with this centre are $3882\,\mathrm{cm}^{-1}$ for the stretch mode of H_{BC}, and 3511 and $1782\,\mathrm{cm}^{-1}$ for the stretch and bending modes of H_{AB}, respectively [21]. However, like isolated H_{BC} in diamond, discussed at the beginning of Sect. 8.2.1, the effective charge for each mode is very small, making the corresponding LVMs difficult to detect [9].

The H_2^* centre in silicon has been proposed from *ab initio* calculations by Chang and Chadi [45, 46] to explain the diffusion features ascribed to molecular hydrogen in silicon. A convincing spectroscopic evidence of the vibrational properties of this centre was provided by Holbech et al. [47]. In this study, the LNT absorption of FZ silicon samples implanted at RT with protons (deuterons) showed four LVMs at 2061.5 (1449.7), 1838.3 (1339.6), 1599.1 (1160.0), and 817.2 (587.7) cm^{-1} that could be related to a new hydrogen-related centre, as all these lines disappeared together after annealing of the samples at $\sim 200^{\circ}\mathrm{C}$ (470 K). The measurements on samples implanted with both protons and deuterons showed new LVMs indicating that the centre contained two non-equivalent hydrogen atoms. Absorption under uniaxial stress showed that the two highest frequency LVMs observed in proton-implanted samples were due to Si–H stretch modes with trigonal symmetry [48], and that the excited state of the $817.2\,\mathrm{cm}^{-1}$ LVM was a doubly degenerate state of a centre with trigonal symmetry [47]. This led to the attribution of the $2061.5\,\mathrm{cm}^{-1}$ LVM to the stretch mode of a Si–H_{BC} bond and the $1838.3\,\mathrm{cm}^{-1}$ one to the stretch mode of a Si $-$ H_{AB} bond. The $817.2\,\mathrm{cm}^{-1}$ LVM was ascribed to the bending mode of the Si$-$H$_{AB}$ bond and the weak LVM at $1599.1\,\mathrm{cm}^{-1}$ to the overtone of this latter LVM.

These attributions were found to be consistent with the results of *ab initio* LDF calculations on a 73-atom cluster, and the frequencies calculated for the different locations of the hydrogen isotopes in the H_2^* dimer are compared with the experimental values in Table 8.1.

A bending mode of Si $-$ H_{BC} is predicted at a frequency significantly lower than that of Si $-$ H_{AB}, but it is not observed either because of its rather small effective charge or because its location, below the Raman frequency of silicon, produces a broadening of the resonant mode [47].

The same kind of measurement as in silicon has been reported in intrinsic germanium crystals irradiated with protons or deuterons at RT, or irradiated at 30 K and stored at RT [22, 49]. In the proton-irradiated germanium samples, four LVMs at 1989, 1774, 1499, and $765\,\mathrm{cm}^{-1}$ were observed at LHeT. In the deuteron-irradiated sample, only LVMs at 1434, 1281, and $1076\,\mathrm{cm}^{-1}$ were reported, together with additional LVMs in samples co-implanted with protons and deuterons (see Table 8.1). The $r(\mathrm{H/D})$ frequency ratio of the stretch modes for the (H_{BC}, H_{AB}) and (D_{BC}, D_{AB}) isotopic combinations deduced from Table 8.1 differs slightly for H_{BC}/D_{BC} (1.3746 and 1.3865 in silicon and in germanium, respectively) and H_{AB}/D_{AB} (1.3722 and 1.3848 in silicon and in germanium, respectively). For the bending mode of antibonded hydrogen in silicon, it is 1.3905.

Table 8.1 Calculated and observed frequencies at LNT (cm^{-1}) of the LVMs of H_2^* in silicon for different locations of the hydrogen isotopes [47]. The measured overtones are given in brackets. The intensities are proportional to the square of the effective charge η (in unit of the electron charge). The last columns give the observed values in germanium [49] and the ratio r of the corresponding frequencies in silicon and germanium

Configuration	Mode	Obs. (Si)	η	Calc. (Si)	Obs. (Ge)	r(Si/Ge)
(H_{BC}, H_{AB})	Stretch H_{BC}	2061.5	0.29	2164.5	1988.8	1.037
	Stretch H_{AB}	1838.3	0.41	1843.7	1773.8	1.036
	Bending H_{AB}	817.2	0.41	1002	765	1.068
		(1599.1)				
	Bending H_{BC}	. . .	0.11	612		
(D_{BC}, D_{AB})	Stretch D_{BC}	1499.7		1548.1	1434.4	1.046
	Stretch D_{AB}	1339.6		1321.4	1280.9	1.046
	Bending D_{AB}	587.7		712		
		(1160.0)				
	Bending D_{BC}	. . .		586		
(H_{BC}, D_{AB})	Stretch H_{BC}	2058.1		2159.3	1984.5	1.037
	Stretch D_{AB}	1342.0		1324.3	1283.3	1.046
	Bending D_{AB}	587.8		712		
	Bending H_{BC}	. . .		612		
(D_{BC}, H_{AB})	Stretch D_{BC}	1488.5		1538.4	1424.6	1.045
	Stretch H_{AB}	1851.5		1855.9	1787.4	1.036
	Bending H_{AB}	817.2		1002		
	Bending D_{BC}	. . .		586		

Table 8.2 Values (cm^{-1} GPa^{-1}) of the piezospectroscopic parameters of the Si − H stretch (s) and bending (b) modes of the trigonal centre H_2^* in silicon and germanium

	A_1		A_2		B		C	
	Si	Ge[c]	Si	Ge[c]	Si	Ge[c]	Si	Ge[c]
H_{BC} (s)	3.74[a]	4.7	4.87[a]	4.1				
H_{AB} (s)	−0.04[a]	2.0	−1.31[a]	−1.50				
H_{AB} (b)	1.35[b]	2.3	0.80[b]	0.1	0.45[b]	0.6	2.51[b]	2.1

[a][48]
[b][47]
[c][49]

A piezospectroscopic study of the LVMs at 1989, 1774, and 765 cm^{-1} has globally yielded the same results as in silicon, giving a trigonal symmetry for the three LVMs. The piezospectroscopic parameters of these LVMs in silicon and germanium are compared in Table 8.2 (see Appendix E for the definition of these parameters).

If H_2^* is generally produced by proton irradiation at RT, the LVMs at 1840 and 2060 cm^{-1} attesting the formation of H_2^* are already observed in silicon implanted with protons at 80 K after annealing at 200 K, due to the thermal dissociation of H_{BC} [23].

In germanium, H_2^* is slightly less stable than in silicon, and the LVMs of this centre anneal near 145°C (430 K).

8.2.3 Hydrogen and Vacancies

8.2.3.1 Group IV Semiconductors

In silicon, vacancies are created by irradiation with high-energy particles. In high-purity silicon, they are immobile up to $\sim$200 K, but above this temperature they start diffusing in the crystal, and they can interact with other point defects, including other vacancies, and with FAs, like the group V donors and O_i. Divacancies can be created as primary defects when the energy of the incident particles is high enough, or as secondary defects when the vacancies become mobile and can coalesce (see Sect. 4.1). The main reason for the interaction of vacancies with hydrogen is that, when neutral, a vacancy (V) includes initially four dangling bonds (unpaired valence electrons of the Si atoms) which interact to form two more stable reconstructed bonds. In crystals where hydrogen has been introduced, at a temperature where isolated hydrogen can diffuse, considering the separation between the Si atoms nns of V or V_2 and the reactivity of H atoms or ions, one can expect the partial or total dissociation of the reconstructed bonds and the formation of one or more Si $-$ H bonds involving the Si atoms nns of V and pointing toward the missing atom. When considering the isolated vacancy, this results in a family of (V, hydrogen) defects which can be noted VH, VH$_2$, VH$_3$, and VH$_4$ when considering the ^{1}H isotope. This is also true, *mutatis mutandis*, for the divacancy (V_2). Vacancies are also produced thermally at high temperature, together with self-interstitial atoms and there are evidences for the interaction of these vacancies with hydrogen in crystals grown in a hydrogen atmosphere.

The evidence for the interaction of vacancies with hydrogen has been obtained from piezospectroscopic results on some hydrogen-related LVMs, from the changes observed in the hydrogen-related LVM spectra in samples co-implanted with protons and deuterons, from eventual correlations with ESR measurements, and from the changes in the hydrogen-related LVMs under thermal annealing.

Under irradiation with high-energy protons or deuterons, vacancies and divacancies are introduced at the same time as hydrogen. However, *in situ* spectroscopic measurements after low-temperature irradiation show that no (V, hydrogen) complexes are formed as primary defects, and that the production of these complexes requires thermal annealing at and above RT. Inversely, when vacancies and divacancies are introduced by high-energy electron irradiation near RT in silicon samples already containing hydrogen, (V, hydrogen) centres are produced without further annealing [50].

There have been many experimental reports of the vibrational absorption of these centres [48, 51–53], to quote early ones, and the modelling of their properties has also been investigated (see for instance [9], and references therein).

It has been pointed out above that in silicon implanted with protons at low temperature, H_{BC} is the only H-related IR-active centre present. Intrinsic defects are also present, but they do not still form complexes with H. Under annealing near 100 K in the dark, or near 60 K under band gap light illumination, new H-related LVMs appear, and their growth is correlated with the mobility onset of the vacancies.

One of these LVMs is at 2039 (1495) cm^{-1} for proton (deuteron) implantation, and as no other LVM is detected after mixed implantation with protons and deuterons, it is deduced that only one H or D atom is involved in the corresponding centre. The temperature threshold for its production and its annealing properties are the same as those of an ESR signal observed in the same samples. Both are attributed to VH, formed by the trapping of a Si vacancy by an H atom. This requires the breaking of one of the V reconstructed bonds and the building of a Si $-$ H bond, which leaves a dangling bond. The stretching of the Si-H bond produces the LVM at 2039 cm^{-1} and the ESR activity is due to the spin on the dangling bond [54]. At low temperature, such a centre has monoclinic-1 symmetry, but the ESR results show that the centre reorients above $\sim$85 K, giving a dynamic trigonal symmetry. The formation kinetics of VH seems to be controlled by the trapping of V^- by H_{BC}^+. Above $\sim$200 K, isolated H becomes mobile and it can be trapped in turn by VH to give VH$_2$, electrically inactive and responsible for two LVMs at 2121 and 2145 cm^{-1}, or by VH$_2$ to give VH$_3$, which is paramagnetic and gives two LVMs at 2155 and 2185 cm^{-1}. The formation kinetics is not clear, however, as the annealing curves show that the signals related to VH, VH$_2$, and VH$_3$ all anneal out near 500 K. Logically, the last complex is VH$_4$, a vacancy fully saturated by H atoms, and it starts forming early (near 250 K) in the annealing process of silicon implanted with protons at low temperature. A single LVM at 2223 cm^{-1} has been related to this electrically inactive defect, and the attribution has been substantiated by piezospectroscopic measurements on this LVM, which have shown that the corresponding centre must have cubic symmetry[4] [55]. Its concentration increases in the same annealing domain where the concentrations of VH, VH$_2$, and VH$_3$ decrease, so that it should also be formed there by the capture of one or several H atoms by partially hydrogenated vacancies. This centre, traditionally labelled VH$_4$ is very stable and it anneals out near 800 K ($\sim$530°C) in proton-implanted silicon. It turns out that VH$_4$ is also the dominant centre in silicon grown in a H$_2$ atmosphere, and this indicates that vacancies produced at high temperature can be stabilized by hydrogen [53, 56]. The cooling down of the crystal to RT is not an equilibrium process and that is the reason for the observation of these centres. A clear Si isotope effect of the LVM of VH$_4$ has been reported [53,56] and it is shown in Fig. 8.3.

The ^{28}Si–^{29}Si IS measured for the T$_2$ mode of VD$_4$ at 1617.5 cm^{-1} is 1.2 cm^{-1}, a value significantly larger than the one for VH$_4$ (0.8 cm^{-1}), and the FWHM of the T$_2$(VD$_4$) LVM is only 0.06 cm^{-1}.

The VH$_4$ centre contains four Si–H bonds, and in ^{nat}Si, the relative abundance of such a centre when the four ^{28}Si–H bonds are replaced by four ^{29}Si–H or ^{30}Si–H bonds is exceedingly small. The only isotopic combination giving relative intensities comparable to the ones of the isotopic components of Fig. 8.3 would result from the isotopic substitution of only one ^{28}Si–H bond by a ^{29}Si–H or ^{30}Si–H bond. It remains

---

[4]A cubic symmetry of the centre associated with this LVM, measured at 2210 cm^{-1} at RT, had also been deduced from the fact that it included several hydrogen atoms while producing only one LVM [53].

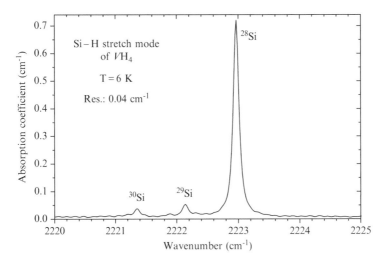

Fig. 8.3 Absorption of the T_2 stretch mode of Si $-$ H of $V\mathrm{H}_4$ in FZ silicon grown in a H_2 atmosphere. The FWHM of the ^{28}Si component is $0.12\,\mathrm{cm}^{-1}$, and the ^{28}Si–^{29}Si and ^{28}Si–^{30}Si ISs, 0.82 and $1.61\,\mathrm{cm}^{-1}$, respectively. The relative intensities of the ^{29}Si and ^{30}Si components with respect to that of ^{28}Si are 0.06 and 0.04, respectively. (Pajot, unpublished)

to calculate if such a substitution can give ISs comparable to the ones observed. On the experimental ground, the measurement of the stretch mode of $V\mathrm{H}_4$ in qmi ^{29}Si or ^{30}Si would bring an answer.

An alternative to $V\mathrm{H}_4$ results from the trapping of four H atoms by a self-interstitial I to give $I\mathrm{H}_4$ (SiH$_4$), possibly located at a vacancy site. This possibility has been considered theoretically [58] and discarded on the ground that the stretch frequency calculated for $I\mathrm{H}_4$ was lower by $\sim\!300\,\mathrm{cm}^{-1}$ than the one observed, but it has been seriously considered from the experimental side as the centre giving the $2223\,\mathrm{cm}^{-1}$ LVM[5]. It is clear that within the $I\mathrm{H}_4$ model, the relative intensities of the ^{29}Si and ^{30}Si components of Fig. 8.3, close to the natural abundances of the three Si isotopes, find a natural explanation as only one Si atom is involved.

From LHeT to RT, the frequency of the T_2 mode of $V\mathrm{H}_4$ decreases from 2223 to $2210\,\mathrm{cm}^{-1}$, and its FWHM reaches $\sim\!5.5\,\mathrm{cm}^{-1}$ at RT [57].

In proton-implanted silicon, an ESR spectrum has been related to $V_2\mathrm{H}$, a divacancy decorated with one H atom [59]. The existence of $V_2\mathrm{H}$ is not unexpected as V_2 is produced by proton implantation. This is further confirmed by the observation, after annealing at $450\,\mathrm{K}$ of proton-implanted silicon, of LVMs at 2466 and $2191\,\mathrm{cm}^{-1}$ attributed to a defect with trigonal symmetry. This defect has been identified with $V_2\mathrm{H}_6$, a divacancy fully saturated with H [22]. This centre has a stability comparable with that of $V\mathrm{H}_4$ since it anneals out near $800\,\mathrm{K}$. In hydrogenated FZ silicon irradiated with high-energy electrons, a LVM at $2072\,\mathrm{cm}^{-1}$ has also been attributed to a stretch mode of $V_2\mathrm{H}_2$ [50].

---

[5]See N. Fukata and M. Suezawa, J. Appl. Phys. 86, 1848–1853 (1999)

Table 8.3 Measured frequencies (cm^{-1}) at LHeT of the LVMs of (vacancy, hydrogen)-related centres in ^{nat}Si [36, 56, 59] and in germanium. The annealing temperatures are indicated for H$^+$- and D$^+$-implanted samples. For the VH$_n$ and VD$_n$ centres, the frequencies calculated *ab initio* are given in brackets [60]

Centre	Symmetry	Mode	Silicon		Germanium[a]	
			Freq.	Anneal. T	Freq.	Anneal. T
VH^0	Monoclinic (C_{1h})<65 K Trigonal >100 K	A"	2038.5 (2248)	~480 K		
VD^0	"	"	1494.6 (1613)	"		
VH$_2$	Orthorhombic-I (C_{2v})	B$_1$	2122.3 (2267)	~470 K	1979.5 (2078)	~400 K
"	"	A$_1$	2145.1 (2316)		1992.6 (2102)	"
VHD	Monoclinic (C_{1h})	A'	2135.4[b] (2292)	"		
"	"	A'	1555.3[b] (1641)	"		
VD$_2$	Orthorhombic-I (C_{2v})	B$_1$	1547.9 (1625)	"	1432.6 (1478)	
"	"	A$_1$	1565.1 (1658)	"	1444.2 (1495)	
VH$_3$	Trigonal (C_{3v})		2155 (2256)	~500 K	2015 (2067)	
"	"		2185 (2318)	"	2025 (2098)	
VH$_4$	Cubic (T_d)	T$_2$	2222.97 (2319)	~770 K	2061.5 (2107)	~590 K
VD$_4$	"	"	1617.53 (1664)		1488.5	"
V_2H$_2$	Monoclinic-I (C_{2h})		2072.5	~570 K		
V_2D$_2$	"		1510.4	"		
V_2H$_6$	Trigonal (C_{3v})	E	2166	~800 K	2014.9	~620 K
"	"	A	2191	"	2024.8	"
V_2D$_6$	"	E	1576	"	1454.3	"
"	"	A	1594	"	1464.8	"

[a][22]
[b][61] in qmi ^{28}Si

With time, there have been changes in the early attributions of the LVMs of the (V, hydrogen) complexes in silicon. In Table 8.3, we have summarized the main spectroscopic properties of some (V, hydrogen) complexes in silicon at the light of the most recent attributions.

It has been suggested that the LVMs observed at LHeT at 2068 and 2073.2 cm^{-1} in proton-implanted silicon could be due to V_3H^0 and V_4H^0, respectively [59], but it is not clear whether the 2073.2 cm^{-1} LVM is the same as the one attributed to V_2H$_2$ or whether there are two distinct LVMs at about the same frequency [50].

Table 8.4 Stretch frequencies (cm^{-1}) at LNT [60] (1st line) and at LHeT after [56] (2nd line) of the VH$_n$D$_m$ configurations with n + m = 4 and n and m $\neq$ 0 in silicon compared to the calculated frequencies given in brackets [60] (Bech Nielsen et al. 1995)

VH$_3$D			VH$_2$D$_2$				VHD$_3$		
Si − H	Si − H	Si − D	Si − H	Si − H	Si − D	Si − D	Si − H	Si − D	Si − D
2250	2224	1620	2244	2225	1628	1615	2236	1636	1616
				2226.6			2237.5		1616.5
(2384)	(2319)	(1677)	(2364)	(2319)	(1690)	(1663)	(2342)	(1705)	(1664)

In addition to the above IR-active stretch mode of VH$_4$ with T$_2$ symmetry, a symmetric A$_1$ breathing mode, IR inactive, but Raman active is expected at a slightly higher frequency. The calculated frequencies of this Raman mode are 2404 and 1721 cm^{-1} for VH$_4$ and VD$_4$, respectively [60]. The Raman scattering lines measured at 2223 and 2257 cm^{-1} at LHeT (2205 and 2234 cm^{-1}, respectively, at RT) in proton-implanted silicon have been attributed to the IR-Raman active T$_2$ mode observed in absorption at 2223 cm^{-1} and to this IR-inactive A$_1$ mode [62].

For the mixed VH$_n$D$_m$ configurations with n + m = 4 and n and m $\neq$ 0, all the Si − H and Si − D modes are IR active. One can note that the frequency of the highest energy one for VH$_3$D is comparable to the one measured for the Raman mode of VH$_4$ (2257 cm^{-1}). The corresponding observed and calculated frequencies are given in Table 8.4.

For the "pure" isotopic combinations VH$_4$ and VD$_4$, the r(H/D) frequency ratios of the T$_2$ mode for silicon and germanium deduced from Table 8.3 are 1.3743 and 1.3850, respectively, and they are comparable to those of the BC stretch modes of H$_2^*$ and D$_2^*$ (1.3746 and 1.3865, respectively) deduced from Table 8.1.

Lifetimes of the excited states of (V, hydrogen) LVM transitions have been measured in silicon samples electron irradiated after annealing at high temperature in a H$_2$ or D$_2$ atmosphere and quenched at RT [63]. They are deduced from the values of the FWHMs of the LVMs measured at LHeT using expression (6.7). The FWHM of T$_2$(VD$_4$) reported in this reference is 0.04 cm^{-1}, a value noticeably smaller than the one reported for an as-grown sample [56]. Contrary to the situation for VH$_4$ and VD$_4$, it is found that for the LVMs attributed to[6] V_2H$_2$ and V_2D$_2$, the FWHM for V_2H$_2$ is 0.02 cm^{-1}, compared to 0.06 cm^{-1} for V_2D$_2$ [63].

In proton-implanted qmi ^{28}Si, ^{29}Si, and ^{30}Si samples, Si ISs of the A$_1$ and B$_1$ Si − H modes of VH$_2$ have been reported. For both modes, they are ~2 cm^{-1} for qmi ^{28}Si/qmi ^{30}Si, and they are well reproduced by a simple vibrational model based on two coupled Morse oscillators [61].

The first overtones of T$_2$(VH$_4$) and T$_2$(VD$_4$) are observed at 4388.5 and 3208.1 cm^{-1}, respectively, with an intensity ratio ~0.01 with respect to the fundamental. Fitting the fundamental and first overtone frequencies with a Morse potential gives for the T$_2$ mode of VH$_4$ and VD$_4$ an anharmonicity parameter x_e

---

[6]In [63], V_2H$_2$ is noted HV.VH$_{(110)}$ to emphasise that the Si − H bonds are located in the same (110) plane in separate halves of the divacancy.

Table 8.5 Frequency differences (cm^{-1}) between the $Si-H$ and $Si-D$ LVMs measured at LHeT in the $V + O + 2$-hydrogen-atom and $V + 2$-hydrogen-atom centres in ^{nat}Si [64]

	$\Delta(Si-H)$		$\Delta(Si-D)$	
$VOH_2 - VH_2$	$+6.4(A_1)$	$+4.1(B_1)$		
$VOD_2 - VD_2$			$+2.3(A_1)$	$+1.8(B_1)$
$VOHD - VHD$	$+5.2(A')$		$+2.0(A')$	

of 0.013 and 0.008, respectively [56]. Effects associated with the anharmonicity of these modes are further discussed in this reference.

The VO centre, whose vibrational properties in relation with oxygen are discussed in Sect. 7.1.1, can trap two H atoms to give VOH_2 (see Sect. 7.1.1.3), and this centre has the same point group symmetry as VH_2. It can be produced by high-energy electron irradiation at RT of CZ silicon where hydrogen has been introduced by high-temperature annealing in a hydrogen atmosphere [64] or by proton implantation [61]. The frequencies of the $Si - H$ and $Si - D$ LVMs of the VOH_2, VOD_2, and VOD centres have been measured in ^{nat}Si samples [64] and in qmi ^{28}Si, ^{29}Si, and ^{30}Si samples [61]. It is found that the frequencies of these LVMs are larger than the ones for the centres where oxygen is absent, and the differences are given in Table 8.5 with reference to Table 8.3.

The increase of the frequency of the $Si-$ hydrogen modes when the O atom is present must be related to the global inward distortion of VO compared to the isolated vacancy. At a difference with VH_4, the FWHMs and intensities of the VH_2 and VOH_2 LVMs do not allow to measure directly the Si IS of the symmetric (A_1) and antisymmetric (B_1) $Si - H$ modes of these centres. However, a $^{28}Si-^{30}Si$ IS of $\sim 2\,cm^{-1}$ for both modes can be estimated from the measurements on qmi ^{28}Si and ^{30}Si samples [61].

In proton-implanted germanium, LVMs associated with $(V,$ hydrogen) complexes have been reported [22]. Their main spectroscopic properties are summarized in Table 8.3.

In CVD diamond, the ESR signature of a $S = 1$ defect with trigonal symmetry, observed and studied between LHeT and RT, has been attributed to a VH defect in the negative charge state [65].

8.2.3.2 Compound Semiconductors

LVMs of hydrogen-related defects have been observed in III–V compounds implanted with protons [66] and with protons and deuterons ([67], and references therein) or grown by the LEC method [68–71]. In the LEC-grown materials, hydrogen comes from the dissociation of the water contained in the encapsulant (see Sect. 1.2.2). In InP, two LVMs, with frequencies at 2316 and $2202\,cm^{-1}$ at LHeT, have been specially investigated and their characteristics are discussed below.

In LEC-grown Fe-doped InP, among five different sharp LVMs near $2300\,cm^{-1}$, the one observed at LHeT at $2315.6\,cm^{-1}$ could not be related to a centre involving a TM [68,69]. In this kind of samples, the FWHM at LHeT of this LVM is $0.03\,cm^{-1}$ [70]. This LVM was reported at $2308\,cm^{-1}$ at RT [72] and at $2315.2\,cm^{-1}$ at

LNT [73] in proton-implanted InP. The piezospectroscopic properties of this LVM have led to its attribution to the P–H stretch mode T_2 of a $V_{In}H_4$ centre with T_d symmetry, similar to VH_4 in silicon, whose spectroscopic parameters A, B, and C (see Table E.3 of Appendix E) are 3.5, −0.06, and 6 $cm^{-1}GPa^{-1}$, respectively [70]. The LVM associated with $V_{In}D_4$ has been reported in deuteron-implanted InP at 1683.4 cm^{-1} at LNT by Fisher et al. [73] and by Tatarkiewicz et al. [74]. In InP samples co-implanted with proton and deuterons, P−H and P−D LVMs close to those of $V_{In}H_4$ and $V_{In}D_4$ have been reported and attributed to mixed structures where H and D atoms are simultaneously present in the centre [75]. In implanted material, $V_{In}H_4$ anneals at ∼550°C(∼800 K), but its complete dissociation in as-grown material requires wafers annealing time between 50 and 100 h at temperatures between 900 and 950°C in a phosphorous atmosphere, and a slow cooling down at RT [76, 77].

As for VH_4 in silicon, there exists an IR-inactive symmetric breathing mode of $V_{In}H_4$. A weak line observed at RT in proton-implanted InP at 2336 cm^{-1} correlates with the LVM stretch mode of $V_{In}H_4$ reported at 2308 cm^{-1} at RT in [72], and it can be attributed to this breathing mode in proton-implanted InP [74]. For $V_{In}H_4$ and $V_{In}D_4$, the calculated frequencies of these breathing modes are 2387.85 and 1713.01 cm^{-1}, respectively [78].

A high-resolution study of the absorption of the LVMs of P − H and P − D in $V_{In}H_nD_m(n + m = 4)$ has been performed on samples from LEC InP crystals grown using an encapsulant wetted with H_2O or D_2O. As these LVMs are very sharp in as-grown InP, the high resolution has allowed to observe the number of modes expected ([71], K. Irmscher, A. Kwasniewski, M. Czupalla, M. Neubert, W. Ulrici, unpublished results, 2007), and this is shown in Fig. 8.4.

The assignment of the lines has allowed a self-consistent comparison of the frequencies of the LVMs observed for the different $V_{In}H_nD_m$ combinations with those from *ab initio* calculations. These results are summarized in Table 8.6.

The overtone of the 2316.6 cm^{-1} LVM has been observed at 4547.59 cm^{-1} with a relative intensity of 0.004 with respect to the fundamental, and a FWHM of 0.04 cm^{-1} [70].

Compared to VH_4 in silicon, the frequency and FWHM of the P−H stretch mode of $V_{In}H_4$ show a modest temperature dependence, with a decrease of the frequency by ∼3 cm^{-1} and an increase of the FWHM to 1 cm^{-1} at 300 K [70].

In a III–V compound, the four bonds of a group III atom are made from the three valence electrons of this atom and from five electrons from the *nn* group V atoms. When a neutral group III atom is ejected, statistically, five electrons remain about V_{III}. In the case of InP, four of these electrons can bind four hydrogen atoms and the fifth one can be considered as a donor electron, making $V_{In}H_4$ an "intrinsic" donor centre. The ionization energy of this donor is not known, but no difference in the 2316 cm^{-1} line shape is observed in n-type InP and in SI InP:Fe [71]. This cannot be considered, however, as a proof that $V_{In}H_4$ retains the same charge state throughout the energy range as it has been shown, for instance, that the isotopic structure of the LVM of the C_{As} acceptor in GaAs is not affected by its charge state (see Sect. 5.3.2.2).

A H-related LVM at 2202 cm^{-1} is observed in proton-implanted InP, and it disappears after annealing at ∼350°C (∼620 K) [67, 73]. In as-grown LEC InP:Fe,

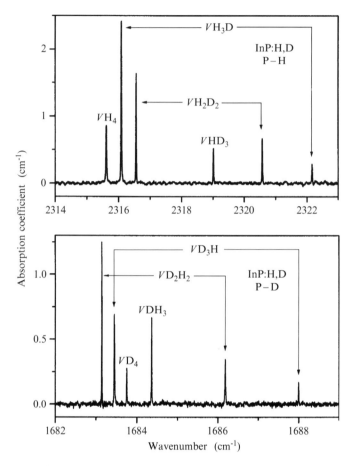

Fig. 8.4 P − H and P − D stretch modes absorption at 7 K for the five different types of $V_{In}H_nD_m$ combinations (n + m = 4) in as-grown InP H- and D-doped samples [71]. With respect to [71], the attributions for VD_4 and the lowest-frequency line of VD_2H_2 have been inverted (K. Irmscher, A. Kwasniewski, M. Czupalla, M. Neubert, W. Ulrici, unpublished results). The unapodized resolution is 0.01 cm^{-1}

depending on the thermal history of the crystal, annealing must be required to observe that LVM [79]. At a difference with proton-implanted InP, in LEC InP:Fe, this LVM, located at 2202.39 cm^{-1} at LHeT, is very sharp, with a FWHM of 0.01 cm^{-1}. When H is replaced by D, the LVM shifts at 1604.00 cm^{-1}, with a FWHM still sharper (0.007 cm^{-1}) [71]. The centre associated with this LVM has a trigonal symmetry and the LVM is assumed to be due to the P − H bond stretching of a single H atom of a $V_{In}H$ centre [70].

No LVM related to the equivalent of $V_{In}H_4$ has been reported in GaAs and GaP. This absence in GaAs has been attributed, on the basis of LDA calculations, to the low stability of the fourth As − H bond when a H atom is added to $V_{Ga}H_3$ [80]. In LEC-grown GaAs, the sharp LVM at 2001.0 cm^{-1}, with a FWHM of 0.03 cm^{-1}, with a D counterpart at 1459.5 cm^{-1}, has been attributed to a centre with trigonal

Table 8.6 Comparison of the frequencies (cm^{-1}) of the P $-$ H and P $-$ D stretch modes of the different $V_{In}H_nD_m$ combinations measured at high resolution at LHeT in InP [71] with those from *ab initio* calculations, given in brackets [78]. With respect to [71], the experimental values for VD_4 and the lowest frequency for VD_2H_2 have been inverted (K. Irmscher, A. Kwasniewski, M. Czupalla, M. Neubert, W. Ulrici, unpublished results). For each combination, the measured FWHMs ($\times 10^{-3}$ cm^{-1}) are given in the columns at the right of the frequencies. For the P $-$ D modes, the last line gives the measured ratio r of the P $-$ H to P $-$ D frequencies

	VH_4		VH_3D		VH_2D_2		VHD_3	
P $-$ H	2315.62	25	2316.08	18	2316.55	15	2319.01	19
	(2356.40)		(2356.32)		(2356.24)		(2364.31)	
			2322.16	19	2320.57	19		
			(2380.18)		(2372.34)			
	VD_4		VD_3H		VD_2H_2		VDH_3	
P $-$ D	1683.75	10	1683.45	15	1683.14	7	1684.36	11
	(1690.82)		(1690.95)		(1691.08)		(1696.43)	
	$r = 1.3753$		$r = 1.3758$		$r = 1.3763$		$r = 1.3768$	
			1687.99	9	1686.18	17		
			(1707.26)		(1701.74)			
			$r = 1.3757$		$r = 1.3762$			

symmetry identified as $V_{Ga}H$ centre [81]. In GaP, LVMs at 2205.9 (H) and 1617.3 (D) cm^{-1} have been tentatively attributed to $V_{Ga}H$ and $V_{Ga}D$, respectively ([13], and references therein). In proton-implanted GaAs and GaP, the LVMs of $V_{Ga}H$ are observed at 2001 and 2204 cm^{-1}, respectively ([82], and references therein).

The frequencies of the stretch and bending modes of $V_{Ga}H_n$ defects (n $\leq$ 4) in w-GaN have been calculated *ab initio* for different charge states [83]. In w-GaN, for the bonding of an H atom to one N atom surrounding V_{Ga}, one must distinguish the three equivalent surrounding N atoms (noted here $N_\perp$) in a plane perpendicular to the c axis and the N atom (noted here $N_{//}$) located along the c axis. For $V_{Ga}H$, $V_{Ga}H_2$, and $V_{Ga}H_3$, this leads to consider two possible configurations called Type I (one H atom bonded to $N_{//}$) and Type II (no H atom bonded to $N_{//}$). The calculated stretch frequencies (between 3100 and 3200 cm^{-1}) are slightly larger for the Type I structures, and the stretch frequencies for $V_{Ga}H_4$ are significantly larger (between $\sim$3400 and 3650 cm^{-1}) [83]. Several LVMs near 3100 cm^{-1} have been observed in w-GaN implanted with protons, and they have been attributed to N $-$ H modes from H-decorated V_{Ga} defects, and they are within the frequency range computed for $V_{Ga}H$, $V_{Ga}H_2$, and $V_{Ga}H_3$ [84]. LVMs observed in w-GaN between $\sim$3150 and 3300 cm^{-1} are also in a frequency range compatible with the stretch mode of some (V_{Ga},H) defects [85].

In hydrogenated ZnO, LVMs are observed at 3312.2 and 3349.6 cm^{-1} at LHeT (the corresponding O–D frequencies are 2486.3 and 2460.7 cm^{-1}, respectively) [86]. They are attributed to O–H bonds, almost respectively parallel and perpendicular to the c-axis of a $V_{Zn}H_2$ centre [87]; this attribution relies heavily on the results of the *ab initio* calculations published in [88].

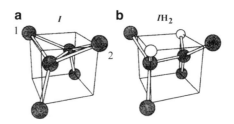

Fig. 8.5 Calculated structure of (**a**) the split $-\langle 110\rangle$ form of the self-interstitial I and (**b**) the hydrogen saturated self-interstitial $I\,H_2$ in silicon (after [89]). The shaded spheres are Si atoms and the white ones H atoms. Copyright 1998 by the American Physical Society

8.2.4 Hydrogen and Self-Interstitials

There is no direct experimental evidence of the isolated self-interstitial (I) in silicon, but there exist many calculations of its possible structures and thermodynamic properties. It is admitted that in silicon, the lowest energy form of the self-interstitial is the split-$\langle 110\rangle$ form, shown in Fig. 8.5a, where the two equivalent atoms forming the defect are fourfold coordinated, while the surrounding atoms 1 and 2 are fivefold coordinated [90].

The stability of the hexagonal form, where the self-interstitial atom is sixfold coordinated has also been studied [91]. Trapping of hydrogen by I has been envisaged, and one of the possible structures is the one where only a H atom is bonded to one fivefold coordinated Si atoms of Fig. 8.5a, in the (110) plane, giving the $I\,H_{\langle 110\rangle}$ structure. In the $I\,H_{2\langle 110\rangle}$ structure shown in Fig. 8.5b, the two Si atoms which were fivefold coordinated in the bare I configuration return to their normal fourfold coordination. The possible bonding of up to four H atoms by a self-interstitial have been envisaged by Gharaibeh et al. [92].

Several LVMs observed in silicon after proton implantation have been discussed in the preceding sections. An absorption feature observed at $1980\,\mathrm{cm}^{-1}$ at RT after such an implantation has been reported [51], and it has been also reported in H-containing silicon after high-energy electron irradiation [93], neutron irradiation [94], and γ-rays irradiation [95]. At LHeT, this feature is resolved into a doublet at 1987 and $1990\,\mathrm{cm}^{-1}$ with components of comparable intensities [50, 89]. These two LVMs are accompanied at lower energies by a pair of LVMs at 743 and $748\,\mathrm{cm}^{-1}$ [89]. The four LVMs anneal out together near $230°\mathrm{C}$ ($\sim500\,\mathrm{K}$), indicating that they are related to the same centre, and co-implantation with protons and deuterons shows that two H atoms are involved in this centre, which was tentatively ascribed to $I\,H_2$. Comparable results have been obtained in proton- and deuteron-implanted germanium [89]. The conjunction of piezospectroscopic measurements on these LVMs and of *ab initio* calculations on possible $I\,H_2$ structures have put the above attribution on firm grounds, yielding for this centre an atomic structure with monoclinic-II symmetry (C_2) very comparable to the one of Fig. 8.5b. When the C_2 symmetry axis is taken along [001], the Si-H or Ge-H bonds are nearly aligned with the [011] and [0$\bar{1}$1] directions. These results are summarized in Table 8.7.

Table 8.7 Measured frequencies (cm^{-1}) of the LVMs of $I H_2$, $I HD$, and $I D_2$ in silicon and germanium at LHeT. The calculated frequencies are given in brackets. Calculations predict two other bend modes with different orientations at lower frequencies [89]

Mode type	Silicon				Germanium			
	$I H_2$	$I D_2$	r(H/D)	$I HD$	$I H_2$	$I D_2$	r(H/D)	$I HD$
Stretch	1990.0[a]	1448.7[a]	1.374	1987.8	1883.5	1359.0	1.386	1882.8
(A)	(2144.7)	(1540.2)	(1.393)	(2143.8)	(2056.7)	(1462.5)	(1.406)	(2055.1)
" (B)	1987.1[a]	1446.5[a]	1.374	1447.3	1881.8	1357.6	1.386	1358.4
	(2142.9)	(1539.9)	(1.391)	(1540.1)	(2053.3)	(1460.1)	(1.406)	(1461.3)
Bend (A)	748.0			745.7	705.5			~703
	(774.7)			(771.4)	(787.4)			(784.9)
" (B)	743.1			(727.3)	700.3			
	(768.1)				(784.7)			

[a][63]

In H-containing silicon irradiated with high-energy electrons or in proton-irradiated silicon annealed at RT, a LVM is observed at $1870\,cm^{-1}$ at LHeT, and this LVM has been specially investigated by Suezawa [50]. Its FWHM is $0.9\,cm^{-1}$ and its D counterpart is at $1364.9\,cm^{-1}$ ($r = 1.370$). The related centre involves two H atoms and it anneals out at 225°C($\sim$500 K). The intensity of the $1870\,cm^{-1}$ LVM is sublinear with the electron dose, and this is comparable to the electron dose dependence for the $2072\,cm^{-1}$ LVM attributed to V_2H_2 (see Table 8.3). For this reason and by comparison of its frequency with that of V_2H_2, it has been ascribed to I_2H_2, a self-di-interstitial having trapped two H atoms [50].

In the preceding presentation, a limiting temperature has generally been indicated above which the hydrogen-related defects dissociate under thermal equilibrium, and this is a thermodynamic property common to complex centres in crystals. Similarly, starting from the low-temperature side, there are thresholds below which a complex defect cannot form because its constituents are immobile. Recently, by *in situ* absorption measurements, it has been shown that VH- and IH-related defects produced "thermodynamically" in proton-implanted silicon annealed at RT could be dissociated by 1.8 MeV ^{3}He implantation near LNT, producing H_{BC} whose presence was detected by the observation of the LVM of H_{BC}^+ at $1998\,cm^{-1}$ [96]. This low-temperature implantation provides evidence of ion-induced dissociation of complex defect resulting in the production of the building blocks of these defects. Vacancies and divacancies were certainly also produced directly by this implantation, and also presumably from the dissociation of V-related centres.

8.3 Hydrogen Molecules

8.3.1 Reminder on Free Hydrogen Molecules

Free H_2 and D_2 molecules are symmetric and their vibration and rotation does not induce any electric dipole moment. They are therefore IR inactive and their rotation-vibration (ro-vibrational) spectrum can only be observed by Raman spectroscopy.[7]

Protons, with spin $1/2$, are fermions. Therefore, H_2 molecules have two possible nuclear spin states: $I = 0$ (para-hydrogen) and $I = 1$ (ortho-hydrogen). At room temperature, under thermodynamic equilibrium, the ratio of ortho- to para-H_2 concentrations is equal to the ratio $3/1$ of the nuclear spin degeneracy's $2I + 1$. The natural ortho to para conversion without catalyst is very slow (days). The total wave function of the molecule has to be antisymmetric with respect to permutation of the nuclei. Para-H_2 with an antisymmetric nuclear spin wave function can only have symmetric rotational wave functions with J even whereas ortho-H_2 with a symmetric nuclear spin wave function can only have antisymmetric rotational wave functions with J odd.

Deuterons, with $I = 1$, are bosons. Therefore, D_2 molecules have three possible nuclear spin states: $I = 0$ and 2 corresponding to ortho-D_2 and $I = 1$ corresponding to para-D_2. At room temperature the ratio of ortho- to para-D_2 concentrations is $2/1$. The total wave function of D_2 has to be symmetric, and therefore ortho-D_2 and para-D_2 can have only symmetric (J even) and antisymmetric (J odd) rotational wave functions, respectively, where J is the rotational quantum number.

The HD molecule is asymmetric, and the restrictions due to the symmetry of H_2 and D_2 do not apply to HD, which behaves as a conventional hetero-nuclear diatomic molecule.

The vibrational frequencies of the hydrogen molecules obtained from Raman scattering measurements in the gas phase at RT are $4161.1, 3632.1$, and $2993.6 \, cm^{-1}$ for H_2, HD, and D_2, respectively [98], giving a $r(H_2/D_2)$ frequency ratio of 1.3900.

8.3.2 Isolated Molecules in GaAs

The first spectroscopic evidence of molecular hydrogen in semiconductors was given by the observation, in plasma-hydrogenated GaAs samples of various origins, of Raman lines associated with hydrogen [99]. Figure 8.6 shows the Raman shifts of three GaAs samples after different plasma treatments.

The shifts of these lines with respect to the energy of the exciting laser are ~ 3930, 3446.5, and $2842.6 \, cm^{-1}$ for H_2, HD, and D_2, respectively, and there is a similitude with what is observed in the gas phase, with a red shift of $\sim 100\text{--}200 \, cm^{-1}$ in GaAs and a $r(H_2/D_2)$ frequency ratio of only 1.382, compared to 1.390 in the gas phase. In III–V compounds, first-principles calculations predict that the H_2 molecule is more stable at a T_{iIII} location than at a T_{iV} location of Fig. 1.5. The calculated frequency for H_2 at a T_{iGa} site in GaAs is $3824 \, cm^{-1}$ ($3750 \, cm^{-1}$ at a T_{iAs} site) compared to the experimental LNT value of $3934 \, cm^{-1}$, reduced to $3912 \, cm^{-1}$ at RT [100]. Thus, this latter line is ascribed to H_2 at a T_{iGa} site.

---

[7]Very weak IR absorption of vibration-rotation transitions of H_2 and D_2 at a pressure of $4 \, MPa$ ($\sim 40 \, bar$) has been reported at RT under an electric field of $7 \, MV \, m^{-1}$ [97].

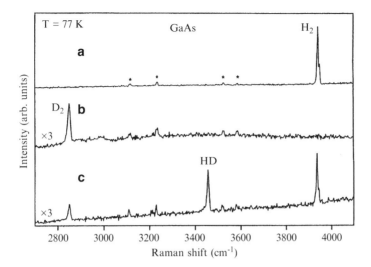

Fig. 8.6 Raman spectra (resolution: $6\,\mathrm{cm}^{-1}$) of three GaAs samples. (**a**) SI LEC GaAs after 8-h H_2-plasma treatment at 254°C, (**b**) Low-temperature (LT) MBE-grown GaAs after 3-h D_2-plasma treatment at 200°C, (**c**) LT GaAs after 8-h 50% H_2/50% D_2 plasma treatment at 800°C. Residual plasma lines of the exciting laser are indicated by stars (after [99]). Copyright 1996 by the American Physical Society

The structure observed in Fig. 8.6 for the H_2-diffused sample is actually a doublet, and in Fig. 3.3 of [99], obtained at a resolution of $2\,\mathrm{cm}^{-1}$, it is clearly resolved in a pair of lines at 3925.9 and $3934.1\,\mathrm{cm}^{-1}$, with an intensity ratio of $\sim 3/1$. The origin of this doublet structure is based on the following: First, the possibility of a rotation of the hydrogen molecule is implicitly assumed, and the vibrational and rotational states are noted v and J, respectively, or $|v,J\rangle$ for a ro-vibrational state.[8] Second, one must take into account the existence of para-hydrogen and of ortho-hydrogen molecules. When considering the situation for H_2, the lowest accessible ro-vibrational state for para-H_2 is $|0,0\rangle$, but for ortho-H_2, this state is symmetry-forbidden and the higher lying state, $|0,1\rangle$ is populated; for D_2, the conclusions are inverted. This situation is represented in Fig. 8.7.

The Raman transitions correspond to a selection rule where $\Delta J = 0$. The rotational constant is smaller in the $v = 1$ vibrational state than in the vibrational ground state because of the ro-vibrational coupling so that the rotational levels are closer in the state $v = 1$. Therefore, the frequency of the $|0,0\rangle \to |1,0\rangle$ transition or Q(0), with the lowest intensity as it corresponds to para-H_2, must be larger than that

---

[8]In vibrational spectroscopy, when considering rovibrational transitions between the ground and excited states of an ordinary molecule, one distinguishes between the sets of $|0, J\rangle \to |1, J+1\rangle$, $|0, J\rangle \to |1, J\rangle$, and $|0, J\rangle \to |1, J-1\rangle$ transitions, called traditionally the P, Q, and R branches. For this reason, the Raman lines, corresponding to $|0, J\rangle \to |1, J\rangle$ transitions, are noted Q(J).

Fig. 8.7 Ro-vibrational transition scheme for ortho-H$_2$ and para-D$_2$ triplet states (left) compared to the ones for para-H$_2$ and ortho-D$_2$ singlet states (right) in GaAs. The double-sided arrows represent the Raman transitions. The levels are also denoted by the *IR*s A$_1$ (singlet) and T$_2$ (triplet) of the T_d point group

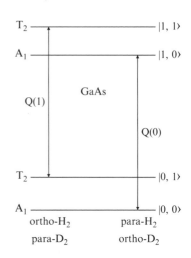

of the $|0,1\rangle \rightarrow |1,1\rangle$ transition (Q(1)), with the highest intensity as it corresponds to ortho-H$_2$. This is what is observed and the lines at 3925.9 and 3934.1 cm^{-1} can then be attributed to ortho-H$_2$ and para-H$_2$, respectively. The para-ortho splitting measured in GaAs for H$_2$ is 8.2 cm^{-1} compared to 5.9 cm^{-1} in the gas phase. It is the difference between the splittings in the ground and excited states. Such a splitting should also be observed for D$_2$ in GaAs, but it should be smaller (it is 2.1 cm^{-1} in the gas phase) because the rotational energies for D$_2$ are smaller. This splitting is certainly smaller than the FWHM of the D$_2$ line at 2842.6 cm^{-1} in GaAs, and this why it is not observed. The Raman line for HD observed at 3446.5 cm^{-1} corresponds to a $|0,0\rangle \rightarrow |1,0\rangle$ transition.

In recent Raman measurements near RT on plasma-hydrogenated GaAs samples, in addition to the H$_2$ line at 3912 cm^{-1}, a weak line at 4043 cm^{-1} and a broad feature at 4112 cm^{-1} were also observed. The line at 4043 cm^{-1} was attributed to the stretching of a H$_2$ molecule located at a T_{iAs} site, and Raman lines corresponding to D$_2$ and HD at this site were also observed at 2920 and 3602 cm^{-1}, respectively [14]. The line at 4112 cm^{-1} was tentatively attributed to the stretching of H$_2$ molecules trapped in voids produced by the plasma treatment near from the surface (see Sect. 8.3.8).

8.3.3 Isolated Molecules in Silicon

Symmetric molecules like H$_2$ embedded in a crystal lattice are subjected to a crystal field that breaks the symmetry of the physical system. Therefore, IR transitions between the ro-vibrational levels of H$_2$ and of D$_2$ become weakly allowed. Actually, the vibration of hydrogen molecules at interstitial tetrahedral sites has been detected in silicon by absorption [101, 102] and Raman spectroscopies [104]. It took, however, nearly a decade and many papers to fully understand the behaviour of the hydrogen molecule in silicon. Only the present understanding is described in the following.

8.3.3.1 Raman Spectroscopy

Results roughly in line with those described in the previous section for GaAs were reported by Leitch et al. [104] in plasma-hydrogenated FZ silicon samples. A LVM was observed at $3601\,cm^{-1}$ at RT ($3618\,cm^{-1}$ at LHeT) in the H_2-treated sample, with a corresponding one at $2622\,cm^{-1}$ at RT in the D_2-treated sample. The frequencies of these LVMs are even lower than those observed in GaAs, but the ratios of the H_2/D_2 frequencies (1.380) are quite comparable, so that they were assigned to the vibration of H_2 and D_2 molecules. A noticeable difference with GaAs was the absence of para–ortho splitting of the H_2 LVM in this pioneering work, but this splitting was reported later for H_2 and D_2 in CZ silicon samples [15], and it is shown in Fig. 8.8 for H_2.

The frequencies measured at 100 K for ortho-H_2 and para-H_2 are 3617 and $3626\,cm^{-1}$, respectively, and the ones for ortho-D_2 and para-D_2, 2645 and $2642\,cm^{-1}$, respectively. For the HD molecule, the splitting of the $|0,0\rangle \rightarrow |1,0\rangle$ and $|0,1\rangle \rightarrow |1,1\rangle$ transitions with frequencies of 3194 and $3189\,cm^{-1}$, respectively, was also observed [15]. Raman measurements in the 170–388 K temperature range have recently allowed to observe the Q(2) transition of interstitial ortho-H_2 [105].

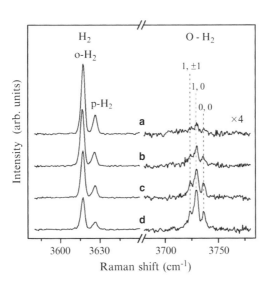

Fig. 8.8 Raman scattering spectra at $\sim$100 K of a CZ silicon sample after exposure to a H_2 plasma. On the left side are shown the ortho-H_2 (o $-$ H_2) Q(1) and para-H_2 (p $-$ H_2) Q(0) transitions of interstitial H_2 split by $9\,cm^{-1}$. The spectra are taken right after hydrogenation (**a**) and after RT annealing for 7.5 (**b**), 19 (**c**), and 77 h (**d**). The lines on the right side are due to transitions of interstitial H_2 perturbed by a nearby interstitial oxygen (O–H_2). Their intensities increase with annealing time at the expense of that of the o-H_2-p-H_2 pair (after [15])

8.3.3.2 Absorption Spectroscopy

In the level scheme of Fig. 8.7, an IR transition between two A_1 levels is symmetry-forbidden. Therefore, para-H_2 and ortho-D_2 molecules at tetrahedral sites cannot be detected by absorption spectroscopy. By comparison, transitions between levels with T_2 symmetry are allowed and the absorption of ortho-H_2 and para-D_2 molecules at tetrahedral sites can be detected. Absorption at LHeT of LVMs at 3618.3 and 2642.5 cm^{-1} in CZ and FZ silicon samples annealed at high temperature in H_2 and D_2 atmospheres was interpreted as due to the presence of isolated H_2 and D_2 molecules, respectively, at interstitial tetrahedral sites [101, 102]. With reference to the preceding conclusion, these LVMs must be attributed to the $|0,1\rangle \rightarrow |1,1\rangle$ transitions of ortho-H_2 and para-D_2. The results of uniaxial-stress experiments on these absorption lines [106, 108] are fully compatible with $T_2 \rightarrow T_2$ transitions [108].

The results on the HD molecule have been the most difficult to interpret. Pritchard et al. [101] reported an LVM of this molecule at 3264.8 cm^{-1} at LHeT. The frequency of this LVM does not correspond to those observed by Raman scattering mentioned in Sect. 8.3.3.1; moreover, it does not scale properly with the frequencies of the transitions observed for the H_2 and D_2 molecules. The puzzle was solved [107] by realizing that within T_d symmetry, an $A_1 \rightarrow T_2$ transition is allowed, and the 3265 cm^{-1} LVM was consistently assigned to the $|0,0\rangle \rightarrow |1, 1\rangle$ transition of the HD molecule at an interstitial tetrahedral site. In measurements at higher temperatures, Chen et al. observed a decrease of the intensity of the 3265 cm^{-1} LVM together with the rising at 75 K of a new LVM at 3191.1 cm^{-1} [107]. The activation energy of 71 ± 4 cm^{-1} deduced from their measurements as a function of temperature agrees well with the frequency difference between the two lines (≈ 74 cm^{-1}). This indicates a thermalization between the initial levels involved in the two transitions. The "hot" line at 3191.1 cm^{-1} must therefore correspond to the $|0,1\rangle \rightarrow |1,1\rangle$ transition, also observed later on by Raman scattering [109]. Subsequent experiments revealed another absorption line of HD at 3425 cm^{-1}, observed at LHeT, and corresponding to a $|0,0\rangle \rightarrow |1, 2\rangle$ transition [110]. All the transitions reported for the HD molecule are summarized in Fig. 8.9.

Experiments under uniaxial stress have confirmed the assignments of the lines observed by absorption spectroscopy [110].

It has been shown that annealing B-doped silicon samples at 1300°C in flowing H_2 for B passivation led to the production of isolated H_2, attested by the observation at LHeT of the 3618 cm^{-1} LVM, together with that of the LVM of the passivating (B, H) centre at 1903 cm^{-1}, presented in Sect. 8.5.1.1 [111]. The interesting point is that annealing the samples at 175°C produced an increase of the intensity of the (B, H) LVM correlated with a decrease of the intensity of the H_2 LVM. From the annealing data, it seems likely that the increase of the (B, H) concentration is due to the capture of a H_2 molecule by a B^- acceptor, with the release of an extra H^+ [111].

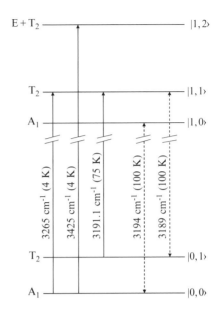

Fig. 8.9 Energy-level diagram of the ro-vibrational transitions of the HD molecule at an interstitial tetrahedral site in silicon. The solid lines correspond to IR absorption and the dashed lines to Raman scattering. The IR and Raman frequencies are taken from [110] and [15], respectively

8.3.3.3 Ortho–Para Conversion

Right after hydrogenation, the low-temperature Raman scattering measurements presented earlier show that the ratio of the ortho- and para-H_2 concentrations is $\sim 3/1$, assuming implicitly that the relevant transitions observed by Raman scattering have comparable oscillator strengths. In silicon, the change with annealing time of this ratio has been used to follow the ortho- to para-H_2 and para- to ortho-H_2 conversions [112]. Absorption spectroscopy has also been used, but with this technique, one measures only the change of the intensity of the ortho-H_2 line at 3618 cm^{-1} [113]. For the ortho- to para-H_2 (o-H_2 to p-H_2) conversion, silicon samples were exposed to a H_2 plasma and then stored at LNT. Either Raman scattering or IR absorption was measured as a function of the storage time at LNT. The result shown in Fig. 8.10 using Raman scattering evidences the decrease of the ortho-H_2 mode and a corresponding increase of the para-H_2 mode.

For the absorption measurements, an exponential decrease of the ortho-H_2 line with a time constant of (229 ± 14) h is observed [113].

The reverse process, para- to ortho-H_2 conversion, has also been investigated [112, 113]. For this kind of experiment, after hydrogenation, the samples are first kept at LNT for a sufficiently long time to achieve an equilibrium population of the two nuclear-spin isomers, comparable to the one represented in spectrum (d) of Fig. 8.10. The Raman or IR measurements are then performed as a function of the annealing time of the samples at RT. In that case, the para-H_2 population decreases and the ortho-H_2 population increases with the annealing time. An exponential process is again observed, with a time constant of 5.6 h [112] and (8.1 ± 0.5) h [113].

Fig. 8.10 Raman scattering at 90 K of the ortho–para-H_2 pair in a CZ silicon sample after exposure to a H_2 plasma and subsequent storage durations at LNT, showing the progressive o-H_2 to p-H_2 conversion. Right after hydrogenation (a) and after storing the sample at LNT for 87 (**b**), 338 (**c**), and 999 h (**d**). Spectra are normalized for the laser powers used for the Raman measurements, the baselines corrected and offset vertically for clarity [112]. Copyright 2007 by the American Physical Society

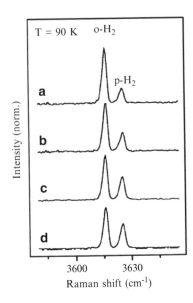

The same kind of absorption experiment was performed by Peng et al. on silicon samples containing D_2 molecules. The time constant of the para-D_2 to ortho-D_2 conversion determined at LNT is (213 ± 72) h, and the one for the inverse conversion at RT is (25 ± 20) h [113].

The physical mechanisms responsible for these conversions are still not clear at the present time.

8.3.4   Isolated Molecules in Germanium

In plasma-hydrogenated germanium, two sharp Raman lines have been reported at 3826 and 3834 cm^{-1} at LNT with a 3826/3834 intensity ratio of about 3/1. By analogy with the situation in GaAs and with the help of modelling, these lines were attributed to the stretching of interstitial ortho-H_2 and para-H_2, respectively, at a T_i site [114]. The para-ortho splitting of D_2 for the Raman line at 2782 cm^{-1}, estimated to 2.5 cm^{-1}, is smaller than the FWHM of the line, and it is not observed. Germanium samples co-diffused with H_2 and D_2 do show the hint of a LVM at 3368 cm^{-1}, but this needs confirmation [114].

8.3.5   Isolated Molecules in ZnO

Zinc oxide (ZnO) has the wurtzite structure. In ZnO samples where the Fermi level is close to the CB, Raman scattering of hydrogen molecules has been detected after annealing at 550°C of samples hydrogenated at 1000°C for 1 h in a sealed quartz ampoule and quenched at RT [40, 43]. The ortho-H_2 and para-H_2 lines are observed near 20 K at 4145 and 4153 cm^{-1}, respectively, with the usual intensity ratio ~3/1. With D_2, the ortho-para splitting cannot be resolved and a single feature at 2895 cm^{-1} is observed. For samples submitted to mixed $H_2 + D_2$ hydrogenation, an additional line at 3626 cm^{-1} due to HD has been reported. These frequencies are very close (to less than 10 cm^{-1}) to those of the hydrogen molecules in the gas phase, given at the end of Sect. 8.3.1.

8.3.6 Isolated Molecules in Gallium Nitride

The absorption spectrum shown in Fig. 8.11 has been observed at LHeT [115] in Mg-doped MOVPE GaN layers using a multitransmission geometry shown in [85].

It mainly consists of two lines at 4090 and 4110 cm^{-1} with FWHMs of about 10 cm^{-1}. These lines are still observed at RT, slightly downshifted in energy (4087 and 4107 cm^{-1}), indicating the vibrational nature of the transitions. As no transitions are observed in the 2000–2100 cm^{-1} range, these lines cannot be attributed to overtones. The Mg-doped layers are known to contain hydrogen; in particular in the (Mg,H) centre responsible for the LVM at 3125 cm^{-1} [85], also observed in this sample. It is therefore natural to attribute the spectrum of Fig. 8.11 to the absorption of isolated H_2 molecules, as H_2 molecules in voids (see Sect. 8.3.8) can be detected only by Raman scattering. As a matter of fact, theoretical investigations [116, 117] predict hydrogen molecules to be stable around the centre of the hexagonal channel of the wurtzite structure; it has also been shown [116] that in this location, a polarization of the molecule occurs that reinforces its infrared activity. In the harmonic approximation, which overestimates the vibration frequencies, the vibration frequency of the molecule in this configuration has been

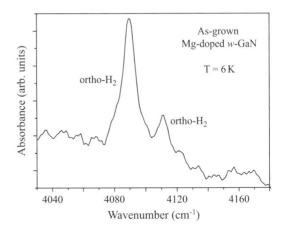

Fig. 8.11 Absorption spectrum of the isolated H_2 molecule in w-GaN. The transitions refer to the level scheme of Fig. 8.12. After [115]

Fig. 8.12 Energy-level
diagram of the ro-vibrational
of the H_2 molecule in GaN.
The energy levels are indexed
by the corresponding *IR*s of
the C_{3v} symmetry group. The
frequencies are indicated in
wavenumbers (cm^{-1}). The
full lines with one arrow
correspond to the transitions
observed in absorption [115].
The broken line with one
arrow is allowed in
absorption and the dotted
lines with two-sided arrows
correspond to the transition
that should be observed by
Raman spectroscopy

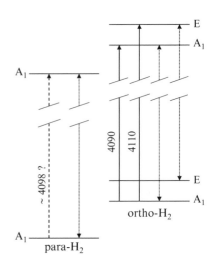

calculated to be $4235\,cm^{-1}$ [83], while calculations taking anharmonicity into
account give a value of $4032\,cm^{-1}$, close to the measured value [117]. Therefore,
the spectrum shown in Fig. 8.11 has been assigned to molecular hydrogen in the
hexagonal channel of the GaN wurtzite structure [115].

The splitting $(20\,cm^{-1})$ between the two transitions is too large for being due to
the para–ortho splitting. A plausible explanation for this splitting is a consequence
of the fact that the symmetry of the potential felt by the molecule is C_{3v}. The $J = 1$
rotational levels are then split by the trigonal field into an A_1 singlet and an E
doublet, giving the energy diagram shown in Fig. 8.12.

The two transitions observed are the two transitions of ortho-hydrogen indicated
by full arrows in Fig. 8.12. This means that the $20\,cm^{-1}$ splitting corresponds to
the splitting of the $J = 1$ rotational level in the $v = 1$ vibrational state. One should
wonder why para-hydrogen is not observed. The transition due to para-hydrogen,
that is indeed allowed, should be about $8\,cm^{-1}$ higher than the $4090\,cm^{-1}$ transition,
i.e. hidden in the high-energy tail of the $4090\,cm^{-1}$ line.

A question, related with Sect. 8.3.5, is why the splitting described just above is
not observed in ZnO? Hydrogen molecules have been observed in ZnO by Raman
scattering only and the Raman selection rules are different from the infrared ones. In
Raman spectroscopy, for the H_2 molecule in GaN, one should observe the transitions
marked by dotted double arrows in Fig. 8.12. If the splittings of the $J = 1$ states
are the same in the $v = 0$ and $v = 1$ vibrational states, the two Raman transitions
of ortho-hydrogen should coincide and one would observe a single line for ortho-
hydrogen together with the line of para-hydrogen. This is probably what happens
in ZnO. In this respect, it would indeed be of interest to investigate also hydrogen
molecules in ZnO by absorption spectroscopy.

8.3.7 Molecules Trapped by Oxygen in Silicon

It has been shown by Pritchard et al. [101] that an absorption peak at $1075\,\mathrm{cm}^{-1}$ observed in CZ Si heated at high temperature in hydrogen gas and quenched down to RT was due to a perturbed bond-centred interstitial $^{16}O_i$ atom. In deuterated samples, the peak shifts to $1076\,\mathrm{cm}^{-1}$ (see Sect. 6.1.1.4). These peaks have been shown to consist of components with relative intensities 3:1 for the hydrogenated samples and 1:2 for the deuterated ones. The investigation of samples heat treated in $H_2 + D_2$ atmosphere led these authors to conclude that the O_i atom was adjacent to a pair of hydrogen atoms.

A hydrogen molecule located in the neighbourhood of an O_i atom experiences a potential with symmetry lower than T_d. This has two consequences for the ro-vibrational properties of the molecule: (1) the $J = 1$ levels split into two sublevels corresponding to $m_J = 0$ and $m_J = \pm 1$, and (2) the selection rules for IR absorption are less restrictive than for molecules in T_d symmetry. LVMs of both para- and ortho-hydrogen can thus be observed in absorption; moreover, there can be two transitions, at low temperature, observable for ortho-H_2 and para-D_2. These transitions have indeed been observed in absorption spectroscopy by Pritchard et al. [101] and Chen et al. [118] for silicon samples heat treated, respectively, in H_2, D_2, and $H_2 + D_2$ atmosphere, and they are correlated with the oxygen-perturbed mode at $1075\,\mathrm{cm}^{-1}$. Chen et al. [118] provided the complete and correct assignment of all of the lines that is indicated by the solid vertical lines of Fig. 8.13.

They also showed that the two components of the 1075 and $1076\,\mathrm{cm}^{-1}$ lines were due to the ortho and para components of H_2 and D_2, respectively.

Hydrogen molecules trapped by interstitial oxygen in silicon have also been investigated by Raman spectroscopy [119]. The three transitions marked by double arrows in Fig. 8.13 have been observed for H_2 molecules. The Raman scattering transitions of O–H_2 at 100 K can clearly be seen in the spectrum at the right side of Fig. 8.8d. The ortho-para conversion at 77 K of H_2 trapped by interstitial oxygen has also been investigated and the same conversion time as the one for "isolated" hydrogen molecules has been found [119]. This indicates that interstitial oxygen does not play a role in the nuclear spin flip process involved in the ortho to para conversion.

8.3.8 Molecules in Platelets and Voids

Platelets are planar extended defects, extending generally perpendicular to threefold axes, that are formed during hydrogenation of elemental semiconductors [120]. In these defects, the crystal lattice is broken and dilated by about 0.3 nm [120]; hydrogen under pressure (about 1 GPa) can fill the voids in these extended defects. In such cases the spectroscopic properties of the molecules are close to those of free

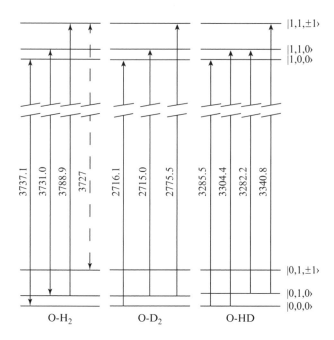

Fig. 8.13 Energy-level diagrams of the ro-vibrational states of the H_2, D_2, and HD molecules trapped by interstitial oxygen in silicon (O–H_2, O–D_2, and O-HD, respectively). The ro-vibrational states are noted $|v, J, m_J\rangle$, and the transition energies are indicated in cm^{-1}. The full lines correspond to the transitions observed in absorption [101, 118]. The full and broken lines with two-sided arrows correspond to those observed by Raman spectroscopy [119]

molecules. Raman spectroscopy is the only appropriate spectroscopic technique for the investigation of hydrogen molecules in these extended defects.

In plasma hydrogenated silicon samples, Fukata et al. [121] reported a rather broad Raman vibrational line at 4158 cm^{-1}, and they showed that it consists of two components at about 4160 and 4130 cm^{-1}, the relative intensities of which depend upon hydrogenation temperature. In addition, they observed a purely rotational transition at 590 cm^{-1}. In deuterated samples, they observed a vibrational peak at 2990 cm^{-1}. Leitch and Weber observed the same vibrational lines and in addition a line at 3629 cm^{-1} in H_2 + D_2 plasma diffused samples [103]. These authors attributed all of these spectroscopic data to H_2, D_2, and HD molecules located within the voids formed by the platelets, which are generated during the plasma treatments of the silicon samples. This is substantiated by the fact that the $r(H_2/D_2)$ frequency ratio for these molecules (1.390) is the same as the one for molecular hydrogen in the gas phase. Hiller et al. [122] have refined these studies. They observed four purely rotational transitions at 353, 587, 815, and 1034 cm^{-1} corresponding to the $\Delta J = 2$ Raman transitions originating at the $J = 0, 1, 2, 3$ states. They also showed that the 4158 cm^{-1} peak consists in fact of four components corresponding to the transitions of ortho- and para-hydrogen for two distinct types of platelets.

For the first type, the transitions are at $4137\,\mathrm{cm}^{-1}$ (ortho) and $4143\,\mathrm{cm}^{-1}$ (para) and for the second one, they are at $4148\,\mathrm{cm}^{-1}$ (ortho) and $4154\,\mathrm{cm}^{-1}$ (para). For both types of platelets, the para–ortho splitting is $6\,\mathrm{cm}^{-1}$, identical to the one of the free hydrogen molecules. Hiller et al. [122] have also investigated the ortho–para conversion of hydrogen molecules in platelets formed in silicon. They found conversion rates much faster than for the molecules at the silicon tetrahedral sites. This indicates that the interaction between neighbouring H_2 is the dominant ortho–para conversion mechanism in platelets; this is consistent with the high hydrogen pressure in the voids suggested by Muto et al. [120].

Platelets are formed in germanium too. The H_2 vibrational mode in germanium platelets has been detected at $4155\,\mathrm{cm}^{-1}$ [123].

In hydrogen plasma diffused GaAs, a rather broad Raman transition has been detected at $4112\,\mathrm{cm}^{-1}$, and this band has been tentatively assigned to H_2 molecules trapped in voids formed by the plasma treatment [14].

8.4 Extrinsic Hydrogen Centres I

In this section and in Sect. 8.5, we discuss the properties of the vibrational spectra due to chemical bonds of extrinsic centres including hydrogen and FAs or hydrogen, FAs, and lattice atoms. The frequencies of the stretch modes of these $X-H$ or $X-D$ bonds depend on the chemical nature of the X atom, but also on the compression or expansion of the bond in the lattice. In Table 8.8 are given as a qualitative reference the measured frequencies of the $X-H$ stretch bonds in some molecules.

In the present section, we do not consider centres related to the passivation of the electrical activity of acceptors and donors. These latter centres are discussed in Sect. 8.5.

Table 8.8 Frequencies (cm^{-1}) of the stretch mode of $X-H$ bonds in some molecules (gas phase unless otherwise specified) and solids. Factor $r(\mathrm{H/D})$ is the ratio of these frequencies with the corresponding ones for the $X-D$ bonds

	Bond	Frequency	$r(\mathrm{H/D})$
$^{11}B_2H_6$	$B-H$	2612	1.3099
$CHCl_3$	$C-H$	3034	1.3416
$CHBr_3$	$C-H$	3017	1.3403
NH_3	$N-H$	3444	1.3432
OH^{-a}	$O-H$	3282.4	1.3459
OH^{-b}	$O-H$	3286.5	1.3442
SiH_4	$Si-H$	2190.6	1.3717
PH_3	$P-H$	2328	1.3710
GeH_3F	$Ge-H$	2121	1.3909
AsH_3	$As-H$	2116	1.3894
$TMP.Ga-H_3$	$Ga-H$	1808	1.3961

[a][124], α-Al_2O_3 at 77 K
[b][125], TiO_2 at LHeT

8.4.1   Isoelectronic Atoms

For substitutional isoelectronic atoms like C in silicon, foreign group III and group V atoms in III–V compounds, and group VI atoms in II–VI compounds, the external electronic configuration is not changed with respect to the lattice atom they replace (antisite locations are ignored), and they are in principle electrically inactive.[9] A general characteristic of the complexes of hydrogen with these atoms seems to be their electrical activity. This can be considered as the inverse of the passivation process, where electrically active atoms or centres are made electrically inactive by interaction with hydrogen.

8.4.1.1   Carbon in Silicon and Germanium

Experimental evidence for low concentrations of H-related acceptors in germanium samples from crystals grown in graphite crucibles under a hydrogen atmosphere was obtained from PTIS measurements. This is a very sensitive method and it has allowed to identify a (C, hydrogen) centre with trigonal symmetry [126]. However, considering the small solubility of carbon in germanium, no LVM related to (C, hydrogen) centres in this material has been detected.

An *ab initio* cluster model has been used to predict and calculate the stabilities and vibrational frequencies of several (C_n, H_m) structures in silicon at a time where only a few experimental results had been published [127]. Later on, most of the publications of experimental results were accompanied by the presentations of calculations of the properties of possible interpretative structural models.

In silicon with $[C_s]$ in the range 10^{16}–10^{17} cm^{-3} hydrogenated by wet chemical etching, DLTS have been ascribed to the donor $0/+$ ($E3$) and acceptor $-/0$ ($H1$) electronic levels of a complex of H with C_s ([128, 129]). The $E3$ donor level has been located at $E_c - 0.15$ eV. The centre giving $E3$ dissociates near 20°C in n-type silicon and the one giving $H1$ near 100°C in p-type silicon, with illumination-dependent effects on the dissociation rates. It has been proposed that these two levels belong to a bistable (C_s, H) centre with trigonal symmetry [129, 130]. With reference to the H_2^* dimer structure of Fig. 8.2, possible configurations of a (C_s, H) complex with trigonal symmetry are (a) $(C_s - H_{BC}Si)$, (b) $(H_{AB} - C_s\,Si)$, (c) $(C_s\,H_{BC} - Si)$, and (d) $(C_s\,Si - H_{AB})$. The results of the calculation for the neutral charge state using LDF theory give (a) for the lowest energy state, followed by (c) [127, 131]. For the positive charge state, $(C_s - H_{BC}Si)$ is also the lowest energy configuration, but it is $(H_{AB}-C_s\,Si)$ for the negative charge state [127]. From these structures, one expects vibrational frequencies $\sim 2900 - 3100$ cm^{-1} for $(C_s - H_{BC}\,Si)^+$ and $(C_s - H_{BC}\,Si)^0$, and of ~ 2600–2800 cm^{-1} for $(H_{AB} - C_s\,Si)^-$, but no IR results have been reported up to now for this defect.

---

[9]Some of them, like the N_{As} or N_P close pairs in GaAs or GaP or O_{VI} in some II–VI compounds display an electrical activity because of the nitrogen or oxygen high electronegativity.

Three kinds of experiments performed between years $\sim$1998 and 2001 have led to the observation of LVMs associated with (C_s, H) complexes. They used FZ silicon samples with [$^{12}C_s$] in the range 1–4 $\times$ 10^{17} cm^{-3} and/or ^{13}C–implanted or ^{13}C–doped FZ samples with comparable or larger concentrations.

A Weakly Bound Centre

In the first kind of experiments, involving defects with only one C atom and one hydrogen atom, hydrogen (H, D, or H + D) was introduced by implantation at low temperature, in the same way as the one used in the study of isolated bond-centred hydrogen, and the experiments were performed *in situ* at LHeT (see Sect. 8.2.1). After annealing above $\sim$100 K, in the ^{12}C–containing samples, Si − H (Si − D) LVMs were observed at 1884.5 cm^{-1} (1362.5 cm^{-1}), with frequencies reduced by $\sim$0.2 cm^{-1} in the ^{13}C–containing samples [132]. Besides the LVM of isolated C_s, weak LVMs at 596 and 691 cm^{-1} were also observed in the (^{12}C, H)-containing samples, shifting to 578 and 641 cm^{-1} in the (^{13}C, H)-containing samples. In the deuterated samples, these C_s–related LVMs showed negative shifts of 0.5 cm^{-1} between the (^{12}C, H)- and (^{12}C, D)-containing samples. All these LVMs were attributed to a complex, labelled CH$_I$ in the original reference, with an annealing temperature of $\sim$220 K. This complex shows a small coupling between the C and H atoms and the stretch frequency excludes direct bonding between them. Weak Si − H (Si − D) LVMs close to those at 1884.5 cm^{-1} (1362.5 cm^{-1}) were also reported, showing the existence of similar complexes with slightly different configurations [132].

In the same reference, possible structures for (C, hydrogen) complexes involving one atom of each kind were investigated by an *ab initio* cluster technique based on LDF pseudopotential theory [132]. The structures considered are noted here[10] ($C_s - H^+_{BC}$ Si), ($H^+_{BC} - $ Si $- C_s$), ($H^+_{BC} - $ Si $- $ Si $- C_s$)$_{(110)}$, in which the four listed atoms are in the same (110) plane, and (H^+_{BC}-Si-Si-C_s)$_{(1\bar{1}0)}$ where the Si–C_s bond is in a plane perpendicular to the (110) plane containing H and the two Si atoms. All the calculated Si − H^+_{BC} and Si − D^+_{BC} stretch frequencies of the relevant structures are smaller than the one calculated for isolated H^+_{BC} and D^+_{BC} in silicon (1852.1 and 1316.2 cm^{-1}, respectively). A plausible structure for the CH$_I$ complex responsible for the LVMs observed at 1985, 661, and 596 cm^{-1} could be then ($H^+_{BC} - $ Si $- C_s$), with calculated frequencies of 1795.0, 642.6, and 574.1 cm^{-1}, respectively, but it is pointed out that the ($H^+_{BC} - $ Si $- $ Si $- C_s$)$_{(110)}$ and ($H^+_{BC} - $ Si $- $ Si $- C_s$)$_{(1\bar{1}0)}$ structures cannot be ruled out completely. The impossibility to observe LVMs which could be attributed to ($C_s - H^+_{BC}$Si) has been explained by the competition with the formation of stronger hydrogen traps like the (V, H) defects [132].

---

[10]In [132], the structures are noted $(C_s H^+_{BC})_1$, $(C_s H^+_{BC})_2$, $(C_s H^+_{BC})_{3a}$, and $(C_s H^+_{BC})_{3b}$, respectively, with reference to Fig. 6 of this paper.

The CH_2^* *Dimers*

In the second kind of experiments, where hydrogen was introduced by annealing the samples at 1250–1350°C in a hydrogen atmosphere, followed by quenching to RT, defects with trigonal symmetry containing one C atom and two hydrogen atoms were investigated ([133], and references therein). In relation with the H_2^* dimer discussed in Sect. 8.2.2, such defects had been anticipated from theoretical results, and two degenerate structures had been proposed: one where C_s replaces Si atom 1 in the H_2^* structure of Fig. 8.2, and another one where it replaces Si atom 2 in the same structure [127]. These structures can be noted ($C_s - H_{BC}Si - H_{AB}$) and ($Si - H_{BC}C_s-H_{AB}$), but they are noted $(CH_2^*)_2$ and $(CH_2^*)_1$, respectively, in [133]. For these centres, with symmetry C_{3v}, there are three stretch modes: $C - H$ (A_1), $Si - H$ (A_1), and $Si - C$ (E), and two bending modes: $Si - H$ (E) and $C - H$ (E) (the corresponding *IR*s are given in brackets). Among several LVMs observed at LHeT in these samples, those at 792.0, 1921.8, and 2752.3 cm^{-1} have been ascribed to $Si - H_{AB}$ bending,[11] $Si - H_{AB}$ stretch, and $C - H_{BC}$ stretch modes of $(CH_2^*)_2$, respectively. Those at 665.3 and 2210.4 cm^{-1} are ascribed to the $Si - C$ bending and $Si - H_{BC}$ stretch modes of $(CH_2^*)_1$, respectively. A LVM has also been reported at $\sim$2688.5 cm^{-1}, with an annealing behaviour similar to that of the 665 and 2210 cm^{-1} LVMs, and it could be due to the $C - H_{AB}$ stretch mode of $(CH_2^*)_1$ [133]. A summary of the measured and calculated frequencies of the LVMs of the $(CH_2^*)_1$ and $(CH_2^*)_2$ centres for the partial and total substitutions of H by D is given in Tables 2 and 1, respectively, in [133], but no measured frequency for the $C - D$ stretch mode is reported. For $(CH_2^*)_1$, whatever the combination, only the stretch mode associated with BC hydrogen and the $Si - C$ mode are observed. There is a rather good agreement between the calculated and measured frequencies of the $Si - C$ mode, but the frequencies calculated for the BC hydrogen mode are smaller by $\sim$40 cm^{-1} than the measured values. The calculated frequency for the $^{12}C - H_{AB}$ stretch mode is 2665 cm^{-1}, close to the LVM observed at 2688.5 cm^{-1}, giving further support to the attribution of this LVM to the $^{12}C - H_{AB}$ stretch mode of $(CH_2^*)_1$.

The comparison of the spectroscopic data of $(CH_2^*)_1$ and $(CH_2^*)_2$ on the one hand and of H_2^* on the other hand shows that the measured frequencies of the $Si-H_{BC}$ and $Si-H_{AB}$ stretch modes in the C-containing centres are larger by $\sim$150 and 48 cm^{-1}, respectively, than the corresponding $Si - H$ modes of H_2^*, indicating more compact centres when C is present.

Piezospectroscopic measurements have been reported on the 792 cm^{-1} LVM of $(CH_2^*)_2$ [134] and on the 2210 cm^{-1} LVM of $(CH_2^*)_1$ [135] which confirm the trigonal symmetry of these centres. The observation of LVMs at 1921.76, 2210.39, and 2752.52 cm^{-1} has been reported in silicon crystals grown in a H_2 atmosphere, with FWHMs of 0.6, 0.1, and 0.1 cm^{-1}, respectively [56], and they correspond to the above-discussed $(CH_2^*)_2$ and $(CH_2^*)_1$ modes. It is worth noting that in the same

---

[11]The overtone of the $Si - H_{AB}$ bending mode has been reported at 1550.2 cm^{-1} [133].

samples, the H_2^* modes (Table 8.1) were not observed. This seems to show that during the non-equilibrium cooling down of the crystal, the $(CH_2^*)_1$ and $(CH_2^*)_2$ centres can form while H_2^* cannot, despite a dissociation temperature of the former centres near 430°C ($\sim$700 K) [134, 135], compared to 470 K for H_2^*.

The Hydrogenated C_s Dicarbon Pair

In the last kind of experiment, C-containing FZ silicon samples were first irradiated at RT with 2 MeV electrons, and subsequently implanted with protons, deuterons, or proton + deuterons at RT [136]. Three LVMs at 2967.4, 911.7, and 654.7 cm^{-1} were observed at LHeT in samples containing ^{12}C and H, which were specially investigated. The intensities of these LVMs increase with annealing temperatures, reaching a maximum at 650°C. They anneal out simultaneously at 750°C, an indication that they are due to the same centre. The vibrational spectra of samples with partial or total ^{13}C and D substitutions showed that the centre contains two equivalent C and two equivalent H atoms and that the 2967.4 cm^{-1} LVM is a $C_s - H$ stretch mode. The results of piezospectroscopic measurements on this latter LVM could be explained by assuming that this LVM is due to the non-degenerate mode of a centre with trigonal symmetry along a <111> axis, while those on the 911.7 cm^{-1} LVM imply a doubly degenerate mode of a trigonal defect [136]. This is clearly demonstrated for a stress along a <111> axis, where one of the stress-induced components corresponds to an orientation of the stress parallel to the trigonal axis of the centre. The two H atoms of this centre are then pushed against each other and the repulsive interaction between the two $C_s - H$ stretch modes increases their frequency with stress as shown in Fig. 8.14. This is also observed for H_2^*.

It is suggested that this centre is formed when diffusing hydrogen atoms are trapped by the dicarbon *nn* pair produced by RT electron irradiation of FZ silicon, discussed in Sect. 7.2.3.

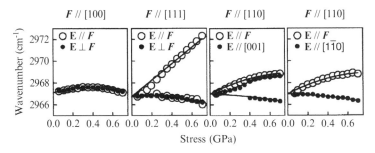

Fig. 8.14 Stress-induced splitting of the 2967.4 cm^{-1} LVM of the hydrogenated dicarbon pair in silicon. The filled (open) circles correspond to the electric vector **E** of the radiation polarized perpendicular (parallel) to the stress. The solid line is a best fit for the frequency shifts using expression (E.1) of Appendix E for a trigonal centre and including terms quadratic with stress [136]. Copyright 2000 by the American Physical Society

Fig. 8.15 Calculated
structure of
$(C_s - H \cdots H - C_s)$ in
silicon. The calculated
distance between the two C
atoms is 346 pm, compared to
a normal Si $\cdots$ Si separation
of 235 pm [136]. Copyright
2000 by the American
Physical Society

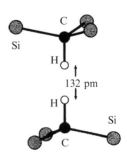

Based on two *nn* C_s atoms and two equivalent H atoms all lying on a trigonal
<111> axis, two possible structures can be considered for the resulting centre: one
where the H atoms are located between the C_s atoms, noted $(C_s - H \cdots H - C_s)$, and
another one with the H atoms antibonded to the C atoms, noted $(H_{AB}-C_s \cdots C_s-H_{AB})$.
Ab initio calculations on a 44-atom Si cluster where the two central atoms are
replaced by two C atoms show that the most stable structure is the one represented
in Fig. 8.15. This structure of the hydrogenated dicarbon pair is in agreement with
the piezospectroscopic results, and this centre must be electrically inactive.

The calculations for this centre, with D_{3d} point group symmetry, gives an A_{2u}
mode at 3216 cm^{-1} involving mainly the $^{12}C_s$ − H stretch mode, one E_u doublet
at 853 cm^{-1} corresponding to the C − H bend mode, and another E_u doublet at
666 cm^{-1} corresponding primarily to the stretching and bending of Si − C bonds.
These three modes can be easily related to the three LVMs observed. There are ten
combinations of partial or complete substitutions of ^{13}C and D to ^{12}C and H for
the $(C_s - H \cdots H - C_s)$ structure. The symmetry of the centre changes from D_{3d}
to C_{3v} for asymmetric substitutions and the number of modes increase. For some
isotopic combinations, a Fermi resonance occurs between the two E_u modes because
of the values of the isotopic shifts [136]. The calculated values of the frequencies of
the $(C_s - H \cdots H - C_s)$ modes for carbon and hydrogen isotopic substitutions are
compared with some values measured at LHeT in Table I of [136].

A LVM at 2967.34 cm^{-1} with a FWHM of 0.5 cm^{-1} measured at LHeT in a FZ
silicon crystal grown in a H$_2$ atmosphere due, from the above presentation, to the
A_{2u} mode of $(^{12}C_s - H \cdots H -^{12} C_s)$, was attributed incorrectly to a NH mode, on
the basis of frequency comparisons with LVMs in GaAs and GaP [56].

8.4.1.2 Nitrogen in III–V Compounds

In the LEC method, the III–V compound crystals are grown under gas pressure, and
most often, the gas used is nitrogen. The encapsulant used is wet boric oxide and the
water contained in the encapsulant is a source of hydrogen. Therefore, LEC-grown
materials are contaminated with both nitrogen and hydrogen, and these atoms can
eventually form complexes.

GaP

In GaP samples where the Fermi level is below $E_v + 0.3$ eV, i.e., in samples in which H diffuses as a proton, no LVMs which could be attributed to nitrogen–hydrogen complexes are detected.

In GaP samples with a Fermi level above $E_v + 0.3$ eV and not too high within the band gap, LVMs at 2885.2, 2054.1, and 1049.8 cm^{-1} were reported at LHeT under TEC and were associated with a ^{14}N–H bond [137, 138]. All these modes are correlated as their intensity ratios are independent of the sample and of the experimental conditions. Experiments under uniaxial stresses [139] have shown that the two LVMs at 2885.5 and 2054.1 cm^{-1} correspond to transitions between singlet states of oscillators aligned along trigonal axes of the samples whereas the mode at 1049.8 cm^{-1} corresponds to a transition between a singlet state and a doublet state of the same type of oscillators. These experiments have also shown no evidence for reorientation of the complexes at RT. These results led first to assume, improperly, that the transitions at 2885.5 and 2054.1 cm^{-1} were due to stretching modes of a complex containing two H atoms. However, experiments performed on samples containing both hydrogen and deuterium proved definitively that the complex contains only one hydrogen atom [13]. The frequencies of these modes and of their deuterated and/or ^{15}N counterparts are given in the second and third columns of Table 8.9. Therefore, the transitions at 2885.5 cm^{-1} and 1049.8 cm^{-1} correspond to the stretch and bending modes of the N−H bond of a (N, H) complex. The observation of a bending mode, together with the absence of reorientation of the complex at RT, suggests that the H is in an antibonding site. The 2054.1 cm^{-1} transition corresponds to an overtone of the mode at 1049.8 cm^{-1}. This overtone is about three times more intense than the fundamental mode, an *a priori* unexpected result.

In GaP samples where the Fermi level is in the upper part of the band gap, under TEC, only one mode is observed at 2729 cm^{-1} [137], with a deuterium counterpart at 2058.5 cm^{-1} [13]. It has to be noted that this mode can also be observed in the samples described in the previous paragraph after either long or complicated sequences of illumination of the samples [137]. Uniaxial-stress experiments have shown that the symmetry of the complex responsible for this mode is lower than trigonal [137, 139].

When illuminating with photons of proper wavelengths samples, containing H, D or both H and D and having Fermi levels above $E_V + 0.3$ eV, including in the upper part of the band gap, the intensities of the LVMs observed under TEC decrease and new LVMs listed in the columns with the heading "Photoinduced" of Table 8.9 appear [13, 137]. Uniaxial stress experiments show that the centre associated with these new LVMs has also trigonal symmetry.

All these experiments prove that the (N, hydrogen) complex exists in three states, two of them, at least, existing at equilibrium. This demonstrates also that this complex is electrically active. The conclusion is that in III–V compounds, complexing an electrically inactive isoelectronic impurity, N_{III}, with hydrogen can create an electrically active centre.

Table 8.9 Frequencies (cm^{-1}) and attributions of the nitrogen–hydrogen LVMs observed at LHeT in GaP and GaAs crystals slightly doped with N under different observation conditions. The indices s, b, and bo indicate stretch, bending, and bending mode overtone, respectively. Superscript n refers to LVMs observed when the Fermi level is in the upper part of the band gap of GaP [13,137,138]

| | GaP | | | | GaAs | | | |
| | Under TEC | | Photoinduced | | Under TEC | | Photoinduced | |
	H	D	H	D	H	D	H	D
(^{14}N–hydrogen)$_s$	2885.5	2161.9	2728.8	2058.5	2947.4	2202.0	2791.8	2102.1
(^{15}N–hydrogen)$_s$	2879.7	2153.4			2941.6		2786.3	
(^{14}N – hydrogen)$_b$	1049.8	771.2	1078.2		1010.1	741.9	1042.6	
(^{14}N – hydrogen)$_{bo}$	2054.1	1519.3	2102.2	1550.9	1984.3	1468.7	2042.5	
(^{15}N – hydrogen)$_{bo}$	2052.4				1983.4			
(^{14}N – hydrogen)$_s^n$	2712.9	2059.5						

Ab initio calculations were performed on this centre, which predicted that the (N, hydrogen) complex contains only one hydrogen atom [140]. They considered only the complex with hydrogen in *AB* location and introduced as a constraint in their calculations a trigonal symmetry for the complex. This constraint is not valid for one of the states of the complex, as mentioned earlier. Dixon et al. [140] found the complex to be electrically active and possibly existing in three different charge states: NH^+, NH^0, and NH^-. When increasing the electron number in the complex, these authors found the stretching mode wave number decreasing and the bending mode wave number increasing. This fits with what is experimentally observed for the two forms of the complex having trigonal symmetry.

GaAs

The situation in GaAs has several similarities with the one in GaP. No complexes are formed when hydrogen is in the H^+ state, i.e., when E_F is below $E_v + 0.5\,eV$. When E_F is above this value, the lines listed in Table 8.9 under the TEC heading are observed [137]. It has to be noted that the line at $2947\,cm^{-1}$ had been previously reported [141] and attributed to an $N - H$ stretch mode. Experiments under uniaxial stress show that the centres responsible for these lines are aligned along trigonal axes and that the line at $1010.1\,cm^{-1}$ is a bending mode of the complex.

Under illumination, the TEC lines decrease and a set of new lines listed in Table 8.9 under the heading "Photoinduced" is observed [137]. Experiments under stress show that the centre revealed by illumination has also trigonal symmetry.

Therefore, the (N, H) complex in GaAs has two states, both with trigonal symmetry. The third state observed in GaP seems to be absent in GaAs.

Calculations have been performed in the frame of the LDA by Kim and Chang [142]. They found that the (nitrogen, hydrogen) complexes should include only one hydrogen atom, in agreement with the experimental results in the samples containing both hydrogen and deuterium [13]. They found that the lowest energy configuration corresponds to bond-centred hydrogen between nitrogen and gallium atoms. This result contrasts with those obtained for GaP [140], but one has to note that the calculated energy difference between bond-centred and antibonding hydrogen configurations is very low [142].

8.4.1.3   III–$V_{1-x}N_x$ Dilute Alloys

$GaAs_{1-x}N_x$ and $GaP_{1-x}N_x$ with nitrogen concentration x in the percent range have raised quite a large interest as the introduction of nitrogen induces a drastic reduction of the band gap of the binary compounds. Moreover, it has been shown that hydrogenation of these alloys restores the GaAs or GaP band gaps [143, 144]. The frequencies of the LVMs observed in these alloys are given in Table 8.10, and

Table 8.10   Frequencies (cm^{-1}) at LHeT of the (N, H) LVMs in $GaP_{1-x}N_x$ and $GaAs_{1-x}N_x$ alloys [145]

	$GaP_{0.993}N_{0.007}$		$GaAs_{0.9913}N_{0.0087}$	
	H	D	H	D
Stretch 1	3302	2381	3195	2376
Stretch 2	2955	2211	2967	2217
Bending-1 overtone	2891	2150	2868	2137
Bending 1	1458	1082	1447	1076
Bending 2	1069		1068	
Bending 3			1057	798

one notes that none of the LVMs observed in the lightly doped samples (frequencies of Table 8.9) is observed in these samples [145].

These modes correspond to the vibrations of two types of N–H bonds in each compound. In samples containing both hydrogen and deuterium, extra lines appear at 2366, 2224, and $1062 \, cm^{-1}$ in $GaAs_{1-x}N_x$ alloys and at 2371 and $2218 \, cm^{-1}$ in $GaP_{1-x}N_x$ alloys. These extra lines have been interpreted as being due to the coupling of the vibrations of the two types of bonds and therefore it is deduced that the same centre contains two non-equivalent weakly coupled N–H bonds [145]. IR absorption measurements under uniaxial stress have been performed on these samples, and they have shown that the centre has C_{1h} symmetry [146]. They have also shown that under a [110] stress applied at temperatures around or above 40 K, the centre could reorient among its equivalent positions; this does not happen when [100] stress is applied, even at room temperature. A barrier height around 100 meV has been deduced from the experiments [146]. A model, in which the two hydrogen atoms are bonded to the same nitrogen atom has been proposed [145, 146].

Alt et al. [147] have presented results, on hydrogenated and deuterated $GaAs_{1-x}N_x$ and $Ga_{1-y}In_yAs_{1-x}N_x$ epilayers, that are difficult to reconcile with those described just above. They investigated a large number of samples and considered carefully the relations between the integrated intensities of the various transitions observed in the samples; they focused on deuterated samples. They found clear correlations between the three modes at 798, 1076, and $2376 \, cm^{-1}$ with constant integrated intensities ratio of about 2.4: 1: 0.26, but they also found that the mode at $2217 \, cm^{-1}$ was not correlated with those three modes, being even absent in several of the investigated samples. It was therefore suggested that there are two different complexes, each of them involving only one hydrogen–nitrogen bond [147]. The first complex would have one stretching mode at $3195 \, cm^{-1}$ (H) or $2376 \, cm^{-1}$ (D) and two bending modes at 1447 and $1057 \, cm^{-1}$ (H) or 1076 and $798 \, cm^{-1}$ (D) whereas the second one would have a stretching mode at $2967 \, cm^{-1}$ (H) or $2217 \, cm^{-1}$ (D) and a bending mode at $1046 \, cm^{-1}$ (H) or $767 \, cm^{-1}$ (D).

8.4.1.4 Discussion on (N, H) Complexes in III–V Compounds and Their Alloys

Several points related to the vibrational results presented earlier are surprising or unclear. One question is why are the centres observed in lightly N-doped materials (Sect. 8.4.1.2) different from those observed in alloys in which the nitrogen concentration is in the percent range (Sect. 8.4.1.3). Another one is how to reconcile the results of [145, 146] with those of [147] in dilute nitrides?

It might be worth reminding results obtained by photoluminescence in nitrogen containing GaP. On the one hand, Singh and Weber [148] reported that in GaP needles weakly doped with acceptor concentrations in the 10^{15}–10^{16} cm^{-3} range, a marked reduction of the intensity and a 25% decrease of the line width of the "isolated" nitrogen excitonic photoluminescence are observed after plasma hydrogenation. On the other hand, Mizuta et al. [149] reported that in more heavily p-type GaP samples ($\sim 2 \times 10^{18}$ cm^{-3}), the "isolated" nitrogen excitonic photoluminescence is strongly observed after ECR hydrogenation. Singh and Weber [148] also report that in VPE grown GaP layers heavily doped with nitrogen ($>10^{18}$ cm^{-3}), hydrogen binds preferentially to the close N–N pairs; the degree of passivation depends then drastically on the pair separation. These behaviours have been explained [150, 151] in terms of the state of the diffusing hydrogen: when the Fermi level is lower than $E_v + 0.3$ eV, hydrogen diffuses as H$^+$ that is trapped by acceptors, including close nitrogen pairs, whereas when the E_F is above $E_v + 0.3$ eV, hydrogen diffuses in an other state that allows the formation of complexes with "isolated" nitrogen. The same situation occurs in GaAs where the "boundary" between the two hydrogen states is at $E_v + 0.5$ eV [153, 154].

At the view of these photoluminescence results, one can understand why different LVMs are observed in lightly nitrogen doped GaP and GaAs and in GaAs$_{1-x}$N$_x$ and GaP$_{1-x}$N$_x$ alloys. In the lightly N-doped samples investigated in Sect. 8.4.1.2, E_F is sufficiently high in the band gap for the hydrogen state making complexes with "isolated nitrogen" to diffuse. Therefore, in these samples the complexes that are described result from the "isolated" nitrogen passivation, involving the binding of a single hydrogen atom to a single N atom. Most likely, the preferential binding of hydrogen to close nitrogen pairs, which has been observed by Singh and Weber [148] in GaP samples containing more than 10^{18} cm^{-3} nitrogen, also occurs in the GaAs$_{1-x}$N$_x$ and GaP$_{1-x}$N$_x$ alloys. This would make complexes resulting from the passivation of close N pairs by hydrogen privileged candidates for being responsible of the local modes described in Sect. 8.4.1.3. This possibility does not seem to have been considered until now in literature. It has to be noted that a complex resulting from the passivation of the closest N pair by a single hydrogen atom would have a C_{1h} symmetry corresponding to the symmetry determined by Wen et al. [146].

The results of [147] could easily be interpreted in the frame of a model of close N pair passivation: the two complexes suggested could correspond to the complexes resulting from the passivation of two different types of pairs by hydrogen. The results of [145] and [146] could also find their place in the frame of a model of passivation of close nitrogen pairs: such pairs could trap two hydrogen atoms that should not necessarily be bonded to the same N atom of the pair; the two N − hydrogen bonds would still be weakly coupled even though the two hydrogen atoms were bonded to the two N partners of the pair. Even though the results of [147] on the one hand and those of [145] and [146] on the other hand are independently compatible with a model of nitrogen pair passivation by hydrogen, they are not compatible with each other.

8.4.2 Interaction with C_i in Silicon

Interstitial C (C_i) represented in Fig. 4.8 has two undercoordinated atoms (C and Si). Between the migration temperatures of H_{BC} and of C_i, ($\sim$200 K and $\sim$300 K, respectively), H could be trapped by either of these two atoms to form a $(C_iH)_{<100>}$Si or $C_i(SiH)_{<100>}$ configuration of the C_iH centre, where the C − Si bond keeps an orientation close to <100>. *Ab initio* calculations have shown that the lowest energy structure is $(C_iH)_{<100>}$Si. Three charge states can exist for C_iH, and for the negative one, a *BC* configuration is predicted for C_i − H, similar to the one of O_i [127].

In Sect. 7.2, the spectroscopic properties of complexes with interstitial atoms produced by irradiation with high-energy particles have been discussed and we extend this discussion to their interaction with hydrogen.

In C-containing hydrogenated FZ silicon and in CZ silicon, after electron or neutron irradiation and annealing in the 400°C–600°C range, or after heat treatment in the 450°C range, a ZPL at 935.1 meV (7542 cm^{-1}) has been observed by photoluminescence [155–158]. This line was first reported with label T [155], and it is known as the T-line. In PL, it is attributed to the recombination of an exciton bound to a neutral centre (IBE), with the electron bound to the centre into a deep-level state, and the hole weakly bound in the Coulomb potential produced by the resulting ion. A high-energy satellite of the T-line, shifted by 1.75 meV (14.1 cm^{-1}) is associated with an excited state of the weakly bound particle [156, 157]. The splitting of this line under a magnetic field and under uniaxial stress has allowed establishing the monoclinic-I symmetry of the related centre, and the use of samples enriched with ^{13}C and deuterium has established that this centre contained carbon and hydrogen.

The frequency of the T-line in ^{nat}Si is 935.11 meV (7542.2 cm^{-1}) and in qmi ^{30}Si, it becomes 936.88 meV (7556.4 cm^{-1}) [158]. This line is relatively sharp, with a FWHM of 0.04 meV ($\sim$0.3 cm^{-1}) at LHeT. The presence of [13]C or [14]C in the sample produces high-frequency satellites shifted by 0.07 or 0.14 meV ($\sim$0.6 or 1.13 cm^{-1}), respectively [156, 159]. Similarly, replacing H by D produces a high-frequency shift of the line by 0.67 meV (5.4 cm^{-1}) [159].

This ZPL is accompanied by vibronic lines at lower energies, noted L_2, L_3, L_4, and L_5, associated with vibrational modes of the centre, and the isotopic structure of line L_5 in samples containing ^{12}C and ^{13}C shows that the centre contains two non-equivalent C atoms [157]. The structural model proposed for this centre can be described starting from the C_i model of Fig. 4.8: the Si_3 and C_3 atoms of the C_i model are replaced by two C atoms relabelled C_3, taken as C_s, and C_4, taken as C_i. This C_i atom binds a H atom along a direction close to $[\bar{1}10]$, producing a small disorientation of the main axis of the defect from [001], but the centre is still labelled $C_s(C_iH)_{<100>}$ [127]. In this configuration, C_i becomes fourfold coordinated and the two C atoms are nearest neighbours, at a difference with the C_iSiC_s structure of Fig. 7.11.

The frequencies of the LVMs associated with $^{12}C_s(^{12}C_iH)$ deduced from the energies of the vibronic lines (in brackets) are: 1056.0 (L_5), 796.0 (L_4), 567.5 (L_3), and 531.5 cm^{-1} (L_2). In ^{13}C-containing samples, the LVM at 1056 cm^{-1} gives additional components at 1038, 1029, and 1011 cm^{-1}. It is thus ascribed to a C − C stretch mode between two non-equivalent C atoms, and this is an *a posteriori* justification of our preceding statement. The calculated frequencies of these LVMs for $^{12}C_s(^{12}C_iH)$ are 1097.8, 743.6, 558.0, and 542.4 cm^{-1}, respectively. The mode at 744 cm^{-1} is attributed from the calculations to a C−Si stretch mode with little amplitude on atom C_4. In addition, the *ab initio* calculations predict also LVMs at 2913.6 and 1180.4 cm^{-1} (the stretch and bending modes of C − H), corresponding to distant vibronic replicas. The comparison between the calculated frequencies and those deduced from the differences between the T-line and its vibronic replicas for the different hydrogen and C isotopic combinations is shown in Table IX of [127].

The interstitial bond-centred O_i configuration of oxygen in silicon, germanium, GaAs and GaP is electrically inactive. O_i does not seem to interact directly with hydrogen, but evidence of indirect interaction was given in the section on molecular hydrogen.

8.5   Extrinsic Hydrogen Centres II

In this section are discussed the vibrational properties of the hydrogen centres passivating the electrical properties of the shallow donors and acceptors in semiconductors. The conditions for the formation of these centres and for their dissociation are reviewed in [8].

8.5.1   *(Acceptor, Hydrogen) Centres*

8.5.1.1   Elemental Semiconductors

As hydrogen diffuses in p-type silicon as a proton, it is expected to interact with negative acceptor (A) ions to form some kind of electrically neutral (A, hydrogen) centre. The vibrational spectroscopy of pairs involving Li^+ and B_s^-, where the pair can be seen as the result of close compensation, has been presented in Sect. 5.3.4.1, and such a mechanism could also be invoked for hydrogen. However, the first IR experiments at RT on hydrogenated B-doped silicon samples showed LVMs near 1870 (1360) cm^{-1} for H (D), which could be attributed to a stretch frequency involving a hydrogen atom, indicating that the interaction between hydrogen and the acceptor was of a different type [160, 161]. A conclusion reached from this result was that the hydrogen atom was bonded to a Si atom and located between the Si and B_s atoms. In this picture, there is no dangling bond, and therefore no electrical activity, in agreement with the passivating property of the centre [160]. Calculations performed on the basis of a linear ($Si \cdots H \cdots acceptor$) model confirmed that in the case of boron, the binding of the H atom was predominantly with the Si atom, while for aluminium, the coupling constants with the Si and Al atoms were comparable [162]. These calculations yielded for ($Si \cdots H \cdots B$) a frequency of 1880 cm^{-1}, in good agreement with experiment, and for ($Si \cdots H \cdots Al$) a predicted frequency of 2220 cm^{-1}, in reasonable agreement with the value of Table 8.11.

Extensive results of the absorption between LHeT and RT of the hydrogen stretch modes related to the passivation of the B, Al, and Ga acceptors in silicon have been reported by Stavola et al. [163]. The frequencies of these modes, which can slightly differ as a function of the hydrogenation method, are given at LHeT in Table 8.11.

The similarity and the values of the frequencies for the (Al, hydrogen) and (Ga, hydrogen) centres confirm a Si – H bonding. At LHeT, the FWHMs of the (B, H) and (B, D) modes are $\sim$4 and 3 cm^{-1}, respectively [164, 165], but larger values have been reported ($\sim$9 and 7.5 cm^{-1}, respectively) in plasma-passivated highly doped samples [163]. A value of the FWHM of 9.2 cm^{-1} has been reported for (Al, H) mode [163]. These values of the FWHMs are too large to allow the observation of the Si IS of a Si – H mode. When measured at increasing temperatures, the (B, H) and (B, D) modes are observed to shift to lower frequencies, reaching 1870 and 1363 cm^{-1}, respectively, at RT. In the case of the (Al, H) and (Ga, H) modes, increasing the temperature reveals also the existence of a low-frequency sideband which can be seen in Fig. 8.16 for (Al, H).

Similar sidebands also occur for the (Al, D) and (Ga, D) centres. They indicate the existence of a low-energy excitation above the vibrational ground state, and its observation when H is replaced by D implies that this excitation involves the hydrogen atom. By extending the Lorentzian fit of Fig. 8.16 to the absorptions

Table 8.11 Frequencies (cm^{-1}) of the stretch modes of the (A, hydrogen) centres in silicon at LHeT [163]. The ratio r of the H-related/D-related frequencies is significantly different for B compared to Al and Ga

Acceptor	B	Al	Ga
(A, H)	1903 (1904.8[a])	2201	2171
(A, D)	1390 (1390.7[a])	1597	1577
r	1.370	1.378	1.377

[a][164]

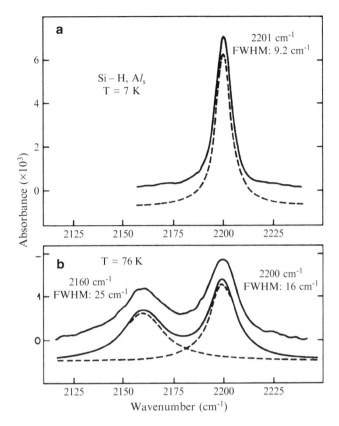

Fig. 8.16 Measured absorption (solid line) of the (A*l*, H) mode in silicon at LHeT (**a**) and at 76 K (**b**) fitted to Lorentzian line shapes (dashed lines). The peak frequencies and FWHMs for the fits are indicated [163]. Copyright 1988 by the American Physical Society

measured at other temperatures, activation energies of 78 and 56 cm^{-1} for the low-energy excitations were deduced for (A*l*, H) and (A*l*, D), respectively, in the vibrational ground state. Because of the larger size of the A*l* and Ga atoms, one possibility to explain this excitation is a low-frequency motion of the hydrogen atom perpendicular to the Si$\cdots$A*l*$_s$ or Si$\cdots$Ga$_s$ axis, similar to the one made possible for O$_i$ in silicon by its static configuration of Fig. 6.1 [163]. Such a configuration would certainly reduce the interaction of the hydrogen atom with the acceptor atom and explain the larger frequency of the Si − H mode in the centres involving A*l* and Ga.

By comparison, results of piezospectroscopic measurements at LHeT on the Si − H, B$_s$ stretch mode confirmed the *BC* location of the H atom, with trigonal symmetry, as can be judged from Fig. 8.17a, where only one component is seen for *F*// [001], for which all the <111>-oriented Si − H bonds are equivalent under stress [166]. More intriguing are the piezospectroscopic results obtained on the

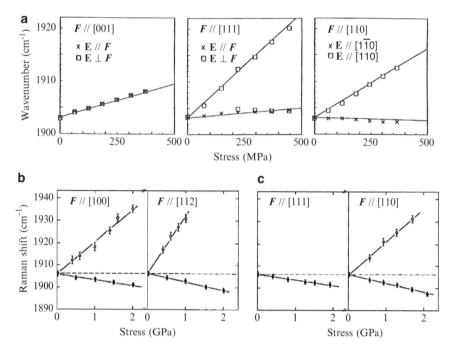

Fig. 8.17 Splitting under uniaxial stress of the Si − H, B_s stretch mode in H-passivated B-doped silicon measured by (**a**) absorption at LHeT [166], and (**b**) and (**c**) Raman scattering at 100 K [167]. Copyright 1988 and 1991 by the American Physical Society

same centre by Raman scattering at 100 K, showing that time two components for $F//$ [001], as seen in Fig. 8.17b.

To explain the non-trigonal symmetry revealed at 100 K, one has to assume some off-axis position of the H atom at higher temperature, with a corresponding lowering of the symmetry. This point is discussed in the paper by Herrero et al. [167].

An estimation of the strength of the interaction between the H and B atoms of the (B, hydrogen) centre was obtained from the B IS of the Si − H, B and Si − D, B stretch modes measured after hydrogenation of silicon samples implanted with ^{10}B and ^{11}B [164]. A small IS of 0.8 cm^{-1} was observed between Si − H, ^{10}B and Si − H, ^{11}B, demonstrating a weak interaction between B and H in this centre. An unexpected result was, however, the increase of this ^{10}B–^{11}B IS to 3.3 cm^{-1} for the D equivalents, as shown in Fig. 8.18.

Besides the Si − hydrogen, B stretch mode, the Raman scattering of a vibrational mode of B_s perturbed by hydrogen has also been reported at 652 (680) cm^{-1} for ^{11}B (^{10}B) in hydrogen-passivated B-doped silicon [167, 168]. This perturbed mode of B_s corresponds to *IR* E of C_{3v}, since its motion is predominantly 2D in the plane perpendicular to the <111> defect axis [169]. The overtone of this mode, is expected at ∼1305 and (1360) cm^{-1} for ^{11}B (^{10}B). While the frequencies of this overtone for ^{11}B and that of the Si − D, B_s stretch mode (1390 cm^{-1}) are relatively

Fig. 8.18 Absorption of the Si— hydrogen stretch mode of the (B, hydrogen) passivating centre in silicon samples implanted with ^{10}B and ^{11}B. It shows the difference between the B isotope shifts for the Si − H, B and Si − D, B stretch modes: (**a**) Si − H, B stretch mode, (**b**) Si − D, B stretch mode [164]. Copyright 1988 with permission from Elsevier

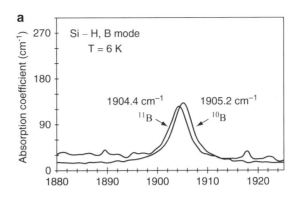

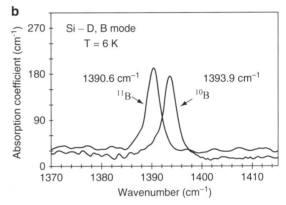

distant, this is not the case for the ^{10}B overtone, and a Fermi resonance, already mentioned in Sect. 5.1, can occur between this latter overtone and the Si − D, B_s stretch mode, with *IR* A_1 of C_{3v}. This is the explanation put forward by Watkins et al. to explain the relatively large ^{10}B/^{11}B IS of the Si − D, B_s stretch mode compared to Si − H, B_s (see Fig. 8.18) [169]. This explanation is very convincing as the absorption of this B_s overtones has indeed been observed for ^{10}B (^{11}B) at 1347.8 (1298.0) cm^{-1} [169]. A self-consistent calculation based on a model of a pair of harmonic oscillators coupled anharmonically has been used to calculate the B IS of the Si − D, B_s stretch mode. The fact that this simple model can be used to fit the experimental results validates the existence of a Fermi resonance between the B_s and Si − D, B modes as an explanation of the anomalous IS observed [169].

The thermal stability of the (A, H) centres in silicon depends on the chemical nature of the acceptor. The (B, H) centre is stable up to ∼160°C [163, 167] while (Ga, H) dissociates around ∼180°C. The most stable centre is (A*l*, H), which starts dissociating near 280°C [163]. A calibration factor of 3×10^{15} cm^{-1} has been derived for the *IA* at LHeT of the LVM of the (B, H) centre at 1903 cm^{-1}, implying an apparent effective charge η of the stretch mode of 1.5*e* [165].

Table 8.12 Frequencies of the LVMs calculated to be associated with the (B_1H_1) centre in ^{12}C diamond [173]. The frequencies in brackets are close to the Raman frequency of diamond, and they are not so reliable as the others. (a) are B-related modes with A'' symmetry, while (b) and (c) are H/D-related modes with A' symmetry

Isotopic combination	Frequency (cm^{-1})		
	(a)	(b)	(c)
$^{11}B - H$	1376	1501	2657
$^{11}B - D$	1376	(1366)	1965
$^{10}B - H$	1399	1511	2668
$^{10}B - D$	1399	(1366)	1984

In B-doped epitaxial diamond films, it has been shown that the electronic acceptor lines of boron in the $2800\,cm^{-1}$ ($350\,meV$) spectral region are suppressed by hydrogenation, showing interaction between boron and hydrogen [170]. Electrical passivation of boron by deuterium has also been reported from capacitance-voltage measurements of synthetic type IIb diamond samples [171, 173]. The atomic structure of (B, H) centres has been investigated by *ab initio* methods. The structure retained is the one with an H atom in a puckered location between B_s and a C atom, with a bonding between the H atom and the B_s and C atoms. This centre is noted B_1H_1 in [172]. In this structure, there is also a weak direct bonding between the B_s and C atoms ([172], and references therein). Vibrational modes of this centre, with C_{1h} symmetry, have been calculated. The one with the lowest frequency is localized on the B_s atom and the other ones correspond to stretch and wag modes of boron-hydrogen. They are given in Table 8.12.

Other structures involving two nn B_s atoms and one or two H atoms (the B_2H_1 and B_2H_2 centres, respectively) have also been considered [173], and their frequencies are given in Tables III and IV of this reference. The passivation of boron by hydrogen in diamond is presented in a review of theoretical models related to diamond doping [172]. Up to now, no vibrational mode associated to (B, hydrogen) centre of any kind has been observed in diamond.

No hydrogen passivation of shallow acceptors has been observed in germanium under usual passivation treatments and this seems to be related to the difficulty for hydrogen to diffuse appreciably in the bulk of this material [174]. The difference between the passivation by hydrogen in silicon and germanium has been discussed by Estreicher and Maric [175].

8.5.1.2   Compound Semiconductors

III–V Compounds

Evidence for the electrical passivation of the electrical activity of shallow acceptors by hydrogen in GaAs was first provided by Johnson et al. [176]. In III–V compounds with sphalerite structure, as a rule, elements of group IIA and IIB atoms Be, Mg,

Zn, and Cd locate on the cation sublattice where they are single acceptors. Similarly, elements of group IVB and VB on the anion sublattice are single acceptors in III–V and II–VI compounds, respectively (see Appendix D). In GaP, AlAs, and GaAs, C sits on the anion sublattice while in InP, it sits preferentially on the cation sublattice, where it acts as a donor. However, evidence of the presence of C on P sites (C_P) in low-temperature MOVPE InP has been obtained by vibrational spectroscopy. For most of these centres, only the vibration of the hydrogen stretch mode is observed, but for some of them, a bending mode can also be observed at lower frequency. Most of these acceptors build passivating centres with hydrogen, in the same way as group III acceptors do in silicon, where the hydrogen atom is bond centred between the acceptor and an *nn* lattice atom. Correlations between frequencies of the different centres and the occasional observation of isotope effects have shown that for electropositive acceptors located on the cation sublattice, the hydrogen atom binds to the more electronegative lattice atom. Inversely, for the more electronegative acceptors on the anion site, the hydrogen atom binds directly to the acceptor. These situations are shown in Fig. 8.19.

These centres display a trigonal C_{3v} symmetry along a <111> axis. A consequence of the binding preferences of hydrogen is that in III–V compounds, for the group IV acceptors which are located in the group V sublattice (we note them A_V), the H stretch mode is $A_V - H_{BC}$. The frequency of this mode is thus expected to be approximately the same in different compounds. Inversely, for the A_{III} acceptors, the H atom is bonded to a group V atom and in a given compound, its frequency will be weakly dependent on the chemical nature of acceptor A_{III}. This can be seen in Table 8.13. In this table, the *BC* index indicating the hydrogen location has been dropped.

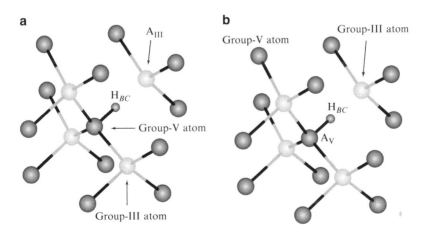

Fig. 8.19 Ball-and-sticks models of the (acceptor, hydrogen) centres in III–V compounds with sphalerite structure. (**a**) Group II acceptor on an atom III site, noted A_{III}. (**b**) Group IV acceptor on an atom V site, noted A_V. These models should also apply, *mutatis mutandis*, to the II–VI compounds with the sphalerite structure

Table 8.13 Measured frequencies ω (cm^{-1}) and FWHM ($\Delta\tilde{\nu}$) (cm^{-1}) at LHeT of the stretch mode of (A$_{III}$, H) and (A$_V$, H) passivating centres in different III–V compounds. Factor r is the ratio of the (X, H) and (X, D) frequencies

	GaAs		AlAs		GaP		InP	
	ω	$\Delta\tilde{\nu}$	ω	$\Delta\tilde{\nu}$	ω	$\Delta\tilde{\nu}$	ω	$\Delta\tilde{\nu}$
(^{12}C$_V$, H)	2635.14‡	0.25	2558.1^{c†}	~8.5	2660.2^d	2.4	2703.3^f	
(^{13}C$_V$, H)	2628.38		2549.7		2652.6		2696.6	
(^{12}C$_V$, D)	1968.55	0.5	1902.6	~5.4	1980.8			
r	1.3386		1.3445		1.3430			
(Si$_V$, H)	2094.7	4.4						
(Si$_V$, D)	1514.4	2.7						
r	1.3832							
(Ge$_V$, H)	2010.3	5.5						
(Ge$_V$, D)	1447.4	3.1						
r	1.3889							
(Be$_{III}$, H)	2036.9^a	3.6			2292.2^c	2.7	2236.49^g	0.43
(Be$_{III}$, D)	1471.2	1.2			1669.8	0.8	1630.85	0.2
r	1.3845				1.3727		1.3714	
(Zn$_{III}$, H)	2146.94^b	1.8			2379.0^c	1.1	2287.71^g	0.23
(Zn$_{III}$, D)	1549.0	0.9			1729.4	0.5	1664.52	0.08
r	1.3860				1.3756		1.3744	
(Cd$_{III}$, H)	2206.72^b	1.3			2434.0^c	1.2	2332.42^g	0.12
(Cd$_{III}$, D)	1591.9	0.3			1768.3	0.6	1695.40	0.10
r	1.3862				1.3765		1.3757	

[a][177]
[b][178]
[c][179]
[d][152]
[e][181]
[f][182]
[g][70]
$\ddagger$Small changes in the frequencies occur for larger values of [C]
$\dagger$In [183], the frequency of (^{12}C$_{As}$, H) in AlAs is given as 2555.4 cm^{-1} for H-plasma passivated sample

In a hydrogenated C-containing AlSb sample, a LVM at 2566.6 cm^{-1} with an extrapolated FWHM ~0.6 cm^{-1} has been reported at LHeT [184]. This LVM, whose frequency is close to the one of (^{12}C, H) in AlAs has been attributed to the stretch mode of (^{12}C$_{Sb}$ − H) [184]. Vibrational modes of (C, hydrogen) centres have been detected by absorption and Raman scattering in C-doped InAs [185].

Calibration factors of the integrated absorption of the stretch modes of (Zn, H) in GaAs and InP for the passivated Zn concentration have been determined [186]. They are $(4.0 \pm 0.2) \times 10^{16}$ cm^{-1} and $(2.0 \pm 0.4) \times 10^{16}$ cm^{-1} for the 2147 and 2288 cm^{-1} LVMs in GaAs and InP, respectively.

The frequency of the stretch mode of the (A_{III}, H) and (A_V, H) centres displays most often a negative shift with temperature. For the C–H complexes, the shift is positive; at RT, it is about $+9\,\mathrm{cm}^{-1}$ for (C_{As}, H) in GaAs and $+4\,\mathrm{cm}^{-1}$ for (C_P, H) in GaP [187]. For InP, the largest negative shift ($\sim 7.5\,\mathrm{cm}^{-1}$ at 140 K) is observed for (Zn_{In}, H) [70].

As a consequence of the trigonal symmetry, no splitting of the hydrogen stretch mode of these (A, hydrogen) centres, whose dipole moments are oriented along $<111>\,C_3$ axes, is expected under a uniaxial stress along a $<100>$ axis (see Fig. 8.17a). This has been checked for (Be_{Ga}, H) in GaAs [188] (C_{As}, H) in GaAs [187] and for (Zn_{In}, H) in InP [70].

In III–V alloys, even if the local symmetry of the (acceptor, hydrogen) centre remains unchanged, the presence of *nn* atoms different from those in the binary compound produces shifts of the vibrational lines of the centre as a function of the modified environments. For instance, the change of the absorption profile of the (C_{As}, H) and (C_{As}, D) stretch modes in hydrogenated $Al_xGa_{1-x}As$ alloys has been investigated for eight values of x between 0.01 and 0.95 by Pritchard et al. [180]. Since the C_{As} − Al bond is shorter and stronger than the C_{As} − Ga bond, it will cost less energy to break a C_{As} − Ga bond to insert hydrogen. As a consequence, for increasing Al concentrations, stretch modes are observed which correspond to combinations with an unchanged C_{As}–H, Ga core, up to $x = 0.33$, containing up to 3 Al atoms *nn* of C_{As}. There is a gap in the values of x investigated, and it is for the following value of x (0.65) that a stretch mode can be attributed to $Al_3 \equiv C_{As}$–H, Al, a configuration similar to the one in AlAs. The frequencies of the (C_{As}, H) and (C_{As}, D) stretch modes observed for the different samples are given in Table 8.14.

While group IIA atom Mg is an interstitial double donor in silicon and germanium, it is a substitutional A_{III} acceptor in III–V compounds. What was thought to be the passivation efficiency of the Mg acceptor by hydrogen in GaAs was found to be small compared to that of other group II acceptors [178], but more efficient passivation was later reported [189]. In hydrogenated (deuterated) Mg-doped GaAs epilayers, the observation at LHeT of a LVM at 2144.0 (1547.0) cm^{-1}

Table 8.14 Frequencies (cm^{-1}) of the (C_{As}, H) and (C_{As}, D) stretch modes with $n = 0, 1, 2, 3$, and 4 Al atoms *nn* of the C_{As} atom measured at LHeT in $Al_xGa_{1-x}As$ alloys for different values of x. The frequencies of the (C_{As}, D) stretch mode are given in brackets (after [180])

$x = 0$	0.01	0.02	0.055	0.24	0.33	0.65	0.95	1	n
						2559	2558	2558	4
						(1906)	(1905)	(1903)	
				2608	2608	2606	2605		3
				(1946)	(1946)	(1944)	(1943)		
		2617	2619	2617	2616	2615			2
		(1952)	(1954)	(1952)	(1952)	(1952)			
	2626	2626	2628	2625	2625				1
	(1960)	(1960)	(1962)	(1960)					
2636	2636	2635	2636						0
(1969)	(1969)	(1969)	(1970)						

with a FWHM of 0.2 (0.1) cm^{-1} has been reported [178, 186]. The frequencies and r factor of 1.3859 of these LVMs led to their attribution to the As $-$ H (As $-$ D) stretch mode of a passivating centre. The 2144 cm^{-1} was, however, not related to Mg passivation because a LVM at the very same frequency was observed, together with the (Zn, H) LVM, in a H-passivated GaAs:Zn sample where Mg was absent (see Fig. 16 of [186]). The increase of the FWHM of the 2144 cm^{-1} LVM with temperature was also found to be moderate compared to the one of the (Zn, H) mode [190]. LVMs with frequencies very close to 2144 cm^{-1} have been later reported [191] in hydrogenated and deuterated $Ga_{1-x}Mn_x$ As alloys with $x = 0.034$ and 0.056, with FWHM displaying the same temperature dependence as the 2144 cm^{-1} LVM. From these results, it seems clear now that the 2144 cm^{-1} LVM must be associated with a (Mn_{Ga}, H) passivating centre, and that its observation in the GaAs:Mg and GaAs:Zn epilayers was due to the presence of Mn in the GaAs substrates. More details on the Mn passivation in GaAs are given in Sect. 8.6. The conclusion is that up to now, no LVM associated to a (Mg, hydrogen) centre has been detected in GaAs.

In hydrogenated InP:Mg, LVMs at 2366.4 cm^{-1} (H) and 1720.3 cm^{-1} (D) have been reported which are not observed in hydrogenated InP:Zn, and they have been attributed to the usual (A_{III}, hydrogen) H_{BC} configuration [190]. The results of *ab initio* calculations of the frequencies of the (Mg, hydrogen) centre in InP in the *AB* and *BC* configurations have given frequencies higher than the experimental ones for both configurations, but the frequencies for the *BC* configuration are closer to the experimental ones [190].

In $w -$ GaN, calculations predicted the H atom to be located antibonded to a N atom *nn* of Mg, with a frequency of 3360 cm^{-1} [192]. The measured stretch frequencies of the (Mg, H) and (Mg, D) centres at LHeT in $w -$ GaN (unless otherwise specified, we drop from now $w-$) are 3125 and 2321 cm^{-1}, respectively, and they correspond to N $-$ H and N $-$ D stretch modes [193]. In this experiment, the conditions were such that only bonds vibrating in a plane perpendicular to the c-axis could be detected.

Experiments with polarized radiation at non-normal incidence on the mode at 3125 cm^{-1} have been performed on epitaxial GaN samples deposited on Al_2O_3 (sapphire) substrates with the c-axis perpendicular to the surface [85]. They confirmed that the (Mg, H) centre was not aligned along the c-axis. It was suggested that the N $-$ H bond associated with the H_{AB} atom of the (Mg, H) centre made an angle of $\sim130°$ with respect to the c-axis [85], but it was shown later that the analysis leading to this value of the angle neglected the reflections at the $Al_2O_3/$GaN interface so that the conclusions derived were incorrect [194]. Further experiments with layers of various thicknesses were performed (L. Siozade, C. Naud, B. Clerjaud, unpublished results, 2003); they confirmed that the value of the angle was dependent upon the layer thickness as suggested by Seager [194], but their results did not agree either with the thickness dependence calculated by Seager [194]. Therefore, in the present state of the art, in GaN:Mg,H, one cannot derive the angle between the N $-$ H bond and the c-axis from experiments performed with polarized radiation. This particular off-axis (OA) configuration of the N $-$ H bond was confirmed by calculations

Table 8.15 Frequencies ω (cm^{-1}) of the LVMs of (C$_{As}$, hydrogen) in AlAs, GaAs and InAs measured by IR absorption or Raman scattering at LHeT (25 K for InAs) and calculated *ab initio*. In [179], the low-frequency modes of the (C, H) centres are denoted X, Y$_1$ and Y$_2$ and the correspondence is indicated in the first column.

Mode			^{12}C-H	^{13}C-H	^{12}C-D		^{13}C-D
			ω	ω	ω	r(H/D)	ω
A$_1$H stretch	AlAs	Exp.[a]	2558.1	2549.7	1902.6	1.3445	1894.4
		Calc.[a]	2885	2877	2111		2100
	GaAs	Exp.[b]	2635.2	2628.5	1968.6	1.3386	1958.3
		Calc.[c]	2950	2942	2154		2144
	InAs	Exp.[d]	2686.5	2678.8	ND		ND
		Calc.[e]	2745	2738	–		–
A$_1$(X)	AlAs	Exp.[a]	487.0	477.2	479.8		471.2
		Calc.[a]	466	453	454		442
	GaAs	Exp.[b]	452.7	437.8	440.2		426.9
		Calc.[c]	456	440	442		428
	InAs	Exp.[d]	393.2	379.7	ND		ND
		Calc.[e]	394	381	–		–
E$^-$ H-like (Y$_2$)	AlAs	Exp.[a]	670.8	652.9	656.6		635.3
		Calc.[a]	740	725	684		662
	GaAs	Exp.[b,†]	738.6	~730	638.6		618.1
		Calc.[c]	888	883	707		693
	InAs	Exp.[d]	ND	ND	ND		ND
		Calc.[e]	689	685	–		–
E$^+$ C-like (Y$_1$)	AlAs	Exp.[a]	ND	ND	ND		ND
		Calc.[a]	559	551	437		436
	GaAs	Exp.[b]	562.6	547.6	466.2		463.8
		Calc.[c]	553	536	495		487
	InAs	Exp.[d]	518.2	502.8	ND		ND
		Calc.[e]	502	486	–		–

[a][179]
[b][200]
[c][198]
[d][185]
[e][199]
†Raman value

[195, 196] and by the interpretation of measurements on the 3125 cm^{-1} mode under hydrostatic pressure [197].

When considering a quasi-molecular model of the (A$_V$ − H$_{BC}$, Group III atom) centre of Fig. 8.19b with C_{3v} symmetry, four modes are expected: a stretch mode of the H bond along the <111> axis, with symmetry A$_1$, a doubly degenerate bending mode of the same bond in a plane perpendicular to <111> axis, with symmetry E$^-$, plus A$_1$ and E modes mainly associated with the vibration of the A$_V$ atom [179]. Beside the stretch mode, other LVMs of the (C$_V$, hydrogen) centres have been observed in GaAs, AlAs and InAs. Their measured frequencies are

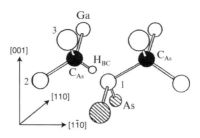

Fig. 8.20 Atomic structure of the $(C_{As}-H, C_{As})^-$ acceptor centre in GaAs whose $C_{As}-H$ stretch mode vibrates at 2688 cm^{-1}. The weak bonding between H_{BC} and Ga atom 1 is not represented (after [203]). The C_{As} atoms, H_{BC}, and Ga atom 1 lie in a (110) plane, and the C_{As} atoms are aligned along the [1$\bar{1}$0] axis

given in Table 8.15, where they are compared with those predicted from *ab initio* calculations. The similarity of the frequencies of the C–H modes in the three compounds show that in InAs, C locates also on an anion (As) site.

In hydrogenated GaAs:Be, in addition to the As − H stretch mode of the (Be$_{Ga}$, hydrogen) centre, LVMs at 555.7 and 553.6 cm^{-1} related to (Be$_{Ga}$, H) and (Be$_{Ga}$, D), respectively, are observed. The small dependence of the frequency of this LVM on the nature of the hydrogen isotope had led to its attribution to the E$^-$ mode of Be$_{Ga}$ weakly perturbed by the hydrogen atom [177].

In highly C-doped GaAs epilayers grown by decomposition of trimethylgallium with either arsine or trimethylarsenic, LVMs at 2643.1, 2650.6, and 2688.4 cm^{-1} were observed at LHeT, with deuterium counterparts at 1974.6, 1980.3, and 2007.7 cm^{-1}, respectively [201]. The (H)/(D) frequency ratios r of these LVMs are 1.3385, 1.3385, and 1.3390, respectively, close to the one (1.3386) of isolated (C_{As}, H) at 2635.2 cm^{-1} discussed before. In samples with a (001)-oriented growth surface, the LVM at 2688 cm^{-1} shows a spontaneous dichroism of the absorption, which is much larger for **E**//[1$\bar{1}$0] than for **E**//[110] [203]. This has been related to the fact that, at the growth surface, the kinetic of layer-by-layer growth is different along the [110] and [$\bar{1}$10] directions. Thus, the surface concentration of defects differs along these directions, and the difference is maintained in the bulk. The 2688 cm^{-1} LVM has been attributed to the C_{As} − H stretch mode of a $(C_{As}$ − H, $C_{As})$ acceptor centre including an *nnn* C_{As} pair and a bond-centred H atom. This particular centre dissociates around 600°C [203]. The proposed structure of this centre is shown in Fig. 8.20.

Ab initio calculations of this centre have shown that the structures where the H atom is located between C_{As} and Ga atom 2 or 3 of Fig. 8.20 are less stable by 0.5 or 0.9 eV, respectively, than the one where it is located between C_{As} and Ga atom 1, as in Fig. 8.20 [204]. Another LVM at 2729 cm^{-1} has been reported in samples where the 2688 cm^{-1} LVM is observed, but it is polarized along the [110] direction [205].

After annealing of samples showing the 2688 cm^{-1} LVM in a H plasma at 320°C, this LVM disappears and two LVMs, showing no polarization effects, appear at 2725 and 2775 cm^{-1}. A subsequent annealing at 450°C removes these new LVMs and the 2688 cm^{-1} LVM reappears. These LVMs were tentatively attributed to complexes involving two C_{As} and two H atoms ([204, 206], and references therein). Dichroism of other (C_{As}, H) complexes has been also reported in GaAs epilayer heavily C-doped [206]. In this reference, a LVM at 576 cm^{-1}, polarized with $E//[110]$ using the orientations[12] defined in Fig. 8.20, is reported to be consistently observed when the 2688 cm^{-1} LVM or its D counterpart is observed. This low-frequency LVM is ascribed to the transverse mode of the unpaired C_{As} atom of the (C_{As} − H, C_{As}) centre. In InAs containing a high C concentration, a LVM at 2757 cm^{-1} has been attributed to a dicarbon-hydrogen pair with a structure similar to the one in GaAs [185].

Four estimates of the calibration factor of the IA at LHeT of the $^{12}C_{As}$ − H stretch mode at 2635.2 cm^{-1} in GaAs have been reported. The most recent one is $(9 \pm 2) \times 10^{15}$ cm^{-1} [207].

II–VI Compounds

In II–VI compounds, evidence for some passivation of the P_{Te} and As_{Te} acceptors in MOCVD as-grown As-doped CdTe and P- and As-doped ZnTe samples came from the observation of H-related LVMs related to the stretching of P_{Te} − H and As_{Te} − H bonds [208]. In this study, reactivation of the passivated As acceptors in CdTe was observed for annealing temperatures starting at 250°C. In ZnSe, N_{Se} is an acceptor with an ionization energy of ∼0.1 eV, about 4% of the band gap value and it can be considered as a shallow acceptor. It has been suspected that the difficulty to achieve large N acceptor concentrations in ZnSe was due to the formation of passivating (N_{Se}, H) centres during the growth-doping process and this has been confirmed by the observation in this material of a N − H stretch mode and of a mode at 783 cm^{-1}, tentatively assigned to a bending mode of N − H [209]. The same effect occurred in ZnSe for the As acceptor, and the observation of the stretch mode of As_{Se} − H has also been reported together with the one of As_{Se} − D when replacing H_2 by D_2 as the carrier gas during the MOCVD growth [211]. In ZnTe, LVMs due to stretch modes of N_{Te} − H and N_{Te} − D bonds have been observed in plasma-hydrogenated and -deuterated ZnTe epitaxial samples [212]. In CdTe, it has been shown that exciton and donor-acceptor N-related luminescence are absent in N-doped samples grown by photon-assisted MBE under atomic hydrogen [213]. In these samples, LVMs at 3210 and 749 cm^{-1}, due to the stretching and bending modes of (N_{Te},H) centres have been observed [213]. The frequencies of the LVMs reported for the stretch modes of the (A_{VI}, hydrogen) centres in some II–VI compounds are given in Table 8.16.

---

[12]In [206], the [110] and [$\bar{1}$10] directions are inverted with respect to those of Fig. 8.20.

Table 8.16 Measured frequencies (cm^{-1}) at LHeT of the stretch mode of (A$_{VI}$, H) passivating centres in some II–VI compounds [214] and references therein, [215]

	N [a,b,c]			P [d]		As [d,e]		
	N − H	N − D	r(H/D)	P − H	P − D	As − H	As − D	r(H/D)
ZnSe	3193.6	2368	1.3488			2165.6	1557.1	1.3908
ZnTe	3346	2489	1.3443	2193		2014		
CdTe	3210					2022		

[a][209] [b][212] [c][213] [d][208] [e][211]

Polarized Raman scattering measurements on the N$_{Se}$ − H bond in ZnSe indicate a C_{3v} trigonal symmetry for the centre, similar to the one for the (A, hydrogen) centres in III–V compounds and in silicon [209].

ZnO can be doped with N, which replaces an O atom, which has been expected to be a potential dopant acceptor in ZnO. Recent first-principles calculations using hybrid functional that describe accurately the band gap have, however, predicted a very large value of the ionization energy of N$_O$ ($\sim$1.3 eV), which seem to show that N$_O$ cannot lead to a practical hole conductivity in ZnO [216]. These theoretical predictions have been experimentally confirmed [217]. As N$_O$ presents interaction with hydrogen similar to those of some acceptors in other materials, they are still presented here. In ZnO, a stretch mode of ^{14}N − H has been reported at 3150.6 cm^{-1} at LHeT, with a FWHM of 2.5 cm^{-1} (at RT, these quantities are 3148 and 8 cm^{-1}, respectively). The use of ZnO samples grown in a mixture of ^{15}NH$_3$ and ^{14}ND$_3$ revealed new LVMs associated with the ^{15}N − H, ^{14}N − D, and ^{15}N − D stretch modes, with ISs of −6.70, −811.0, and −820.7 cm^{-1}, respectively, with respect to the frequency of the ^{14}N − H stretch mode at LHeT. The absence of other LVMs indicates that these complexes include only one nitrogen and one hydrogen atom, and they were ascribed to the (N$_O$, hydrogen) complex [218]. The r-factor obtained for ^{14}N is 1.3466 and it is comparable to the ones of Table 8.16 for ZnSe and ZnTe. Anharmonicity of the N − H stretch mode allowed the observation of an overtone at 6132.9 and 6120.7 cm^{-1} for ^{14}N − H and ^{15}N − H, respectively.

ZnO has the wurtzite structure, with an anisotropy axis (the c-axis) shown in Fig. B.5 of Appendix B. From the RT absorption measurements of the N − H stretch mode with the **k** vector of the radiation along the c-axis, the polarized absorption of this mode was found to be independent of the orientation of the electric vector **E** in the plane perpendicular to the c-axis. This indicates that the N − H bonds or dipoles are distributed uniformly around the c-axis. The angle θ of the N − H dipoles with the c-axis can be estimated from the ratio of the intensities of the mode integrated over the polarization angle for the **k** vector perpendicular and parallel to the c-axis, and a value of 114° is obtained for θ [218]. This situation bears some resemblance with the one met for the above-discussed (Mg, H) centre in w − GaN.

The dissociation of this (N, H) centre occurs under annealing in the 700–800°C temperature range with a corresponding increase of the acceptors concentration [218].

8.5.1.3 Stress-induced Reorientation

Because of the interest in their dynamic properties, in relation with their diffusion coefficients, and also because of their relatively simple atomic structure, stress-induced dichroism of the vibrational modes of the (A, H) centres has been used to determine their reorientation energies in the crystal. Stress-induced dichroism has been outlined in Sects. 6.1.1.1 and 7.1.1.1 in relation with O_i and VO in silicon, but we recall here in more details the principles of the method.

The orientational degeneracy of a defect along different directions of the crystal allows its reorientation from a direction to an equivalent one at a temperature comparable to the activation energy for reorientation E_r, but statistically, the equipartition of the populations of the different orientations is preserved. The reorientation energy E_r can be obtained from the thermal relaxation of the stress-induced dichroism. This is based on the fact that, for some directions of a uniaxial stress, the atomic energy of one (or more) orientation of the defect becomes lower than that of the other ones. As a consequence, maintaining under such stresses for some time a sample containing this defect at a temperature where thermal jumps can occur at a reasonable rate produces an increase of the population of the orientation(s) with the lowest energy with respect to stress. The resulting difference must reflect on the absorption coefficients $K_{//}$ or $K_\perp$ of LVMs or electronic absorption lines of the defect when measured with the electric vector of the radiation $\mathbf{E}$ parallel ($K_{//}$) or perpendicular ($K_\perp$) to the aligning stress. This is the principle of the stress-induced linear dichroism of the absorption of the LVMs of these defects. The sign of the difference $K_\perp - K_{//}$ depends on the relative populations and on the orientation of the dipole associated with the transition considered (see Sect. 6.1.1.3). The normalized absorption difference is the dichroic ratio taken as:

$$D_R = (K_\perp - K_{//})/(K_\perp + K_{//}). \tag{8.1}$$

In most cases, the decay with time of dichroism follows a first-order kinetics as a function of the annealing time t, expressed as:

$$D_R = D_{R0} \exp(-k_d t), \tag{8.2}$$

where the rate of decay k_d of dichroism at temperature T associated with the thermal reorientation of the defect is proportional to the jump rate k_r of the centre from an orientation to the other. This jump rate is expressed as a function of an attempt frequency ν_0 and of the activation energy for the reorientation E_r by:

$$k_r = \nu_0 \exp(-E_r/k_B T). \tag{8.3}$$

The rate of jump k_r from one specific orientation to another is proportional to k_d, with a proportionality factor depending on the configuration of the defect in the lattice. In the case of the (A, H) centres, the dipole associated with the H stretch mode is along a $<111>$ axis and there are four equivalent orientations of this dipole

Fig. 8.21 Isothermal decay of the stress-induced dichroic ratio of the absorption of the Si − H and Si − D stretch modes of the (B, H) and (B, D) centres, respectively, in silicon.[13] The inset shows the initial dichroism measured at 65 K, where the polarization is taken with respect to the [110] axis (after [220])

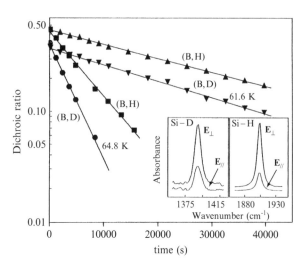

in the lattice. It can then be shown [188] that in that case, k_d is related to the reorientation rate k_r of the centre by $k_d = 4\,k_r$. For the measurement of dichroism, the stress is usually applied along the [110] axis and $K_\perp$ measured for $\mathbf{E}//[001]$. The aligning stress is of course released at the temperature of the measurement.

The reorientation parameters can be obtained from the decay of the dichroic ratio measured by isothermal or isochronal annealing. Isothermal annealing has been used for the study of the (B, H) and (B, D) centres in silicon, and it is shown in Fig. 8.21. The inset of Fig. 8.21 shows the effect on the polarized absorption of the Si − H and Si − D stretch modes of the cooling down to 65 K of a silicon sample containing both (B, H) and (B, D) centres with a stress of 450 MPa applied along the [110] axis. The ratio of the absorptions for the electric vector of the radiation $\mathbf{E}$ perpendicular ($\mathbf{E}_\perp$) and parallel ($\mathbf{E}_{//}$) to the aligning stress is ∼2.6 and 3.6 for Si − H and Si − D, respectively. The straight lines on this figure are a fit on a logarithmic scale of the isothermal decay of the dichroic ratio of the stretch modes for two not too different temperatures. From expression (8.2), the slopes are proportional to $-k_d$.

A plot of $\ln(k_r)$ vs T^{-1} is shown in Fig. 8.22. The value of E_r/k_d is deduced from the slope of the line fitted to the data assuming that the reorientation kinetics follows a first-order law.

The purpose of these figures is to show how a reorientation energy can be obtained from the time decay of the stress-induced dichroism. The physical interpretation of the dependence of the jump rate on T^{-1} is not as simple as it could appear from Fig 8.22, which corresponds to a very narrow temperature range; moreover, the fact that the jump rate of the heavier D isotope is larger than the

---

[13]It is assumed that at the temperatures chosen, the loss of dichroism during the time of the spectroscopic measurement can be neglected.

Fig. 8.22 H and D jump rates vs. T^{-1} between equivalent orientations of the (B, H) and (B, D) centres in silicon derived from the time decay of the dichroism. The activation energies corresponding to the solid lines are indicated. In the inset, the H jump from a *BC* site to an equivalent one is materialized for the (B, H) centre. The temperature range is ∼57 to 71 K (after [220]). Copyright 1994 by the American Physical Society

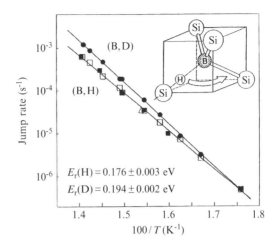

$E_r(H) = 0.176 \pm 0.003$ eV

$E_r(D) = 0.194 \pm 0.002$ eV

one for H is surprising. Jump rates and reorientation energies of (B, H) have been determined at 125 K by internal friction measurements [219]. Their comparison with the above results show that the reorientation of the (B, hydrogen) centre in silicon is governed by thermally assisted tunnelling [220].

Instead of the isothermal annealing method described earlier, an isochronal annealing method has also been used. The population difference between the different orientations is produced by the same method, but the base temperature at which stress is released is generally LHeT, where the initial dichroic ratio is measured. The sample is then warmed at the first annealing temperature, where it is kept for a given time, and cooled down again at LHeT where the new dichroic ratio is measured. This procedure is then repeated for the same annealing time (isochronal annealing) at increasing temperatures till complete loss of dichroism.

This method has been used to follow the decay of D_R for the (As − H, Be$_{Ga}$) stretch mode in GaAs as a function of the annealing temperature, and the result is shown in Fig. 8.23.

In this experiment, the sample was initially cooled from 180 to 15 K under a stress of 250 MPa along [110], and a dichroic ratio of ∼0.15 measured at 15 K without stress. The sample was then warmed at the first annealing temperature (63 K) for 20 min, cooled down to 15 K, where D_R was remeasured, and the procedure repeated for increasing annealing temperatures. The data points of Fig. 8.23 are fit to $D_R = D_0 \exp(-4 k_r t)$ where k_r is given by expression (8.3), and the solid line of this figure corresponds to $E_r = 0.374$ eV, assuming a value of 10^{12} s^{-1} for ν_0.

When considering a positive population difference $n_a - n_b$ of the vibrational ground states of (A, H) centres for orientations a and b produced by stress-induced alignment, the difference between the ground state energies E_a and E_b of these orientations, assumed to follow a Maxwell–Boltzmann statistics is:

$$n_a/n_b = \exp[(E_b - E_a)/k_B T], \tag{8.4}$$

Fig. 8.23 Decay of the stress-induced dichroism of the 2037 cm^{-1} LVM of (As − H, Be$_{Ga}$) in GaAs as a function of isochronal (20 min) annealing temperatures. The solid line corresponds to the fit described in the text and the activation energy for reorientation with the attempt frequency used is indicated (after [188])

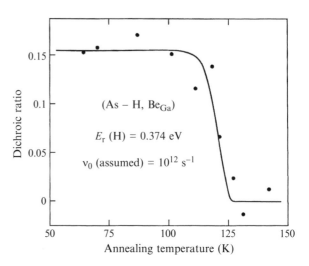

where T is the temperature where the stress alignment has been performed. For the (A, H) centres with trigonal symmetry, for the ground state, only the piezospectroscopic parameter A_2, noted here A_2' to avoid confusion with the transition piezospectroscopic parameter A_2 of Table E.2, is required to describe this stress-induced ground state energy difference. With the aligning stress along [110] used in the above-described experiments, the n_a/n_b ratio is given by the ratio $K_\perp/K_{//}$ of the absorption coefficients measured for the electric vector $\mathbf{E}$ of the radiation perpendicular and parallel to the aligning stress. In terms of the magnitude $\sigma_{[110]}$ of the stress, A_2' is given by:

$$A_2' = (E_b - E_a)/2\sigma_{[110]}. \qquad (8.5)$$

From expressions (8.4) and (8.5), one finds that the slope of a plot of $\ln(K_\perp/K_{//})$ vs. $\sigma_{[110]}$ is $2A_2'/k_B T$ [221].

Parameters of some (A, H) centres in silicon and III–V compounds related to their static and dynamic behaviour under stress are given in Table 8.17.

The lower value of the reorientation energy of (Si$_{As}$, H) in GaAs compared to (C$_{As}$, H) can be attributed to the extra elastic energy of the stressed Si$_{As}$ − H bond compared to the C$_{As}$ − H bond.

8.5.2 (Donor, Hydrogen) Centres

The report on the passivation by hydrogen of the Si donor in GaAs by Chevallier et al. [224] was followed by a more general paper where was reported the passivation of the group IV and group VI donors in GaAs [225]. The first observation of LVMs associated to (Donor, H) centres in GaAs was reported in GaAs in 1987 [226] and

Table 8.17 Piezospectroscopic parameters A_1 and A_2 (cm^{-1}/GPa) of the hydrogen stretch mode, piezospectroscopic parameter A'_2 (meV/GPa) of the vibrational ground state, and activation energy for the reorientation E_r (eV) of some (A, hydrogen) centres in different semiconductors. A compressive stress is taken as positive

Material	Bond	A_1	A_2	A'_2	E_r
Si:B, H	Si $-$ H	12.7[a]	13.5[a]	11.6[b]	0.18[c]
GaAs:Be, H	As $-$ H	5.0[b]	6.5[b]	9.3[b]	0.37[d]
GaAs:Si, H	Si$_{As}$ $-$ H	7.0[e]	8.4[e]	7.8[e]	0.26[e]
GaAs:C, H	C$_{As}$ $-$ H	2.4[b]	2.5[b]	8.2[b]	0.50[f]
GaP:C, H	C$_P$ $-$ H	4.3[g]	4.1[g]		0.35[h]
InP:Zn, H	P $-$ H	4.1[i]	5.1[i]		0.8[j]
InP:Zn, D	P $-$ D	3.1[i]	3.7[i]		

[a][166]
[b][222]
[c][205]
[d][188]
[e][202]
[f][187]
[g][139]
[h][150]
[i][70]
[j][223]

the one in silicon in 1988 [227]. In silicon, the dissociation energy (1.1–1.2 eV) of the (Donor, H) centres is lower than the one (1.3–1.4 eV) of the (Acceptor, H) centres. In GaAs, the (Si$_{Ga}$, H) passivating centre dissociates by thermal annealing near 400°C. As a rule, with the exception of the (Si$_{Ga}$, H) centre in GaP, the frequencies of the stretch modes of the (Donor, H) centres in semiconductors are significantly lower than those of the (A, H) centres.

8.5.2.1 Compound Semiconductors

Donors on Cation Sites

The passivation by hydrogen of group IV donors has only been observed in GaAs, where it is very efficient. The vibrational spectroscopy of the (Donor, hydrogen)-related modes in these materials is fully described in [13]. LVMs associated with Si$_{III}$ have been reported in GaAs, AlAs, and GaP [228–230], and with Sn$_{Ga}$ in GaAs [231].

The observation at LHeT, for H passivation of the Si donor in GaAs and AlAs, of stretch modes at 1717.25 and 1608.95 cm^{-1}, respectively, and of bending modes at 896.82 and 891.06 cm^{-1}, respectively, indicates that in these compounds, H is more probably not located at a *BC* site but at an *AB* site. The bonding of hydrogen to the Si atom is inferred from a Si IS of the stretch mode observed in GaAs. The observation of this Si IS is made possible because the FWHMs of the Si$_{Ga}$–H and Si$_{Ga}$–D stretch modes are only ~0.2 cm^{-1}, compared to ~0.4 and 0.5 cm^{-1} for the $^{28}\text{Si}_{\text{Ga}}$–H/$^{29}\text{Si}_{\text{Ga}}$–H and $^{28}\text{Si}_{\text{Ga}}$ – D/$^{29}\text{Si}_{\text{Ga}}$ – D ISs, respectively. The LVM of the Si_{Ga} atom shifts from 384 cm^{-1} when isolated (see Sect. 5.3.5) to 409.95 and 409.45 cm^{-1} when perturbed by H and D, respectively [228].

The frequency of the stretch mode decreases logically when the mass of the donor atom increases (1327.80 cm^{-1} for (Sn_{Ga}, H) in GaAs [231]). Uniaxial stress measurements on the Si_{Ga}–H mode are consistent with a trigonal symmetry of this bond in the lattice [82]. From the measured frequency of the $^{28}\text{Si}_{\text{Ga}}$–D stretch mode, the r-factor for this AB mode is 1.3764, compared to 1.3831 for the BC mode of Si_{As}–H of the (Acceptor, H) centre (see Table 8.13). This can be attributed to a higher anharmonicity of the AB mode as the harmonic value is 1.3904.

In GaAs, the intensity of the Si_{Ga}–H bending mode is larger than that of the stretch mode by a factor of $\sim$3, but in AlAs, this is the intensity of the Si_{Al}–H stretch mode which is the largest [229].

The absorption of the Si–H bending mode has been reported in H-passivated n-type $\text{Al}_{0.2}\text{Ga}_{0.8}$As : Si. The observation of several LVMs in the 880–900 cm^{-1} range shown in Fig. 8.24 is attributed to Si–H pairs differing in the chemical nature of the group III atom nnn in the Si atom [232].

A value of the position of the electronic level in the band gap of GaAs above which hydrogen is negatively charged has been estimated from the relationship between the observation of the LVM of Si_{Ga}–H at 1717 cm^{-1} and the position of E_{F} in LEC GaAs containing residual Si donors [139]. This LVM is detected only in samples where the electron free absorption is also detected. Since the Si_{Ga}–H centres can be formed only when H is negatively charged, an approximate value of $E_{\text{c}} - 0.1$ eV is obtained for the level above which hydrogen is negatively charged.

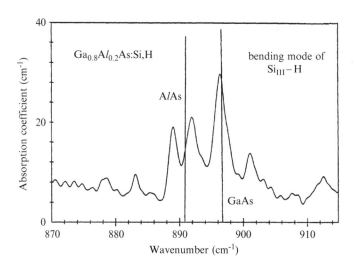

Fig. 8.24 Absorption at LHeT of Si–H bending modes in an n-type Si-doped $\text{Al}_{0.2}\text{Ga}_{0.8}$As sample showing the perturbation of this mode by mixed (Al, Ga) configurations nnn of Si_{Al} and Si_{Ga}. The frequencies of Si–H in the pure materials are indicated by bars (after [232])

The results of calculation of the (Si_{III}, H) structure give the one with H_{AB} as the more stable one, and the calculated frequencies are in good agreement with experiment [229, 233].

In as-grown LEC GaP doped with silicon, a LVM at 2175.1 cm^{-1} with a FWHM of 0.4 cm^{-1} has been reported and assigned to the (Si_{Ga}, H) centre, but the high frequency of this LVM suggests that it is due to the stretch mode of a P–H bond [230].

Piezospectroscopic measurements give a trigonal symmetry for this centre. Hence, a configuration with the H atom bond centred between the Si atom and one of the nn P atoms has been proposed for this centre in GaP [230]. However, this BC location of H could be questioned, as the (donor, hydrogen) complexes form by the interaction of H$^-$ with a positively charged donor. Therefore, hydrogen should somehow retain a negatively charged character and avoid the BC site, a location where the negative charge density is large, and an AB location would be more likely, as for the other (donor, hydrogen) complexes. However, an argument against the AB location is that no bending mode which could be related to the P $-$ H bond of this (Si_{Ga}, H) centre is observed in GaP, but this absence could still be explained within the frame of an AB location: in Pauling's electronegativity scale, hydrogen and phosphorous have the same electronegativity, and this means that the extra electron brought by H$^-$ is not strictly localized around hydrogen, but shared between H and the P atom to which it is bonded. Now, the force constant of the bending mode is mostly due to the repulsive Coulombic interaction between hydrogen and the three Ga $-$ P bonds into which it is encaged. If the negative charge around hydrogen is small, the force constant and the associated bending frequency are weak, and the multiphonon absorption bands can prevent the observation of this bending mode. All together, the structure of the (Si_{Ga}, H) centre in GaP cannot be considered as firmly established, and the help of theory would be welcomed for a definite understanding of this complex.

Donors on Anion Sites

The chalcogen atoms (S, Se, and Te) are donors on the group V sublattice. They are common n-type dopants in III–V compounds, and their passivation by hydrogen in GaAs was demonstrated by Pearton et al. [225]. The LVMs of the (Chalcogen, H) complexes measured at LHeT in GaAs can be attributed to stretch modes near 1500 cm^{-1} and to bending modes near 780 cm^{-1}, showing both a small frequency dependence on the chemical nature of the chalcogen and they are characterized by rather small FWHMs ($\sim$0.1 cm^{-1}) [234–236]. The near independence of these modes on the nature of the chalcogen atom is further confirmed by the $r(H/D)$ ratios for the stretch and bending modes, as can be seen from Table 8.18.

From these results, the H atom of these centres is expected to form a Ga–H bond with a Ga atom nn of the chalcogen donor atom, where the H atom sits in an AB site. However, no $^{69}Ga/^{71}Ga$ IS is observed, and in the present case, a Ga IS of 0.23 cm^{-1} is calculated within the harmonic approximation. However, from the small values of the FWHMs of the chalcogen-related stretch modes, the experimental Ga IS is deduced to be smaller than 0.05 cm^{-1} [13]. Piezospectroscopic measurement on the

Table 8.18 Frequencies and FMHMs (cm^{-1}) at LHeT of the stretch and bending modes of the H bonds of the chalcogen donor centres in GaAs. The intensity ratio of the bending to stretch modes is $\sim$7 [235, 236]

Donor	Mode	(Donor, H) Ga $-$ H$_{AB}$		(Donor, D) Ga $-$ D$_{AB}$		
		Frequency	FWHM	Frequency	FWHM	r(H/D)
S	Stretching	1512.3	0.07	1088.4	0.4	1.3895
	Bending	780.6	0.05	556.1	0.3	1.4037
Se	Stretching	1507.5	0.08	1084.8	0.7	1.3897
	Bending	778.0	0.08	554.3	0.6	1.4036
Te	Stretching	1499.9	0.13			
	Bending	771.8	0.09	550.0	0.6	1.4033

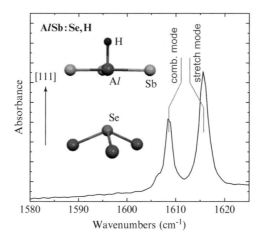

Fig. 8.25 Absorption spectrum at LHeT of hydrogenated A*l*Sb:Se in the vicinity of the stretch mode of (Se$_{Sb}$, H) showing the combination mode. The proximity of the two LVMs implies a resonant interaction between them. The FWHMs of the stretch mode and of the combination mode are 2.2 and 1.5 cm^{-1}, respectively. The inset shows the structure of the centre with trigonal symmetry [237]. Copyright 2009 by the American Physical Society

stretch mode of the (Se, H) and (Te, H) centres in GaAs indicate a trigonal symmetry of these centres [236].

The observation of H- and D-related LVMs has also been reported at LHeT in hydrogen-passivated Se- and Te-doped A*l*Sb [210]. In the hydrogenated Te-doped samples, only two LVMs at 1599.0 and 665.0 cm^{-1} are observed at LHeT, attributed to the stretch and bending modes of (Te$_{Sb}$, H), respectively, with (Te$_{Sb}$, D) counterparts [237]. In the Se-doped samples, LVMs are also observed near 1600 and 665 cm^{-1}, but seven other H-related LVMs were also observed, two near 1600 cm^{-1} and five between $\sim$690 and 1330 cm^{-1} [210]. The (Se$_{Sb}$, H) spectrum near 1600 cm^{-1} is displayed in Fig. 8.25.

Besides the stretch and bending modes of the A*l*–H bond, the centre represented in the inset of Fig. 8.25 has also low-frequency longitudinal in-phase A$_1$ and E transverse vibrational motions of Al–H. These latter modes are somewhat similar to the X and Y$_1$ modes of the (Acceptor, hydrogen) centres of Table 8.15, and in the present case, they mainly involve the motion of the Al atom. Their frequencies, together with the ones of the stretch and bending modes of Al–H and Al–D have been calculated by a first-principle DFT method [237]. Within the harmonic approximation, the frequencies of these longitudinal (A$_1$) and transverse (E) modes are near 170 and 270 cm^{-1}, respectively, practically independent of the mass of the hydrogen isotope. The frequencies of these modes lie in the phonon gap of AlSb, so that they can be in principle detected, but they have not been looked for [237]. The overtone of the Al–H bending mode is split into two components A$_1$ (Γ_1) and E (Γ_3) with frequencies in the 1330 cm^{-1} range, observed in the (Se$_{Sb}$, H) spectrum. The LVM on the low-energy side of the LVM stretch mode of Al–H has been attributed to a combination mode of the E overtone with the low-frequency E mode, resulting in a frequency $\sim$1611 cm^{-1}. Given its frequency, this combination mode interacts with the Al–H stretch mode, and this has been demonstrated spectroscopically by measurements under hydrostatic pressure [238]. The calculated frequencies of the (Se$_{Sb}$, hydrogen) centre taking into account anharmonicity are compared in Table 8.19 with the ones measured at LHeT.

The frequencies of the stretch and bending modes of Al–H for the (S$_{Sb}$, H) centre calculated *ab initio* are 1665.4 and 640.1 cm^{-1}, respectively. A comparison with

Table 8.19 Comparison of the frequencies (cm^{-1}) of vibrational Al–H modes of the centres passivating the Se$_{Sb}$ and Te$_{Sb}$ donors in AlSb measured at LHeT [210] with the values calculated taking into account anharmonicity and from *ab initio* calculations in the harmonic approximation (values in brackets) [237]. The attributions are those of [237]. The integrated absorptions (*IA*s) are normalized to that of the stretch mode (after [210])

| (Se, hydrogen) | Al–H | | | Al–D | | | |
	Meas.	Calc.	IA	Meas.	Calc.	IA	r(H/D)
Stretch	1615.7	1599.1 (1655)	1	1173.4	1196.4 (1188)	1	1.3769
Overtone (E)/transverse (E) combination	1608.6		0.42				
Bending overtone (E)	1333.0	1383.5	0.067	957.4	971.2	0.014	
" (A$_1$)	1315.8	1382.7	0.10	948.4	970.9	0.046	
Bending	665.7	691.7 (641.2)	3.33	478.3	485.6 (453)	3.7	1.3918
Transverse (E)		(269)			(268)		
Longitudinal (A$_1$)		(168)			(165)		
(Te, hydrogen)	Meas.	Calc.		Meas.	Calc.		r(H/D)
Stretch	1599.0	(1648.5)		1164.4			1.3732
Bending	665.0	(639.2)		478.2			1.3906

those of Table 8.18 for (Se$_{Sb}$, H) and (Te$_{Sb}$, H) confirms that they are practically independent on the chemical nature of the donor atom.

LVMs with smaller intensities that may correspond to unidentified impurities in the AlSb sample (ref. 25 of [237]) were reported at 692.2, 992.6, and 1031.8 cm^{-1} in the (Se$_{Sb}$, H) spectrum [210].

Taking for the transverse mode E an experimental value of 278 cm^{-1} independent from the hydrogen isotope, the frequency of the combination mode of the (Se, D) centre corresponding to the one at 1609 cm^{-1} for (Se, H) should be $\sim$1235 cm^{-1}, far above the frequency of the Al–D stretch mode, so that no detectable interaction should occur between these two modes [237].

The observation at LHeT of LVMs at 2204.3, 2209.6, and 2217.1 cm^{-1} has been reported in H-passivated n-type GaP samples doped with S, Se, and Te, respectively, where the chalcogen atom replaces a P atom [239]. S being a residual impurity in GaP, the 2204.3 cm^{-1} LVM had been reported a long time ago without a specific attribution [69]. The similarity of the frequencies, their values, and the absence of an IS despite the sharpness of these LVMs (their FWHMs are $\sim$0.2 cm^{-1}) has led to their attributions to P $-$ H stretch modes. The results of piezospectroscopic measurement on this P $-$ H stretch mode in Te-doped GaP have precluded a trigonal symmetry for the (Te$_P$, H) centre. The conclusion is that the centre displays a monoclinic-I (C_{1h}) symmetry, where the P atom of the P $-$ H bond and the Te$_P$ atom are *nnn*s in the same (110) plane.

For stresses along the <001>, <111>, and <110> directions, the stretch mode splits into 2, 3, and 4 components, respectively. The relative intensities of these components for E parallel and perpendicular to the stress depend on the actual angle between the direction of the P $-$ H bond and the <001> direction[14] [13]. This angle should be 54.7° when the P $-$ H bond is along the <111> axis, but the experimental results show that this bond is slightly off-axis, being 35, 40, and 40° for (S, H), (Se, H), and (Te, H), respectively. The structure of the complexes is not fully understood. Originally, a near *BC* structure was proposed, but it suffers the same drawback as for the (Si$_{Ga}$, H) centre in GaP (see previous subsection), as H complexing with a donor should be negatively charged and stay in region with low electron densities, i.e. near *AB* locations, as shown in Fig. 8.26b.

Positively charged donors should attract negatively charged hydrogen. This should favour an increase of the angle between the P $-$ H bond and the <001> direction when hydrogen is at the AB_1 location and a decrease of this angle at the AB_2 and AB_3 locations. This angle being experimentally determined below 54.7°, location AB_2 or AB_3 could be the most probable, but one has to be careful as lattice relaxations have not been included in the figure. Here again, a theoretical investigation would improve the full understanding of the structure of these complexes.

LEC-grown GaAs contains B concentrations in the 2×10^{17} cm^{-3} range because of the boron oxide encapsulant. Two lines at 2383.2 and 2392.8 cm^{-1} have

---

[14]Similar conclusions had been reached before for different centres [48,240].

Fig. 8.26 Schematic representations in the {110} plane of possible atomic structures of the complexes of chalcogen donors with hydrogen in GaP. In these representations, the lattice relaxation is not included: (a) originally proposed structure with H located near bond centre [239], (b) possible configurations of antibonded H

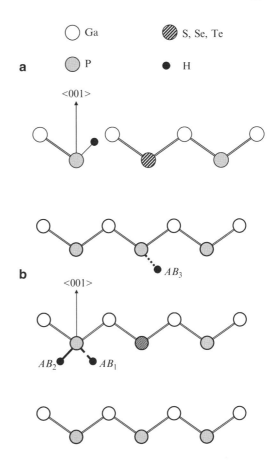

been reported at LHeT in S-doped LEC GaAs samples, with FWHMs of 0.4 and 0.6 cm^{-1}, respectively. These lines can still be observed at RT, indicating a vibrational origin, and the ratio of the integrated absorptions IA_{2382}/IA_{2393} is 4 ± 0.4 [241]. They have been attributed to the stretch mode of $^{11}B - H$ and $^{10}B - H$ bonds, respectively as their intensity ratio is equal to the natural isotopic abundance ratio of ^{11}B and ^{10}B, and as their frequency difference (9.6 cm^{-1}) fits the IS calculated (9.9 cm^{-1}) for a "free" $^{10}B - H$ pseudo molecule with a frequency of 2392.8 cm^{-1}. A sharp, but weak, LVM observed at 897.6 cm^{-1} was attributed to the bending mode of the $^{11}B - H$ bond. In these S-doped LEC samples, the Ga $-$ H stretch mode at 1515 cm^{-1} of the trigonal (S, H) centre reported in [236] was not observed. Inversely, the two above-described LVMs were not observed in SI and p-type GaAs.

The uniaxial stress measurements on the 2382 cm^{-1} LVM indicated a non-trigonal symmetry for the $^{11}B - H$ bond. This led Ulrici and Clerjaud to ascribe them to a special kind of S passivation in LEC GaAs [241]. In this material, instead of a trigonal ($S_{As} - Ga - H_{AB}$) centre, the passivating centre is ($S_{As} - B_{Ga} - H$) complex where the H atom is antibonded to a B_{Ga} atom nn of S_{As}. The difference between the two centres comes from the non-trigonal symmetry of the latter one. It is explained by the result of polarized light measurements similar to the ones described earlier for the (Te, H) passivating centre in GaP. They show that instead of making with the <001> axis of Fig. 8.26 an angle θ of 54.7° as when antibonded along a trigonal axis, the B $-$ H bond makes an angle of 70° with this axis [241].

8.5.2.2 Silicon

Under H_2 and D_2 plasma treatments, the electrical activity of n-type silicon is found to be reduced. The absorption of LVMs attributed to (Donor, hydrogen) centres resulting from this passivation in P-, As-, and Sb-doped samples was reported [227]. For each chemical donor, three LVMs are observed, and the dependence of their frequencies on the chemical nature of the donor is found to be small. A trigonal symmetry of these (Donor, hydrogen) centres in silicon has been implicitly assumed.

For (P, H), (As, H), and (Sb, H), the highest frequencies, in the $1650–1670\,\mathrm{cm}^{-1}$ range, are comparable to the one of the stretch mode of the (Si_{Ga}, H) passivating centre in n-type GaAs, and they were logically ascribed to the stretch mode of a Si–H bond with A_1 symmetry. The lowest frequency, near $810\,\mathrm{cm}^{-1}$ was ascribed to the bending mode of this bond, indicating an antibonding location of the H atom. The antibonding location could also be inferred from the already signalled small dependence of the frequencies on the chemical nature of the donor. The frequency of the third LVM, relatively close to that of the stretch mode, has been a matter of debate, and several assignments have been proposed ([242], and references therein). Now, the overtone of the bending mode with symmetry E is split into two components: A_1 (Γ_1) and E (Γ_3) with frequencies in the $1620\,\mathrm{cm}^{-1}$ range. It has been finally recognized that this third mode was due to the A_1 overtone of the bending mode in Fermi resonance with the stretch mode [242]. Expressions for the measured energies of the coupled stretch mode and bending overtone as a function of the decoupled energies of these modes and of coupling parameters have been calculated by Zheng and Stavola. The determination of the coupling parameters necessitates the comparison of the experimental results for the (Donor, H) and (Donor, D) centres. The observed spectrum for the hydrogenated Sb donor is shown in Fig. 8.27.

The ratio of the integrated intensities of the stretch mode and of the overtone depends on the resonance conditions and it has been estimated self-consistently from the anharmonic interaction for the different centres investigated. For (Sb, H) and (Sb, D), it is 1.2 and 2.6, respectively, compared with experimental values of 1.3 and 2.8, respectively [242].

As for the (Donor, hydrogen) centres with trigonal symmetry in III–V compounds, a second overtone of the bending mode is also expected. This second overtone has indeed been observed for the (Donor, H) centres, and it can be observed near $1620\,\mathrm{cm}^{-1}$ in Fig. 8.27b [242]. The frequencies of the different modes observed for the different group V donors are listed in Table 8.20.

These LVMs are relatively broad compared to those observed in compound semiconductors: at LHeT, the FWHMs of the H and D stretch modes are $\sim$10

Fig. 8.27 Absorption at LHeT of the stretch (s) mode and of the A_1 overtone of the bending mode (bo) of the (Sb, hydrogen) centre in silicon showing the effect of the Fermi resonance. The vertical bars give the calculated positions of the decoupled modes $^d\omega_{bo}$ and $^d\omega_s$ in the absence of resonance, and their spacing (cm^{-1}) is indicated (after [242]). Copyright 1996 by the American Physical Society

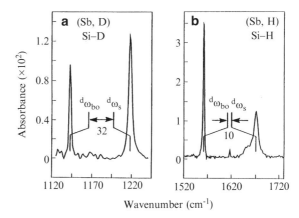

Table 8.20 Frequencies (cm^{-1}) and relative integrated absorptions (*IA*s) of the (Donor, H) and (Donor, D) stretch, bending, and bending overtone modes measured at LHeT in H- and D-plasma-passivated n-type silicon samples containing P, As, and Sb (after [242]). The corresponding *IR*s for a trigonal centre are indicated. The difference in the values of r(H/D) for the stretch mode is due to the perturbations of the frequencies due to the above-discussed resonance effect

Donor	Mode	(Donor, H) (Si–H$_{AB}$)		(Donor, D) (Si–D$_{AB}$)		r(H/D)
		Frequency	*IA*	Frequency	*IA*	
P	Stretch (A_1)	1645.5	0.92	1215.5	0.98	1.3538
	Bend. overtone (E)	1615.5	0.04			
	" (A_1)	1555.4	0.62	1141.5	0.42	
	Bending (E)	809.5	1	584.7	1	1.3845
As	Stretch (A_1)	1661.4	0.98	1222.0	1.29	1.3596
	Bend. overtone (E)	1617.0	0.02			
	" (A_1)	1561.0	0.61	1143.1	0.39	
	Bending (E)	809.9	1	584.4	1	1.3859
Sb	Stretch (A_1)	1670.6	0.85	1218.1	1.04	1.3715
	Bend. overtone	1616.7	0.02			
	"	1561.8	0.65	1142.7	0.43	
	Bending (E)	809.5	1	584.4	1	1.3852

and $\sim$5 cm^{-1}, respectively, while those of the bending modes are $\sim$2 cm^{-1} for both H and D [227].

The annealing of the (P, H) centres starts near 370 K ($\sim$100°C) and it is completed near 430 K ($\sim$150°C). The As and Sb centres are slightly more stable, but they do not survive above about 470 K ($\sim$200°C) [227].

In silicon, hydrogen can also form complexes with the chalcogen double donors and produce shallow H-related thermal donors (see, for instance, [8], and references therein, and Sect. 6.4.2 of [243]). However, no LVMs related to these centres have been reported, presumably because their concentration is too small.

8.6   Hydrogen and Transition Metals (TMs)

8.6.1   Complexes with TMs in Silicon

The study of the interaction of hydrogen with TMs, which are fast-diffusing contaminants in silicon, has long been of interest [1]. In Pt-containing FZ silicon samples annealed in hydrogen at 1250°C, combined ESR and IR absorption studies allowed to identify a Pt-related centre containing two hydrogen atoms [244, 245]. This centre is a double acceptor that can exist in three different charge states, one of them being paramagnetic. The two related electronic levels are located near mid-gap and 0.1 eV below the *CB* minimum. Every charge state of the PtH_2, PtHD, and PtD_2 configurations has two stretch modes due to symmetric and antisymmetric motions of the hydrogen atoms. The frequencies of these Pt-related LVMs are listed in Table 8.21.

Some of these LVMs have also been detected by internal multireflection spectroscopy in samples hydrogenated by wet chemical etching and annealed at 200°C [246]. ESR and piezospectroscopic measurements showed that in the paramagnetic state PtH_2^-, the complex has C_{2v} symmetry. Two possible configurations of a PtH_2 complex with this symmetry are presented in Fig. 8.28.

The complex can reorient among energetically equivalent configurations above 40 K. If, on the one hand, an analysis of the components of the hyperfine tensor [245] favours the "antibonding" configuration of Fig. 8.28b, on the other hand, a reorientation of this configuration at temperatures as low as 40 K is quite unexpected. Theoretical calculations [247] also favour the antibonding situation. Even though the literature favours the antibonding configuration of Fig. 8.28b, the actual configuration of the defect might still be a subject of debate.

Two charge states of a PtH complex have also been detected by LVM spectroscopy [246]. The neutral one is characterized by a LVM at 1880.7 cm^{-1} at LHeT, and the negative one, observed once the centre has trapped an electron, by a LVM

Table 8.21 Frequencies (cm^{-1}) at LHeT of the LVMs noted A, B, and C associated with the three charge states of the complex of Pt with two hydrogen atoms. The paramagnetic charge state corresponds to the -1 charge state and the diamagnetic charge states to 0 and -2. The stretch modes for the antisymmetric and symmetric motions are indexed 1 and 2, respectively [245]

	-2	-1	0
PtH_2	1889.6 (A1$_H$)	1888.2 (B1$_H$)	1873.1 (C1$_H$)
	1898.0 (A2$_H$)	1901.6 (B2$_H$)	1891.9 (C2$_H$)
PtHD	1893.9 (A3$_H$)	1894.6 (B3$_H$)	1880.3 (C3$_H$)
	1367.5 (A3$_D$)	1366.9 (B3$_D$)	1361.0 (C3$_D$)
PtD_2	1363.3 (A1$_D$)	1362.5 (B1$_D$)	1352.4 (C1$_D$)
		1370.7 (B2$_D$)	1365.2 (C2$_D$)

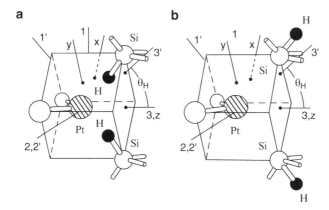

Fig. 8.28 Tentative models for the PtH$_2$ complex in silicon. In (**a**), the two weak Si–H bonds point toward the Pt atom. In (**b**), the H atoms point away from the Pt atom. Although the Si atoms are represented on their lattice site, there may be substantial relaxations that are not shown [245]. Copyright 1995 by the American Physical Society

at 1897.2 cm^{-1}. The related deep acceptor level is in the range 0.43–0.55 eV below the *CB* minimum [246].

In Au-diffused FZ silicon samples annealed in H$_2$ and/or D$_2$ atmospheres at 1250°C, two types of gold–hydrogen complexes have been reported [248]. Pairs of LVMs with C_{2v} symmetry, observed near 1800 cm^{-1}, have been attributed to the symmetric and antisymmetric stretch modes of a single charge state of AuH$_2$, AuD$_2$, and AuHD complexes, with the symmetric mode at a slightly larger frequency than the antisymmetric one. Their configuration is assumed to be almost identical to the one of PtH$_2$ [248].

The LVMs of three charge states of the AuH and AuD complexes, with frequencies in the same spectral range as the AuH$_2$ complexes, have also been identified, and the results of uniaxial stress measurements are consistent with a C_{1h} symmetry or lower for the complexes with a single hydrogen atom [248].

Theoretical calculations have been performed on these centres. Their results favour configurations where the hydrogen atom(s) is (are) bonded to Si atoms *nn*(s) of the Au atom, and located in "antibonding" locations [249].

The LVMs of two charge states of (Fe,Au,H) and (Fe,Au,D) complexes have also been detected in these samples, where iron was a contaminant, with frequencies comparable to those of the (Au,H) and (Au, D) complexes. For these (Fe, Au, hydrogen) complexes, the results of experiments under uniaxial stress indicate a C_{1h} symmetry [248].

8.6.2   Copper–Hydrogen Complexes in ZnO

The LVM of copper–hydrogen complexes in ZnO was among the first hydrogen-related LVMs reported in semiconductors [250]. At RT, a mode at 3234 (2402) cm^{-1} in ZnO samples containing Cu and H (D) was reported [250]; in samples containing both H and D, only these two lines were observed, indicating the presence of only one hydrogen atom in the centre. At LHeT, the H-related LVM was reported at 3194 cm^{-1} (3192 cm^{-1} in the subsequent works). The absorption coefficient for light polarization perpendicular to the c-axis was found to be about twice larger than for polarization parallel to the c-axis. It was deduced from these spectroscopic measurements that this LVM was due to a (Cu$_{Zn}$, H/D) complex, with the H/D atom bonded to an O atom, in a BC location in one of the three Cu$_{Zn}$ − O bonds not oriented along the c-axis. A more complete analysis allowed deducing an angle of 116.6° between the c-axis and the O–H bond [250]. In subsequent works, the same group [251, 252] reported a great number of LVMs in more heavily doped ZnO samples corresponding to a large variety of (Cu, hydrogen) complexes, some containing more than one H/D atom.

Lavrov and Weber performed piezospectroscopic measurements on the LVM at 3192 cm^{-1} [253]. When the stress is applied along the c-axis, no splitting is observed whereas when the stress is applied along [1$\bar{2}$10] or [10$\bar{1}$0] axes, the 3192 cm^{-1} line splits into two components. This is fully consistent with the microscopic structure suggested by Mollwo et al. [250]. The surprising fact is that the shift of the 3192 cm^{-1} line when the stress is applied along the c-axis is a strongly non-linear function of the applied stress. At the same time, Lavrov and Weber [253] also showed that the application of stresses along the c-axis induces the appearance of two new modes at 3229 and 3263 cm^{-1}. In deuterium-doped samples, the main line is at 2379 cm^{-1} and those appearing under stress at 2412 and 2429 cm^{-1}. These results evidence the stress-induced mixing of the excited state of the 3192 cm^{-1} transition with two other states that breaks forbidding selection rules. Experiments performed around 15 K show that the ground state of the transition is also influenced by two nearby states. These results allowed Lavrov and Weber to establish the energy level diagram for this complex. It is shown in Fig. 8.29. The origin of the low energy excitations is not clear yet. At zero stress, the only transition observed at low temperature is g$_0$ → e$_0$ (line 0). Line 1 (g$_1$ → e$_1$) has been observed at elevated temperatures [251]. The energies of the low-frequency excitations are deduced from the extrapolations at zero stress of the high-stress results.

Stress-induced dichroism experiments have been performed on the (Cu, H) and (Cu, D) LVMs at 3192 and 2379 cm^{-1}. They have shown that a stress applied at RT along the [10$\bar{1}$0] axis of ZnO results in an alignment of the (Cu, H/D) complex. Once frozen under stress at low temperature, the complexes can reorient statistically for annealing temperatures above 155 K, giving for the thermally activated reorientation (see Sect. 8.5.1.3 for the actual procedure) an activation energy about 0.5 eV [254].

Lavrov et al. [255] have shown that two modes at 3346.9 and 3373.9 cm^{-1}, already reported by Gärtner and Mollwo [251, 252] were due to a (Cu,H$_2$) complex.

Fig. 8.29 Energy level diagram for the (Cu, H/D) complex in ZnO. The frequencies of the transitions are given in wavenumbers (cm^{-1}). The dashed lines denote transitions forbidden in the electric dipole approximation. They appear under stress and the frequencies indicated are extrapolations at zero stress. The values in brackets are for (Cu, D) [253]. Copyright Wiley-VCH Verlag GmbH & Co. KGa. Reproduced with permission

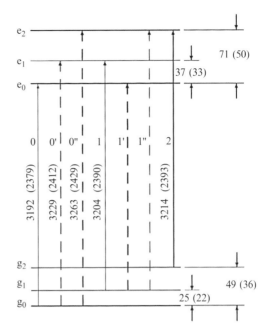

For the CuD$_2$ complex, the corresponding modes are at 2485.3 and 2504.7 cm^{-1} and for the CuHD complex, they are at 2494.8 and 3361.4 cm^{-1}.

Piezospectroscopic measurements have shown this complex of Cu with two hydrogen atoms consists in a substitutional Cu$_{Zn}$ atom with two adjacent *BC* H/D atoms, each of which is bound to O atoms in the basal plane (perpendicular to the *c*-axis) of the defect.

Measurements of stress-induced alignment of the O–H and O–D bonds of the (Cu, H$_2$) and (Cu, D$_2$) complexes have also been made; they show that the defect can be aligned under stress applied along [10$\bar{1}$0]. The reorientation temperature for the (Cu, H$_2$) and (Cu, D$_2$) complexes are 87 and 116 K, respectively, and this corresponds to activation energies for the reorientation of 0.29 and 0.39 eV, respectively.

Ab initio calculations have been performed on the copper–hydrogen complexes in ZnO [88]. They agree with the configuration deduced from the experiment for the (Cu, H) complex with one H atom, but not for the one with two H atoms.

8.6.3 TM-Hydrogen Complexes in III–V Compounds

LVMs of complexes between TMs and hydrogen have been observed in InP and GaAs [68, 69]. The most documented cases concern the Mn-hydrogen complexes. In InP, the frequency of the LVM associated with the (Mn, H) complex is 2272 cm^{-1}, and it can be seen as a very weak feature in Fig. 4.29. Experiments under uniaxial stresses performed on this LVM reveal no splitting for a stress along a <001> axis whereas a splitting into two components is observed for a stress along a <110> axis [256]. This is the piezospectroscopic signature of a complex aligned along trigonal axes of the InP lattice.

The frequency of the (Mn, H) complex indicates the vibration of a P − H bond. The similarity with the one of the P − H bond of the (Zn, H) passivating complex in InP ($2288\,cm^{-1}$) indicates that the H atom of the (Mn, H) complex is bond centred between the Mn and P atoms, in a structure identical to the one of the H-shallow acceptor complexes (see Fig. 8.19b). The shift of the transition under <001> stress is strongly non-linear [256]. As for the copper–hydrogen complex in ZnO discussed in the previous subsection, this probably indicates the interaction of one of the states involved in the transition with an energetically nearby state that has not been directly detected. It has also been shown [256] that the complex does not reorient at temperatures up to 100°C. This implies a reorientation barrier height larger than 1.2 eV.

The LVM associated with the (Mn, H) complex in GaAs has been observed in Mn-doped epitaxial layers grown by MBE [191, 257]; it has been shown [191] that it is in fact the $2144\,cm^{-1}$ mode first reported by Pajot et al. [258] attributed, at that time, to the complex of a H atom with an unknown acceptor. Its D analogue is at $1547\,cm^{-1}$. It is suggested [191] that the microscopic structure of the complex is identical to the one in InP discussed in the previous paragraph. It has to be noted that the passivation of manganese by hydrogen in epilayers doped with manganese in the percent range suppresses the ferromagnetism in the layers [259, 260].

The detection of the LVM due to the complexing of Mn with H in GaP has been quite delicate [261, 262]. The reason for this difficulty is the following: the complexes form according to the reaction: $A^- + H^+ \rightarrow$ (A, H) where A means acceptor. Therefore, to form (A, H) complexes, A^- and H^+ must be present simultaneously. This is what happens, for instance, in InP or GaAs where the Mn acceptor level is clearly below the level at which hydrogen passes from H^+ to an other state. In the case of GaP, the Mn acceptor level is deeper than in GaAs and InP and is slightly above the level at which hydrogen passes from H^+ to an other state. Hopefully, the complexes form during the growth of the samples at rather high temperatures, where the Fermi functions are not fully step-like. Therefore, if the samples have a Fermi level in between the level at which hydrogen changes of state and the Mn acceptor level, a weak overlap exists between the H^+ and A^- distributions and a small fraction of (Mn, H) complexes can form. The stretch mode of the (Mn, H) complex in GaP has been observed at $2369.4\,cm^{-1}$ at LHeT and the one of (Mn, D) at $1724.2\,cm^{-1}$ [261, 262]. Here again, the proximity of these frequencies with those of the (Zn, H) and (Zn, D) LVMs (2379 and $1728\,cm^{-1}$, respectively) suggests that the microscopic structure of the complexes is the same as for the complexes of shallow acceptors with hydrogen/deuterium. In GaMnP alloys in which the Mn concentration is in the percent range, the Mn acceptors form a band that could extend below the level at which hydrogen changes of state, and therefore, the formation of (Mn, H) complexes is easier. That is why, in these alloys, the introduction of hydrogen can also destroy the ferromagnetism of the layers [263].

Nevertheless, the LVMs due to the (Mn, H) complexes cannot be detected in these materials [263].

8.7 Other H-Related Centres

8.7.1 Hydrogen-related LVMs in Diamond

The existence of vibrational modes related to hydrogen-related centres in diamond has been predicted from first-principles calculations [9], and this has been anticipated by the observation of several hydrogen-related LVMs in natural and synthetic type I diamonds. In this section, we limit ourselves almost uniquely to a H-related LVM observed at $3107 \, \text{cm}^{-1}$ ($3.22 \, \mu\text{m}$) in nearly all the type Ia diamonds, but whose attribution is still tentative. A thorough presentation of the IR vibrational lines of hydrogen in natural diamond can be found in [264]. A commented list of H-related LVMs in natural and synthetic diamond is given in [265]. The origin of the $2828 \, \text{cm}^{-1}$ LVM observed in CVD diamond is discussed in [266].

Type Ia natural diamonds contain nitrogen mainly in the form of (N, V) centres (see Sect. 1.2.1). These diamonds consistently display in their vibrational spectrum a series of absorption lines dominated by a LVM at $3107 \, \text{cm}^{-1}$, first reported by Charette [267]. A ^{13}C counterpart of this LVM is located at $3098 \, \text{cm}^{-1}$, but no N IS has been reported for this mode [268, 269]. For this reason, this LVM has been attributed to the stretch mode (ω_s) of a C $-$ H bond. The $3107 \, \text{cm}^{-1}$ LVM has also been observed in synthetic diamonds when the growth conditions allowed the formation of (N, V) centres [269].

LVMs due to the bending mode (ω_b), bending overtones ($n\omega_b$), stretch overtone ($m\omega_s$), and combination modes ($\omega_s + k\omega_b$) of the stretch mode of this C $-$ H bond have been reported [269, 270], and references therein). The results are summarized in Table 8.22.

Table 8.22 Frequencies (cm^{-1}), FWHM (cm^{-1}), relative intensities, and assignments of the IR lines associated with the absorption at $3107 \, \text{cm}^{-1}$ of a C–H bond observed in natural type Ia diamonds (room temperature values). The positions, FWHMs, and assignments are from [270], and the relative intensities from [269, 271]

Freq.	FWHM	Relative peak height	Assignment
1405	1.5	400	ω_b
2786	2.5	34	$2\omega_b$
3107	2.9	1000	ω_s
4169	4.4	4	$3\omega_b$
4499	4.8	29	$\omega_s + \omega_b$
5555	–	~0.1	$4\omega_b$
5889	–	~0.6	$\omega_s + 2\omega_b$
6070	–	~1.2	$2\omega_s$

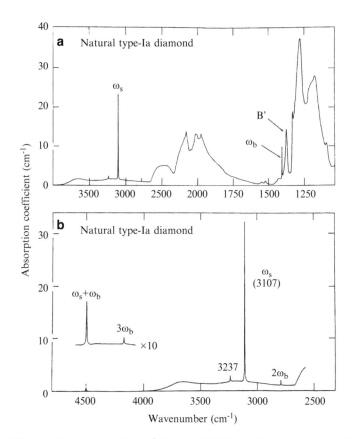

Fig. 8.30 RT absorption of the $3107 \, \text{cm}^{-1}$ H-related LVM and of other associated LVMs in a natural diamond sample. The assignments are those of Table 8.21. Spectrum (**a**) was recorded with a commercial double beam spectrometer, with automatic change of the wavenumber scale by a factor of 2 at $2000 \, \text{cm}^{-1}$, and a region of the sample $8 \times 2 \, \text{mm}$ was used. For spectrum (**b**), a region $2 \, \text{mm} \times 20 \, \mu\text{m}$ was used, and the specimen inhomogeneity explains the absorption difference with respect to (**a**). After [270]. Band B' is due to the presence of platelets in the sample and the line at $3237 \, \text{cm}^{-1}$ tentatively attributed to a N–H vibration [269]

In a synthetic diamond grown from ^{13}C and estimated from the published spectra [269] to contain about 90% of ^{13}C, the frequencies of the $^{12}\text{C} - \text{H}$ and $^{13}\text{C} - \text{H}$ stretch modes are found to be the same as in natural diamond. For the bending mode, however, which is relatively close to the Raman frequency of diamond, the frequencies for $^{12}\text{C} - \text{H}$ and $^{13}\text{C} - \text{H}$ decrease to 1396 and 1391 cm^{-1}, respectively, compared to $1405 \, \text{cm}^{-1}$ in natural diamond, because of the decrease of the Raman frequency by about $50 \, \text{cm}^{-1}$ in the quasi-^{13}C diamond [269].

The frequencies and the FWHMs of these modes present a very small variation with temperature between LHeT and RT ([270], and references therein). The most intense transitions of the centre associated with this $\text{C} - \text{H}$ bond are displayed in Fig. 8.30a, b.

The exact nature of the centre associated with the $3107\,\text{cm}^{-1}$ LVM remains elusive. If it seems clear that the $3107\,\text{cm}^{-1}$ LVM does not involve an $N - H$ bond, it has been pointed out that this did not necessarily mean that N was absent in the centre itself. A good counter-example is the (Donor, hydrogen) centres in silicon, where the LVM observed is due to a $Si - H_{AB}$ stretch mode, and a comparable structure, involving a distant N atom has been suggested for the $3107\,\text{cm}^{-1}$ centre in diamond [272]. In the spectrum of a natural type IaB diamond sample, the $3107\,\text{cm}^{-1}$ LVM has been reported to disappear after annealing for 12 h at $\sim 1900°\text{C}$ under a pressure of $\sim 7\,\text{GPa}$, but these annealing conditions had practically no effect on a natural type IA/B diamond [273].

A relatively weak LVM at $3237\,\text{cm}^{-1}$ (reported also at $3236\,\text{cm}^{-1}$ at RT) showing no C IS has also been observed in all these diamonds [269]. It does not correlate to the intensities of the LVMs of the $3107\,\text{cm}^{-1}$ family, and it has been suggested that it could be due to an $N - H$ stretch mode [268]. In natural diamond samples where the $3236\,\text{cm}^{-1}$ line is more intense, it has been reported to be consistently associated with another line at $4703\,\text{cm}^{-1}$, with a peak intensity weaker by a factor of ~ 70 [264].

In diamonds with type Ib characteristics, the $3107\,\text{cm}^{-1}$ LVM is very weak or absent. In these samples, three LVMs at 3343, 3372, and $3394\,\text{cm}^{-1}$ are prominent and they have been tentatively assigned to $N - H$ stretch modes [268].

8.7.2 (O, H) Centres in GaAs and GaP

8.7.2.1 The 3300 cm^{-1} Line in GaAs

In LEC SI GaAs samples containing O_i, the absorption at LHeT between ~ 2950 and $3460\,\text{cm}^{-1}$ of several sharp vibrational lines was consistently observed under TEC [141, 274]. The most intense one, located at $3300.0\,\text{cm}^{-1}$, with a FWHM of $0.13\,\text{cm}^{-1}$, displays a weak satellite at $3289.5\,\text{cm}^{-1}$, with a relative intensity of 0.003. This satellite has been attributed to the ^{18}O isotopic component of the $3300\,\text{cm}^{-1}$ line, assumed to be due to the stretch mode of a $^{16}\text{O–H}$ bond. The frequency of the D counterpart of the $3300\,\text{cm}^{-1}$ line was reported at $2455.0\,\text{cm}^{-1}$ [274], and it can be checked that the $^{16}\text{O}/^{18}\text{O}$ IS and a value of $r(\text{H/D})$ of 1.346 are consistent with the O–H attribution. This is confirmed by their similarities with the values of Table 8.8 for (O–H)$^-$ bonds in oxide crystal.

The measurements of the splitting of the $3300\,\text{cm}^{-1}$ line (line 14 of Table I of [141]) under a uniaxial stress exclude a trigonal symmetry and predict distinct orientations under stress of this (O–H) centre, and its possibility to reorient at LHeT [141]. The frequency of the $3300\,\text{cm}^{-1}$ line shows an *increase* with temperature, which amounts to $2.8\,\text{cm}^{-1}$ at 210 K [275]. This indicates a rather weak coupling of this bond to the lattice. A low-temperature illumination through an interference filter[15] centred at $1.25\,eV$ ($\sim 10000\,cm^{-1}$) produces a decrease of the intensity of the $3300\,cm^{-1}$ line and the rise of a new line at $3296.4\,cm^{-1}$ (line 13 of Table I of [141]), with a FWHM of $0.5\,cm^{-1}$. The same is observed for the O–D mode at $2455\,cm^{-1}$ and the new line appears at $2451.6\,cm^{-1}$ [13].

It has been suggested that the centre giving the $3300\,cm^{-1}$ line (noted $(O–H)^A$ in [274]) was an OH radical centred in an As vacancy V_{As} [141]. It would be stabilized in this location by a rearrangement of the orbitals of the Ga atoms nn of V_{As} and this seems to be comforted by the small anharmonicity of the stretch mode. The piezospectroscopic effects can be explained by the stress-induced rearrangement of the Ga orbitals and a location of OH along <111> axes has been proposed. The trapping of a photohole by V_{As} would reduce the electronic potential felt by the OH radical, and could allow for hindered rotation of OH among equivalent orientations, and this could possibly account for the fourfold increase of the FWHMs of the photoinduced lines at $3296\,cm^{-1}$ [141].

Isochronal annealing experiments have shown that this OH centre decays in the 550–650°C temperature region [13].

8.7.2.2   Other (O,H) Centres in GaAs and GaP

In the same LEC SI GaAs samples, H-related LVMs at 3108.0 and $3235.1\,cm^{-1}$ were observed at LHeT (lines 7 and 11, respectively, of Table I of [141, 274], and their equivalent in GaP were reported at 3106.0 and $3250.9\,cm^{-1}$, respectively [274]. The FWHMs of the LVMs at $3108/3106\,cm^{-1}$ are $\sim 0.06–0.09\,cm^{-1}$ while those of the LVMs at $3235/3250\,cm^{-1}$ are $\sim 0.3–0.5\,cm^{-1}$. Only one D-related set at $2308/2307\,cm^{-1}$ corresponding to the $3108/3106\,cm^{-1}$ set is observed in samples containing both H and D. This differs from the $3235/3250\,cm^{-1}$ set for which two close D-related LVMs (2400.1 and $2403.3\,cm^{-1}$ in GaAs and 2414.4 and $2415.7\,cm^{-1}$ in GaP) are observed in this case. The $r(H/D)$ ratios of these LVMs are between 1.346 and 1.348. The corresponding centres, assumed to involve oxygen and to be the same in GaAs and GaP, are labelled $(O–H)^I$ and $(O–H)^{II}$ for the 3108/3106 and $3235/3250\,cm^{-1}$ sets, respectively [274].

In GaAs, the intensity of a LVM at $1043\,cm^{-1}$ is correlated to the one of the $3235\,cm^{-1}$ LVM of $(O–H)^{II}$, and it should be a bending mode of an O–H bond. In GaAs samples containing both H and D, this bending mode is detected at $1043.0\,cm^{-1}$, but two bending modes of O–D bonds are observed at 777.5 and $792.4\,cm^{-1}$. It is suggested that in the $(O–H)^{II}$ centre, two H atoms on equivalent sites are bonded to an O atom. In samples containing H and D, the two D-related lines correspond to a centre with two D atoms or only one D atom. The observation of unresolved H-related LVMs for this centre has been attributed to the weakness

---

[15]The process is faster by using white light and saturation is reached after an illumination time of only 10 min, compared to 150 min. with the filter.

of the interaction between the two O–H stretch modes of the H atoms on equivalent sites.

The frequencies of the O–H stretch modes of the $(O–H)^{II}$ centre in GaAs and GaP display a negative shift for increasing temperature, almost linear between 80 and 300 K (-0.018 and $-0.015\,\text{cm}^{-1}\,\text{K}^{-1}$ for GaP and GaAs, respectively) [274]. At a difference with the $3300\,\text{cm}^{-1}$ LVM discussed earlier, this indicates that the O atom of the $(O–H)^{II}$ centre interacts "reasonably" with the lattice. Considering the existence of O_P in GaP and postulating the existence of O_{As} in GaAs, it would be interesting to calculate the stability of a structure resulting from the bonding of two H atoms to O_s by the breaking of two weak O_s − Ga bonds and the reconstruction of a bond between the Ga dangling bonds. It must be pointed out that in GaAs, O_{As} should be associated with off-centre O_{As} discussed in Sect. 7.1.3, and the formation of the centre would only require the bonding of two H atoms to the off-centre O_{As} atom. Because of the ionicity of GaAs and of the magnitude of the distortion of the O atom from the substitutional site, this situation would be different from the one in silicon, where, in the VOH_2 centre, the two H atoms bind to Si atoms (see Fig. 7.5). The kind of bonding of two H atoms to the O_{As} atom suggested for GaAs and GaP would certainly still reduce the interaction between the O atom and the two Ga atoms, leaving as a limiting case, something looking like a lattice vacancy occupied by a water molecule.

The frequency of the stretch mode of the O–H bond of $(O–H)^I$ centre in GaAs shifts by $-0.2\,\text{cm}^{-1}$ between LHeT and 200 K [275]. Under illumination with 1.2 eV photons, the intensities of the 2308 and $3108\,\text{cm}^{-1}$ stretch modes of $(O–D)^I$ and $(O–H)^I$ decrease and new lines appear at 2314.7 and $3117.6\,\text{cm}^{-1}$, with FWHMs of 0.3 and $0.5\,\text{cm}^{-1}$, respectively. This bears some resemblance with the behaviour of the stretch modes of $(O–H)^A$ discussed in the preceding section, but there is presently no model of the $(O–H)^I$ centre in GaAs and GaP.

8.7.3 "Hidden" Hydrogen in InP

In semiconductors, it can be difficult to detect molecular hydrogen by infrared absorption, because of the small value of the effective charge associated to this molecule in crystals. For this reason, it has sometimes been termed "hidden" hydrogen, but it has been seen above that it could generally be detected in absorption, even if Raman scattering is more sensitive.

In LEC InP, there exists no direct evidence of the presence of molecular hydrogen. Now, in LEC SI InP:Fe samples where the $V_{In}H_4$ LVM at $2316\,\text{cm}^{-1}$ is observed, together with the electronic lines of Fe^{2+} between 2800 and $3200\,\text{cm}^{-1}$, intriguing spectroscopic results have been obtained [276].[16] When these InP:Fe samples are illuminated during their cooling down from RT to LHeT (actually

---

[16]In this reference, the contents of pp. 852 and 853 have been inverted.

between $\sim$80 and 20 K) at a typical cooling rate of $\sim$2 K/min with photons between $\sim$1 and 1.3 eV (IDCD procedure), two new absorption features noted B(road) and S(harp) are observed at 2103 and 2220 cm^{-1}, respectively, together with an increase of the intensity of the Fe^{2+} electronic lines. No other photoinduced change is observed between 450 and 4600 cm^{-1}. It must be pointed out that the IDCD procedure is mandatory to observe the changes and that illumination performed at LHeT produces no change with respect to thermal equilibrium. The above photoinduced features can be bleached out by heating the samples to 80 K, and the process can be started again.

At LHeT, the FWHM of B is $\sim$3 cm^{-1} and the S feature can be resolved into three components at $\sim$2220.4, $\sim$2220.5, and 2220.69 cm^{-1}, with FWHMs of $\sim$0.07 cm^{-1}. At LHeT, these features are metastable as they are observed with external illumination turned-off, but S can be bleached out by illumination with 0.7 eV photons, indicating that B and S are associated with different centres [276].

The frequency of the S feature is within the range of the P $-$ H stretch modes frequencies in InP. The one of B is lower, but these two features have been none the less attributed to vibrational modes.

The presence of Fe^{3+} in the InP samples showing these photoinduced effects was detected by ESR. The photoinduced increase of the intensity of the Fe^{2+} lines arises from transitions from the VB to Fe^{3+}, leaving holes that can be trapped by different centres. The occurrence of features B and S can be explained by assuming that at not too low temperatures, IR-inactive H-related centres turn into metastable IR-active states when trapping a hole, and this ad hoc mechanism can explain the origin of the S and B features. It is very difficult, however, to ascertain if molecular hydrogen is involved in these processes.

8.7.4   Unidentified H-Related LVMs

In this chapter, we have discussed what seemed to us the H-related vibrational absorption of the most important and reasonably identified centres in diamond and in semiconductors. Many other H-related LVMs have been reported that are not mentioned here [13, 22, 56, 265, 277], and their frequencies can be found in these references.

Recently, some light has been shed on the origin of LVMs reported without identification in FZ silicon with a high carbon concentration [56], and more specifically those at 2183.2, 2214.5, and 2826.9 cm^{-1}. These LVMs, and several other ones, have been observed in hydrogenated and deuterated microcrystalline silicon samples annealed above $\sim$500°C [278]. The result of mixed hydrogenation and deuteration showed that the centre associated with these three LVMs contained several H atoms. The annealing conditions after which these LVMs were observed coupled to theoretical calculations established that they were due to the vibrational modes of a VH_4 centre (see Sect. 8.2.3.1) where one H atom is bonded to a C atom (VH_3, HC) centre. The frequencies of Si–H, Si–D, C–H, and C–D modes of this centre observed in (hydrogenated + deuterated) samples are also given in this reference ([278], and references therein).

References

1. S.J. Pearton, J.W. Corbett, M. Stavola, *Hydrogen in Crystalline Semiconductors* (Springer, Berlin, 1992)
2. J.I. Pankove, N.M. Johnson (eds.), *Semiconductors and Semimetals. Hydrogen in Semiconductors*, vol. 34 (Academic, San Diego, 1991)
3. S. Pearton (ed.), *Hydrogen in Compound Semiconductors*. Mater. Sci. Forum **148–149** (1994)
4. N. Nickel (ed.), *Semiconductors and Semimetals. Hydrogen in Semiconductors II*, vol. 61 (Academic, San Diego, 1999)
5. J. Chevallier, Hydrogen in crystalline semiconductors. Ann. Rev. Mater. Sci. **18**, 219–256 (1988)
6. S.M. Myers, M.I. Baskes, H.K. Birnbaum, J.W. Corbett, G.G. DeLeo, S.K. Estreicher, E.E. Haller, P. Jena, N.M. Johnson, R. Kircheim, S.J. Pearton, M.J. Stavola, Hydrogen interactions with defects in crystalline solids. Rev. Mod. Phys. **64**, 559–617 (1992)
7. S.K. Estreicher, Hydrogen-related defects in crystalline semiconductors: a theorist's perspective. Mater. Sci. Eng. R **14**, 319–412 (1995)
8. J. Chevallier, B. Pajot, Interaction of hydrogen with impurities and defects in semiconductors. Diff. Defect Data B (Solid State Phenom.) **85–86**, 203–283 (2002)
9. J.P. Goss, Theory of hydrogen in diamond. J. Phys. Cond. Matter **15**, R551–R580 (2003)
10. J.I. Pankove, D.E. Carlson, J.E. Berkeyheiser, R.O. Wance, Neutralization of shallow acceptor levels in silicon by atomic hydrogen. Phys. Rev. Lett. **51**, 2224–2225 (1983)
11. C.T. Sah, J.Y.C. Sun, J.J. Tzou, Deactivation of the boron acceptor in silicon by hydrogen. Appl. Phys. Lett. **43**, 204–206 (1983)
12. M.D. McCluskey, Local vibrational modes of impurities in semiconductors. J. Appl. Phys. **87**, 3593–3617 (2000)
13. W. Ulrici, Hydrogen-impurity complexes in III-V semiconductors. Rep. Prog. Phys. **67**, 2233–2286 (2004)
14. E.V. Lavrov, J. Weber, Hydrogen molecules in GaAs. Physica B **340–342**, 329–332 (2003)
15. M. Hiller, E.V. Lavrov, J. Weber, Raman scattering study of H_2 in Si. Phys. Rev. B **74**, 235214/1–9 (2006)
16. M. Hiller, E.V. Lavrov, J. Weber, Raman spectroscopy of hydrogen molecules in germanium. Physica B **376–377**, 142–145 (2006)
17. M.K. Weldon, V.E. Marsico, Y.J. Chabal, A. Agarwal, D.J. Eaglesham, J. Sapjeta, W.L. Brown, D.C. Jacobson, Y. Caudano, S.B. Christman, E.E. Chaban, On the mechanism of the hydrogen-induced exfoliation of silicon. J. Vac. Sci. Technol. B **15**, 1065–1073 (1997)
18. E.V. Lavrov, J. Weber, Structural properties of hydrogen-induced platelets in silicon: a Raman scattering study. Physica B **308–310**, 151–154 (2001)
19. F. Fuchs, C. Wild, K. Schwarz, W. Müller-Sebert, P. Koidl, Hydrogen induced vibrational and electronic transitions in chemical vapor deposited diamond, identified by isotopic substitution. Appl. Phys. Lett. **66**, 177–179 (1995)
20. S.K. Estreicher, M.A. Roberson, D.M. Maric, Hydrogen and hydrogen dimers in c-C, Si, Ge, and α-Sn. Phys. Rev. B **50**, 17018–17027 (1994)
21. J.P. Goss, R. Jones, M.I. Heggie, C.P. Ewels, P.R. Briddon, S. Öberg, Theory of hydrogen in diamond. Phys. Rev. B **65**, 115207/1–13 (2002)
22. M. Budde, Doctoral thesis, University of Aarhus, Denmark (1998)

23. H.J. Stein, Vacancies and the chemical trapping of hydrogen in silicon. Phys. Rev. Lett. **43**, 1030–1033 (1979)

24. Yu.V. Gorelkinskii, N.N. Nevinnyi, Electron paramagnetic resonance of hydrogen in silicon. Physica B **170**, 155–167 (1991)

25. Yu.V. Gorelkinskii, Electron paramagnetic resonance studies of hydrogen and hydrogen-related defects in crystalline silicon. Semiconductors and Semimetals, **61**, 25–81 (1999)

26. M. Budde, B. Bech Nielsen, C. Parks Cheney, N.H. Tolk, L.C. Feldman, Local vibrational modes of isolated hydrogen in germanium. Phys. Rev. Lett. **85**, 2965–2968 (2000)

27. K. Bonde Nielsen, B. Bech Nielsen, J. Hansen, E. Andersen, J.U. Andersen, Bond-centered hydrogen in silicon studied by *in situ* deep-level transient spectroscopy. Phys. Rev. B **60**, 1716–1728 (1999)

28. A. Balsas, V.J.B. Torres, J. Coutinho, R. Jones, B. Hourahine, P.R. Briddon, M. Barroso, Vibrational properties of elemental hydrogen centres in Si, Ge and dilute SiGe alloys. J. Phys. Cond. Matter **17**, S2155–S2164 (2005)

29. M. Budde, G. Lüpke, C. Parks Cheney, N.H. Tolk, L.C. Feldman, Vibrational lifetime in bond-center hydrogen in crystalline silicon. Phys. Rev. Lett. **85**, 1452–1455 (2000)

30. R.N. Pereira, T. Ohya, K.M. Itoh, B. Bech Nielsen, Local vibrational modes of bond-centered H in ^{28}Si, ^{29}Si, and ^{30}Si crystals. Physica B **340–342**, 697–700 (2003)

31. R.N. Pereira, B. Bech Nielsen, J. Coutinho, V.J.B. Torres, R. Jones, T. Ohya, K.M. Itoh, P.R. Briddon, Anharmonicity and lattice coupling of bond-centered hydrogen and interstitial oxygen defects in monoisotopic silicon crystals. Phys. Rev. B **72**, 115212/1–13 (2005)

32. S.V.S.N. Rao, S.K. Dixit, G. Lüpke, N.H. Tolk, L.C. Feldman, Effect of energetic ions on the stability of bond-center hydrogen in silicon. Phys. Rev. B **75**, 235202/1–6 (2007)

33. C.G. Van de Walle, Theory of interstitial hydrogen and muonium in crystalline semiconductors. Semiconductors and Semimetals **34**, 585–622 (1991)

34. R.N. Pereira, B. Bech Nielsen, L. Dobaczewski, A.R. Peaker, N.V. Abrosimov, Local modes of bond-centered hydrogen in Si:Ge and Ge: Si. Phys. Rev. B **71**, 195201/1–13 (2005)

35. R.N. Pereira, B. Bech Nielsen, J. Coutinho, Local modes of hydrogen defects in Si:Ge and Ge:Si. Physica B **376–377**, 22–27 (2006)

36. M. Budde, C. Parks Cheney, G. Lüpke, N.H. Tolk, L.C. Feldman, Vibrational dynamics of bond-center hydrogen in crystalline silicon. Phys. Rev. B **63**, 195203/1–18 (2001)

37. C. Parks Cheney, M. Budde, G. Lüpke, L.C. Feldman, N.H. Tolk, Vibrational dynamics of isolated hydrogen in germanium. Phys. Rev. B **65**, 035214/1–9 (2002)

38. R.N. Pereira, B. Bech Nielsen, M. Stavola, S.K. Sanati M Estreicher, M. Mizuta, Local vibrational modes of hydrogen in GaN: observation and theory. Physica B **376–377**, 464–467 (2006)

39. E. Mollwo, Die Wirkung von Wasserstoff auf die Leifähigkeit und Lumineszenz von Zincoxydkristallen. Z. Physik **138**, 478–488 (1954)

40. E.V. Lavrov, F. Herklotz, J. Weber, Identification of two hydrogen donors in ZnO. Phys. Rev. B **79**, 165210/1–13 (2009)

41. K. Ip, M.E. Overberg, Y.W. Heo, D.P. Norton, S.J. Pearton, C.E. Stutz, B. Luo, F. Ren, D.C. Look, J.M. Zavada, Hydrogen incorporation and diffusivity in plasma-exposed bulk ZnO. Appl. Phys. Lett. **82**, 385–387 (2003)

42. M.G. Wardle, J.P. Goss, P.R. Briddon, First-principles study of the diffusion of hydrogen in ZnO. Phys. Rev. Lett. **96**, 205504/1–4 (2006)

43. E.V. Lavrov, F. Herklotz, J. Weber, Identification of hydrogen molecules in ZnO. Phys. Rev. Lett. **102**, 185502/1–4 (2009)

44. P. Briddon, R. Jones, G.M.S. Lister, Hydrogen in diamond. J. Phys. C **11**, L1027–L1031 (1988)

45. K.J. Chang, D.J. Chadi, Diatomic-hydrogen-complex diffusion and self-trapping in crystalline silicon. Phys. Rev. Lett. **62**, 937–940 (1989)

46. K.J. Chang, D.J. Chadi, Hydrogen bonding and diffusion in crystalline silicon. Phys. Rev. B **40**, 11644–11653 (1989)

47. J.D. Holbech, B. Bech Nielsen, R. Jones, P. Sitch, S. Öberg, H_2* defect in crystalline silicon. Phys. Rev. Lett. **71**, 875–878 (1993)

48. B. Bech Nielsen, H.G. Grimmeiss, Effect of uniaxial stress on local vibrational modes of hydrogen in ion-implanted silicon. Phys. Rev. B **40**, 12403–12415 (1989)

49. M. Budde, B. Bech Nielsen, R. Jones, J. Goss, S. Öberg, Local modes of the H_2^* dimer in germanium. Phys. Rev. B **54**, 5485–5494 (1996)

50. M. Suezawa, Formation of defect complexes by electron-irradiation of hydrogenated crystalline silicon. Phys. Rev. B **63**, 035201/1–7 (2000)

51. H.J. Stein, Bonding and thermal stability of implanted hydrogen in silicon. J. Electron. Mater. **4**, 159–174 (1975)

52. B.N. Mukashev, K.H. Nussupov, M.F. Tamendarov, V.V. Frolov, On the identification of the vibrational spectra of hydrogen-implanted crystalline silicon. Phys. Lett. A **87**, 376–380 (1982)

53. G.R. Bai, M.W. Qi, M.W. Xie, T.S. Shi, The isotope study of the Si − H absorption peaks in the FZ − Si grown in hydrogen atmosphere. Solid State Commun. **56**, 277–281 (1985)

54. B. Bech Nielsen, P. Johannesen, S. Stallinga, K. Bonde Nielsen, J.R. Byberg, Identification of the silicon vacancy containing a single hydrogen atom by EPR. Phys. Rev. Lett. **79**, 1507–1510 (1997)

55. B. Bech Nielsen, J. Olajos, H.G. Grimmeiss, Hydrogen-related center with tetrahedral symmetry in ion-implanted silicon. Phys. Rev. B **39**, 3330–3336 (1989)

56. B. Pajot, B. Clerjaud, Z.J. Xu, High-frequency hydrogen-related infrared modes in silicon grown in a hydrogen atmosphere. Phys. Rev. B **59**, 7500–7506 (1999)

57. J.M. Chen, L.M. Xie, G.R. Bai, Y.G. Zhao, Temperature effects of several intense IR absorption peaks in NTD Si:H. Phys. Stat. Sol. B **153**, 107–114 (1989)

58. B. Hourahine, R. Jones, S. Öberg, P.R. Briddon, Self-interstitial-hydrogen complexes in silicon. Phys. Rev. B **59**, 15729–15732 (1999)

59. P. Stallinga, P. Johannesen, S. Herstrøm, K. Bonde Nielsen, B. Bech Nielsen, J.R. Byberg, Electron paramagnetic resonance study of hydrogen-vacancy defects in crystalline silicon. Phys. Rev. B **58**, 3842–3852 (1998)

60. B. Bech Nielsen, L. Hoffmann, M. Budde, R. Jones, J. Goss, S. Öberg, H interacting with intrinsic defects in Si. Mater. Sci. Forum **196–201**, 933–938 (1995)

61. T. Ohya, K.M. Itoh, R.N. Pereira, B. Bech Nielsen, Host isotope effect on the local vibration modes of VH_2 and VOH_2 defects in isotopically enriched ^{28}Si, ^{29}Si and ^{30}Si single crystals. Jpn. J. Appl. Phys. **44**, 7309–7313 (2005)

62. E.V. Lavrov, J. Weber, L. Huang, B. Bech Nielsen, Vacancy-hydrogen defects in silicon studied by Raman spectroscopy. Phys. Rev. B **64**, 035204/1–5 (2001)

63. M. Budde, G. Lüpke, E. Chen, X. Zhang, N.H. Tolk, L.C. Feldman, E. Tarhan, A.K. Ramdas, M. Stavola, Lifetimes of hydrogen and deuterium related vibrational modes in silicon. Phys. Rev. Lett. **87**, 145501/1–4 (2001)

64. V.P. Markevich, L.I. Murin, M. Suezawa, J.L. Lindström, J. Coutinho, R. Jones, P.R. Briddon, S. Öberg, Observation and theory of the $V − O–H_2$ complex in silicon. Phys. Rev. B **61**, 12964–12969 (2000)

65. C. Glover, M.E. Newton, P.M. Martineau, S. Quinn, D.J. Twitchen, Hydrogen incorporation in diamond: the vacancy-hydrogen complex. Phys. Rev. Lett. **92**, 135502/1–4 (2004)

66. C. Ascheron, C. Bauer, H. Sobotta, V. Riede, Investigations of hydrogen implanted GaP single crystals by means of particle induced γ-spectroscopy, infrared spectroscopy, and Rutherford backscattering channelling technique. Phys. Stat. Sol. A **89**, 549–557 (1985)

67. D.W. Fisher, M.O. Manasreh, D.N. Talwar, G. Matous, Isochronal annealing of local vibrational modes in proton- and deuteron-implanted InP. J. Appl. Phys. **73**, 78–83 (1993)

68. B. Clerjaud, D. Côte, C. Naud, Evidence for hydrogen-transition metal complexes in as-grown indium phosphide. J. Cryst. Growth **83**, 190–193 (1987)

69. B. Clerjaud, D. Côte, C. Naud, Evidence for complexes of hydrogen with deep-level defects in bulk III-V materials. Phys. Rev. Lett. **58**, 1755–1757 (1987)

70. R. Darwich, B. Pajot, B. Rose, D. Robein, B. Theys, R. Rahbi, C. Porte, F. Gendron, Experimental study of the hydrogen complexes in InP. Phys. Rev. B **48**, 17776–17790 (1993)
71. W. Ulrici, A. Kwasniewski, M. Czupalla, M. Neubert, Phosphorus-hydrogen complexes in LEC-grown InP. Phys. Stat. Sol. B **242**, 873–880 (2005)
72. V. Riede, H. Sobotta, H. Neumann, C. Ascheron, C. Nellmeijer, A. Schindler, Depth dependence of localized mode absorptions in proton-implanted InP. Phys. Stat. Sol. A **116**, K147–K152 (1989)
73. D.W. Fisher, M.O. Manasreh, G. Matous, Local mode spectroscopy of proton- and deuteron-implanted InP. J. Appl. Phys. **71**, 4805–4808 (1992)
74. J. Tatarkiewicz, B. Clerjaud, D. Côte, F. Gendron, A.M. Hennel, Local modes of vibration in proton- and deuteron-implanted InP. Appl. Phys. Lett. **53**, 382–384 (1988)
75. F.X. Zach, E.E. Haller, D. Gabbe, G. Iseler, G.G. Bryant, D.F. Bliss, Electrical properties of the hydrogen defect in InP and the microscopic structure of the 2316 cm^{-1} hydrogen related line. J. Electron. Mater. **25**, 331–335 (1996)
76. A. Zappettini, R. Fornari, R. Capelleti, Electrical and optical properties of semi-insulating InP obtained by wafer and ingot annealing. Mater. Sci. Eng. B **45**, 147–151 (1997)
77. Y.W. Zhao, X.L. Xu, M. Gong, S. Fung, C.D. Beling, X.D. Chen, N.F. Sun, T.N. Sun, S.L. Lin, G.Y. Yang, X.B. Buo, Y.Z. Sun, L. Wang, Q.Y. Zheng, Z.H. Zhou, J. Chen, Formation of P_{In} defect in annealed liquid-encapsulated Czochralski InP. Appl. Phys. Lett. **72**, 2126–2128 (1998)
78. C.P. Ewels, S. Öberg, B. Jones, B. Pajot, P.R. Briddon, Vacancy- and acceptor-H complexes in InP. Semicond. Sci. Technol. **11**, 502–507 (1996)
79. B. Pajot, Electrical and optical properties of hydrogen-containing indium phosphide, *InP and Related Compounds: Materials, Applications and Devices*. (*Optoelectronic Properties of Semiconductors and Superlattices*), vol. 9 ed. by M.O. Manasreh (Gordon and Breach, New York 2000), Chapt. 5
80. A. Amore Bonapasta, M. Capizzi, Hydrogen as a deep impurity in semiconductors and its interaction with deep centers in III-V compounds. Defect Diff. Forum **157–159**:133–174 (1998)
81. B. Clerjaud, M. Krause, C. Porte, W. Ulrici, Study of the hydrogen related complexes in GaAs under uniaxial stress, *Proceedings of the 19th International Conference on the Physics of Semiconductors* ed. by W. Zawadzki (Institute of Physics, Polish Academy of Sciences 1988) pp. 1175–1178
82. B. Pajot, B. Clerjaud, J. Chevallier, Vibrational properties of hydrogen in compound semiconductors. Physica B **170**, 371–382 (1991)
83. A.F. Wright, Interaction of hydrogen with gallium vacancies in wurtzite GaN. J. Appl. Phys. **90**, 1164–1169 (2001)
84. M.G. Weinstein, C.Y. Song, M. Stavola, S.J. Pearton, R.G. Wilson, R.J. Shul, K.P. Kileen, M.J. Ludowise, Hydrogen-decorated lattice defects in proton-implanted GaN. Appl. Phys. Lett. **72**, 1703–1705 (1998)
85. B. Clerjaud, D. Côte, A. Lebkiri, C. Naud, J.M. Baranowski, K. Pakula, D. Wasik, T. Suski, Infrared spectroscopy of Mg-H local vibrational mode in GaN with polarized light. Phys. Rev. B **61**, 8238–8241 (2000)
86. E.V. Lavrov, J. Weber, F. Börnert, C.G. Van de Walle, R. Helbig, Hydrogen-related defects in ZnO studied by infrared absorption spectroscopy. Phys. Rev. B **66**, 165205/1–7 (2002)
87. F. Herklotz, E.V. Lavrov, Vl. Kolkovsky, J. Weber, M. Stavola, Charge states of a hydrogen defect with a local vibrational mode at 3326 cm^{-1} in ZnO. Phys. Rev. B **82**, 115206/1-10 (2010)
88. M.G. Wardle, J.P. Goss, P.R. Briddon, Theory of Fe, Co, Ni, Cu, and their complexes with hydrogen in ZnO. Phys. Rev. B **72**, 155108/1–13 (2005)
89. M. Budde, B. Bech Nielsen, P. Leary, J. Goss, R. Jones, P.R. Briddon, S. Öberg, S.J. Breuer, Identification of the hydrogen-saturated self-interstitials in silicon and germanium. Phys. Rev. B **57**, 4397–4412 (1998)

90. C.G. Van de Walle, J. Neugebauer, Hydrogen interactions with self-interstitials in silicon. Phys. Rev. B **52**, R14320–R14323 (1995)

91. R.J. Needs, First-principles calculations of self-interstitial defect structures and diffusion paths in silicon. J. Phys Cond. Matter **11**, 10437–10450 (1999)

92. M. Gharaibeh, S.K. Estreicher, P.A. Fedders, P. Ordejon, Self-interstitial-hydrogen complexes in Si. Phys. Rev. B **64**, 235211/1–7 (2001)

93. T.S. Shi, G.R. Bai, M.W. Qi, J.K. Zhou, IR studies of the electron-irradiated silicon crystal grown in hydrogen atmosphere. Mater. Sci. Forum **10–12**, 597–602 (1986)

94. L.M. Xie, M.W. Qi, J.M. Chen, The nature of several intense Si − H infrared stretching peaks in the neutron-transmutation-doped Si − H system. J. Phys. Condens. Matter **3**, 8519–8528 (1991)

95. Y.C. Du, Y.F. Zhang, X.T. Meng, H.Y. Shen, Infrared absorption bands of Si–H centers in γ-ray irradiated FZ-Si grown in hydrogen atmosphere and their identification. Sci. Sinica A **30**, 176–185 (1987)

96. S.V.S.N. Rao, S.K. Dixit, G. Lüpke, N.H. Tolk, L.C. Feldman, Reconfiguration and dissociation of bonded hydrogen in silicon by energetic ions. Phys. Rev. B **83**, 045204/1–6 (2011)

97. R.W. Terhune, C.W. Peters, Electric field induced vibration rotation spectrum of H_2 and D_2. J. Mol. Spectr. **3**, 138–147 (1959)

98. B.P. Stoicheff, High resolution Raman spectroscopy of gases: IX. Spectra of H_2, HD, and D_2. Can. J. Phys. **35**, 730–741 (1957)

99. J. Vetterhöffer, J. Wagner, J. Weber, Isolated hydrogen molecule in GaAs. Phys. Rev. Lett. **77**, 5409–5412 (1996)

100. C.G. Van de Walle, Energetics and vibrational frequencies of interstitial H_2 molecules in semiconductors. Phys. Rev. Lett. **80**, 2177–2180 (1998)

101. R.E. Pritchard, M.J. Ashwin, J.H. Tucker, R.C. Newman, E.C. Lightowlers, M.J. Binns, S.A. McQuaid, R. Falster, Interaction of hydrogen molecules with bond-centered interstitial oxygen and another defect center in silicon. Phys. Rev. B **56**, 13118–13125 (1997)

102. R.E. Pritchard, M.J. Ashwin, J.H. Tucker, R.C. Newman, Isolated interstitial hydrogen molecules in hydrogenated crystalline silicon. Phys. Rev. B **57**, R15048–R15051 (1998)

103. A.W.R. Leitch, V. Alex, J. Weber, H_2 molecules in c-Si after plasma treatment. Solid State Commun. **105**, 215–219 (1998)

104. A.W.R. Leitch, V. Alex, J. Weber, Raman spectroscopy of hydrogen molecules in crystalline silicon. Phys. Rev. Lett. **81**, 421–424 (1998)

105. S. Koch, E.V. Lavrov, J. Weber, Rovibrational states of interstitial H_2 in Si. Phys. Rev. B **83**, 233203/1–4 (2011)

106. J.A. Zhou, M. Stavola, Symmetry of molecular H_2 in Si from a uniaxial stress study of the $3618.4\,cm^{-1}$ vibrational line. Phys. Rev. Lett. **83**, 1351–1354 (1999)

107. E.E. Chen, M. Stavola, W.B. Fowler, P. Walters, Key to understanding interstitial H_2 in silicon. Phys. Rev. Lett. **88**, 105507/1–4 (2002)

108. E.E. Chen, M. Stavola, W.B. Fowler, J.A. Zhou, Rotation of molecular hydrogen in Si: unambiguous identification of ortho-H_2 and para-H_2. Phys. Rev. Lett. **88**, 245503/1–4 (2002)

109. E.V. Lavrov, J. Weber, Ortho and para interstitial H_2 in silicon. Phys. Rev. Lett. **89**, 215501/1–4 (2002)

110. G.A. Shi, M. Stavola, W.B. Fowler, E.E. Chen, Rotational-vibrational transitions of interstitial HD in Si. Phys. Rev. B **72**, 085207/1–6 (2005)

111. R.E. Pritchard, J.H. Tucker, R.C. Newman, E.C. Lightowlers, Hydrogen molecules in boron-doped silicon. Semicond. Sci. Technol. **14**, 77–80 (1999)

112. M. Hiller, E.V. Lavrov, J. Weber, Ortho-para conversion of interstitial H_2 in Si. Phys. Rev. Lett. **98**, 055504/1–4 (2007)

113. C. Peng, M. Stavola, W.B. Fowler, M. Lockwood, Ortho-para transition of interstitial H_2 and D_2 in Si. Phys. Rev. B **80**, 125207/1–13 (2009)

114. M. Hiller, E.V. Lavrov, J. Weber, B. Hourahine, R. Jones, P.R. Briddon, Interstitial H_2 in germanium by Raman scattering and *ab initio* calculations. Phys. Rev. B **72**, 153201/1–4 (2005)

115. B. Clerjaud, D. Côte, C. Naud, R. Bouanani-Rahbi, D. Wasik, K. Pakula, J.M. Baranowski, T. Suski, E. Litwin-Staszewska, M. Bockowski, I. Grzegory, The role of oxygen and hydrogen in GaN. Physica B **308–310**, 117–121 (2001)
116. A.F. Wright, Influence of the crystal structure on the lattice site and formation energies of hydrogen in wurtzite and zinc-blende GaN. Phys. Rev. B **60**, R5101–R5104 (1999)
117. S. Limpijumnong, C.G. Van de Walle, Stability, diffusivity, and vibrational properties of monoatomic and molecular hydrogen in wurtzite GaN. Phys. Rev. B **68**, 235203/1–10 (2003)
118. E.E. Chen, M. Stavola, W.B. Fowler, Ortho and para O-H_2 complexes in silicon. Phys. Rev. B **65**, 245208/1–9 (2002)
119. M. Hiller, E.V. Lavrov, J. Weber, A Raman study of H_2 trapped near O in Si. Physica B **401–402**, 97–100 (2007)
120. S. Muto, S. Takeda, M. Hirata, Hydrogen-induced platelets in silicon studied by transmission electron microscopy. Phil. Mag. A **72**, 1057–1074 (1995)
121. N. Fukata, S. Sasaki, K. Murakami, K. Ishioka, K.G. Nakamura, M. Kitajima, S. Fujimura, J. Kikuchi, H. Haneda, Hydrogen molecules and hydrogen-related defects in crystalline silicon. Phys. Rev. B **56**, 6642–6647 (1997)
122. M. Hiller, E.V. Lavrov, J. Weber, Raman scattering study of H_2 trapped within {111}-oriented platelets in Si. Phys. Rev. B **80**, 045306/1–8 (2009)
123. M. Hiller, E.V. Lavrov, J. Weber, Hydrogen-induced platelets in Ge determined by Raman scattering. Phys. Rev. B **71**, 045208/1–5 (2005)
124. H. Engstrom, J.B. Bates, J.C. Wang, M.M. Abraham, Infrared spectra of hydrogen isotopes in $\alpha - Al_2O_3$. Phys. Rev. B **21**, 1520–1526 (1980)
125. J.B. Bates, R.A. Perkins, Infrared spectral properties of hydrogen, deuterium, and tritium in TiO_2. Phys. Rev. B **16**, 3713–3722 (1977)
126. J.M. Kahn, R.E. McMurray, E.E. Haller, L.M. Falicov, Trigonal hydrogen-related acceptor complexes in germanium. Phys. Rev. B **36**, 8001–8014 (1987)
127. P. Leary, R. Jones, S. Öberg, Interaction of hydrogen with substitutional and interstitial carbon in silicon. Phys. Rev. B **57**, 3887–3899 (1998)
128. A. Endrös, Charge-state-dependent hydrogen-carbon-related deep donor in silicon. Phys. Rev. Lett. **63**, 70–73 (1989)
129. Y. Kamiura, M. Tsutsue, M. Yayashi, Y. Yamashita, F. Hashimoto, Stability and defect reaction of two hydrogen-carbon complexes in silicon. Mater. Sci. Forum **196–201**, 903–908 (1995)
130. Y. Kamiura, N. Ishiga, S. Ohyama, Y. Yamashita, Structure and charge-state-dependent instability of a hydrogen-carbon complex in silicon. Mater. Sci. Forum **258–263**, 247–252 (1997)
131. C. Kaneta, H. Katayama-Yoshida, Atomic configurations and electronic states of carbon-hydrogen complex in silicon, in *Proc. 22th Internat. Conf. Phys. Semicond.*, ed. by D.J. Lockwood (World Scientific, Singapore, 1995), pp 2215–2218
132. L. Hoffmann, E.V. Lavrov, B. Bech Nielsen, B. Hourahine, R. Jones, S. Öberg, P.R. Briddon, Weakly bound carbon-hydrogen complex in silicon. Phys. Rev. B **61**, 16659–16666 (2000)
133. V.P. Markevich, B. Hourahine, R.C. Newman, R. Jones, M. Kleverman, J.L. Lindström, L.I. Murin, M. Suezawa, S. Öberg, P.R. Briddon, Stable hydrogen pair trapped at carbon impurities in silicon. Defect Diff. Forum **221–223**, 1–9 (2003)
134. V.P. Markevich, L.I. Murin, J. Hermansson, M. Kleverman, J.L. Lindström, N. Fukata, M. Suezawa, $C_s - H_2^*$ defect in crystalline silicon. Physica B **302–303**, 220–226 (2001)
135. B. Hourahine, R. Jones, S. Öberg, P.R. Briddon, V.P. Markevich, R.C. Newman, J. Hermansson, M. Kleverman, J.L. Lindström, L.I. Murin, N. Fukata, M. Suezawa, Evidence for H_2^* trapped by carbon impurities in silicon. Physica B **308–310**, 197–201 (2001)
136. E.V. Lavrov, L. Hoffmann, B. Bech Nielsen, B. Hourahine, R. Jones, S. Öberg, P.R. Briddon, Combined infrared absorption and modeling study of a dicarbon-dihydrogen defect in silicon. Phys. Rev. B **62**, 12859–12867 (2000)

137. B. Clerjaud, D. Côte, W.S. Hahn, A. Lebkiri, W. Ulrici, D. Wasik, On the way to the investigation of hydrogen in GaN: hydrogen in nitrogen-doped GaP and GaAs. Phys. Stat. Sol. A **159**, 121–131 (1997)

138. B. Clerjaud, D. Côte, W.S. Hahn, A. Lebkiri, W. Ulrici, D. Wasik, Nitrogen-dihydrogen complex in GaP. Phys. Rev. Lett. **77**, 4930–4933 (1996)

139. W.S. Hahn, Doctoral thesis, Université Pierre et Marie Curie, Paris (1994)

140. P. Dixon, D. Richardson, R. Jones, C.D. Latham, S. Öberg, V.J.B. Torres, P.R. Briddon, Nitrogen-hydrogen defects in GaP. Phys. Stat. Sol. B **210**, 321–326 (1998)

141. B. Pajot, C.Y. Song, OH bonds in gallium arsenide grown by the liquid-encapsulated Czochralski crystal-growth method. Phys. Rev. B **45**, 6484–6491 (1992)

142. Y.S. Kim, K.J. Chang, Nitrogen-monohydride versus nitrogen-dihydride complexes in GaAs and GaAs$_{1-x}$N$_x$ alloys. Phys. Rev. B **66**, 073313/1-4 (2002)

143. A. Polimeni, G.B. Höger von Högersthal, M. Bissiri, M. Capizzi, A. Frova, M. Fischer, M. Reinhardt, A. Forchel, Role of hydrogen in III-N-V compound semiconductors. Semicond. Sci. Technol. **17**, 797–802 (2002)

144. I.A. Buyanova, M. Izadifard, I.G. Ivanov, J. Birch, W.M. Chen, M. Felici, A. Polimeni, M. Capizzi, Y.G. Hong, H.P. Xin, C.W. Tu, Direct experimental evidence of unusual effects of hydrogen on the electronic and vibrational properties of GaN$_x$P$_{1-x}$ alloys: a proof of a general property of diluted nitrides. Phys. Rev. B **70**, 245215/1-4 (2004)

145. S. Kleekajai, F. Jiang, K. Colon, M. Stavola, W.B. Fowler, K.R. Martin, A. Polimeni, M. Capizzi, Y.G. Hong, H.P. Xin, C.W. Tu, G. Bais, S. Rubini, F. Martelli, Vibrational properties of the H-N-H complex in dilute III-N-V alloys: Infrared spectroscopy and density functional theory. Phys. Rev. B **77**, 085213/1-9 (2008)

146. L. Wen, F. Bekisli, M. Stavola, W.B. Fowler, R. Trotta, A. Polimeni, M. Capizzi, S. Rubini, F. Martelli, Detailed structure of the H-N-H center of GaN$_x$P$_{1-x}$ revealed by vibrational spectroscopy under uniaxial stress. Phys. Rev. B **81**, 233201/1-4 (2010)

147. H.Ch. Alt, P. Messerer, K. Köhler, H. Riechert, H- and D-related mid-infrared absorption bands in Ga$_{1-x}$In$_x$As$_{1-x}$N$_x$ epitaxial layers. Phys. Stat. Sol. B **246**, 200–205 (2009)

148. M. Singh, J. Weber, Shallow impurity neutralization in GaP by atomic hydrogen. Appl. Phys. Lett. **54**, 424–426 (1989)

149. M. Mizuta, Y. Mochizuki, N. Takadoh, K. Asakawa, Hydrogen passivation of impurities in GaP as studied by photoluminescence spectroscopy. J. Appl. Phys. **66**, 891–895 (1989)

150. B. Clerjaud, D. Côte, W.S. Hahn, Hydrogen in crystalline gallium phosphide. Mater. Sci. Forum **148–149**, 281–294 (1994)

151. B. Clerjaud, D. Côte, W.S. Hahn, D. Wasik, W. Ulrici, Donor level of interstitial hydrogen in GaP. Appl. Phys. Lett. **60**, 2374–2376 (1993)

152. B. Clerjaud, D. Côte, W.S. Hahn, Carbon-hydrogen complex in GaP. Appl. Phys. Lett. **58**, 1860–1862 (1991)

153. B. Clerjaud, F. Gendron, M. Krause, W. Ulrici, Electronic level of interstitial hydrogen in GaAs. Phys. Rev. Lett. **65**, 1800–1803 (1990)

154. B. Clerjaud, F. Gendron, M. Krause, Donor level of interstitial hydrogen in GaAs. Modern Phys. Lett. B **5**, 877–880 (1991)

155. N.S. Minaev, A.V. Mudryi, Thermally-induced defects in silicon containing oxygen and carbon. Phys. Stat. Sol. A **68**, 561–565 (1981)

156. E. Irion, N. Bürger, K. Thonke, R. Sauer, The defect luminescence spectrum at 0.9351 eV in carbon-doped heat-treated or irradiated silicon. J. Phys. C **18**, 5069–5082 (1985)

157. A.N. Safonov, E.C. Lightowlers, G. Davies, P. Leary, R. Jones, S. Öberg, Interstitial-carbon hydrogen interaction in silicon. Phys. Rev. Lett. **77**, 4812–4815 (1996)

158. S. Hayama, G. Davies, J. Tan, J. Coutinho, R. Jones, K.M. Itoh, Lattice isotope effects on optical transitions in silicon. Phys. Rev. B **70**, 035202/1-9 (2004)

159. A.N. Safonov, E.C. Lightowlers, Hydrogen related optical centres in radiation damaged silicon. Mater. Sci. Forum **143–147**, 903–908 (1994)

160. J.I. Pankove, P.J. Zanzucchi, C.W. Magee, G. Lucovsky, Hydrogen localization near boron in silicon. Appl. Phys. Lett. **46**, 421–423 (1985)

161. N.M. Johnson, Mechanism for hydrogen compensation of shallow acceptor impurities in single-crystal silicon. Phys. Rev. B **31**, 5525–5528 (1985)

162. G.G. DeLeo, W.B. Fowler, Hydrogen-acceptor pairs in silicon: pairing effect on the hydrogen vibrational frequency. Phys. Rev. B **31**, 6861–6864 (1985)

163. M. Stavola, S.J. Pearton, J. Lopata, W.C. Dautremont-Smith, Vibrational spectroscopy of acceptor-hydrogen complexes in silicon: Evidence for low-frequency excitations. Phys. Rev. B **37**, 8313–8318 (1988)

164. B. Pajot, A. Chari, M. Aucouturier, M. Astier, A. Chantre, Experimental evidence for boron-hydrogen interaction in boron-doped silicon passivated with hydrogen. Solid State Commun. **67**, 855–858 (1988)

165. S.A. McQuaid, R.C. Newman, J.H. Tucker, E.C. Lightowlers, R.A.A. Kubiak, M. Goulding, Concentration of atomic hydrogen diffused into silicon in the temperature range 900–1300°C. Appl. Phys. Lett. **58**, 2933–2935 (1991)

166. K. Bergman, M. Stavola, S.J. Pearton, T. Hayes, Structure of acceptor-hydrogen and donor-hydrogen complexes in silicon from uniaxial stress studies. Phys. Rev. B **38**, 9643–9648 (1988)

167. C.P. Herrero, M. Stutzmann, A. Breitschwerdt, Boron-hydrogen complexes in crystalline silicon. Phys. Rev. B **43**, 1555–1575 (1991)

168. C.P. Herrero, M. Stutzmann, Microscopic structure of boron-hydrogen complexes in crystalline silicon. Phys. Rev. B **38**, 12668–12671 (1988)

169. G.D. Watkins, W.B. Fowler, M. Stavola, G.G. DeLeo, D.M. Kozuch, S.J. Pearton, J. Lopata, Identification of a Fermi resonance for a defect in silicon: deuterium-boron pair. Phys. Rev. Lett. **64**, 467–470 (1990)

170. J. Chevallier, B. Theys, A. Lusson, C. Grattepain, A. Deneuville, E. Gheeraert, Hydrogen-boron interactions in p-type diamond. Phys. Rev. B **58**, 7966–7969 (1998)

171. R. Zeisel, C.E. Nebel, M. Stutzmann, Passivation of boron in diamond by deuterium. Appl. Phys. Lett. **74**, 1875–1876 (1999)

172. J.P. Goss, R.J. Eyre, P.R. Briddon, Theoretical models for doping diamond for semiconductor applications. Phys. Stat. Sol. B **245**, 1679–1700 (2008)

173. J.P. Goss, P.R. Briddon, R. Jones, Z. Teukam, D. Ballutaud, F. Jomard, J. Chevallier, M. Bernard, A. Deneuville, Deep hydrogen traps in heavily B-doped diamond. Phys. Rev. B **68**, 235209/1–10 (2003)

174. M. Stutzmann, J.-B Chevrier, C.P. Herrero, A. Breitschwerdt, A comparison of hydrogen incorporation and effusion in doped crystalline silicon, germanium, and gallium arsenide. Appl. Phys. A **53**, 47–53 (1991)

175. S.K. Estreicher, D.M. Maric, What is so strange about hydrogen interaction in germanium? Phys. Rev. Lett. **70**, 3963–3966 (1993)

176. N.M. Johnson, R.D. Burnham, R.A. Street, R.L. Thornton, Hydrogen passivation of shallow-acceptor impurities in p-type GaAs. Phys. Rev. B **33**, 1102–1105 (1986)

177. P.S. Nandhra, R.C. Newman, R. Murray, B. Pajot, J. Chevallier, R.B. Beall, J.J. Harris, The passivation of Be acceptors in GaAs by exposure to a hydrogen plasma. Semicond. Sci. Technol. **3**, 356–360 (1988)

178. R. Rahbi, B. Pajot, J. Chevallier, A. Marbeuf, R.C. Logan, M. Gavand, Hydrogen diffusion and acceptor passivation in p-type GaAs. J. Appl. Phys. **73**, 1723–1731 (1992)

179. R.E. Pritchard, B.R. Davidson, R.C. Newman, T.J. Bullough, T.B. Joyce, R. Jones, S. Öberg, The structure and vibrational modes of $H - C_{As}$ pairs in passivated AlAs grown by chemical beam epitaxy. Semicond. Sci. Technol. **9**, 140–149 (1994)

180. R.E. Pritchard, R.C. Newman, J. Wagner, F. Fuchs, R. Jones, S. Öberg, Bonding of $H - C_{As}$ pairs in $Al_x Ga_{1-x}$ As alloys. Phys. Rev. B **50**, 10628–10636 (1994)

181. M.D. McCluskey, E.E. Haller, J. Walker, N.M. Johnson, Vibrational spectroscopy of group-II-acceptor-hydrogen complexes in GaP. Phys. Rev. B **52**, 11859–11864 (1995)

182. B.R. Davidson, R.C. Newman, C.C. Button, Vibrational modes of carbon acceptors and hydrogen-carbon pairs in semi-insulating InP doped using CCl_4. Phys. Rev. B **58**, 15609–15613 (1998)

183. B.R. Davidson, R.C. Newman, R.E. Pritchard, D.A. Robbie, A. Fischer, K. Ploog, Infrared and Raman studies of carbon impurities in highly doped MBE AlAs:C. Mater. Sci. Forum **143–147**, 247–252 (1994)

184. M.D. McCluskey, E.E. Haller, P. Becla, Carbon acceptors and carbon-hydrogen complexes in AlSb. Phys. Rev. B **65**, 045201/1–4 (2001)

185. S. Najmi, X.K. Chen, A. Yang, M. Steger, M.L.W. Thewalt, S.P. Watkins, Local vibrational mode study of carbon-doped InAs. Phys. Rev. B **74**, 113202/1–4 (2006)

186. J. Chevallier, B. Clerjaud, B. Pajot, Neutralization of defects and dopants in III-V. Semiconductors and semimetals **34**, 447–510 (1991)

187. B. Clerjaud, D. Côte, F. Gendron, W.S. Hahn, M. Krause, C. Porte, W. Ulrici, Carbon-hydrogen interaction in III-V compounds. Mater. Sci. Forum **83–87**, 563–568 (1992)

188. M. Stavola, S.J. Pearton, J. Lopata, C.R. Abernathy, K. Bergman, Structure and dynamics of the Be-H complex in GaAs. Phys. Rev. B **39**, 8051–8054 (1989)

189. M.C. Wagener, J.R. Botha, A.W.R. Leitch, Passivation and thermal reactivation of Mg acceptors in p-type GaAs. Phys. Rev. B **62**, 15315–15318 (2000)

190. R. Bouanani-Rahbi, B. Pajot, C. Ewels, S. Öberg, J. Goss, R. Jones, Y. Nissim, B. Theys, C. Blaauw, Is H passivating the Mg acceptor bond-centred in InP:Mg and antibonded on GaAs:Mg? in *Shallow-Level Centers in Semiconductors*, ed. by C.A.J. Ammerlaan, B. Pajot (World Scientific, Singapore, 1997), pp. 171–178

191. R. Bouanani-Rahbi, B. Clerjaud, B. Theys, A. Lemaitre, F. Jomard, Neutralization of manganese by hydrogen in GaAs. Physica B **340–342**, 284–287 (2003)

192. J. Neugebauer, C.G. Van de Walle, Hydrogen in GaN: Novel aspects of a common impurity. Phys. Rev. Lett. **75**, 4452–4455 (1995)

193. W. Götz, N.M. Johnson, D.P. Bour, M.D. McCluskey, E.E. Haller, Local vibrational modes of the Mg-H acceptor complex in GaN. Appl. Phys. Lett. **69**, 3725–3727 (1996)

194. C.H. Seager, Comment on "Infrared spectroscopy of Mg-H local vibrational mode in GaN with polarized light". Phys. Rev. B **67**, 037201/1–2 (2003)

195. C.J. Fall, R. Jones, P.R. Briddon, S. Öberg, Electronic and vibrational properties of Mg- and O-related complexes in GaN. Mater. Sci. Eng. B **82**, 88–90 (2001)

196. S. Limpijumnong, J.E. Northrup, C.G. Van de Walle, Entropy-driven stabilization of a novel configuration for acceptor-hydrogen complexes in GaN. Phys. Rev. Lett. **87**, 205505/1–4 (2001)

197. M.D. McCluskey, K.K. Zhuraviev, M. Kneissl, W. Wong, D. Treat, S. Limpijumnong, C.G. Van de Walle, N.M. Johnson Vibrational spectroscopy of GaN:Mg under pressure. Mat. Res. Soc. Symp. Proc. **693**, 23–28 (2002)

198. R. Jones, J. Goss, C. Ewels, S. Öberg, *Ab initio* calculations of anharmonicity of the C-H stretch mode in HCN and GaAs. Phys. Rev. B **50**, 8378–8388 (1994)

199. V.J.B. Torres, J. Coutinho, P.R. Briddon, *Ab-initio* modeling of carbon and carbon-hydrogen defects in InAs. Physica B **401–402**, 275 (2007)

200. J. Wagner, K.H. Bachem, B.R. Davidson, R.C. Newman, T.J. Bullough, T.B. Joyce, Dynamics of the H-C$_{AS}$ complex in GaAs determined from Raman measurements. Phys. Rev. B **51**, 4150–4158 (1995)

201. D.M. Kozuch, M. Stavola, S.J. Pearton, C.R. Abernathy, W.S. Hobson, Passivation of carbon-doped GaAs layers by hydrogen introduced by annealing and growth ambients. J Appl. Phys. **73**, 3716–3724 (1993)

202. D.M. Kozuch, M. Stavola, S.J. Spector, S.J. Pearton, J. Lopata, Symmetry, stress alignment, and reorientation of the Si$_{As}$ − H complex in GaAs. Phys. Rev. B **48**, 8751–8756 (1993)

203. Y.M. Cheng, M. Stavola, C.R. Abernathy, S.J. Pearton, W.S. Hobson, Aligned defect complex containing carbon and hydrogen in as-grown GaAs epitaxial layers. Phys. Rev. B **49**, 2469–2476 (1994)

204. J.P. Goss, R. Jones, S. Öberg, P.R. Briddon, (C$_{As}$)$_2$-hydrogen defects in GaAs: a first-principle study. Phys. Rev. B **55**, 15576–15580 (1997)

205. M. Stavola, J.F. Zheng, Y.M. Cheng, C.R. Abernathy, S.J. Pearton, Novel properties of hydrogen-containing complexes revealed by their hydrogen vibrations. Mater. Sci. Forum **196–201**, 809–816 (1995)

206. B.R. Davidson, R.C. Newman, H. Fushimi, K. Wada, H. Yokoyama, N. Inoue, Aligned carbon-hydrogen complexes in GaAs formed by the decomposition of trimethylgallium during metalorganic vapor phase epitaxy and atomic layer epitaxy. J. Appl. Phys. **81**, 7255–7260 (1997)

207. B.R. Davidson, R.C. Newman, T.B. Joyce, T.J. Bullough, A calibration of the $H-C_{As}$ stretch mode in GaAs. Semicond. Sci. Technol. **11**, 455–457 (1996)

208. L. Svob, Y. Marfaing, B. Clerjaud, D. Côte, D. Ballutaud, B. Theys, D. Druilhe, W. Kuhn, H. Stanzl, W. Gebhardt, Incorporation and interaction of hydrogen with acceptor impurities in II-VI semiconductor compounds. Mater. Sci. Forum **143–147**, 447–452 (1994)

209. J.A. Wolk, J.W. Ager III, K.J. Duxstad, E.E. Haller, N.R. Taskar, D.R. Dorman, D.J. Olego, Local vibrational mode spectroscopy of nitrogen-hydrogen complex in ZnSe. Appl. Phys. Lett. **63**, 2756–2758 (1993)

210. M.D. McCluskey, E.E. Haller, W. Walukiewicz, P. Becla, Hydrogen passivation of Se and Te in AlSb. Phys. Rev. B **53**, 16297–16301 (1996)

211. M.D. McCluskey, E.E. Haller, F.X. Zach, E.D. Bourret-Courchesne, Vibrational spectroscopy of arsenic-hydrogen complexes in ZnSe. Appl. Phys. Lett. **68**, 3476–3478 (1996)

212. H. Pelletier, A. Lusson, B. Theys, J. Chevallier, N. Magnéa, Spectroscopic evidence of the formation of N − H and N − D complexes in plasma hydrogenated and deuterated ZnTe:N layers. Appl. Phys. Lett. **73**, 28–30 (1998)

213. Z. Yu, S.L. Buczkowski, M.C. Petcu, N.C. Giles, T.H. Myers, Hydrogenation of undoped and nitrogen-doped CdTe grown by molecular beam epitaxy. Appl. Phys. Lett. **68**, 529–531 (1996)

214. B. Clerjaud, D. Côte, A. Lebkiri, Acceptor neutralisation by hydrogen in GaN and wide band gap II-VI materials, Phys. Stat. Sol. B **210**, 497–506 (1998)

215. Z. Yu, S.L. Buczkowski, L.S. Hirsch, T.H. Myers, An infrared absorption investigation of hydrogen, deuterium and nitrogen in ZnSe grown by molecular beam epitaxy. J. Appl. Phys. **80**, 6425–6428 (1996)

216. J.L. Lyons, A. Janotti, C.G. Van de Walle, Why nitrogen cannot lead to p-type conductivity in ZnO. Appl. Phys. Lett. **95**, 252105/1–3 (2009)

217. M.C. Tarun, M. Zafar Iqbal, M.D. McCluskey, Nitrogen is a deep acceptor in ZnO. AIP Advances **1**, 022105/1–7 (2011)

218. S.J. Jokela, M.D. McCluskey, Structure and stability of N − H complexes in single-crystal ZnO. J. Appl. Phys. **107**, 113536/1–5 (2010)

219. G. Cannelli, R. Cantelli, M. Capizzi, C. Coluzza, F. Cordere, A. Frova, A. Lo Presti, Reorientation of the B-H complex in silicon by anelastic relaxation measurements. Phys. Rev. B **44**, 11486–11489 (1991)

220. Y.M. Cheng, M. Stavola, Non-Arrhenius reorientation kinetics for the B-H complex in Si: Evidence for thermally assisted tunnelling. Phys. Rev. Lett. **73**, 3419–3422 (1994)

221. M. Stavola, Vibrational spectroscopy of dopant-hydrogen complexes in III-V semiconductors. Mater. Sci. Forum **148–149**, 251–280 (1994)

222. I.A. Veloarisoa, M. Stavola, Y.M. Cheng, S. Uftring, G.D. Watkins, S.J. Pearton, C.R. Abernathy, J. Lopata, Ground-state energy shift of acceptor-hydrogen complexes in Si and GaAs under uniaxial stress. Phys. Rev. B **47**, 16237–16241 (1993)

223. R. Darwich, B. Pajot, C.Y. Song, B. Rose, B. Theys, C. Porte, Symmetry and reorientation of zinc-hydrogen complex in indium phosphide, in *Proc. 20th Conf. Physics of Semicond.*, ed. by E.M. Anastassakis, J.D. Joannopoulos (World Scientific, Singapore 1990), pp.791–794

224. J. Chevallier, W.C. Dautremont-Smith, C.W. Tu, S.J. Pearton, Donor neutralization in GaAs(Si) by atomic hydrogenation. Appl Phys. Lett. **47**, 108–110 (1985)

225. S.J. Pearton, W.C. Dautremont-Smith, J. Chevallier, C.W. Tu, K.D. Cummings, Hydrogenation of shallow-donor levels in GaAs. J. Appl. Phys. **59**, 2821–2827 (1986)

226. A. Jalil, J. Chevallier, J.C. Pesant, R. Mostefaoui, B. Pajot, P. Murawala, R. Azoulay, Infrared spectroscopic evidence of silicon related hydrogen complexes in hydrogenated n-type GaAs doped with silicon. Appl. Phys. Lett. **50**, 438–441 (1987)

227. K. Bergman, M. Stavola, S.J. Pearton, J. Lopata, Donor-hydrogen complexes in passivated silicon. Phys. Rev. B **37**, 2770–2773 (1988)

228. B. Pajot, R.C. Newman, R. Murray, A. Jalil, J. Chevallier, R. Azoulay, High-resolution infrared study of the neutralization of silicon donors in gallium arsenide. Phys. Rev. B **37**, 4188–4195 (1988)

229. E. Tuncel, H. Sigg, E. Meier, L. Pavesi, P. Gianozzi, D. Martin, F. Morier-Genoud, F.K. Reinhart, Effects of hydrogen in Si-doped $AlAs$. Mater. Sci. Forum **83–87**, 635–640 (1992)

230. W. Ulrici, B. Clerjaud, D. Côte, Hydrogen passivation of the Si_{Ga} donor in GaP. Phys. Stat. Sol. B **235**, 102–106 (2003)

231. D.M. Kozuch, M. Stavola, S.J. Pearton, C.R. Abernathy, J. Lopata, Sn-H complex in hydrogen-passivated GaAs. in *Impurities, Defects, and Diffusion in Semiconductors: bulk and layered structures*, Mat. Res. Soc. Symp. Proc. **163**. eds. D.J. Wolford, J. Bernholc, E.E. Haller (MRS, Pittsburgh, 1990), p. 477–482

232. B. Pajot, Hydrogen passivation of shallow donors and acceptors in GaAs. Inst. Phys. Conf. Ser. No. 95 (Institute of Physics Publishing, Bristol, 1988), pp. 437–446

233. P.R. Briddon, R. Jones, *Ab initio* calculations on the passivation of shallow impurities in GaAs. Phys. Rev. Lett. **64**, 2535–2538 (1990)

234. R. Rahbi, B. Theys, R. Jones, B. Pajot, S. Öberg, K. Somogyi, M.L. Fille, J. Chevallier, Neutralization of group VI donors by hydrogen in gallium arsenide. Solid State Commun. **91**, 187–190 (1994)

235. J. Vetterhöffer, J.H. Svensson, J. Weber, A.W.R. Leitch, J.R. Botha, Local hydrogen vibrational modes in GaAs doped with S, Se, and Te. Phys. Rev. B **50**, 2708–2710 (1994)

236. J. Vetterhöffer, J. Weber, Hydrogen passivation of shallow donors S, Se, and Te in GaAs. Phys. Rev. B **53**, 12835–12844 (1996)

237. M.D. McCluskey, Resonant interaction between hydrogen vibrational modes in $AlSb$:Se. Phys. Rev. Lett. **102**, 135502/1–4 (2009)

238. M.D. McCluskey, E.E. Haller, W. Walukiewicz, P. Becla, Anti-crossing behavior of local vibrational modes in $AlSb$. Solid State Commun. **106**, 587–590 (1998)

239. B. Clerjaud, D. Côte, W. Ulrici, Complexes of group VI donors with hydrogen in GaP. Physica. B **273–274**, 803–806 (1999)

240. G. Davies, E.C. Lightowlers, M. Stavola, K. Bergman, B. Svensson, The 3942-cm^{-1} optical band in irradiated silicon. Phys. Rev. B **35**, 2755–2766 (1987)

241. W. Ulrici, B. Clerjaud, Evidence of a sulfur-boron-hydrogen complex in GaAs grown by the liquid encapsulation Czochralski technique. Phys. Rev. B **70**, 205214/1–5 (2004)

242. J.F. Zheng, M. Stavola, Correct assignment of the hydrogen vibrations of the donor-hydrogen complexes in Si: a new example of Fermi resonance. Phys. Rev. Lett. **76**, 1154–1157 (1996)

243. B. Pajot, *Optical Absorption of Impurities and Defects in Semiconducting Crystals I. Hydrogen-like Centres* (Springer, Berlin 2010)

244. P.M. Williams, G.D. Watkins, S. Uftring, M. Stavola, Structure-sensitive spectroscopy of transition-metal-hydrogen complexes in silicon. Phys. Rev. Lett. **70**, 3816–3819 (1993)

245. S.J. Uftring, M. Stavola, P.M. Williams, G.D. Watkins, Microscopic structure and multiple charge states of a PtH_2 complex in Si. Phys. Rev. B **51**, 9612–9621 (1995)

246. M.G. Weinstein, M. Stavola, K.L. Stavola, S.J. Uftring, J. Weber, J.U. Sachse, H. Lemke, Pt–H complexes in Si: complementary studies by vibrational and capacitance spectroscopies. Phys. Rev. B **65**, 035206/1–10 (2001)

247. A. Resende, R. Jones, S. Öberg, P.R. Briddon, The structural properties of transition-metal hydrogen complexes in silicon. Mater. Sci. Engin. B **58**, 146–148 (1999)

248. M.J. Evans, M. Stavola, M.G. Weinstein, S.J. Uftring, Vibrational spectroscopy of defect complexes containing Au and H in Si. Mater. Sci. Eng. B **58**, 118–125 (1999)

249. A. Resende, R. Jones, S. Öberg, P.R. Briddon, Calculations of electrical levels of deep centers: application to Au-H and Ag-H defects in silicon. Phys. Rev. Lett. **82**, 2111–2114 (1999)
250. E. Mollwo, G. Müller, D. Zwingel, Optical and paramagnetic properties of ZnO-crystals simultaneously doped with copper and hydrogen. Solid State Commun. **15**, 1475–1479 (1974)
251. F.G. Gärtner, E. Mollwo, IR absorption of OH and OD centres and OH/OH, OD/OD, and OH/OD complexes in Cu-doped ZnO single crystals. I. Experimental results. Phys. Stat. Sol. B **89**, 381–388 (1978)
252. F.G. Gärtner, E. Mollwo, IR absorption of OH and OD centres and OH/OH, OD/OD, and OH/OD complexes in Cu-doped ZnO single crystals. II. Discussion and quantitative interpretations of models. Phys. Stat. Sol. B **90**, 33–44 (1978)
253. E.V. Lavrov, J. Weber, Uniaxial stresss study of the Cu-H complex in ZnO. Phys. Stat. Sol. B **243**, 2657–2664 (2006)
254. F. Börrnert, E.V. Lavrov, J. Weber, Hydrogen motion in the Cu-H complex in ZnO. Phys. Rev. B **75**, 205202/1–5. Erratum: Phys. Rev. B **76**, 119903/1 (2007)
255. E.V. Lavrov, J. Weber, F. Börnert, Copper dihydrogen complex in ZnO. Phys. Rev. B **77**, 155209/1-10 (2008)
256. B. Clerjaud, D. Côte, A. Lebkiri, A. Mari, Investigation of the manganese-hydrogen complex in InP. Mater. Sci. Forum **196–201**, 975–980 (1995)
257. M.S. Brandt, S.T.B. Goennenwein, T.A. Wassner, F. Kohl, H. Huebl, T. Graf, M. Stutzmann, A. Koeder, W. Schoch, A. Waag, Passivation of Mn acceptors in GaMnAs. Appl. Phys. Lett. **84**, 2277–2279 (2004)
258. B. Pajot, A. Jalil, J. Chevallier, R. Azoulay, Spectroscopic evidence for the hydrogen passivation of zinc acceptors in gallium arsenide. Semicond. Sci. Technol. **2**, 305–307 (1987)
259. S.T.B. Goennenwein, T.A. Wassner, H. Huebl, M.S. Brandt, J.B. Philipp, M. Opel, R. Gross, A. Koeder, W. Schoch, W. Waag, Hydrogen control of ferromagnetism in a dilute magnetic semiconductor. Phys. Rev. Lett. **92**, 227202/1–4 (2004)
260. A. Lemaitre, L. Thevenard, M. Viret, L. Largeau, O. Mauguin, B. Theys, F. Bernardot, R. Bouanani-Rahbi, B. Clerjaud, F. Jomard, Tuning the ferromagnetic properties of hydrogenated GaMnAs. in *Proc. 27th Internat. Conf. Phys. Semicond.*, ed by J. Menendez, C.G. Van de Walle. AIP Conf. Proc. **772**, 363–364 (2005)
261. B. Clerjaud, D. Wasik, R. Bouanani-Rahbi, G. Strzelecka, A. Hruban, M. Piersa, M. Kaminska, Manganese-hydrogen complex in GaP. Physica B **401–402**, 258–261 (2007)
262. B. Clerjaud, D. Wasik, R. Bouanani-Rahbi, G. Strzelecka, A. Hruban, M. Kaminska, On the formation of complexes between Mn acceptors and hydrogen in GaP and GaMnP alloys. J. Appl. Phys. **103**, 123507/1–5 (2008)
263. C. Bihler, M. Kraus, M.S. Brandt, S.T.B. Goennenwein, M. Opel, M.A. Scarpulla, R. Farschi, D.M. Estrada, O. Dubon, Suppression of hole-mediated ferromagnetism in Ga$_{1-x}$Mn$_x$P by hydrogen. J. Appl. Phys. **104**, 013908/1–5 (2008)
264. E. Fritsch, T. Hainschwang, L. Massi, B. Rondeau, Hydrogen-related optical centers in natural diamond: An update. New Diam. Frontier Carbon Technol. **17**, 63–89 (2007)
265. A.M. Zaitsev, *Optical Properties of Diamond. A Data Handbook* (Springer, Berlin, 2001)
266. C.J. Tang, A.J. Neves, L. Rino, A.J.S. Fernandes, The 2828 cm^{-1} C − H related IR vibrations in CVD diamond. Diam. Relat. Mater. **13**, 958–964 (2004)
267. J.J. Charette, Essai de classification des bandes d'absorption infrarouge du diamant. Physica **27**, 1061–1073 (1961)
268. G.S. Woods, A.T. Collins, Infrared absorption spectra of hydrogen complexes in type I diamonds. J. Phys. Chem. Solids **44**, 471–475 (1983)
269. F. De Weerdt, Yu.N. Pal'Yanov, A.T. Collins, Absorption spectra of hydrogen in ^{13}C diamond produced by high pressure, high temperature synthesis. J. Phys. Cond. Matt. **15**, 3163–3170 (2003)
270. G. Davies, A.T. Collins, P. Spear, Sharp infra-red absorption lines in diamond. Solid State Commun. **49**, 433–436 (1984)

271. E. Fritsch, K. Scarrat, A.T. Collins, Optical properties of diamonds with an unusually high hydrogen content, *New Diamond Science and Technology: Proc. 2nd Internat. Conf., Washinton, DC, September 23–27 1990*, eds. R. Messier, J.T. Glass, J.E. Butler, R. Roy (Materials Research Society, Pittsburgh PA, 1991) pp. 671–676

272. J. Chevallier, F. Jomard, Z. Teukam, S. Koizumi, H. Kanda, Y. Sato, A. Deneuville, Hydrogen in n-type diamond. Diam. Relat. Mater. **11**, 1566–1571 (2002)

273. F. De Weerdt, I.G. Kupriyanov, Report on the influence of HPHT annealing on the $3107\,cm^{-1}$ hydrogen related absorption peak in natural type Ia diamonds. Diam. Relat. Mater. **11**, 714–715 (2002)

274. W. Ulrici, M. Jurisch, Vibrational absorption of hydrogen bonded to interstitial oxygen in GaAs and GaP. Phys. Stat. Sol. B **233**, 263–269 (2002)

275. C.Y. Song, Doctoral thesis, Université Paris 7 (1992)

276. B. Pajot, C.Y. Song, R. Darwich, F. Gendron, C. Ewels, Photo-induced changes of hydrogen bonding in semi-insulating iron-doped indium phosphide. Solid State Commun. **95**, 851–854 (1995)

277. J. Tatarkiewicz, A. Breitschwerdt, A. Witowski, Hydrogen vibrations in CdS. Phys. Rev. B **39**, 3889–3891 (1989)

278. C. Peng, H. Zhang, M. Stavola, W.B. Fowler, B. Esham, S.K. Estreicher, A. Docaj, L. Carnel, M. Seacrist, Microscopic structure of a VH_4 center trapped by C in Si. Phys. Rev. B **84**, 195205/1–7 (2011)

Appendix A
Energy Units Used in Spectroscopy and Solid-State Physics

The energy of an electron accelerated by a potential of one volt is one electron volt (eV), a quantity of the order of magnitude of the energies at the atomic scale. The infrared spectroscopists prefer the wavenumber (the number of wavelengths λ per unit length, usually noted $\tilde{\nu}$, but often noted ω in the book), specially when dealing with vibrational energies. It is commonly expressed in reciprocal centimetre (cm^{-1}). The phonon frequencies are often evaluated in terahertz. The absolute temperature is often used to measure energy in statistical mechanics. The correspondence with macroscopic energies is provided by multiplying the energies in eV by the Avogadro constant N_A and evaluating the result in kJ mol^{-1} ($1J = 6.24151 \times 10^{18}$ eV).

The correspondences between the eV and these units are given below. It is derived from $E = eV = hc\tilde{\nu} = h\nu = k_B T = hc/\lambda$ (the Boltzmann constant is noted k_B).

E (eV)	$\tilde{\nu}$ (cm^{-1})	ν (THz)	K (kelvin)	kJ mol^{-1}	λ (μm)
1	8065.545	241.7992	11604.50	96.48534	1.239842
1.239842×10^{-4}	1	0.0299792	1.438781	0.0119627	10000
0.004135667	33.35641	1	47.99237	0.399030	299.792
8.61734×10^{-5}	0.695036	0.0208366	1	0.00831444	143878
0.0103643	83.5935	2.50608	120.273	1	119.627
1.239842	10000	299.792	14387.81	119.627	1

In the book, 1 cm^{-1} is taken as 0.1239842 meV. In the visible and UV regions of the spectrum, the nanometre (nm) wavelength unit is used (1 Å $= 0.1$ nm). In the IR region of the spectrum, the μm wavelength unit is mostly used above 2500 nm and below 1 mm.

B. Pajot and B. Clerjaud, *Optical Absorption of Impurities and Defects in Semiconducting Crystals*, Springer Series in Solid-State Sciences 169, DOI 10.1007/978-3-642-18018-7, © Springer-Verlag Berlin Heidelberg 2013

465

A.1   Values of Selected Physical Constants Recommended by CODATA (2006)

Except for the value for c, $\mu_0 = 4\pi \times 10^{-7}$, and ε_0, taken as exact, all the physical constants are rounded.

Speed c of light in vacuum (m s^{-1}):	299792458
Magnetic constant μ_0 (N A^{-2}):	$12.566370614\ldots \times 10^{-7}$
Electric constant $\varepsilon_0 = 1/\mu_0 c^2$ (F m^{-1}):	$8.854187817\ldots \times 10^{-12}$
Electron charge e (C):	$1.602176487\,(10) \times 10^{-19}$
Planck constant h (J s):	$6.62606896\,(33) \times 10^{-34}$
Planck constant h (eV s):	$4.13566733\,(10) \times 10^{-15}$
Planck constant over 2π $\hbar$ (J s):	$1.054571628\,(53) \times 10^{-34}$
Planck constant over 2π $\hbar$ (eV s):	$6.58211899\,(16) \times 10^{-16}$
Boltzmann constant k_B(J K^{-1}):	$1.3806505\,(24) \times 10^{-23}$
Boltzmann constant k_B(eV K^{-1}):	$8.617343\,(15) \times 10^{-5}$
Fine structure constant $\alpha = e^2/4\pi\varepsilon_0\hbar c$	$7.297352533\,(27) \times 10^{-3}$
Bohr radius a_0 (m) $= 4\pi\varepsilon_0\hbar^2/m_e e^2$	$0.529177208\,(59) \times 10^{-10}$
Rydberg constant R_∞(m^{-1}) $= \alpha^2 m_e c/2h$	$10973731.568527\,(73)$
hcR_∞ (eV) $= m_e e^4/(8\varepsilon_0 h^2)$	$13.60569193\,(34)$
Avogadro constant N_A(atom mol^{-1}):	$6.02214179(30) \times 10^{23}$
Electron mass m_e (kg)	$9.10938215\,(45) \times 10^{-31}$
Atomic mass constant $m_u = \frac{1}{12}m(^{12}C)$ (kg)	$1.660538782 \times 10^{-27}$
Bohr magneton $\mu_B = e\hbar/2m_e$(JT^{-1})	$927.400915\,(23) \times 10^{-26}$
Bohr magneton $\mu_B = e\hbar/2m_e$(eVT^{-1})	$5.7883817555 \times 10^{-5}$

The above expressions for the Rydberg constant and the Bohr radius are given in SI units. In cgs units, where the electron charge is 4.803204×10^{-10} esu (statcoulomb), R_∞(cm^{-1}) is $2\pi^2 m_e e^4/h^3 c$ and the Bohr radius a_0 (cm) is $\hbar^2/m_e e^2$. The atomic mass unit u is defined to be equal to m_u.

Appendix B
Bravais Lattices, Symmetry, and Crystals

Three-dimensional (3D) space can be filled without voids or overlapping by identical prismatic cells with well-defined symmetries, and their types are limited to seven. These unit cells can be defined by the lengths of three primitive vectors $\mathbf{a}_1$, $\mathbf{a}_2$, and $\mathbf{a}_3$ and by the angles α, β, and γ between these vectors. They generate the seven simple crystal systems or classes, defined by the sets of all points taken from a given origin of these cells, that are defined by vectors

$$\mathbf{R} = n_1\mathbf{a}_1 + n_2\mathbf{a}_2 + n_3\mathbf{a}_3 \tag{B.1}$$

where n_1, n_2, and n_3 are integers. Table B.1 enumerates these crystal systems and their geometric characteristics.

Table B.1 The seven 3D simple crystal systems

System	Restrictions for vectors lengths and angles
Triclinic	$a_1 \neq a_2 \neq a_3$
	$\alpha \neq \beta \neq \gamma$
Monoclinic	$a_1 \neq a_2 \neq a_3$
	$\alpha = \gamma = 90° \neq \beta$
Orthorhombic or rhombic	$a_1 \neq a_2 \neq a_3$
	$\alpha = \beta = \gamma = 90°$
Tetragonal	$a_1 = a_2 \neq a_3$
	$\alpha = \beta = \gamma = 90°$
Hexagonal	$a_1 = a_2 \neq a_3$
	$\alpha = \beta = 90°$, $\gamma = 120°$
Trigonal	$a_1 = a_2 = a_3$
	$\alpha = \beta = \gamma \neq 90°$
Cubic (isometric)	$a_1 = a_2 = a_3$
	$\alpha = \beta = \gamma = 90°$

The conditions on the primitive vectors of the unit cells and on their orientations are indicated. Angle γ is taken as the one between $\mathbf{a}_1$ and $\mathbf{a}_2$

B. Pajot and B. Clerjaud, *Optical Absorption of Impurities and Defects in Semiconducting Crystals*, Springer Series in Solid-State Sciences 169, DOI 10.1007/978-3-642-18018-7, © Springer-Verlag Berlin Heidelberg 2013

The other crystal lattices can be generated by adding to some of the above-defined cells extra high-symmetry points by the so-called centring method. Table B.2 shows the new systems added to the simple crystal lattices (noted s, or P, for primitive) and the numbers of lattice points in each conventional unit cell. The body-centred lattices are noted bc or I (for German *Innenzentrierte*), the face-centred, fc or F, and the side-centred or base-centred lattices are noted C (an extra atom at the centre of the base). These 14 lattice systems are known as the Bravais lattices (noted here *BLs*). A representation of their unit cells can be found in the textbook by Kittel [1].

A primitive cell of a *BL* is a cell of minimum volume that contains only one lattice point, so that the whole lattice can be generated by all the translations of this cell. This definition allows for different primitive cells for the same *BL*, but their volumes must be the same. The parallelepiped defined by the three primitive vectors a_1, a_2, and a_3 of a simple *BL* is a primitive cell of this lattice.

The conventional unit cell showing the symmetry of the hexagonal system is that of a right prism, whose height is usually noted c, with a regular hexagon as a base. This cell contains three lattice points, hence three primitive cells consisting in a right prism with a base made of a rhomb with one 120° angle.

The unit cells of the simple P systems are primitive cells. Primitive cells are not unique and most do not have the *BL* symmetry, but it is possible to construct a primitive cell with the symmetry of the *BL*. The recipe is to connect a given lattice point to its nearest neighbours by straight lines and to intersect these lines at mid-point by perpendicular planes. The inner volume defined by these planes is the volume of the primitive cell and it is known as the Wigner–Seitz cell. In particular, the Wigner–Seitz cell for the hexagonal system is a hexagonal prism whose volume is that of the hexagonal unit cell.

Real crystal lattices are made from atoms, atomic or molecular entities associated with lattice points of the *BLs* or of their combinations. For instance, when they are centred at the lattice points of a fcc *BL*, entities of two same atoms lying along the diagonal of the unit cell of this *BL* and separated by one quarter of this diagonal generate the diamond structure (when the two atoms are different, the structure generated is that of sphalerite, also called zinc-blende).

Table B.2 Number of lattice points in the unit cells of the 14 3D Bravais lattices

System	Simple (P)	Body-centred	Face-centred	Base-centred
Triclinic	1	–	–	–
Monoclinic	1	–	–	2
Orthorhombic	1	2	4	2
Tetragonal	1	2	–	–
Hexagonal	1	–	–	–
Trigonal	1	–	–	–
Cubic	1 (sc)	2 (bcc)	4 (fcc)	–

B.1   The Reciprocal Lattice

When dealing with the interactions of crystals with particles that can display wave-like properties, like photons, phonons, or electrons, it is useful to introduce a reciprocal lattice associated with the real (or direct) crystal lattice. Let us consider a set of vectors $\mathbf{R}$ constituting a given 3D *BL* and a plane wave $e^{i\mathbf{k}\cdot\mathbf{r}}$. For special choices of $\mathbf{k}$, it can be shown that $\mathbf{k}$ can also display the periodicity of a *BL*, known as the reciprocal lattice of the direct *BL*. For all $\mathbf{R}$ of the direct *BL*, the set of all wave vectors $\mathbf{G}$ belonging to the reciprocal lattice verify the relation

$$e^{i\mathbf{G}\cdot(\mathbf{r}+\mathbf{R})} = e^{i\mathbf{G}\cdot\mathbf{r}} \tag{B.2}$$

for any $\mathbf{r}$. The reciprocal lattice can thus be defined as the set of wave vectors $\mathbf{G}$ satisfying

$$e^{i\mathbf{G}\cdot\mathbf{R}} = 1 \tag{B.3}$$

The reciprocal lattice of a *BL* whose primitive unit cell is defined by three vectors $\mathbf{a}_1$, $\mathbf{a}_2$, and $\mathbf{a}_3$ is generated by three primitive vectors

$$\mathbf{b}_1 = 2\pi\frac{\mathbf{a}_2 \wedge \mathbf{a}_3}{v}, \quad \mathbf{b}_2 = 2\pi\frac{\mathbf{a}_3 \wedge \mathbf{a}_1}{v}, \quad \mathbf{b}_3 = 2\pi\frac{\mathbf{a}_1 \wedge \mathbf{a}_2}{v} \tag{B.4}$$

where $v = \mathbf{a}_1\cdot(\mathbf{a}_2 \wedge \mathbf{a}_3)$ is the volume of the primitive unit cell of the direct lattice (the notation $\mathbf{u} \wedge \mathbf{v}$ denotes the vector product of vectors $\mathbf{u}$ and $\mathbf{v}$).

It is clear that the $\mathbf{a}_i$ and $\mathbf{b}_j$ satisfy condition B.3 as $\mathbf{a}_i \cdot \mathbf{b}_j = 2\pi\delta_{ij}$, where δ_{ij} is the Kronecker symbol (0 if $i \neq j$, 1 if $i = j$). Similarly, it can be checked that for any vector $\mathbf{G} = m_1\mathbf{b}_1 + m_2\mathbf{b}_2 + m_3\mathbf{b}_3$ (m_1, m_2, and m_3 being integers) of the lattice generated by the $\mathbf{b}_j$, condition B.3 is met when $\mathbf{R}$ is a vector of the direct lattice.

It can be also checked by using expressions B.4 that the reciprocal lattice of the reciprocal lattice is the original direct lattice and that the volume of the primitive unit cell of the reciprocal lattice is $(2\pi)^3$. The Wigner–Seitz primitive cell of the reciprocal lattice is known as the first Brillouin zone (BZ) of the reciprocal lattice. As an example, the reciprocal lattice of the fcc *BL* with conventional cubic unit cell of side a is the corresponding bcc *BL* with a conventional cubic unit cell of side $4\pi/a$, and by applying twice the construction of a reciprocal lattice, it is seen that the reciprocal lattice of the bcc *BL* is the corresponding fcc *BL*. The angular correspondence is not a general rule, however, and the reciprocal lattice of the hexagonal *BL* is another hexagonal *BL* rotated through 30° about the *c*-axis of the direct lattice. A general account on the symmetries of the Wigner–Seitz cells for the different *BL*s can be found in the review by Koster [2] and it can be easily extrapolated to the first BZs.

B.2   Lattice Planes and Miller Indices

Let us start with a few definitions. A lattice plane of a given 3D *BL* contains at least three noncollinear lattice points and this plane forms a 2D *BL*. A family of lattice planes of a 3D *BL* is a set of parallel equally spaced lattice planes separated by the minimum distance d between planes, and this set contains all the points of the *BL*. The resolution of a given 3D *BL* into a family of lattice planes is not unique, but for any family of lattice planes of a direct *BL*, there are vectors of the reciprocal lattice that are perpendicular to the direct lattice planes. Inversely, for any reciprocal lattice vector $\mathbf{G}$, there is a family of planes of the direct lattice normal to $\mathbf{G}$ and separated by a distance d, where $2\pi/d$ is the length of the shortest reciprocal lattice vector parallel to $\mathbf{G}$. A proof of these two assertions can be found in [3].

As one generally uses a vector normal to a lattice plane to specify its orientation, one can as well use a reciprocal lattice vector. This allows to define the Miller indices of a lattice plane as the coordinates of the shortest reciprocal lattice vector normal to that plane, with respect to a specified set of direct lattice vectors. These indices are integers with no common factor other than 1. A plane with Miller indices h, k, l is thus normal to the reciprocal lattice vector $\mathbf{G} = h\mathbf{b}_1 + k\mathbf{b}_2 + l\mathbf{b}_3$ and it is contained in a continuous plane $\mathbf{G} \cdot \mathbf{r} = $ constant. This plane intersects the primitive vectors $\mathbf{a}_i$ of the direct lattice at the points of coordinates $x_1\mathbf{a}_1$, $x_2\mathbf{a}_2$, and $x_3\mathbf{a}_3$, where the x_i must satisfy separately $\mathbf{G}.x_i\mathbf{a}_i = $ constant. Since $\mathbf{G} \cdot \mathbf{a}_1$, $\mathbf{G} \cdot \mathbf{a}_2$, and $\mathbf{G} \cdot \mathbf{a}_3$ are equal to h, k, and l, respectively, the x_i are inversely proportional to the Miller indices of the plane. When the plane is parallel to a given axis, the corresponding x value is taken for infinity and the corresponding Miller index taken equal to zero.

Lattice planes are specified by giving their Miller indices in parentheses: $(h\ k\ l)$. For instance, in the cubic system, the Miller indices of a plane intersecting the $\mathbf{a}_1$, $\mathbf{a}_2$, and $\mathbf{a}_3$ axes at 3, -1, and 2, respectively, will be $(2\bar{6}3)$ and the plane will be noted $(2\bar{6}3)$. The corresponding normal direction in the direct lattice is noted $[2\bar{6}3]$. The body diagonal of the unit cell of the cubic lattice lies in a [111] direction and more generally, the lattice point $n_1\mathbf{a}_1 + n_2\mathbf{a}_2 + n_3\mathbf{a}_3$ lies in the direction $[n_3\ n_2\ n_3]$ from the origin. For symmetry reasons, there exists equivalent families of planes in nontriclinic crystals, and the equivalent planes are noted collectively $\{u\ v\ w\}$. For instance, in the cubic lattice, the (100), (010), and (001) planes are noted $\{100\}$. Similarly, the [100], [010], [001], [$\bar{1}$00], [0$\bar{1}$0], and [00$\bar{1}$] directions are collectively noted $\langle 100 \rangle$.

In fcc and bcc lattices, there are no cubic primitive cells whereas in simple cubic (sc) system, the reciprocal lattice is also sc and the Miller indices of a family of lattice planes represent the coordinates of a vector normal to the planes in the usual Cartesian coordinates. As the lattice planes of a fcc cubic lattice or a bcc cubic lattice are parallel to those of a sc lattice, it has then been fixed as a rule to define the lattice planes of the fcc and bcc cubic lattices as if they were sc lattices with orthogonal primitive vectors of the reciprocal lattice.

The lattice planes of the hexagonal structures can be defined by three coplanar basis vectors $\mathbf{a}_1, \mathbf{a}_2$, and $\mathbf{a}_3$ at $120°$ from one another and such as $\mathbf{a}_1 + \mathbf{a}_2 + \mathbf{a}_3 = 0$

and by axis c perpendicular to these vectors. The Miller indices of a plane for these structures is written $(h\,k\,i\,l)$ where h, k, and i are the reciprocals of the intercept of the plane with $\mathbf{a}_1$, $\mathbf{a}_2$, and $\mathbf{a}_3$ and l the reciprocal of the intercept in the c direction. The indices h, k, and i are not linearly independent and their sum must be zero. The first BZ of the hexagonal BL is shown in Fig. B.1.

B.3   A Toolbox for Symmetry Groups

B.3.1   The Abstract Groups

A presentation of the optical spectroscopy of impurity centres in crystals requires some understanding of group theory and we provide here basic definitions. Specific answers to many questions on group theory and to its applications in solid-state physics and spectroscopy can be found in [4]. Among other properties [5], the abstract finite groups are characterized by (1) their order, i.e. the number of elements they contain; (2) a closed combination law within the group such that the application of this law to any two elements of the group still yields an element of the group. The order of application is important because for any two group elements G and P, the element resulting from GP is usually different from that resulting from PG, where multiplication is used as the combination law. When GP gives the same result as PG

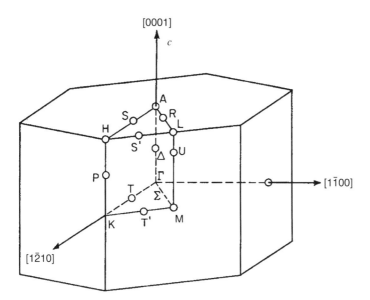

Fig. B.1 First Brillouin zone of the hexagonal BL. The *letters* correspond to critical points (*CPs*) of the BZ with specific symmetries

whatever G and P, the group is said to be abelian; (3) the existence of an identity element (noted E) such as, for any G belonging to the group, GE or EG yields G; and (4) the existence for any element G of an inverse element G^{-1} belonging to the group and such that $G\ G^{-1} = G^{-1}G = E$. Two elements A and B of a group are said to be conjugate if $A = G\ B\ G^{-1}$, from which $B = G^{-1}AG$ is readily derived. A set of mutually conjugate elements of a group constitutes a class of the group and any element of the group appears only in one class (E is a class by itself). A subset of a given group displaying general group properties with the same multiplication law as the initial group is called a subgroup of this group.

B.3.2 The Symmetry Point Groups

The symmetry point groups are a particular category of finite groups whose elements are spatial symmetry operations and 32 of them, derived from the symmetries of the BLs, are known as the crystallographic point groups. A BL or any entity left invariant under all the symmetry operations of a given point group is said to belong to this point group. Two or three of these point groups can sometimes show a one-to-one correspondence between their elements, with the same formal multiplication tables, despite the fact that the symmetry operations are spatially different. Such groups, which correspond to the same abstract group, are said to be isomorphous. There are two notations for the point groups: the international one, also known as the Hermann-Mauguin notation, mainly used by crystallographers, and the one based on the Schönflies notation (used here), mainly used in molecular and semiconductor physics. The correspondence between the two notations is given in Tables 7.2 and 7.3 of [3].

A short description of the 32 crystallographic point groups is given below. They are:

- The pure rotation groups $C_n (n = 1, 2, 3, 4, 6)$ containing only rotations $2\pi k/n$ about an axis (k is an integer between 1 and n). The rotations made clockwise are noted C_n, and those made counter-clockwise C_n^{-1} (they are obviously the inverse of each other), except for $n = 2$ where the two rotations yield the same result. For the C_n groups, these two rotations are unilateral (not equivalent) and they form two distinct classes. When k differs from n and from n/2 ($C_n^{n/2} = C_2$), the rotations C_n^k belong to different classes. For instance, the different classes of the C_6 group are E, C_6, $C_6^2 = C_3$, $C_6^3 = C_2$, $C_6^4 \equiv C_3^{-1}$, and $C_6^5 \equiv C_6^{t1}$. The C_n groups are called cyclic as the C_n operation repeated n times (C_n^n) gives E. The only element of group C_1 is E. Other point groups derived from C_n are:

- The S_{2n} groups ($n = 1, 2, 3$), with additional rotations π/n about the main axis, followed by a reflection through a plane perpendicular to the main axis (S_{2n} or S_{2n0}^{-1} rotation-reflections). For $n = 1$, this corresponds to inversion I. The S_n operations are called improper rotations, by comparison with the proper rotations

C_n. The only element of group S_2 (besides E) is I so that this group is also noted C_i.

- The C_{nh} groups (n $= 1, 2, 3, 4, 6$), with additional S_n rotation-reflections and a symmetry reflection σ_h through a plane perpendicular to the main axis, plus I for n even. A symmetry reflection is another kind of improper rotation (rotation-inversion) resulting from a rotation C_2 followed by inversion (IC_2);
- The C_{nv} groups (n $= 2, 3, 4, 6$), with additional reflections through n symmetry planes containing the main axis (one kind, σ_v, for n $= 3$, two kinds, σ_v and σ'_v, for n even);
- The D_n groups (n $= 2, 3, 4, 6$), with additional rotations of an angle π through n axes perpendicular to the main axis (one kind, C'_2, for n $= 3$, two kinds, C'_2 and C''_2 for n even). For the groups including these additional rotations, the C_n and C_n^{-1} rotations about the main axis are equivalent (bilateral) and they belong to the same class of symmetry operations.
- The D_{nh} groups (n $= 2, 3, 4, 6$), derived from the D_n groups by adding reflections through n symmetry planes containing the main axis and the C'_2 axes (one kind, σ_v, for n = 3, two kinds, σ_v and σ'_v, for n even), a reflection σ_h through a plane containing the C'_2 axes, plus I for n even;
- the D_{nd} groups (n $= 2, 3$), derived from the D_n groups by adding reflections through n symmetry planes containing the main axis and midway of the C'_2 axes (one kind, σ_v, for D_{3d}, two kinds, σ_v and σ'_v, for D_{2d}), plus I for D_{3d}.

The five other point groups are known as the cubic point groups. They are groups T and O, including all the proper rotational symmetries of the tetrahedron and of the cube, respectively, group T_h, derived from T by adding a centre of symmetry, and finally groups T_d and O_h, including all the rotational symmetry transformations of the tetrahedron and of the cube, respectively.

As already said, the above point groups are derived from the symmetries of the BLs. They cannot therefore include groups with C_5 rotational symmetry, like the C_5 group and the groups derived from it. The icosahedral point group, sometimes noted Y, contains fifteen C_2, ten C_3, and six C_5 axes. It displays the rotational symmetries of the regular icosahedron and dodecahedron, the two other regular polyhedra (platonic solids) besides the tetrahedron, the cube, and the octahedron. The I_h point group, also often referred to as the icosahedral point group, is derived from Y by the addition of a centre of symmetry and it is the point group with the largest number of symmetry elements (120). I_h is the symmetry point group attributed to fullerene (C_{60}), whose structure possesses regular hexagonal and pentagonal faces. C_5 rotational symmetry can also be found in some quasicrystals (for a review, see [6]).

B.3.3   Representations and Basis Functions

A set of matrices transforming under the multiplication laws of a group constitutes a representation of this group. When this set is in the diagonal form and that it can be reduced into subsets that cannot be further reduced (we assume the reader is familiar with matrix algebra), these subsets form irreducible representations (*IR*s) of this group. When the initial set cannot be reduced, it is already an *IR* of the group. There are as many *IR*s of a group as the number of classes of this group. The sums of the diagonal elements of the diagonalized matrices are the characters of the *IR*s and they are the same for all the elements of a given class. As the identity *E* is a class by itself, the characters of the *IR*s corresponding to *E* are simply the dimensions of the *IR*s. Most of the group characters are real numbers, but some of them can also be imaginary (for instance, in group C_4) or complex (for instance, in group C_3). The character tables of the 32 crystallographic point groups can be found in [7] and in [5]. A function or a set of functions that transforms under the symmetry operations *R* of a group through the set of matrices corresponding to a given representation forms a basis for this representation (actually, the basis functions are used to determine the representations). Among the *IR*s, there is always a unit representation, 1D, whose characters are 1, whatever the class.

In the notation of [8], the *IR*s are noted by capital letters eventually with indices and/or primes, the convention being to label by A or B the 1D *IR*s, by E (not to be confused with the identity operation *E*), the 2D ones, and by T the 3D ones. In the notation of Bethe [9] used by Koster [7], the *IR*s are simply noted Γ_i ($i = 1, 2, 3,$ etc.), eventually with + or − exponents.

The symmetry operations considered up to now are supposed to apply on components x, y, z of polar vectors (like those of a force or of an electric field), that change sign under inversion symmetry, or on components S_x, S_y, S_z of axial vectors, or pseudo-vectors (like the angular momentum or the magnetic field) that do not change sign under inversion. It is possible to calculate the characters of the 3D matrix representations associated with the components of polar and axial vectors for the different symmetry operations of the 32 point groups and the corresponding list [10] is given in Table B.3.

Table B.3 Characters of the representations spanned by polar and axial vectors for the different symmetry operations of the crystallographic point groups

Symmetry operation	E	C_2	C_3	C_4	C_6	I	σ	S_3	S_4	S_6
Polar vector	3	−1	0	1	2	−3	1	−2	−1	0
Axial vector	3	1	0	1	2	3	−1	2	1	0

This table can be used to determine the representation for a polar or axial vector in a given symmetry group. In some cases, these representations are irreducible, as for the T_d group, but for the others, they are reducible and the character table of the *IR*s of the group must be used for the reductions into *IR*s.

As an example, the character tables of the C_{6v} and D_{3d} symmetry point groups are given in Tables B.4 and B.5. C_{6v} is the symmetry point group of the crystals with wurtzite structure, and D_{3d} is the point group of the linear O_i structure in silicon discussed in Sect. 6.1.1.1.

In Table B.4, z is taken as the sixfold axis and the six reflection planes intersect on the z-axis and make 60° angles with one another.

Table B.4 Character table for the C_{6v} point group. In the header, the numbers before the symmetry operations are the number of elements of a symmetry class

Symmetry classes: IR s of C_{6v}	E	C_2	$2C_3$	$2C_6$	$3\sigma_d$	$3\sigma_v$	Basis functions
Γ_1, A_1	1	1	1	1	1	1	$z; z^2 + y^2; z^2$
Γ_2, A_2	1	1	1	1	-1	-1	S_z
Γ_3, B_2	1	-1	1	-1	1	-1	$x^3 - 3xy^2$
Γ_4, B_1	1	-1	1	-1	-1	1	$y^3 - 3yx^2$
Γ_5, E_1	2	-2	-1	1	0	0	x, y
Γ_6, E_2	2	2	-1	-1	0	0	$x^2 - y^2, xy$

The first IR indicated corresponds to the notation of [2] and the second one to that of [8]

Table B.5 Characters table for the D_{3d} point group

Symmetry classes: IR s of D_{3d}	E	$2C_3$	$3C_2'$	I	$2S_6$	$3\sigma_d$	Basis functions
Γ_1^+, A_{1g}	1	1	1	1	1	1	R
Γ_2^+, A_{2g}	1	1	-1	1	1	-1	S_x
Γ_3^+, E_g	2	-1	0	2	-1	0	$S_x - iS_y, -(S_x + iS_y)$
Γ_1^-, A_{1u}	1	1	1	-1	-1	-1	zS_z
Γ_2^-, A_{2u}	1	1	-1	-1	-1	1	z
Γ_3^-, E_u	2	-1	0	2	1	0	$(x - iy), -(x + iy)$

The C_2' axes are perpendicular to the C_3 axis. The numbers before the symmetry operations is the number of elements of a symmetry class. The first IR indicated corresponds to Koster's notation [7] and the second one to that of Mulliken [8]. R is any function going into itself under all proper and improper rotations

In Table B.5 z is taken as the threefold axis. The three twofold axes C_2' are perpendicular to C_3 and they make 120° angles with respect to one another. The three reflection planes σ_d contain the z-axis and are each perpendicular to one of the C_2' axes.

Now, to go further and to provide conceptual tools that will be used in the interpretation of the electronic spectra of impurities in crystals, a new group has to be introduced, the 3D rotation group, noted here $R^+(3)$, which is the group of all the rotations through any angle about any axis. $R^+(3)$ is an infinite group and its IRs and their basis functions are intimately related to the quantum-mechanical properties of the total angular momentum of an electron in a free atom. In the one-electron approximation, quantum mechanics tells us that the energy level of an electron whose eigenvalue of angular momentum j is $(2j + 1)$-fold degenerate. This level is associated with $(2j + 1)$ eigenfunctions differing in the value m of the z-component of the angular momentum, running from j to $-j$. For integral values of j, these eigenfunctions are the spherical harmonics

$$Y_{lm} = N_{lm} P_{l|m|}(\cos \theta)e^{im\varphi}$$

where θ and ϕ are the spherical polar coordinates, $P_{l|m|}$ an associated Legendre polynomial, and N_{lm} a normalizing factor. When the electron spin is included, j

can take integral and half-integral values so that the degeneracy is 1, 2, 3, etc
.... From the quantum-mechanical analogy between the operators of an infinitely
small rotation and angular momentum, it can be shown that the value of j can
be used to label the $(2j + 1)$-dimensional IRs of $R^+(3)$, noted $D^{(j)}$. The unit
representation of $R^+(3)$ is $D^{(0)}$ and the components of an axial vector transform
as IR $D^{(1)}$ of $R^+(3)$. Under rotation by angle ϕ about a given axis, the basis
functions of IR j are multiplied by $e^{im\phi}$. For half-integral values of j, it is
seen that the rotation of 2π about an arbitrary axis does not correspond to the
unit element E for $R^+(3)$ as the basis functions change sign, but to a new
element of the group, usually noted $\bar{E}$ (notations $\hat{E}$ and Q are also found) and
such that $\bar{E}\bar{E} = E$. This can be translated to point groups when studying the
symmetry properties of electronic systems with half-integral values of the angular
momentum in crystals. In that case, besides $\bar{E}$, one has to introduce for the
point group new classes of symmetry operations, noted here generically $\bar{R}$ with
respect to the usual ones, such that $R\bar{E} = \bar{R}$, and they lead to a two-valued
representation of the group, referred to as the double group in this particular
case. For instance, a $\bar{C}_n$ class corresponds to $C_n\bar{E}$. The number of classes of the
double group is larger than that of the original group, but not always twice as
large.

The tables of characters of a point group are very useful to determine the splitting
of a degenerate electronic or vibrational energy level in a crystal field of a given
symmetry. They also allow to determine if a transition between two levels associated
with different IRs is IR-allowed, Raman-allowed, or forbidden. This is facilitated by
a table of multiplication of the IRs for the different symmetry point groups, like the
ones given in [7]. For the double groups, the characters of the IRs not involving spin
are the same for the R and $\bar{R}$ symmetry operations. For those involving spin, the
characters are different, unless the R and $\bar{R}$ operations belong to the same class.

B.3.4   The Symmetry Space Groups

The global symmetry of a crystal is specified not only by a spatial invariance with
respect to the proper and improper rotations defined by the elements of its point
group, but also by the translation operations[1] by vectors $\mathbf{t}_n$ defined by relation B.1.
The primitive translation vectors are defined by the lattice points of the primitive
cells of the different BLs and they constitute an invariant symmetry group. The
symmetry space group of a crystal contains elements combining the operations of
the point (or rotation) group and of the translation group of the crystal. The number
of symmetry space groups is finite and equal to 230 in 3D. The translation group of
operations is a subgroup of the space group of the crystal. When this group contains

---

[1]For a crystal of finite size, translation symmetry necessitates proper consideration of boundary
conditions [5].

only the primitive translations of the *BL*, the rotation group is also a subgroup of the space group of the crystal, which is then called symmorphic or simple space group. There are 73 such space groups in 3D. The translation groups of the other space groups (157 in 3D) contain vectors that are not primitive vectors of the *BL*s and the rotation groups associated with these space groups are not subgroups of these space groups. These latter space groups are called nonsymmorphic [2].

We consider here a few particular space groups. The fcc *BL* is generated by three primitive translation vectors making equal angles with one another. The unit cell contains four lattice points. If one lattice point is at the corner of the cube, the three primitive translations extend from this point to the centre of the faces of the cube adjacent to this corner. They can be taken as:

$$\mathbf{t}_1 = (t/\sqrt{2})(\mathbf{i} + \mathbf{j})$$
$$\mathbf{t}_2 = (t/\sqrt{2})(\mathbf{i} + \mathbf{k}) \tag{B.5}$$
$$\mathbf{t}_3 = (t/\sqrt{2})(\mathbf{j} + \mathbf{k})$$

where $\mathbf{i}$, $\mathbf{j}$, and $\mathbf{k}$ are unit vectors along the edges of the cube and t the length of the translations. The combinations of these primitive translations with the T_d and O_h point groups result in the symmorphic T_d^2 and O_h^5 space groups[2] (noted $F\bar{4}3m$ and $Fm\bar{3}m$, respectively, in the international notation). T_d^2 is the space groups of sphalerite (cubic ZnS), a crystal structure shared by several III–V and II–VI compounds, and O_h^5 the space group of sodium chloride and calcium fluoride. When adding to the fcc primitive translations (B.5) the nonprimitive translation $1/4(\mathbf{t}_1 + \mathbf{t}_2 + \mathbf{t}_3)$ and combining with O_h, the space group generated is O_h^7 ($Fd\bar{3}m$). By construction, this space group is not symmorphic and it generates the diamond structure.

The BZ of the fcc *BL*, associated with space group O_h^5 is shown in Fig. B.2, where the Miller indices of the main symmetry axes are indicated. The critical points Δ, Λ, and Σ are general points inside the BZ on the indicated axes.

The BZ of the O_h^7 and T_d^2 space groups have the same geometry, but the point group symmetries associated with the *CP*s can differ. These symmetries are given in Table B.6.

Table B.6 Point group symmetries associated with the critical points of the BZ of the fcc *BL* for different space groups

CPs:	Γ	Δ	Λ	Σ	X	L	K	W
Space group			Point group symmetries					
O_h^5 and O_h^7	O_h	C_{4v}	C_{3v}	C_{2v}	D_{4h}	D_{3d}	C_{2v}	D_{2d}
T_d^2	T_d	C_{2v}	C_{3v}	C_{2v}	D_{2d}	C_{3v}	C_{2v}	S_4

---

[2]These notations simply mean that T_d^2 was the second space group including T_d and O_h^5 the fifth space group including O_h derived by Schönflies.

Fig. B.2 First Brillouin zone
of the fcc *BL*, showing the
critical points. Its geometry is
the same as that of the
Wigner–Seitz primitive cell
of the bcc *BL*

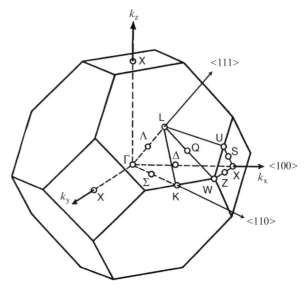

The combination of the primitive translation vectors of the hexagonal *BL* and of a
nonprimitive translation vector to be defined later with the C_{6v} rotation group results
in the C_{6v}^4 space group ($P6_3mc$). This space group is the one of wurtzite (hexagonal
ZnS) to which belong the III–V nitrides and several II–VI compounds. The BZ of
the hexagonal *BL* is shown in Fig. B.1. The point groups along the $\Gamma - \Delta - A$,
K-P-H, and M-U-L axes of the BZ for the wurtzite structure are C_{6v}, C_{3v}, and C_{2v},
respectively [11].

B.4   Some Crystal Structures

B.4.1   Cubic Structures

The cubic structure is found in many crystals, but with different arrangements of the
atoms. The simplest ones are the NaCl and the CsCl structures. The NaCl structure is
the superposition of two identical fcc Bravais sublattices shifted by 1/2 of the edges
of their unit cell; one Na$^+$ (Cl$^-$) ion has 6 Cl$^-$ (Na$^+$) *nn*s along $\langle 100 \rangle$ directions.
The CsCl structure is the superposition of two identical sc sublattices translated by
1/2 of the diagonal of their unit cell; one Cs$^+$ (Cl$^-$) ion has 8 Cl$^-$ (Cs$^+$) *nn*s of the
other sublattice along $\langle 111 \rangle$ directions. The symmorphic space group of CsCl is O_h^1
($Pm3m$).

The fluorite (CaF$_2$) lattice is the superposition of a fcc sublattice of Ca^{++} ions
with a sc sublattice of F$^-$ ions. The lengths of the edges of the unit cells of the
Ca^{++} and F$^-$ sublattices are in the ratio of 2 to 1, respectively, and the F$^-$ unit

cell is shifted by 1/4 along the diagonal of the Ca^{++} cubic cell. The CaF_2 lattice is thus made of unit cells containing each four Ca^{++} ions and eight F^- ions. Crystals with the same atomic arrangement as fluorite, but where the more electronegative element is exchanged with the more positive one of fluorite, like Mg_2 Si, are said to have the antifluorite structure. This Mg-based family of crystals has semiconductor properties.

The diamond structure and the cubic ZnS (sphalerite or zinc-blende) structure can be seen as the superposition of two identical fcc Bravais sublattices translated by one quarter of the diagonal of their unit cell. In these structures, each atom is bonded to its four *nn*s in a regular tetrahedral configuration (see Fig. B.3a,b). In the diamond structure, the atoms of the two sublattices are the same and the associated rotational symmetry is the one of the fcc structure, O_h, or m3m in the international notation, which includes inversion symmetry. In the sphalerite structure, as the two atoms are different, there is no more inversion symmetry and the point group symmetry is T_d or $\overline{4}3$ m. There must be no confusion with the site symmetry of a substitutional impurity, which is T_d for both structures. A partial list of crystals with these structures is given in Appendix C.

With reference to the fluorite structure, the diamond and sphalerite structures can be seen as the superposition of a Ca^{++}-like fcc sublattice to a F^--like sc sublattice where half of the atoms have been removed, in order to yield the second fcc sublattice of the diamond and sphalerite structures.

In the cubic cuprite (Cu_2O) structure, the Cu sites form a fcc sublattice shifted by 1/4 of the body diagonal from the bcc lattice formed by the O sites, as shown in Fig. B.4.

With reference to the fluorite/antifluorite structure, the cuprite structure can be seen as the superposition of a cation-like (Cu^+) fcc sublattice to an anion-like (O^{2-}) sc sublattice with lattice parameters in a 2/1 ratio, where 3/4 of the sites are vacant, in order to yield a bcc sublattice. This results in a sc lattice structure which bears similitude with the diamond structure. When the origin of the unit cell is an O site,

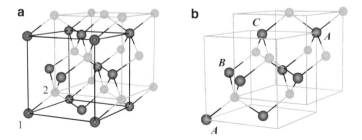

Fig. B.3 (a) Ball and stick model of the sphalerite structure showing the two interpenetrating fcc unit cells. Each cell contains only one type of atom. The displacement between the two cells is materialized by the bond between atoms 1 and 2. (b) Same cells as in (a) showing atoms bonding along a privileged $\langle 111 \rangle$ axis of the crystal. Along this axis, the period of the crystal is the diagonal of one unit cell and it contains three stacks of atoms of one type (the **ABC** stacking period of sphalerite). The atoms not involved in the oriented bonding have been omitted for clarity

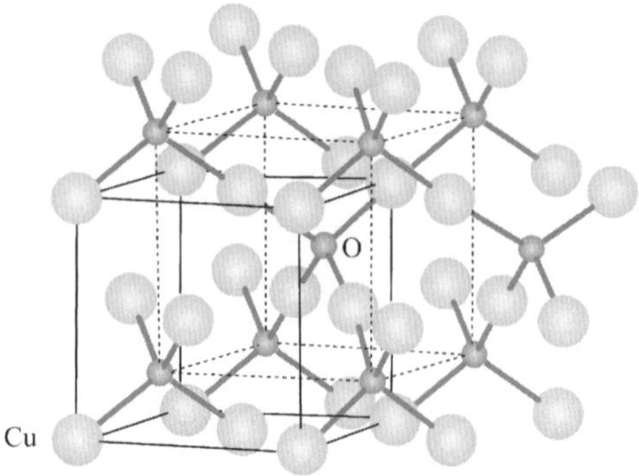

Fig. B.4 Ball and stick model of the cuprite structure. The fcc unit cell of the Cu sublattice is outlined by *solid lines* and the bcc unit cell of the O sublattice by *dashed lines*. These cells both contain four Cu and two O atoms, and they represent the Cu_2O stoichiometry

Fig. B.5 Ball and stick model of the wurtzite structure. Along the c direction, the period of the crystal is equal to the height c of the cell and it contains two stacks of atoms of one type (the *AB* stacking period of wurtzite). To better appreciate the symmetry, the limits of one hexagonal subunit have been outlined

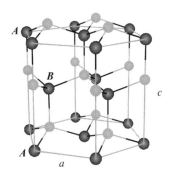

this cell contains 2 O sites and 4 Cu sites, with a centre of inversion about each Cu site, and it is described by the O_h point group. The BZ for this structure is a cube, and the associated space group is O_h^4.

B.4.2   Hexagonal Structures

The hexagonal closed-packed (hcp) structure can be viewed as two interpenetrating hexagonal *BL*s where one is shifted vertically along the c-axis by half of the height c of the hexagonal unit cell and horizontally so that the points of one hexagonal lattice lie directly above the centres of the triangles formed by the points of the other one. The hcp structure is the same as that of a close-packed stack of identical spheres. If the radius of these spheres is a, the distance $c/2$ between the first and second layers is $\sqrt{2/3}a$ and this packing condition determines the ratio 1.633 between the side a of the hexagon and the height c of the unit in the hcp structure. The wurtzite structure (so-called after the hexagonal allotropic form of ZnS) is the superposition of two hcp sublattices whose unit cells are shifted by $5c/8$ along the height of the cell (the c-axis) and it is shown in Fig. B.5.

The symmetry point group associated with this structure is C_{6v} (6 mm) and it is derived from the corresponding space group. The symmetry difference between wurtzite and sphalerite leads naturally to environment differences: in sphalerite, an atom has 24 closer third *nn* atoms and 12 more distant third *nn* atoms. In wurtzite there are four categories of third *nn* atoms: only one-third *nn* atom is at a distance only slightly larger than the *nns*, along the same *c*-axis as the reference atom. The three other categories contain 6, 6, and 12 atoms. The real crystals with wurtzite structure do show a small crystal distortion along the *c*-axis so that the ratio c/a between the height of the unit cell and the side of the regular hexagonal base differs from the ideal value 1.6333. This produces a small increase of the *nn* and *nnn* distances for orientations along or predominantly along the *c*-axis.

Many IIA-sulphides and -oxides as well as IIIA-nitrides crystallize in the wurtzite form, but some of them (ZnS, of course, but also CdS, GaN, and others) can also be found in the sphalerite form. When this occurs, the wurtzite and sphalerite forms are often designated as the α- and β-forms, respectively. SiC can adopt the wurtzite form ($2H$-SiC) or less frequently the sphalerite form ($3C$-SiC), with a notable difference in the band gap (3.3 or 2.3 eV, respectively), but when grown by

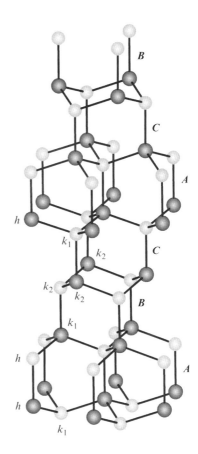

Fig. B.6 Unit cell of the $6H$-SiC polytype showing the **ABCACB** stacki ng sequence and the different sites (see text)

vapour-phase epitaxy, SiC is usually obtained in the form of polytypes with stacking periods different from those of $3C$-SiC and $2H$-SiC. One of the most common variety is $6H$-SiC, whose stacking period along the c-axis is **ABCACB**. Its unit cell is shown in Fig. B.6.

In the $6H$-SiC polytype, there are three different sites: an hexagonal one noted h, and two cubic ones noted k_1 and k_2. As shown in Fig. B.6, an h site is surrounded by three k_1 and one h nn sites, a k_1 site by three h and one k_2 nns sites, and a k_2 site by three k_2 and one k_1 nns sites. Very small carbon crystals with the hexagonal structure of Fig. B.5 have been found in some meteorites and this form of carbon is called lonsdaleite.

References

1. C. Kittel, *Introduction to Solid State Physics*, 7th edn. (Wiley, New York, 1996)
2. G.F. Koster, Space groups and their representations, in *Solid State Physics Advances in Research and Application*, vol. 5, ed. by F. Seitz , D. Turnbull (Academic, New York, 1957), pp. 173–256
3. N.W. Ashcroft, N.D. Mermin, *Solid State Physics* (Saunders College, Philadelphia, 1976)
4. M.S. Dresselhaus, G. Dresselhaus, A. Jorio, *Group Theory: Application to the Physics of Condensed Matter* (Springer, New York, 2008)
5. V. Heine, *Group Theory in Quantum Mechanics* (Pergamon Press, Oxford, 1960)
6. J.W. Cahn, Quasicrystals. J. Res. Natl. Inst. Stand. Technol. **106**, 975–982 (2001)
7. G.F. Koster, J.O. Dimmock, R.G. Wheeler, H. Statz, *Properties of the Thirty-Two Point Groups* (M.I.T. Press, Cambridge, 1963)
8. R.S. Mulliken, Electronic structures of polyatomic molecules and valence. IV. Electronic states, quantum theory of the double bond. Phys. Rev. **43**, 279–302 (1933)
9. H. Bethe, *Termaufspaltung in Kristallen*. Ann.der Phys. **3**, 133–208 (1933)
10. M. Lax, *Symmetry Principles in Solid State and Molecular Physics* (Dover, Mineola, 1974)
11. L. Patrick, D.R. Hamilton, W.J. Choyke, Growth, luminescence, selection rules, and lattice sums of SiC with wurtzite structure. Phys. Rev. **143**, 526–536 (1966)
12. L.B. Bouckaert, R. Smoluchowski, E. Wigner, Theory of the Brillouin zones and symmetry properties of wave functions in crystals. Phys. Rev. **50**, 58–67 (1936)

Appendix C
Optical Band Gaps and Crystal Structures of Some Insulators and Semiconductors

Band-gap energies E_g (eV) of some insulators and semiconductors with direct (D) or indirect (I) band gaps at RT and, when known, at LHeT. For the uniaxial crystals, the band gaps for $\mathbf{E}\|c$ and $\mathbf{E}\perp c$ slightly differ and the value given is an average. The equivalent of the RT value of E_g is expressed in wavelength (λ) in the last column. It is close to the high-frequency limit of transparency for pure and non-diffusing materials. The visible range extends from 400 to about 750 nm ($\sim$3.1 to about 1.7 eV).

For the crystal structures, c, h, and hcp stand for cubic, hexagonal, and hexagonal close-packed, respectively. The names of the crystals used as references are bold-faced.

Material	Crystal structure	E_g (RT)	E_g (LHeT)	λ(nm)
c-BN	c (sphalerite)	6.4I		194
AlN	hcp (wurtzite)	6.2D		200
C (lonsdaleite)	hcp (wurtzite)	$\sim$5.5		$\sim$220
$\mathbf{C_{diam}}$ **(diamond)**	$\mathbf{C_{diam}}$ **(c)**	5.475 I	5.487[a]	226
h-BN	h 2D	5.2I		238
α-**ZnS (wurtzite)**	**Wurtzite (hcp)**	3.8D		325
β-**ZnS (sphalerite)**	**Sphalerite (c)**	3.734D	3.835	332
CuCl	c (Sphalerite)	3.4 D	3.399	$\sim$365
w-GaN or α-GaN	hcp (wurtzite)	$\sim$3.4D	3.50	$\sim$365
ZnO	hcp (wurtzite)	3.4D	3.44	365
c-GaN or β-GaN	c (sphalerite)	3.30D	3.41	376
2H-SiC or α-SiC	hcp (wurtzite)	3.3 I	3.33	376
4H-SiC	Polytype	3.23 I	3.27	384
6H-SiC	Polytype	2.86 I	3.03	434
ZnSe	c (sphalerite)	2.722D	2.825	460
β-CdS	c (sphalerite)	2.50D		496
α-CdS	hcp (wurtzite)	2.49D		498
AlP	c (sphalerite)	2.45 I	2.505	506

(continued)

B. Pajot and B. Clerjaud, *Optical Absorption of Impurities and Defects in Semiconducting Crystals*, Springer Series in Solid-State Sciences 169, DOI 10.1007/978-3-642-18018-7, © Springer-Verlag Berlin Heidelberg 2013

(continued)

Material	Crystal structure	E_g (RT)	E_g (LHeT)	λ(nm)
3C-SiC or β-SiC	c (sphalerite)	$\sim$2.3I	2.41	$\sim$540
ZnTe	c (sphalerite)	2.290D	2.394	544
GaP	c (sphalerite)	2.272I	2.350	546
Cu$_2$O	c	$\sim$2.1D	2.174	$\sim$590
AlAs	c (sphalerite)	2.15 I	2.229	577
GaSe	quasi-2D	$\sim$2D		$\sim$620
CdSe	hcp (wurtzite)	1.714 D	1.829^a	709
AlSb	c (sphalerite)	1.62 I	1.686	765
CdTe	c (sphalerite)	1.526D	1.607	812
GaAs	c (sphalerite)	1.424D	1.519	873
InP	c (sphalerite)	1.344D	1.424	923
Si	c (C_{diam})	1.124 I	1.1700	1101
InN	hcp (wurtzite)	$\sim$0.8D		$\sim$1550
GaSb	c (sphalerite)	0.727D	0.811	1705
Ge	c (C_{diam})	0.670 I	0.7447	1851
Mg$_2$Si	c (antifluorite)	$\sim$0.6I	0.77	$\sim$2070
Mg$_2$Ge	c (antifluorite)	0.54 I	0.74	$\sim$2480
PbS	c (NaCl)	0.41D	0.29	$\sim$3350
InAs	c (sphalerite)	0.354D	0.418	$\sim$3440
PbTe	c (NaCl)	0.29D	0.190	$\sim$4280
PbSe	c (NaCl)	0.26D	0.165	$\sim$4770
SnTe	c (NaCl)	0.19 D		$\sim$6320
Mg$_2$Sn	c (antifluorite)	0.18I		$\sim$6900
InSb	c (sphalerite)	0.18D	0.2344	$\sim$6900

^{a}At LNT

Appendix D
Table of Isotopes

An asterisk denotes a radioactive isotope whose lifetime is indicated in the natural abundance column. When a stable element has several radioactive isotopes, a few ones have been chosen for their interest in life or different applications. The isotopes with the longest lifetimes and at least one with a nonzero nuclear spin I are also indicated. The electronic configuration of an element with atomic number Z is given in *italics* in the name and symbol box. When relevant, the old group label notation of the periodic table is indicated in square brackets in this box. Radioactive elements like ^{40}K, ^{115}In or ^{238}U, with lifetimes larger than 10^9 years, are considered as stable.

Name and symbol	Z	Number of nucleons	Natural abundance (%)	Average mass (amu)	I
Hydrogen (H or ^{1}H)	1	1	(99.985)	1.008	1/2
Deuterium (D or ^{2}H)		2	(0.0148)		1
Tritium* (T or ^{3}H) $1s$		3*	12.32 y		1/2
Helium (He) $1s^2$	2	3	(0.000138)	4.003	1/2
		4	(99.999862)		0
Lithium (Li) [IA]	3	6	(7.6)	6.941	1
[He]$2s$		7	(92.4)		3/2
Beryllium (Be) [IIA]	4	9	(100)	9.012	3/2
[He]$2s^2$		10*	1.52×10^6 y		0
Boron (B) [IIIB]	5	10	(19.8)	10.81	3
[He]$2s^2 2p$		11	(80.2)		3/2
Carbon (C) [IVB]	6	12	(98.93)	12.01	0
[He]$2s^2 2p^2$		13	(1.07)		1/2
		14*	5715 y		0
Nitrogen (N) [VB]	7	14	(99.632)	14.01	1
[He]$2s^2 2p^3$		15	(0.368)		1/2
Oxygen (O) [VIB]	8	16	(99.757)	16.00	0
[He]$2s^2 2p^4$		17	(0.038)		5/2
		18	(0.205)		0

(continued)

B. Pajot and B. Clerjaud, *Optical Absorption of Impurities and Defects in Semiconducting Crystals*, Springer Series in Solid-State Sciences 169, DOI 10.1007/978-3-642-18018-7, © Springer-Verlag Berlin Heidelberg 2013

(continued)

Name and symbol	Z	Number of nucleons	Natural abundance (%)	Average mass (amu)	I
Fluorine (F) [VIIB]	9	18*	1.83 h	19.00	1
$[He]2s^2 2p^5$		19	(100)		1/2
Neon (Ne)	10	20	(90.48)	20.18	0
$[He]2s^2 2p^6$		21	(0.27)		3/2
		22	(9.25)		0
Sodium (Na) [IA]	11	22*	2.605 y	22.99	3
$[Ne]3s$		23	(100)		3/2
Magnesium (Mg) [IIA]	12	24	(78.99)	24.31	0
$[Ne]3s^2$		25	(10.00)		5/2
		26	(11.01)		0
Aluminium (Al) [IIIB]	13	26*	7.1×10^5 y	26.98	5
$[Ne]3s^2 3p$		27	(100)		5/2
Silicon (Si) [IVB]	14	28	(92.23)	28.086	0
$[Ne]3s^2 3p^2$		29	(4.67)		1/2
		30	(3.10)		0
		31*	2.62 h		0
Phosphorus (P) [VB]	15	31	(100)	30.97	1/2
$[Ne]3s^2 3p^3$		32*	14.28 d		1
Sulphur (S) [VIB]	16	32	(94.93)	32.07	0
$[Ne]3s^2 3p^4$		33	(0.76)		3/2
		34	(4.29)		0
		36	(0.02)		0
Chlorine (Cl) [VIIB]	17	35	(75.78)	35.45	3/2
$[Ne]3s^2 3p^5$		36*	301000 y		0
		37	(24.22)		3/2
Argon (Ar)	18	36	(0.337)	39.95	0
$[Ne]3s^2 3p^6$		38	(0.063)		0
		39*	268 y		7/2
		40	(99.600)		0
Potassium (K) [IA]	19	39	(93.26)	39.10	3/2
$[Ar]4s$		40	(0.012)		4
		41	(6.73)		3/2
Calcium (Ca) [IIA]	20	40	(96.941)	40.08	0
$[Ar]4s^2$		41*	102000 y		7/2
		42	(0.647)		0
		43	(0.135)		7/2
		44	(2.086)		0
		46	(0.004)		0
		48	(0.187)		0
Scandium (Sc)	21	45	(100)	44.96	7/2
$[Ar]3d4s^2$		46*	83.81 d		4
Titanium (Ti)	22	44*	67 y	47.87	0
$[Ar]3d^2 4s^2$		46	(8.25)		0
		47	(7.44)		5/2

(continued)

(continued)

Name and symbol	Z	Number of nucleons	Natural abundance (%)	Average mass (amu)	I
		48	(73.72)		0
		49	(5.41)		7/2
		50	(5.18)		0
Vanadium (V)	23	50	(0.25)	50.94	6
$[Ar]3d^34s^2$		52	(99.75)		7/2
Chromium (Cr)	24	50	(4.35)	52.00	0
$[Ar]3d^54s$		51*	27.7 d		7/2
		52	(83.79)		0
		53	(9.50)		3/2
		54	(2.36)		0
Manganese (Mn)	25	53*	3.7×10^6 y	54.94	7/2
$[Ar]3d^54s^2$		55	(100)		5/2
Iron (Fe)	26	54	(5.85)	55.85	0
$[Ar]3d^64s^2$		56	(91.75)		0
		57	(2.12)		1/2
		58	(0.28)		0
		60*	1.5×10^6 y		0
Cobalt (Co)	27	58*	70.9 d	58.93	2
$[Ar]3d^74s^2$		59	(100)		7/2
		60*	5.271 y		5
Nickel (Ni)	28	58	(68.08)	58.69	0
$[Ar]3d^84s^2$		59*	76000 y		3/2
		60	(26.22)		0
		61	(1.14)		3/2
		62	(3.63)		0
		64	(0.93)		0
Copper (Cu) [IB]	29	63	(69.17)	63.55	3/2
$[Ar]3d^{10}4s$		64*	12.701 h		1
		65	(30.83)		3/2
		66*	5.09 m		1
Zinc (Zn) [IIB]	30	64	(48.63)	65.41	0
$[Ar]3d^{10}4s^2$		65*	243.8 d		5/2
		66	(27.90)		0
		67	(4.10)		5/2
		68	(18.75)		0
		70	(0.62)		0
Gallium (Ga) [IIIB]	31	69	(60.11)	69.72	3/2
$[Ar]3d^{10}4s^24p$		70*	21.1 m		1
		71	(39.89)		3/2
		72*	14.10 h		3
Germanium (Ge) [IVB]	32	70	(20.38)	72.64	0
$[Ar]3d^{10}4s^24p^2$		71*	11.2 d		1/2
		72	(27.31)		0
		73	(7.76)		9/2
		74	(36.72)		0

(continued)

(continued)

Name and symbol	Z	Number of nucleons	Natural abundance (%)	Average mass (amu)	I
		75*	1.38 h		1/2
		76	(7.83)		0
		77*	11.30 d		7/2
Arsenic (As) [VB] $[Ar]3d^{10}4s^24p^3$	33	75	(100)	74.92	3/2
		76*	26.3 h		2
Selenium (Se) [VIB] $[Ar]3d^{10}4s^24p^4$	34	74	(0.89)	78.96	0
		76	(9.37)		0
		77	(7.63)		1/2
		78	(23.77)		0
		79*	65000 y		7/2
		80	(49.61)		0
		82	(8.73)		0
Bromine (Br) [VIIB] $[Ar]3d^{10}4s^24p^5$	35	77*	2.376 d	79.90	3/2
		79	(50.69)		3/2
		81	(49.31)		3/2
Krypton (Kr) $[Ar]3d^{10}4s^24p^6$	36	78	(0.35)	83.80	0
		80	(2.28)		0
		82	(11.58)		0
		83	(11.49)		9/2
		84	(57.00)		0
		85*	10.73 y		9/2
		86	(17.30)		0
Rubidium (Rb) [IA] $[Kr]\,5s$	37	83*	86.2 d	85.47	5/5
		85	(72.17)		5/2
		87	(27.83)		3/2
Strontium (Sr) [IIA] $[Kr]\,5s^2$	38	84	(0.56)	87.62	0
		86	(9.86)		0
		87	(7.0)		9/2
		88	(82.58)		0
		90*	29.1 y		0
Yttrium (Y) $[Kr]4d\,5s^2$	39	89	(100)	88.91	1/2
Zirconium (Zr) $[Kr]\,4d^25s^2$	40	90	(51.45)	91.22	0
		91	(11.22)		5/2
		92	(17.15)		0
		94	(17.38)		0
		96	(2.80)		0
Niobium (Nb) $[Kr]\,4d^45s$	41	92*	3.7×10^7 y	92.91	7
		93	(100)		9/2
		94*	24000 y		6
Molybdenum (Mo) $[Kr]4d^55s$	42	92	(14.84)	95.94	0
		93*	3500 y		5/2
		94	(9.25)		0
		95	(15.92)		5/2

(continued)

(continued)

Name and symbol	Z	Number of nucleons	Natural abundance (%)	Average mass (amu)	I
		96	(16.68)		0
		97	(9.55)		5/2
		98	(24.13)		0
		100	(9.63)		0
Technetium* (Tc) [Kr] $4d^5 5s^2$	43	97*	2.6×10^6 y		9/2
		98*	4.2×10^6 y		6
		99*	213000 y		9/2
Ruthenium (Ru) [Kr]$4d^7 5s$	44	96	(5.52)	101.1	5/2
		98	(1.88)		0
		99	(12.70)		0
		100	(12.60)		0
		101	(17.00)		5/2
		102	(31.60)		0
		104	(18.70)		0
Rhodium (Rh) [Kr]$4d^8 5s$	45	101*	3.5 y	102.9	1/2
		102*	2.9 y		6
		103	(100)		1/2
Palladium (Pd) [Kr]$4d^{10}$	46	102	(1.02)	106.4	0
		104	(11.14)		0
		105	(22.33)		5/2
		106	(27.33)		0
		107*	6.5×10^6 y		5/2
		108	(26.46)		0
		110	(11.72)		0
Silver (Ag) [IB] [Kr]$4d^{10} 5s$	47	105*	41.3 d	107.9	1/2
		107	(51.83)		1/2
		109	(48.17)		1/2
Cadmium (Cd) [IIB] [Kr] $4d^{10} 5s^2$	48	106	(1.25)	112.4	0
		108	(0.89)		0
		110	(12.49)		0
		111	(12.80)		1/2
		112	(24.13)		0
		113	(12.22)		1/2
		114	(28.73)		0
		116	(7.49)		0
Indium (In) [IIIB] [Kr]$4d^{10} 5s^2 5p$	49	111*	2.805 d	114.8	9/2
		113	(4.3)		9/2
		115	(95.7)		9/2
Tin (Sn) [IVB] [Kr] $4d^{10} 5s^2 5p^2$	50	112	(1.0)	118.7	0
		114	(0.7)		0
		115	(0.4)		1/2
		116	(14.7)		0
		117	(7.7)		1/2
		118	(24.3)		0

(continued)

(continued)

Name and symbol	Z	Number of nucleons	Natural abundance (%)	Average mass (amu)	I
		119	(8.6)		1/2
		120	(32.4)		0
		122	(4.6)		0
		124	(5.6)		0
Antimony (Sb) [VB] $[Kr]4d^{10}5s^25p^3$	51	121	(57.3)	121.8	5/2
		122*	2.72 d		2
		123	(42.7)		7/2
		124*	60.30 d		3
Tellurium (Te) [VIB] $[Kr]\,4d^{10}5s^25p^4$	52	119*	16 h	127.6	1/2
		120	(0.09)		0
		122	(2.55)		0
		123	(0.89)		1/2
		124	(4.74)		0
		125	(7.07)		1/2
		126	(18.84)		0
		128	(31.74)		0
		130	(34.08)		0
Iodine (I) [VIIB] $[Kr]\,4d^{10}5s^25p^5$	53	127	(100)	126.9	5/2
		129*	1.7×10^7 y		7/2
		131*	8.04 d		7/2
Xenon (Xe) $[Kr]\,d^{10}5s^25p^6$	54	124	(0.10)	131.3	0
		126	(0.09)		0
		127*	3.64 d		1/2
		128	(1.91)		0
		129	(26.40)		1/2
		130	(4.10)		0
		131	(21.20)		3/2
		132	(26.90)		0
		134	(10.40)		0
		136	(8.90)		0
Caesium (Cs) [IA] $[Xe]\,6s$	55	133	(100)	132.9	7/2
		134*	2.065 y		4
		135*	2.3×10^6 y		7/2
		137*	30.2 y		7/2
Barium (Ba) [IIA] $[Xe]\,6s^2$	56	130	(0.106)	137.3	0
		132	(0.101)		0
		133*	10.53 y		1/2
		134	(2.417)		3/2
		135	(6.592)		0
		136	(7.854)		3/2
		137	(11.23)		0
		138	(71.70)		0

(continued)

(continued)

Name and symbol	Z	Number of nucleons	Natural abundance (%)	Average mass (amu)	I
Lanthanum (La)	57	137*	60000 y	138.9	7/2
[Xe] $5d\,6s^2$		138	(0.09)		5
		139	(99.91)		7/2
Cerium (Ce)	58	136	(0.19)	140.1	0
[Xe] $4f\,5d\,6s^2$		138	(0.25)		0
		139*	137.6 d		3/2
		140	(88.48)		0
		142	(11.08)		0
Praseodymium (Pr)	59	141	(100)	140.9	5/2
[Xe] $4f^36s^2$					
Neodymium (Nd)	60	142	(27.13)	144.2	0
[Xe]$4f^46s^2$		143	(12.18)		7/2
		144	(23.80)		0
		145	(8.30)		7/2
		146	(17.19)		0
		148	(5.76)		0
		150	(5.64)		0
Promethium* (Pm)	61	145*	17.7 y		5/2
[Xe] $4f^56s^2$		146*	5.53 y		3
		147*	2.62 y		7/2
Samarium (Sm)	62	144	(3.1)	150.4	0
[Xe] $4f^66s^2$		146*	1.03×10^8 y		0
		147	(15.0)		7/2
		148	(11.2)		0
		149	(13.8)		7/2
		150	(7.4)		0
		151*	90 y		5/2
		152	(26.8)		0
		154	(22.8)		0
Europium (Eu)	63	151	(47.8)	152.0	5/2
[Xe] $4f^76s^2$		152*	13.5 y		3
		153	(52.2)		5/2
		154*	8.59 y		3
		155*	4.76 y		5/2
Gadolinium (Gd)	64	152	(0.20)	157.3	0
[Xe] $4f^75d\,6s^2$		154	(2.18)		0
		155	(14.80)		3/2
		156	(20.47)		0
		157	(15.65)		3/2
		158	(24.84)		0
		160	(21.86)		0
Terbium (Tb)	65	157*	110 y	158.9	3/2
[Xe] $4f^85d\,6s^2$		158*	180 y		3
		159	(100)		3/2

(continued)

(continued)

Name and symbol	Z	Number of nucleons	Natural abundance (%)	Average mass (amu)	I
Dysprosium (Dy)	66	156	(0.06)	162.5	0
$[Xe]\,4f^9 5d\,6s^2$		158	(0.10)		0
		160	(2.34)		0
		161	(18.90)		5/2
		162	(25.50)		0
		163	(24.90)		5/2
		164	(28.20)		0
Holmium (Ho)	67	163*	162.9 y		7/2
$[Xe]\,4f^{10}5d\,6s^2$		165	(100)	164.93	7/2
Erbium (Er)	68	162	(0.14)	167.3	0
$[Xe]\,4f^{11}5d\,6s^2$		164	(1.61)		0
		166	(33.60)		0
		167	(22.95)		7/2
		168	(26.80)		0
		170	(14.90)		0
Thulium (Tm)	69	169	(100)	168.93	1/2
$[Xe]\,4f^{12}5d\,6s^2$		171*	1.92 y		1/2
Ytterbium (Yb)	70	168	(0.13)	173.0	0
$[Xe]\,4f^{13}5d\,6s^2$		170	(3.05)		0
		171	(14.30)		1/2
		172	(21.90)		0
		173	(16.12		5/2
		174	(30.80)		0
		176	(12.70)		0
Lutecium (Lu)	71	173*	1.37 y	175.0	7/2
$[Xe]\,4f^{14}5d\,6s^2$		174*	3.3 y		1
		175	(97.41)		7/2
		176	(2.59)		7
Hafnium (Hf)	72	174	(0.16)	178.5	0
$[Xe]\,4f^{14}5d^2 6s^2$		176	(5.20)		0
		177	(18.60)		7/2
		178	(27.10)		0
		179	(13.74)		9/2
		180	(35.20)		0
Tantalum (Ta)	73	180	(0.012)	180.9	0
$[Xe]\,4f^{14}5d^3 6s^2$		181	(99.988)		7/2
Tungsten (W)	74	180	(0.13)	183.9	0
$[Xe]\,4f^{14}5d^4 6s^2$		182	(26.30)		0
		183	(14.30)		1/2
		184	(30.67)		0
		186	(28.60)		0
Rhenium (Re)	75	185	(37.4)	186.2	5/2
$[Xe]\,4f^{14}5d^5 6s^2$		187	(62.6)		5/2

(continued)

(continued)

Name and symbol	Z	Number of nucleons	Natural abundance (%)	Average mass (amu)	I
Osmium (Os) $[Xe]\,4f^{14}5d^66s^2$	76	184	(0.02)	190.2	0
		186	(1.58)		0
		187	(1.6)		1/2
		188	(13.3)		0
		189	(16.1)		3/2
		190	(26.4)		0
		192	(41.0)		0
Iridium (Ir) $[Xe]\,4f^{14}5d^76s^2$	77	191	(37.3)	192.2	3/2
		193	(62.7)		3/2
Platinum (Pt) $[Xe]\,4f^{14}5d^96s$	78	190	(0.01)	195.1	0
		192	(0.79)		0
		193*	60 y		1/2
		194	(32.90)		0
		195	(33.80)		1/2
		196	(25.30)		0
Gold (Au) [IB] $[Xe]\,4f^{14}5d^{10}6s$	79	193*	17.62 h	197.0	3/2
		198	(7.20)		0
		195*	186.12 d		3/2
		197	(100)		3/2
Mercury (Hg) [IIB] $[Xe]\,4f^{14}5d^{10}6s^2$	80	193*	3.80 h	200.6	3/2
		195*	9.5 h		1/2
		196	(0.15)		0
		198	(10.10)		0
		199	(17.00)		1/2
		200	(23.10)		0
		201	(13.20)		3/2
		202	(29.65)		0
		204	(6.80)		0
Thallium (Tl) [IIIB] $[Xe]\,4f^{14}5d^{10}6s^26p$	81	201	(29.524)	204.4	1/2
		204*	3.78 y		2
		205	(70.476)		1/2
Lead (Pb) [IVB] $[Xe]\,4f^{14}5d^{10}6s^26p^2$	82	204	(1.4)	207.2	0
		206	(24.1)		0
		207	(22.1)		1/2
		208	(52.4)		0
Bismuth (Bi) [VB] $[Xe]4f^{14}5d^{10}6s^26p^3$	83	207*	35 y	208.98	9/2
		209	(100)		9/2
Polonium* (Po) $[Xe]4f^{14}5d^{10}6s^26p^4$	84	207*	2.898 y		1/2
		209*	102 y		0
		210*	138.38 d		0
Astatine* (At) $[Xe]$ $4f^{14}5d^{10}6s^26p^5$	85	210*	8.1 h		5
		211*	7.2 h		9/2
Radon* (Rn) $[Xe]4f^{14}5d^{10}6s^26p^6$	86	211*	14.6 h		1/2
		222*	3.824 d		0

(continued)

(continued)

Name and symbol	Z	Number of nucleons	Natural abundance (%)	Average mass (amu)	I
Thorium (Th) $[Rn]6d^27s^2$	90	229*	7900 y	232.04	5/2
		232	(100)		0
Protactinium* (Pa) $[Rn]5f^26d7s^2$	91	231*	32000 y		3/2
Uranium (U) $[Rn]5f^36d7s^2$	92	233*	2.45×10^5 y	238.03	5/2
		234*	(0.0055) 2.45×10^5y		0
		235*	(0.7200) 7.04×10^8 y		7/2
		238	(99.2745)		0

Appendix E
Uniaxial Stress and Orientational Degeneracy

The atoms, impurity complexes, and localized defects in crystals are characterized by definite atomic symmetries which belong to one of the 32-point group symmetries. The symmetry of a centre in the crystal induces a kind of degeneracy called orientational degeneracy, which is the number of equivalent orientations it can take in the crystal. For instance, in a cubic crystal with the diamond structure, this orientational degeneracy is four when the centre is oriented along a bond axis.

For centres only known by their vibrational spectrum, the knowledge of this orientational degeneracy can help to determine their atomic structure. It can be determined by applying to the sample containing the centre uniaxial stresses along chosen symmetry directions, which reduces this degeneracy in a controllable way. These stresses generally produce a splitting of the vibrational lines, and from symmetry arguments, the orientational degeneracy can be determined, helping to deduce the atomic structure of the centre.

Centres with different symmetries are characterized by a piezospectroscopic tensor A whose components are determined by symmetry considerations [1, 2]. The shift in energy Δ with stress of a transition of a centre referenced in an orthogonal system is:

$$\Delta = A_{xx}\sigma_{xx} + A_{yy}\sigma_{yy} + A_{zz}\sigma_{zz} + 2(A_{xy}\sigma_{xy} + A_{yz}\sigma_{yz} + A_{zx}\sigma_{zx}) \qquad (E.1)$$

where the A_{ij} are the components of the piezospectroscopic tensor, which depend on the orientational degeneracy of the centre in the crystal. These components are expressed in energy per unit stress. The components σ_{ij} of the stress tensor are defined by:

$$\sigma_{ij} = \mathbf{n}_i\,\mathbf{n}_j\,T \qquad (E.2)$$

where $\mathbf{n}_i$ and $\mathbf{n}_j$ are the direction cosines of the stress in the reference frame and T the magnitude of the compressional stress.

The values of Δ for the different orientations of noncubic centres are given in Table 3 of the paper by Kaplyanskii [1]. The shift of the centre of gravity of the split components is independent from the direction of stress and it is equal to one-third

B. Pajot and B. Clerjaud, *Optical Absorption of Impurities and Defects in Semiconducting Crystals*, Springer Series in Solid-State Sciences 169, DOI 10.1007/978-3-642-18018-7, © Springer-Verlag Berlin Heidelberg 2013

of the shift for a hydrostatic stress of the same magnitude. The stress splitting of doubly degenerate states of centres with tetragonal and trigonal symmetry has been treated by Hughes and Runcinman [3].

The number of independent components A_{ij} in expression (E.1) depends on the symmetry of the centre and the values of these components on the interaction of the centre with the crystal.

A piezospectroscopic tensor usually noted B, related to strain instead of stress can alternatively be defined [1, 3, 4], and its components are expressed in energy, as strain is dimensionless.

Orientational degeneracy depends on the symmetries of the centre and of the crystal in which it is embedded. To go further, it is necessary to give a classification of the possible symmetries of centres in crystals. This has been done for cubic crystals by Kaplyanskii [1] to study their stress splitting, but it can be extended to other kinds of perturbations [5].

In cubic crystals, one can distinguish six kinds of centres whose symmetries are those of the polyhedra shown in Fig. E.1. In this figure are also indicated the components A_{ij} of the piezospectroscopic tensor, symmetric with respect to the main diagonal, defined by expression (E.1).

In Table E.1, these polyhedra are defined by their edges a_1, a_2, and a_3 (along the z [001] axis) and the angles α, β, and γ between these edges (angle γ taken as the one between a_1 and a_2), and the triclinic centres have been added.

More generally, in a diamond-type crystal, the orientational degeneracy R of a centre whose symmetry corresponds to a subgroup of order g of the diamond point group is $R = G/g$, where G is the order (48) of the full cubic group O_h. This orientational degeneracy, is shared by the electronic and vibrational lines associated with the centre, in addition to their possible intrinsic degeneracies. Under

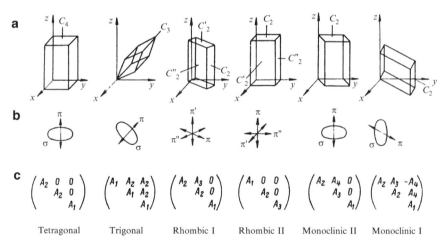

Fig. E.1 (a) Polyhedra having the symmetries of the different noncubic centres in a cubic crystal. (b) Dipole oscillators of the centres. (c) Corresponding piezospectroscopic tensors. C_2' and C_2'' indicate also axes perpendicular to the symmetry planes σ_v' and σ_v (after [1,2]). The type of centres are those of Table E.1

Table E.1 Main symmetry characteristics of the noncubic centres in cubic crystals. The notations of Fig. E.1 are indicated first for the type of centre. Columns 2 and 3 refer to the polyhedra of Fig. E.1. Possible point-group symmetries for the related centres are given in the last column

Type of centre	Restrictions on the lengths of the edges and angles	Symmetry axes	Point groups
Tetragonal	$a_1 = a_2 \neq a_3$ $\alpha = \beta = \gamma = 90°$	C_4 along [001]	$D_{4h}, D_4, C_{4v},$ $D_{2d}, C_{4h}, C_4,$ S_4
Trigonal	$a_1 = a_2 = a_3$ $\alpha = \beta = \gamma \neq$ $90° < 120°$	C_3 along [111]	$D_{3d}, D_3, C_{3v},$ C_{3i}, C_3
Rhombic I	$a_1 \neq a_2 \neq a_3$ $\alpha = \beta = \gamma = 90°$	C_2 along [110] C_2' along [001] C_2'' along [1$\bar{1}$0]	D_{2h}, D_2, C_{2v}
Rhombic II	$a_1 \neq a_2 \neq a_3$ $\alpha = \beta = \gamma = 90°$	C_2, C_2', C_2'' along $\langle 100 \rangle$ directions	D_{2h}, D_2, C_{2v}
Monoclinic II	$a_1 \neq a_2 \neq a_3$ $\alpha = \beta = 90° \neq \gamma$	C_2 or a normal to σ_v along [001]	C_{2h}, C_{1h}, C_s
Monoclinic I	$a_1 \neq a_2 \neq a_3$ $\alpha = \beta = 90° \neq \gamma$	C_2 or a normal to along [110]	C_{2h}, C_{1h}, C_s
Triclinic	No restriction		C_1

nonisotropic perturbations such as oriented stresses and magnetic or electric fields, orientational degeneracy can be partially or totally lifted by defining sub-families of centres with the same orientational degeneracy R_i with respect to stress and a line splitting is observed. Similar orientational degeneracies also occur in noncubic crystals. For a few noncubic symmetries there can exist however high-symmetry orientations along which stress has no effect on the initial orientational degeneracy.

The splitting under stress of nondegenerate electronic and of vibrational transitions of impurity or defect centres depend almost exclusively on the symmetry of these centres through their orientational degeneracy, which has been defined before.

We consider in detail here the case of centres with trigonal symmetry in a cubic crystal. A trigonal centre has a C_3 symmetry axis along a $\langle 111 \rangle$ direction and it displays a fourfold degeneracy. Its spectroscopic tensor A_{trig} has two independent components A_1 and A_2.

The reference frames chosen for the definition of the direction cosines n_j of stress depend on the sub-family of centres considered. The stress-induced splittings for transitions between nondegenerate electronic or vibrational levels, labelled here $A \rightarrow A$, and for transitions to the doubly degenerate E level ($A \rightarrow E$ transitions) of trigonal centres in cubic crystals are expressed in terms of the piezospectroscopic parameters A_1, A_2, B, and C, and they are given in Table E.2. For the E level, the

Table E.2 Piezospectroscopic characteristic of transitions from the ground state to singly degenerate (A) or doubly degenerate (E) excited state of a trigonal centre in a cubic crystal

Stress along	R_i	n_x, n_y, n_z	LD	Δ_i	Intensities for E//σ : E⊥σ	
					π dipoles	σ dipoles
[100]	4	1, 0, 0	A	A_1	1:1	1:1
			E	$A_1 - 2B$	4:1	
				$A_1 + 2B$	0:3	
[111]	1	$\frac{1}{\sqrt{3}}, \frac{1}{\sqrt{3}}, \frac{1}{\sqrt{3}}$	A	$A_1 + 2A_2$	3:0	0:3
	3	$\frac{1}{\sqrt{3}}, -\frac{1}{\sqrt{3}}, \frac{1}{\sqrt{3}}$		$A_1 - 2A_2/3$	1:4	8:5
				$A_1 + 2A_2$	0:1	
			E	$A_1 - 2A_2/3 - 4C/3$	8/3:1/6	
				$A_1 - 2A_2/3 + 4C/3$	0:3/2	

					Intensities for E// [110] : [001] : [1$\bar{1}$0]		
[110]	2	$\frac{1}{\sqrt{2}}, \frac{1}{\sqrt{2}}, 0$	A	$A_1 + A_2$	2:1: 0	1:2: 3
	2	$\frac{1}{\sqrt{2}}, -\frac{1}{\sqrt{2}}, 0$		$A_1 - A_2$	0:1: 2	3:2: 1
				$A_1 + A_2 + C - B$	0:0: 3	
			E	$A_1 + A_2 - C + B$	1:2: 0	
				$A_1 - A_2 + C + B$	0:2: 1	
				$A_1 - A_2 - C - B$	3:0: 0	

R_i denotes the residual orientational degeneracies under stress. The direction cosines n_x, n_y, and n_z of stress in the reference frames associated with each subfamily are the ones given by Kaplyanskii [1] and the shift Δ_i per unit stress is calculated using expression (E.1). LD is the level degeneracy of the excited state. The relative intensities are indicated for the orientation of the electric vector **E** of radiation parallel or perpendicular to stress σ

additional piezospectroscopic components B and C correspond to the electronic or vibrational degeneracy [3].

When the components of the stress tensor are referred to the mutually orthogonal principal axes of the trigonal defect (e.g. [111], [1$\bar{1}$0], and [11$\bar{2}$]), the stress tensor takes the diagonal form

$$\Delta = A_{xx}\sigma_{xx} + A_{yy}\sigma_{yy} + A_{zz}\sigma_{zz}$$

where $A_{xx} = A_1 + 2A_2$, $A_{yy} = A_{zz} = A_1 - A_2$.

The shift of the centre of gravity Δ_{cg} of the split components is independent of the orientation of stress and it is given by:

$$\Delta_{cg} = \frac{1}{R} \sum_{i=1}^{R_i} \Delta_i$$

where the sum is taken on the sub-families for one of the stress orientations. It can be checked that for the trigonal symmetry, Δ_{cg} is A_1 [1, 2]. It corresponds to one-third of the response of the line to a hydrostatic pressure.

Table E.3 Piezospectroscopic characteristic of transitions from the ground state to the triply degenerate excited state of a cubic centre in a cubic crystal Δ is the shift per unit stress [2]

Stress direction	Δ	Intensities for $E//\sigma : E\perp\sigma$	
[100]	$A - B$	0:1	
	$A + 2B$	1:0	
[111]	$A - C/3$	0:1	
	$A + 2C/3$	1:0	
		$k// [1\bar{1}0]$	$k// [001]$
[110]	$A + B/2 + C/2$	1:0	1:0
	$A + B/2 - C/2$	0:0	0:1
	$A - B$	0:1	0:0

It must be noted that in centres with orientational degeneracy, a uniaxial stress produces also a shift of the ground-state energies of the different configurations. It can be also described by expression (E.1), but with piezospectroscopic parameters A_{ij} different from those for a vibrational transition. At a temperature where the centres can reorient under a uniaxial stress to minimize their energy, the reorientation produces a population difference, directly related to the energy difference. When cooling a sample under stress at low temperature and releasing the applied stress, the population difference remains frozen. It can then be probed by the difference observed in the low-temperature absorption of transitions from the ground state when the electric vector of the radiation E is perpendicular and parallel to the aligning stress. For trigonal centres, the spectroscopic method to measure the piezospectroscopic parameter A_2 for the ground state is described at the end of Sect. 8.5.1.3.

A substitutional atom in a cubic crystal with the diamond or sphalerite structure has T_d symmetry and no orientational degeneracy. The stress-induced splitting of a triply degenerate mode can be expressed as a function of the pezospectroscopic parameters A, B, and C. They are given in Table E.3.

References

1. A.A. Kaplyanskii, Noncubic centres in cubic crystals and their piezospectroscopic investigation. Opt. Spectrosc. (USSR) **16**, 329–337 (1964)
2. A.A. Kaplyanskii, Opt. Spectrosc. (USSR) **16**, 1031–1042 (1964)
3. A.E. Hughes, W.A. Runcinman, Uniaxial stress splitting of doubly degenerate states of tetragonal and trigonal centres in cubic crystals. Proc. Phys. Soc. **90**, 827–838 (1967)
4. G.D. Watkins, Oxygen-related defects in silicon: studies using stress-induced alignment, in *Early Stages of Oxygen Precipitation in Silicon. NATO ASI Series. 3 High Technology*, vol. 17, ed. by R. Jones (Kluwer, Dordrecht, 1996), pp. 1–39
5. A.A. Kaplyanskii, Noncubic centers in cubic crystals and their spectra in external fields. J. Phys. Colloque C4 **28** (Suppl. 8–9), C4–39 (1967)

Index

0.79 eV line in silicon, *see C* line in silicon
0.97 eV line in silicon, *see G* line
2DLFM, *see* oxygen (insterstitial), silicon
3H line in diamond, *see* diamond, 3H line
3942 cm^{-1}-line defect in silicon, 137

v_3 mode of interstitial oxygen
  germanium, 283, 286, 288, 301
  silicon, 245

A aggregate or *A* centre in diamond, *see*
          nitrogen, diamond, N$_{2s}$ pair
A centre in silicon, *see V* O centre, silicon
A LVM in GaAs, *see V*O, gallium arsenide
A and *B* LVMs in silicon (O$_i$, C$_s$), 273, 278
ab initio calculations, 194
absorption
  coefficient, 43
    interband, 65
  cross-section, 90
  integrated -, 89
  intrinsic -, 3, 43
  phonons-, *see* phonon, 92
  saturated, 92
(acceptor, hydrogen) centres
  compound semiconductors, 418
  elemental semiconductors, 413
    piezospectroscopy, 415
    reorientation, 427
aluminium
  III–V and II–VI compounds, 214
  group-IV crystals, 212
amphoteric
  atom, 23, 28
  behaviour, 12

anharmonic interaction, 193
anharmonicity, 191
  electrical, 193
  parameters, 192
anion site, 12, 228, 419
antibonding state, 329, 352
antibonding (*AB*) site or location, 13, 14
antisite atom, 9, 12, 152, 193, 210, 220
apex angle, 14, 244, 248, 285, 306, 338
A$_{1g}$ symmetric mode, 249
A$_{2u}$ antisymmetric mode, 249

B aggregate or *B* centre, *see* nitrogen, *V*N$_4$
B LVM in GaAs, *see V* O, gallium arsenide
B′ LVM in GaAs, *see V* O, gallium arsenide
B(1) centre in GaAs (B$_{Ga}$, As$_i$), 362
B(2) centre in GaAs, *see* B$_{As}$ in GaAs
band gap (E$_g$), 1, 58
  confinement dependence, 73
  isotopic dependence, 72
  isotopic mass coefficient (IMC)
    compound materials, 73
    germanium, 73
    silicon, 73
  magnetic field dependence, 72
  pressure dependence, 70
  temperature dependence, 67
  values, 64, 483
band-structure parameters, 61
beam splitter (BS), 95
  spectral range, 97
bending mode, 369, 423
beryllium, 197
biexciton, 79, 80
  radiative lifetime, 81
bistable centre, 18

B. Pajot and B. Clerjaud, *Optical Absorption of Impurities and Defects in Semiconducting*   501
Crystals, Springer Series in Solid-State Sciences 169,
DOI 10.1007/978-3-642-18018-7, © Springer-Verlag Berlin Heidelberg 2013

bolometer, 98
bond-centred hydrogen (H_{BC}), *see* hydrogen (bond-centred)
bond-centred (BC) location, 13, 244
bond reconstruction, 10, 117
boron (substitutional) B_s, 210
 III–V compounds, 212
 diamond, 210, 211
 GaAs
  B_{As}, 213
  B_{Ga}, 212
  BX centres, 214
 germanium, 210, 212
 $15R$-SiC, B-related gap mode, 212
 silicon, 210, 211
  B_s $I B_s$, 360
  B_s pair or P centre, 211
  overtone, 211, 212
boron - donor pairs, 214
 III–V compounds, 217
 germanium, 217
 silicon, 214
  (B, Li) pair, 215
boron interstitial in silicon (B_i or BI), 359
 B_i pair, 359, 360
 Si-G28 ESR spectrum $(BI)^0$, 359
bound exciton, *see* exciton
Brewster's incidence, 44, 45, 201
Bridgman growth method, 7
Brillouin zone (BZ), 45, 469, 471, 478
 folded -, 48
broadening of spectral lines
 homogeneous, 261
 inhomogeneous, 91, 103
 instrumental, 89, 251
Burstein-Moss effect, 74

C absorption or centre in diamond, *see* nitrogen in diamond, N_s
C line in silicon, *see* (C_i O_i) centre in silicon
C(1) centre in GaP, 357
 apex angle, 357
C(1) LVM in GaAs, 357
C(1) LVMs in silicon, (C_i), 51, 348
C(2) centre in silicon, 347
C(3) centre and LVMs in silicon, 353
C(4) LVMs in silicon, 355
C(5) LVM in silicon, 348
C–H modes
 III–V compounds, 418, 420, 423
 diamond, 445
 silicon, 402, 404, 405, 412
CH_2^* dimers in silicon, 404

C_i, *see* carbon (interstitial)
C-related centres
 C N interstitial pair in GaAs and GaP, 315
 CO interstitial pair in GaAs, 316
C_i I C_s, *see* C_i Si C_s in silicon
(C_i O_i) centre in silicon, 327, 353
 C-line, 115, 134
  qmi ^{30}Si, 135
  sidebands, 136
 ($I C_i$ O_i) centre, 355
 precursors, 355
 Si-G15 ESR spectrum $(C_i O_i)^+$, 353
 thermal stability, 135
C_i Si C_s in silicon, 131, 348
 bistability, 348
 C_i S_i C_s^*, 348
 (C_i Si C_s)$_A$, 133, 348, 349
  Si-G11 ESR spectrum $(C_i$ Si $C_s)_{A+}$, 133, 349
  Si-G17 ESR spectrum $(C_i$ Si $C_s)_{A-}$, 349
 (C_i Si C_s)$_B$, 133, 349
  G-line, 131, 348
  Si-SL7 ESR spectrum $(C_i$ Si $C_s)_{B-}$, 349
 thermal stability, 134
C_s(C_iH) in silicon, 412
calcium, 198
calibration factor, 90, 190
 B_{Ga} in GaAs, 214
 (B, H) in silicon, 417
 C_{As} in GaAs, 206
 (C_{As}, H) in GaAs, 425
 C_s in silicon, 134, 203
 divacancy in silicon, 124
 EL2, 156
 H (isolated) in ZnO, 375
 N_{As} in GaAs, 227
 N_i-N_i split pair in silicon, 315
 N_s in diamond, 222, 223
 N_{2s} in diamond, 224
 O_i in GaAs, 307
 O_i in germanium, 300
 O_i in silicon, 268
 Si in GaAs, 219
 $V N_4$ in diamond, 224
 $V O$ in GaAs, 346
 $V O$ in silicon, 332
 (Zn_{III}, H) in GaAs and InP, 420
carbon (interstitial)
 diamond, *see* self-interstitial atom
 silicon
  C_i, 130, 347
  (C_iSn) defect in silicon, 131, 348
  $I C_i$, 347

Si-G12 ESR spectrum, $(C_i)^+$, 130
Si-L6 ESR spectrum, $(C_i)^-$, 130
carbon (substitutional)
  GaAs, 203
    ab initio calculations, 208
    anharmonicity, 206
    charge-state effect, 205
    FWHM, 205
    host-isotope shift, 204
    hydrogen passivation, 210, 419, 423
    overtone, 206
  germanium, 203
    solubility, 203
  other III–V compounds, 207
    C_{In} gap mode in InP, 207
    hydrogen passivation, 420
  silicon, 201
    contamination, 6
    detection limit, 203
    diffusion coefficient, 21
    lattice contraction coefficient, 17
    overtone, 201, 202
    in qmi -, 201, 202
    SiGe alloys, 202
    solubility, 20, 203
cation site, 12, 196, 200
channelled spectrum, 44, 101
chromium, 5, 158
  II–VI compounds, 159, 161, 200
  III–V compounds, 158, 160
closed shells ions, 2
cobalt, 165, 172, 200
combination mode, 191
compensation, 24
  ratio, 24
  self -, 25, 196
contamination, 6
copper in GaAs, 174
corner-cube geometry, 91, 95, 96
corundum, 4, 5, 104
critical point (*CP*) of the Brillouin zone, 49,
    471, 477, 478
cuprite, *see* cuprous oxide
cuprous oxide (Cu_2O), 60, 77, 479
  yellow excitonic series, 77, 78
Czochralski (CZ) growth method, 6

dangling bond, 117, 131, 139, 151, 336, 380
deep centres, 4, 23, 25, 26, 113
defect
  elementary -, 9
  extrinsic -, 115

intrinsic -, 115
radiation -, 10, 115
defect molecule, 194
degeneracy
  electronic, 35, 57, 117
  orientational, 14, 34, 114, 427
  spin, 72
density of states (DoS), 2, 24
  1-phonon-, 49, 55, 262
  2-phonon -, 202, 212, 262
  3-phonon -, 261
diamond;, 47, 61
  3H line, 140, 148
  5RL line, 140
  A aggregate or *A* centre, *see* nitrogen, N_{2s}
    pair
  ABC -, 224
  B aggregate or *B* centre, *see* nitrogen, VN_4
  B' band, 224, 446
  brown, 143, 148, 149
  ^{13}C-, 48, 73, 139–141, 148
  C-absorption (N_s), 222
  C-*R2* ESR spectrum (I^0), 139
  Dresden Green -, 5, 141
  GR1, *see* vacancy, diamond
  graphitization, 138
  *H*1a centre, *see* nitrogen, N_{2i} [100] pair
  mass dependence of E_g, 72, 73
  N3 centre, *see* nitrogen, VN_3
  ND1 line, *see V*
  Ni-related centre, 5, 8
  plastic deformation, 5, 138
  platelets, 146, 224, 446
  R11 line, *see* self-interstitial
  type-I, 4, 137, 144
  type-Ia, 4
  type-IaA, 145, 146
  type-IaA/B, 146
  type-IaB, 146, 148, 224
  type-Ib, 4, 145
  type-II, 4, 5, 137, 139
  type IIa, 5, 139
  type IIb, 5
  (*V*,Si) centres, 150
    C-KUL1 ESR spectrum, 151
    C-KUL8, ESR spectrum, 151
diamond anvil cell (DAC), 107
dicarbon pair
  in III–V compounds, 208
  C_s-C_s in silicon, 203, 352
    hydrogenated, 405
    Si-GGA-2 ESR spectrum, 353
dichroic ratio, 263, 300, 330, 427–429
dichroism, 93

dielectric constant, 22, 43
  relative -, 50
    compound semiconductors, 51
    elemental semiconductors, 50
diffusion coefficient, 20
  temperature dependence, 21
  transition metals, 28
dipole moment, 45, 48, 263
  electric, 51, 118, 189, 193
    π dipole, 245, 263, 298
    σ dipole, 245, 263
  magnetic -, 34
  quenching, 229
  second-order, 52
dislocation loops, 9
divacancy (V_2)
  diamond, 143
  silicon, 116
    absorption, 118, 122
    photoconversion, 121
    piezospectroscopy, 120
    qmi ^{30}Si, 119
    reorientation (atomic), 118
    reorientation (electronic), 117, 119
    thermal stability, 116
    V_2 hydrogen centres, 381
(donor, hydrogen) centres, 430
Drude optical absorption, 3

E-centre, see $(V,$ donor$)$ in silicon
effective charge, 191
effective mass, 22
  longitudinal -, 57
  tensor, 56
  theory (EMT), 22
  transverse -, 57
EL2, 152
  charge states, 152, 153
  metastable state EL2*, 153–155
  optical absorption, 153
  paramagnetism, 155
  photoinduced metastability, 152
  symmetry, 154
electron-attractive potential, 30
electron-hole
  condensation, 80
  radiative recombination, 67
electron-hole droplet (EHD), see electron-hole
    condensation
electron irradiation, 10, 115
electron-phonon interaction, 69, 73
emissivity, 90, 105
encapsulant, 6, 210, 340

ENDOR, 35, 116, 146
energy gap, see band gap
excitation spectroscopy, 91
excitation spectrum, 123
exciton
  binding energy, 74
  bound - (BE), 32, 134
    localization energy, 32, 78
  effective Bohr radius, 75
  effective Rydberg, 75
  free - (FE), 30, 67, 75
  Frenkel -, 79
  isoelectronic bound - (IBE), 30
  lifetime, 74, 78, 79
  Mott-Wannier -, 74, 79
  orthoexciton, 77
  paraexciton, 77
  recombination, 32
excitonic gap, 74, 76, 77
  IMC, 77
excitonic molecule, see biexciton
extended x-ray absorption fine structure
    (EXAFS), 17
extinction coefficient, 43

Fano resonance, 154, 175, 193, 197
Faraday configuration, 109
Fermi level, 22, 362
Fermi resonance, 193, 417, 438
first principles calculations, see ab initio
    fluence
float-zone (FZ) growth method, 7
fluence, 31
fluorite, 479
forbidden transition, 35, 57, 60, 66, 77, 130,
    139, 170, 173, 254, 391, 394
foreign molecule, 33
Fourier-transform spectroscopy (FTS), 94, 95
  photometric accuracy improvement, 97
Franck-Condon
  energy, 24
  shift, 23, 26
free exciton, see exciton
free hole, 2
full width at half maximum (FWHM), 89

gallium arsenide (GaAs), 47, 51, 64
  O_i absorption, see oxygen interstitial
gallium phosphide (GaP), 47, 61
  camel's back, 64
  O_i absorption, see oxygen interstitial

gap mode, 32, 190
  II–VI compounds, 199, 201
  A*l*As, 221
  GaP, 171, 213, 221, 229
germanium, 47, 63
  multiphonon spectra, 55
  O_i absorption, *see* oxygen (interstitial)
  qmi, 77
GeSi alloys, 63, 217, 266
G-line, 131
  piezospectroscopic parameters, 133
  qmi ^{30}Si, 133
GR1, *see* vacancy, diamond
GR2 - GR8 system, *see* vacancy, diamond
Green's functions
  calculations, 214
  technique, 194

H 1a in diamond, *see* nitrogen, N_{2i}[100]
H2 and H3 in diamond, *see* nitrogen, VN_2
H4 in diamond, *see* nitrogen, V_2N_4
H_{AB}, *see* hydrogen antibonded
Hall coefficient, 25
harmonic
  frequency, 192
  oscillator, 191
  potential, 51, 191
H_{BC}, *see* hydrogen bond-centred
helium-like centre, 23
Hexavacancy, *see* V_6
higher-order bands (HOBs), 125
high pressure, high temperature (HPHT)
  growth method, 8
high spin/low spin configuration, 165
hole-attractive potential, 30
hole burning, 91, 147, 154
Huang-Rhys diagram, 23
Huang-Rhys factor, 113, 145, 162
Hubbard correlation energy, 26
Humble pairing model, 16, 140, 355
hydrogen antibonded (H_{AB}), 369
  silicon, 377
    (donor, hydrogen) centres, 438
    overtone, 377
hydrogen bond-centred (H_{BC}), 369
  diamond, 370
  germanium, 373
  silicon, 371
    ab initio calculations, 372
    (acceptor, hydrogen) centres, 413
    donor level, 371
    piezospectroscopic parameters, 371
    qmi crystals, 372

Si-AA9 ESR centre, 371
  stability, 371, 373
  vibrational lifetime, 373
Si:Ge alloys, 374
hydrogen dimer (H_2^*) in silicon and
  germanium, 376
  piezospectrocopic parameters, 378
hydrogen in diamond, 445
hydrogen ("isolated") in *w*-GaN, 375
hydrogen ("isolated") in ZnO, 375
  diffusion, 375
  n-type conductivity, 375
  thermal stability, 376
hydrogen-like atom, 22
hydrogen molecule, 16, 33, 389
  gallium arsenide, 390
  gallium nitride, 397
  gaseous, 390
  germanium, 396
  ortho- and para-hydrogen, 390, 392–401
  platelets and voids, 399
  rotational transition, 400
  silicon, 392
    trapped by O_i, 399
  ZnO, 396
hydrogen (off-axis), 416, 422, 436
hydrostatic stress, 70, 107, 210, 423, 435, 496
hyperfine structure and tensor, 35, 116, 440

infrared activity, 3
in situ measurements, 105, 125, 370, 373–375,
    379, 389, 403
insulator, 1, 2
interaction mass, 248, 253, 293, 338, 339
interferogram, 94, 96, 97, 102
interstitialcy, 130, 243, 347, 356
interstitial hydrogen, *see* hydrogen, (bond-
    centred) and (antibonded)
interstitial locations, 13, 243
interstitial oxygen, *see* oxygen (interstitial)
iron, 8, 21, 158, 200
irreducible representation (*IR*), 37, 47, 57, 58,
    474, 476
isoelectronic atom, 29
isotope shift, 33, 166, 190, 195
    ^{14}C values, 134, 202, 347, 348, 354, 412
    electronic, 129, 132, 135, 149, 166
    host -, 195, 204, 206, 213, 227, 229, 250
isotopic disorder, 262
isotopic mass coefficient (IMC), 73, 77

Jahn-Teller (J-T) distortion, 17, 117, 127, 147,
    151, 157

labelling of centres, 35, 140, 147, 383, 403
Landau energy, 72
lattice
    contraction or expansion, 17
        coefficient, 17
    distortion, 12, 16, 30, 202, 214
    relaxation, 18, 23, 27, 145, 166, 169, 229,
        437
LEC growth method, 6
lifetime, 91, 92, 117, 120, 156, 213, 260, 298
linear atomic chain, 193, 228
linewidth
    homogeneous, 91, 260
    inhomogeneous, 91
Li-related defects in silicon, 128
    Q ZPL, 129
lithium
    interstitial (Li_i), 195
    pairing with acceptors, 216
    pairing with donors, 196
    substitutional acceptor, 195
localized vibrational mode (LVM), 33, 189
Luttinger parameters, 62, 64

magnesium
    II–VI compounds, 199
    in hydrogenated III-V compounds, 421,
        422
magnetic circular dichroism (MCD), 93, 155
magnetic semiconductors, 28
manganese, 160, 200
    III–V compounds, 173, 174
    H passivation in GaAs, 422, 424, 443, 444
metal ions notation, 29
metal-to-insulator transition (MIT), 3, 81
metastability, 16, 18, 28, 152, 275, 301, 345
molecular model
    XYX nonlinear molecule, 245
        apex angle, 244
        isotope shifts, 248
molecular spectroscopy, 51, 189, 194, 245
monoisotopic crystal, see qmi
                (quasimonoisotopic) crystal
Morse potential, 192
Mott insulator, 2

N(1) mode in GaP, 315
N3 centre in diamond, see nitrogen, VN_3
N9 spectrum in diamond, see nitrogen, VN_4
ND1 line, see diamond, V^-
(N, H) centres in III-V compounds, 406, 411
(N, H) centres in III-V_{1-x} N_x diluted alloys,
        409

natural radioactivity, 5
negative-U behaviour, 26
negative-U centre, 27, 28, 30, 340, 349
neutron irradiation, 10, 119
nickel
    isotope effects in GaAs, 166
    in ZnS, 169
niobium in GaAs, 164
nitrogen
    diamond, 4, 137
        C-P1 ESR spectrum (N_s^0), 145, 221
        C-P2 ESR spectrum (VN_3), 147
        C-W15 ESR spectrum (NV^-), 147
        C-W24 ESR spectrum (N_{2s}^+), 146
        N_{2i}[100] pair, 308–310
        N_s, 145, 221
        N_{2s} pair (A aggregate), 4, 146, 223
        NV centre, 147
        split pair, 15
        VN_2, 148
        V_2N_4, 148
        VN_3, 146
        VN_4 (B aggregate), 4, 146, 224
    germanium
        N_s, 225
        split pair N_i-N_i, 311
    N_{As} in GaAs, 226
        N_{As}–V_{Ga} centre, 227
    N_s in GaP and ZnSe, 226
    SiC, 224
        N_C gap mode in $4H$-SiC, 224
    silicon
        ab $initio$ calculations, 225
        isolated interstitial (ab $initio$
            calculations), 311
        NNO centre, 312
        N_s, 225
        Si-SL5 ESR spectrum (N_s), 224
        solubility, 315
        split pair N_i–N_i, 15, 310

O–H modes
    in GaAs and GaP, 448
    in ZnO, 375
off-centre O in GaAs, see VO in GaAs
O_i in GaAs, 306
    apex angle, 306, 307
    low-frequency motion, 307
    phonon spectroscopy, 307
    piezospectroscopic parameters, 307
    stress-induced dichroism, 307
O_i in GaP, 308

O_i in germanium, 282
  ab initio calculations, 267
  apex angle, 285, 291
  asymmetric isotopic substitution, 287
  combination band, 296, 300
  dichroism, 300
  dimer, 300
  frequencies, 283, 289, 290, 292, 294, 296
  interaction mass, 293
  isotope effects
    Ge, 289, 290, 292, 296
    O, 283, 291, 292, 294
    qmi effects, 288, 290, 292
  lifetimes and linewidths, 298
  phonon spectroscopy, 285, 292, 296, 298
  piezospectroscopic parameters, 298
  relative intensities, 288, 289
  reorientation energy, 300
  rotational isotope effect, 292
  rotational levels, 295, 296
  rotational motion, 284, 285
  sideband spectra, 293
O_i in silicon, 14, 244
  2DLFM, 245, 247, 256, 259, 265, 266
  9 μm absorption band, 245, 269
  A_{1g} symmetric mode, 249, 256, 257, 259,
      265, 269
  A_{2u} antisymmetric mode, 249, 250, 254,
      259
  ab initio calculations, 249, 253, 258
  apex angle, 249, 253, 258
  combination band, 256, 257
  dichroic ratio, 264
  diffusion coefficient, 21, 263
  dimer, 271
    skewed configuration, 271
    staggered configuration, 271, 273
  equilibrium position, 246, 250, 265
  E_u mode, 254
  frequencies, 250, 254, 258
  hydrostatic pressure effects, 255, 265
  intensities (relative), 268
  interacting with
    C_s (O_i, C_s), 278
    C_i, *see* carbon (interstitial), silicon
    hydrogen, 279, 399
    lithium, 282
    nitrogen, 312
    self-interstitial (I), 282
    tin, 266
  interaction mass, 253
  isotope effects
    O, 247, 250, 252, 269
    qmi effects, 253, 255, 258

Si, 247, 250, 252, 253
  lattice expansion coefficient, 17
  lifetimes and linewidths, 260, 262
  linear model, 249
  linewidths, 260
  low-frequency excitation (LFE), 247
    LFE0, 259
    LFE1, 250, 259
  phonon spectroscopy, 246
  piezospectroscopic parameters, 264
  radial distribution, 271
  radial oscillator model, 249
  relative intensities, 251, 254
  reorientation energy, 263
  Si-Ge alloys, 266
  temperature effects
    frequency shift, 251, 260
    thermalization, 251
  trimer, 273
orientational degeneracy, *see* degeneracy,
      orientational
oscillator
  1D, 191
  3D, 192
overtone, 191, 192, 194
oxygen (interstitial), *see* O_i
oxygen precipitation
  in germanium, 278, 305
  in silicon, 276
oxygen (substitutional)
  II–VI compounds, 229
  GaN, 229
  GaP, 217, 229, 325
oxygen-vacancy centre, *see* VO

P centre in silicon, *see* B_s pair
P line in silicon, 134, 356
  qmi ^{30}Si, 135
  sidebands, 136, 356
  thermal stability, 135
pairing of impurities, 14, 196, 211, 214
passivation of dopants, 25, 413
phonon
  2TA overtone, 52
  absorption
    multiphonon, 43, 51
    one-phonon, 43, 48
  acoustic branches, 45
  anharmonic coupling, 51
  difference process, 52, 54
  energies at zone centre, 47
  momentum-conserving -, 32, 66, 70
  occupation number, 54, 66

optic branches, 45
qmi effects, 48, 54
Raman frequency, 32, 48
summation process, 52, 54
wave vector, 46
phonon gap, 32, 47, 193
phonon spectroscopy, 93
phosphorus
  in III–V compounds, 227
  in silicon and germanium, 227
photoconductivity, 93, 142, 145, 152, 153
photoionization spectrum, 31
photonic up-conversion, 170
photo-thermal ionization spectroscopy, 93
photothreshold, 29
piezospectroscopic parameters, 497
piezospectroscopic techniques, 105
piezospectroscopic tensor, 496
platelets, 370, 399
polarization
  p -, 44
  s -, 44
polarizer (wire grid), 100
polyvacancies clusters, 126
pressure-transmitting media, 108
pseudo-acceptor, 30
pseudo-donor, 30
pseudomolecule, 195, 204, 306
pseudopotentials, 194
PTIS, *see* photothermal ionization
        spectroscopy
puckered configuration, 244, 245
pump-probe method, 92

Q lines, B_i-related LVMs in silicon, 360
quasimonoistopic, *see* qmi
qmi crystal, 4
quasi-rotational mode, 33
quasi-substitutional oxygen, *see* VO

R LVMs in silicon, 359
R11 line in diamond, *see* self-interstitial
Racah parameters, 159
Raman frequency, *see* phonon
Raman scattering, 31, 33
reflectivity, 43
refractive index, 43, 50
reorientation energy, 14
resistivity, 3
  intrinsic, 3
resonant mode, 32, 189

rotation-tunnelling motion, 14
ruby, 5

S LVMs in silicon, 359
sapphire, 5, 104
saturated absorption, 92
scattering mirror, 100
segregation coefficient, 19, 190
selenium in ZnS, 229
self-interstitial atom (I), 9
  in diamond, 139, 316
    R11 line (I^0), 139
  (IH_2) in germanium, 388
  in silicon, 115
    hydrogenated (IH_2), 388
semiconductor
  direct-band gap -, 57
  extrinsic, 3
  indirect-band gap -, 57
  intrinsic, 2
semi-insulating, 2
semimetal, 1
shallow centre, 22
silent mode, 47
silicon (crystal), 47, 54, 61
  lattice distortion, 17
  O_i absorption, *see* oxygen (interstitial)
  qmi, multiphonon absorption, 54
  qmi, $O(\Gamma)$ phonon, 48
silicon (impurity), 217
  III–V compounds, 218
    gap mode of Si_{Al} in $AlAs$, 221
    Si_{III}–Si_V pairs in GaAs, 219
    (Si, X) and (Si, Y) centres in GaAs, 219
  diamond, 149
  germanium, 217
silicon carbide (SiC), 13, 47, 48, 51, 61, 481
  $4H$ polytype, 2
  $6H$-polytype, 481
  precipitation in silicon, 20, 203
solubility, 18
  equilibrium, 19
  non-equilibrium, 19
spectral resolution, 89
spin-orbit (s-o) coupling, 57, 58, 151
spin-orbit (s-o) splitting, 57–59
split interstitial, *see* interstitialcy
stacking sequence, 14, 479, 481
stress-induced dichroism, *see* appropriate
        impurity or centre
stress-induced reorientation method, 263, 427
stretch mode, 189, 369

sulphur in II–VI compounds, 231
sulphur in GaP, 229

T line in silicon, $C_s(C_iH)$, 412
Tanabe-Sugano diagram, 159, 163
TH5 line, *see* divacancy in diamond
thermal donors in germanium, 9, 301–305
    metastability, 301
    O isotope effect, 304
thermal donors in silicon, 9, 271–276
    metastability, 275
    overcoordinated oxygen, 274
    shallow - (STDs), 9
time domain spectroscopy, 92
time-resolved transient spectroscopy, 374
titanium, 200
TPA, *see* two-photon absorption
transition metals (TMs), 158, 200
    charge transfer bands, 169
    contamination, 7
    in diamond, 8, 165
    hydrogen-like transitions, 171
    LVMs in II–VI compounds, 200
    (TM, hydrogen) centres, 440, 443
transmittance, 44
transmutation doping, 8
tunable lasers, 99
two-photon absorption (TPA), 78, 92

up-conversion, *see* photonic up-conversion

V (vacancy), 9, 10
    diamond, 140
        C-S1 ESR spectrum (V^-), 143
        GR1 line (V^0), 141
        GR2-GR8 system, 141
        ND1 (V^-), 143
        thermal stability, 143
    silicon, 10
        negative-U (V^+), 30
        Si-G2 ESR spectrum, V^-, 30
(V, donor) in silicon, 126
(V, hydrogen) centres, 379
    in GaN, 387
    in germanium, 382, 384
    in silicon, 379, 382
    in ZnO, 387
V_6 in silicon, 11, 126
$V H_4$
    germanium, 382, 384
    InP, 384

ab initio calculations, 385
    breathing modes, 385
    overtone, 385
    piezospectroscopic parameters, 385
    stretch mode, 385
silicon, 380
    breathing mode, 383
    interacting with C, 450
    isotope effect, 380
    overtone, 383
    stretch mode, 380–382
VO
    gallium arsenide (OV_{As}), 340
        A LVM, 340–346
        abnormal intensity ratio, 345
        As_{Ga} antisite, 346
        B LVM, 340–346
        B' LVM, 340–344
        charge states, 344
        negative-U behaviour, 340
        photoconversion, 340
        piezospectroscopic parameters, 343
    germanium, 337
        ab-initio calculations, 339
        apex angle, 338
        charge states, 337
        Ge isotope effect, 338
        interaction mass, 338
    silicon, 10, 325
        A-centre, 326
        apex angle, 332
        ENDOR transitions, 328
        perturbation by foreign atoms, 334
        piezospectroscopic parameters, 331
        reorientation energy, 330
        Si-A ESR spectrum, (VO), 36, 326
        stress-induced dichroism, 330
        VOH_2, 335, 384
V_nO_m centres in silicon, 332
    Si-A14 ESR spectrum $(V_2O)^0$, 333
valence band (VB), 58
    heavy-hole -, 61
    light-hole -, 61
    parameters, 61, 64
vanadium, 29, 162, 165, 200
Vegard's law, 16
vertical-gradient freeze growth method, 7
vibrational spectroscopy, 33
vibronic
    coupling, 114, 162, 165
    replica, *see* sidebands
    sidebands, 32, 114, 130
    transition, *see* sidebands

virtual crystal approximation (VCA),
    48
voids, 9, 225, 311
Voigt configuration, 109

wag mode, *see* bending mode
work function, *see* photothreshold

X ZPL in silicon, 136

Zeeman term, 34
zero-phonon line (ZPL), 31, 113
zero-point motion, 72
ZnO, 9, 47, 51, 66
    hydrogen shallow donor, 375

Printed by Printforce, the Netherlands